microbe

microbe

Moselio Schaechter
Department of Biology
San Diego State University
and
Division of Biological Sciences
University of California, San Diego
San Diego, California

John L. Ingraham
Section of Microbiology
Division of Biological Sciences
University of California, Davis
Davis, California

Frederick C. Neidhardt
Department of Microbiology and Immunology
University of Michigan
Ann Arbor, Michigan

WASHINGTON, D.C.

Address editorial correspondence to ASM Press, 1752 N St. NW, Washington, DC 20036-2904, USA

Send orders to ASM Press, P.O. Box 605, Herndon, VA 20172, USA
Phone: (800) 546-2416 or (703) 661-1593
Fax: (703) 661-1501
E-mail: books@asmusa.org
Online: estore.asm.org

Library of Congress Cataloging-in-Publication Data

Schaechter, Moselio.
Microbe / Moselio Schaechter, John L. Ingraham, Frederick C. Neidhardt.
p. cm.
Includes bibliographical references and index.
ISBN-13: 978-1-55581-320-8 (pbk.)
ISBN-10: 1-55581-320-8 (pbk.)
1. Microbiology. I. Ingraham, John L. II. Neidhardt, Frederick C. (Frederick Carl), 1931- III. Title.

QR41.2.S33 2006
579—dc22

2005016180

10 9 8 7 6 5 4 3 2 1

Illustrations: Patrick Lane, ScEYEnce Studios
Cover and interior design: Susan Brown Schmidler

Cover image: copyright Gregor Reid (London, Ontario, Canada); created with the assistance of Mark Neysmith

This book is dedicated to

Edith Koppel
*Marjorie Ingraham**
Germaine Chipault

Though not in a position to share the unique pleasures of team writing, these patient companions of the authors were not spared its demanding aspects. Each provided support, encouragement, and understanding—skills these three spouses had long ago perfected.

M.S.
J.L.I.
F.C.N.

*Deceased 20 July 2005

Contents

Preface

The microbial world is in surprising ways the only world we humans inhabit. Our intention in writing this textbook has been to introduce you to the world of microbes, its importance in all things human and, beyond that, its importance in maintaining the well-being of our planet.

Our working plan has been to emphasize concepts, not facts, although facts have a tendency to creep in. We like to tell stories about the way microbes are put together, what they must do to grow and survive, and how they interact with all other living things. Our approach is far from encyclopedic, although this is a fairly thick book because there is so much to tell about microbes.

We know that a number of you will enter the health professions and that you may be particularly eager to read about microbes as disease agents. We share this interest, as we have taught in medical schools and have written textbooks with a medical orientation. But we think that knowledge of all aspects of microbial life and activities is necessary as preparation for dealing with them as pathogens. Should you wish to anticipate what is coming in this area, turn to chapter 22 of the book and read it first. It requires little background and can be read early in a microbiology course.

Not wanting to overload this book with information, we turn you to the wealth of material available on the Internet. In order to make the choices palatable and not overwhelm you with too many interesting websites, we have selected a few especially relevant ones for your attention, and we refer to them in the text. See "Study Aids," the last item in the front section of this book, for more information.

Acknowledgments

We thank the following individuals for their comments, suggestions, answers to queries, and in some cases reviews of entire chapters—all given generously and in a spirit of helpfulness: Jennifer Antonucci, Douglas Bartlett, Paul Baumann, Mya Breitbart, Veronica Casas, Mike Cleary,

Kathleen Collins, Dean Dawson, Wesley Dunnick, Cary Engleberg, Jack Fellman, Joshua Fierer, Margo Haygood, Gary Huffnagle, Mark Hildebrand, John Howieson, Philip King, Denise Kirschner, Roberto Kolter, Sydney Kustu, David Lipson, Stanley Maloy, Jack Meeks, Hal Mensch, Mylene Mozafarzadeh, Cristián Orrego, Joe Pogliano, Kit Pogliano, Malini Raghavan, Monica Riley, Forest Rohwer, Dirk Schüler, Anca Segall, Deborah Spector, David Stollar, David Thorley-Lawson, and Joseph Viznet.

Special thanks go to Terry Beveridge for supplying a number of electron micrographs.

We thank members of the 2003 bacteriology class at the University of California, San Diego, who read an early draft and made incisive and useful comments: Cathy Chang, Collin Chang, Jason Chen, Jennifer Han, Jennifer Hsieh, Jennifer Kim, Jennifer Lee, Neal Mehta, and Ilya Monosov.

Special appreciation is extended to those individuals—experienced teachers of undergraduate microbiology all—who read the entire text and offered corrections and guidance from their individual perspectives: Rod Anderson, Robert Bender, Gary Ogden, James Russell, and Amy Cheng Vollmer. Their thoughtful and substantive comments were extremely helpful.

Finally, we are grateful to the staff of ASM Press and the freelancers who worked on this book for their skill, expertise, and patience. Jeff Holtmeier (Director) assembled a crackerjack team consisting of Kenneth April (production editor), Elizabeth McGillicuddy (copy editor), Kathleen Vickers (proofreader), Patrick Lane (artist), and Susan Schmidler (cover and interior designer). With good humor and total dedication, they provided what it takes to turn a manuscript into a book.

About the Authors

The three of us have interacted and collaborated for many years. In fact, this is our fifth joint book title. Among these works are a textbook, *Physiology of the Bacterial Cell,* and a reference book, Escherichia coli *and* Salmonella: *Cellular and Molecular Biology.* We can all vouch for the pleasure of working with people of like mind and how it has deepened our friendship. We have shared other experiences as well; each of us, for example, has served as president of the American Society for Microbiology.

Moselio Schaechter (middle) was born in Italy, lived in Ecuador as a teen, and obtained his Ph.D. at the University of Pennsylvania. He spent most of his academic life at Tufts University School of Medicine and moved to San Diego in 1995. His research interests involved aspects of bacterial physiology, including growth rate regulation, membrane biology, and chromosome transactions. His e-mail address is mschaech@sunstroke.sdsu.edu.

John L. Ingraham (right) is a lifelong Californian, having been born, raised, and educated in Berkeley and having spent his academic career at the University of California, Davis. He, too, studied aspects of bacterial physiology, including growth at low temperature, the malolactic fermentation, fusel oil formation, pyrimidine metabolism, and denitrification. His e-mail address is jingrah1@earthlink.net.

Frederick C. Neidhardt (left) was born in Philadelphia, majored in biology at Kenyon College in Ohio, and received his Ph.D. at Harvard University. He held academic posts at Harvard, Purdue University, and the University of Michigan. His research focused on catabolite repression, growth rate regulation, aminoacyl-tRNA synthetases, and heat shock and other global cell networks. His e-mail address is fcneid@umich.edu.

Study Aids

The Website

Readers of this textbook will have easy and direct access to sites on the Internet. To enhance the learning experience, a website associated with this book has been set up at http://www.microbebook.org. This website will contain links to other sites related to items discussed in the book as well as to the online study guide described below. We suggest that readers designate www.microbebook.org as a favorite or bookmark. Once the site is opened, the reader will be able to open links with a single click. In the text, the linked sites are indicated by sidebars (or, in one instance, in a table) with numbers corresponding to links on the website.

The linked sites are selected for various purposes, the main ones being to fulfill the reader's desire for more information and to provide animation or video clips. These websites have been chosen for stability and will be monitored and updated.

The Online Study Guide

Another study aid is the BrainX Digital Tutoring System. *Microbe* comes with online access to this powerful online study guide. This system incorporates the latest research on the neurobiology of learning. Simply read a chapter of the book to gain a good understanding of the material presented in the chapter. Then go online and use the tutoring system to reinforce the information until it is locked in your long-term memory.

To get started, go to http://www.microbebook.org. There you will be able to link to the site where you can open your own account on the BrainX Digital Tutoring System and use the study guide for *Microbe*.

If you bought this book new, you will be asked to enter the authentication number printed on the inside front cover to get free access to the BrainX Digital Tutoring System and the study guide for *Microbe*. This number can be used only once, by the original purchaser of this textbook. If you bought *Microbe* as a used book, please go to www.microbebook.org and follow the instructions for purchasing a pass code for the online study guide.

section I microbial activity

chapter 1 the world of microbes

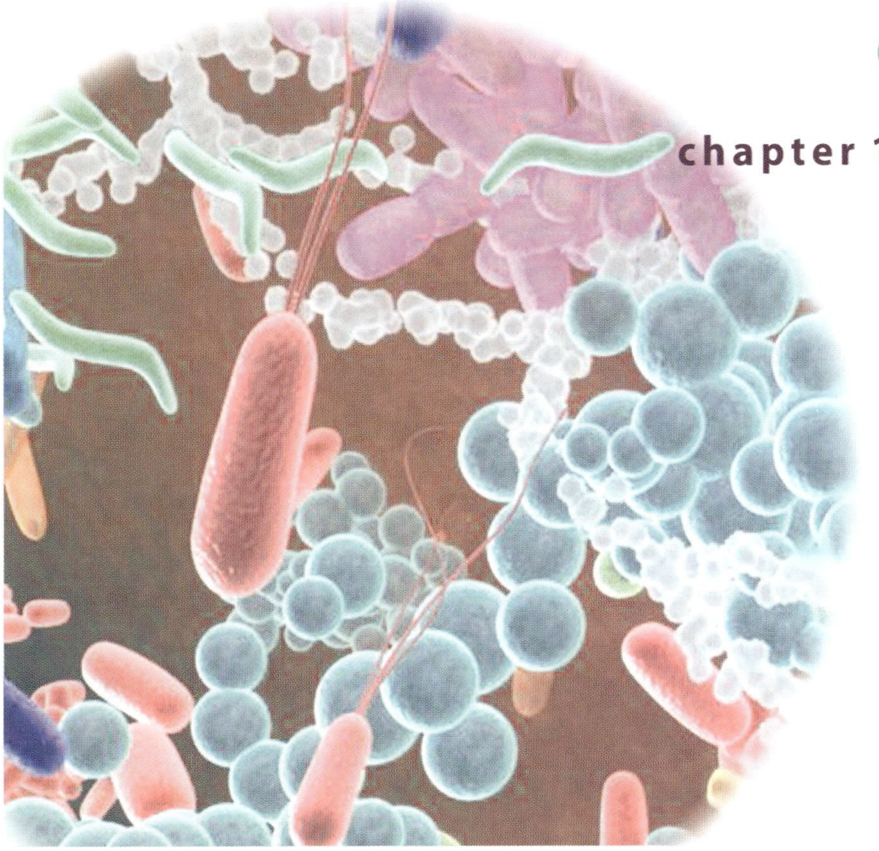

chapter 1 the world of microbes

INTRODUCTION

1.1 *General background to the chapter*

Most living things are microbes, organisms not visible to the unaided eye; the organisms that are visible are a small minority. Microbes interact with all members of the living world and much of the inanimate world. The impact of this interaction is truly staggering (Table 1.1), and for this reason, the study of microbes transcends microbiology and impinges on all of biology. We believe that it is not possible to study any branch of biology or the earth sciences without giving serious consideration to the activities of microbes. They have great impact on the lives of humans (Table 1.2). In this chapter, we will introduce you to the microbes' more notable characteristics.

WHAT IS A MICROBE?

Traditionally, microbes have been described as free-living organisms so small (less than about 100 micrometers [μm]) they are visible only under the microscope, but some microbes are large enough to be seen with the naked eye. The smallest bacteria are barely 0.2 μm long, but giant bacteria and protozoa can be 1 millimeter (mm) in length or even longer. Microbes are either **prokaryotes** (cells lacking a true nucleus) or **eukaryotes** (cells with a true nucleus) (Fig. 1.1). Eukaryotic microbes, other than algae and fungi, are collectively called **protists**. A complicating factor is that some microbes are not free-living; some are obligate intracellular parasites, such as the leprosy bacillus (*Mycobacterium leprae*). A second complication is that some groups are especially hard to delineate, in part because they have large relatives. For example, yeasts are certainly microbes, but mushrooms are not, yet both are fungi.

Table 1.1 Characteristics of microbes

Microbes:
Are the source of all life forms
Are more phylogenetically diverse than plants and animals
Are enormously abundant
Grow in virtually every place on earth where there is liquid water
Carry out transformations of matter essential for life
Transform the geosphere
Affect the climate
Participate in countless symbiotic relationships with animals, plants, and other microbes
Cause disease
Influence the behavior of animals and plants

Table 1.2 Human use of microbes

Carrying out chemical activities of major industrial importance
Engineering for production of useful proteins, e.g., vaccines
Enhancing food production and preservation
Providing vital public health measures, such as sewage disposal
Bioremediation of polluted sites
Malevolent intent (biological warfare, bioterrorism)

Table 1.3 Bacteria and Archaea

The two differ in chemical properties of cell wall and membranes.
Bacteria are sensitive to antibiotics; archaea are not sensitive to many of them.
The archaeal protein- and nucleic acid-synthesizing enzymes resemble those of eukaryotes; not so the bacterial ones.
Bacteria include animal and plant pathogens; archaea do not.
Typical bacteria are *Escherichia coli*, staph (*Staphylococcus*), and strep (*Streptococcus*).
Typical archaea are extreme thermophiles and methane producers.

Finally, what about the **viruses?** Are they microbes? Not really, because they lack one key quality of living cells, the capacity to maintain their bodily integrity. Viruses reproduce and mutate, as do cellular organisms, but they disassemble during their reproductive cycle. The major constituents of viruses are nucleic acids and proteins that are made separately in their host cells and only later are reassembled into progeny viral particles. Viruses do not have the capacity to make proteins (they have no ribosomes) and rely on the host to supply all the building blocks needed for biosynthesis of their macromolecules. Outside a host cell, a viral particle is inert and incapable of reproduction. Viral biology is discussed in Chapter 17.

The prokaryotes comprise two separate but very large groups, the **Bacteria** and the **Archaea**. As we will see in Chapter 2, the two have generally the same body plan: they lack nuclei and membrane-bound organelles, such as mitochondria or chloroplasts (Table 1.3). Molecular methods based on ribosomal ribonucleic acid (RNA) sequencing indicate that they comprise two distinct evolutionary groups. The branch point of their descent is ancient, and its details are controversial. According to many investigators, the Archaea are actually closer to the eukaryotes than to the Bacteria (Fig. 1.2). Here, the Bacteria form one branch, and the Archaea and the eukaryotes form the other. Other researchers propose yet other schemes. The term prokaryote does not have phylogenetic meaning and should be limited to describing a cell body type (mainly, the lack of a true nucleus). What must be noted, however, is that Bacteria and

Figure 1.1 Microbes. Microbes include both prokaryotes and eukaryotes.

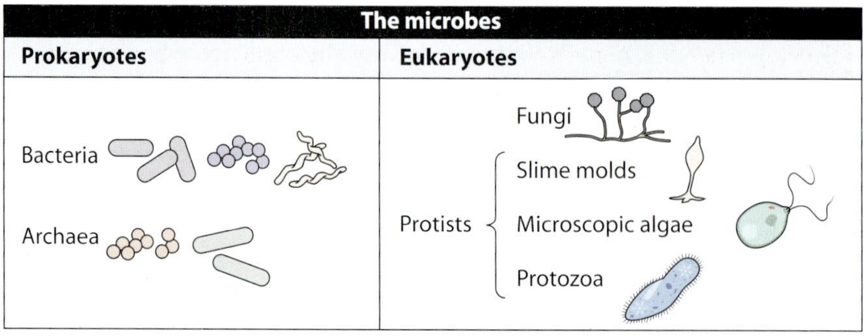

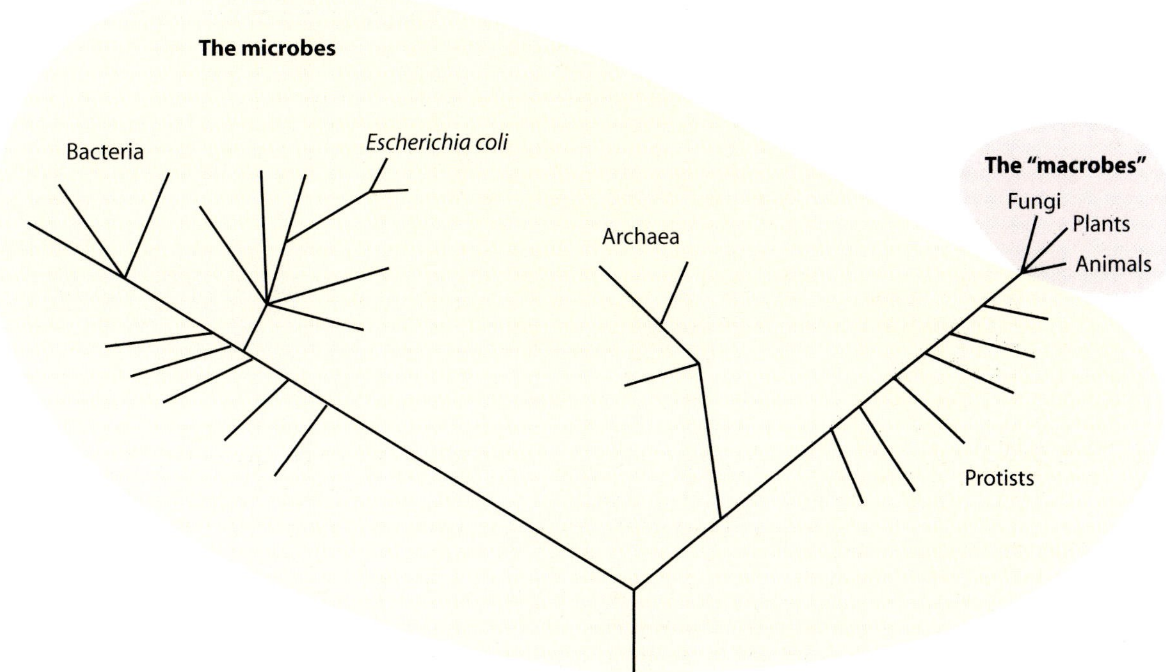

Figure 1.2 Phylogenetic relationships in the living world based on small ribosomal RNA homology. All living organisms fall into one of three **domains,** the Bacteria and Archaea **(prokaryotes)** and the Eukarya **(eukaryotes),** each divided into a number of phyla. Note that the number of domains in the prokaryotic world is greater than that in the eukaryotic one. Further, the number of phyla of organisms that cannot be seen with the naked eye (the microbes) is far greater than that of those that can be seen (the "macrobes"). Animals, plants, and fungi are perched on a small branch of this "tree of life."

Archaea have *greater phylogenetic diversity than all eukaryotes combined.*

HAVING A LONG PAST

Any meaningful discussion of evolution must include the microbes. Life started with microbes, and such unicellular organisms had the planet to themselves for about 80% of the time that life has existed on Earth (Fig. 1.3). We do not know the nature or location of the "primordial ooze" (or what Charles Darwin called a "warm little pond") where life began, but we do know that microbes were the first cellular life forms to arise and thrive. Being so ancient, microbes have had a very long time to evolve and to develop the basic metabolic mechanisms that made all other life possible. Through evolution, microbes have come to occupy a great variety of ecological niches, including some that seem improbable from a human point of view. Microbes grow in the frozen tundra, in waters whose temperature is over the boiling point (at high pressure), in strong acid and alkali, and in concentrated brine.

The microbial fossil record is scant, but together with genomic information, it is sufficient to tell us that microbes diversified early on into a great variety of shapes and life styles. Prokaryotes were alone on Earth

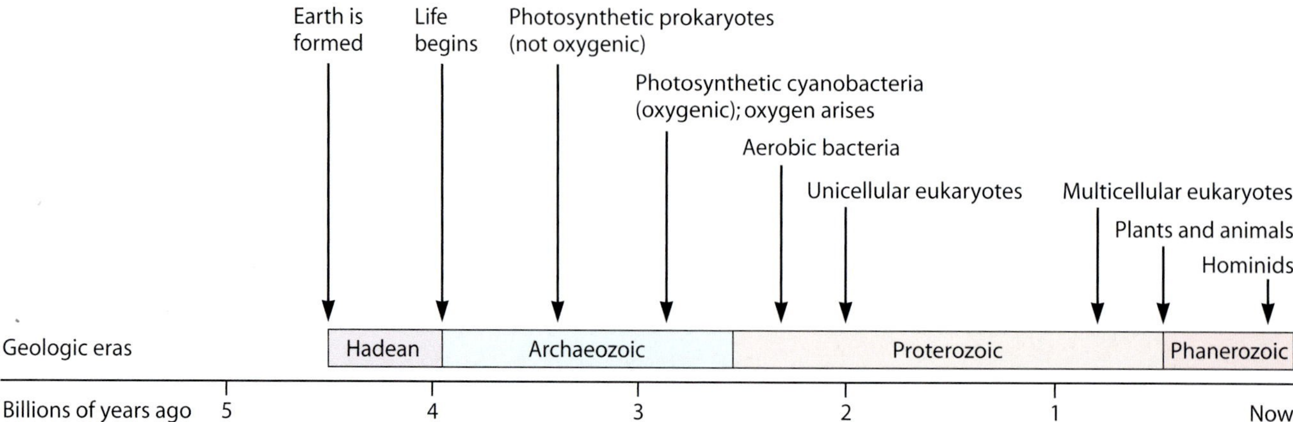

Figure 1.3 Time scale of life on Earth. The arrows indicate the approximate times when different life forms arose. Note that the Phanerozoic includes three main eras, Paleozoic (540 million to 248 million years ago), when relatively simple animals and plants arose; Mesozoic (248 million to 65 million years ago), the age of dinosaurs, birds, and fish; and Cenozoic (65 million years ago until now), the era of mammals.

for some 2 billion years (about half the time there has been life on Earth), after which eukaryotic microbes arose. Multicellular organisms did not arise until some 750 million years ago. As we will see later in this book, *some microbes are the only forms of life that are self-sufficient, that is, only they can exist indefinitely without any other living things.*

Prokaryotes are the ancestors of all other life forms. All eukaryotes, from simple yeasts and algae to humans, arose from prokaryotic progenitors. A special aspect of this development is seen in the organelles of eukaryotic cells—mitochondria and chloroplasts. The genomes of these organelles, which were acquired by the ingestion of early microbes, although reduced, are unmistakably microbial.

BEING SMALL

Microbes are small: the volume of a representative bacterium is about 1 μm^3, roughly 1/1,000 that of human cells. Some bacteria are considerably larger, others smaller, than this average. One gargantuan species (*Epulopiscium fishelsoni*) that lives in the gut of the surgeonfish is astonishingly large: it is over 0.5 mm in length and about 1/10 mm^3 in volume, and it is visible with the naked eye. The range in volume of the smallest to the largest known bacteria is well over 1 millionfold (Fig. 1.4). It follows that size alone is not a good way to distinguish prokaryotes from eukaryotes. Some marine algae (e.g., *Ostreococcus tauri*) are about 1 μm in diameter; they are the smallest known eukaryotes, well within the size range of most prokaryotes. However, it is true that most prokaryotic cells are smaller than most eukaryotic cells.

To picture how small a small bacterium really is, let us imagine that one which is slightly over 1.5 μm in length and a human being were each enlarged 1 millionfold. The bacterium would now have the approximate size of a human being; the human being would stretch all the way from New York to Chicago. In terms of volume, it takes about 10^{17} bacteria, or 100,000 terabacteria, to occupy the volume of a nonenlarged adult human. Smaller volume means a greater surface-to-volume ratio: the

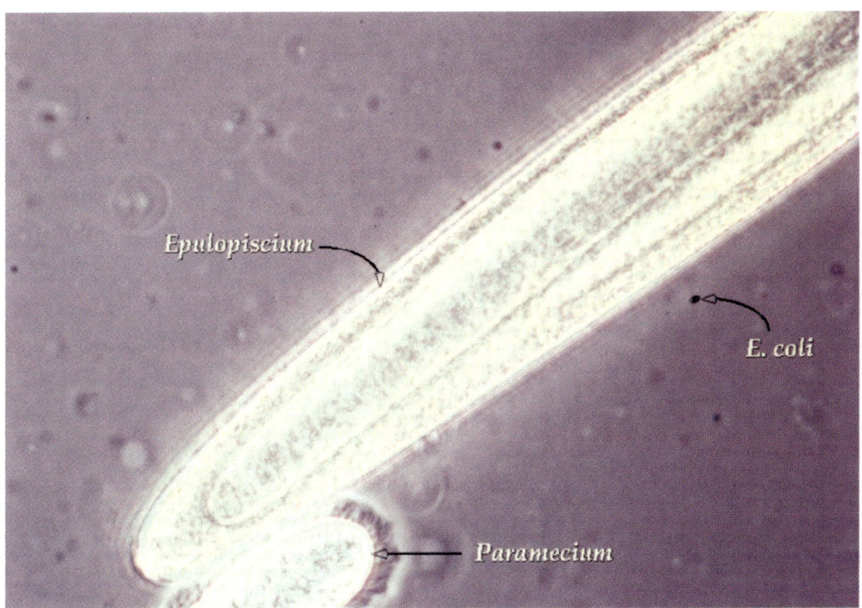

Figure 1.4 How big is big? One of the largest prokaryotes known *(E. fishelsoni)* greatly exceeds an average bacterium *(Escherichia coli)*, and even a large unicellular eukaryote *(Paramecium)*, in size.

ratio of a typical bacterium is 1,000 times greater than that of a typical animal cell, which allows higher rates of uptake of nutrients from the environment.

Simple microscopic observation of bacteria is not very revealing. Most of them look like unassuming little sausages or small spheres without distinctive markings. Despite this simplicity, microbiologists who are used to observing a given microbial species find that its size and shape are quite distinctive. This is useful, because one can assess the relative state of health of the organisms just by looking at them under the microscope. Distressed bacteria in cultures often look misshapen—they are unevenly swollen, elongated, and irregular—whereas healthy ones are typically uniform in shape.

To get a sense of prokaryotic complexity, let us imagine a fanciful voyage, where we have shrunk some 10 millionfold and are able to fit inside a typical-size bacterium. What would we see? What would the surroundings feel like to the touch? The largest objects would be ribosomes, the size of beach balls, and they would almost completely fill the field of vision. Also visible would be a long strand of deoxyribonucleic acid (DNA), tightly folded on itself like a skein of wool. We would find the medium to be thick and gummy.

An analogous voyage into an animal or plant cell would show not only that it is larger than a bacterium but also that it has a different body plan. Here, the interior is divided into compartments bounded by membranes: the nucleus, mitochondria, Golgi apparatus, endoplasmic reticulum, and, in plants, chloroplasts. These structures, although spatially secluded, communicate with each other via metabolic messages. In most bacteria, by contrast, there are few real compartments, and there is a greater premium on the efficient use of space and energy.

Microbes often compensate for the inherent vulnerability of smallness by forming aggregates of one type or by living in communities together with other microbes. Such masses of cells are often visible to the naked eye. They are not just lumps of cells: individuals often become metabolically and morphologically differentiated, together resembling in some respects a multicellular organism. In some instances, they become distinct and often beautiful structures (see Fig. 14.5). Individual cells communicate with one another within such structures.

BEING MANY

As small as individual microbes are, their total collective mass on earth is staggering. Microbes are nearly ubiquitous, and they are found practically anywhere there is free liquid water. In the oceans alone, there are nearly 1 million bacteria per ml of water, for a total of at least 10^{29} cells in all the world's oceans. It has been estimated that the total biomass of microbes on this planet is nearly as great as that of all plant life and, according to some calculations, greater. Accordingly, the amount of carbon in prokaryotic cells is nearly as large as, or by some calculations even larger than, that in all plants on earth, terrestrial and marine. Also, because prokaryotes contain proportionally more nitrogen and phosphorus than plants, they are the largest reservoir of these elements in the biological world.

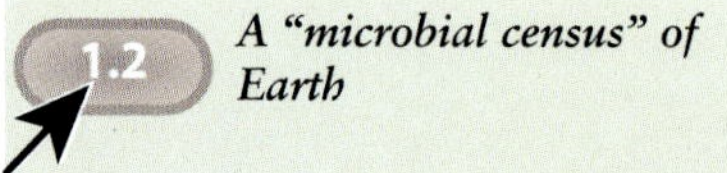

A "microbial census" of Earth

Numbers, large and small, matter. Some diseases, such as bacterial dysentery, are acquired by swallowing just a few bacterial cells, but close to 1 trillion (one terabacterium) of them weigh scarcely a gram and are approximately the volume of a sugar cube. For the dysentery bacilli, this number is sufficient to infect not only all humans but also all other susceptible vertebrates on Earth with plenty to spare. How much fecal contamination would it take to pollute a large body of water and make even a few swallows of its water infectious? Not much. The lower intestines are so teeming with microbes that the dry weight of vertebrate feces is one-third to one-half bacteria. Such crowding is not unusual in nature. The space between teeth and gums (the gingival crevice that dental hygienists probe for pockets), for example, contains wall-to-wall bacteria. Put another way, there are more bacteria in and on our bodies than all our human cells.

GROWING AND PERSISTING

Under optimal conditions in the laboratory, some bacteria can divide once every 20 minutes (producing 8 by 1 hour, 64 in the next hour, 128 in the next, and so forth). Some bacteria grow even more rapidly. One can calculate the number produced in a given time starting with a single one by using the equation $N = 2^g$, where N is the number of bacteria and g is the number of generations (doublings). It takes only 20 generations (or 6 hours 40 minutes) for one bacterial cell to become over 1 billion bacteria. (For the calculations, see "The Law of Growth" in Chapter 4.) In less than 2 days, this single bacterium would have multiplied to a volume much greater than that of the earth. So why doesn't this happen? The answer is that the population runs out of food, and even if food were plentiful, the metabolic activities of the bacteria would result in the accumulation of growth-inhibiting toxic products.

The high surface-to-volume ratio of a bacterial cell results in high metabolic rates. Rapidly growing bacteria process their nutrients at rates

10 to 1,000 times higher per gram than do animal cells. There are significant consequences of such rapid chemical turnover, some of which are easy to grasp. How long does it take for the body of an animal to decompose, for bread to rise, or for a person to get sick with an infectious agent? Hours? Days?

On the other hand, some microbes grow exceedingly slowly even under optimal conditions. Fast growth may be required for success in some environments but not in others. The tubercle bacillus *(Mycobacterium tuberculosis)*, which causes tuberculosis, divides only about once every 24 hours. Certain microbes living in barren habitats grow even more slowly than the tubercle bacillus, doubling perhaps once a month, once a year, or even less often. Those who have visited the deserts in the western United States may remember seeing sun-baked rocks "painted" with bright colors. These are not the actual colors of the rocks but just those of a thin surface patina known as "desert varnish," the layer that Native Americans chipped away to create their stunning petroglyphs.

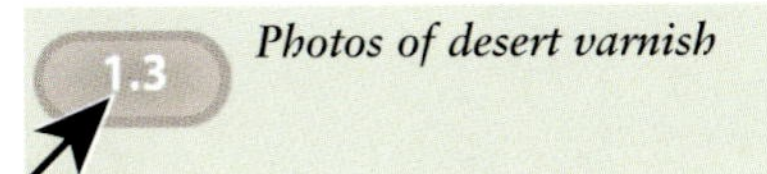

Photos of desert varnish

Desert varnish consists of a thin veneer of iron and manganese oxides, deposited over eons by the actions of resident bacteria. The organisms are shielded from the sun's lethal ultraviolet radiation by clay particles. Imagine how infrequently such bacteria grow and divide. Precise figures are not available, but years or centuries do not seem unreasonable guesses. Growth and division under such circumstances are not continuous and are likely to take place in short but infrequent spurts. However slow the growth, microbes make a huge difference over geological times.

The lesson we might gain from these beautiful painted rocks is that bacteria do not have to grow very fast or divide very often to *survive* for a long period of time. This suggests that microbes have formidable mechanisms to adapt to harsh, unfavorable conditions, such as dryness, high temperatures, radiation, and high osmotic pressure No wonder the microbial world is such an integral component of our planet.

COLONIZING EVERY NICHE AND MAKING EARTH INHABITABLE

Where there is life, there are microbes. No niche on Earth that supports life is without microbes. This includes environments considered exceedingly harsh. Credibility is strained here: microbes not only survive but actually grow under extreme conditions, such as temperatures above that of boiling water, pH 1.0, pressures of thousands of atmospheres, and the salinity of concentrated brine (Table 1.4). The current world champion among **thermophiles** (as organisms that grow only at high temperatures are called) grows even at 121°C, a temperature attainable only at high pressures, such as those found at the bottom of the ocean. To grow such organisms in the laboratory requires a pressure cooker or an autoclave. This might suggest that autoclaves may be ineffective to sterilize microbiological media because they do not kill these heat-loving organisms. However, most microbiological work is done at much lower temperatures. In the unlikely event that such extreme thermophiles were present, they would not grow at the temperatures used for usual experiments and

Table 1.4 The known extremes of life

Some microbes *survive*:
5 megarads of gamma radiation (ca. 10,000 times what would kill a human)
Very high pressures (ca. 8,000 atmospheres, or 117,000 pounds per square inch
Some microbes *grow at*:
Extremes of pH (0 to 11.4)
Extreme temperatures (−15 to 121°C)
High hydrostatic pressures (ca. 1,300 atmospheres, or 18,500 pounds per square inch)
High osmotic pressures (5.2 M NaCl)

therefore would not be perceived. Equally improbably, a bottle of 1.0 normal (N) hydrochloric acid left on a shelf long enough may develop some fuzzy balls—the growth of molds in the liquid. Clearly, we have an anthropocentric view of life: we think that these organisms are bizarre, and we call them **extremophiles**. However, from the perspective of these organisms, pH 7, a temperature of 37°C, and low salinity are not simply uncomfortable but are actually incompatible with life. The notion of what is extreme, like beauty, is in the eye of the beholder.

Being able to grow at high temperatures implies that the biological molecules of a thermophile must be endowed with greater stability than was usually thought possible. Whatever the mechanism, thermal stability has uses in research and industry. For example, some laundry detergents, which are meant to work at a high temperature, contain heat-stable hydrolases obtained from thermophilic bacteria. Another example is the widely used **polymerase chain reaction** (**PCR;** the same method used for forensic testing of DNA), which depends on alternating high and low temperatures. The heat-resistant enzyme originally used was derived from a thermophilic organism growing in a hot spring in Yellowstone National Park.

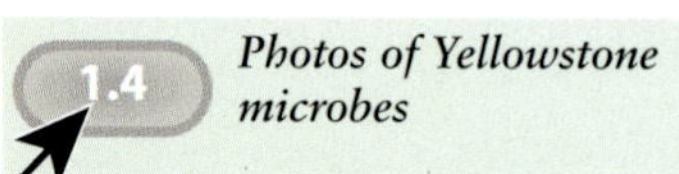

Photos of Yellowstone microbes

Not only do microbes have the ability to live under harsh physical conditions, they can also obtain metabolic energy in unexpected ways. Many of the microbes on this planet are photosynthetic; in fact, they invented photosynthesis. Others have evolved ways to derive energy not from light but from chemical bonds in inorganic compounds. Thus, microbes oxidize molecular hydrogen, reduced iron, sulfides, and other intractable-sounding but energy-rich compounds. These organisms derive their cell carbon from CO_2 in much the same way that CO_2 is "fixed" in photosynthesis. However, they do not rely on the energy of the sun. Such metabolic skills permit life in dark places that lack a supply of organic nutrients. Indeed, such bacteria are the life-sustaining force in deep-sea vents in the ocean and in cracks and fissures in rocks. So large are these communities that they are thought to outweigh all the oxygen-using organisms above ground. Thus, microbes occupy more than just the Earth's thin surface layers. The extraordinary metabolic and ecological repertoire of such microbes has led to speculation that analogous life forms may even exist in other places in our solar system, such as Mars and one of Jupiter's moons, Europa.

The abundance of microbes on Earth has direct effects on the physical and chemical qualities of the environment. By their intense biochemical activities, microbes carry out major changes in the biosphere. To sustain life, such essential elements as carbon, oxygen, and nitrogen must be recycled. The oxygen of our atmosphere, for example, is constantly replenished by the activity of living things and was originally produced by photosynthetic bacteria. To this day, at least half of the oxygen in the air is the result of microbial photosynthesis; the rest originates from plant photosynthesis. Were it not for the breaking down of organic material by microbes, carbon would accumulate in dead plant matter and not be available in sufficient quantities as carbon dioxide for plant growth. Besides chemically synthesized nitrates used for fertilizers, prokaryotic microbes are the only significant source of utilizable nitrogen, essential for all of life. It can be estimated that if this microbial activity were to cease abruptly, plants would run out of usable nitrogen in about 1 week. Sometimes, not enough attention is paid to such microbial activities. When scientists sought to build a closed ecosystem in Arizona (Biosphere 2), an acceptable self-contained environment could not be sustained, in good part because the behavior of microbes in the soil was not sufficiently taken into account. As a result of microbial respiration, the oxygen decreased to levels lower than expected. On a small scale, this illustrates the impact of microbes on our environment and their essential role in making Earth habitable.

Different microbes utilize a vast repertoire of organic substances for nutrition and growth. It has been said that for nearly any naturally occurring organic compound, there is at least one kind of microbe that can break it down, as long as water is available. This is particularly useful for the purpose of cleaning up environmental pollutants. The technique of fostering the activity of microbes in contaminated sites is known as **bioremediation** (see Chapter 23). It was used successfully in the cleanup of the 1989 Alaska oil spill from the tanker *Exxon Valdez* and has continued to find other important applications to the present day.

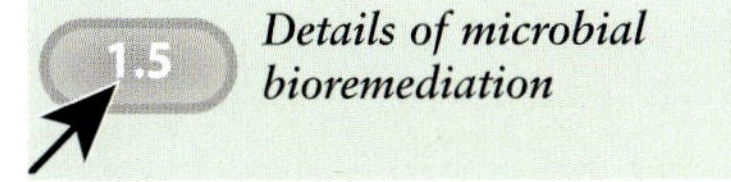

Details of microbial bioremediation

The presence of microbes in so many environments makes it difficult for life to originate anew at the present time. The chemicals needed to make a cell would not accumulate long enough for a cell to be assembled. The microbes in the environment would simply eat them up.

Some microbes are "picky eaters" and utilize only a narrow spectrum of compounds as sources of carbon and energy. Many of the bacteria associated with the human body fall into this category, possibly because they have adapted to a rich and relatively constant environment. Other microbes have the ability to utilize an amazing collection of compounds. Certain species of bacteria called *Pseudomonas* can metabolize hundreds of organic compounds. Not surprisingly, these are species found in a variety of natural environments. *Pseudomonas* species are abundant in soil, plant roots, bodies of water, and animals. They also have lots of genes and, accordingly, have some of the largest microbial genomes.

SHAPING THE PLANET

Taking into account the metabolic versatility of microbes, it should not require a leap of faith to realize that they play a major role in the geology and meteorology of this planet. For example, many large caverns owe their existence to microbes that oxidize hydrogen sulfide to produce sulfuric acid. Deposits of calcium carbonate are converted by microbial sources of sulfuric acid into the more soluble calcium sulfate, which is dissolved away by water streams. This process is relatively slow, but the activity of the bacteria is steady and unremitting. Given time, these microbial miners excavate giant holes in the ground.

Microbes also make rocks. Limestone accounts for about 10% of all rocks on Earth and is largely made by microbial members of the marine plankton called coccolithophores. Each year, these organisms make 1.5 million tons of calcite, the main constituent of limestone. Analogously, diatoms account for the huge deposits of silica-containing rocks. Microbes are also involved in the weathering of rocks, the process whereby rocks wear and break apart. The surfaces of rocks were commonly thought to harbor a few microbes, but microbes are present in large amounts in the crevices of rocks going down several kilometers from the surface, biomass that is thought to exceed that on the surface of the earth.

The geology of the planet is, of course, greatly influenced by the weather and the cycling of the major gases in the atmosphere, nitrogen and oxygen. This cycling is due in large extent to microbial action. Free molecular oxygen first appeared in the atmosphere around 2.5 billion years ago, and Bacteria account for a large proportion of the oxygen produced by photosynthesis. If eukaryotic members of the microbial world, such as the diatoms, are included, the preponderance of photosynthesis is indeed microbial in origin. The main greenhouse gas, carbon dioxide, is both produced and consumed by microbes.

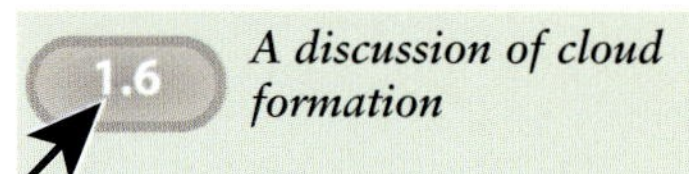

A discussion of cloud formation

Even cloud formation can have microbial origins. Marine algae and bacteria combine to produce a volatile compound, dimethyl sulfide, to the tune of some 50 million tons annually. Dimethyl sulfide turns into sulfate in the atmosphere, and sulfate acts as a nucleating agent for water vapor to become water droplets. The result is clouds. Note the feedback loop: the greater the cloud cover, the less light there is for photosynthesis, and the production of cloud-forming sulfur compounds declines. In time, this leads to fewer clouds, and the cycle begins again. Microbes indeed shape this planet.

MAKING A LIVING

All free-living organisms have constant needs that fall into three general categories: **nutrition** (intermittent availability of food), **occupancy** (the need to remain in a certain habitat), and **resistance** to damaging agents.

The example of the microbes in the human gut is illuminating. In most people, the relative number of different kinds of bacteria is more or less stable, although not every human has the same species ("to each his own"). Equally surprising is the fact that some species are abundant, up to 10 billion cells per gram, while others are perhaps 1 million-fold less frequent. Why don't the dominant species completely outgrow the others? The answer seems to be different niches. On a macro scale, the human gut is a tube that has at least two distinct compartments, the **lumen** (the central space within the tube) and the epithelial layer that acts as a wall. These regions are distinct because the liquid is not evenly stirred near the lining and its surface is coated by viscous mucus.

Each compartment presents a distinct challenge to microbes. In the lumen, species that do not thrive will be outgrown by those that do. Here, the premium is on rapid utilization of the available nutrients and therefore on rapid growth. Additionally, this is a moving environment, and bacteria are constantly being flushed away. Survival in the lumen depends not only on making large numbers of cells but also on the ability to respond rapidly to a changing environment. Although food is abundant in the intestinal lumen, a bacterium must compete with other species and also with its own kin. Thus, microbes must be able to make efficient use of their resources and, of equal importance, to adapt rapidly to changing nutritional conditions, a life of feast and famine. Such organisms must strike a balance between being efficient and being adaptable.

Other bacteria have evolved a different strategy: they adhere to the intestinal surface. These will not be readily washed away and remain in place even if they do not grow very fast. Here, the premium is on occupancy, not just on rapid growth. Surface occupancy is a big issue in the microbial world. Every particle of soil, every rock in the seas, or the epithelial surfaces of animals are at one time or another colonized by microbes. Only rarely does one encounter surfaces where there is no resident population. An example is the lower reaches of our respiratory tract. Here, powerful host defense mechanisms prevent bacteria from sticking. The epithelium of our respiratory tree is covered with an **elevator**, a set of cilia that are constantly beating upward, removing particles from the surface and carrying them toward the outside of the body. When this ciliary epithelium is impaired, as in certain viral infections, bacterial bronchitis or pneumonia almost always follows. As may be expected, microbes, including many pathogens, have evolved ways to attach to surfaces. Many make sticky substances that help them adhere; others have special appendages that recognize specific sites on the surfaces of cells. These microbial molecular devices are called **adhesins**. Colonization, therefore, is a critical topic in clinical and environmental microbiology.

COOPERATING FOR COMPLEX ENDEAVORS

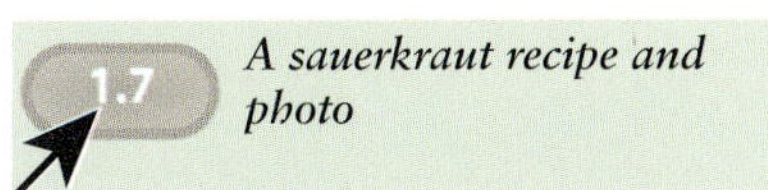

A sauerkraut recipe and photo

Consider sauerkraut. It may not be everyone's favorite food, but it illustrates a linked chain of microbial events. Sauerkraut is the result of a microbial fermentation of cabbage that takes place in a highly reproducible manner. All it takes is to dice the cabbage, add salt, shield it from the air, and presto, a few weeks later, the final product is ready for consumption. It occurs every time with nearly unfailing precision, yet it requires the activities of dozens of different microbes working in an assembly line (an ecological succession). This is what happens. The addition of salt causes the cabbage cells to plasmolyze and release a large amount of soluble substances—sugars, amino acids, and vitamins—that serve as nutrients for bacteria and yeasts. Some of these microbes make enzymes that soften the cabbage and change its texture, and other enzymes convert sugars into organic acids, mainly lactic, acetic, and propionic. At the end of this first fermentation, other bacteria take over and convert remaining substrates into the compounds that give sauerkraut its characteristic aroma. This second fermentation imparts the flavor to the finished product. By this time, the pH has dropped precipitously to about 3.5, but this is not low enough to inhibit other bacteria. A third fermentation is then carried out by another bacterial population that is quite acid resistant, resulting in the formation of even more lactic acid, with the pH dropping to about 2.0. Now the sauerkraut has the desired flavor, and the low pH stops further microbial growth (as long as air is excluded). It can be stored for a long time. Three distinct bacterial populations have been at work, the first to prepare the food for the next, the second to impart the flavor, and the third to make it into a preserved food.

What have we learned? The cabbage itself must have been the source of many of the organisms involved, suggesting that even fresh and unblemished cabbage carries diverse microbes. Some microbes may have come from the hands of the handlers, the vessel used, or other environmental sites, but the reproducibility of sauerkraut making suggests that the main source of the **inoculum** (the initial cells) is the cabbage itself. As the environment changes in the various steps, new microbes proliferate to occupy what becomes a new ecological niche. We learned that even where the microbial biota is highly complex, only some of its members may be active at any particular time. Later, when the environment changes, others become active.

The linking of microbial activities is not uncommon. For example, mammals do not produce enzymes that break down cellulose. Instead, cattle, sheep, deer, and other ruminants harbor cellulose-digesting microbes in a large fermentation chamber called the rumen. The animal provides the microbes with a suitable habitat for growth; the microbes reciprocate by breaking down the cellulose into volatile fatty acids, which are used by the animal as a major energy source. Such interactions are part of the grand recycling system of all organic material on Earth. No animal or plant ever decomposes through the activities of a single microbial species. Food chains and interactive populations are the norm.

In nature, microbes live in communities that may consist of hundreds of different species. How they coexist is difficult to determine. We know that interactions between species include food competition, food chains, predation, and the excretion of antimicrobial metabolites and antibiotics. At first glance, it may appear that the coexistence of hundred of species is too complex for us to comprehend. However, trying to understand it is not much different from asking how hundreds of different plant species coexist in a given plot of a tropical forest. In neither case do we have simple answers to important questions.

In most natural habitats, bacteria reside on surfaces in complex structures called **biofilms**. Examples are corrosive growth on metal pipes, plaque on teeth, or the membranous scum on an old-fashioned pickle barrel. Biofilms are typically several cell layers thick and may be visible to the naked eye. They may contain a single species but are more often made up of several species. Biofilms that are thicker than a certain size have special structural architecture that ensures the exchange of substrates and waste products. Such biofilms are complex pieces of engineering, developing channels to ensure the proximity of fresh liquids. In this manner, biofilm-making bacteria overcome the disadvantage of being in the depth of a segregated structure where they would be limited in their ability to feed and exchange substances with the environment.

CONCLUSIONS

Microbes are small, ubiquitous, abundant, adaptable, and necessary for the continued existence of other life forms. They play a vital role in shaping this planet. Thus, their study is essential for all who seek to understand life on Earth. In the next two chapters, we deal with the body plan of the prokaryotic cell in anticipation of discussing their physiology, genetics, ecology, and evolution.

STUDY QUESTIONS

1. Prepare an outline for a lecture to high school seniors on the importance of microbes on this planet.

2. Prepare an outline for a lecture to high school seniors on the importance of microbes for human beings.

3. Bacteria range in volume over a millionfold. Discuss some of the consequences of being much larger or much smaller than the average.

4. In what ways have Bacteria and Archaea influenced the evolution of eukaryotes in the past? How do they influence it in the present?

section II

structure and function

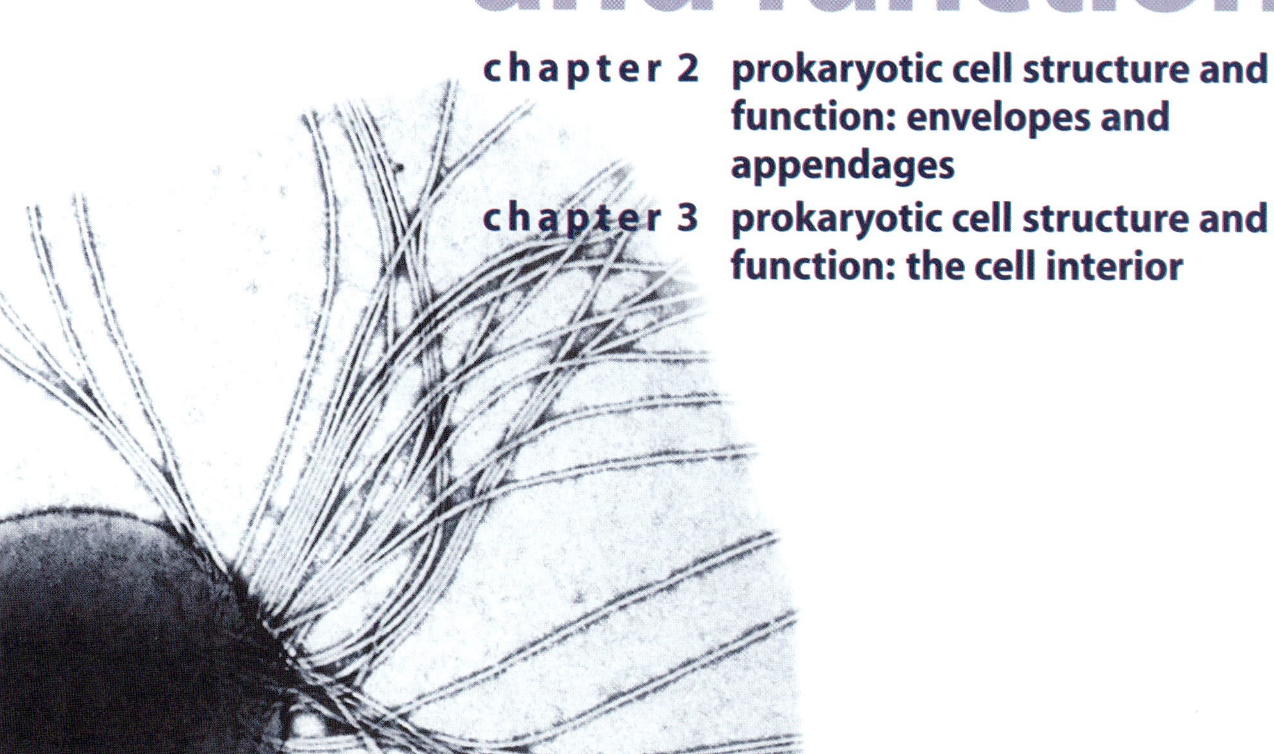

chapter 2 prokaryotic cell structure and function: envelopes and appendages

PROKARYOTIC CELLS

The microbial world spans the largest cleft in cell types within the living world: that between prokaryotes and eukaryotes. Prokaryotic cells lack nuclei, mitochondria, and chloroplasts (Fig. 2.1). They do not carry out **endocytosis** or **pinocytosis**; that is, they are incapable of ingesting particles or liquid droplets. Prokaryotes and eukaryotes also differ in important biochemical details, such as the structures of their ribosomes and the compositions of their lipids. Furthermore, most prokaryotes are haploid and usually have a single kind of circular **chromosome** accompanied by extrachromosomal elements, called **plasmids**; most eukaryotes have a diploid phase and several different linear chromosomes. This chapter and the next deal with prokaryotic cell structures only. The cell structures of selected eukaryotic microbes are discussed in Chapter 16.

The cells of Bacteria and Archaea may be small, but they are quite complex in organization. Prokaryotic cells are not just tiny nondescript balls or sausages; rather, some of their proteins and DNA are located at specific sites in the cell. In addition, they make use of elaborate structures for their morphogenesis and cell division, including proteins equivalent to those that are involved in the eukaryotic cytoskeleton. A number of proteins, including some involved in chemotaxis (the movement of bacteria in response to chemical stimuli), are located at only one of the poles (see Chapter 13). It can thus be said that bacteria have a "nose" that senses attractants and repellents. Examples that illustrate the sophisticated anatomy of the prokaryotic cells recur and will be found throughout this book.

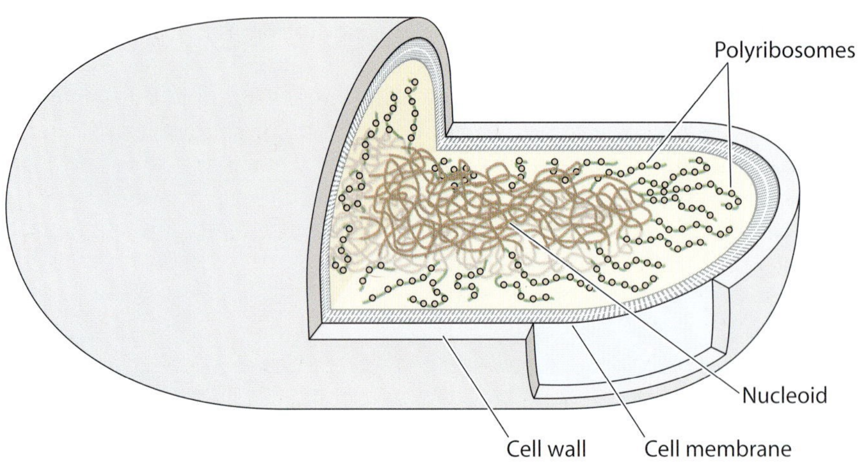

Figure 2.1 Ultrastructure of a characteristic prokaryotic cell. Inside the cell membrane, the DNA is wrapped into a mass called the nucleoid. Several ribosomes are attached to a single messenger RNA (mRNA) molecule (making polyribosomes). Note that these form while mRNA molecules are being transcribed. Outside the cell membrane is a cell wall and, typically, a layer called the extracellular matrix (not shown).

MICROSCOPES

The importance that microbiologists place on visualizing microbes has waxed and waned over the years. At its beginning, microbiology was strictly an observational science, and microscopists led the way. Antonie van Leeuwenhoek—possibly the first microbiologist—observed the presence of *Giardia* cells in his stools during a bout of an intestinal illness. This finding depended entirely on microscopic observation.

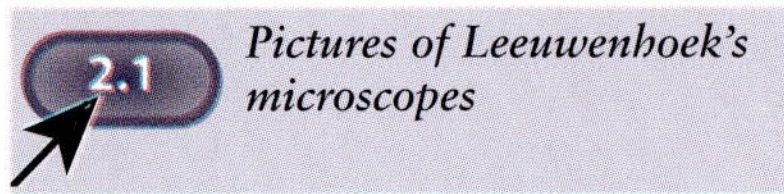

Pictures of Leeuwenhoek's microscopes

Throughout the "golden age of medical microbiology" in the late 19th century, the light microscope remained the microbiologist's most valuable tool. By then, it had become possible to overcome the inherent difficulty of observing prokaryotic cells: they lack color and therefore do not generate sufficient contrast to be seen clearly by light microscopy. Dyes came to be used singly and in combination to stain and thereby increase the contrast of prokaryotic cells, making them clearly visible. Stains that exploited unique properties of microbial cell surfaces opened the way to major achievements. Robert Koch, a 19th-century German microbiologist, discovered the cause of tuberculosis by using a "differential" stain that enabled him to see tubercle bacilli within host tissues. Christian Gram developed a differential staining procedure that allowed him to sort bacteria into two groups (gram-positive and gram-negative bacteria [see below]).

Prokaryotic cells approach the limit of resolution of light microscopes. Under optimal conditions, a light microscope has a **resolving power** (the minimum distance that permits two points to be perceived as separate) of about 0.22 micrometers (also called microns) (μm), i.e., one-third the width of a typical prokaryotic cell. Therefore, only the largest intracellular structures of prokaryotes are revealed by light microscopy. It is not surprising that the most fundamental and obvious difference in cell types—the eukaryotic-prokaryotic distinction—was discovered only after suitable electron microscopes became available in the late 1930s. The superior resolving power of electron microscopes reveals internal structures not seen with an optical microscope.

The development of the **phase-contrast microscope** and its cousin, the **Nomarski differential interference contrast microscope**, made it possible to visualize unstained living cells, active and undistorted by fixation. By generating interference between light passing through and around cells, these microscopes produce contrast by exploiting the high refractive index of microbial cells.

In time, powerful tools (including heavy-metal staining, shadow casting, freeze-fracture, and freeze-etching) were developed to study the internal structure of prokaryotic cells by using the electron microscope. Light microscopy has also benefited greatly from the development of novel fluorescence-based techniques. At present, microscopy has again become an integral part of microbiological research.

Newer microscopes

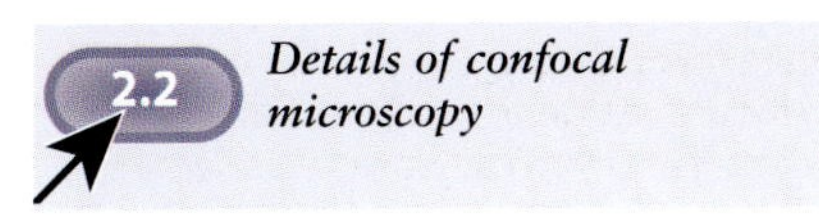

The light microscope does not give a three-dimensional (3-D) picture of an object. Very sharp 3-D images of microscopic objects can be obtained with two other kinds of microscopes, the **confocal microscope** and the **deconvolution microscope.** Both of them create "optical slices" by focusing on different layers of an object. These slices can be assembled to produce a 3-D image of the specimen. In general terms, this resembles the way radiologists use computed-tomography (CT) scans to obtain 3-D images of organs of patients. The two microscopes differ in how they eliminate the blur from other planes of focus. The confocal microscope uses a system of pinholes placed before a collection aperture, which block out all light from the sample except for the light originating directly from the plane of focus. In **deconvolution microscopy,** a software-based process is used for this purpose. Here, optical slices are collected through an object (as in confocal microscopy). However, instead of eliminating out-of-focus light by mechanical means, it is removed from each focal plane by a computer that uses so-called nearest-neighbor algorithms. For an example, see Fig. 2.6.

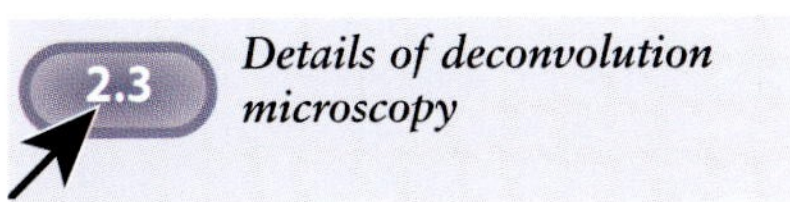

Scanned-proximity probe microscopes

Scanned-proximity probe microscopes can resolve and visualize single atoms and molecules without using optical means. Unlike traditional microscopes, they do not use lenses but sense the height of an object by scanning it with a very thin probe. Thus, they are not limited by the laws of diffraction, and their limit of resolution is set by the size of the probe. The probe can be sharpened to a point as small as a single atom and held very close to the surface of the specimen, at the distance of about one atom's diameter. The measurement is then converted into an electrical signal that generates an image. The most commonly used kinds of scanned-proximity probe microscopes are the **scanning tunneling microscope** and the **atomic force microscope.**

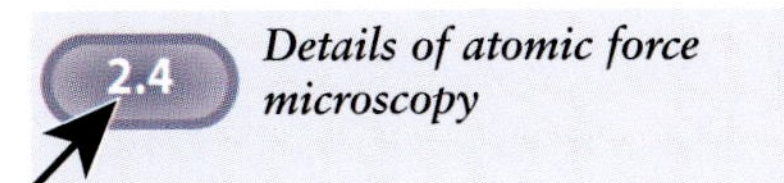

Molecular stains

Traditional stains are used to increase contrast and also to identify or highlight certain parts of cells, e.g., spores or flagella. Molecular stains are more discriminating and can be used to visualize specific types of molecules.

Fluorochromes. Specific cells or cellular components can be localized by the use of selective antibodies or DNA probes that have been modified to carry a fluorescent chemical called a **fluorochrome.** Such tagged components then light up under a fluorescence microscope to reveal their locations in cells or tissues.

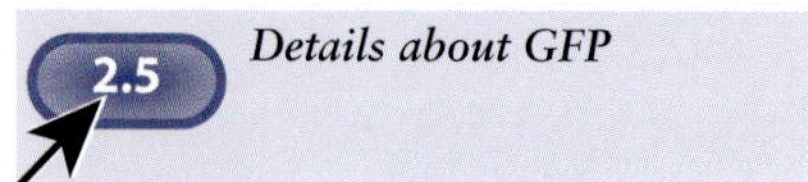

Fluorescent proteins. The location of a protein can also be determined in living cells by fusing it to a **fluorescent protein.** Such an artificial protein can be made by genetically fusing the **gene** for the protein being examined to a gene encoding a fluorescent protein. The original gene for such proteins was derived from a jellyfish, and its product is known as the **green fluorescent protein (GFP).** Several modified forms of the protein, e.g., the **yellow fluorescent protein (YFP)** and the **cyan fluorescent protein (CFP;** cyan is a greenish-blue color), have some advantages over GFP and are increasingly being used. An important advantage of using fluorescent-protein fusions is that they can be used with living cells; thus, the fate of a given protein can be examined over time in the same cell.

PROKARYOTES HAVE COMPLEX ENVELOPES AND APPENDAGES

Microbial anatomy tells us a great deal about how microbes cope with environmental challenges. Free-living, small unicellular organisms such as bacteria confront nutritionally varied conditions, withstand perilous physical and chemical challenges, and, commonly, must have the ability to attach to surfaces. The solutions to such challenges are clearly revealed in the body plans of the Bacteria and Archaea. Nowhere is this better illustrated than in the *composition and structure of surface layers*, which is where microbes contact and communicate with their inanimate environment and with other cells, including those of a host they might have invaded. Surface components often determine whether an organism can survive in a particular environment.

Microbial cells are surrounded by **complex envelope layers** and **appendages** that differ in composition among the major groups and, more subtly, even between strains of individual species. These structures protect the organisms from hostile environments, such as extreme osmolarity, harsh chemicals, and even antibiotics. We will first describe the structural properties of bacterial structures and later contrast them with those of the Archaea.

THE CELL MEMBRANE

Like all other cells, the Bacteria possess a **cell membrane** consisting of the usual lipid bilayer. We will discuss below the substantially different way that the Archaea have of making their membranes. *In either case, the cell membrane is the true boundary between the cell interior and the environment.* Typically, for bacteria, the cell membrane consists of phospholipids and upward of 200 different kinds of proteins. The latter account for approximately 70% of the mass of the membrane, a considerably higher proportion than in mammalian cell membranes.

The cell membranes of bacteria are busy places. They assume functions that in eukaryotic cells are distributed among the **plasma membrane** and the membranes of intracellular organelles (Table 2.1). Its most critical function is the uptake of **substrates** from the medium. Basically, the cell membrane is an *osmotic and solute barrier containing specific transport systems and electron transport chains.* The cell membrane contains

Table 2.1 Some functions of prokaryotic and eukaryotic cell membranes and organelles

Function	Structure: Prokaryotes	Structure: Eukaryotes[a]
Osmotic barrier	Cytoplasmic membrane	Cytoplasmic membrane
Transport of solutes	Cytoplasmic membrane	Mainly cytoplasmic membrane
Respiratory electron transport	Cytoplasmic membrane	Mitochondrial membrane
Protein synthesis	Polyribosomes in cytoplasm	Polyribosomes on endoplasmic reticulum
Synthesis of lipids	Cytoplasmic membrane	Smooth endoplasmic reticulum, Golgi apparatus
Synthesis of wall polymers	Cytoplasmic membrane	Golgi apparatus in plants
Protein secretion	Cytoplasmic membrane	Endoplasmic reticulum and secretory vesicles
Photosynthesis	Various membrane types, some continuous with, others independent of the cytoplasmic membrane	Chloroplast membranes

[a] Main sites only.

specific proteins that bring about in a great number of ways the entry of most metabolites. Many of these **transport mechanisms** can concentrate their specific substrates within the cell to be as much as 10^5 times higher than in the surrounding medium (this is discussed in detail in Chapter 6).

HOW THE CELL MEMBRANE IS PROTECTED

Like all biological membranes, the bacterial cell membrane can be disrupted by detergents and other **amphipathic** compounds (those containing both polar and nonpolar domains). Also, the membrane must be mechanically stabilized to withstand high intracellular osmotic pressures. Most bacteria solve these problems by surrounding the membrane with a tough saclike structure. There are several variations on this theme, generating four architectural styles: **gram positive, gram negative** (Fig. 2.2), **acid-fast,** and that of mycoplasmas.

The Gram stain (named after the early Danish microbiologist who devised it) divides most bacteria into two groups, gram positive and gram negative. The gram-negative species are more numerous than the gram-positive species. However, many human- and animal-pathogenic microbes are gram positive. The Gram stain depends on the ability of certain bacteria (the gram-positive bacteria) to retain a complex of a purple dye and iodine after a brief alcohol wash (Table 2.2). Gram-negative bacteria do not retain the dye and become translucent, but they can then be counterstained with a red dye. So, after being stained, gram-positive bacteria look purple under the microscope, and gram-negative bacteria look red. The distinction between these two groups turns out to reflect fundamental differences in their cellular envelopes.

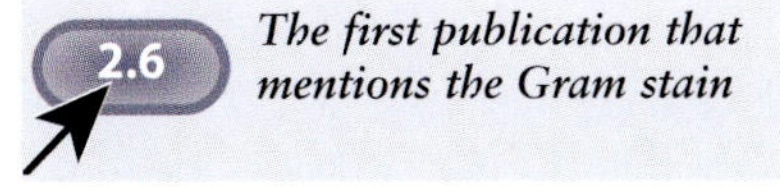

The first publication that mentions the Gram stain

Table 2.2 The Gram stain

1. Stain with crystal violet (purple); wash.
2. Mordant (bind) the dye with potassium iodide; wash.
3. Flush with alcohol; wash. Gram-negative bacteria are decolorized; gram-positive bacteria remain purple.
4. Counterstain with safranin (red); wash. Gram-negative bacteria become red; gram-positive bacteria remain purple.

The gram-positive solution

Gram-positive bacteria, e.g., staph (*Staphylococcus*) and strep (*Streptococcus*), have a thick **cell wall** that protects the membrane from the high internal turgor pressure. The major constituent of the wall is a

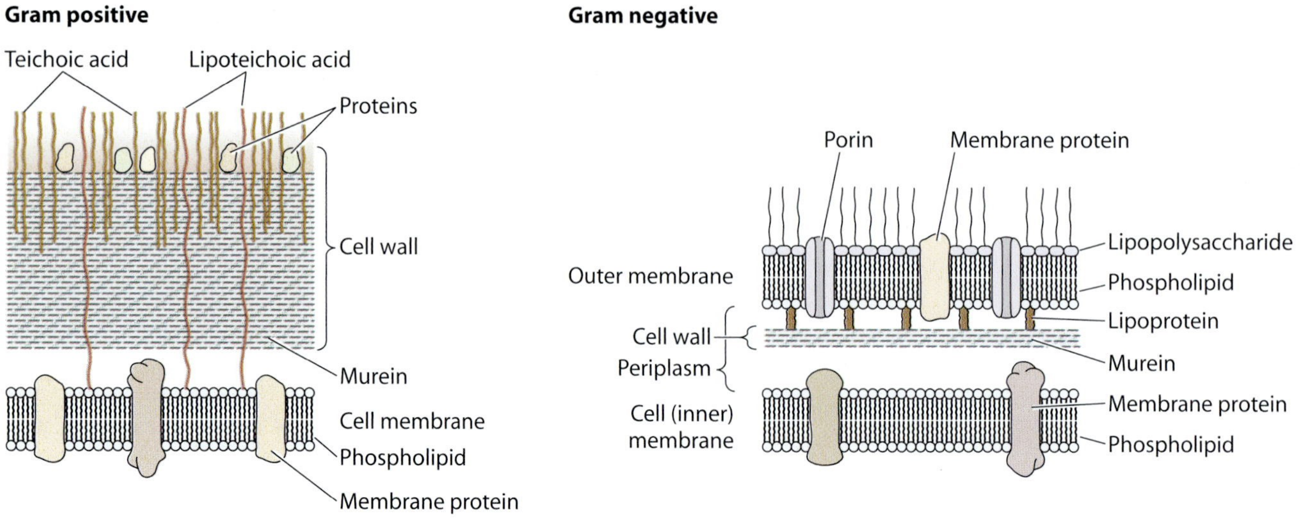

Figure 2.2 Envelopes of gram-positive and gram-negative bacteria. Gram-positive bacteria have a thick cell wall made up of many layers of murein. Several constituents (teichoic acids, lipoteichoic acids, and protein) protrude from this layer. Gram-negative bacteria have a thin cell wall and an outer membrane. The outer leaflet of the outer membrane is made up of LPS. The space between the two membranes is called the periplasm.

peptidoglycan, a complex polymer of sugars and amino acids. The particular peptidoglycan found in bacteria is called **murein** (Fig. 2.3). Besides protecting the membrane, murein is also responsible for *shape* and *rigidity.*

Murein is a tough polymer that is *unique to bacteria.* This fact alone should evoke the thought that it might be a good target for antibiotics. Indeed, this is the case: many antibiotics, e.g., the entire class of β-lactams, which includes penicillin, work by inhibiting murein synthesis. Murein is a fabric of long strands cross-linked to one another (Fig. 2.3), resembling chain mail (the armor used in the Middle Ages). The strands are composed of glycan (sugar) chains linked by short peptides. The overall structure of murein is similar in all species, but it differs in chemical details (Fig. 2.4). This two-dimensional (2-D) polymeric fabric is wrapped around the length and girth of the bacterium to form an intact sac (called a **sacculus**) that defines its size and shape. Depending on the shape of the sacculus, bacteria may have the appearance of a **bacillus** (rod; plural, bacilli), **coccus** (sphere; plural, cocci), or **spirillum** (spring, or helix; plural, spirilla), or a number of other, less frequently encountered shapes, including triangles and squares. When emptied, the murein sac retains the size and shape of the cell from which it is derived (Fig. 2.5). The arrangement of the various layers in a gram-positive bacterium is shown in Fig. 2.2.

Murein is not the only determinant of bacterial cell shape. Many rod-shaped bacteria contain an actin-like protein that forms fibrous spirals under the cell membrane (Fig. 2.6). **Mutant** strains lacking this protein become spherical. Interestingly, this protein is found in most rod-shaped bacteria, but not in cocci. The existence of this and of another protein, FtsZ (a homolog of another cytoskeletal protein, tubulin), suggests that the eukaryotic cytoskeleton may have a bacterial ancestry. Bacterial and eukaryotic actin molecules differ greatly in amino acid composition (with

Figure 2.3 Structure of murein. Murein consists of glycan chains of alternating units of *N*-acetylglucosamine and *N*-acetylmuramic acid to which a short peptide is connected. Some of these peptides are cross-linked to the peptides of another glycan chain, making a 2-D network. Murein has several unusual components, including D-amino acids and diaminopimelic acid.

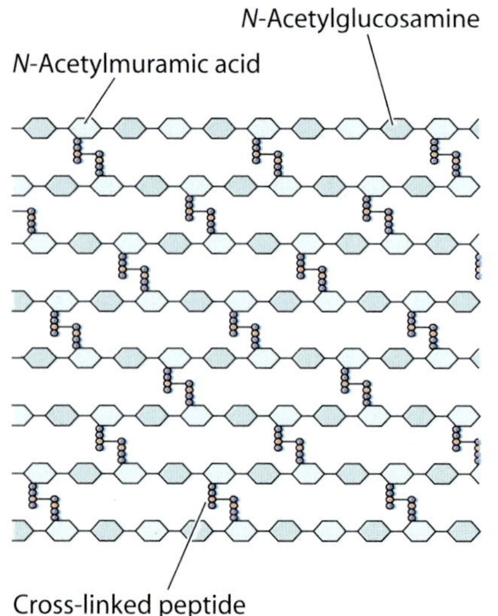

Gram positive

[—GlcNAc — MurNAc —]$_n$
L-Ala
D-Glu
L-Lys — $(Gly)_5$ — D-Ala
D-Ala L-Lys
D-Glu
L-Ala
[—GlcNAc — MurNAc —]$_n$

Gram negative

[—GlcNAc — MurNAc —]$_n$
L-Ala
D-Glu
DAP —— D-Ala
D-Ala DAP
D-Glu
L-Ala
[—GlcNAc — MurNAc —]$_n$

Figure 2.4 Murein composition. Typically, in the murein of gram-positive bacteria, peptide cross-links are between a lysine and a D-alanine residue. In gram-negative bacteria, the cross-links are between diaminopimelic acid (DAP) and D-alanine.

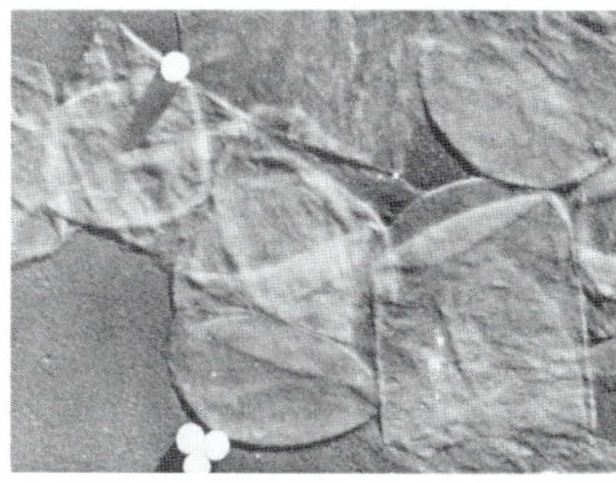

Figure 2.5 Murein sacculi. The electron micrograph shows flattened murein sacculi from *E. coli.* In 3-D, the structure would have the same dimensions as the cell from which it is derived. The white spheres are latex beads 0.25 μm in diameter, included to show the scale.

sequence identity of only about 15%) but have nearly identical 3-D structures, suggesting that this is what has been worth conserving.

Examples of bacterial proteins that have eukaryotic cytoskeleton counterparts are listed in Table 2.3.

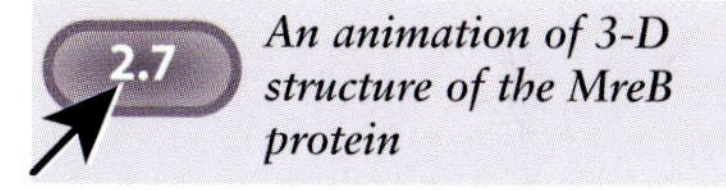

2.7
An animation of 3-D structure of the MreB protein

How does the cell wall contribute to the defense of the cell membrane? The rigid murein sacculus allows bacteria to survive in media with osmotic pressure less than that of their cytoplasm. Without a rigid corset-like structure to push against, the membrane would burst and the cells would lyse. A simple experiment demonstrates both the protective and shape-determining roles of murein. The murein layer can be hydrolyzed by **lysozyme,** an enzyme present in many mammalian tissues and secretions, including tears. If the environmental osmotic pressure is low, the cells will lyse. If it is high, lysozyme-treated bacteria do not lyse but change their shape. If the original cells were rod shaped, they will become spherical (Fig. 2.7). Such structures are commonly called **spheroplasts** when they are found in gram-negative bacteria or **protoplasts** when they are found in gram-positive bacteria.

The cell wall of gram-positive bacteria, consisting of several layers of murein, forms a thick barrier that hinders the passage of hydrophobic compounds because its phosphates, sugars, and charged amino acids are highly polar. Thus, the cells are surrounded with a thick *hydrophilic layer*. Consequently, gram-positive bacteria can generally withstand noxious hydrophobic compounds, including the bile salts found in our intestines.

Table 2.3 Bacterial cytoskeletal proteins

Name	Known function	Eukaryotic counterpart
FtsZ	Division septum formation	Tubulin
MreB	Cell shape in rods	F-actin
ParM	Plasmid segregation	F-actin
Crescentin	Cell curvature	Intermediate filaments

Figure 2.6 Helical structure made by protein MreB. The gene for the actin-like MreB protein in *E. coli* was fused to that for a fluorescent protein. The protein could then be localized within the cells under a fluorescence microscope. The images were sharpened by a process called deconvolution microscopy (see the text). Image 1 was taken before deconvolution, and image 2 was taken after deconvolution. Images 3 and 4 are turned 180° with respect to each other. Image 5 shows the outline of the cell.

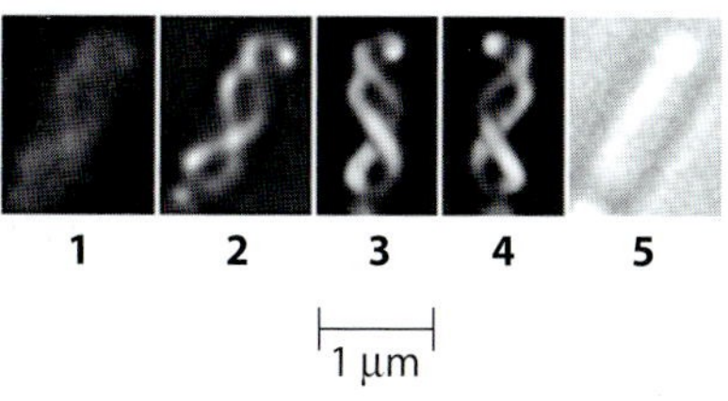

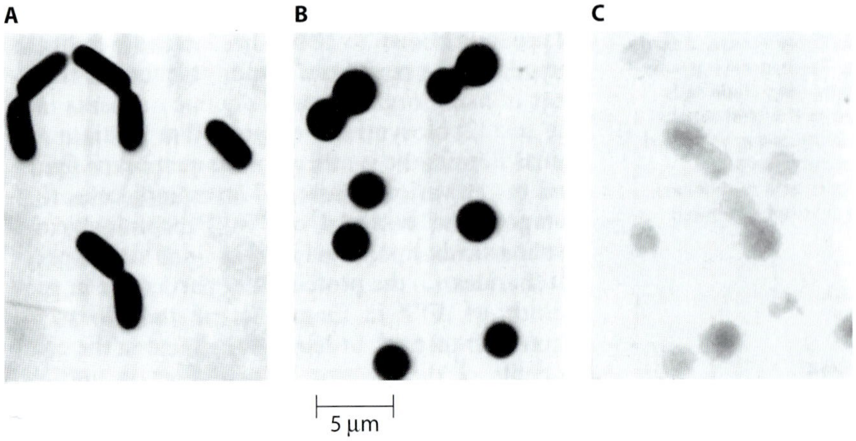

Figure 2.7 Protoplasts of a member of the genus *Bacillus* (*Bacillus megaterium*). **(A)** Intact cells. **(B)** Protoplasts formed by the action of lysozyme in the presence of 0.5 M sucrose. **(C)** Ghosts (i.e., empty membranes) formed after osmotic lysis of protoplasts in a hypotonic medium.

Gram-positive walls contain other unique polymers, **teichoic acids,** which are chains of a sugar alcohol, either ribitol or glycerol, linked by **phosphodiester bonds** (Fig. 2.8). Teichoic acids play a role in pathogenesis by promoting the adherence of organisms such as streptococci to tissues of the host.

The gram-negative solution

Gram-negative bacteria have evolved a radically different mechanism to protect their cell membranes. Instead of a thick murein cell wall, they have a thin murein layer surrounded by a unique **outer membrane** (Fig. 2.2). The outer membrane is chemically distinct from all other biological membranes and is especially resistant to harmful chemicals. Although it is a bilayered structure, its outer leaflet contains a distinctive component. This is a **lipopolysaccharide (LPS),** a complex molecule that, like murein, is found only in prokaryotic cells. The inner leaflet of the outer membrane, on the other hand, is composed of the customary phospholipids that comprise both layers of the cell membranes of all bacteria and eukaryotes.

LPS consists of three parts (Fig. 2.9):

- A lipid, called **lipid A,** embeds LPS in the outer leaflet of the membrane. Lipid A is an unusual glycolipid composed of a disaccharide with attached short-chain fatty acids and phosphate groups. Lipid A causes fever and shock in vertebrates when released from the outer membrane. For this reason, LPS is also known as **endotoxin.**

Figure 2.8 Teichoic acid structure. The repeating units of two kinds of teichoic acids, ribitol and glycerol, are shown. The lengths of chains are variable among species.

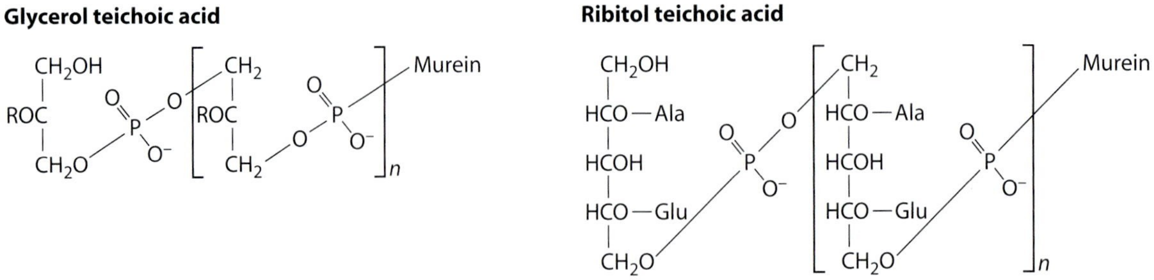

- Attached to lipid A is a short series of sugars, the **core,** whose composition is relatively constant among gram-negative bacteria and which includes two characteristic sugars, keto-deoxyoctanoic acid and a heptose.
- The **O antigen** is a long carbohydrate chain (up to 40 sugar residues in length) bound to the core. The hydrophilic carbohydrate chains of the O antigen swathe the bacterial surface and exclude hydrophobic compounds. The importance of the O-antigen chains is illustrated by mutants deficient in their biosynthesis. Mutants that make little O antigen or merely shortened chains become sensitive to bile salts and antibiotics. As the name implies, the *O antigen is highly immunogenic;* it elicits a strong antibody response when introduced into a vertebrate animal. O antigens are highly varied and differ among species and even in strains within a species (an example is the "O157" carried by the *Escherichia coli* O157:H7 strain, which causes severe infections after ingestion of contaminated hamburger meat). Because of the O-antigens' variability, antibodies against one type will not protect an individual against infection from bacteria containing another kind. In addition, O antigens are toxic, which accounts for some of the **virulence** (disease-causing properties) of certain gram-negative bacteria.

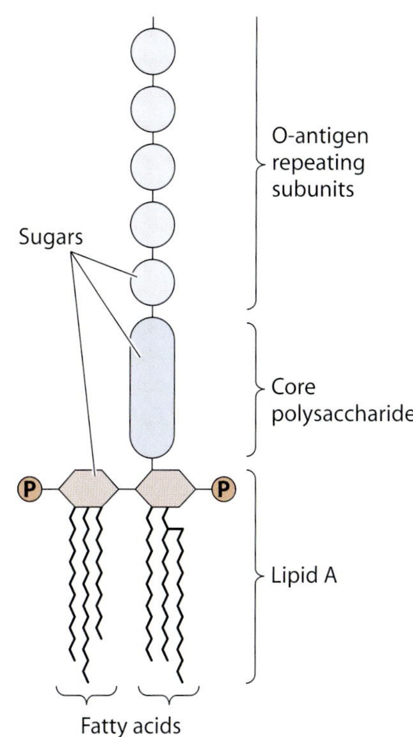

Figure 2.9 LPS structure. LPS consist of three parts: lipid A, a phosphorylated disaccharide to which fatty acids are attached; a core polysaccharide region; and the O antigen, made up of repeating sugars. This region is highly variable among strains and is the main reason for antigenic variety among gram-negative bacteria.

To exclude deleterious *hydrophobic* compounds, gram-negative bacteria, like gram-positive bacteria, rely on the presence of external hydrophilic polysaccharides. Only their locations differ: the murein sacculus in the case of gram-positive bacteria and the outer membrane in the case of gram-negative bacteria. How do essential *hydrophilic* nutrients pass through the two membranes? The inner membrane of gram-negative bacteria, like the cytoplasmic membranes of all cells, is endowed with special transport mechanisms of considerable complexity (see Chapter 6). But what about passage through the outer membrane? To duplicate the transport systems of the inner membrane seems not only wasteful but also counterproductive for the purpose of protecting the inner membrane. In addition, the outer membrane lacks an energy source, as found in the **cytoplasmic membrane.** The gram-negative bacteria have again found an interesting solution to this problem: the outer membrane contains **special channels** that nonspecifically permit the diffusion of many hydrophilic compounds, such as sugars, amino acids, and certain ions. These channels are formed by triads of protein molecules; they are aptly called **porins.** The porin channels are narrow, just large enough to permit the entry of compounds of up to 600 to 700 daltons, about the size of a trisaccharide (Fig. 2.10).

Certain hydrophilic compounds necessary for survival are larger than 600 to 700 daltons. These molecules include vitamin B_{12}, sugars larger than trisaccharides, and iron in the form of chelates. Such compounds cross the outer membrane by separate, specific transport mechanisms that utilize proteins especially designed to translocate each of the compounds. Thus, the outer membrane *allows the passage of small hydrophilic nutrients; excludes hydrophobic compounds, large or small; and allows the entry of certain larger hydrophilic molecules by specific dedicated mechanisms.*

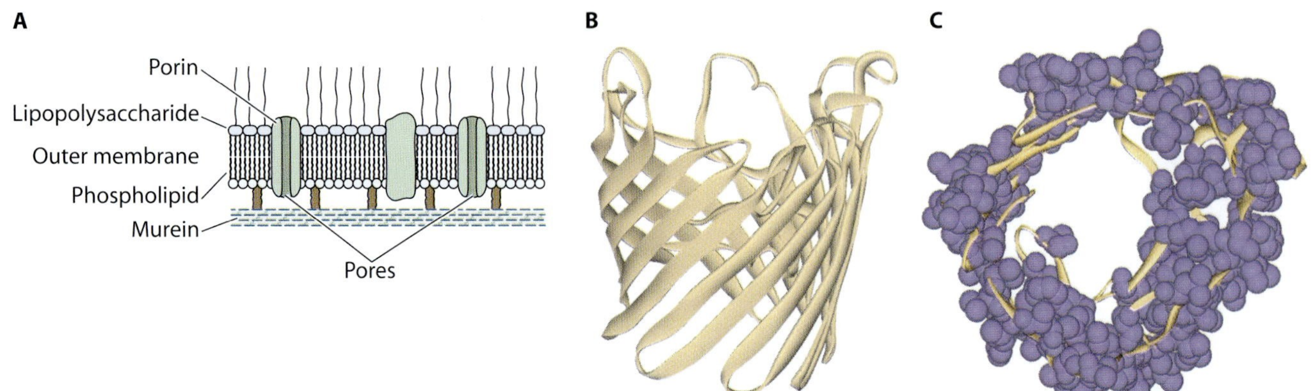

Figure 2.10 Porins. (A) The outer membrane of a gram-negative bacterium shows pores made by porin molecules. **(B)** The 3-D structure of a porin (side view). Porins form a 16-strand β-barrel inserted in the outer membrane. **(C)** Top view of a porin showing the hydrophilic residues making a water-filled channel.

The two membranes of gram-negative bacteria create a compartment between them called the **periplasm** (Fig. 2.2). This compartment contains the murein layer and a gel-like solution of murein precursors and proteins that assist in nutrition. These include degradative enzymes, such as phosphatases, **nucleases,** and **proteases,** that break down molecules to digestible size. In addition, the periplasm contains so-called **binding proteins,** which have high binding affinity for sugars and amino acids and help soak them up from the medium. The periplasm also contains enzymes, called the **β-lactamases,** that inactivate antibiotics, such as penicillins and cephalosporins. The gram-positive bacteria do not have a periplasmic compartment, but they secrete similar enzymes into the environment.

The murein layer is not a separate structure floating in the periplasm but is bound to several proteins in the outer membrane, one of which is a lipoprotein.

The acid-fast solution

Some bacteria, notably the tubercle bacillus (*Mycobacterium tuberculosis*) and its relatives, have developed yet another way to protect their cell membranes from environmental challenges. Their cell walls contain large amounts of **waxes,** complex branched hydrocarbons (60 to 90 carbons long). X-ray diffraction studies revealed that the waxes, known as **mycolic acids,** are arranged in two layers, with their hydrophobic tails directed toward the space between them. These mycolic acid layers are exceptionally thick and are covalently attached to underlying layers of complex sugars and the murein cell wall. Thus, mycolic acids form a highly ordered lipid bilayer membrane, not just a disorganized waxy layer. Proteins are embedded in this layer, where they form water-filled pores through which nutrients and certain drugs can pass slowly, resembling in this regard the porins in the outer membranes of gram-negative bacteria.

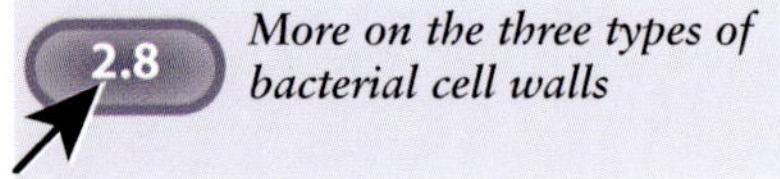

2.8
More on the three types of bacterial cell walls

With such a robust hydrophobic cover, these bacteria are impervious to many harsh chemicals, including disinfectants and strong acids. If a dye is introduced into these cells by brief heating or transitory treatment with detergents, it cannot be removed by dilute hydrochloric acid, as in all other bacteria. These organisms are therefore called **acid-fast,** meaning acid resistant (Table 2.4). The waxy coat is interlaced with murein, polysaccharides, and lipids. These organisms resist not only caustic chemicals but also white blood cells. All this is at a cost: these organisms grow very slowly, probably because the rate of nutrient uptake is limited by their waxy covering. In fact, their permeability for hydrophilic molecules is 100- to 1,000-fold lower than for *E. coli,* a gram-negative bacterium. Some acid-fast bacteria, such as the human tubercle bacillus, divide only once every 24 hours.

Table 2.4 The acid-fast stain

1. Stain with hot fuchsin (red); wash.
2. Decolorize with acid alcohol; wash. Only acid-fast bacteria remain red.
3. Counterstain with methylene blue; wash. All other material becomes blue.

Crystalline surface layers

Some Bacteria and Archaea have **crystalline surface layers,** yet another variation on the theme of the organization of the bacterial envelope. These layers have protein subunits arranged in crystalline arrays that may be square, hexagonal, or oblique (Fig. 2.11). These structures are called **S-layers,** and they are located on the outermost cell surface. S-layers are present on a large number of species that embrace all the major groups of bacteria. In some Archaea, they are the only layer external to the cell membrane. In the gram-positive bacteria, the S-layer is external to the murein wall; in the gram-negative bacteria, it is external to the outer membrane. In both gram-positive and gram-negative organisms, this structure is sometimes several molecules thick.

S-layers are usually made of a single kind of protein molecule, sometimes with carbohydrates attached. The isolated molecules are capable of self-assembly, that is, they make sheets similar or identical to those present on the cells. Think of an S-layer as a kind of lipidless membrane that consists exclusively of proteins or glycoproteins. As can be expected for proteins that are exposed to the environment, S-layer proteins are quite resistant to proteolytic enzymes and protein-denaturing agents.

The function of S-layers is probably protective. In some organisms (e.g., the intestinal pathogen *Campylobacter*), S-layers help prevent phagocytosis. S-layers guard bacteria from infection by bacterial viruses (bacteriophages, or phages) or, paradoxically, they promote infection by other phages by functioning as receptors. S-layers also participate in the adherence of certain bacteria to surfaces.

Biotechnologists foresee several potential uses for S-layers. The periodic structure of the layers can be exploited for creating regular arrays of

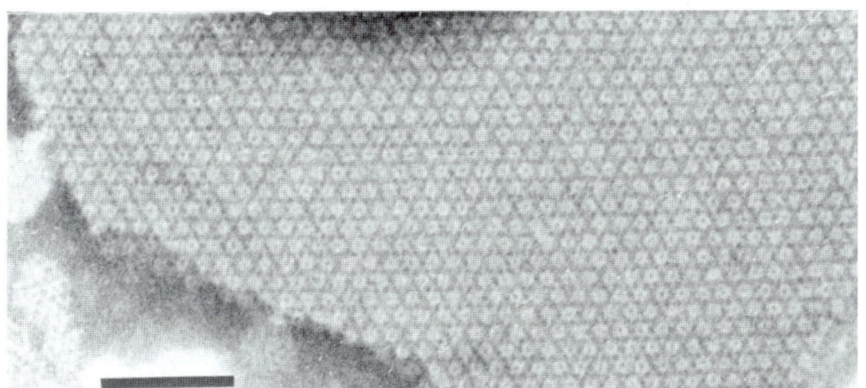

Figure 2.11 The S-layer of Archaea and Bacteria. A fragment of an S-layer shows the regular arrangement of the subunits as seen under an electron microscope. Bar = 100 nm.

molecules and particles useful in manufacturing materials for electronics, photonics, and magnetics. Bacteria can be engineered to make a fusion molecule consisting of the protein in question plus the portion of the S-protein involved in its secretion but lacking the portion that anchors it to the envelope. Such recombinant bacteria can make large amounts of soluble fusion proteins that can be readily purified and used for industrial purposes.

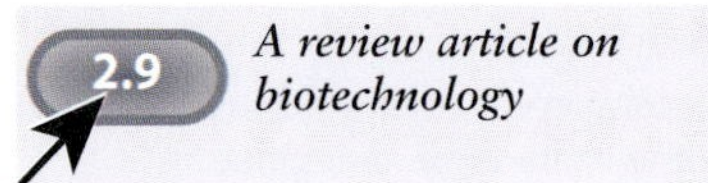

A review article on biotechnology

Bacteria without cell walls: the mycoplasmas

There are exceptions to the universal use of murein by bacteria. The **mycoplasmas** are *cell wall-lacking bacteria* with *no murein.* There are also wall-less Archaea, although their structure and physiology are less well studied. Lacking a target for penicillins, mycoplasmas are resistant to these drugs, as well as their relatives, the cephalosporins. Some, like *Mycoplasma pneumoniae,* an agent of pneumonia, contain **sterols** in their membranes, which is the rule in eukaryotic cells but unusual among prokaryotes. Genomic analysis places the mycoplasmas close to the gram-positive bacteria, from which they may well have been derived.

One might think that mycoplasmas would be quite feeble, yet they cope well without a rigid cell wall. Although quite delicate in culture, they have a remarkable ability to persist in the human body and even in harsh environments. Such toughness is also seen in archaeal wall-less equivalents of the mycoplasmas, e.g., species of *Thermoplasma,* a heat-loving and acid-loving genus. These organisms are certainly no weaklings: one strain was isolated from a slow-burning coal mine in Indiana.

We have mentioned above that when the cell wall of rod-shaped bacteria is removed, the cells become round (**spheroplasts**). Spheroplasts can grow in culture and make colonies resembling those of mycoplasmas. The difference between spheroplast cultures (at one time called L forms) and mycoplasmas is that when the murein is allowed to reform, the spheroplasts revert to their original bacterial shape, but mycoplasmas never do.

Mycoplasmas are not the only bacteria that lack murein. This unusual property is shared by the Planctomycetes, a highly distinctive and phylogenetically divergent phylum of bacteria found in terrestrial and aquatic environments. These organisms are protected by a protein S-layer.

Cell envelopes of the Archaea

The cell envelopes of the Archaea differ from those of the Bacteria, which points to the **evolutionary distance** between the two groups of prokaryotes. Archaeal cell membranes contain unique lipids, isoprenoids, rather than fatty acids, linked to glycerol by an ether rather than an ester linkage. Some of these lipids have no phosphate groups, and therefore, they are not phospholipids. In some species, these molecules are about the length of a phospholipid, and each forms one half of the membrane lipid bilayer with their hydrophilic isoprenoid moieties interacting in the middle of the membrane. Other species have a surprising architecture: their membranes are made of **lipid monolayers,** not the customary bilayers. These lipids are much longer (about twice as long as a phospholipid) and double headed, with glycerol ethers at both ends. A single molecule orients itself just as phospholipids do: polar groups are on the outer surface with a nonpolar hydrocarbon chain in the interior (Fig. 2.12). Many Archaea grow at very high temperatures (see Chapter 4), and their special lipids contribute to adaptation to such conditions. Indeed, single-

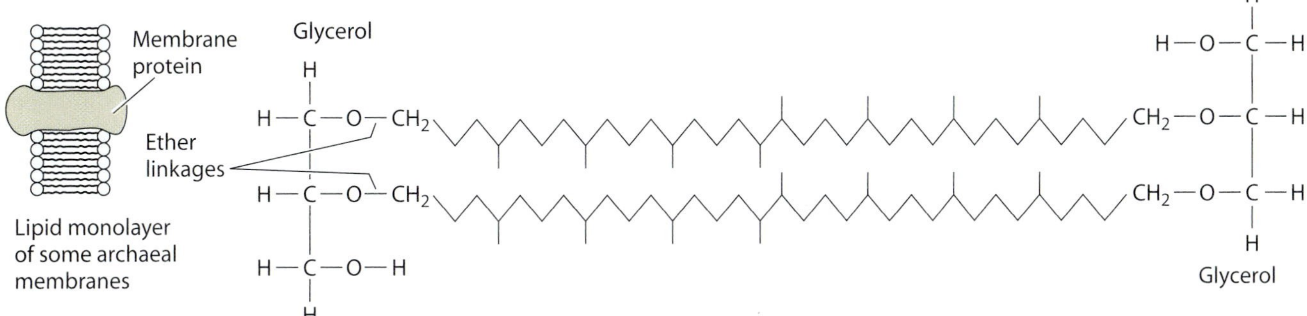

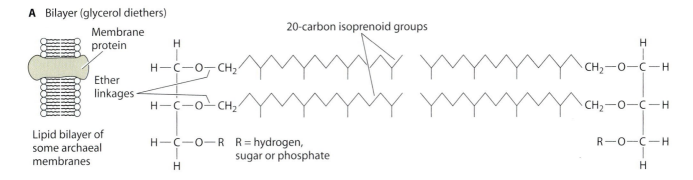

Figure 2.12 Two types of archaeal membranes. (A) Bilayers of isoprenoids linked to glycerol by ether bonds, the same arrangement as in phospholipid membranes. **(B)** Monolayers of isoprenoids that are twice as long as those in panel A, linked to glycerol by an ether bond.

headed isoprenoid membranes are more stable in vitro to high temperature, high salt, and low pH than phospholipid membranes, and the double-headed ones are even more stable.

The Archaea do not have cell walls like the Bacteria. Some have a simple S-layer that is more sensitive to detergents than the bacterial S-layers. In many species, the S-layer is made up of glycoproteins. Some Archaea have a rigid cell wall made up of polysaccharides or a peptidoglycan called **pseudomurein,** which makes up their sacculus. It differs from murein by having L-amino acids rather than D-amino acids and disaccharide units with an α-1,3 rather than a β-1,4 linkage. Archaea that have a pseudomurein cell wall are gram positive. Species that lack cell walls are more common among Archaea than Bacteria. Surprisingly, these include some that live in harsh, hot, acid environments.

CAPSULES, FLAGELLA, AND PILI: HOW PROKARYOTES COPE IN CERTAIN ENVIRONMENTS

The morphological diversity of prokaryotes is not limited to walls and membranes. Some bacteria and archaea have yet other exterior structures: **capsules, flagella,** and **pili.** These components are not always present; they are important for survival under certain circumstances.

Capsules and slime layers

Many bacteria, gram positive and gram negative alike, are surrounded by a coat of slime. When the slime is tenacious and remains attached to the cells, it is called a capsule (Fig. 2.13). If it is looser, it is known as a **slime layer**, but this is not a sharp distinction. In most cases, this layer is made up of a high-molecular-weight polysaccharide, either heteropolymeric (i.e., containing more than one kind of monosaccharide) or homopolymeric. In some bacteria, the layer is not a polysaccharide but a polymer of amino acids, e.g., glutamate in the anthrax bacillus (*Bacillus anthracis*). The synthesis of this material depends on the environment and is not required at all times. Organisms that lack capsules or slime layers grow well, at least under laboratory conditions, but these layers are essential for survival in certain natural environments.

The slime layer is frequently a major determinant of a bacterial cell's ability to colonize a niche. It may also play a role in protection from desiccation. This is exemplified in an organism involved in initiating tooth decay (caries), *Streptococcus mutans,* which splits sucrose into glucose and fructose and converts glucose into a sticky polymer called dextran, which helps the organism adhere to the enamel of teeth and form plaque. A slime layer also enables bacteria to adhere to surfaces in the environment (e.g., rocks or plants in bodies of water) and to make biofilms.

The capsule is also a *major line of bacterial defense against phagocytosis* because it hinders the uptake of the bacteria by phagocytic cells. Many bacteria that must travel through the blood to reach target organs are indeed encapsulated. Examples are the agents of bacterial meningitis, such as the meningococci (*Neisseria meningitidis*) and *Haemophilus influenzae.*

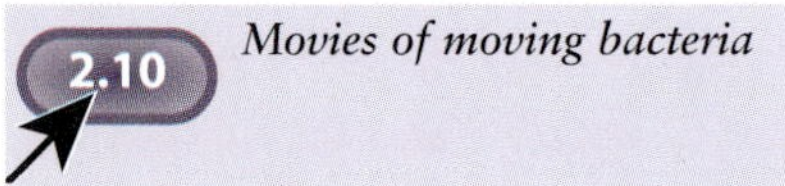

Flagella

Many bacteria and archaea are able to move actively through liquids and across wet surfaces. There are several ways of doing this, the most common being by the action of helical filaments called flagella (singular, flagellum). *Flagella rotate* and thus act as propellers. This contrasts with

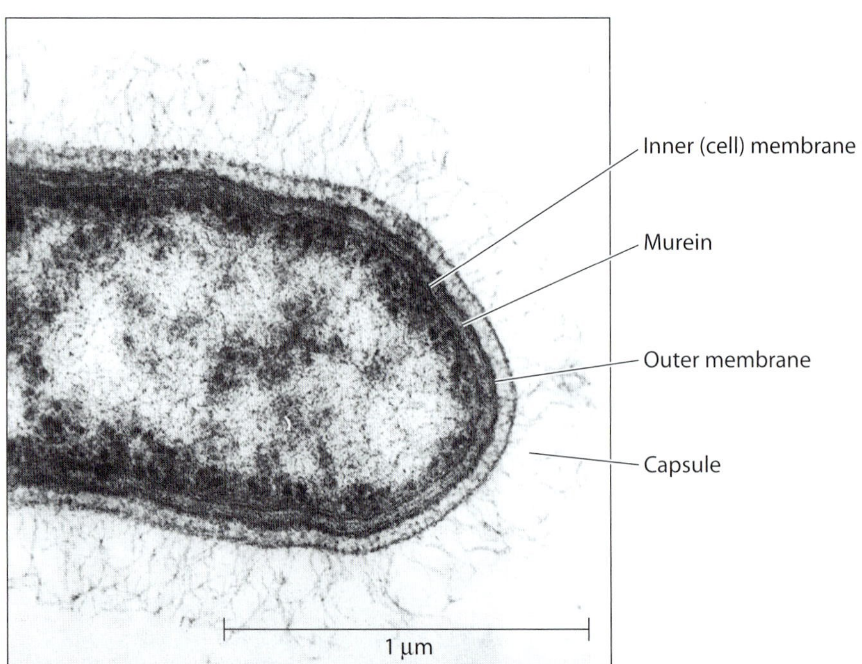

Figure 2.13 Bacterial capsule. The capsule is the fuzzy material surrounding the cell envelopes in this electron micrograph thin section. Note that its thickness is about one-fourth of the cell's diameter. Some bacteria have considerably thicker capsules.

eukaryotic flagella, which move in a 2-D, whiplike motion. Each prokaryotic flagellum *is fitted with a motor* that turns it, a fact that should fascinate all of us, especially those interested in nanotechnology. Can you think of other natural examples of motors in biology, and on such a small scale?

Many successful pathogens are motile, which probably aids their spread in the environment and possibly in the bodies of their hosts. Depending on the species, a single cell may have one or many flagella (Fig. 2.14). In some, the flagella are located at the ends of the cells (polar), and in others, they are at random points around the periphery (peritrichous, or "hairy all over"). This distinction is useful in taxonomy and in diagnostic microbiology. Flagella are not the only filamentous structures protruding through the surface layers of bacteria. There are others, such as pili (also called **fimbriae**) and fibrils (Fig. 2.15).

Flagella are composed of three parts (Fig. 2.16): a long helical **filament** connected via a **hook** to a **basal body.** This is a complex structure that anchors the flagellum to the cell envelope and contains the flagellum-turning motor.

The filament is a helical structure that may extend 5 to 10 μm into the medium, several times the length of the cell. It is hollow. In most bacteria, the filament is composed of a single protein, **flagellin.** Flagellins are highly antigenic and sometimes elicit specific immune responses to infections. Mechanically speaking, flagellins are strikingly rigid, about 100 times stiffer than the F actin of muscle. Flagellin molecules aggregate spontaneously to form the structure characteristic of the flagellar filament, which is an example of morphogenesis on the molecular level. In vitro, isolated filaments can be dissociated into a flagellin solution. When the ionic environment is properly adjusted, the flagellin molecules can then reform into filaments indistinguishable from the natural product.

The hook is a short curved structure, slightly larger in diameter than the filament, that connects the filament to the cell. It appears to act as a "universal joint" between the motor in the basal structure and the filament. The molecular structure of the hook is also relatively simple. Like the filament, it is made up of a single type of protein.

The basal body is composed of some 15 proteins that aggregate to form a rod to which four rings are attached (in the case of gram-negative bacteria) (Fig. 2.16). The rings act as bushings, or stators, anchoring the structure in the various layers of the cell envelope while allowing the rod

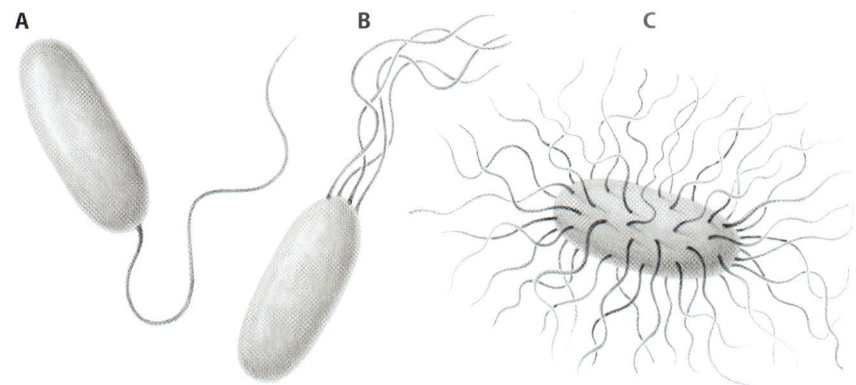

Figure 2.14 Arrangements of flagella in some types of bacteria. (A) Single polar flagellum. **(B)** Multiple polar flagella. **(C)** Flagella arising from all around the cell (peritrichous).

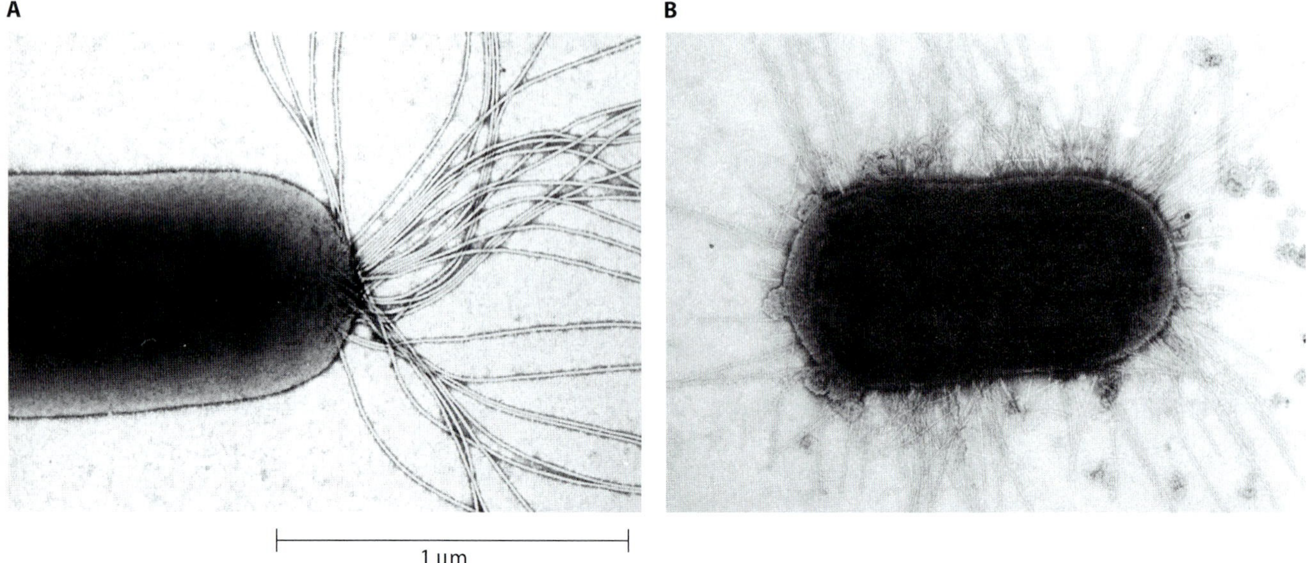

Figure 2.15 Flagella and pili seen under the electron microscope. (A) Flagella inserted at the pole of a bacterium. In other organisms, flagella are inserted all over the cell or a single flagellum may be inserted at a cell's pole. **(B)** Pili surrounding a gram-negative bacterium.

(the rotor) to rotate. It is not yet known how the rod is physically retained on the cell surface. As shown in Fig. 2.16, each ring corresponds to a specific layer of the gram-negative cell envelope.

As might be expected, gram-positive and gram-negative bacteria have different flagellar basal bodies. Gram-positive bacteria have only two rings, one imbedded in their cell membrane and the other associated with the teichoic acid component of their wall.

In helical bacteria known as spirochetes, a group that includes the agents of syphilis and Lyme disease, the flagella do not extend into the environment. They are contained within the periplasm, attached near both of the cell poles (Fig. 2.17). When the filaments rotate, the cell turns in a corkscrew-like manner. Spirochetes can move in media so viscous that it impairs the motility of externally flagellated bacteria. It is thought that motility is essential for spirochetes to be able to colonize the intestine.

How are flagella made? The answer is, stepwise. First, the basal body is assembled and inserted into the cell envelope. Then the hook is added, and finally, the filament is assembled progressively by the addition of new flagellin subunits to its growing tip. One may guess that the addition of new molecules is from the end nearest the cell. Not so. Flagella grow from the tip that protrudes from the cell. Conveniently, flagella have a hollow central channel through which the new flagellin molecules are extruded. When it reaches the tip, each molecule spontaneously condenses with its predecessors, and thus the filament elongates (Fig. 2.18). With time, the process slows down. This, together with mechanical breakage, may explain why flagella are limited in length.

The assembly of flagella is a regulated process. Picture what would happen if flagellin molecules were to accumulate in the cell. If they were assembled intracellularly into filaments, these might well interfere with central physiological processes (the bacteria would get "indigestion"). In fact, *synthesis of flagellin molecules is coupled to their utilization in fla-*

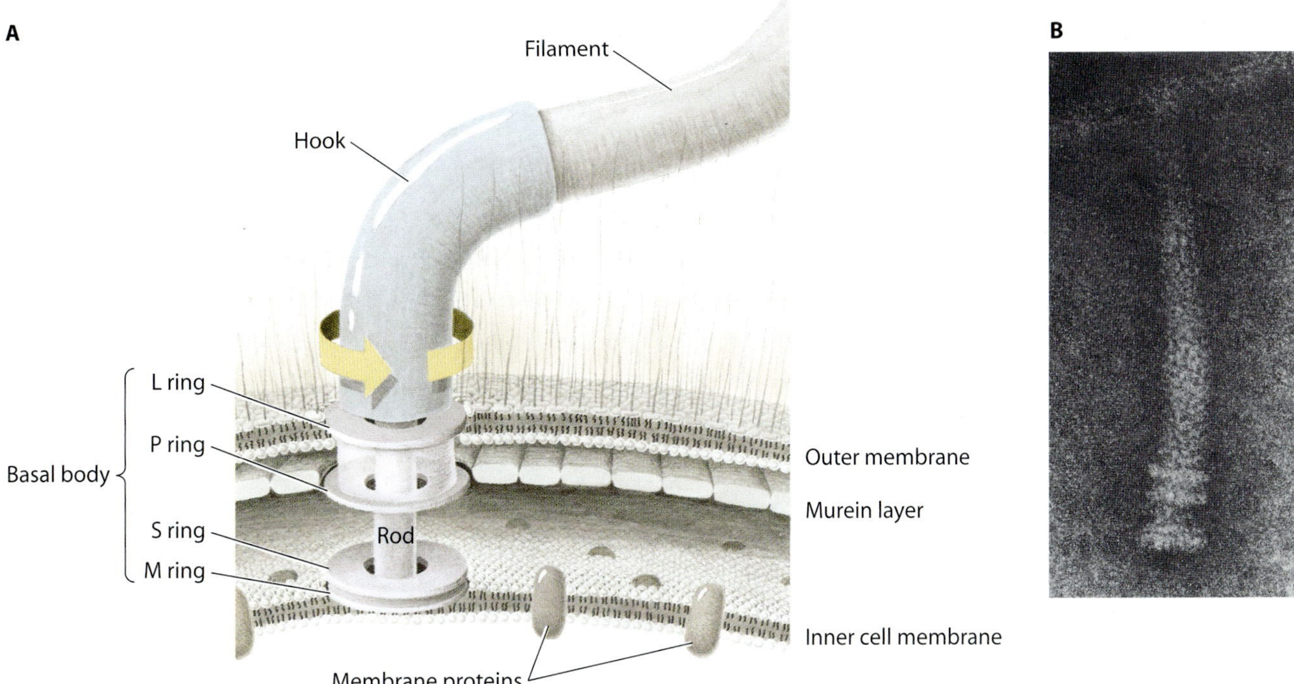

Figure 2.16 Structure of flagella. **(A)** Insertion of a flagellum into the double membrane of gram-negative bacteria. Flagella of gram-positive bacteria lack the outer two rings. **(B)** Electron micrograph of the portion of a flagellum of *E. coli* closest to the cell.

gellar assembly. In *E. coli*, this coupling takes place via an inhibitor of flagellin synthesis. This inhibitor prevents the accumulation of intracellular flagellin molecules and their assembly into a filament. However, once the basal body is inserted into the envelope, the *inhibitor is secreted from the cytoplasm into the medium.* Thus, assembly of the flagella leads to the loss of the inhibitor, allowing the making of new flagellin molecules.

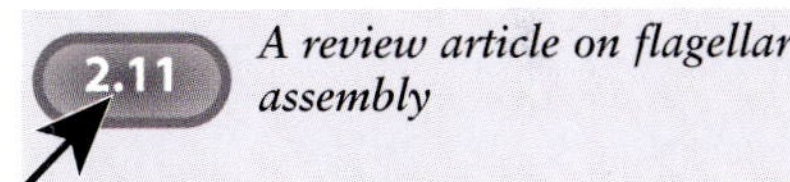

A review article on flagellar assembly

The molecular circuitry that brings about chemotaxis, and related, seemingly purposeful movement is described in detail in Chapter 13.

Pili

Pili (singular, pilus; also called fimbriae) are structures involved in strikingly diverse physiological functions, such as attachment of bacteria to host cells and other surfaces, the transfer of proteins and nucleic acids to other cells, and motility. Not all pili do all these things—most are specialized and restricted in their functions. Pili consist of straight protein filaments, thinner and often shorter than flagella (although some are

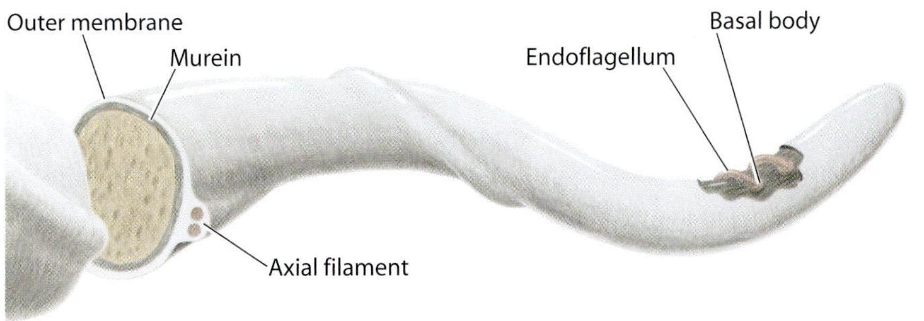

Figure 2.17 Arrangement of the axial flagella within the periplasm.

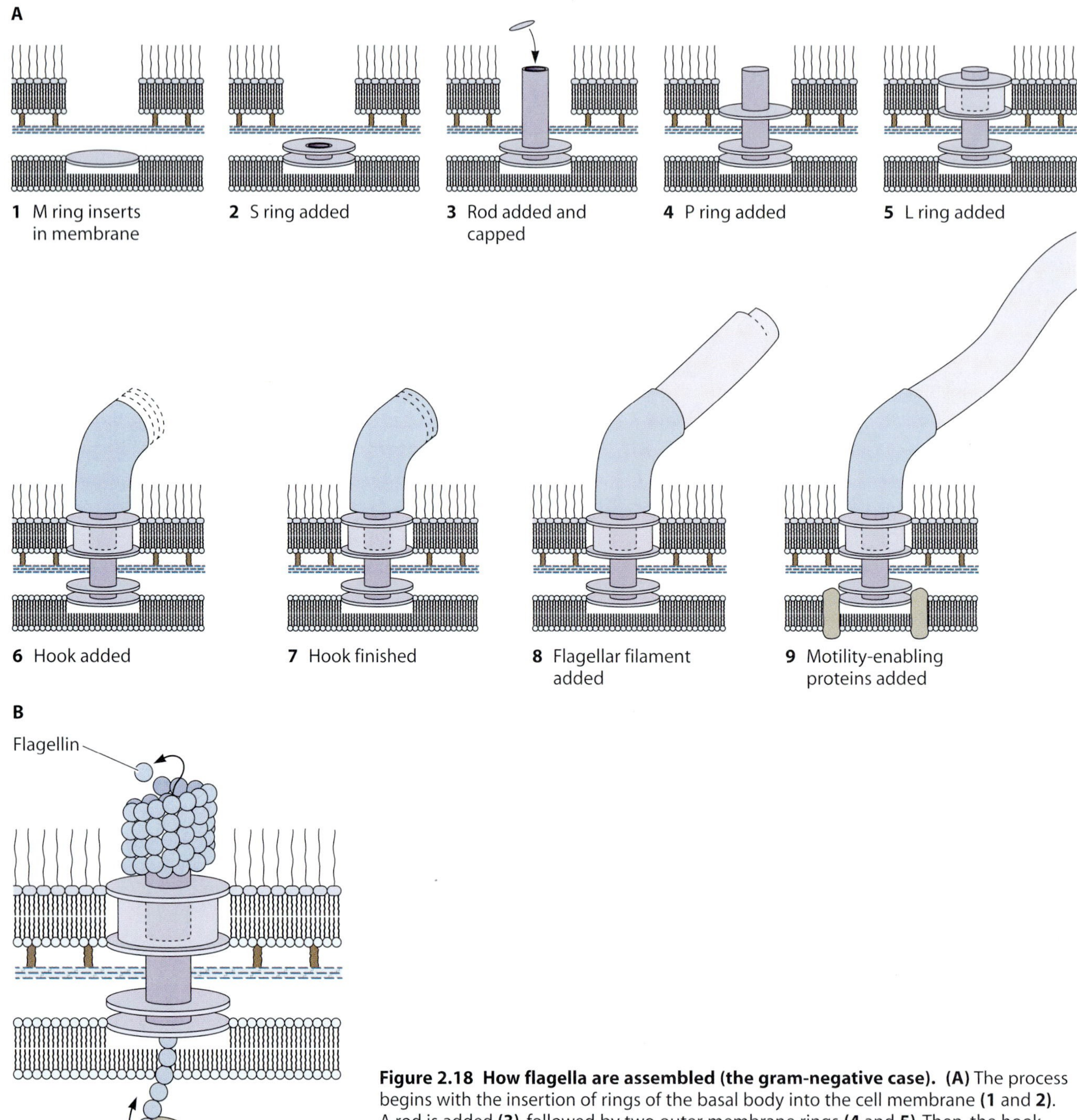

Figure 2.18 How flagella are assembled (the gram-negative case). (A) The process begins with the insertion of rings of the basal body into the cell membrane **(1** and **2)**. A rod is added **(3)**, followed by two outer membrane rings **(4** and **5)**. Then, the hook **(6** and **7)** and the filament (**8**) are added. The last step is the addition of motility proteins in the cell membrane **(9)**. **(B)** Growth of the flagellar filament (step 8 in panel A) by extrusion of flagellin subunits through the core of the flagellum.

longer). They are distributed, sometimes in large numbers, over the surfaces of some bacteria (Fig. 2.14). In some species, they occur in tufts at one or both poles of the cell.

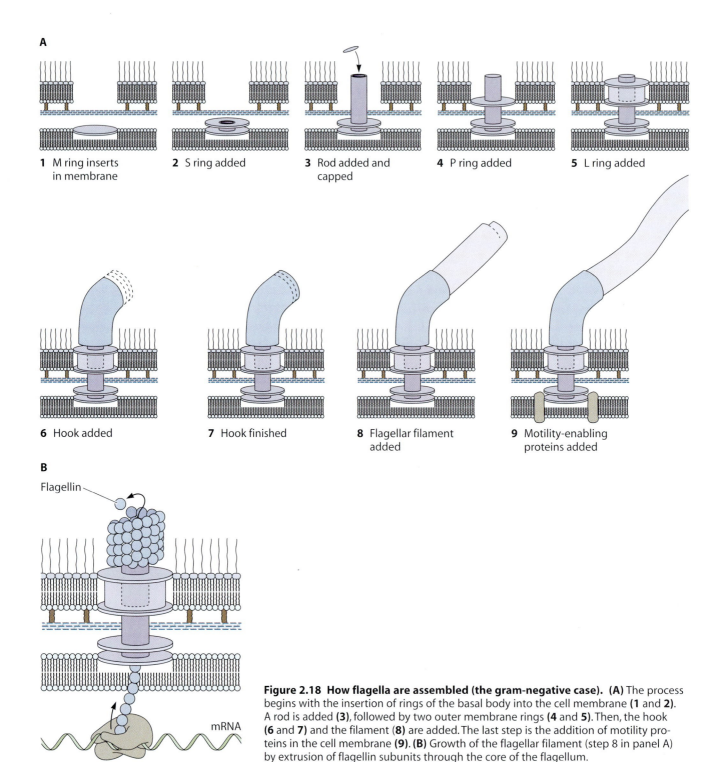

Virtually all gram-negative bacteria possess pili, but relatively few gram-positive bacteria do. In the human body, pili allow bacteria to adhere to mucosal surfaces (e.g., in the gastrointestinal and urinary tracts). Sticking is mediated by **adhesins,** specific proteins at the tips or sides of the pili. Pili also have other roles in disease. Like the capsules, pili inhibit the phagocytic ability of white blood cells. They are also antigenic and elicit an immune response in the host. The proteins of pili, **pilins,** are highly changeable, permitting some bacteria to put on a succession of disguises that enable them to outflank the immune system. For example, gonococci, the cause of gonorrhea, have a large number of genes that code for variants of pilin. Each version of pilin is antigenically distinct and elicits the formation of different antibodies by the host. In the presence of antibodies to one type of pilin, there is rapid selection for strains that have switched to the synthesis of another antigenic type of pilin. Thus, in this quick-change scenario, they keep one step ahead of the host immune response. It is easy to see why vaccines against gonococci containing pilins have failed so far.

Motility via pili is completely different from flagellar motion. Pili are straight rods and do not rotate. Their tips strongly adhere to surfaces at a distance from the cells. Pili then depolymerize from the inner end, thus retracting inside the cell. In this respect, they resemble grappling hooks (reminiscent of those used by ninjas). The result is that, during repeated rounds of this process, the bacterium moves in the direction of the adhering tip. This kind of surface motility is called **twitching** and is quite widespread among piliated bacteria (see "Twitching motility" in Chapter 13).

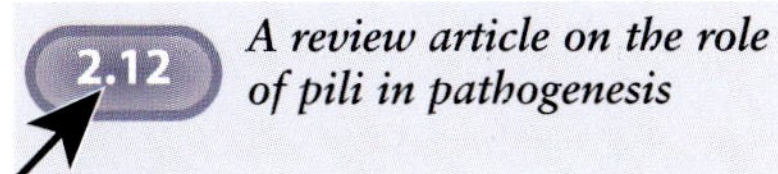

A review article on the role of pili in pathogenesis

How are pili made? Unlike flagella, they grow from the inside of the cell outward. Although pili have a central channel, it is too small for pilins (the pilus proteins) to pass through.

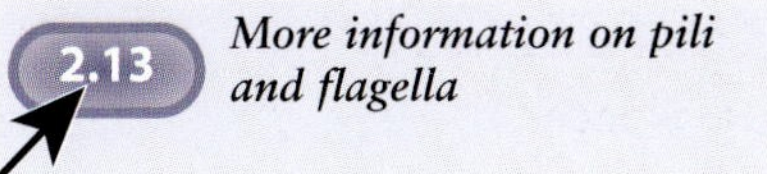

More information on pili and flagella

In conclusion, as small as bacteria and archaea are, they are endowed with a sophisticated set of cell envelopes that play a major role in their survival in the environment. The wall, capsule, and outer membrane of the bacterial cell contain molecules and residues—odd sugars, D-amino acids, and unique lipids—not found elsewhere in the biological world. Bacterial flagella and pili have no direct counterparts in eukaryotes. This makes "microbial dermatology" a specialized area of study.

STUDY QUESTIONS

1. How do the gram-positive and the gram-negative bacteria protect their cytoplasmic membranes from harmful chemicals?

2. Do the differences in composition of the envelopes of the Archaea and the Bacteria reflect their ecological distribution?

3. How do hydrophilic compounds with a molecular mass of 700 daltons or less enter gram-negative cells?

4. How do acid-fast bacteria withstand membrane-damaging chemicals?

5. What are crystalline S-layers, and how widespread are they among prokaryotes?

6. What role do prokaryotic capsules and slime layers play in the survival of bacteria?

7. What are the main features of the structure and function of bacterial flagella? How do they differ between gram-positive and gram-negative bacteria?

8. For what functions are bacterial pili (fimbriae) used? What is their structure, and how are they assembled?

chapter 3

prokaryotic cell structure and function: the cell interior

GENERAL OBSERVATIONS

We saw in Chapter 2 that the surface structures of prokaryotes are architecturally and biochemically unique. What about the cell interior? Is it also different from that of eukaryotic cells? The answer, which we will explore in this chapter, is yes. Here, we examine the two major components inside the prokaryotic cell envelope: the **nucleoid** and the **cytoplasm**. That there are *only* two says a lot. Mitochondria, nucleus, Golgi apparatus, endoplasmic reticulum, plastids, mitotic apparatus—these elements, which are almost universal among eukaryotic cells, are absent from Bacteria and Archaea. The first impulse is to decide that there may be little of interest to be said about the interiors of prokaryotes. Quite the contrary. Though complexity of the eukaryotic sort has not been selected for during bacterial evolution, the very streamlining of the cell's interior has enabled the development of some unique and fascinating structural features (Table 3.1).

We shall focus on the functional consequences of the unusual nature of the bacterial nucleoid and cytoplasm and then briefly review some of the less prominent interior structures of certain bacteria. Much of what we say also applies to archaea.

THE NUCLEOID

Perhaps the most fundamental difference between prokaryotes and eukaryotes is the cytological organization of their genomes. Prokaryotes have no true nuclei; instead they package their DNA in a structure known as the nucleoid. The nucleoids of most prokaryotes contain a single circular chromosome, which is a largely unadorned DNA molecule. Nucleoids look like irregular blobs of DNA that are separated from the cytoplasm but, with

Table 3.1 What's inside the cell membrane of prokaryotes

Structure or component	Functions
Nucleoid	Repository of genome Transcription
Cytosol	
Polyribosomes	Protein synthesis
Enzymes	Metabolism
Regulatory proteins	Regulation of gene expression
Metabolites, precursors, energy compounds, salts	Participate in metabolism
Vesicles (in some only)	
Gas vesicles	Cell buoyancy
Photosynthetic vesicles	Photosynthesis
Chemosynthetic vesicles	Chemosynthesis
Carboxysomes	Enhance CO_2 fixation in heterotrophs?
Enterosomes	Metabolism of propanediol, others
Storage granules	Store energy-rich compounds:
Acidocalcisomes	Polyphosphates, polyhydroxyalkanoates
Others	Glycogen-like compounds, sulfur compounds
Other structures	
Magnetosomes	Involved in directional orientation with respect to magnetic field

some exceptions, without the benefit of a membrane (Fig. 3.1). The exceptions are seen in the planctomycetes, a distinct group of bacteria we encountered in Chapter 2. In these organisms, the nucleoid is surrounded by a double membrane reminiscent of that of eukaryotic nuclei (Fig. 3.2). The distinction between prokaryotic nucleoids and eukaryotic nuclei that still holds is that prokaryotes have no eukaryotic-type mitotic apparatus.

Prokaryotes (as well as eukaryotic cells and viruses) must solve a demanding topological problem in organizing their long, thin DNA. If stretched out, the DNA molecule of *Escherichia coli* would be about 1,000 times the length of the cell. If this chromosome were magnified in thickness to the size of common spaghetti, its length would be equivalent to 200 average platefuls, with each "spaghetto" linked end to end to the one on the next plate. The problem of how to package such a long DNA molecule in a small space is to *tightly fold it into a nucleoid*. The physical state of the DNA molecule in the nucleoid is somewhat mysterious, because in the test tube, a solution 1/100 the concentration in the cell is a solid gel! Clearly, proteins and counterions play important roles in keeping the DNA molecule folded.

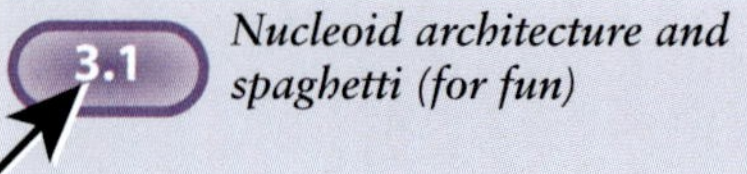

Nucleoid architecture and spaghetti (for fun)

Some information about the folding of DNA into a compact structure comes from in vitro studies. When bacteria are broken by mechanical means, the **lysate** is highly viscous because the nucleoid "explodes," in part due to electrostatic repulsion of the highly (negatively) charged strands of DNA. However, if the ionic strength of counterions, such as magnesium, is high (e.g., 0.1 M), the nucleoids remain condensed and the lysate is not viscous. Thus, in living cells, the ionic environment inside and around the nucleoids works together with DNA-binding proteins to induce a special (albeit unknown) state of condensation of the DNA. This condensation excludes from the nucleoid the numerous **ribosomes** and polyribosomes of the cell. For this reason, electron micrographs of sections

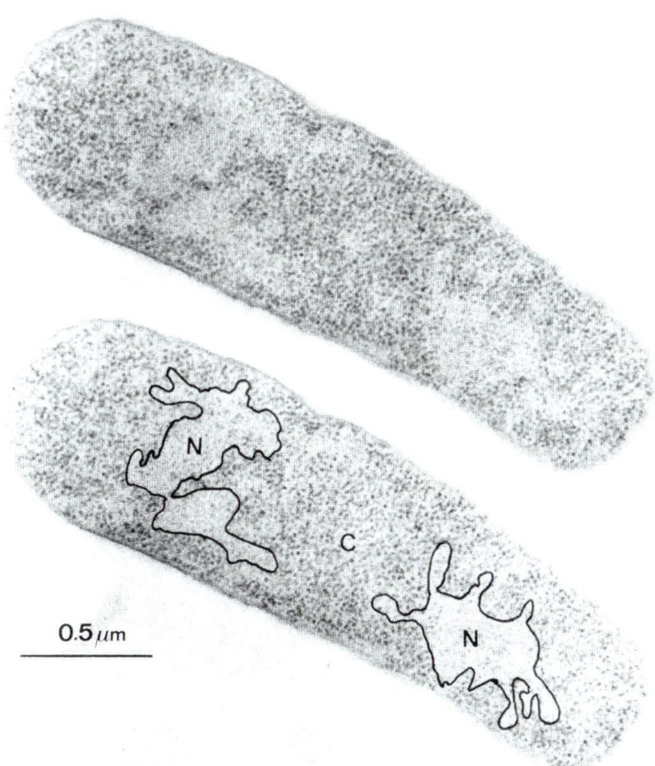

Figure 3.1 Electron micrograph of a thin section through an *E. coli* cell. The clear irregular area corresponding to the nucleoid (N) has been outlined in the lower picture. The granular appearance of the rest of the cell, the cytoplasm (C), reflects its high content of ribosomes.

Figure 3.2 The prokaryotic nuclear envelope. An electron micrograph of a thin section of the planctomycete *Gemmata obscuriglobus* shows a nucleoid (N) surrounded by a nuclear envelope (NE) consisting of two membranes. The dark granules are the ribosomes, found both within and outside this structure.

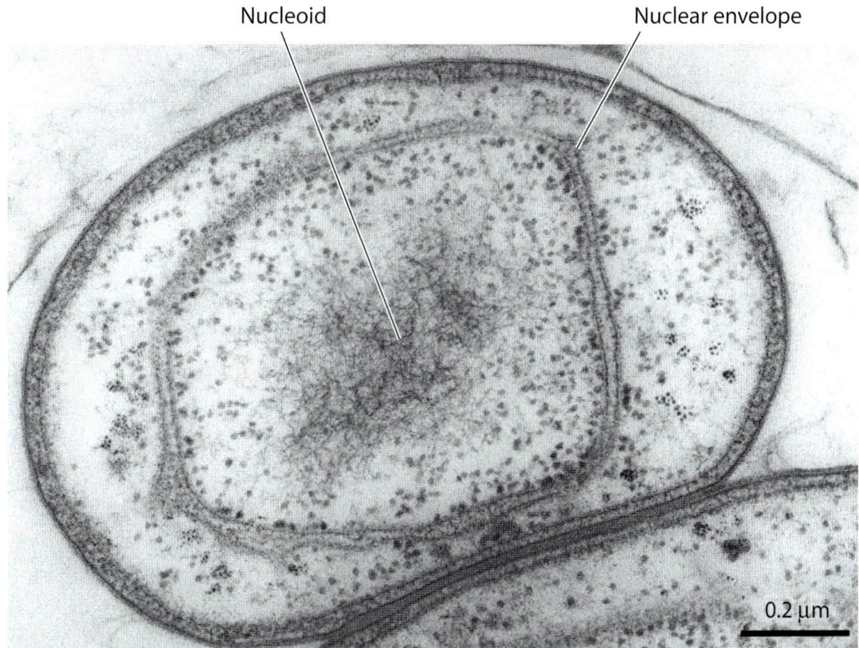

of prokaryotic cells (Fig. 3.1) display a fibrous-appearing nucleoid (due to the tightly folded DNA fiber) and a granular cytoplasm (due to the large numbers of ribosomes). Note that DNA in eukaryotic metaphase chromosomes or within viral particles can be even more highly condensed than this.

Since nucleoid DNA is highly condensed, *transcription of the DNA into RNA takes place only at the nucleoid-cytoplasm interface*. This fact was demonstrated by the following experiment. Cells were exposed for a short time to a radioactive precursor of RNA, namely, tritium-labeled uridine, for a short time. After ultrathin sections of such cells were overlaid with photographic emulsion, some time was allowed for the tritium atoms to disintegrate, and the photographic grains were developed, the preparations were examined with an electron microscope. Photographic grains were seen only at the surfaces of nucleoids, not within them, demonstrating that RNA polymerase molecules work only at the nucleoid surface. This finding further suggests that DNA is not static but writhes, snakelike, from the interior to the nucleoid surface and back. Such movement would give each gene the opportunity to be transcribed. This image brings up a whole new subject: the nucleoid-cytoplasm interface makes possible a largely prokaryotic feature, the tight coupling of transcription and translation. Protein synthesis begins as ribosomes attach to messenger RNA (mRNA) molecules while they are still elongating (the planctomycetes even have ribosomes within their "nuclear membrane"). Transcription and translation occur simultaneously, with transcription of DNA coupled with simultaneous translation of the growing mRNA chains into proteins (see Chapter 8).

In most bacteria, the **genome** consists of a *single circular molecule* of double-stranded DNA. Circularity of the chromosome solves one problem: cells contain exonucleases that chew up the ends of linear DNA molecules. Having no ends means being immune to these enzymes. In eukaryotes, the chromosomes are linear and the ends are protected by specialized structures, the telomeres. There are exceptions to the rule of circularity, because some bacteria have linear chromosomes. Here, the ends are protected either by double-stranded hairpins or by binding of specific proteins at the ends of the chromosomes. Examples of linear chromosomes are found in disparate bacterial species, such as the agent of Lyme disease *(Borrelia burgdorferi)* and various antibiotic-producing soil bacteria *(Streptomyces)*. Why are some bacterial chromosomes circular and others linear? We do not yet know.

Being circular endows the bacterial chromosome with another property: it allows it to be **supercoiled**, that is, it has twists of a higher order (like the annoying twists of a coiled telephone cord). As a consequence, the chromosomes are energetically activated, or if you wish, spring-loaded. This lowers the energy barrier for strand separation, which is needed for both replication of the DNA and its transcription into mRNA. Supercoiling is thought to be achieved by the balance of two topoisomerases, enzymes that change the topology of the DNA. One of these, **DNA gyrase**, introduces supercoils into circular DNA; its action is counteracted by a second enzyme, **topoisomerase I**, which relaxes the supercoils by making single-strand breaks. The circular chromosome is not a single supercoiled molecule but is divided into individual domains, each

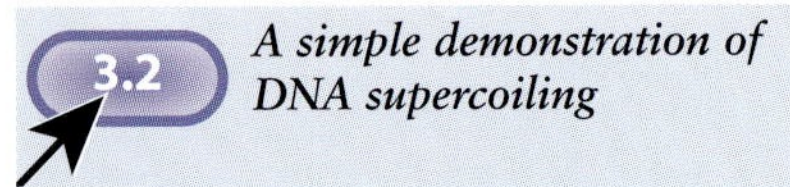

A simple demonstration of DNA supercoiling

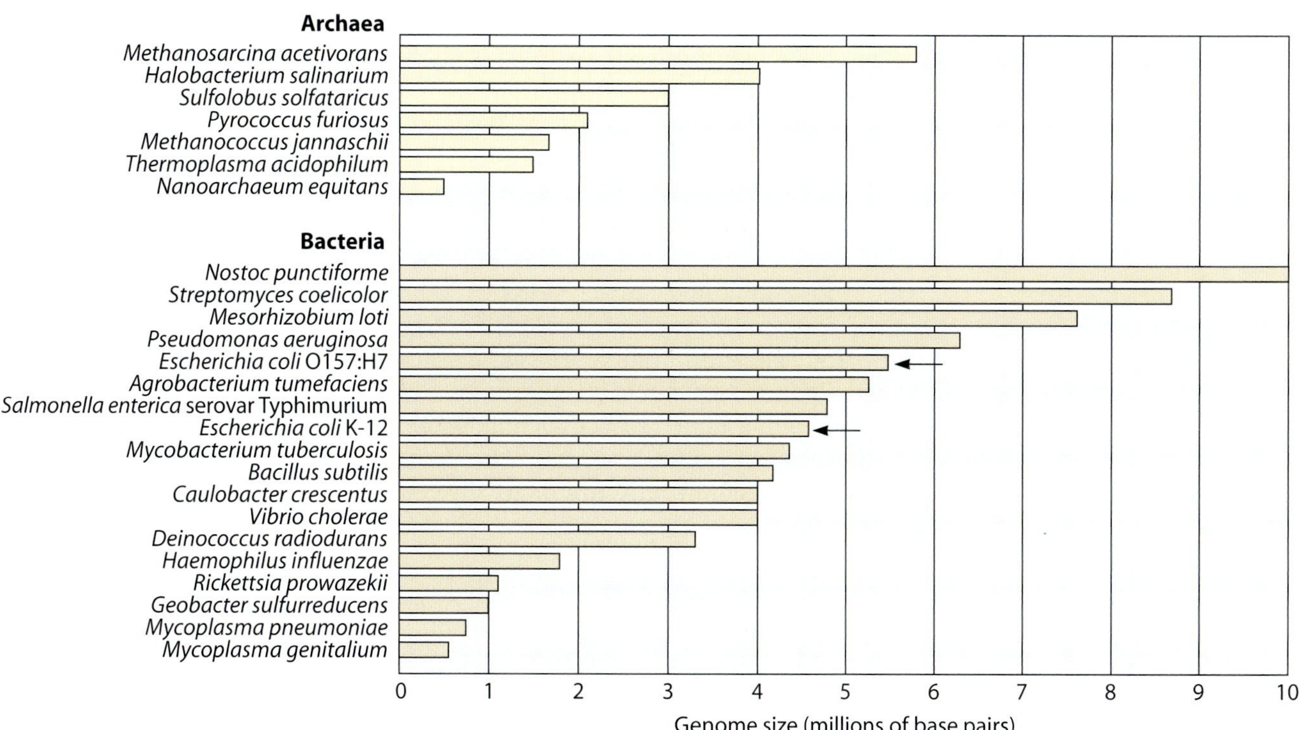

Figure 3.3 Genome sizes of archaea and bacteria. The overall range in genome sizes is about 20-fold. Note that the genomes of two strains of *E. coli*, O157:H7 and K-12 (arrows), differ by about 20%, or 1 million base pairs.

of which is individually supercoiled. Supercoiled molecules become "relaxed" by making single-strand breaks, but it takes several such cuts to relax the whole chromosome. This finding suggests considerable organization within the nucleoid.

The sizes of prokaryotic genomes vary considerably (Fig. 3.3). The smallest known consist of some 580,000 base pairs, and the largest contain almost 10 million base pairs. For comparison, eukaryotic chromosomes vary from 2.9 million to over 4 billion base pairs. In general, the size of a prokaryote's genome reflects the number of functions, and hence the number of genes, it needs to prosper in its particular range of habitats. This subject is taken up in greater detail later (see Chapter 13).

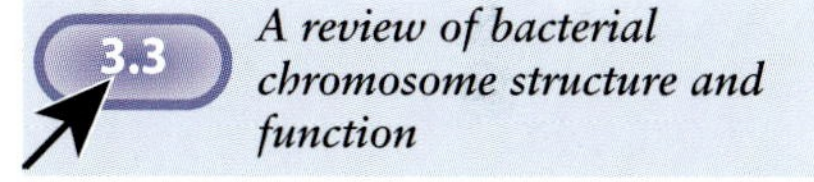

3.3
A review of bacterial chromosome structure and function

Many, perhaps most, bacteria have a single chromosome, or in genetic parlance, a single **linkage group**, but there are exceptions to this. The cholera bacillus has two dissimilar chromosomes, some rhizobia (nitrogen-fixing bacteria associated with the nodules of legumes) have three, and yet others have four. We do not know why this is so, but the question arises as to how such chromosomes are apportioned, one each, to progeny cells. The mechanism that ensures partitioning of chromosomes is discussed in Chapter 9.

In addition to chromosomes, many bacteria carry extrachromosomal DNA genetic elements called **plasmids** (see Chapter 10). Some plasmids are very large and stably inherited; therefore, telling them apart from a chromosome becomes a matter of definition. Plasmids are defined as genetic elements encoding dispensable functions: a bacterium that loses a plasmid remains viable, but one that loses a chromosome does not.

Although most bacteria have a single discrete chromosome, they may have more than one copy of it. In bacteria, the number of nucleoids, and therefore the number of chromosomes, can be readily altered by growth conditions. Rapidly growing bacteria have more nucleoids per cell than slowly growing ones. The process can be quickly reversed: upon decreasing the growth rate, the cells return to having a single nucleoid. Therefore, it is not correct to speak of this state as being akin to eukaryotic ploidy, which is a more stable condition. Thus, bacteria can be haploid and, at times, multinucleated.

How does a molecule of the physical complexity of DNA replicate, segregate, and become transcribed into mRNA? Clearly, it is not easy to visualize how such an enormous "plate of spaghetti" can keep from becoming hopelessly tangled as it fulfills its many functions. The implications of nucleoid structure and supercoiling for DNA replication, segregation, and transcription are discussed in Chapters 8 and 9.

THE CYTOPLASM

For many years (even after the advent of the electron microscope), a common notion was that the bacterial cytoplasm was a bag of soluble enzymes and other molecular machines, such as the ribosomes. Accordingly, the interesting structures of prokaryotes were thought to be on the outside of the cell, protecting a rather plain inside. This view has been contradicted by the development of techniques that augment electron and fluorescence microscopy. In Chapter 1, we imagined a trip inside a bacterial cell and tried to envisage what we would see. Here, we will deal in more detail with physical properties of the bacterial interior.

The interior of a prokaryotic cell is crowded with macromolecules. The cytoplasm is teeming with the metabolic machinery of the cell, most of which is devoted to the manufacture of proteins. We have already noted that the granular appearance of the cytoplasm is the result of its large concentration of ribosomes (20,000 to 200,000 in a single *E. coli* cell, depending on its growth rate). A typical cell consists (by wet weight) of approximately 20% protein, 7 to 20% RNA (depending on growth conditions), and 2% DNA. Soluble precursors and inorganic ions add 1 to 2%. Typically, the concentration of all macromolecules is about 30%, which is three times the protein concentration of a chicken's egg white. Moreover, macromolecules bind water, which increases their effective molecular volume. By any comparison, the concentration of macromolecules is very high, which makes the interior of a bacterium quite viscous. As a consequence, chemical reactions within a bacterial cell take place in an environment totally different from what is common in test tube studies, and extrapolations from in vitro biochemistry to the living cell are risky. In particular, neither enzyme kinetics nor rates of macromolecular movement can be inferred from test tube measurements. Although the cytoplasm is visibly separated from the nucleoid, the two are intimate. As we have mentioned, transcription of DNA into mRNA takes place at the nucleoid-cytoplasm interphase. mRNA molecules do not wait to be finished to become loaded with ribosomes but do so soon after they begin to be made. As transcription continues and the mRNA becomes longer, more and more ribosomes become attached, forming what are known as **polyribosomes,** or **polysomes.** Thus, in the cytoplasm of growing bacteria, few ribosomes are floating freely: most are attached to mRNA and are engaged in protein synthesis.

Specialized physical techniques (e.g., fluorescence recovery after photobleaching [FRAP]) permit one to determine the rates of diffusion of a fluorescent protein, such as green fluorescent protein, cloned into cells.

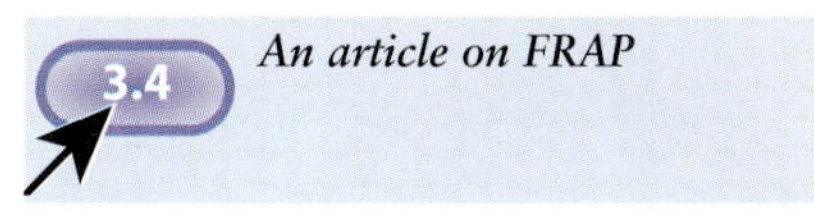

Such measurements show that green fluorescent protein diffuses about 10 times more slowly in *E. coli* than in water and about 4 times more slowly than in the eukaryotic cytoplasm or mitochondria. However, the diffusion rate is still relatively high. A small protein can travel the length of a typical bacterium (1 micrometer [μm]) in about 100 milliseconds, which is reasonably fast for some biological responses, although slow for chemical reactions. This diffusion rate is high enough to account for the rate at which genes are turned on by effectors made elsewhere in the cell, but not by much. It is possible, therefore, that the diffusion rates of certain molecules may limit some intracellular transactions.

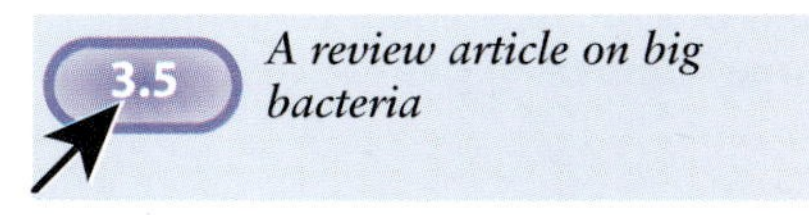

Cell size matters here: the bigger the bacterium, the greater is the impact of diffusion. In bacteria that exceed 1 μm in their smallest dimension, diffusion may well be growth rate limiting. At the known extreme of prokaryotic cell size is a gargantuan bacterium, *Epulopiscium fishelsoni*, found in the intestines of certain fish, which can reach 0.5 mm in length and 80 μm in width (Fig. 3.4). It is about the size of a period on this page and thus can be seen with the naked eye. *E. fishelsoni* cannot yet be cultured in the laboratory, so we do not know how fast it grows and how it fulfills its biosynthetic requirements, nor do we know how nutrients and metabolites are transported in and out of the interiors of these cells. Note that size alone may not mean that diffusion presents a problem. Another mammoth bacterium, the spherical *Thiomargarita namibiensis*, is between 300 and 700 μm in diameter (Fig. 3.4). However, most of *T. namibiensis* consists of an internal vacuole filled with concentrated nitrate (>0.1 M), which is used for the oxidation of sulfide (see Chapter 18 for details). The relevant cytoplasm of this organism is a shell only 1 to 2 μm thick.

Figure 3.4 Two gargantuan bacteria. **(A)** The world's largest known prokaryote, *T. namibiensis*. The white bodies are sulfur granules made by the oxidation of hydrogen sulfide. Most of the cell consists of a vacuole filled with nitrate. The cytoplasm is a thin shell, about 1 μm thick, in the same range as that of common bacteria. **(B)** *E. fishelsoni*, the world's biggest bacterium, if the vacuole of *T. namibiensis* is excluded. Note the difference in scale. On either scale, an ordinary bacterium would be smaller than the period at the end of this sentence.

A

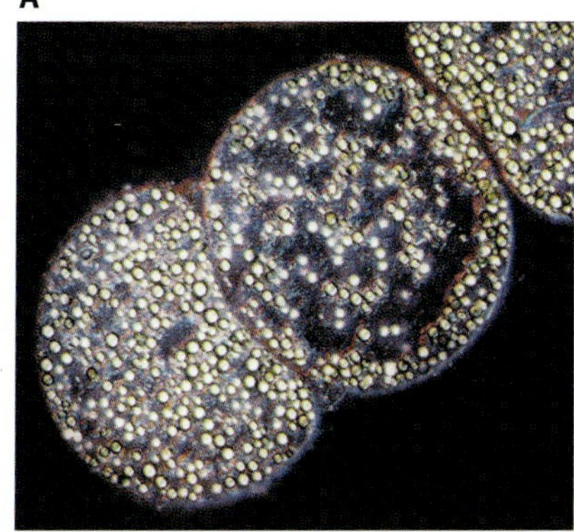

B

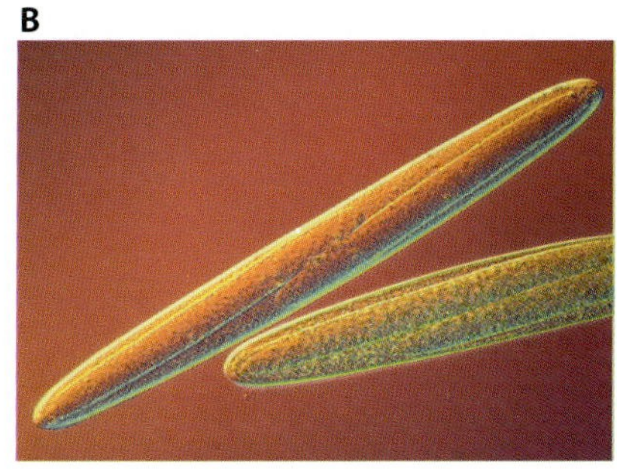

INCLUSIONS AND VESICLES

We have mainly dealt with "ordinary" bacteria and archaea, but there is a wonderland of prokaryotic shapes, forms, and structures, and some of this variety can be seen within the cell's interior. Some prokaryotes have internal vesicles that are involved in various aspects of the cell's physiology. Some are bound by lipidless protein membranes, while others are bound by bilayered lipid-protein membranes. What distinguishes these structures from mitochondria and chloroplasts is their lack of DNA.

Gas vesicles

In bodies of water, photosynthetic bacteria have a problem of buoyancy. The greatest amount of light is, of course, near the surface of the water. These bacteria, though light in weight, would eventually sink unless provided with buoyancy. Many such organisms are made buoyant by internal gas vesicles, which are structures filled by a gas similar to that of the environment (Fig. 3.5). These vesicles are surrounded by a shell *composed entirely of protein*; thus, they are not typical biological membranes.

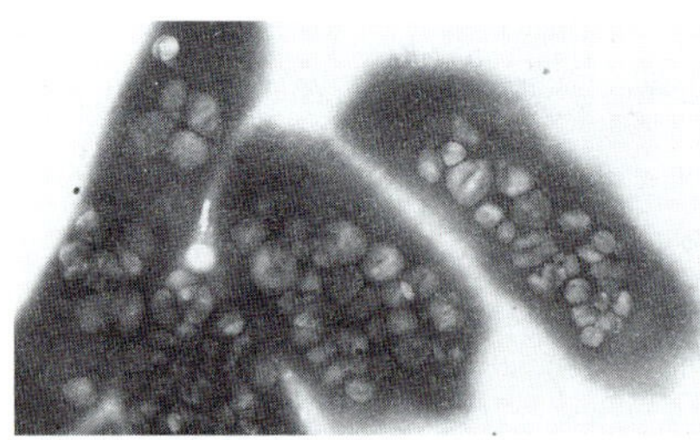

Figure 3.5 Negatively stained empty cells of a *Halobacterium* containing large numbers of gas vesicles.

The protein shell is freely permeable to gases, but not to water. The vesicles do not keep their shape because they are inflated by the gas but because the protein shell is quite rigid. Gas vesicles collapse when cells are subjected to a sudden increase in hydrostatic pressure.

Gas vesicles are characteristic of aquatic photosynthetic bacteria belonging to a large group known as the **blue-green bacteria**, or **cyanobacteria**. Not all use light of the same wavelength for photosynthesis, and thus, they may favor different locations in the water. The gas vesicles inflate or deflate according to controls that are geared to placing the cells at a position in the water column having light of the proper wavelength and intensity.

Organelles for photosynthesis and chemosynthesis

Photosynthesis takes place in highly organized spaces. In plants, it is carried out in intricate multicompartment assemblies, the **chloroplasts.** One of these compartments consists of the **thylakoids,** which is where light energy is captured and converted into chemical energy. Thylakoids are stacks of membranous sacs that greatly increase the surface area on which the photosynthetic reactions take place. In a fluid compartment surrounding the thylakoids, the **stroma,** energy is used to carry out the conversion of carbon dioxide into sugars (the Calvin cycle).

Photosynthetic bacteria have corresponding structures, but they differ in organization. Photosynthetic bacteria have various kinds of photosynthetic membranous structures, some continuous with the cell membrane and some not. Similarly, bacteria that obtain their energy from oxidation of inorganic material, such as reduced iron, sulfide, or hydrogen (the **chemoautotrophs**), also have amplified membrane structures where these processes take place. The special metabolism of these species of bacteria is described in Chapter 6.

Carboxysomes

Autotrophic bacteria (see Chapter 6), those that fix carbon dioxide to make their biochemical building blocks, often contain yet another kind of structure, the **carboxysomes** (Fig. 3.6). These are polyhedral bodies surrounded by a protein shell (not a lipid membrane) containing the key enzyme in carbon fixation, **RuBisCo (ribulose bisphosphate carboxy-**

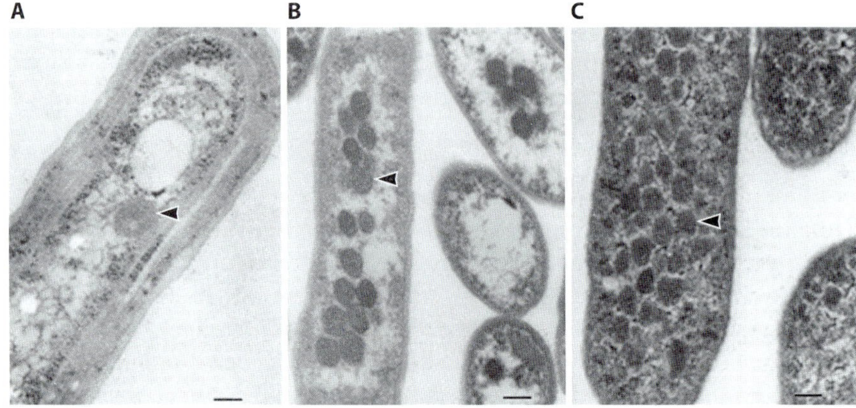

Figure 3.6 Various bacterial vesicles. **(A)** Carboxysomes in a photosynthetic cyanobacterium, *Synechococcus.* **(B)** Carboxysomes in a sulfide-reducing autotroph, *Halothiobacillus.* **(C)** Enterosomes in a heterotroph, *Salmonella.* The arrowheads point to typical structures. Bars, 100 nm.

lase). The structured environment in the carboxysomes may enhance the catalytic power of this enzyme, but how this takes place is not known.

Enterosomes

Surprisingly, structures similar to carboxysomes are also found in heterotrophic bacteria, such as *Salmonella* and *E. coli*, where they have been called **enterosomes** (because they were first found in enteric bacteria). Under the electron microscope, enterosomes appear to be similar to carboxysomes. They do not contain RuBisCo (Fig. 3.6), but some of their other proteins are homologous to proteins in the carboxysomes. Enterosomes contain enzymes required for the metabolism of certain substrates, such as propanediol or ethanolamine. Indeed, these structures are found only when the organisms are grown on these substrates. One may ask what role such substrates have and why complex enzymatic arrays are needed to metabolize them. Propanediol, it turns out, is a metabolite of fucose, a sugar found on the intestinal wall of mammals that can be degraded by intestinal bacteria. Enterosomes are relevant to *Salmonella* pathogenesis, because certain nonvirulent mutants have proved to be defective in propanediol metabolism.

The reason for the existence of enterosomes is not well understood. It has been proposed that they serve to cope with a toxic intermediate in propanediol metabolism, propanaldehyde. They must be of considerable importance, judging by the fact that they are retained by these bacteria even though their synthesis requires a considerable expenditure of energy.

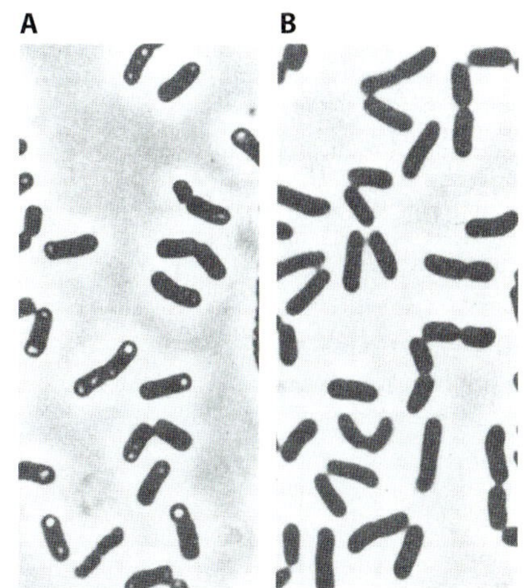

Figure 3.7 Storage granules in a member of the genus Bacillus *(Bacillus megaterium*; phase-contrast microscopy). (A) Cells growing at a high concentration of glucose and acetate. The light areas are granules of polyhydroxyalkane. **(B)** Cell from the same culture after having been incubated for 24 hours in the absence of a carbon source.

Storage granules and others

Under the microscope, some bacteria are seen to have refractile bodies in their cytoplasm (Fig. 3.7). These so-called **inclusion bodies** serve to store sulfur, phosphate (in several chemical forms, including polymeric pyrophosphates), calcium, or organic polymers, such as glycogen. In some cases, inclusion bodies are surrounded by a membrane, in which case they are commonly called volutin granules or **acidocalcisomes**, and are similar to some structures found in eukaryotic cells.

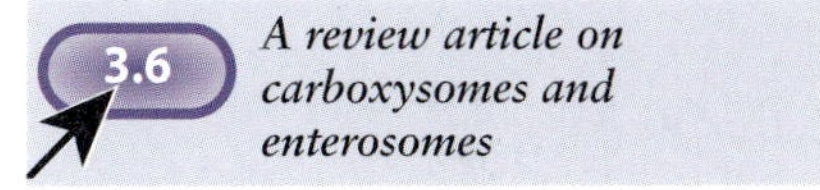

A wide variety of bacteria make storage compounds called **polyhydroxyalkanoates**, which are polymers that have industrial uses. Some of these compounds make stiff and brittle plastics; others make rubberlike substances. Because they are inherently biodegradable, they are regarded as an attractive source of nonpolluting plastics.

Some motile aquatic bacteria possess **magnetosomes**, which are membrane-bound crystals of magnetite or other iron-containing substances that function as tiny magnets. Such bacteria can orient themselves by responding to the magnetic fields of the earth (see Chapter 13).

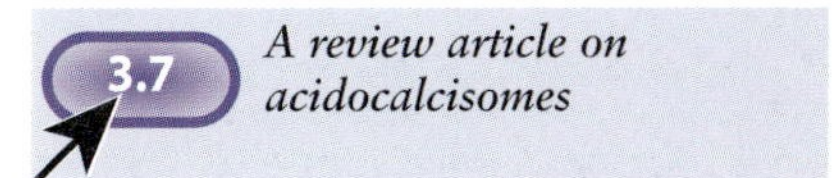

CONCLUSIONS

Prokaryotic cells generally lack a nuclear membrane, and their genomic DNA forms an irregular mass called the nucleoid. Some bacteria and many archaea contain vesicles that are surrounded either by proteins or by protein-lipid membranes. Although prokaryotic cells lack the organelles that characterize their eukaryotic counterparts, their interiors are surprisingly complex, densely packed with metabolic machinery, and exceedingly viscous. In the next chapter, we examine some characteristics of microbial growth and reproduction, preparatory to learning how this is all brought about.

STUDY QUESTIONS

1. Contrast the structure of the cytoplasm of a typical prokaryotic cell and that of a typical eukaryotic cell.

2. How does the bacterial nucleoid differ from a true nucleus?

3. What difference does it make that the DNA of the bacterial nucleoid is circular?

4. How do some bacteria change their buoyant densities?

5. What "extra" organelles are found in some prokaryotic cells? What are their known functions?

section III growth

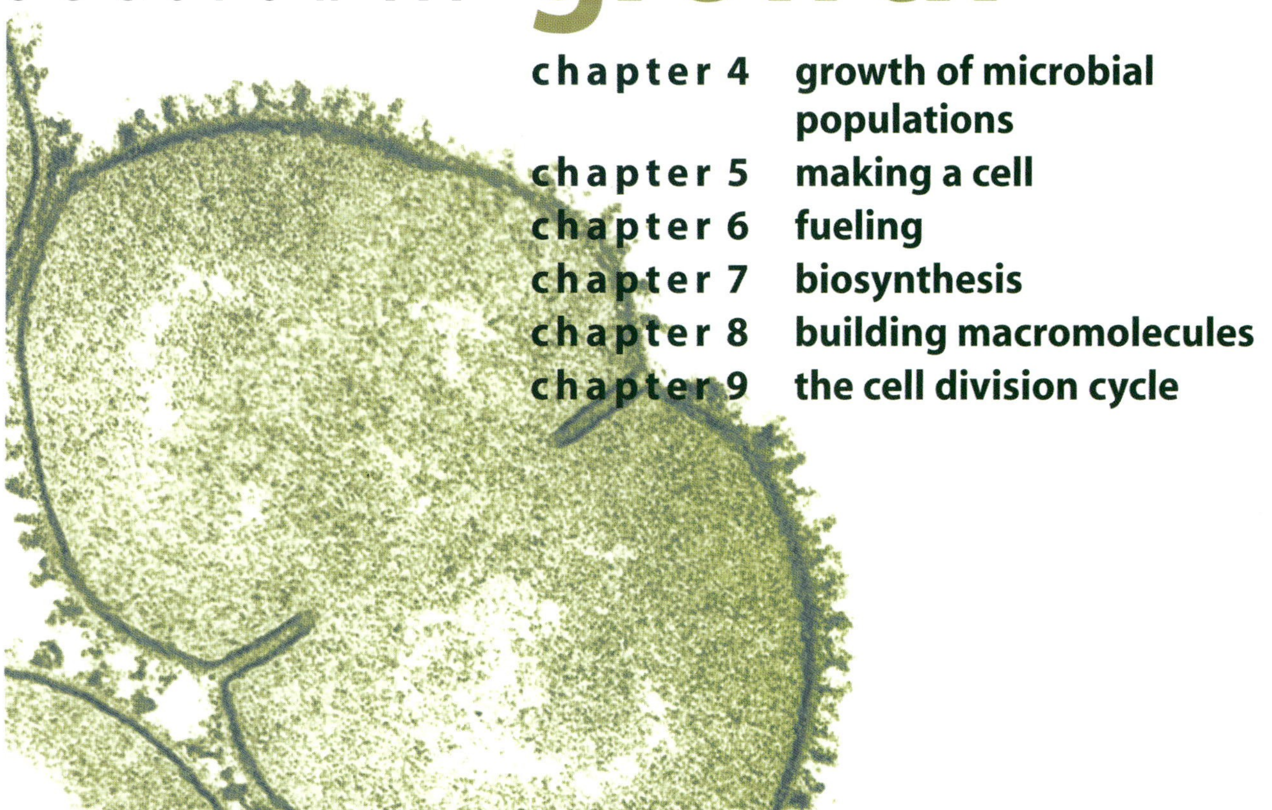

chapter 4 growth of microbial populations

INTRODUCTION

Most prokaryotes do not have an obligatory life cycle in the sense that more complex organisms do. As long as the environment is propitious, most prokaryotes reproduce continuously. Some bacteria undergo developmental changes, such as spore formation, but bacterial development is seldom the result of the inexorable program of **differentiation** seen in higher plants or animals. Thus, when bacteria are not actually growing, they are poised to do so.

4.1 *A movie of growing bacteria*

There are exceptions: some bacteria have developmental cycles that are obligatory. An example is the bacterium *Caulobacter*, which differentiates into two different kinds of cells at every cell division (see Chapter 14). Most other bacteria and practically all archaea have a more repetitive life style: each cell grows and eventually divides into two equal **progeny cells**, which then carry on like their progenitor. For such organisms, each generation closely resembles the previous one.

It is customary to use the term "growth" to indicate an increase in the number of cells in a population, as well as in the size of an individual. Although some bacteria grow quite slowly, requiring days or more for one cell to become two new ones, rapid growth is characteristic of many species, particularly those that have to compete in the environment for scarce or only intermittently available nutrients. Growth as a response provides the most comprehensive picture of how a free-living microbe is faring. Is it thriving, or is it being held back? Is it being challenged by changing environmental stimuli, or is it enjoying a relatively unperturbed existence? Growth is the most global indicator of a microbe's status, and it therefore deserves our attention.

There is an evolutionary premium on efficient growth. Most microbes do not live in isolation but must compete for food with their neighbors. Being able to utilize nutrients rapidly imparts a strong selective advantage. No wonder bacteria have evolved to grow at astounding speeds. The entire process of synthesizing a single *Escherichia coli* cell takes only about 20 minutes at 37°C in rich media, and *E. coli* is not even the speed champion. A bacterium isolated from salt marshes, *Vibrio natriegens*, can double in less than 10 minutes! Keep in mind, however, that fast growth is not the only available survival strategy. For example, adhering to surfaces allows the long-range survival of organisms that do not compete well for food but are able to stay put in a particular environmental niche.

To introduce the basic aspects of bacterial growth measurements, we will consider a straightforward question that can be readily answered in the laboratory: how fast does a given bacterial species grow in different culture media? To find out the answer experimentally, we first need to learn something about the culture media used in the laboratory. One might think that microbiologists would tend to use culture media that resemble the natural environment of the organism. This is true to some extent, especially for bacteria that are **photoautotrophic** (obtain their energy from light) or **chemoautotrophic** (obtain their energy from the oxidation of inorganic compounds) (see Chapter 6). But often, the medium is chosen for practical reasons, such as how fast colonies appear on agar plates, how easy it is to manufacture, or how much it costs. In research laboratories, other characteristics become important, such as knowing the precise chemical composition of a medium and being able to reproduce that composition readily.

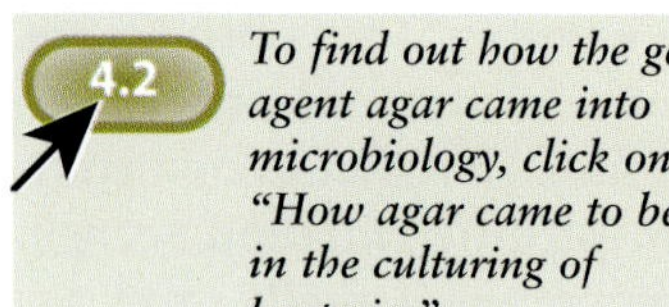

To find out how the gelling agent agar came into microbiology, click on "How agar came to be used in the culturing of bacteria."

In diagnostic laboratories, the behavior in culture of organisms isolated from a patient's sample often helps to determine their identity. For example, the ability to ferment lactose is a characteristic used to distinguish between certain intestinal bacterial pathogens and nonpathogens. One can simply use a **differential medium** that contains lactose and a pH indicator dye: lactose fermenters produce acid that changes the color of the medium around their colonies on agar plates, non-lactose fermenters do not. Other media are designed to let some organisms grow and not others. An example of such a **selective medium** is one that contains 5% sodium chloride, which doesn't hinder the growth of staphylococci but inhibits many other bacteria that might be present in a clinical sample. Also, the shape and color of a bacterial **colony** often suggest the identity of the organism or at least allow us to tell what it is not.

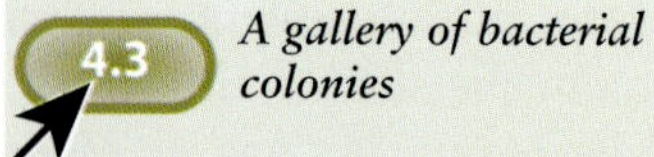

A gallery of bacterial colonies

To return to our experiment, many different culture media could be used to support the growth of a given species. If we wanted to grow *E. coli* or some other familiar bacterium, we could choose a rich medium called **nutrient broth**, which contains meat extract (similar to what is in beef broth bouillons) and another ill-defined constituent, peptone (which is a meat digest), which supply the organism with sugars, amino acids, purines, pyrimidines, vitamins, and other nutrients. Obviously, the precise chemical composition of such a medium is not known. At the other end of the spectrum, we could choose a fully defined synthetic **minimal medium**, one consisting only of inorganic salts with glucose as the sole carbon source (Table 4.1). We could increase the complexity of this medium by adding a number of nutrients, e.g., amino acids or vitamins, alone or in combination. Before using any culture medium, we must sterilize it

Table 4.1 A minimal medium known as MOPS medium for the growth of *E. coli*[a]

Constituent	Concentration (mM)
Potassium phosphate (as K_2HPO_4)	1.32
Ammonium chloride	9.52
Magnesium chloride	0.523
Potassium sulfate	0.276
Ferrous sulfate	0.01
Calcium chloride	5×10^{-4}
Sodium chloride	50
MOPS (a buffer)	40
Molybdenum, boron, cobalt, copper, manganese, zinc	Trace amt
Carbon source (e.g., 0.2% glucose)	

[a]MOPS, morpholinepropanesulfonic acid. To make a **rich medium** that supports high growth rates, the following supplements are added: all 20 amino acids, purines and pyrimidines, and several vitamins. Note that minimal media are totally defined and hence reproducible. In addition, they permit the use of isotopes (^{14}C, ^{35}S, and ^{32}P) and organic precursors for efficient labeling of cell components.

to destroy all the organisms that may be present, usually by using an **autoclave** (a fancy pressure cooker). Note that this procedure may not get rid of all forms of life because extreme thermophiles (see "Temperature" below) may withstand the treatment. Such organisms are found only in particular environments and in any case would not grow at 37°C, which is the temperature we would use in this experiment. Another sterilization technique for liquid media is via passage through a membrane filter that retains cellular microbes, although most viruses pass through. This method provides a means to prepare sterile media that contain heat-labile components.

We can now proceed with the experiment by choosing, for example, three different media: no. 1, nutrient broth; no. 2, minimal medium with glucose as the carbon source; and no. 3, the same as no. 2 with 10 amino acids added. We can proceed to add the organisms to each of these media. This process is called **inoculation**, and it usually involves the transfer of a small number of cells from a previously prepared pure culture (called the **inoculum**). We can now place the culture at 37°C in a shaker apparatus to provide it with oxygen (our *E. coli* likes it that way) and follow growth as the increase in number or mass of the bacteria versus time. For a typical laboratory strain of *E. coli*, the values of these parameters will double every 20 minutes in medium 1, every 50 minutes in medium 2, and every 35 minutes in medium 3. Nothing could be simpler, and yet this experiment contains several hidden truths and brings up a number of questions that are best left for Chapter 12. But first, how was the growth rate determined?

HOW TO MEASURE GROWTH OF A BACTERIAL CULTURE

There are various ways to measure the growth rate of a culture. One has the choice of following the rates of increase of the number of cells or of any cell constituent, disappearance of nutrients, or accumulation of metabolites. The most widely used measurement is that of **turbidity**,

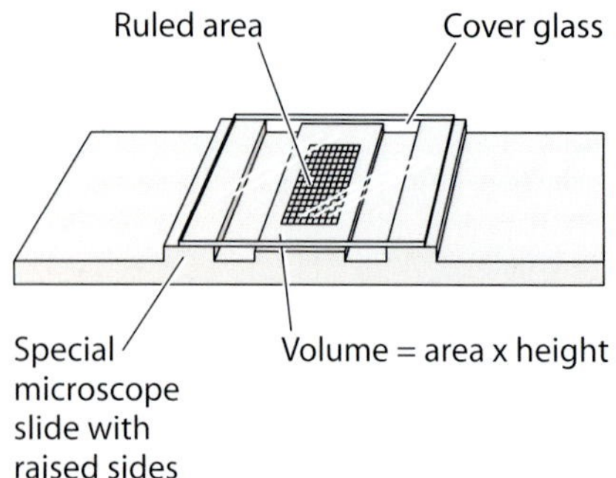

Figure 4.1 Chamber for counting cells under the microscope (also known as a hemocytometer). The ruled portion defines an area, and the space between the raised cover glass and the slide defines the height. Counting the particles over a certain area gives an estimate of their number per unit volume.

Figure 4.2 Electronic particle counter. A dilute suspension of particles, e.g., bacteria, in a buffer is placed in the outer beaker. A certain volume of liquid is sucked through a hole on the side of the inner tube, and conductivity between the two electrodes is measured. To count bacteria effectively, the hole in the inner tube can be as large as 30 micrometers (μm).

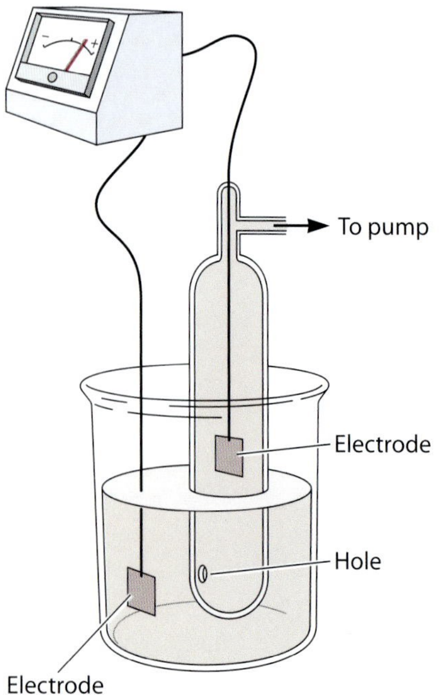

which can be determined nearly instantaneously using a colorimeter or a spectrophotometer. This measurement is not only convenient, it is accurate. Turbidimetric measurement depends on the ability of bacterial cells to scatter light (not on their light absorption, because they are virtually transparent). There are instances, however, when such a measurement is not practical, e.g., when the medium itself is cloudy. However, for the purpose of determining the growth rate in the experiment presented above, turbidimetry would be the method of choice. If attention is paid to the chemical composition of the medium and the temperature, this measurement of growth rate can be reproducible and accurate, easily to within an error of 5%.

Sometimes one wishes to determine the number of *living cells in a population* rather than the total bacterial biomass. Counting living cells relies on the ability of a single bacterium to produce a colony on a petri dish containing a suitable culture medium solidified with agar. To carry out such a **viable count**, one dilutes the sample, spreads a measured sample of the dilution on the surface of the agar, incubates the petri dish for an appropriate time, and counts the number of colonies that develop. The technique has potential for errors that in some cases are difficult to avoid. For example, if the bacteria clump together, the number of colonies reveals the number of clumps, not the number of individual cells.

Total-particle counts can be carried out using a microscope or an electronic particle counter. Under the microscope, cells in a sample can be counted with a **counting chamber** (Fig. 4.1), which is a glass slide with a central depression of known depth (for bacteria, usually 0.02 mm), the bottom of which is ruled into squares of known area. The depression is filled with a bacterial suspension, covered with a coverslip, and allowed to stand until the cells have settled to the bottom (so they will all be in the same plane of focus). The number of cells per unit area (here, proportional to the volume) reveals the total number of cells in the suspension. An analogous device used for counting blood and other animal cells is called a **hemocytometer**.

Electronic counting is based on the principle that a bacterial cell conducts less electricity than a saline solution. An electronic particle counter consists of two chambers containing a saline solution to which a sample of a bacterial suspension has been added. The chambers are connected through a small hole (Fig. 4.2). Every time a bacterium passes through the hole, it decreases the electrical conductivity, and the number of changes is tallied electronically. Cell counters are often used for counting other particles, for example, white blood cells or animal cells in culture. Note that these machines are different from flow cytometers, which are normally used to determine the distribution of some property (e.g., DNA content per cell) in a population.

WHEN SHOULD THE GROWTH RATE BE DETERMINED?

The changes in the turbidity of a typical culture are plotted in Fig. 4.3. Note that the culture grew over a period of many hours but had stopped growing by 24 hours. It was then said to be in the **stationary phase** of growth. What made growth stop? The cells in the culture used up nutrients and secreted waste products. Eventually, some essential nutrient became exhausted or some waste product reached an inhibitory concentration.

If a culture that has been in the stationary phase for some time is transferred to fresh medium, it may not start growing right away. Growth is delayed for a period of time, the length of which depends on the organism, the particular medium, and the length of time the culture has been in stationary phase. This period is called the **lag phase.** In spore-forming bacteria, such as *Bacillus subtilis*, as well as others that form resting structures (see Chapter 14), the lag phase has special attributes. Because these structures form during the stationary phase, subsequent growth of the culture in fresh medium can resume only after the lengthy process of spore **germination.** *E. coli* also undergoes physiological rearrangements during the stationary phase, although the changes are less obvious than in sporulation (see Chapter 13). Once growth resumes at a steady rate, the culture is said to be in the **exponential phase.** Here, the growth rate is constant, which (to answer the question in the subheading) is when the measurement should be carried out.

A culture in the exponential phase grows at a constant rate as long as the concentration of substrate being consumed does not fall below a utilizable level. The growth rate remains constant and decreases to zero when the substrate is depleted. Therefore, the growth rate changes only when the substrate concentration falls below a certain threshold value, usually in the micromolar range. The relationship between growth rate and substrate concentration approximates first-order kinetics (Fig. 4.4), which resembles the relationship between the velocity of an enzyme-catalyzed reaction and the concentration of the substrate for that reaction, as described by the Michaelis-Menten equation.

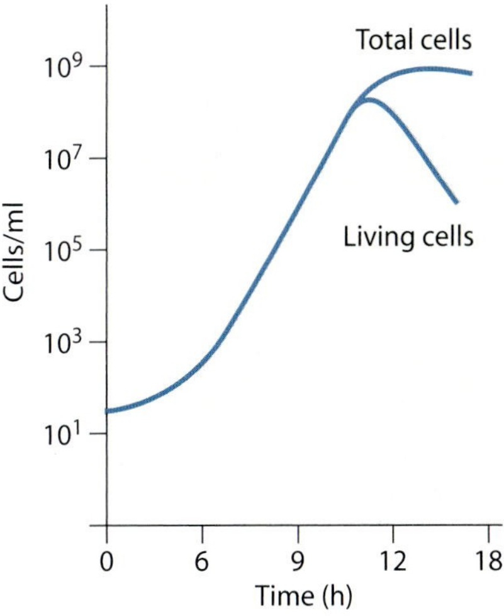

Figure 4.3 A typical bacterial growth curve. Measurements were carried out for the total number of cells and for the number of viable cells. Note that the two are the same during much of the growth of the organisms, indicating that **viability** was nearly 100%. However, as growth slowed down, some cells died. They may still be intact, however, and thus contribute to the total number. Not all bacteria do this. Some lyse as their growth slows down.

THE LAW OF GROWTH

Have you ever heard the saying "When cells divide, they multiply?" At each division, one bacterium becomes two, these become four, etc. The series 2, 4, 8, 16, ..., 2^n describes the increase in cell numbers, where n is

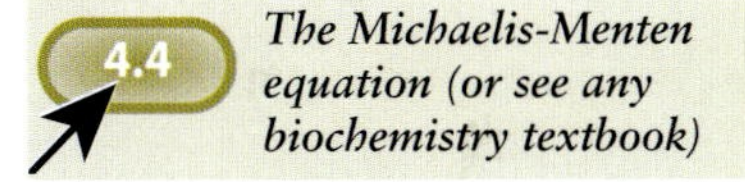

The Michaelis-Menten equation (or see any biochemistry textbook)

Figure 4.4 Change in growth rate as a function of concentration of an essential limiting substrate. The growth rate is constant over a wide range of substrate concentrations but is lower when the substrate is scarce. For measurements to be carried out in this region, the bacterial cultures must be highly dilute, or else the cells would readily consume the small amount of substrate and thus alter its concentration over the time of the measurement. For a perspective, the concentration of glucose that results in the half-maximal growth rate (marked by the horizontal and vertical lines) in *E. coli* is 10^{-6} M.

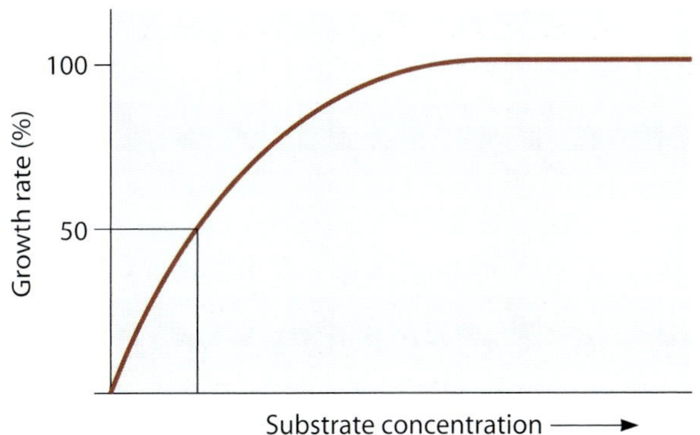

the number of **generations** (1, 2, 3, 4, etc.). The number of cells, N, after n generations beginning with N_0 cells, can be calculated from the following equations:

$$N = 2^n N_0 \tag{1}$$

and

$$\log N = n \times \log_{10} 2 + \log N_0 \tag{2}$$

If the numbers of cells at the beginning and at any particular time thereafter are known, the number of intervening generations can be calculated from the equation:

$$n = \frac{\log N - \log N_0}{0.301} \tag{3}$$

For example, if we start with 1,000 bacteria/ml, how long will it take for the population to reach 1 billion cells/ml? From equation 3, n is equal to $(\log 10^9 - \log 10^3)/0.301$; thus, n is equal to 6/0.301, or about 20 generations.

If we know the time interval over which this increase occurred, we can calculate how much time (t) it takes for the cells to double in number (one **generation time** [g]) from the equation $g = t/n$. If the time were 10 hours, g would be 10/20, or 30 minutes.

So far, we have described growth in steps of cell doubling. Calculating the rate of growth for *any time interval* requires a mathematical digression (which will give you a chance to use what you learned in a calculus course, perhaps for the first time).

In mathematical terms,

$$dN/dt = kN \tag{4}$$

where dN is the change in the number (N) of cells per unit volume over time (t), and k is a proportionality constant called the **specific growth rate**. The units of k are reciprocal time, usually expressed as hours^{-1}.

The steps from equation 4 to equation 7 are for mathematical aficionados. Integrating equation 4 yields:

$$N/N_0 = e^{kt} \tag{5}$$

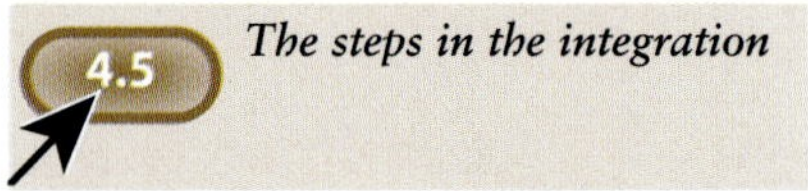

Taking the logarithms of both sides,

$$\log N/N_0 = \log N - \log N_0 = \log(e)kt \tag{6}$$

and converting to log base 10,

$$k = 2.303\,(\log N - \log N_0)/t \tag{7}$$

The specific growth rate (k) of the culture describes how fast a particular bacterium grows in a particular environment. For example, if a culture contains 10^3 cells at time zero and 10^8 cells 6 hours later, the specific growth rate is $(8 - 3)2.303/6$, or $k = 1.92\ \text{hours}^{-1}$. From equation 7, we can also determine the relationship between k and g. Because g is the time for N_0 to double, i.e., for 1 to become 2, we see that $k = 2.303(0.301 - 0)/g$, or $k = 0.693/g$.

Even without mathematics, it should be reasonably obvious that *the number of cells or of any of their components in a growing culture increases at a rate proportional to the number of cells present at any particular time*. The more cells there are, the more cells are made in a particular interval of time. Thus, **exponential growth (logarithmic growth)** mimics an autocatalytic, or first-order, chemical reaction. The fundamental law of growth expressed in equation 5 is the same as that of compound interest (which Benjamin Franklin is said to have called the "eighth wonder of the world").

BALANCED GROWTH

As long as a culture is in the exponential phase, all cell constituents increase by the same proportion over the same interval of time. In other words, the time it takes to double the number of cells in the culture is the same as the time it takes to double the amount of DNA or the number of ribosomes, individual enzyme molecules, and so on. As an example of how handy this can be, if on a single occasion you determined that the average DNA content of a single *E. coli* cell is 26 femtograms (fg) in a culture that is doubling every 40 minutes, you would know *without any further measurement* that the average rate of synthesis of DNA is 0.65 fg/bacterium/minute. This condition is called **balanced growth**, and whereas it can be approximated over a considerable time in the laboratory, it does not usually persist for long in the environment. Under most natural conditions, bacteria alternate between periods of growth and nongrowth, that is, they are constantly entering into and coming out of the stationary phase.

In the laboratory, using cultures in balanced growth (growing exponentially) has some obvious advantages. This is the *only phase of growth of a culture that is readily reproducible* and therefore can be replicated on different occasions and in different laboratories. In the same medium with a suitable gas phase and under the same conditions of temperature, osmotic pressure, and pH, a culture will behave the same way, day after day. The organisms will then be in the *same physiological state*. In the alternative condition, where growth is unbalanced, the properties of the organisms will vary as conditions vary. With cultures in balanced growth, it does not matter when the culture is sampled (as long as the time of sampling is recorded and related to some growth measurement).

Balanced growth also suggests that the *mean* cell size remains constant, a condition that might at first glance appear paradoxical, because as they grow, individual cells increase in size and eventually divide. Indeed, balanced growth refers to the *average behavior of cells in a population*, not to that of individual cells.

When we go from theory to practice, terms related to balanced growth, such as "unchanging environment" and "steady state," must be used as approximations only. They really mean "presumed unchanging environment" or "approximate steady state." This point is a subtle one. *Some cell properties change early during the growth of a culture, long before the increase in mass slows down.* Some of these changes are often imperceptible, which means that balanced growth conditions can be approximated only at low cell densities. This fact is sometimes inconvenient,

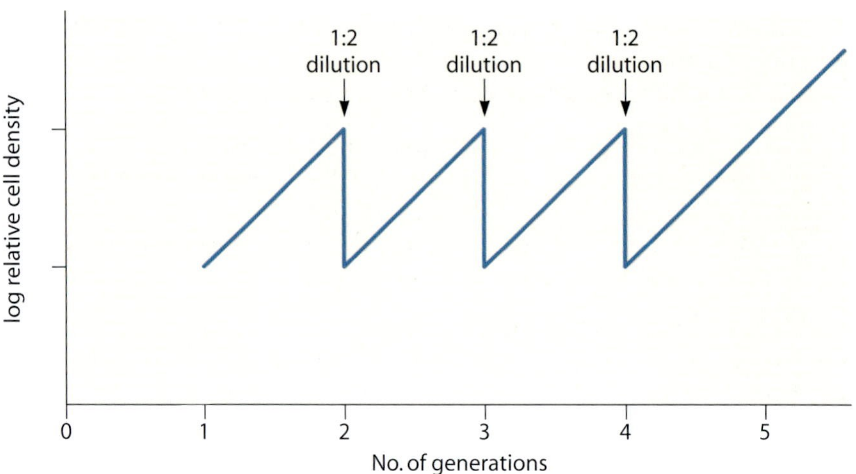

Figure 4.5 How to keep a culture growing exponentially. A culture will grow exponentially, i.e., be in balanced growth, as long as it does not exceed a certain cell density. One way to achieve this condition is to dilute the culture with the same medium whenever it reaches a certain cell density. In this example, a twofold dilution was carried out after each generation. Although tedious, this procedure can keep a culture in a steady state, in principle forever.

because certain biochemical measurements require a fairly large sample, but it holds true nonetheless.

Reference is often made in the literature to cultures harvested in "early" or "late" **logarithmic (log) phase,** or even in "mid-log" phase; these imprecise statements are disturbing because they give the impression that no great care was taken to ensure that cultures are in balanced growth. Therefore, there is no assurance that growth was unchanging at the time of sampling. Unless growth is monitored throughout a physiological experiment, the results may not be reproducible, whether in one's own or somebody else's laboratory. *In fact, a physiological experiment done with a poorly characterized culture is all but useless.*

CONTINUOUS CULTURE

In the laboratory, a culture can be maintained in balanced growth by diluting it at set intervals with fresh medium (Fig. 4.5). If this is done frequently (e.g., once every few doubling times), the culture will remain in the exponential phase of growth. As long as the operator is willing to carry out such dilutions, the cells will grow in an unrestricted manner. However, making periodic dilutions over a long period of time can be a tedious business. Such tedium can be avoided by automation. In order to maintain cultures in a state of balanced growth over a range of growth rates, one can use an apparatus called a **continuous-culture device,** or **chemostat.**

A sketch of a chemostat is shown in Fig. 4.6. In principle, a chemostat is a simple apparatus consisting of a culture vessel kept at the desired temperature that can be aerated for the growth of aerobes. Fresh medium is added at a rate set by a metering pump or a valve that regulates the flow rate. The volume is kept constant by removing medium through an overflow device at the same rate as fresh medium is added. Vigorous mixing is employed to ensure that the fresh medium rapidly equilibrates with the contents of the culture vessel. For a chemostat to function properly, the bacterial density should not exceed that which allows balanced growth in a batch culture. This condition is achieved by *making an essential nutrient limiting.* In practice, this is done by reducing the concentra-

Figure 4.6 Continuous-culture device, or chemostat. Fresh medium from a reservoir is fed at a constant rate into a culture vessel by using a metering pump. The volume of the culture is maintained constant by the removal of excess through an overflow at the same rate as new medium is added. Not shown is a device for mixing the culture vessel contents. The size of the culture vessel may vary from a few milliliters to thousands of gallons.

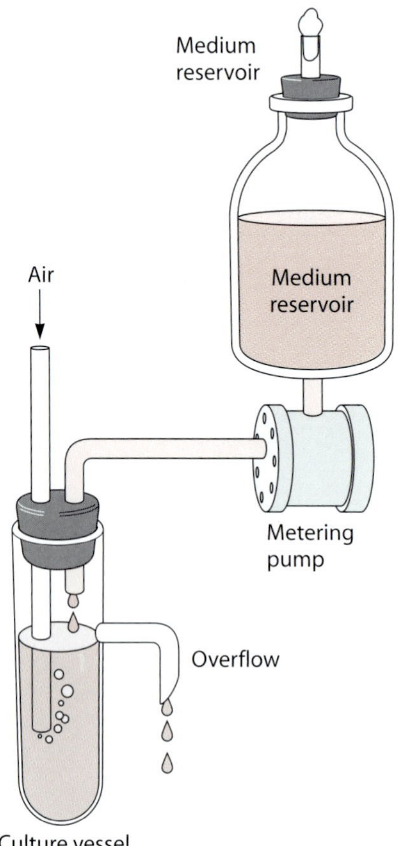

tion of some essential nutrient, such as glucose, ammonia, phosphate, or a required amino acid.

The important properties of a chemostat are as follows.

1. The rate of addition of fresh medium (per volume) determines the growth rate in the culture vessel (up to the maximum possible for that organism in that particular medium).
2. The density of bacteria in the culture vessel is constant and is determined by the concentration of a limiting nutrient.

As might be expected, the chemostat has proven useful in the laboratory for studies of bacterial **mutagenesis** and evolution. The device also has applications in large-scale industrial fermentations.

To illustrate how the chemostat is a self-correcting system, let us imagine increasing the rate of addition of fresh medium. The rate of outflow will increase, and cells will be lost at a greater rate than they are formed. Consequently, the density of cells in the vessel will decrease, utilizing the limiting nutrient at a lower rate, and the concentration of nutrient in the vessel will increase. The growth rate will then increase until it matches the rate of loss of cells through the outflow. If the rate of addition of fresh medium were to decrease, the opposite series of events would ensue. Thus, the growth rate of the culture adjusts to match the rate of loss from the vessel, and the culture density remains constant.

HOW IS THE PHYSIOLOGY OF THE CELLS AFFECTED BY THE GROWTH RATE?

All organisms grow faster and do better when provided with a good diet, but the effects on bacteria are especially dramatic: under otherwise constant environmental conditions (pH, temperature, and osmotic strength), the growth rate of a bacterial culture can vary greatly. Many bacteria are particularly well adapted to exploit their nutritional environment and to convert it into their own special form of selective advantage—a high growth rate. To accomplish this remarkable feat, the makeup of a bacterial cell changes profoundly with the growth rate as it is imposed by the nutritional properties of the medium: both the macromolecular composition and cell size change with the growth rate. The bases of some of these changes are obvious. For example, if a bacterium is to grow faster, it needs more protein-synthesizing capacity. Unused machinery is a disadvantageous expenditure of energy, so to grow at the maximum rate in a particular medium, a bacterial cell must contain an optimal amount of the protein-synthesizing machinery. More or less of it would decrease the growth rate.

The growth rate that a particular medium supports, not its specific composition, determines the **physiological state** (cell size and macromolecular composition) of cells growing in it. For example, the physiological states of cells are the same even in different media, as long as the media support the same growth rate. Thus, changes in physiological state are *nutrition mediated but not nutrition specific*. Temperature, pH, osmotic strength, and hydrostatic pressure also affect the growth rate, but changes in these nonnutritional parameters do not greatly affect the overall macromolecular composition and cell size. For example, a change in

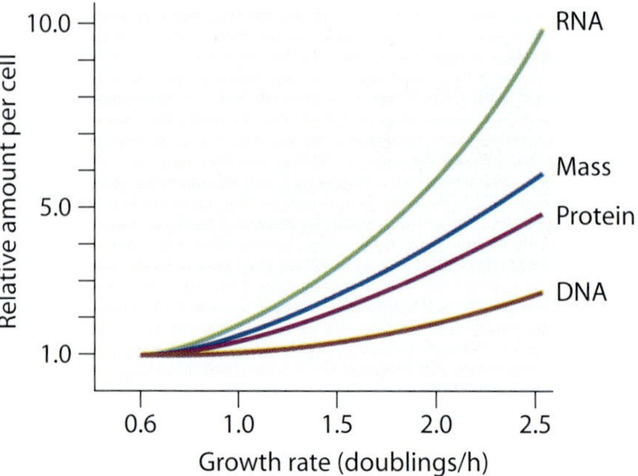

Figure 4.7 Compositions of *E. coli* cultures growing at different rates. Individual cultures were grown at the same temperature in a variety of media, each supporting a different growth rate. The relative amounts of various constituents are plotted as values *per cell.*

incubation temperature that would change the growth rate severalfold will change the intracellular levels of specific proteins (see Chapter 12), but it will have an almost imperceptible effect on the fraction of cell mass that is protein, RNA, or DNA. Temperature changes affect the rates of all cellular processes by similar factors; hence, no modification of the cell's overall macromolecular composition is needed to adjust to the new environment.

How does the growth rate affect macromolecule composition? As shown in Fig. 4.7, cell mass, protein, RNA (mainly in ribosomes), and DNA change with the growth rate: as the cell grows faster, it becomes bigger and contains more of each component. These are major changes: the average cell in a culture of *E. coli* growing at 2.5 doublings per hour contains more than 10 times as many ribosomes as do cells from a culture growing at 0.6 doubling per hour. Since there must be a place to put this extra RNA, fast-growing cells are larger. Figure 4.8A shows how much larger fast-growing cells are than slow-growing ones. Here, cells growing with a doubling time of 22 minutes were mixed with some cells growing with a doubling time of 72 minutes. It is hard to believe they are the same species!

Individual cell constituents change in concentration to different degrees with the growth rate. RNA increases much faster than mass, protein increases somewhat more slowly, and DNA increases even more slowly. To illustrate this, let us plot the ribosome content as a *percentage of total mass rather than on a per-cell basis* (Fig. 4.8B). Note that above 0.5 doublings/hour, the proportion of the cell mass comprised of ribosomes is a linear function of the growth rate. In other words, the faster cells grow, the more ribosomes they must have in order to fulfill their need to synthesize a larger amount of protein. Such a linear relationship suggests that *each ribosome* (and its accompanying components of the protein-synthesis apparatus) *synthesizes proteins at the same rate in fast- and slow-growing cells.* Note that this is an efficient form of regulation. Consider an alternative in which cells possess about the same amount of ribosomes regardless of growth rate. If this were the case, either all ribosomes in slow-growing cells would have to work at a lower rate or a fraction of them would function at a high rate, with the rest being idle. Either alternative imposes a great energy demand. Thus, instead of wasting energy on making inefficient ribosomes, *the cells optimize their numbers.*

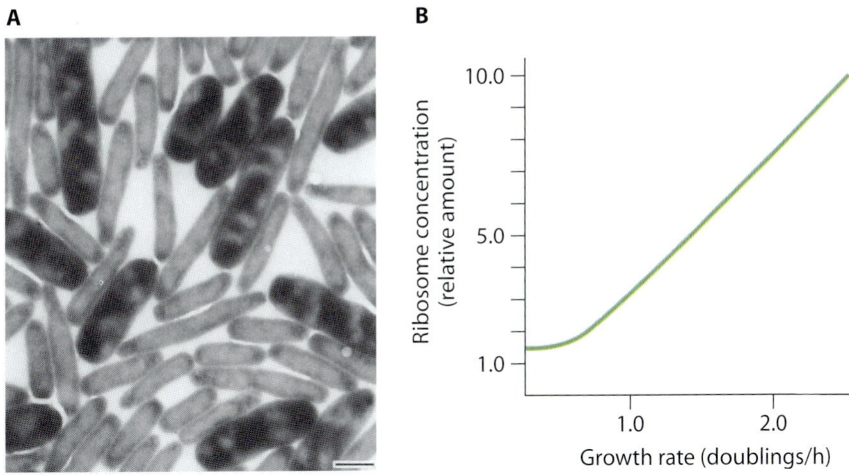

Figure 4.8 Change in cell size and ribosome content with growth rate. (A) Electron micrograph of a mixture of *E. coli* cells grown in different media. The large cells grew at 2.3 doublings per hour, and the smaller ones grew at 0.8 doubling per hour. Bar, 1 micrometer. **(B)** Cultures were grown as for Fig. 4.5, and their ribosome contents were expressed as proportions of the cell mass. Because the sizes of cells vary with the growth rate, this plot appears different from than in Fig. 4.7. Note that the relationship between the ribosome concentration and the growth rate deviates from linear at low rates. This suggests that, no matter how slowly the cells grow, they must retain a minimum number of ribosomes, possibly to be able to resume growth rapidly when nutritional conditions improve.

In Chapter 9, we will see that cells use the same strategy to make their DNA. There again, the rate of DNA synthesis is nearly invariant.

The graph in Fig. 4.8B makes another point. In very slowly growing cultures, the ribosome content deviates from linearity and cells have “extra” ribosomes. By the argument above, this would seem to be wasteful. Indeed, making these ribosomes is not energetically efficient. However, consider that cultures in a poor medium will, at some later time, be in a rich medium. In order to compete with other organisms, these cells must crank up their protein synthesis as rapidly as possible. If they had no ribosomes at all (as if the linear relationship in the graph were to extrapolate to zero growth rate), they would not be able to resume growth under the new conditions. Having some ribosomes permits them to resume growth.

How cells adjust the proportion of their resources that goes to building ribosomes at different growth rates has been a central question in bacterial physiology. What we know of the answer is described in the section “Global response networks” in Chapter 13.

EFFECTS OF TEMPERATURE, HYDROSTATIC PRESSURE, OSMOTIC STRENGTH, AND pH

The cells within large animals exist in a constant, near-optimal environment; they are bathed in an isotonic solution that is maintained at optimal pH and temperature. In contrast, microbes are in intimate contact with an environment that can change abruptly and become threatening, so we should not be surprised to learn that most microbes have evolved

Table 4.2 Temperature responses of microbes

Class	Typically grows at:
Psychrophile	5°C
Facultative	Maximum at 20°C or above
Obligate	Maximum <20°C
Mesophile	37°C
Thermophile	50 to 70°C
Eurythermophile	Broad range of temperatures; can grow at 37°C
Stenothermophile	Narrow range of temperatures; cannot grow at 37°C
Hyperthermophile	>70°C

an array of mechanisms to cope with environmental stresses and changes. The ecology of a few prokaryotes, called **extremophiles**, is remarkable in another respect. Various members of this group have evolved not only to tolerate but also to thrive in environments at startling extremes of temperature (Table 4.2), pH, hydrostatic pressure, or concentrations of salt. Some even tolerate more than one of these environmental challenges. The theme of microbial adaptations to change, including those imposed by the environment, is discussed in detail in Chapter 13. Here, we present an overview of how microbes interact with their environment.

Temperature

It is easy to summarize how temperature affects microbial growth: microbes can grow wherever there is liquid water, regardless of the ambient temperature. The range of temperatures for liquid water on Earth, and hence the collective growth range of microbes, extends well beyond the freezing and boiling points of pure water at sea level. Water is kept liquid below the freezing point by dissolved solutes and above the boiling point by high pressure. Consequently, some microbes can grow in high-salt environments at −15°C and probably lower. It is for this reason that household freezers are set at temperatures below −15°C. But as those who have had hams spoil after some weeks or months in the freezer know, even such low temperatures do not ensure that salty food will remain palatable and safe. Note that the very slow growth that occurs at these temperatures tests the investigator's patience and raises doubts about whether transient rises in temperature may account for observed growth. At the other end of the temperature scale, there are microbes that can grow at well above the temperature of boiling water, which exists under high pressure in the depths of the ocean or in fissures of deep rocks.

There is definite specialization among microbes with respect to growth at temperature extremes. Broadly speaking, bacteria and fungi dominate the lowest temperatures and archaea dominate the highest, but all microbes specialize in one particular range of temperature. Few prokaryotes, if any, grow over a range greater than 40 Celsius degrees, and many are restricted to considerably narrower ranges.

Effect on growth rate

Regardless of the particular temperature and range that a microbe prefers, a generalization about the effect of temperature on growth is possible. Over a certain range (the **"normal"** range), the temperature affects the growth rate the same way it does the rate of a chemical reaction. The growth rate obeys the Arrhenius equation, which describes the logarithm of a chemical reaction rate as a linear function of the reciprocal of absolute temperature (Fig. 4.9).

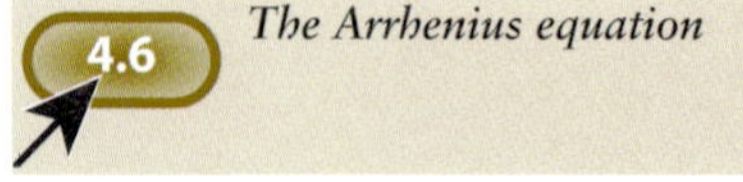

The Arrhenius equation

At both the high- and low-temperature ends of the normal range, growth becomes increasingly slower than is predicted by the Arrhenius equation. At the high end, this decline in the growth rate passes through a maximum (the **optimum growth temperature**) and then falls precipitously to zero (the **maximum growth temperature**). At the low end, the decline is slower, but eventually the slope of the curve becomes vertical at the **minimum growth temperature**. The general shape of this curve is the same for all microbes, although its location on the temperature scale and the slope of the linear region are variable.

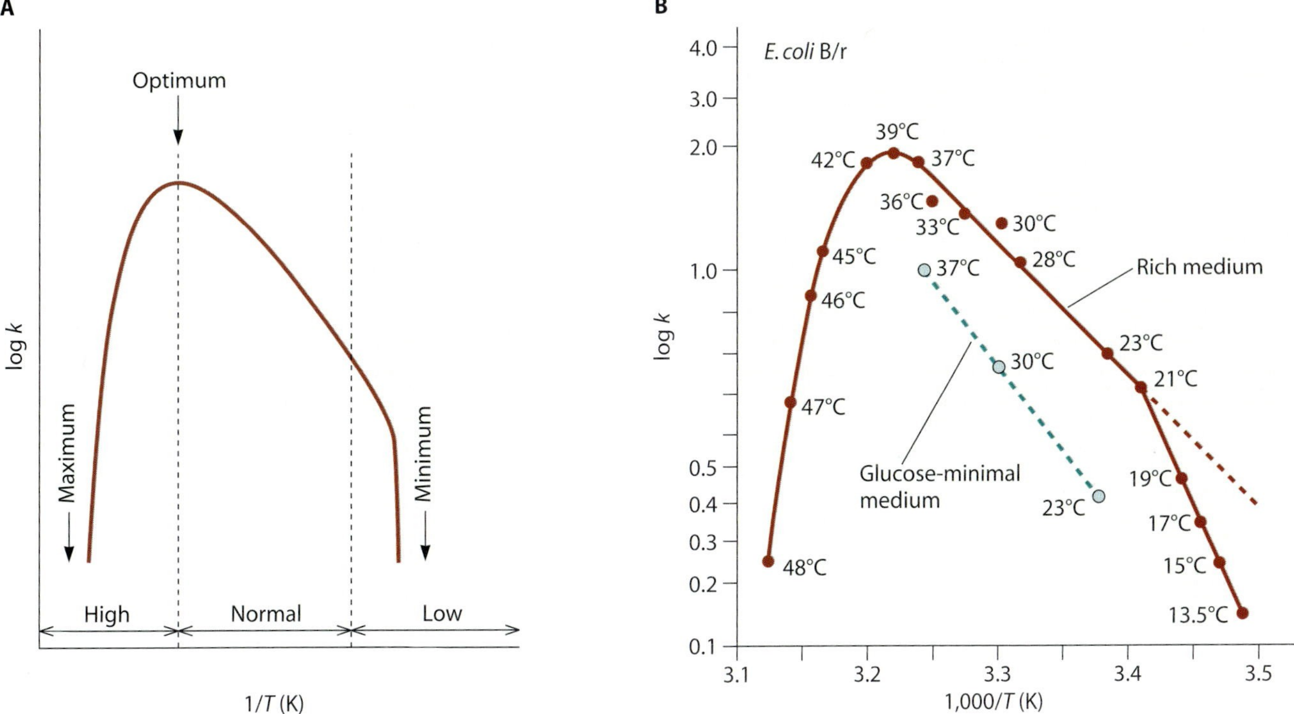

Figure 4.9 Effect of temperature on bacterial growth. (A) General form of an Arrhenius plot. Temperature (*T*) as the reciprocal of absolute temperature (Kelvin [K]) is plotted against the logarithm of the growth rate as *k*. **(B)** Specific values of the growth rate of *E. coli*. Note how it grows more slowly in minimal medium than it does in a rich medium, but the slope of the Arrhenius plot in the normal range is the same for both media. The minimum temperature of growth for *E. coli* is 7.8°C.

Classifying the temperature responses of microbes

Given the wide variation of various microbes' responses to temperature, it is a challenge to assign them to neat, well-defined classes. Still, microbiologists have enjoyed trying, although they have not always been able to agree on precise definitions. One way to classify temperature responses is by the microbe's optimum temperature for growth: a microbe is a **psychrophile** if it is low, a **mesophile** if it is moderate, a **thermophile** if it is high, and a **hyperthermophile** if it is very high. These are useful, widely used terms, but they do not tell us about maximum and minimum temperatures of growth or the extent of the growth range. The champion psychrophile at present is *Psychromonas ingrahamii*, which actually grows at −12°C.

Growth limits at temperature extremes

We might ask why the growth rate declines and eventually stops at the limits of a microbe's growth range. The reason for cessation at **high temperature** is relatively straightforward: some vital cellular component, most likely a protein, becomes thermally inactivated. Many mutations (**heat-sensitive mutations**) decrease that microbe's maximum temperature of growth by diminishing the thermal stability of an essential protein.

The existence of the extreme thermophiles defies ordinary "cool" biochemistry and invites a reexamination of the molecular properties of proteins, nucleic acids, and lipids. In ordinary organisms, these molecules are unstable at temperatures that are far lower than those required for the growth of extremophilic Bacteria and Archaea. Thermophiles, therefore, have adapted to high temperature by evolving constituents with unique biochemical properties. Enzymes isolated from such organisms are active at temperatures close to those for optimal growth and are often inactive at room temperature. Heat resistance is genetically encoded: when cloned in mesophilic hosts, these enzymes generally retain their thermal properties. Such proteins do reveal massive departures from the usual in their overall amino acid compositions. There is evidence, however, that their thermostability depends on their tertiary structure, as influenced by hydrogen bonding, hydrophobic internal packing, and salt bridges. Accordingly, proteins of thermophiles have a larger proportion of charged amino acids, such as glutamic acid and arginine, on their molecular surfaces, which allows additional ionic bonds. In addition, hyperthermophiles manufacture heat shock proteins, which protect other proteins from heat denaturation. Some heat shock proteins are also **chaperones,** proteins that facilitate the folding of other proteins or assembly of multiprotein complexes. Up to 80% of the dry weight of the archaeon *Pyrodictium occultum*, when grown at 108°C, is a specific chaperone. It seems likely, then, that thermal resistance is based partly on inherent properties of the proteins plus their protection by chaperones.

Also puzzling is how cellular nucleic acids maintain their integrity at high temperatures. In solution, DNA melts (its two strands separate) at temperatures well below 80°C, so some mechanism must operate to keep this from happening. Among the factors involved may be the fact that thermophiles have both thermoprotective DNA-binding proteins and relatively high magnesium concentrations that stabilize the molecules by neutralizing their phosphates. In addition, thermophiles have a unique enzyme, a reverse gyrase, which is the only topoisomerase known to introduce positive supercoils into DNA. It is thought that such supercoiling contributes to the thermostability of the DNA. It seems that no single mechanism is responsible for the stability of the proteins and nucleic acids in hyperthermophiles. Rather, increased thermostability is probably due to a number of fairly subtle alterations that do not follow obvious rules.

What about the stability of the thermophiles' lipids? In some of the Archaea, a domain with many thermophiles, membranes are not composed of phospholipid bilayers but rather are isoprenoids linked to glycerol by highly stable ether bonds (rather than ester bonds as in ordinary lipids) (see Fig. 2.12). In some cases, pairs of these molecules form the membrane; in others, single double-headed molecules span the membrane as monolayers. These unique molecular properties suggest that archaeal membranes are well adapted to high temperatures, because ethers are more stable than esters and monolayers are probably more stable than bilayers. A problem with this notion is that nonthermophilic Archaea also have this unusual architecture. A reasonable guess is that the Archaea may have first arisen in a high-temperature environment and later evolved into forms that grow at lower temperatures. This leads to the much-debated notion that life itself may have originated at high temperatures, characteristic of Earth's early periods. Supporting this notion is the fact that the most ancient lineages in both the Archaea and Bacteria are thermophilic (Fig. 4.10).

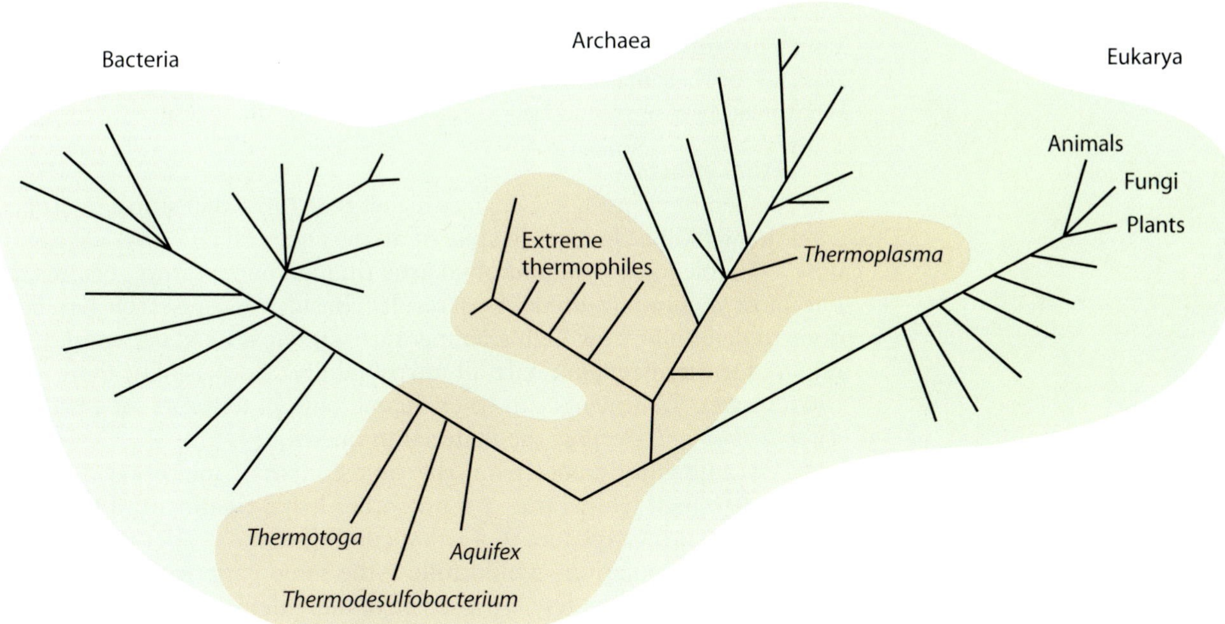

Figure 4.10 Phylogeny of thermophiles. Bacteria and Archaea that can grow at high temperatures (>60°C) are found in relatively few groups (shaded) that tend to be clustered in the older branches of the phylogenetic tree.

Thermostable proteins from several thermophiles have found industrial uses. For example, heat-stable proteases are widely used in laundry detergents. Not only do they have a longer shelf life, they also have higher activity and thus can be used in smaller amounts. In the laboratory, heat-stable **DNA polymerases** simplify the polymerase chain reaction (PCR) technology used to amplify minute amounts of DNA. The PCR requires alternating the reaction mixture between high and low temperatures: using a heat-stable polymerase makes it unnecessary to add fresh enzyme after each heating cycle.

The existence of a minimum temperature of growth, which is actually a precise value and not just a test of the investigator's patience, has a more complex explanation. Chemical reactions proceed more slowly as the temperature is decreased, but they do not stop. Why should this stop microbial growth? The only chemical forces that weaken at low temperature are hydrophobic interactions. The reason is the change in the properties of solvating water, which interferes with these interactions. Weakened hydrophobic interactions cause conformational changes in proteins, some of which change activity and preclude growth. For example, the activity of **allosteric proteins** (those that change conformation upon being bound by ligands [see Chapter 12]) is modulated by their conformational change. If they become too sensitive to inhibition by their effector, as can happen at low temperature, they cease to function. Conformational changes can also interfere with the assembly of proteins into complex structures, such as ribosomes. In fact, much evidence indicates that initiation of protein synthesis on ribosomes is the vital cold-sensitive step for *E. coli.*

As we will see later on (in Chapter 12), at the limits of their growth range, microbes synthesize special proteins and adjust the fatty acid composition of their membranes, thereby extending the growth range beyond what it otherwise would be.

Lethal effects

Temperatures just a few degrees higher than those that stop a microbe's growth may kill it. Indeed, in spite of all the chemical and physical agents now available to control microbial growth, high temperature remains the one most commonly used. As a result, considerable research has been done to determine how high a temperature and how long an exposure are required to **sterilize** (eliminate all microbial life). Surprisingly, there is no clear answer. Treatments can be evaluated only in terms of the *probability that they will sterilize* the material in question.

This ambiguity comes from the time course of microbial death on exposure to high temperature (or any other lethal treatment). One might anticipate that on exposure to a lethal condition, the death of individuals in microbial populations would follow the same pattern as the death of plants or animals. The few most sensitive individuals would die first, followed by the great majority with average sensitivity, and finally, by the most resistant few. But that's not the way microbes die. Instead of a **normally distributed** (according to a bell-shaped curve) range of sensitivities, a microbe's chance of dying in any period of time—immediately or after prolonged exposure—is constant. If 90% of the population dies in the first 5 minutes, 90% of the remaining population will die in the next 5 minutes, and so on. Put mathematically, the logarithm of the number of survivors is a linear function of time of exposure (Fig. 4.11). The curve describes **single-hit kinetics**—the response one would expect if paintballs were fired randomly at a constant rate into a crowd and if each hit were individually visible. We might ask what the heat equivalents of the paintball and the single target are. The answers are by no means clear.

In order to determine how prolonged a treatment is required to eliminate a particular microbial population, we need to know three things: (i) the **decimal reduction time,** or ***D* value** (the time necessary to decrease the

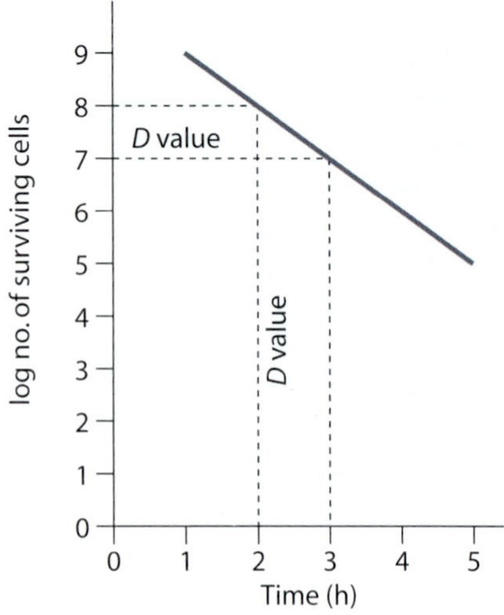

Figure 4.11 Lethal effect of temperature on microbes. The number of surviving cells in the population declines 90% (1 log unit) for each *D* value. In this case, the *D* value is 60 minutes.

viable population by 1 log unit, i.e., to kill, e.g., 90% of the survivors), for that particular microbe at that treatment temperature; (ii) the size of the population; and (iii) the assurance we require that all cells will be killed.

Low temperature itself is not lethal to microbes, but sudden shifts of growing cultures to low temperature can kill. The phenomenon is called **cold shock.** For example, if a culture of *E. coli* growing rapidly at 37°C is suddenly shifted to 5°C, 90% of the population will die. Freezing also kills microbes, but that's not a consequence of low temperature; instead, it is the transient exposure to high osmotic strength that the microbes undergo as the suspending medium freezes that kills them.

Hydrostatic pressure

High hydrostatic pressures, such as those encountered at great depths in the ocean, can crush and kill plants and animals, but not microbes. Most microbes are protected from being crushed to death because their outer cell barrier, including the cell membrane, is freely permeable to water. Thus, the pressure within a microbial cell rapidly balances the external pressure. Hydrostatic pressure does affect microbes at the molecular level by resisting increases and favoring decreases in molecular volume. In spite of this, many ordinary microbes withstand extraordinary pressure. For example, *E. coli* thrives at pressures up to 300 atmospheres (about 4,400 pounds per square inch), and some other microbes can withstand about four times that pressure (see Table 1.4). Some prokaryotes, called **barophiles,** grow faster at increased pressure than they do at 1 atmosphere. Some barophiles are obligate; they can grow only when exposed to pressures higher than 1 atmosphere. Yeast is a curious exception. The growth of brewer's yeast, *Saccharomyces cerevisiae*, stops at 8 atmospheres, the pressure that a sturdy champagne bottle can withstand.

A probable explanation for different responses of microbes to high hydrostatic pressure is the difference in molecular volume between ground and activated states of their enzyme-catalyzed reactions. If the volume of the molecule in the activated state is larger, the reaction will be inhibited by increased pressure; if the volume is smaller, the rate will be enhanced by or possibly even be dependent on high pressure.

Osmotic pressure

We humans added salt to our food in order to preserve it long before we knew that it inhibited the growth of the microbes that spoil food. In the 19th century, salt pork was the sailor's staple on long voyages; it was also the westward pioneer's major sustenance in winter. High concentrations of salt completely stop the growth of microbes likely to be present in food, but some other microbes can tolerate such environments. The range of salt concentrations over which microbes can thrive runs the gamut from distilled water (with nutrients being slowly supplied from the air) to over 5.2 M sodium chloride, which is close to its saturation point. Archaea, such as *Halobacterium* (as an aside, the "*-bacterium*" in this genus name indicates that it was named before the Archaea were discovered), grow at the highest concentrations of salt, but some bacteria, for example, *Halomonas*, are not far behind.

An environment with high osmotic pressure presents a microbe with two challenges: (i) it decreases the activity of water and hence its availability to the cells, effectively drying them out, and (ii) it diminishes the cell's turgor pressure by decreasing the difference in osmotic pressure between the inside and the outside. Most prokaryotes (with the exception of wall-less species) are dependent on a high turgor pressure, presumably to expand their cell walls as they grow. When exposed to a high-osmotic external environment, they maintain turgor by increasing the solute concentration of their cytoplasm, either by pumping solutes in or by synthesizing more of them. Such compensatory solutes include potassium ion, glutamate, proline, choline, betaine (trimethylglycine), and the sugar trehalose. As the internal concentration of solutes rises, the microbe faces another problem, namely, inactivation of its proteins. However, certain solutes, called **compatible solutes** (which include proline, choline, and betaine), are less damaging than most others and may even be protective. Trehalose has a special status. It protects membranes from salt-induced and other forms of drying. Baker's yeast can survive desiccation and remain active only if it contains high concentrations of trehalose. (This strategy is not confined to microbes but is also used by a remarkable small soil-dwelling animal called the water bear, which contains high concentrations of trehalose and can withstand complete desiccation and return to life when rehydrated.)

pH

Most microbial species grow over a somewhat narrow range of pH, but collectively, they span a remarkably broad pH range. **Acidophiles** are abundant in the leachings of mines at pH 1.0, and **alkaliphiles** grow in soda lakes in deserts of the American West at pH 11.5. In general, fungi tend to prefer a slightly acid pH and bacteria a slightly alkaline one.

Prokaryotes can grow over a wider range of pH than their proteins can tolerate. They deal with this dilemma in a special way. Instead of adapting to the unfavorable pH as they do to unfavorable temperature, they resist it by pumping protons in or out of their cells, thereby maintaining a nearly constant internal pH. *E. coli*, for example, can grow at its full rate between pH 6.0 and 8.0 by maintaining an intracellular pH close to 7.8. Prokaryotes can adapt to and survive exposure to extreme values of pH. A pH of 5.7 is lethal to an unadapted culture of *Salmonella enterica* serovar Typhimurium, but growing it at pH 5.8 for one generation renders it 100- to 1,000-fold more resistant to subsequent exposure to pH 3.3.

CONCLUSIONS

We stand in awe both of the ability of microbes to survive conditions that to humans appear truly extreme and of their facility to adapt to changes in these conditions. The continued existence of a bacterium or an archaeon is predicated on growth and survival. Growth lends itself to a higher degree of formalism than nongrowth and can be expressed by simple mathematical equations. In the next four chapters, we will examine the biochemical activities that lead to the acquisition of energy and biochemical building blocks and their conversion into new cells.

STUDY QUESTIONS

1. How can one estimate the number of bacteria in a population? Do the various techniques provide the same information?

2. How can one measure the rate of growth of a bacterial culture?

3. Why is "balanced growth" an important concept?

4. What goes on during the growth curve of a bacterial culture?

5. Explain how a continuous-culture device (chemostat) works. What is the point of using such a device?

6. Discuss the ranges of temperature, pH, and atmospheric and osmotic pressure that are compatible with the growth of bacteria and archaea.

chapter 5 making a cell

INTRODUCTION

Earth is truly a planet of microbes. Microbial populations are so widespread that they can be said to *define the extent of the biosphere*: where there is life, there are microbes; where there are no microbes, there is no life.

Clearly, the domination of Earth by microbes is related to their ability to reproduce. What does microbial reproduction involve? In this and the succeeding four chapters, we shall examine its biochemical aspects. In doing so, we shall probe the most fundamental property of all life: **autocatalysis** (in biology, the ability to self-replicate). We shall discover that much of what we know about the biochemical aspects of self-reproduction has been learned from the study of microbial systems.

This chapter will construct a coherent framework to organize the many details of biochemistry. In a sense, it will provide a **metabolic logic** permitting us to appreciate the roles of the myriad of individual biochemical details that result collectively in the continuation of microbial life on Earth.

GROWTH METABOLISM: MAKING LIFE FROM NONLIFE

Life looks different from nonlife. Nonliving things may be chemically active on a small scale or geologically active on a grand scale, but they take no organized role in events; by and large they appear inert. Living things, on the other hand, may breathe, move, respond to changes in their environment, modify their surroundings, and—their hallmark and universal feature—reproduce themselves. These attributes of living organisms are the result of highly organized chemical reactions collectively called **metabolism.**

Living systems are improbable arrays of improbable molecules. Neither the individual macromolecules—proteins, carbohydrates, nucleic acids, and lipids—nor the complex cell structures made from them are likely to accumulate spontaneously, at least under present Earth conditions. (If any should form, the microbes already present would rapidly consume them.) Rather, cells utilize information embedded in their existing structures to guide reactions of synthesis and assembly that result in the production of new cells. This reproduction of living systems is made possible by four overarching principles (Table 5.1). (i) *Specific catalysts*—enzymes—accelerate otherwise extremely slow reactions. (ii) A design strategy called *reaction coupling* makes individual chemical processes that are necessary for life energetically favorable by coupling them with other favorable reactions, particularly the hydrolysis of high-energy bonds (bonds that can release large amounts of free energy), such as **adenosine triphosphate (ATP)** to adenosine diphosphate (ADP) plus inorganic phosphate. In this sense, ATP can be said to drive otherwise-unfavorable reactions. (iii) *Harvesting energy* to make ATP occurs by means of organic, inorganic, or photochemical oxidation-reduction (redox) reactions. (iv) *Biological membranes transduce energy* into different forms, enabling the energy harvested from oxidation to generate ion gradients that can be used either directly to perform work (such as turning flagella) or to generate the universal energy currency, ATP. Enzyme catalysis, energetically coupled reactions, and acquisition of chemical energy and its transduction into ATP distinguish life from nonlife and enable organisms to create order out of disorder.

Metabolism is the sum of chemical processes in a living system. Many (though by no means all) of the several thousand reactions in a microbial cell directly contribute to the formation of new cells. This ensemble of

Table 5.1 Chemical processes that form the basis of all cellular metabolism

Chemical process	Function
Enzyme-mediated catalysis	Catalysts accelerate chemical reactions, in both the forward and reverse directions, by bringing reactant molecules together and by lowering the activation energy needed for the reaction. Most biocatalysts are proteins, but some, called ribozymes, are RNAs.
Reaction coupling	As an example, when the conversion of molecule A to molecule B is energetically unfavorable, it can be driven by coupling the reaction to a highly favorable one, such as the hydrolysis of ATP to ADP + inorganic phosphate. Thus, A might be adenylylated to form A-AMP, which would more readily form B.
Energy harvesting by redox reactions Organic substrates Inorganic substrates Photochemical reactions	Oxidation (the removal of electrons or of protons plus electrons from organic or inorganic molecules) is used to accumulate energy in a metabolically usable form, such as a proton pool or ATP. Alternatively, the energy of sunlight can be used to raise electrons to a higher level, from which they can generate a proton pool or ATP by returning to their original energy level.
Use of membranes to form gradients of charge and chemical concentration	Biological membranes make it possible to transduce energy, whether harvested by chemical or photochemical oxidation, into metabolically useful forms. Phospholipid membranes of Bacteria and isoprenoid membranes of Archaea, and the special proteins they contain, generate ionic gradients from the harvested energy, and these in turn are used to form ATP or to perform work directly.

reactions we call **growth metabolism.** We shall be focusing on the growth metabolism of microbes in this and the next three chapters. The **non-growth reactions** of metabolism are those responsible for such vital cellular activities as maintenance of intracellular metabolite pools and turgor, repair of cellular structures, secretion, motility, and other responses to environmental stress. These processes will be discussed in Chapter 13.

In this chapter, we shall pass through some territory undoubtedly familiar to you from general biology and biochemistry. The reason is that many details of metabolism are common to all organisms. In fact, much of what we know of metabolism (particularly biosynthesis) in plants and animals has been learned through research on microbes, because they are easier to study. Bacteria and yeasts are especially easy to grow, manipulate genetically, label with isotopes, and extract for enzyme analysis. Consequently, most of the pioneering studies of metabolic pathways were carried out with microbes.

We shall also encounter aspects of life's chemistry that are *unique to microbes*. In doing so, we shall come to appreciate that the biochemical potential of life on this planet is defined to a large extent by the diversity of microbial metabolic processes.

FRAMEWORK OF GROWTH METABOLISM

As we saw in Chapter 4, microbial cells are masters of growth. Though some grow no faster than cells of humans or other complex organisms (which typically require 12 to 24 hours to reproduce), many bacteria are capable of rapid growth and reproduction. We have already noted (see Chapters 2 and 3) that the structural design of prokaryotic cells can lend itself to speedy reproduction. This structural streamlining is reflected in more rapid metabolism. The metabolic rates of bacteria are generally at least an order of magnitude higher than those of plant or animal cells, in keeping with the fact that some microbes can double in just minutes.

This remarkable feat of reproduction involves a very large number of chemical reactions. *How many* reactions can be estimated from the biochemical knowledge accumulated largely over the past half-century, augmented now by **informatic analysis** (computer-assisted use of genome sequence information) of the genomes of many microbes. Although the genomes of hundreds of microbes have been sequenced, the fundamental biochemical, genetic, and physiological knowledge gathered for *Escherichia coli* far exceeds that for any other organism. Hence, here and elsewhere in this presentation of metabolism we shall rely on the detailed information available from over a half-century of concentrated study of this bacterium. This coli-centricity is also justified by the fact that in genome size and metabolic complexity, *E. coli* is rather average. Table 5.2 summarizes the numbers of genes of *E. coli* involved in various cell functions. Nearly half of the over 4,200 genes encode enzymes with known functions. Many are involved in making protoplasm from glucose and inorganic salts.

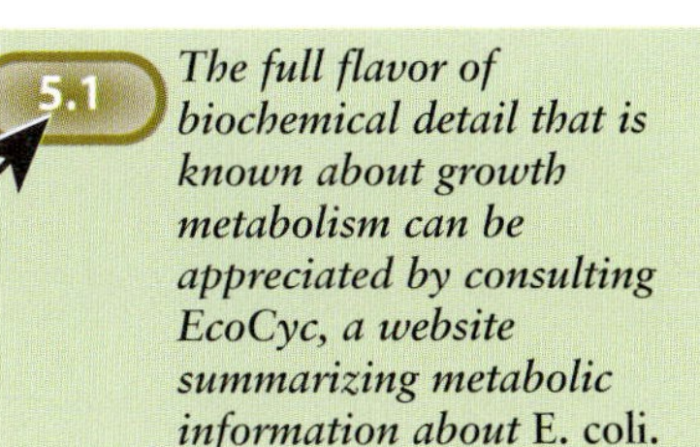

The full flavor of biochemical detail that is known about growth metabolism can be appreciated by consulting EcoCyc, a website summarizing metabolic information about E. coli.

The lists of known metabolic genes and the enzymes they encode may well overwhelm you. This clearly calls for some organizing framework. Such a framework can take the form of a flowchart that proceeds from food materials through the various steps of growth metabolism and ends with a new cell. A framework of this sort is presented in Fig. 5.1.

Table 5.2 Gene products of *Escherichia coli* associated with various metabolic processes

Functional category	No. of genes
Metabolism of small molecules	
Degradation and energy metabolism	316
Central intermediary metabolism	78
Broad regulatory function	51
Biosynthesis	
Amino acids and polyamines	60
Purines, pyrimidines, nucleosides, and nucleotides	98
Fatty acids	26
Metabolism of macromolecules	
Synthesis and modification	406
Degradation	69
Cell envelopes	168
Cell processes	
Transport	253
Other, e.g., cell division, chemotaxis, mobility, osmotic adaptation, detoxification, and cell killing	118
Miscellaneous	107
Total	1,894

Although reaction flowcharts are familiar in chemistry, a word of caution is in order here. This flowchart is unlike the ones usually encountered. First, in organic chemistry, one usually expects different products if different starting materials are used. In this flowchart, however, *one can change the starting materials and still end up with essentially the same product*. In the case of a heterotroph (see below), it does not matter whether the organic nutrient is glucose, succinate, maltose, or any other usable food material, the end product is always a cell of that heterotroph.

Second, the successful production of a living cell, as depicted in Fig. 5.1, depends not only on the chemistry of the metabolic reactions, but also on the cooperative interplay of those reactions. The flowchart omits the myriad feedback loops and other control devices that, operating in unison, result in the orderly production of a living cell. The chart would never "flow" to its conclusion without these controlling forces. Later (see Chapter 12), we shall address the central question of metabolic physiology—how these hundreds of reactions are made to function cooperatively in a grand synthesis, the everyday miracle of making a new cell.

How does the framework in Fig. 5.1 help us organize the myriad reactions of growth metabolism? We should begin at the end, that is, at the right side of the flowchart, where the cell that is to be made is found, and proceed stepwise back to the food that is the starting material for this organic synthesis. Oddly enough, proceeding in this fashion makes conceptual and biological sense, because *each step dictates what must be accomplished in the preceding step*.

Making two from one

Once a cell has successfully doubled in mass and size, having made every cellular compound and structure, the now-enlarged cell must divide into

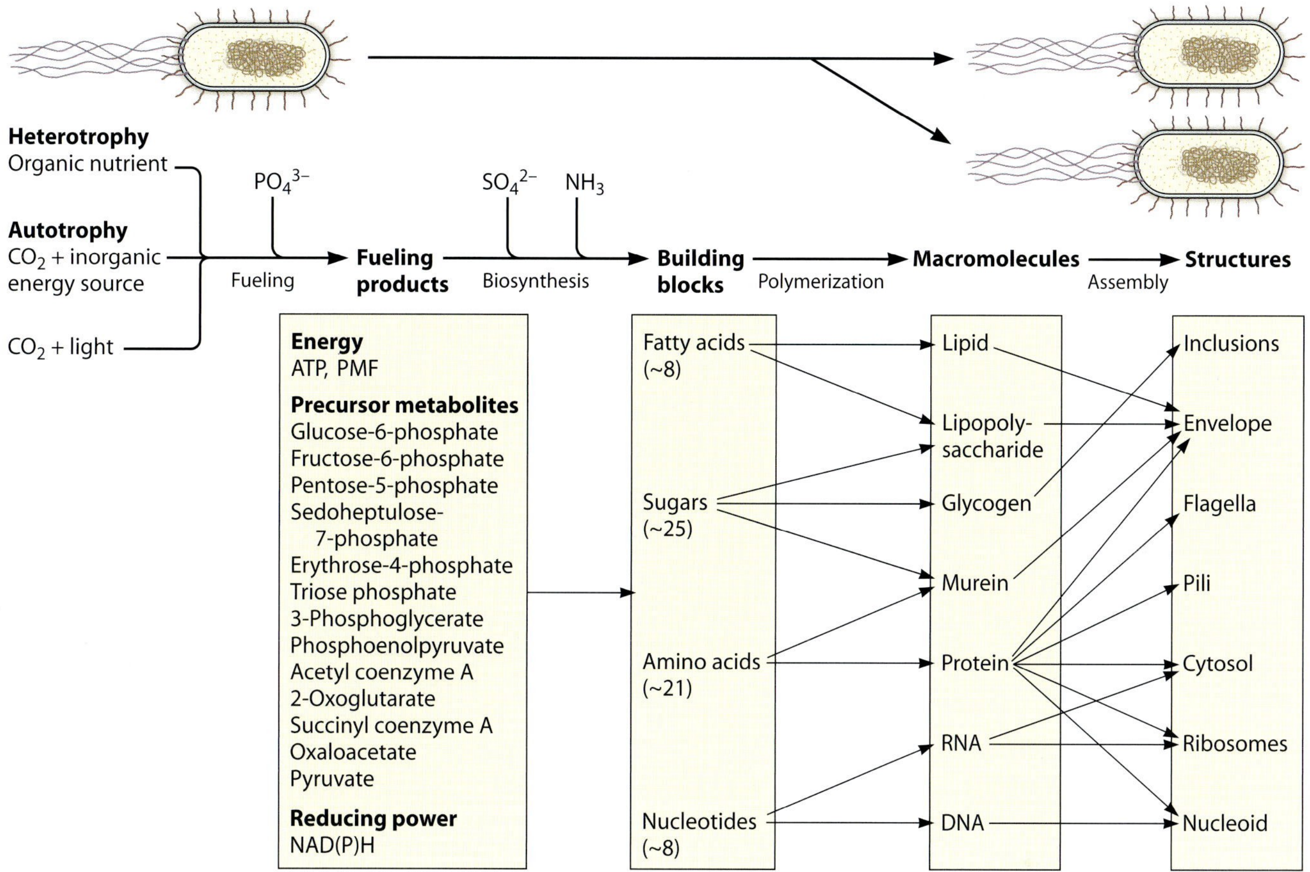

Figure 5.1 Framework of bacterial growth metabolism leading to the production of two cells from one. The diagram illustrates the biochemical flow that converts organic substrates (heterotrophy) or CO_2 (autotrophy) into the structures of a bacterial cell through the sequential processes of fueling, biosynthesis, polymerization, and assembly.

two living units. The process of division (usually by **binary fission**) is far from simple, as you might imagine, and we shall devote an entire chapter (Chapter 9) to it.

Assembling cell structures

Accurate reproduction dictates that each of the new cells have the same structures as their progenitor, that is, each must have (i) a complex envelope, including, for some, a flagellar apparatus and other appendages; (ii) a nucleoid; and (iii) a cytoplasm rich in polyribosomes (multiple ribosomes bound to individual messenger RNA molecules) and enzymatic machinery. **Assembly reactions** form each of the cell's structures out of macromolecules (Fig. 5.1). In some cases (ribosomes or flagellar filaments, for example), structures seem almost to self-assemble as if by condensation, while in other cases (the outer membrane of gram-negative bacteria, for example), it is clear that more elaborate processes have evolved to guide the formation of the finished structure. The assembly of some structures depends on enzyme-catalyzed reactions, while that of others does not. Assembly creates structures out of molecules, but it also involves translocating (moving) molecules (proteins in particular) from their points of manufacture to their ultimate locations. In Chapter 8, we shall examine the mechanisms that have evolved to translocate macromolecules and assemble microbial structures.

Making macromolecules

Based on an analysis of microbial structure (see Chapters 2 and 3), one can list the chemical compositions of the major cellular structures: proteins, nucleic acids, carbohydrates, and lipids, plus hybrids of these, such as lipoproteins, lipopolysaccharides, and murein (Table 5.3 and Fig. 5.2). The quantities of these macromolecular components, all of which must be made de novo rather than acquired from the environment, vary among different bacteria and under different growth conditions. Table 5.3 summarizes approximate values for one strain of *E. coli* growing in one well-defined environment. (Note: here, and occasionally elsewhere, we use specific data from *E. coli*, not because it is more important than other bacteria, but simply because more is known about it.) These values differ significantly from microbe to microbe; the Archaea, to cite an extreme example, lack murein. Macromolecules are formed by the **polymerization** of their **building blocks:** amino acids, **nucleotides,** fatty acids, sugars, and a score of related compounds (Fig. 5.1). (Note that lipids are included with the macromolecules, and fatty acids with the monomeric building blocks, simply for convenience; lipids are not sufficiently large to be in a class with proteins and nucleic acids, and fatty acids are not polymerized to make lipids.) **Polymerization reactions** require large amounts of energy in the form of ATP. The incorporation of an amino acid into a growing polypeptide chain requires four **high-energy phosphate bonds.** The addition of a nucleotide to growing nucleic acid chains expends two high-energy bonds per residue. In addition, energy is needed for **proofreading,** the correction of errors, the subsequent modification of the completed macromolecule chains, the folding of proteins into their mature forms, and the coiling of DNA.

Table 5.3 Overall composition of an average *Escherichia coli* cell

Substance	% of total dry wt
Macromolecules	
Protein	55.0
RNA	20.4
23S RNA	10.6
16S RNA	5.5
5S RNA	0.4
Transfer RNA (4S)	2.9
Messenger RNA	0.8
Miscellaneous small RNAs	0.2
Phospholipid	9.1
Lipopolysaccharide	3.4
DNA	3.1
Murein	2.5
Glycogen and other storage material	2.5
Total macromolecules	**96.1**
Small molecules	
Metabolites, building blocks, vitamins, etc.	2.9
Inorganic ions	1.0
Total small molecules	**3.9**

Synthesizing building blocks

The **building blocks** for polymerization reactions are amino acids, nucleotides, fatty acids, sugars, and other molecules used by the cell to make large molecules. If these building blocks are not available from the environment, they must be made de novo by **biosynthetic reactions** (Fig. 5.1). All these biosynthetic pathways begin with one or more of just 13 compounds, the common **precursor metabolites** listed in Fig. 5.1. Building blocks (for reasons we shall see presently) are usually at a more reduced state than the precursor metabolites from which they are made; they are also generally larger and more complex. Consequently, large amounts of reduced NADPH (nicotinamide adenine dinucleotide phosphate, the most common form of reducing power in biosynthetic pathways) and of energy (largely as ATP) are consumed in their synthesis.

The cost of biosynthesis in reducing power and energy is significant, and thus it is not surprising that bacteria that habituate nutrient-rich environments, including those that live in or on animal hosts, have lost much of their biosynthetic capability, thereby avoiding carrying genes that make unnecessary and costly enzymes.

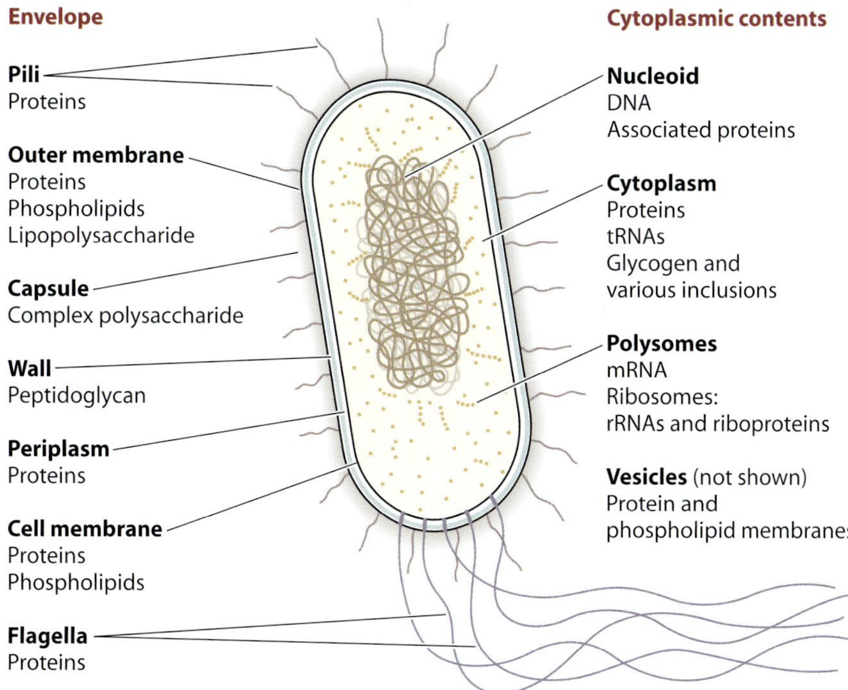

Figure 5.2 Macromolecular components assembled into a bacterial cell. The distribution of macromolecules forming each of the major cellular structures or compartments is indicated, using a gram-negative cell as an example.

Fueling

Getting the precursor metabolites, energy, and reducing power needed for biosynthesis is the function of the **fueling reactions** (Fig. 5.1). However, the burden on fueling is greater than simply providing for biosynthesis, because the energy demands of the cell go well beyond the manufacture of building blocks. Even a nongrowing cell is engaged in myriad activities that require energy. Table 5.4 summarizes the major activities, growth related or not, that consume energy. The quantities of energy needed for these activities vary with environmental conditions and from organism to organism, but they can be very large. The greatest demand for energy, by far, is for growth. The growth requirements for energy (ATP) and reducing power (NADPH) can be calculated in a particular instance from knowing (i) the composition of a cell under specified growth conditions and (ii) details of its pathways of biosynthesis and polymerization (Table 5.5). How is this prodigious amount of chemical energy obtained? All of it is derived from organic, inorganic, or photochemical *redox reactions*. Animals and fungi use exclusively organic redox reactions, plants use photochemical redox reactions, and protists use both organic and photochemical redox reactions, but prokaryotes use all three types of redox reactions: organic, inorganic, and photochemical.

Cell division, the assembly of cell structures, the polymerization of macromolecules, and the biosynthesis of building blocks all share one marvelously simple feature: *they are fundamentally the same in all prokaryotes*. Not so for fueling. Here, microbes display a variety and versatility not seen in the rest of the living world. Consider these two points: (i) *prokaryotes as a group can use as a sole carbon source any organic*

Table 5.4 Some cellular activities requiring energy

Cellular activity
Growth related
Entry of nutrients
Biosynthesis of building blocks
Polymerization of macromolecules
Modification and transport of macromolecules
Assembly of cell structures
Cell division
Growth independent
Motility
Secretion of proteins and other substances
Maintenance of metabolite pools
Maintenance of turgor pressure
Maintenance of cellular pH
Repair of cell structures
Sensing the surroundings
Communication among cells

compound on Earth (save for a few man-made plastics), as well as inorganic carbon (such as CO_2), and (ii) they can obtain energy and reducing power by oxidizing either inorganic or organic compounds or by harvesting energy from light. Some of these fueling modes are unexpected and seem weird if one's biochemical horizon is limited to the metabolic activities of plants and animals.

So diverse is fueling in the prokaryotic world that it has been necessary to invent descriptive terms to organize prokaryotes into broad classes. *The first distinction used in this metabolic classification is the nature of the carbon source.* Those microbes that obtain their carbon from organic compounds are called **heterotrophs;** those using CO_2 as the major carbon source are called **autotrophs.** (These terms are also applied to plants, animals, fungi, and protists.)

The second distinction is the source of energy and reducing power. If chemical sources are used, the prefix "**chemo-**" is added to heterotroph or autotroph. If light is used, the prefix "**photo-**" is used. Chemo-

Table 5.5 Energy and reducing power used in making macromolecules

Macromolecular component to be made from glucose	Energy cost (μmol of ~P[a]/g of cells)	Reducing power cost (μmol of NADPH/g of cells)
Protein		
Synthesis of amino acids	7,287	
Polymerization, mRNA, other steps	21,970	
Total	29,257	11,253
RNA		
Synthesis of ribonucleotides	6,540	
Polymerization, modification, other steps	256	
Total	6,796	427
Phospholipid		
Total	2,578	5,270
DNA		
Synthesis of deoxyribonucleotides	1,090	
Polymerization, methylation, coiling	137	
Total	1,227	200
Lipopolysaccharide		
Synthesis of nucleotide conjugates	470	
Total	470	564
Murein		
Synthesis of amino acids and sugars	248	
Cross-linking	138	
Total	386	193
Glycogen		
Synthesis of ADP-glucose	154	
Total	154	

[a]~P, high-energy phosphate bonds.

Table 5.6 Patterns of fueling reactions among microbes

Class	Source of carbon	Source of energy
Heterotrophs[a]		
Chemoheterotrophs	Organic compound	Organic compound
Photoheterotrophs	Organic compound	Light
Autotrophs		
Chemoautotrophs[b]	CO_2	Inorganic compound
Photoautotrophs	CO_2	Light

[a]Organotrophs.
[b]Lithotrophs.

autotrophs utilize *inorganic* molecules (including H_2, CO, NH_3, NO_2, H_2S, S, $S_2O_3^{2-}$, and Fe^{2+}) as their source of energy and reducing power. These autotrophs are called **lithotrophs** (stone eaters), though it is also common to employ the terms "lithotroph" and "autotroph" as synonyms. (Some microbiologists use the newer term **organotroph** in place of heterotroph. Yes, it is confusing!) The definitions used by us are summarized in Table 5.6. Details of the unique fueling reactions of these classes of microbes are presented in Chapter 6.

Fueling in *heterotrophs* begins with the entry of the organic substances that serve as sources of carbon and energy. In many cases, entry requires an expenditure of energy in order to concentrate certain nutrients present in dilute form in the environment. Fueling in *autotrophs* also begins with the entry of the sources of carbon and energy, though in many cases these are just CO_2 and light or an inorganic compound. Whatever the food source, it must ultimately be converted by heterotrophs *and* autotrophs alike into the 13 precursor metabolites (unless these are themselves available as nutrients). This conversion is achieved ultimately through the familiar central pathways of metabolism, which are reviewed in the next chapter.

Conversion of energy, if it is not derived from light, into a form useful to the cell involves the oxidation of the food molecules, i.e., the removal of electrons from inorganic compounds or of hydrogen atoms (electrons and protons) from organic substrates. The same enzymatic pathways that convert the food substrates into the precursor metabolites generate ATP. That is the reason that precursor metabolites are generally more oxidized than the average building block and that some of the reducing power formed in the **fueling pathways** must be used in biosynthesis. Microbes, as well as all other living things, use oxidation reactions to generate ATP either by **substrate level phosphorylation** or by **proton motive force** established commonly by means of **electron transport.** In the latter case, electrons move from carrier to carrier within the cell membrane down an electrochemical gradient to produce a proton (H^+) gradient across the membrane; this proton gradient is then used to make ATP from ADP. These modes of energy transduction will be examined in Chapter 6.

Where microbes live is related quite directly to their metabolic prowess, and in the fueling domain of metabolism, one sees a clear matchup of environmental characteristics with microbial life style. That is to say, organisms that use photochemistry to obtain energy live in the light, those that use oxidation of inorganic forms of sulfur, iron, or other elements, live where those elements abound, and so on.

GLOBAL EFFECTS OF GROWTH METABOLISM

The fame (or infamy) of microbes as agents of disease has tended to overshadow the vital role of these organisms in nature. The chemical activities of microbes may well be the most underappreciated aspect of life on Earth. Earth's biosystem is *based* on the metabolism of microbes. To put it bluntly, were microbes to vanish, all life on Earth would soon cease.

Earth's chemical cycles

Microbes are responsible for molding the chemistry of our Earth and for keeping it in balance. Each of the major elements of living matter—carbon, nitrogen, sulfur, and phosphorus—exists in nature in its own cyclic process in which it is constantly consumed and replenished (see "Biogeochemical Cycles" in Chapter 18). Microbes are indispensable participants in these geochemical cycles, because microbial metabolism brings about interconversions of matter that are vital in keeping the pools of the major elements in balance. The several roles played by microbes in these geochemical cycles are necessary to keep Earth habitable for all other organisms.

How can microbes have such major global effects on the environment? Several factors are involved. Microbes exist throughout the biosphere in huge numbers, accounting for much of the Earth's biomass. They have high rates of metabolism and can metabolize any naturally occurring organic compound, and as a group, they are adept at inorganic as well as organic chemical conversions. Although heterotrophic bacteria are more familiar to most microbiologists than are autotrophs, it is possible that autotrophy is globally the dominant mode of metabolism, with enormous implications for the operation of each of the geochemical cycles.

In Chapter 18, we shall examine the microbial role in each of the major cycles.

Bioremediation

The collective ability of microbes as a group to degrade any naturally occurring organic compound (and most man-made ones), to oxidize or reduce many inorganic compounds, and to thrive in diverse environmental niches has taken on a special significance in recent decades. Industrial manufacture and use of tens of thousands of chemicals, massive transporting and consumption of coal and petroleum, and military and civilian use of nuclear energy have together led to a highly threatening pollution of Earth's environment. All regions of the biosphere—surface water and groundwater, soil, and the atmosphere—have been contaminated by toxic chemicals. Concern over this crisis has led to the development of the strategy called **bioremediation,** in which living microbes are used to assist in restoring clean water and soil environments, cleanup of hazardous waste, and maintaining a sustainable balance of Earth's geochemical and physical cycles.

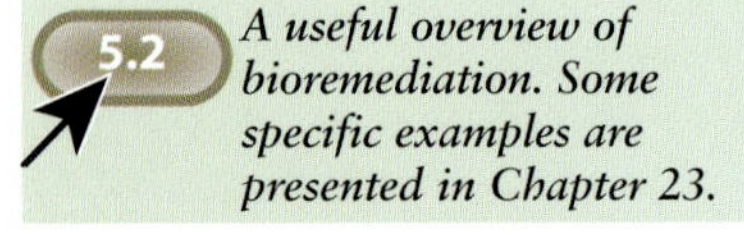

A useful overview of bioremediation. Some specific examples are presented in Chapter 23.

The field of bioremediation is large and expanding. It includes measures to enhance microbial degradation of pesticides and hazardous compounds in soil and groundwater, removal of excess nutrients from streams and lakes, concentration and removal of toxic minerals from groundwater, concentration of uranium and other radioactive materials for removal from mine wastewater and stores of military wastes, cleanup of oil spills—the list of processes under development or already in use is impressive.

SUMMARY AND PLAN

The intent of this chapter has been to present a conceptual framework to aid our understanding of how the thousands of individual metabolic reactions produce a new cell. The next four chapters will take us along the route from food to a new living cell. Though we shall not dwell on the biochemical details of the individual reactions and pathways (many can be viewed on the *EcoCyc* website), a review of the overall features and special principles of each stage of metabolism (fueling, biosynthesis, polymerization, and assembly) will be central to our purpose. This attention is appropriate because the place of the microbe in Earth's biosphere is defined by the biochemical transformations brought about by these dominating but largely invisible creatures. Our focus will be on the uniqueness of the prokaryotic way of doing things.

One must not come away from this presentation of a framework for metabolism with the notion that all microbes have identical metabolic activities and capabilities. Far from it; microbes are successful colonizers of Earth precisely because each group has evolved specialized metabolic programs to deal with specific environmental circumstances. The startling array of fueling reactions, for example, is possible because of the enormous variety of often quite specialized microbes.

STUDY QUESTIONS

1. Life on this planet is made possible by (i) protein enzyme catalysis, (ii) reaction coupling, (iii) energy harvesting, and (iv) membrane-mediated interconversion of different forms of energy (see "Growth Metabolism: Making Life from Nonlife"). Invent and describe the features of a different life form (on an imagined planet) that might not need one or another of these processes.

2. What is the basis for saying that the extent of Earth's biosphere is defined by the distribution of prokaryotic microbes (Bacteria and Archaea)?

3. The broad processes of metabolism (fueling, biosynthesis, polymerization, assembly, and cell division) are interrelated by the fact that each requires as starting material the products of the preceding step. For each of these phases of metabolism, list the kinds of starting materials and products. Could any of the phases be dispensable? If so, under what circumstances? (Hint: consider the effect of nutrition.)

chapter 6 fueling

OVERVIEW OF FUELING REACTIONS

The metabolic route to making a new cell starts with fueling. Here, we examine how microbes obtain both the metabolites and energy needed for biosynthesis and the additional energy needed for growth and a myriad of other cellular activities. As we have noted, the ensemble of reactions supplying these needs—the 13 precursor metabolites, reducing power, and energy—are the **fueling reactions** of the cell.

In large measure, diversity of fueling holds the key to microbial domination of Earth's biosphere and is the basis of the dependence of all life forms on the chemical activities of microbes. Microbes, particularly the prokaryotes, have evolved an almost unbelievable variety of ways to get energy and make metabolites. This capability enables them to grow in profusion in many unexpected locations and even to alter the geochemistry of Earth's land and waters. It is in fueling that prokaryotes least resemble eukaryotes metabolically, and it is fueling that distinguishes the two major metabolic types of prokaryotes: heterotrophs and autotrophs.

For heterotrophic microbes (those using organic sources of carbon) (Fig. 6.1), fueling reactions can be conveniently sorted into three stages: **entry processes**, which move *organic and inorganic food substrates* from the environment into the cell; **feeder pathways**, which convert food substrates once in the cell into one or another metabolite of central metabolism; and **central metabolism**, a group of pathways, common to the metabolisms of most cells, which produce all the precursor metabolites. These three aspects of heterotrophic fueling are diagrammed in Fig. 6.1.

For autotrophic microbes (those using CO_2 as the main source of carbon) (Fig. 6.2), one can speak of the same stages, but with some unique wrinkles. Autotrophic CO_2 fixation and the generation of energy from light or inorganic chemicals are the defining processes of autotrophy;

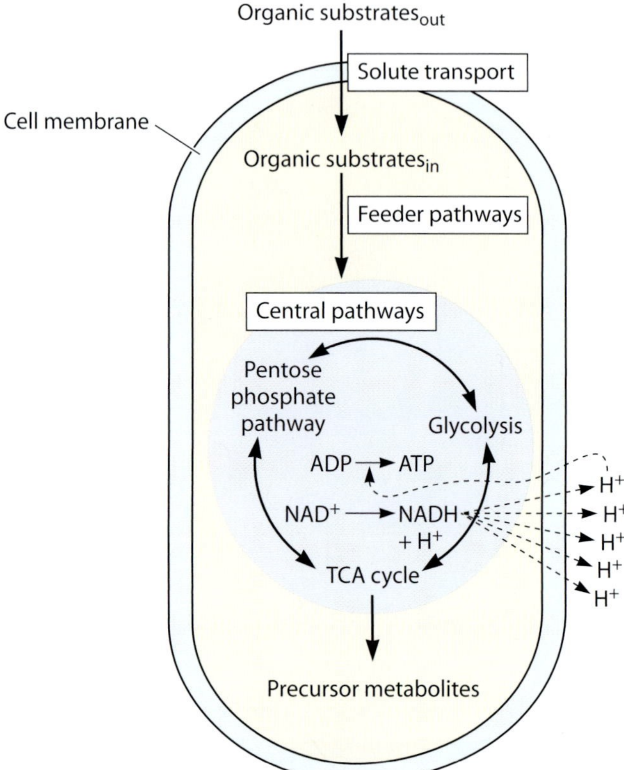

Figure 6.1 Fueling processes in heterotrophs. The diagram shows the three components of fueling: *solute transport*, which enables the cell to acquire substrates; *feeder pathways*, which transform the substrates into one or another of the metabolites in the central pathways; and the three *central pathways* (glycolytic, pentose phosphate, and TCA pathways) of metabolism. The three products of fueling (ATP, NADH, and precursor metabolites) and the generation of ATP by proton motive force are indicated.

pathways of central metabolism are used solely to provide the precursor metabolites, and the feeder pathways are designed to lead from the products of CO_2 fixation to central metabolism. Autotrophic fueling deserves a separate diagram (Fig. 6.2).

We shall turn our attention first to the energy–reducing-power aspect of fueling and will deal later with how the precursor metabolites are made.

GETTING ENERGY AND REDUCING POWER

Driving force and its generation

At the most fundamental level, all energy in living systems is derived from the movement of electrons down an energy gradient to produce two forms of **usable energy**: (i) the high-energy phosphoryl (~P) bonds of ATP (and similar molecules, including other nucleoside triphosphates) and (ii) transmembrane ion gradients. Both forms are necessary: ATP, because most of metabolism is driven by the device of coupling individual reactions to ATP hydrolysis (see "Growth Metabolism: Making Life from Nonlife" in Chapter 5), and transmembrane ion gradients, because

A Chemoautotrophs

Inorganic ions$_{out}$
CO_2
Cell membrane
CO_2 fixation
Electron flow
Inorganic ions$_{in}$
Feeder pathways
H^+
H^+
H^+
H^+
Reductant
ADP
H^+
ATP
NADH
NAD^+
Oxidant
Central pathways
Pentose phosphate pathway
Glycolysis
TCA cycle
Precursor metabolites

B Photoautotrophs

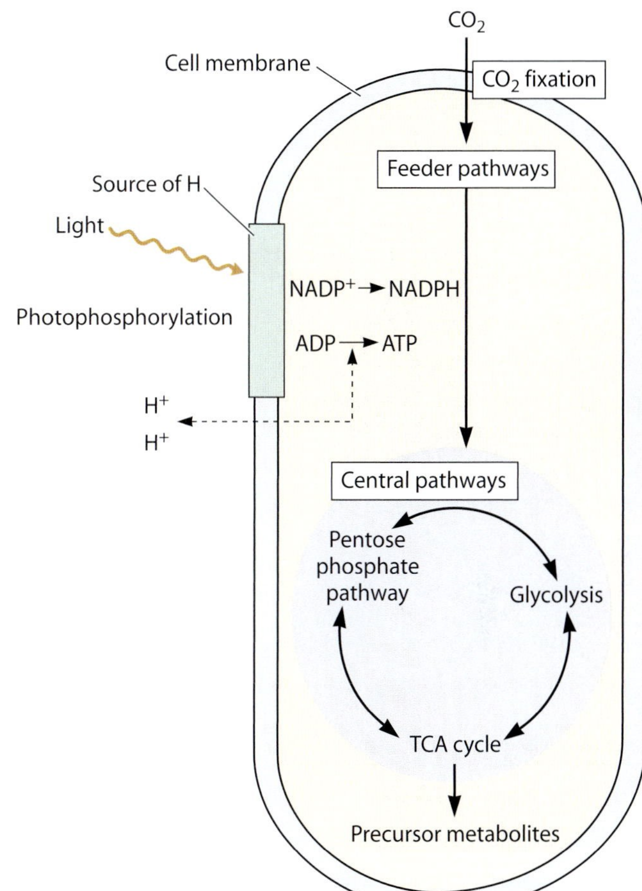

Figure 6.2 Fueling processes in autotrophs. The defining feature of autotrophy, CO_2 fixation, is shown in relation to the feeder pathways and the three familiar central metabolic pathways. **(A)** Chemoautotrophic fueling. Inorganic redox reactions leading to the generation of ATP by proton and electron transport on membrane carriers are schematically indicated as a distinguishing feature of chemoautotrophy. **(B)** Phototrophic fueling. Harvesting light energy to generate ATP and reducing power is schematically indicated as the distinguishing feature of phototrophy.

they drive many cellular processes, such as turning flagella and transporting certain nutrients into the cell. As we shall presently see, these two forms of usable energy are interconvertible; an ion gradient can be used to generate ATP, and ATP can be used to establish an ion gradient.

The electrons are derived from oxidation reactions, many of which involve dehydrogenation (the removal of hydrogen atoms from organic molecules). The hydrogen atoms (H is a proton plus an electron) are transferred to either of two related **coenzymes,** NAD^+ (nicotinamide adenine dinucleotide) or $NADP^+$ (nicotinamide adenine dinucleotide phosphate). Their reduced forms, NADH and NADPH, are ready sources of protons and electrons for many metabolic reactions. NADH participates in most fueling reactions, and NADPH is used commonly in biosynthesis. Together, NADH and NADPH are said to represent the **reducing power** of the cell, just as ATP is said to represent the cell's energy. In heterotrophs,

NAD(P)H (as we refer noncommittally to the pair of coenzymes) functions more or less as a shuttle between fueling and biosynthesis. In autotrophs, however, specific means must be used to form them. We shall examine the rationale for the existence of two coenzymes rather than one when we discuss biosynthesis in Chapter 7.

Reducing power and energy are equivalent in cell biology because they are *interconvertible* through forward and reverse electron transport, as we shall see in "Substrate level phosphorylation and fermentation" below. Given their interconvertibility, we shall refer to energy and reducing power collectively as the **driving force** of the cell.

The biological world has two means to make ATP: (i) substrate level phosphorylation and (ii) harvesting a transmembrane ion gradient to make ATP via a membrane-bound ATP synthase. All organisms employ both modes of energy generation, but prokaryotes stand out for their remarkable creativity in evolving *different versions* of these two energy-harvesting processes (Table 6.1). As we have noted, this inventiveness accounts for the remarkable ability of prokaryotes to exploit all of Earth's life-sustaining habitats at the price of a huge variety of energy-yielding reactions, including some that yield only minuscule quantities of energy compared to that needed to make a molecule of ATP.

This subject clearly merits some close attention. We shall examine the two energy-harvesting mechanisms one at a time.

Table 6.1 Comparison of energy metabolism in prokaryotes and eukaryotes

Major energy-generating mechanism	Prokaryotes	Eukaryotes
Substrate level phosphorylation		
Fermentation: use of the protons and electrons removed from one organic substrate in oxidation reactions to reduce a second organic molecule using only NADH as intermediary; ATP is formed by substrate level phosphorylation	Common among anaerobes and facultative bacteria, where it can serve as the sole source of ATP; great variety of substrates and by-products	In plants and animals, present only incidental to central pathway function; in animal tissues, it serves as a means to survive transient anerobiosis. Can serve as sole ATP source only in the microbial eukaryotes, notably yeast.
Transmembrane ion gradient		
Respiration: electrons from a reductant are transferred by membrane-based carriers (electron transport chain) to oxidants, forming a transmembrane ion gradient useful in ATP formation and other energy-requiring processes	Enormous variety of organic and inorganic electron donors (reductants) and acceptors (oxidants); forms basis of chemosynthetic autotrophy; ion gradient may be other than protons	Major mode of energy formation in animals, but restricted to organic reductants and to O_2 as oxidant
Photosynthesis: light energy activates an electron from chlorophyll to flow down an electron transport chain to its original ground-state chlorophyll or to $NADP^+$, generating a transmembrane ion gradient	Several varieties, differing in whether it is cyclic or noncyclic and in the source of H and electrons; forms basis of photosynthetic autotrophy	One form only (oxygenic), with H_2O as electron and hydrogen source
Enzyme ion pumps: membrane proteins not part of an electron transport chain translocate ions across the cell membrane	Common among a large variety of bacterial species	Not known to occur
Scalar reactions: enzymes that are part of or separate from an electron transport chain create a transmembrane ion gradient by consuming or producing ions, not by moving them	Common among a large variety of bacterial species	Not known to occur

Substrate level phosphorylation and fermentation

The simplest and neatest trick to generate ATP is **substrate level phosphorylation.** At first sight, this may seem a strange name to apply to a process that generates ATP from ADP, but closer inspection reveals the rationale. In this process, an organic substrate first becomes phosphorylated with inorganic phosphate (PO_4^{3-}) in a reaction that requires no energy input. The resulting phosphoryl group on the substrate is not attached by a high-energy bond, so the crucial step comes next: the phosphorylated substrate is oxidized, and the energy that would otherwise be released in the reaction as heat is trapped. The originally **low-energy phosphoryl bond** is now a **high-energy phosphoryl bond** and can be transferred to ADP to form ATP. In this manner, the phosphorylation of ADP by a metabolic substrate makes ATP, hence the term *substrate level phosphorylation.*

An instructive example is the reaction in the **glycolytic pathway** (see Fig. 6.13) whereby 3-phosphoglyceraldehyde is *both* phosphorylated *and* oxidized (removal of 2H) by the enzyme phosphoglyceraldehyde dehydrogenase, generating a high-energy phosphoryl bond in a single reaction:

$$\text{3-phosphoglyceraldehyde} + NAD^+ + P_i \longrightarrow \text{1,3-bisphosphoglycerate} + NADH + H^+$$

$$\begin{array}{l} HC{=}O \\ \quad | \\ HC{-}OH \\ \quad | \\ H_2C{-}O{-}PO_3H_2 \end{array} \qquad\qquad \begin{array}{l} \quad O \\ \quad \| \\ \quad C{-}O \sim PO_3H_2 \\ \quad | \\ HC{-}OH \\ \quad | \\ H_2C{-}O{-}PO_3H_2 \end{array}$$

The acyl phosphate group on carbon 1 of the product is attached by a high-energy bond, and it can be used to generate ATP in the immediately following reaction catalyzed by 3-phosphoglycerate kinase:

$$\text{1,3-bisphosphoglycerate} + ADP + NADH + H^+ \longrightarrow \text{3-phosphoglycerate} + ATP + NAD^+$$

$$\begin{array}{l} \quad O \\ \quad \| \\ \quad C{-}O \sim PO_3H_2 \\ \quad | \\ HC{-}OH \\ \quad | \\ H_2C{-}O{-}PO_3H_2 \end{array} \qquad\qquad \begin{array}{l} \quad O \\ \quad \| \\ \quad C{-}O^- \\ \quad | \\ HC{-}OH \\ \quad | \\ H_2C{-}O{-}PO_3H_2 \end{array}$$

Subsequent reactions in the glycolytic pathway convert 3-phosphoglycerate to 2-phosphoglycerate, which can then be dehydrated by the enzyme enolase to form phosphoenol pyruvate, in which the phosphoryl group is also attached by a high-energy bond. Phosphoenol pyruvate can then perform substrate level phosphorylation on ADP. The net result of the original phosphorylation and dehydrogenation of 3-phosphoglyceraldehyde is a net gain of two ADP-to-ATP transformations per molecule of glucose converted to two molecules of pyruvate (Fig. 6.3).

Substrate level phosphorylation is an integral event in the operation of the glycolytic pathway and the tricarboxylic acid (TCA) cycle. At first glance, there would seem to be nothing uniquely microbial about it, but there is. Unlike plants and animals, many microbes can live indefinitely

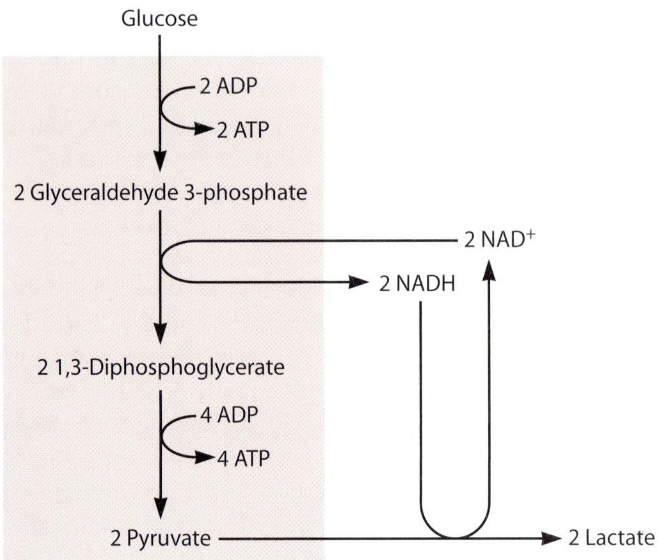

Figure 6.3 Homolactic fermentation. A shortened version of the glycolytic pathway combined with the reaction catalyzed by lactic dehydrogenase is shown to emphasize the reduction and reoxidation of NAD^+, which permits sustained ATP generation from glucose utilization.

using *solely* this mode of ATP generation in a process called **fermentation.** Fermentation is the use of the protons and electrons removed from one substrate in oxidation reactions to reduce a second organic molecule, usually a product of the metabolism of the first, using only NADH as an intermediary carrier. Why is the reduction of the second organic molecule important? Unless something is done to regenerate NAD^+, **glycolysis** will stop. This is a major issue, because large quantities of substrate must be used if sufficient ATP to satisfy the energy needs of the cell is to be made exclusively by substrate level phosphorylation. To use glycolysis as an example, the cell gains only 2 moles of ATP (or two high-energy bonds [~P]) for every mole of glucose metabolized to pyruvate. This same amount of glucose metabolism also generates 2 moles of NADH from NAD^+. The crucial step in fermentation, therefore, is to remove the hydrogen atoms from NADH, reoxidizing it to NAD^+.

It is clear that the production of reducing power (NADH) for biosynthesis is no problem for microbes fueled by fermentation—their problem is the opposite: *getting rid* of reducing power, i.e., regenerating oxidized NAD. Of the large quantities of substrate that must be consumed for ATP generation, very little is used to synthesize cell material; most of it can be recovered in the form of the fermentation end products that serve as the final sink for all the excess hydrogen atoms from NADH. What are these fermentation products? In many bacteria, the pathways of fermentation involve pyruvate as an oxidized intermediate. Figure 6.4 illustrates fermentative end products formed from pyruvate by various bacteria. However, prokaryotes have been enormously inventive in devising fermentation paths, and a huge array of substrates and reduced end products are known to be employed by different organisms (Table 6.2).

In the fermentations illustrated in Fig. 6.4, NADH oxidation is accomplished by dehydrogenases (e.g., lactate or ethanol dehydrogenase),

Table 6.2 Examples of microbial fermentations

Substrate	Product(s)	Organism(s)
Glucose and related sugars	Lactate, ethanol, acetate, butyrate, butanediol, and other products (Fig. 6.4)	*Lactobacillus* spp. and many other organisms
Amino acid (alanine and others)	Acetate (from glycine)	*Clostridium sporogenes*
H_2 and CO	Acetate	*Clostridium aceticum*
Oxalate	Formate + CO_2	*Oxalobacter formigenes*
Malonate	Acetate + CO_2	*Malonomonas rubra*
Lactate	Propionate + acetate	*Clostridium proponicum*
Pyruvate	Acetate	*Desulfotomaculum thermobenzoicum*

but some bacteria have an enzyme called hydrogenase that can oxidize NADH and other reduced **cofactors** to produce hydrogen gas. This reaction is thermodynamically feasible only if the concentration of the resulting hydrogen gas is very low. Efficient removal of hydrogen is accomplished in some natural environments by the scavenging activities of certain other bacteria (called methanogens and sulfate reducers). This process of interspecies hydrogen transfer allows the bacteria to increase their ATP production by using additional kinase (e.g., acetate and butyrate kinase) reactions.

Fermentations have been used by humans throughout history to produce beverages (the whole array of alcoholic products made from grain or fruit) and to process food for preservation and flavor (for example, pickled fruits and vegetables). Readers may recall from Chapter 1 the description of sauerkraut production—a classic example of a practical fermentation, which is in fact a series of sequential fermentations. But not all fermentations are pleasing to human senses. Among many odoriferous products are butyric acid, methane, and amines such as cadaverine. Cadaverine is aptly named (from "cadaver"), because it is one of the products of decaying flesh (which is largely a fermentation process).

Because it is based on the use of organic substrates (sugars, amino acids, compounds from the central pathways of metabolism, and the like), the fermentative mode of fueling is restricted to *heterotrophs*.

Figure 6.4 Fermentation products from pyruvate in various microbes. The NADH produced during glycolysis is used to reduce pyruvate and its derivatives, forming fermentation products distinctive for each organism.

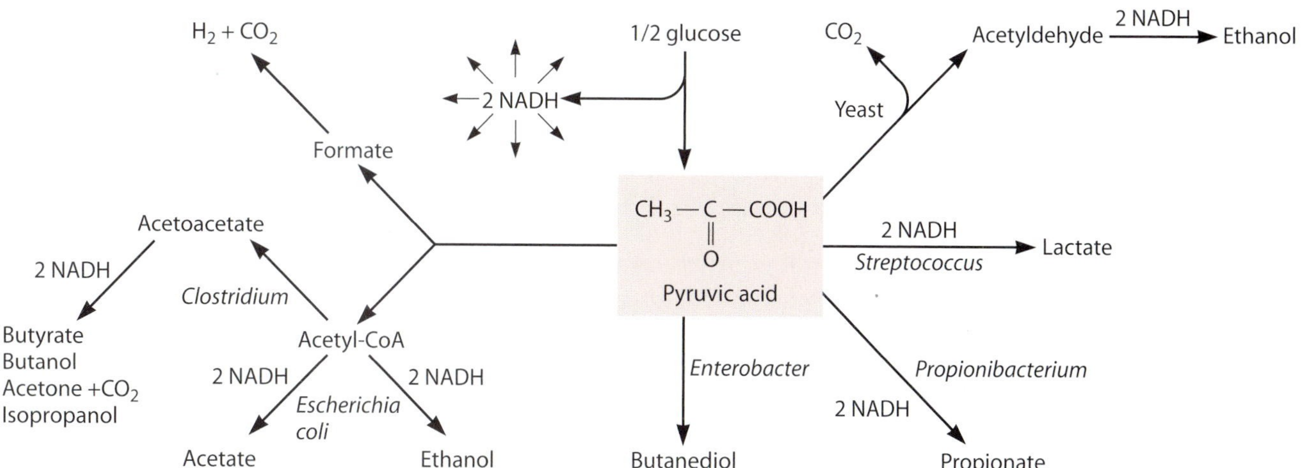

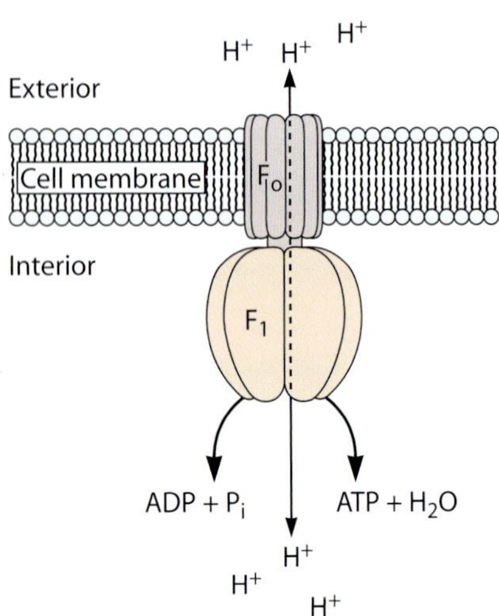

Figure 6.5 Membrane ATP synthase, or F_1F_o ATPase. This multi-subunit enzyme consists of two large complexes. One, F_o, forms a proton channel through the membrane; the other, F_1, is on the inner surface of the membrane. Passage of three or four protons from the exterior through F_o of the energized membrane drives F_1 to form one ATP molecule from ADP. The reaction is poised to proceed in either direction, depending on the ratio of ATP to membrane potential: ATP hydrolysis can produce a proton gradient when the ratio is insufficient, and proton entry can produce ATP when the ATP concentration is too low for cellular functions (hence the alternative names for this complex).

Transmembrane ion gradients

Although substrate level phosphorylation in the process of fermentation is very common among microbes, most microbial energy generation involves the use of an ion gradient across the cell membrane. In the case of a *proton* gradient, the membrane is said to be energized by **proton motive force.** Before surveying the vast array of prokaryotic means used to create these gradients, note should be made of how gradients can produce ATP. Within the cell membrane is an enzyme complex, **F_1F_o synthase,** originally called ATPase (and it still goes by that name) (Fig. 6.5). Its two parts function to interconvert the electrochemical energy of the gradient and the chemical energy of ATP. The F_o portion of the enzyme is a hydrophobic protein; it traverses the membrane and conducts the ions through it to the F_1 component, which is located on the interior surface of the membrane. The F_1 component catalyzes the reaction ATP → ADP + inorganic phosphate (P_i), but, importantly, it can also use the energy of ion flow through F_o to drive the reverse reaction, ADP + P_i → ATP. Estimates indicate that the passage of three or four protons into the cell interior is sufficient to generate one molecule of ATP. (This process occasionally goes by an old name, **oxidative phosphorylation,** meant to distinguish it from substrate level phosphorylation.) The same F_1F_o ATP synthase can work in the opposite direction, hydrolyzing ATP and driving the secretion of protons. One might fairly ask what determines in which direction the system will work. The answer is simple: it depends on the internal concentration of ATP and the magnitude of the electrochemical gradient across the membrane. A low ATP concentration and a steep gradient promote the entry of protons and generation of ATP; a high internal ATP concentration and a shallow gradient lead to export of protons at the expense of ATP. In fact, the hydrolysis of ATP by membrane ATPase/ATP synthase is the *only* way some bacteria have of establishing a transmembrane ion gradient.

Although the most common gradient used by bacteria to generate ATP is that of protons, many bacteria live in alkaline or high-sodium environments (e.g., marine bacteria and those that inhabit the rumen of cattle). Here, a proton gradient cannot be established because the protons would react with the excess of hydroxyl ions. The solution microbes have evolved is to employ a gradient of another cation, commonly Na^+ (in which case the ion gradient is said to constitute a **sodium motive force**).

Besides synthesis of ATP, ion gradients can be used for other purposes, including (i) secondary active transport (ion-coupled transport) across the cell membrane, (ii) maintaining the cell's turgor (water pressure), (iii) maintaining the cell's interior pH, (iv) turning flagella (see Chapters 2 and 14), and (v) driving a reverse flow of electrons through the respiratory chain to reduce NAD^+ when the supply of NADH is inadequate.

Given the importance of transmembrane ion gradients, their generation is clearly central to life processes. Displaying the diversity for which they are noted, *prokaryotes have evolved no fewer than four ways to generate these gradients: (i) respiration, (ii) photosynthesis, (iii) enzyme pumps, and (iv)* ***scalar reactions*** (Table 6.1).

Respiration

Like fermentation, **respiration** is a process by which electrons are passed from an **electron donor (reductant)** to a terminal **electron accep-**

tor (**oxidant**). However, in respiration, the electrons reach the oxidant via a number of membrane-bound electron carriers that function as a transport chain, handing off the electrons from one to another in steps down the electrochemical gradient between the reductant and the oxidant (Fig. 6.6). Such cascades create a proton gradient by (i) using some of the energy to pump protons out of the cell, (ii) exporting protons from the cell during electron-to-hydrogen transfers, or (iii) scalar reactions (defined in "Scalar reactions" below) that consume protons in the cell interior (NADH dehydrogenase does this) or produce them outside the cell (formate dehydrogenase does this).

Collectively, these carriers are called the **electron transfer system** (**ETS**). The oxidation-reduction (redox) potentials of the components of the chain (flavoproteins, quinones, cytochromes, and other electron carriers) are poised in such a manner that each member can be reduced by the reduced form of the preceding member. Thus, reducing power (as hydrogen atoms or electrons) can flow through the chain of carrier molecules to O_2 or some other terminal electron acceptor (NO_3^- and fumarate are examples). This process results in the transfer of hydrogen atoms (H) from an organic compound to some terminal electron acceptor. Some members of the chain carry hydrogen atoms; others carry only electrons. Reduction occurs by transfer of reducing power from a hydrogen carrier to an electron carrier, releasing a proton (H^+). When hydrogen carriers and electron carriers alternate in the chain, the flow of reducing power leads to expulsion of protons from the cell. This transfer of protons across the membrane is brought about in different ways by the different members of the ETS. Some act vectorially as proton pumps expelling the protons; some (the quinones), which act as reservoirs of electrons, can

Figure 6.6 Overview of electron transport system in respiration. Electrons removed from substrates (reductants) travel down a redox gradient from carrier to carrier in a membrane, forming a proton (H^+) gradient, which can be used to form ATP from ADP.

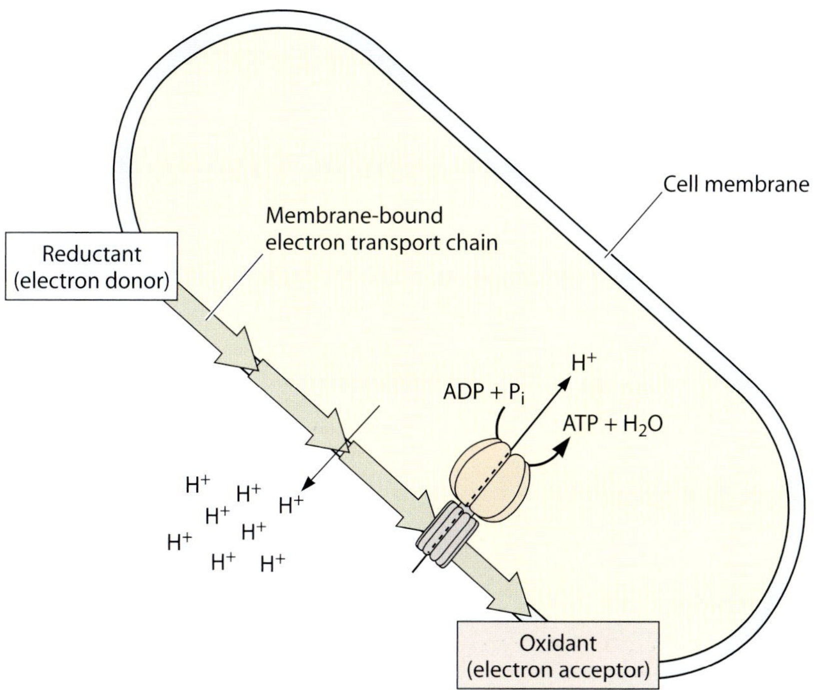

pick up protons from the cell interior and release them externally while passing on electrons to the next system member, a cytochrome. Overall, protons are taken from the cytosol by reduction of a hydrogen carrier on the inside of the membrane and are released outside the cell by reduction of an electron carrier on the outside of the membrane. Variations on this theme are almost endless. Prokaryotes are quite adaptable in electron transfer, using carriers with appropriate electrochemical characteristics under different environmental redox conditions (Fig. 6.7).

Besides its enormous variety, one further feature distinguishes prokaryotic respiration. Respiration in eukaryotes occurs within the *mitochondrial membrane* complex, whereas in prokaryotes, the *cell membrane itself* is the site of respiration. Thus, the whole bacterial cell acts as though it were a mitochondrion. This fact becomes less surprising in view of the evidence that mitochondria are highly evolved progeny of bacterial cells that were engulfed at some early time by a primitive eukaryote (see "Mitochondria, chloroplasts, and the origin of eukaryotic cells" in Chapter 19).

The energy benefit of respiration is obvious: the increased generation of ATP over what can be accomplished by substrate level phosphorylation in fermentation is tremendous. One mole of glucose yields only 2 to 4 moles of ATP by substrate level phosphorylation in most fermentations but 30-some moles (depending on the species and the conditions of growth) by respiration if the glucose were to be completely respired.

The term "respiration" is used regardless of whether O_2 is the final electron and proton acceptor. When O_2 is the acceptor (resulting in the reduction of O_2 to water), it is referred to as **aerobic respiration**, and when some other molecule is the final acceptor, it is called **anaerobic res-**

Figure 6.7 Examples of electron transport systems operative in *E. coli* under different conditions. (A) High oxygen tension. Protons and electrons (e^-) are removed from reduced NAD by NADH dehydrogenase I. The protons are secreted, and the electrons are passed to a quinone pool (Q), from which they can be used by a cytochrome oxidase (Cyo oxidase) to reduce molecular oxygen to water, accompanied by the secretion of four more protons. **(B)** Low oxygen tension. The same NADH dehydrogenase and quinone are used, but the terminal reaction is catalyzed by a different oxidase (Cyd oxidase), which does not secrete protons. **(C)** Absence of oxygen. Oxidation of lactate to pyruvate involves the removal of two electrons by lactate dehydrogenase; the electrons are then used to reduce menaquinone (MK) to MKH_2, which in turn allows the reduction of fumarate to succinate. This process is called anaerobic respiration.

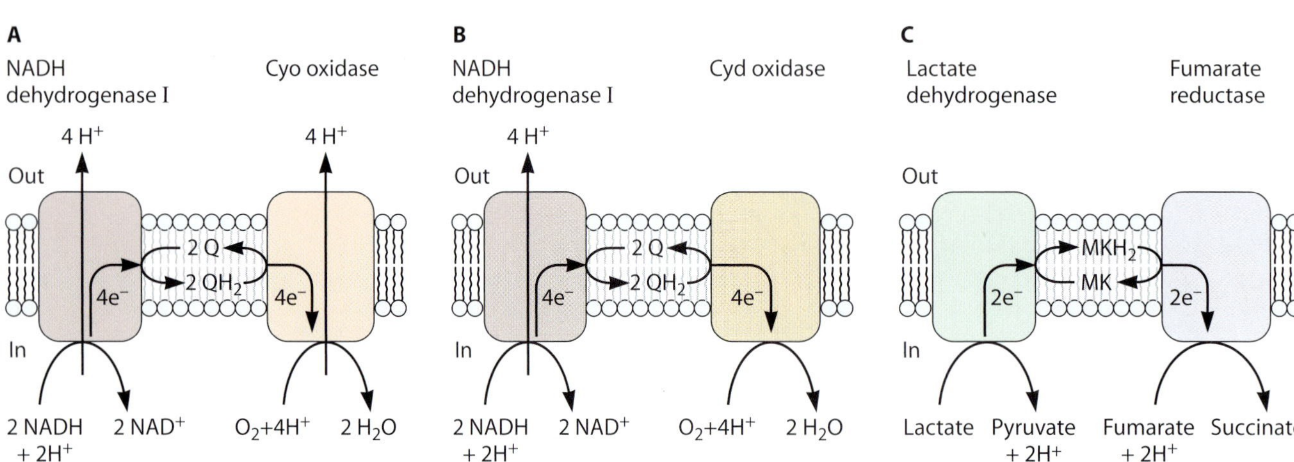

Table 6.3 Compounds that can serve as electron acceptors in anaerobic respiration, replacing oxygen

Organic compounds	Inorganic compounds
Fumarate	Nitrate (NO_3^-)
Dimethylsulfoxide (DMSO)	Nitrite (NO_2^-)
Trimethylamine *N*-oxide (TMAO)	Nitrous oxide (N_2O)
	Chlorate (ClO_3^-)
	Perchlorate (ClO_4^-)
	Manganic ion (Mn^{4+})
	Ferric ion (Fe^{3+})
	Gold (Au^{3+})
	Selenate (SeO_4^{2-})
	Arsenate (AsO_4^{3-})
	Sulfate (SO_4^{2-})
	Sulfur (S^0)

piration (Fig. 6.7C). As a group, microbes are adept at using a variety of compounds, some organic and many inorganic, as terminal electron acceptors in anaerobic respiratory chains (Table 6.3). Many parts of our planet are anaerobic. Anoxic conditions are common beneath the land's surface, yet soil, lakes, streams, and the oceans contain nitrate and sulfate, sometimes in great abundance. These vast ecosystems are colonized by bacteria particularly suited to gain all the energetic advantage of anaerobic respiration. The consequences for life on this planet are striking. For example, the use of nitrate as an acceptor, which yields nitrite (NO_2^-), followed in turn by reduction to nitric oxide (NO), nitrous oxide (N_2O), and finally dinitrogen gas (N_2), accounts for most of the nitrogen in the atmosphere of Earth (see Chapter 18). Just as fermentation products can be smelled, so can some products of anaerobic respiration, e.g., the unpleasant odor of rotting eggs (due to H_2S) and of dead ocean fish (due to trimethylamine).

The CO_2-fixing pathways by which autotrophs make precursor metabolites do not yield energy or reducing power. Instead, they use vast amounts of both. **Chemoautotrophs** obtain these essential driving forces by oxidizing inorganic compounds, and **photoautotrophs** obtain them by harvesting energy from light. Many regard respiration in chemoautotrophs (chemolithotrophs) as the supreme achievement of prokaryotic metabolism (though students of photosynthesis might argue the point). Certainly in the fueling area, chemoautotrophy can hardly be surpassed for expanding the niches prokaryotes can colonize. On the other hand, there is nothing special about the way chemoautotrophs obtain ATP and reducing power. They employ the same respiratory mechanisms—passing electrons through a membrane-embedded electron transport chain and using the proton motive force so generated to make ATP by admitting some protons back through a transmembrane ATPase/ATP synthase—as chemoheterotrophs do. Chemoautotrophs obtain reducing power (NADPH) in a similar manner—using proton motive force to reverse the flow of electrons in the ETS to reduce $NADP^+$. *The uniqueness of chemoautotrophs lies in their ability to feed the electron transport chain from inorganic sources* (Table 6.4). Most of these organisms carry out aerobic respiration, but many are capable of various forms of anaerobic respiration as well.

Table 6.4 Inorganic compounds that can serve as electron donors for chemoautotrophs

Compound	Biological processes
NH_4^+	Ammonia-oxidizing bacteria, by producing nitrite, mediate the first step of nitrification, the process, occurring principally in soil, by which ammonia is rapidly converted to nitrate (see Chapter 18).
NO_2^-	Nitrate-oxidizing bacteria, by producing nitrate, mediate the second step of nitrification.
H_2	The ability to oxidize hydrogen for chemoautotrophic growth is widespread among bacteria—both gram positive and gram negative—and archaea. Most such prokaryotes are also capable of chemoheterotrophic growth, i.e., they are facultative chemoautotrophs, sometimes referred to as mixotrophs.
Fe^{2+}	Ferrous-ion-oxidizing bacteria—called iron bacteria—are widespread in acid aquatic environments, where Fe^{2+} can persist without being spontaneously oxidized. They are found in streams, bogs, and some marine environments. They are noticeable in such locations by the clumps of iron oxide that they form.
Mn^{2+}	Most iron bacteria are also capable of chemoautotrophic growth at the expense of manganous ion.
H_2S, S^0, $S_2O_3^{2-}$	Chemoautotrophs that oxidize reduced sulfur compounds are also widespread in nature. They are the primary producers for the elaborate biological communities that exist near hydrothermal vents in the ocean floors. Because H_2S is spontaneously oxidized by O_2, bacteria that utilize it are located in regions where their sources of H_2S and O_2 diffuse in from opposite directions.

Many bacteria have come to be known by the food they eat. Some of the more common of these informal names of bacteria are listed in Table 6.5. Incidentally, you may have noticed that anaerobic respiration and chemoautotrophic fueling are complementary variations of aerobic respiration: anaerobic respiration differs with respect to the *terminal electron acceptor* of an electron transport chain, and chemoautotrophic fueling differs with respect to the *primary electron donor* to a chain.

Photosynthesis

In **photosynthesis**, light energy activates an electron from chlorophyll to flow back down through an electron transport chain to the original ground-state chlorophyll or to $NADP^+$. If the electrons flow to $NADP^+$, more must be continually supplied to chlorophyll; the sources can be water or sulfur compounds. Among prokaryotes, photosynthesis is restricted to the Bacteria, and nearly all are autotrophs (Table 6.6). Some halophilic Archaea have a primitive mechanism for capturing light energy (see Chapter 16), but it is adequate to supply energy only for motility, not autotrophic growth. Some bacteria are photoheterotrophs, deriving precursor metabolites from organic compounds and the driving force from photosynthesis.

One group of photoautotrophic bacteria, the cyanobacteria, utilize the same **oxygenic** (oxygen-producing) **system of photosynthesis** that you may be familiar with in plants and algae, i.e., they employ chlorophyll *a*, as well as photosystems I and II. It is not really a coincidence. Chloroplasts, the organelles where photosynthesis occurs in plants and algae, are the modern evolutionary products of a cyanobacterium that was captured by a primitive eukaryote. The oxygen produced in oxygenic photosynthesis is the by-product of stripping electrons from water.

Table 6.5 Informal names of some prokaryotes based on fueling activities

Informal name	Distinguishing characteristic	Representative genus
Methanogens	Autotrophic archaea forming methane as a by-product of using H_2, CO_2, formate, methylamine, or similar molecules as energy source	*Methanobacterium*
Hydrogen bacteria	Autotrophs using hydrogen as sole energy source	*Pseudomonas*
Nitrifiers	Autotrophs using ammonia or nitrite as sole energy source	*Nitrosomonas*
Denitrifiers	Organisms using oxidized forms of nitrogen as acceptors in anaerobic respiration, thereby generating N_2	*Paracoccus*
Sulfur bacteria	Autotrophs utilizing sulfur, hydrogen sulfide, or thiosulfate as sole energy source, either aerobically or anaerobically	*Sulfolobus*
Iron bacteria	Autotrophs oxidizing ferrous to ferric iron as a source of energy	*Thiobacillus*
Phototrophs	Autotrophs utilizing light energy	*Rhodobacter*
Methylotrophs	Autotrophs utilizing methane, methanol, methylamine, or formate as sole energy source	*Hyphomicrobium*

Other bacterial groups carry out **anoxygenic** (not oxygen-producing) **photosynthesis** because they utilize a compound other than water as a supply of electrons to generate reducing power. A generalized reaction (formulated by C. B. van Niel) for all photosynthesis is as follows:

$$H_2A + CO_2 \rightarrow A + \text{carbohydrate}$$

where A can be O, S, or other elements. The general characteristics of these bacterial groups are summarized in Table 6.6.

Table 6.6 Properties of photosynthetic prokaryotes

Group	Electron donor	Chlorophyll type(s)	Example(s)
Oxygenic			
Cyanobacteria	H_2O	Chlorophyll *a* and phycobilins	*Synechococcus* spp., *Oscillatoria* spp., and *Nostoc* spp.
Anoxygenic			
Purple nonsulfur bacteria	Many substrates, including H_2, alcohols, organic acids, and Fe^{2+}	Bacteriochlorophylls *a* and *b*	*Rhodobacter* spp. and *Rhodopseudomonas* spp.
Purple sulfur bacteria	Reduced sulfur compounds, (H_2 and also certain acids	Bacteriochlorophylls *a* and *b*	*Cromatium spp.*
Green sulfur bacteria	Reduced sulfur compounds (H_2S and $S_2O_3^{2-}$)	Bacteriochlorophylls *c*, *d*, and *e*	*Chlorobium* spp.
Heliobacteria	Lactate, other organic, acids	Bacteriochlorophyll *g*	*Heliobacillus* spp.

Figure 6.8 Cyclic photophosphorylation in bacteria. Bacteriochlorophyll (Bchl), at the expense of light energy, is raised to an activated state (Bchl*), which is readily oxidized by passing electrons (e^-) through an electron transport chain composed in part of bacteriopheophytin (Bpheo) and quinones to cytochrome c_2, which reduces the oxidized bacteriochlorophyll. The cyclic flow of electrons—from Bchl and back again—is indicated by the arrows. E'_0 is the electrochemical potential.

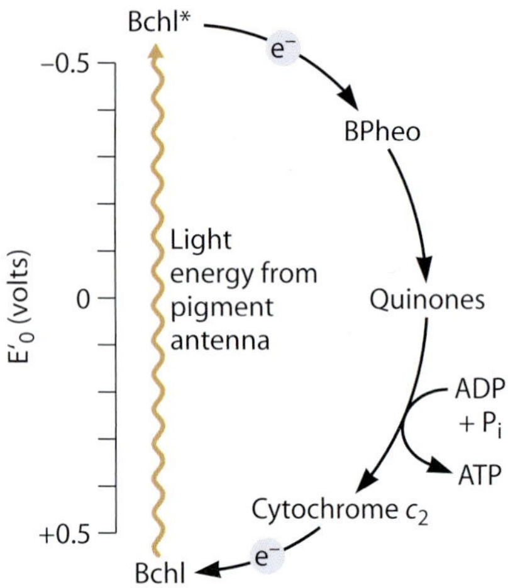

Those interested in a full taste of microbial metabolic diversity might be impressed to learn how some bacteria cross the classification lines of fueling reactions. Anoxygenic photosynthesis is an anaerobic process, but the bacteria called purple nonsulfur bacteria are, in addition, capable of *chemoheterotrophic growth in the presence of oxygen.* They, like the hydrogen bacteria (bacteria that use hydrogen as their source of energy), are **facultative autotrophs**. Moreover, as Table 6.6 suggests, the purple nonsulfur bacteria are also capable of *photoheterotrophic* growth. The purple sulfur bacteria can also cross the classification lines of fueling reactions. If conditions are microaerophilic, they grow as chemoautotrophs by oxidizing H_2 or H_2S.

Photosynthesis generates proton motive force (which can be used for ATP synthesis) in a manner basically similar to respiration in that a flow of electrons down an electrochemical gradient leads to the extrusion of protons across the cell membrane. In the case of photosynthesis, the electrons start at a high electrochemical potential because they are in light-activated **chlorophyll** molecules. Light energy (photons, measured as $h\nu$, where h is the Planck constant and ν is the frequency of the radiation in hertz), captured by **pigment antenna**, is therefore the driving force. Chlorophyll is raised from its ground state to its activated state, which is easily oxidized and donates electrons to the beginning of the transport chain. The electrons are either returned to ground-state chlorophyll (**cyclic photophosphorylation**) (Fig. 6.8) or transferred to $NADP^+$ (**noncyclic photophosphorylation**). [Cyclic photophosphorylation generates ATP, but no reducing power is stored as NAD(P)H. Members of the various groups of **phototrophs** solve this problem in different ways, two of which are shown in Fig. 6.9.]

Enzyme pumps

Enzyme pumps (Table 6.1) are certain membrane proteins that are not part of an electron transport chain but that nevertheless function independently to pump protons or other ions across the membrane, creating a gradient that can be used for the usual energy-requiring processes. Sometimes these reactions augment a gradient produced by other means. There is a light-driven enzyme of *Halobacterium halobium* (an archaeon) that pumps protons out of the cell, and *Klebsiella aerogenes* (a bacterium) has an oxalacetate decarboxylase that pumps sodium ions outside. In other cases, ion pumps do the whole job, as with the membrane-bound **methyltransferases** and heterodisulfide reductases of methanogens (Table 6.5), which pump out sodium ions.

Scalar reactions

Scalar reactions (Table 6.1) have an intriguing name and an equally interesting role in energy metabolism. At first sight, the definition seems strange: *a scalar reaction is one in which the substrates and products are in the same location or compartment.* (This definition is meant to contrast with a **vectorial reaction**, in which one or more products are moved to a separate compartment as a result of the reaction.) In the context of energy metabolism, scalar reactions are those that directly or indirectly create a transmembrane ion gradient *without moving the ions.* Some scalar reactions are associated with electron transport systems, but others exist separately. They function independently to supply part or all of

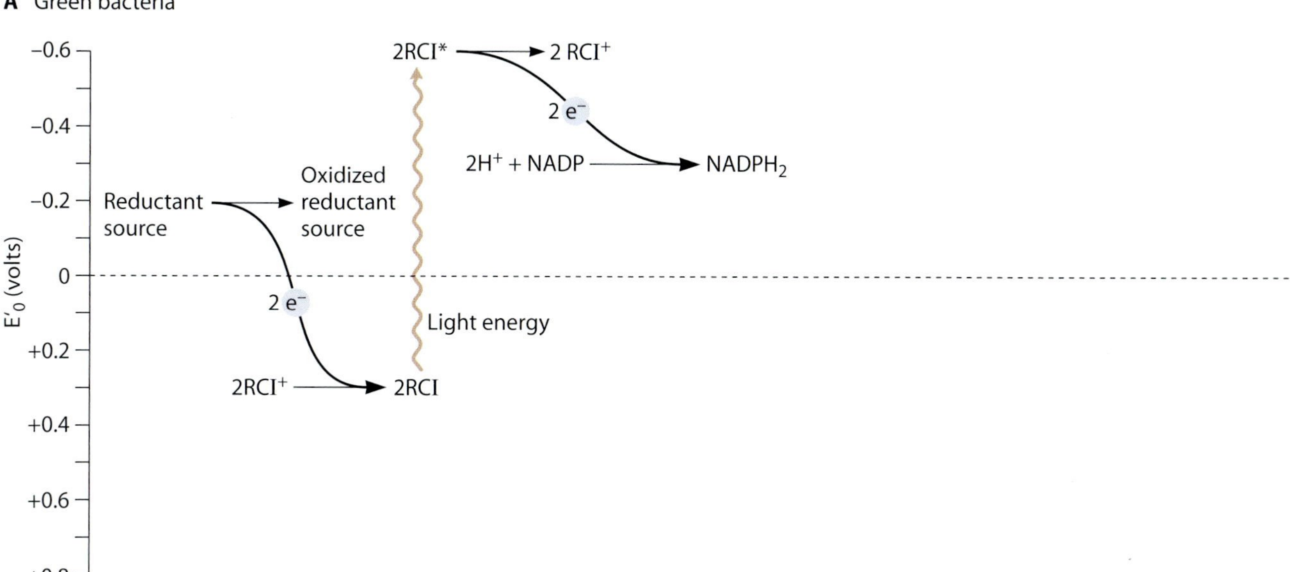

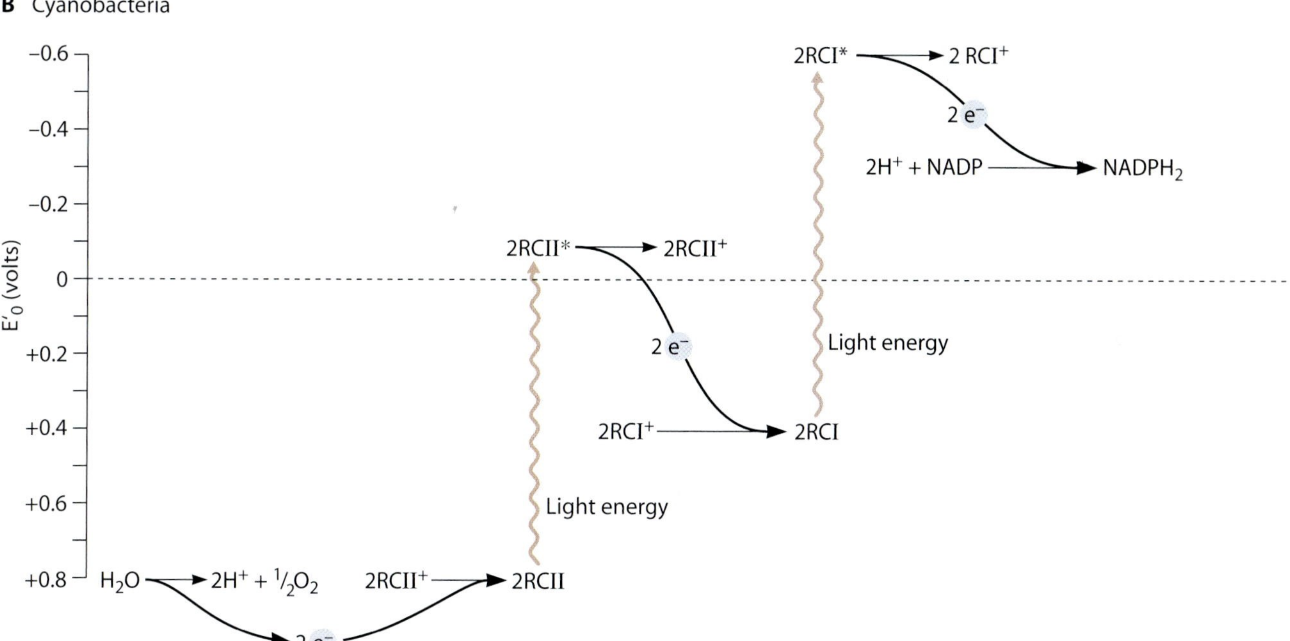

Figure 6.9 Generation of reducing power in noncyclic photophosphorylation. **(A)** In green bacteria, electrons (e^-) from the reductant source pass to oxidized bacteriochlorophyll in reaction center I (RCI^+), thereby reducing it (RCI). Light energy converts the bacteriochlorophyll in RCI to its activated form, RCI^+, which is readily oxidized, donating its electron to a transport chain, which leads to reduction of $NADP^+$ to NADPH. **(B)** In cyanobacteria (as in algae and plants), two sequential photoexcitations of chlorophyll—first in RCII and then in RCI—raise the reducing power of electrons in water sufficiently to reduce $NADP^+$. The thick arrows represent electron transport chains.

a cell's energy in the oddest ways. *Oxalobacter formigenes*, for example, obtains all its energy by decarboxylating oxalate to yield formate. This reaction consumes one proton, thereby creating a proton gradient across the cell membrane, enabling it to make ATP. In still another example, malolactic organisms, which acidify and enrich the complexity of red wines (see Chapter 23), augment their energy by a scalar reaction involving taking malate into the cell and decarboxylating it to lactate. The lactate is still in the cell, but it is then secreted in a subsequent step, taking a proton along with it in a process known as proton symport (see below), thereby establishing a proton gradient. These examples demonstrate that the evolution of organisms under the pressure of surviving and competing in specialized environments is a powerful force with creative outcomes.

Oxygen and life

At many points, our discussion of energy metabolism has hinged on the presence or absence of molecular oxygen. The relationship of living organisms to oxygen is complex, since oxygen and its metabolic derivatives are extraordinarily toxic to cells. Consequently, cell interiors contain very low concentrations of oxygen, or none at all. Still, O_2 is exceedingly effective as a terminal electron acceptor in biological **redox reactions,** whether the energy source is organic or inorganic.

The ability of an organism to withstand the toxic effects of oxygen depends on two factors: whether the enzymes of the cell are intrinsically sensitive to oxygen and whether the cell is able to break down two highly toxic metabolic products of oxygen, hydrogen peroxide (H_2O_2) and the superoxide anion (O_2^-). Depending on these two factors, there is an enormous spectrum of bacterial responses to oxygen. At one extreme are the strict anaerobes, which are so exquisitely sensitive to oxygen that brief exposure to ambient air is lethal, and at the other end of the spectrum are cells (like those of our bodies) that are completely addicted to oxygen. Organisms that can tolerate oxygen generally have protective enzymes, including the following:

- **catalase,** which converts H_2O_2 to H_2O and O_2
- **superoxide dismutase,** which converts superoxide (O_2^-) to O_2 and H_2O_2, which is subsequently degraded by catalase

Table 6.7 defines the categories of responses to oxygen and gives examples of bacteria demonstrating them.

MAKING PRECURSOR METABOLITES: HETEROTROPHY

Acquiring nutrients

The fundamental role of the cell membrane is to contain the cell. Keeping cell components, including low-molecular-weight metabolites, from diffusing into the environment is essential for life, yet substances must get into the cell and many others must be excreted. It will come as no surprise, therefore, that a considerable fraction of the cell's genome encodes structures that deal with solute transport.

The rapid growth of microbes has a natural corollary: food must enter them at high rates. However, we know that the survival of microbes also depends on excluding substances that are potentially harmful to vital cell machinery. Rapid entry of food, therefore, must occur in a *highly selective fashion.* In addition, most microbes grow rapidly even in highly dilute nutrient solutions that permit only low rates of diffusion into cells. Thus, mechanisms for concentrating food molecules inside the cell are

Table 6.7 Microbes classified by response to oxygen

Class	Synonym	Growth in air	Growth without oxygen	Possession of catalase and superoxide dismutase	Description	Example(s)
Aerobe	Strict aerobe	Yes	No	Yes	Requires oxygen; cannot ferment	*Mycobacterium tuberculosis, Pseudomonas aeruginosa, Bacillus subtilis*
Anaerobe	Strict anaerobe	No	Yes	No	Killed by oxygen; ferments in absence of O_2	*Clostridium botulinum, Bacteroides melaninogenicus*
Facultative		Yes	Yes	Yes	Respires with O_2; ferments or uses anaerobic respiration in absence of O_2	*Escherichia coli, Shigella dysenteriae, Staphylococcus aureus*
Indifferent	Aerotolerant anaerobe	Yes	Yes	Yes	Ferments in presence or absence of O_2	*Streptococcus pneumoniae, Streptococcus pyogenes*
Microaerophilic		Slight	Yes	Small amounts	Grows best at low O_2 concentration; can grow without O_2	*Campylobacter jejuni*

necessary. As we shall now see, these three challenges—fast entry, selective admission, and the need to concentrate food molecules—have been successfully met by microbes.

Given the chemical variety of molecules that must be ingested and the fundamentally different architectures of gram-positive and gram-negative envelopes, it will come as no surprise that a number of different means of **solute transport** have evolved to enable high-speed, selective entry of nutrients. Much of the variety of transport mechanisms is related to the cell membrane. We shall examine these, but first we should deal with the outer membrane (OM) of gram-negative cells.

Transport through the OM of gram-negative bacteria

The thick murein lattice of gram-positive cells is readily permeable to water and solutes, and thus it is the cell membrane that provides the first hydrophobic barrier. In gram-negative bacteria, on the other hand, solutes first encounter the hydrophobic OM. This structure would present an impenetrable barrier for water and dissolved ions were it not for the porin-formed channels it contains (see Chapter 2). These pores traverse the OM, allowing water molecules and hydrophilic solutes to diffuse readily into the periplasm, as long as they are no larger than 600 to 700 daltons. Hydrophobic compounds cannot pass through the pores of the OM. This means that sugars, amino acids, and most ions can diffuse into the periplasm, but for hydrophobic molecules and larger hydrophilic ones, special structures are needed to provide selective transport through the OM. Substrate concentrations must be lower in the periplasm than in the external medium, since the OM has no means to pump solutes rapidly through the pores. Exceptions exist. For example, pores that admit the large vitamin B_{12} molecule appear to be energized, though the mechanism seems mysterious since there is no obvious source of ATP in the OM.

Also, there are pores that preferentially admit other large molecules, such as disaccharides and the organic iron chelates that enable the cell to scavenge this vital mineral.

Transport through the cell membrane

A great variety of mechanisms have evolved to admit solutes selectively and rapidly into the cell, presumably because no single process would serve for all solutes in all environmental circumstances. These mechanisms are of three types: **passive transport, active transport,** and **group translocation.**

1. **Passive transport** relies on diffusion, uses no energy, and hence operates only when the solute is at higher concentration outside than inside the cell. **Simple diffusion** accounts for the entry of very few nutrients. Dissolved oxygen, carbon dioxide, and water itself just about complete the list. Simple diffusion provides neither speed nor selectivity. **Facilitated diffusion**, like simple diffusion, requires no energy input, and so the net movement of solute molecules still depends on a downward concentration gradient. Facilitated diffusion is, however, selective. **Channel proteins** form selective channels that facilitate the passage of specific molecules. Facilitated diffusion is common in eukaryotic microbes (this is how yeast cells obtain sugar), but it is rare in prokaryotes. *Escherichia coli* and other enteric bacteria transport glycerol by facilitated diffusion, and some species of *Zymomonas* and *Streptococcus* transport glucose in this manner.
2. **Active transport** mediates the entry of virtually all nutrients. There are a variety of such mechanisms, illustrated and summarized in Table 6.8 and Fig. 6.10. All use energy to pump molecules into the cell at high rates, often against a concentration gradient. *Many nutrients are concentrated more than a thousandfold as a result of active transport.* These mechanisms are of two sorts, depending on the immediate source of energy employed: ion-coupled transport and ATP-binding **cassette** (ABC) transport.
   - **Ion-coupled transport** is driven by the electrochemical gradient (proton or sodium motive force) established across the cell membrane either by electron transport or by hydrolysis of ATP by membrane-bound ATPases. In essence, this is a preload scheme; the energy investment in priming the membrane is made before it is to be used. Hence, this transport mechanism is sometimes referred to as **secondary active transport**, with the prior export of protons considered the primary event. Proton or sodium motive force is used in three different but related ways (**symport**, **antiport**, and **uniport**) to drive solutes into the cell (Table 6.8 and Fig. 6.10). Ion-coupled transport is particularly common among aerobic organisms, which have an easier time generating an ion motive force than do anaerobes.
   - **ABC transport** employs ATP directly to pump solutes into the cell. Specific **binding proteins**, located in the periplasm of gram-negative cells or attached to the outer surface of the cell membrane of gram-positive cells, confer specificity by carrying

Table 6.8 Examples of active transport of solutes in prokaryotes

Mechanism	Energy source	Components	Process	Distribution and function
Ion-coupled transport	Proton motive force	Membrane: specific transmembrane proteins	Symport: transport of proton accompanied by transport of neutral solute or negative ion Antiport: two like-charged ions transported simultaneously in opposite directios Uniport: single molecule transported driven by electrochemical gradient	Common among aerobes Responsible for transport of some sugars, amino acids, and many inorganic ions
Group translocation (PTS system)	Phosphoenolpyruvate (PEP)	Cytoplasm: enzyme I and histidine protein (HPr) Membrane: enzymes II (specific proteins for different substrates), which have multiple subunits	The phosphoryl group of PEP is transferred sequentially from enyzme I to HPr to a subunit of the appropriate enzyme II, which phosphorylates the incoming solute, usually a sugar; another subunit of enzyme II forms the translocation channel.	Common for sugar transport in anaerobes, but not restricted to them
ABC transport	ATP (or acetyl-PO_4)	Periplasm (G^-) or membrane (G^+): set of specific binding proteins for different substrates Membrane: two channel proteins, two ATPases	Cassette consisting of solute bound to its specific binding protein docks at the membrane channel, and solute transport is driven by hydrolysis of ATP.	Many bacteria contain large numbers of these cassettes for amino acids and other organic nutrients.
Iron siderophore transport	ATP	Envelope: group of eight or more proteins that span all layers of the envelope Secreted siderophore (a chelator of Fe^{3+})	The siderophore (enterochelin is one of *Escherichia coli*'s) is made and secreted; after binding Fe^{3+}, it is transported across all layers of the envelope by the eight-protein envelope complex.	Common among many pathogens and others living in Fe-poor environments

their particular ligand to a protein complex on the membrane. Hydrolysis of ATP is then triggered, and the energy is used to open the membrane pore and allow unidirectional movement of the substrate into the cell. This mode of transport is extremely common among many bacteria; *E coli*, for example, has many dozens of specific binding proteins, which handle nearly half of the substrates transported by the organism.

3. A **phosphotransferase system** (PTS) accomplishes transport by chemically modifying the solute, which arrives in the cell as a different molecule. The rationale of this group translocation process is that energy is expended not for the transport process but rather to form an intracellular derivative that is membrane impermeable and thus trapped within the cell. (Note that in a strict sense group translocation is not active transport because no concentration gradient is involved.) The bacterial PTS (the only known group translocation process) results in the phosphorylation of the substrate as it traverses the cell membrane. Phosphoenolpyruvate is one of the common sources of the high-energy phosphate group (Fig. 6.11). This means of transport is responsible in many bacteria for the entry of a variety of different sugars. Observant readers will note that transport of glucose by the PTS can be regarded as a free gift energetically; since glycolysis commences with the phos-

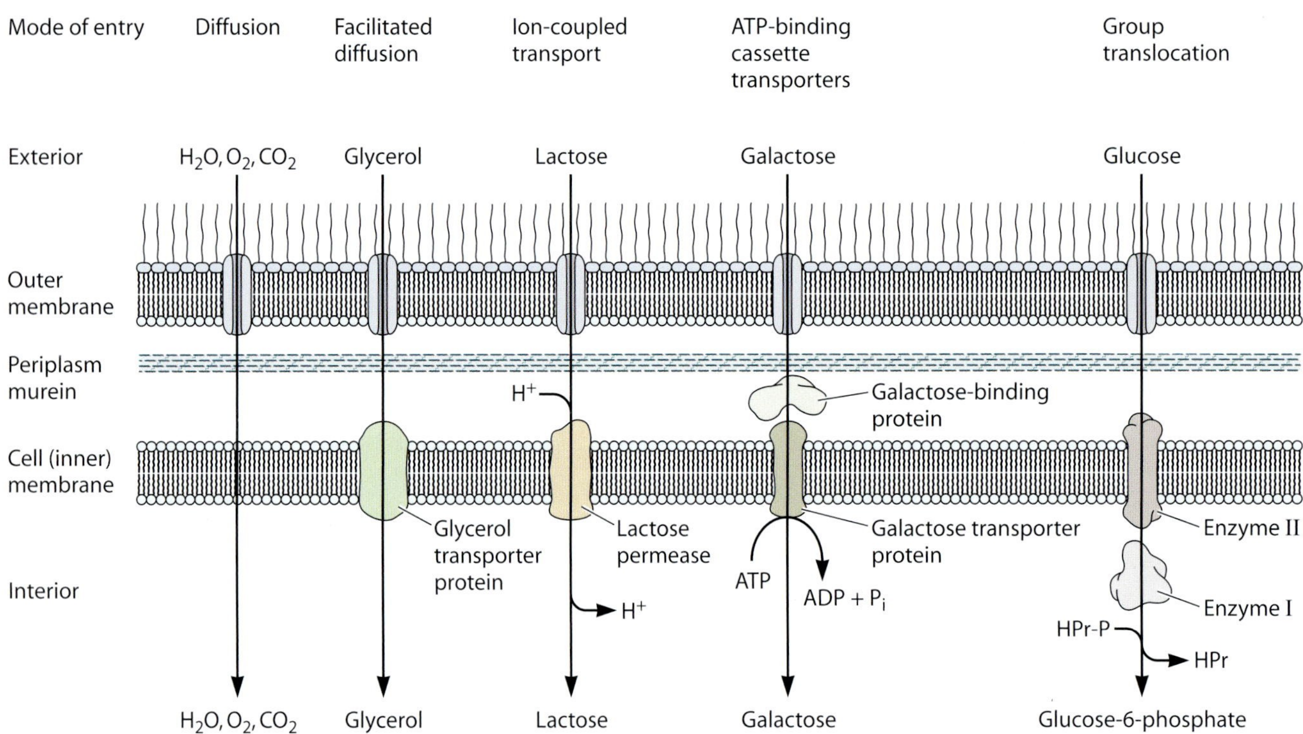

Figure 6.10 Varieties of solute transport processes. The various modes are illustrated by carbohydrate transport in *E. coli*.

phorylation of glucose to glucose-6-phosphate, the expenditure of a ~P would have had to be made anyway. For this reason, the PTS is particularly common among energetically challenged microbes, such as those living in the absence of oxygen. We shall encounter the PTS again in two different contexts, as a part of the mechanism of the **global control system** (systems that regulate many genes with a large variety of related functions are said to be global) called catabolite repression (see Chapter 12) and as a participant in chemotaxis (see Chapter 14).

Special transport processes supplement these main mechanisms. Perhaps the most prominent is the manner in which many bacteria, including important pathogens, steal iron (Fe) from their mammalian hosts (Fig. 6.12). Fe is as much an essential nutrient for bacteria as it is for eukaryotic organisms, yet the internal compartments of animals contain virtually no free Fe; it is sequestered in complexes with such proteins as **transferrin** and **lactoferrin**. Bacteria solve this dilemma by secreting powerful chelators of Fe called **siderophores**. An example from *E. coli* is a siderophore called **enterochelin**. This molecule is synthesized and secreted. After binding Fe, the enterochelin-Fe complex is actively transported back into the cell by the cooperative actions of a group of eight proteins that collectively span the OM, periplasm, and inner membrane of the bacterium. It is an ATP-driven process.

Finally, it must be emphasized that the active transport of substances out of the cell (certain end products and toxic materials) is extremely important and occupies a significant fraction of the prokaryotic cell's membrane machinery.

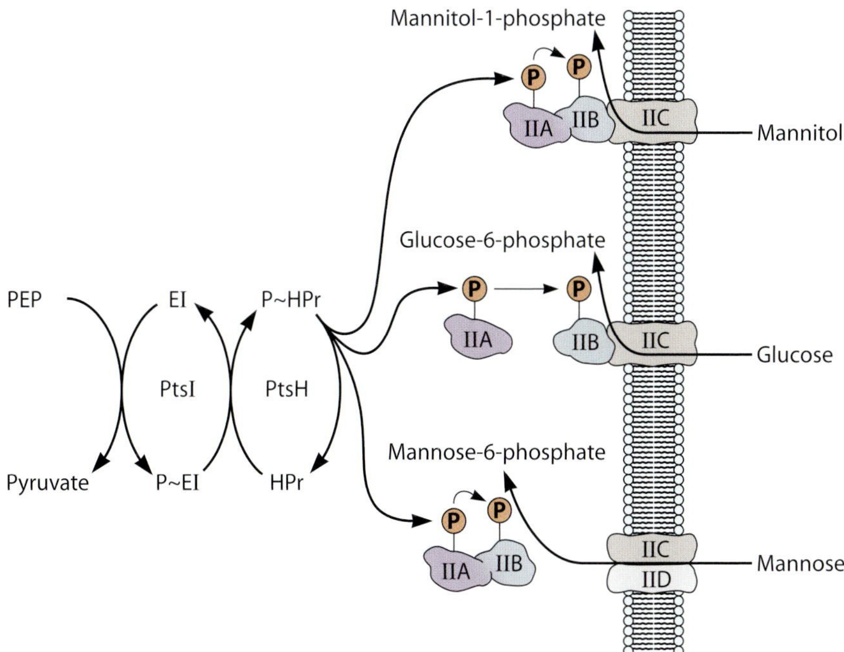

Figure 6.11 PTSs for three sugars in *E. coli.* EI and HPr are general proteins shared by all systems and serve to phosphorylate the sugar-specific transport components. IIA to IID, proteins specific for a given sugar.

Feeder pathways

Once imported, carbon compounds that will serve as the basis for making precursor metabolites and generating driving force need to be introduced into the **central pathways** of fueling reactions. The variety of substances that can be substrates for growth is enormous. For *Salmonella*, an experimental exploration made decades ago identified over 80 organic compounds that can serve as the sole source of carbon and energy (Table 6.9). For organisms that inhabit the soil (*Pseudomonas* species, for example), the list of potential substrates is even larger and includes many compounds that play critical roles in the cycling of organic substances in nature. **Feeder pathways** convert this enormous variety of substances into one or another of the metabolites of the central pathways, from which

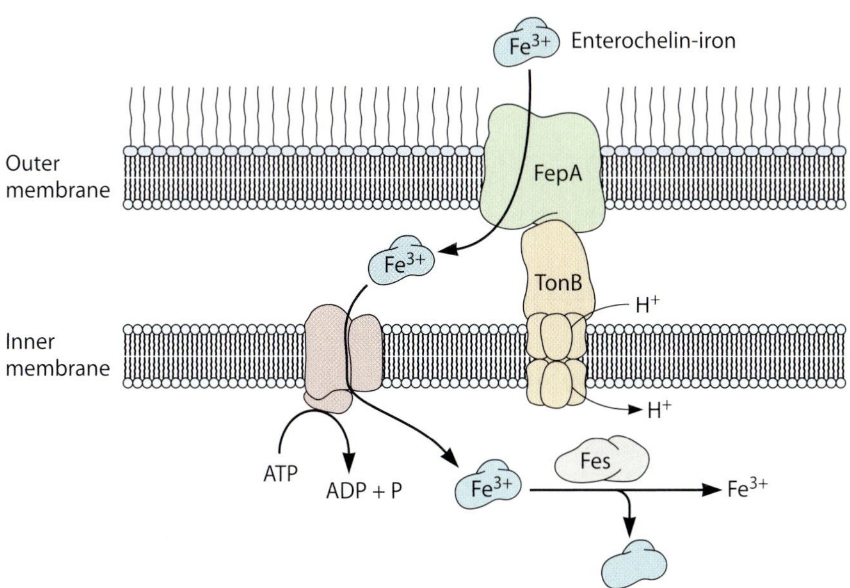

Figure 6.12 Enterochelin iron transport system of *E. coli.* The enterochelin molecule with its bound Fe passes through the FepA protein of the OM in a process requiring proton motive force energy transmitted by the TonB complex, which extends through the energized cell membrane. Next, the complex passes across the cell membrane through other Fep proteins, driven by ATP hydrolysis. An esterase (Fes) cleaves the enterochelin, releasing the Fe.

Table 6.9 Compounds that serve as a sole source of carbon for *Salmonella*

Acetic acid	L-Fucose	*meso*-Inositol	L-Proline
N-Acetyl-D-glucosamine	Fumaric acid	DL-Isocitric acid	Propionic acid
N-Acetyl-D-mannosamine	Galactaric acid	α-Ketoglutaric acid	Pyruvic acid
Adenosine	Galactitol	L-Lactic acid	L-Rhamnose
L-Alanine	D-Galactose	Lauric acid	D-Ribose
L-Arabinose	D-Glucaric acid	L-Lyxose	D-Serine
DL-Citramalic acid	D-Gluconolactone	L-Malic acid	L-Serine
Citric acid	D-Gluconic acid	D-Maltose	D-Sorbitol
L-Cysteine	D-Glucosamine	D-Mannitol	Succinic acid
Cytidine	D-Glucose	D-Mannose	*meso*-Tartaric acid
2-Deoxyadenosine	D-Glucose-6-phosphate	D-Mannosamine	Thymidine
2-Deoxycytidine	D-Glucuronic acid	Melibiose	D-Trehalose
2-Deoxyguanosine	D-Glucuronolactone	L-Methionyl-L-alanine	Tricarballylic acid
2-Deoxy-D-ribose	DL-Glyceric acid	6-Methylaminopurine	Tridecanoic acid
Deoxyuridine	Glycerol	α-Methyl-D-galactoside	Uridine
Dihydroxyacetone	α-Glycerophosphate	Myristic acid	D-Xylose
D-Erythrose	Glycolic acid	Oleic acid	
D-Fructose	Guanosine	Oxalacetic acid	
D-Fructose-6-phosphate	Inosine	3-Phosphoglyceric acid	

the cell can make the 13 precursor metabolites and driving force needed for biosynthesis. Some feeder pathways consist of a single reaction (the conversion of sucrose to two molecules of hexose, for example), but others consist of many sequential enzyme-catalyzed steps, and some make considerable ATP and reducing power, although no precursor metabolites.

The key principle to keep in mind is that organic compounds imported as food for fueling must be converted, by however many sequential reactions it takes, to derivatives that can enter at least one of the three interconnected pathways of central metabolism, which we discuss below.

Central metabolism

Central metabolism consists of three common and a number of species-specific auxiliary pathways of fueling.

Common pathways of central metabolism

Because they are universally distributed among cellular organisms and together account for the synthesis of all 13 precursor metabolites, three interconnected metabolic pathways have a special status. Every biochemistry course will introduce the principle features of the three pathways that constitute central metabolism. The outline of these pathways, shown in Fig. 6.13, highlights the key fueling products of the three pathways.

1. **Glycolysis pathway.** In the glycolytic pathway (known also as the **Embden-Meyerhof-Parnas [EMP] pathway**, for its early discoverers), glucose is phosphorylated and then split into two molecules of triosephosphate, eventually yielding two molecules of pyruvate, a precursor metabolite. Along the way, three other precursor metabolites, reducing power in the form of two molecules of NADH, and a net of two molecules of ATP are formed.

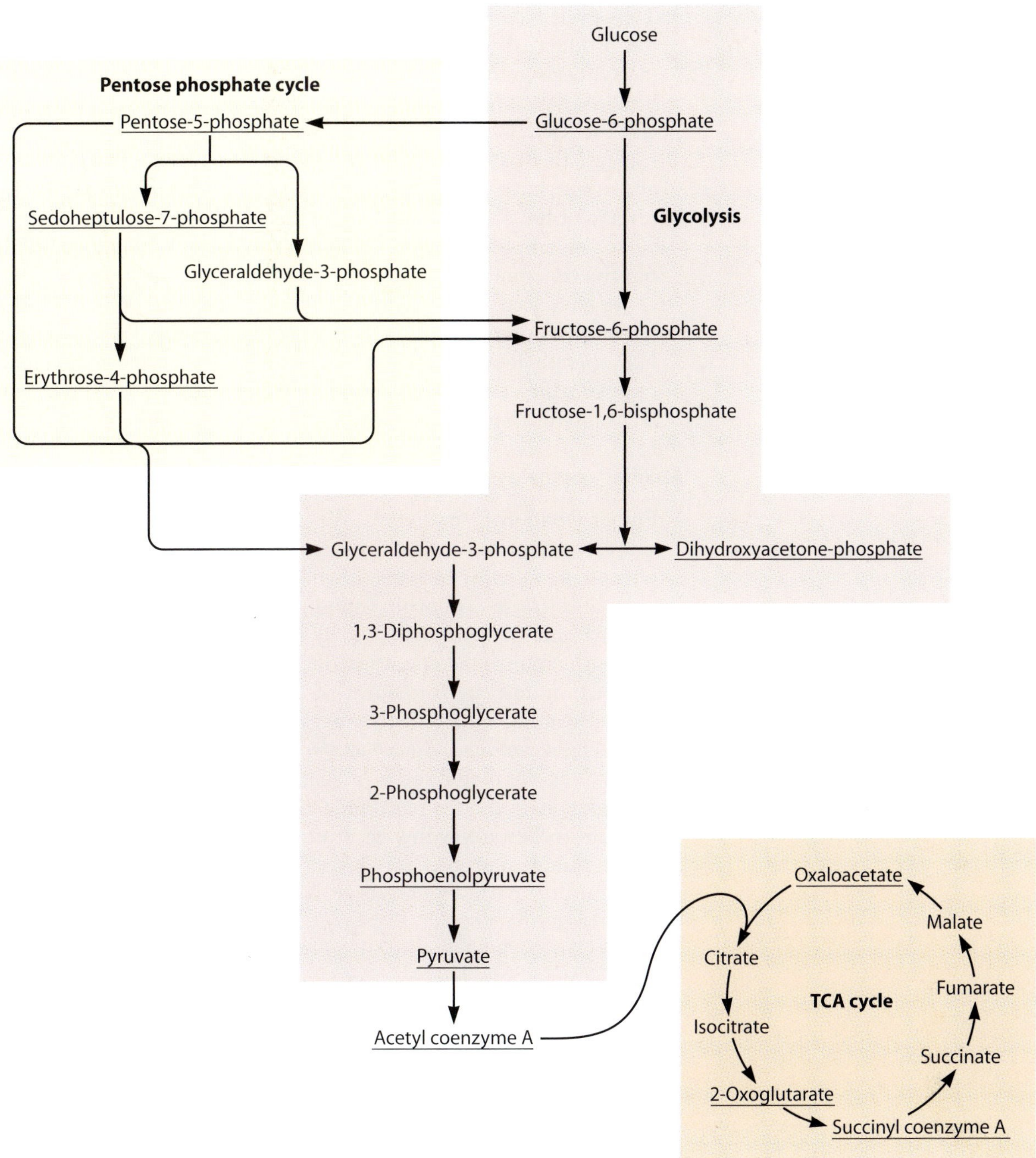

Figure 6.13 Central fueling pathways. The 13 precursor metabolites are underlined.

2. **Pentose phosphate pathway.** Sometimes called the hexose monophosphate shunt, this pathway appears to be an alternate route for converting glucose-6-phosphate to triosephosphate, but it plays the vital role of supplying two additional precursor metabolites. It is also responsible for forming two molecules of reducing power in the form of NADPH, the form most commonly used for biosynthesis.

3. **TCA cycle.** The TCA cycle (or **Krebs cycle**) is fed from pyruvate, the product of glycolysis, through a linker precursor metabolite (acetyl-coenzyme A). This cycle can form three precursor metabolites and four units of reducing power while producing two molecules of CO_2, although under many, perhaps even most, circumstances, this pathway does not function as a cycle. Many strict anaerobes have only a portion of the cycle, functioning in the reverse of the normally depicted direction. By this means, the precursor metabolites needed for biosynthesis can be produced without the concomitant production of NADH, the disposal of which, as we have seen, is a challenge in the absence of oxygen. Common **facultative organisms** (such as *E. coli*) may possess the entire cycle but operate it anaerobically as two separate arms working in opposing directions. This flexibility is described in "Diversity and flexibility of central metabolism" below.

Auxiliary pathways of central metabolism

1. **Entner-Doudoroff pathway.** The Entner-Doudoroff pathway is simply an alternative link between an intermediate of the pentose phosphate pathway (6-phosphogluconate) and two compounds of glycolysis (triose-3-phosphate and pyruvate) (Fig. 6.14). It is mediated by only two enzymes (a dehydratase and an aldolase). Use of this path requires the operation of the glycolytic path for both ATP production and the formation of precursor metabolites, and yet the pathway is widely distributed among diverse bacteria, and in some (*Pseudomonas*, *Zymomonas*, and *Erwinia* species) it appears to serve as the major route of sugar metabolism. *E. coli* metabolizes gluconate by this path. One particular value of this pathway appears to be that pathways from a number of sugar acids feed into its single intermediate (2-keto-3-deoxy-6-phosphogluconate), and thus, the pathway can serve as a collector of metabolites from other feeder pathways.
2. **Glyoxylate cycle.** A modification of the TCA cycle, the glyoxylate cycle is unique to bacteria, protozoa, and plants (Fig. 6.15). The pathway is actually a bypass, or short-circuiting, of two of the

Figure 6.14 The Entner-Doudoroff auxiliary pathway.

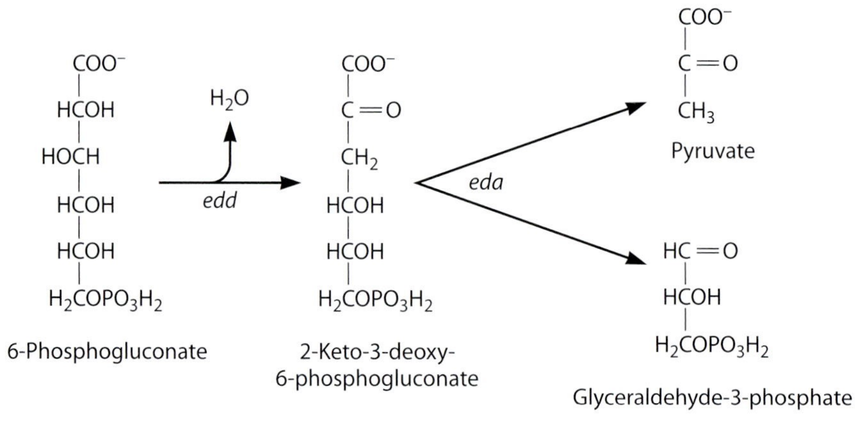

COO^- – HCOH – HOCH – HCOH – HCOH – $H_2COPO_3H_2$
6-Phosphogluconate

→ *edd* (H_2O released) →

COO^- – C=O – CH_2 – HCOH – HCOH – $H_2COPO_3H_2$
2-Keto-3-deoxy-6-phosphogluconate

→ *eda* →

COO^- – C=O – CH_3
Pyruvate

HC=O – HCOH – $H_2COPO_3H_2$
Glyceraldehyde-3-phosphate

decarboxylation reactions of part of the TCA cycle, enabling the organisms to grow on acetate (or fatty acids converted into acetate).

3. **Fermentation pathways.** You will recall that many anaerobic heterotrophs can generate all the ATP they need by substrate level phosphorylation, but only if they find some way to reoxidize NADH to NAD^+. This feat, called fermentation, is achieved by passing most of the hydrogen atoms from NADH to pyruvate or some compound derived from pyruvate. These short pathways (many are shown in Fig. 6.4) are properly considered auxiliary **fueling pathways** of central metabolism.

Diversity and flexibility of central metabolism

Not every biochemistry course deals with the peculiar adeptness with which microbes manipulate these pathways. Bacteria as a group have long been noted for their ability to grow without oxygen and to grow on two-carbon (or even on one-carbon) compounds and other nonsugars, but there are many other unique adaptations of the central pathways evolved by individual microbes to permit growth under various ecological circumstances. How, for example, can cells acquire the fueling products (precursor metabolites, reducing power, and energy) if they are provided only with a fatty acid, or a pentose, or malate, for example, instead of glucose as a sole source of carbon and energy?

The answer is that the central pathways, depending on environmental circumstances, must operate in *either the forward or reverse direction*, because the precursor metabolites must be generated no matter where a feeder pathway pours metabolites into a particular central pathway.

The modifications in the operation of central pathways that enable many prokaryotes to grow aerobically, anaerobically, and in situations requiring **gluconeogenesis** (synthesis of hexose from one-, two-, three-, or four-carbon substrates) are illustrated in Fig. 6.16. This flexibility of

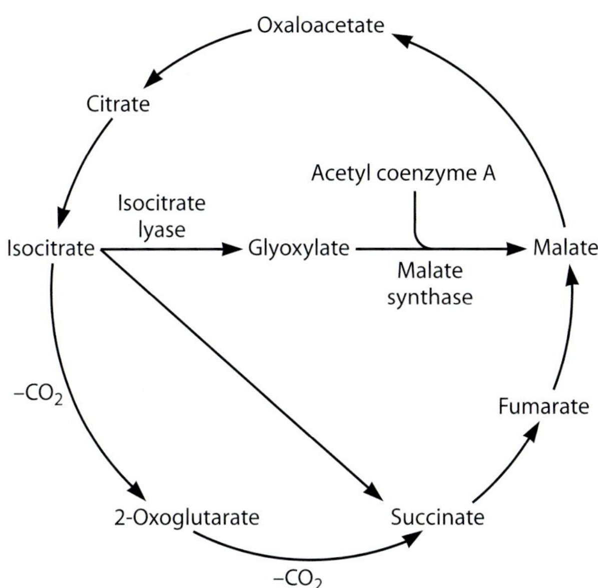

Figure 6.15 Glyoxylate shunt.

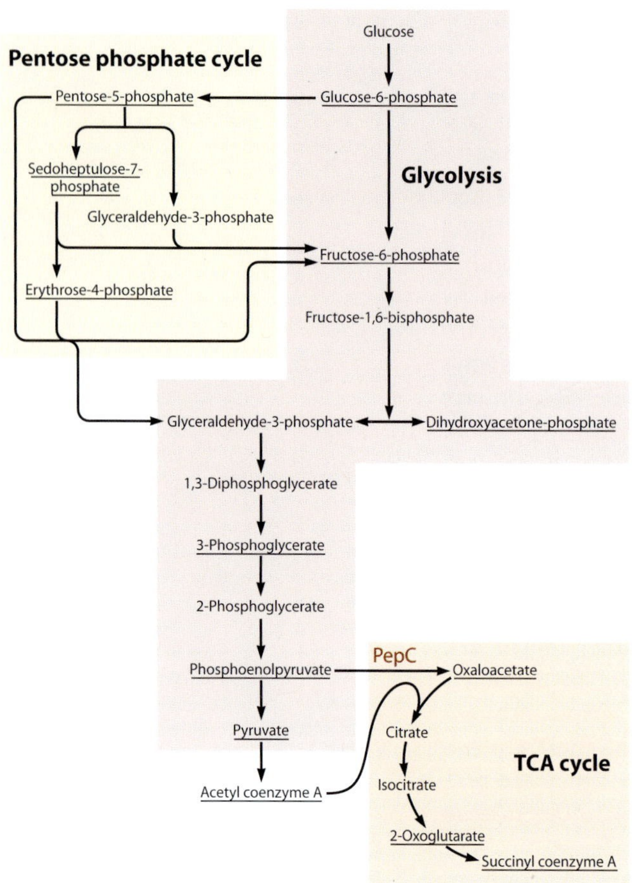

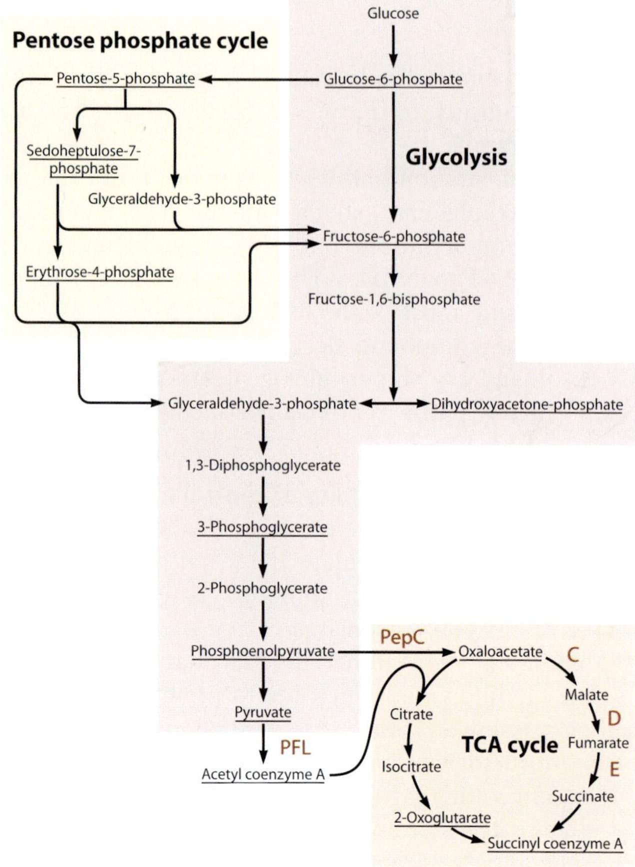

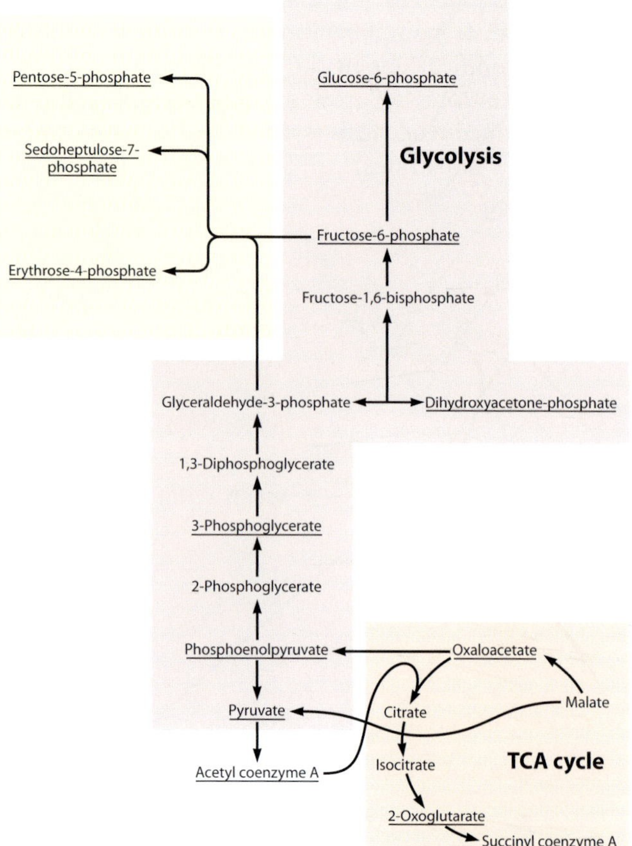

Figure 6.16 Modes of function of central fueling pathways. The examples refer to a facultative heterotroph, such as *E. coli.* The 13 precursor metabolites are underlined. **(A)** Aerobic growth on glucose. PepC (phosphoenolpyruvate carboxylase) forms oxaloacetate by carboxylation of phosphoenolpyruvate, the major replenishing reaction of the TCA pathway. Components of the TCA cycle function almost exclusively to provide three precursor metabolites, not as an energy-generating cycle. **(B)** Anaerobic (fermentative) growth on glucose. PepC is the major replenishing reaction of the TCA pathway. PFL is pyruvate-formate lyase, which replaces the aerobic pyruvate dehydrogenase in catalyzing this reaction. The reactions marked C, D, and E are also catalyzed by enzymes that replace the aerobic enzyme. **(C)** Aerobic growth on malate. Malate is converted to pyruvate by either of two malate enzymes. Conversion of pyruvate to phosphoenolpyruvate is catalyzed by a supplemental enzyme, phosphoenolpyruvate synthase. Likewise, fructose-6-phosphate is formed by the supplemental enzyme fructose-1,6-bisphosphatase. Also, an anaerobic branch of the pentose phosphate cycle is called into play.

operation demands that at some key steps the same chemical reaction must be catalyzed by two separate enzymes, subject to separate controlling factors, so as to achieve the direction of metabolite flux that is appropriate in a given circumstance. How this control is achieved is dealt with in Chapter 12.

MAKING PRECURSOR METABOLITES: AUTOTROPHY

The defining characteristic of autotrophs, plant or microbial, is their ability to fix (convert into organic compounds) sufficient CO_2 to supply carbon for their precursor metabolites, and hence all their cellular constituents, and to do so rapidly enough to permit growth at a reasonable rate. This is a difficult and energetically expensive metabolic process.

The Calvin cycle

The overwhelming majority of autotrophs fix CO_2 via **ribulose bisphosphate carboxylase (RuBisCo)**, a remarkable enzyme that catalyzes the addition of CO_2 to the five-carbon sugar phosphate **ribulose bisphosphate**, producing a six-carbon intermediate that splits spontaneously, yielding two molecules of 3-phosphoglycerate. The **Calvin cycle** (also known as the reductive pentose phosphate pathway), of which RuBisCo is an integral part, regenerates another molecule of CO_2-accepting ribulose-1,5-bisphosphate. Along the way, glyceraldehyde-3-phosphate (triosephosphate) is an intermediate. Because triosephosphate is also an intermediate of the glycolytic pathway, it can flow into the interconnecting pathways of central metabolism, thereby generating the precursor metabolites. Thus, the Calvin cycle, along with the pathways of central metabolism, is sufficient to synthesize all precursor metabolites from CO_2.

RuBisCo, the functional heart of this critical pathway, plays a unique role in nature. It is responsible for the overwhelming majority of the primary production of organic material on the planet. It is also the most abundant enzyme in nature, partly because of its central role, but also because it is not a very good enzyme. On a molecular basis, it is rather inefficient (with a low turnover number), and it has difficulty distinguishing between CO_2 and O_2. It favors CO_2 over O_2 by a factor of 100. However, because of the overwhelmingly higher concentration of O_2 present in the atmosphere, RuBisCo fixes one molecule of O_2 for every three molecules of CO_2. Each time it fixes a molecule of O_2, it produces a useless and toxic product (phosphoglycolate) that must be eliminated at a significant cost in metabolic energy. Some organisms (for example, C-4 plants, such as crabgrass) have evolved mechanisms to increase the intracellular concentration of CO_2, thereby increasing the efficiency of RuBisCo. Speculation about why a better RuBisCo has not evolved is abundant; indeed, there have even been attempts, so far unsuccessful, at **genetic engineering** to give nature a hand, but there are no convincing answers. Possibly, RuBisCo is just faced with catalyzing a difficult chemical reaction.

Other CO_2-fixing cycles

Although the Calvin cycle is the most widespread route by which autotrophs fix CO_2, it is not unique. Two other CO_2-fixing pathways

Table 6.10 Pathways of CO_2 fixation other than the Calvin cycle

Pathway	Key reaction	Distribution
Reductive TCA cycle	Fixes CO_2 by running the TCA cycle backwards; instead of releasing two molecules of CO_2, it fixes two.	*Chlorobium* and certain other photoautotrophic bacteria; some hydrogen-oxidizing bacteria
Hydroxypropionate pathway	Fixes CO_2 via two reactions in which acetyl-CoA and propionyl-CoA are CO_2 acceptors; in a cyclic series of reactions, acetyl-CoA is regenerated.	*Chloroflexus*, a photoautotrophic bacterium
Acetyl-CoA pathway[a]	Fixes CO_2 in a fermentation in which H_2 is used to reduce CO_2 and ultimately form acetate	*Desulfobacterium autotrophicum*, certain clostridia, and other acetogenic bacteria and archaea

[a]CoA, coenzyme A.

exist among autotrophs: the **reductive TCA cycle** and the **hydroxypropionate pathway**. Each of these is restricted to a relatively small group of prokaryotes. Additionally, a fourth route exists in several unrelated anaerobic bacteria and archaea that make acetate from CO_2 and H_2 (Table 6.10). The existence of four pathways for CO_2 fixation illustrates a fact often encountered in microbial metabolism. Natural selection has led to the development of multiple ways to achieve what seems like the same end. This seeming redundancy may reflect our incomplete understanding of the advantages of particular pathways in specific ecological niches.

All of these pathways for CO_2 fixation produce one or another metabolite of the central pathways or some compound easily converted to a central metabolite.

SUMMARY

In this chapter, we have examined how nutrients enter the cell and provide the substrates for pathways that produce reducing power and the 13 precursor metabolites needed for biosynthesis, along with the considerable amounts of ATP needed for growth and other activities.

The following points bear repeated emphasis.

- The flexibility of the central pathways. These pathways are reversible, depending on the substrates provided by the environment, and yet all lead to the generation of all the energy needed by the cell.
- The enormous number of feeder paths. Consequently, microbes play an essential role in recycling *all* living things and *all* their products.
- The manner of coping with the toxic effects of oxygen. Freedom from oxygen addiction is demonstrated by the anaerobes.
- The cleverness of utilizing the cell envelope as an energy transducer.
- The exotic variety of organic and inorganic substances that serve for one organism or another as sources of energy.
- The preponderance of the autotrophic mode of fueling on a global basis.

The last point deserves special attention. All of life's energy is obtained by oxidation of reduced compounds, be it familiar ones, such as glucose, or, in the case of microbes, an unexpected collection of virtually every organic compound on earth, as well as most inorganic compounds. The list strains the imagination, as it includes hydrogen, ferrous iron, sulfide, ammonia, and many others. Microbes seem to "eat" anything that has energy, and therefore, they are the ultimate scavengers for energy on the planet. Oxidation requires that something else become reduced, and here again, the microbial world continues to surprise us. Animals and plants use oxygen, but microbes use, among others, sulfate, nitrate, and ferrous iron. Since we refer to the taking up of oxygen as breathing, microbes, by analogy, breathe rocks! All this alters our conventional view of energy metabolism and expands it to include esoteric-sounding biochemical reactions, but the point is that these processes take place on a wide scale and are essential in the cycles of matter in nature. Without this strange biochemistry, life on Earth could not be sustained.

STUDY QUESTIONS

1. Explain the assertion that microbial domination of Earth's biosphere is related in large measure to the diversity of microbial fueling reactions.

2. What is the justification for combining energy and reducing power into a single entity, driving force?

3. Cellular generation of ATP occurs by two distinct mechanisms, substrate level phosphorylation and transmembrane ion gradients. Explain how each of these mechanisms employs movement of electrons down an energy gradient.

4. ATP generation by transmembrane ion gradients involves two steps: establishing the gradient, and using its energy to produce ATP. What mechanisms have microbes evolved for each of these steps?

5. How is energy generated by photosynthesis?

6. Plants are autotrophs, and animals are heterotrophs. Some prokaryotes are autotrophs, some are heterotrophs, and some can employ either mode of fueling. Beyond this, what justifies the statement that prokaryotes exhibit a greater diversity of fueling modes than do any other living things?

7. What accounts for the variety of solute entry reactions employed by prokaryotes? That is, are all the entry processes equivalent in what they accomplish and what they cost in energy?

8. Most of the reactions and pathways of central metabolism are shared by autotrophs and heterotrophs alike. This being the case, comment on the assertion that central metabolism does not possess the great variety seen in energy-generating steps of fueling.

chapter 7 biosynthesis

SOME GENERAL OBSERVATIONS

Biosynthesis and nutrition

Whether growing in the nutritionally sparse environment of a mountain stream or the rich environment of chicken broth, a microbial cell must present to its polymerization machinery the same complete array of building blocks (Table 7.1). What the cell cannot make it must acquire preformed from its environment. *Biosynthetic capability and nutritional requirements are reciprocal: the greater a cell's biosynthetic capabilities, the fewer its nutritional requirements.* Counter to what might be one's first thought, a cell with *fewer* growth requirements is, in a metabolic sense, *more* complex.

Most of the building blocks can be transported into cells, albeit at some energy cost. Some species of bacteria rely exclusively on an exogenous source of one or more of these substances for use as essential nutrients. Some of them, called **fastidious** bacteria, seem to require just about every building block. *Leuconostoc citrovorum* (a dairy bacterium) requires 19 amino acids, two purines, one pyrimidine, and eight vitamins; these bacteria are harder to please than mammals. At the other end of the nutritional spectrum, many species are complete synthetic chemists, i.e., they possess a complete array of biosynthetic pathways and thus can synthesize every one of the building blocks. Between these extremes exists just about every intermediate state of nutritional competency. Not surprisingly, bacteria that grow in environments rich in organic matter have tended to lose unneeded pathways as a result of natural selection. A hint of the variety of nutritional types of bacteria is given in Table 7.2.

Bacterial studies and biosynthetic pathways

With few exceptions, all the biosynthetic reactions that generate building blocks are known, and with few exceptions, they are the same in all

Table 7.1 Major building blocks needed to produce a typical gram-negative bacterium[a]

Protein amino acids	DNA nucleotides	Glycogen monomers
Alanine	dATP	Glucose
Arginine	dGTP	
Asparagine	dCTP	**Polyamine units**
Aspartate	dTTP	Ornithine
Cysteine		
Glutamate	**Lipid components**	**Coenzymes**
Glutamine	Glycerol phosphate	NAD^+
Glycine	Serine	$NADP^+$
Histidine	Fatty acids (several)	CoA
Isoleucine		CoQ
Leucine	**LPS components**	Bactoprenoid
Lysine	UDP-glucose	Tetrahydrofolate
Methionine	CDP-ethanolamine	Cyanocobalamine
Phenylalanine	Hydroxymyristic acid	Pyridoxal phosphate
Proline	Fatty acid	Nicotinic acid
Selenomethionine	CMP-KDO	Other coenzymes
Serine	NDP-heptose	
Threonine	TDP-glucosamine	**Prosthetic groups**
Tryptophan		FMN
Tyrosine	**Murein monomers**	FAD
Valine	UDP-*N*-acetylglucosamine	Biotin
	UDP-*N*-acetylmuramic acid	Cytochromes
RNA nucleotides	Alanine	Lipoic acid
ATP	Diaminopimelate	Thiamine pyrophosphate
GTP	Glutamate	
CTP		
UTP		

[a]GTP, guanosine triphosphate; UTP, uridine triphosphate; dTTP, deoxyribosylthymine triphosphate; UDP, uridine diphosphate; CDP, cytidine diphosphate; CMP, cytidine monophosphate; KDO, 2-keto-3-deoxyoctulosonic acid; NDP, nucleoside diphosphate; TDP, ribosylthymine diphosphate; FMN, flavin mononucleotide; FAD, flavin adenine dinucleotide; LPS, lipopolysaccharide.

organisms in which they occur. In the mid-20th century, isolation of mutants of **prototrophs** (bacteria having no nutritional requirements for building blocks) that had picked up one or another nutritional requirement provided precisely the means to discover the pathways of biosynthesis. Such **auxotrophic mutants**, largely of the enteric bacteria *Escherichia coli* and *Salmonella*, provided clues because they are able to grow on intermediates after the blocked reaction, and when starved for the building block, they spill into the environment large quantities of the intermediate before the blocked reaction. These intermediates, in turn, could be shown to feed mutants blocked at prior steps. In vivo use of isotope-labeled metabolites and in vitro measurement of enzyme activities led to the identification of the enzyme missing in each mutant. One by one, the steps responsible for making each building block were worked out. This achievement in microbiology carried an extra bonus for all of biology: the pathways discovered in bacterial studies proved (as hoped) to be largely copied by other bacteria, microbes, plants, and animals.

The concept of a biosynthetic pathway

If one lays out on a very large sheet the structures (or even just the names) of all 75 to 100 building blocks (in which we include enzyme cofactors) and then adds for each of them the enzymatically catalyzed reactions leading to their synthesis, the resulting metabolic map is overwhelming;

Table 7.2 Nutritional needs of two heterotrophic bacteria

Organism	Inorganic	Trace[a]	Carbon/energy source	Required nutrients[b]
	Medium components			
			Organic	
Escherichia coli	K^+, NH_4^+, Mg^{2+}, Fe^{2+}, Cl^-, SO_4^{2-}, PO_4^{3-}	Mn^{2+}, Mo^{6+}, Cu^{2+}, Co^{2+}, Zn^{3+}, B^{3+}	Glucose	None
Streptococcus agalactiae[c]	Same as above	Same as above	Same as above	Alanine, arginine, aspartic acid, asparagine, cystine, cysteine, glutamic acid, glycine, glutamine, histidine, leucine, lysine, isoleucine, methionine, phenylalanine, proline, serine, tryptophan, tyrosine, valine, nicotinic acid, pantothenic acid, pyridoxal, thiamine, riboflavin, biotin, folic acid, adenine, guanine, xanthine, uracil

[a]These substances are usually present in sufficient amounts as contaminants of glassware, distilled water, and the major components of the medium.

[b]In many cases, only the natural enantiomorph (e.g., the L form of an amino acid) will suffice.

[c]Data from N. P. Willett and G. E. Morse, *J. Bacteriol.* **91:**2245–2250, 1966.

there are hundreds of enzymatically catalyzed reactions on the chart. A computer works better for a display of this size and complexity, and you can see biosynthetic metabolism displayed in this way at the EcoCyc website, also mentioned in Chapter 5.

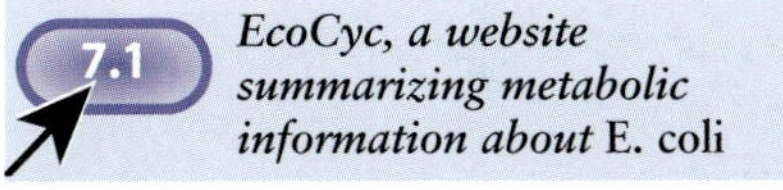

EcoCyc, a website summarizing metabolic information about E. coli

On inspection, however, a simplifying organization emerges from the maze of biosynthetic reactions (Fig. 7.1). It can be seen that the nearly 100 building blocks are made from only 13 **precursor metabolites**—compounds that we identified as intermediates of fueling pathways in Chapter 5. Thus, the overall pattern of the biosynthetic chart is one of parallel and branching pathways emerging from a mere baker's dozen of familiar starting compounds.

On further inspection, other features are evident.

1. The **building blocks**—end products of each pathway—are in general more reduced than the precursor metabolites, and therefore the pathways consume reducing power in the form of the reduced coenzyme nicotinamide adenine dinucleotide phosphate (NADPH). This is an ideal point to be reminded that fueling reactions in general (the pentose phosphate pathway is an exception) produce reducing power in the form of reduced nicotinamide adenine dinucleotide (NADH) and biosynthetic pathways utilize it largely in the form of NADPH. The existence of this pair of seemingly equivalent coenzymes makes it possible for the individual members of the pair to be set at different values of **reductive poise** (the ratio of the reduced form to the oxidized form). Thus, cells maintain a high oxidized nicotinamide adenine dinucleotide (NAD^+)/NADH ratio (low reductive poise), which facilitates the role of this coenzyme in *accepting* hydrogens from redox reactions in fueling paths. The NADPH/oxidized nicotinamide adenine dinucleotide phosphate ($NADP^+$) ratio is high (high reductive poise), which facilitates its role in *donating* hydrogens in biosynthetic pathways. The relative poises between the two forms of reducing power must be adjusted. They are set largely by enzymes called **transhydrogenases.**
2. Most of the pathways utilize the energy in ATP to drive the thermodynamically favorable flow of material from precursor to building block. (The other source of energy, transmembrane ion

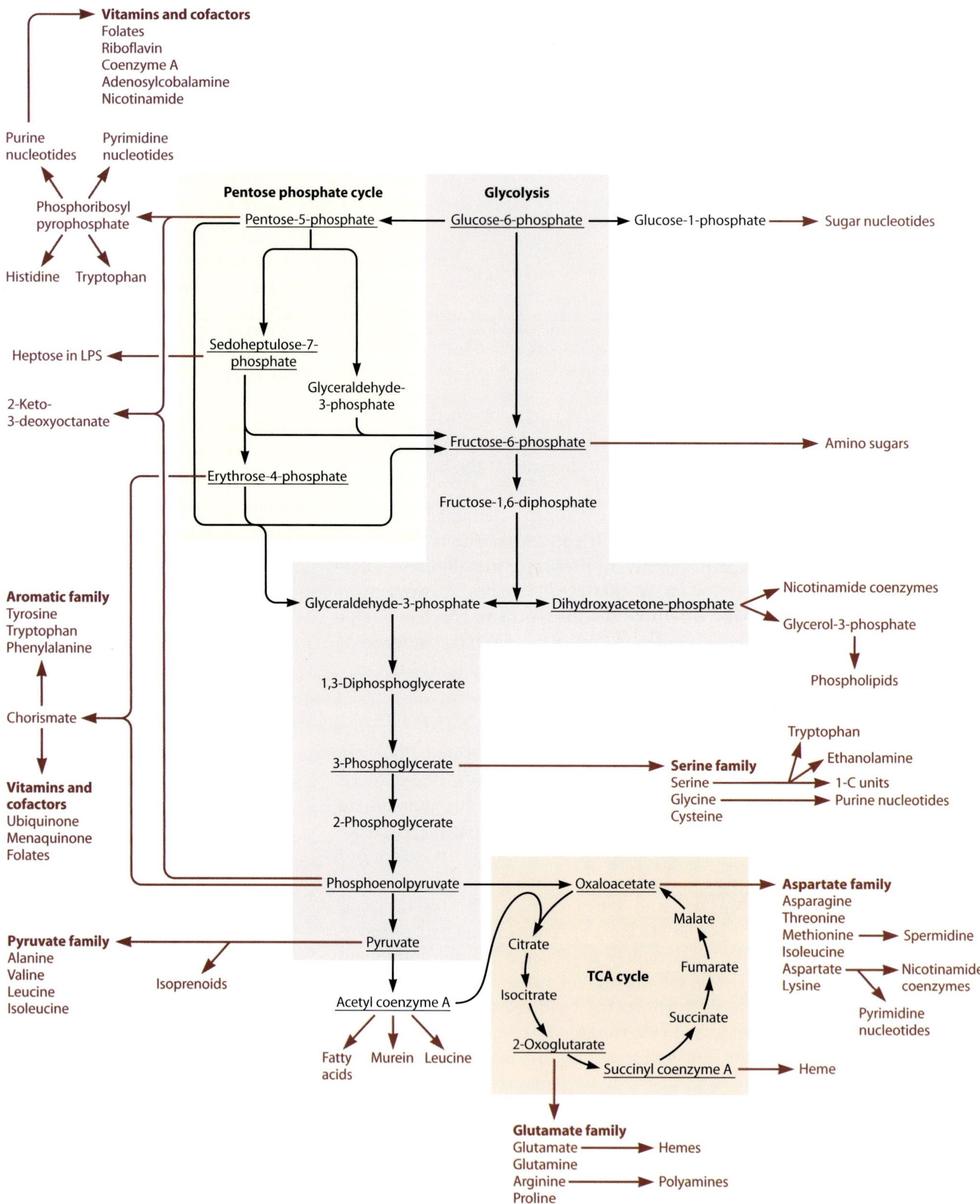

Figure 7.1 Paths from central metabolism to biosynthetic end products. The 13 precursor metabolites are underlined. The biosynthetic end products are shown in red.

gradients, is not directly useful for these exclusively intracellular reactions, though of course, in many bacteria, these gradients make the ATP.)

3. Less evident in this metabolic map is one of the most critical features of biosynthetic pathways. Enzymes that appear at key reactions in each pathway are **allosteric**, meaning they can assume two forms, one catalytically active and one inactive. The form chosen at any one time is determined by the binding or lack of binding of a particular molecule, the **controlling ligand**, which is a metabolite that is a significant measure of the need for the products of the given enzyme. The ligand is bound at a specific site on the enzyme protein that is distinct from its catalytically **active site**. Such ligands are usually low-molecular-weight metabolites. If the ligand is bound, the enzyme is forced into a form that is inactive. (Yes, it also works the other way: the binding of a ligand *activates* some allosteric enzymes.) **Allostery** provides a means by which the activity of an enzyme can be modified by compounds not even remotely resembling its substrates or products. *A common feature of biosynthetic pathways is that the end product, a building block, is an allosteric ligand for one or more enzymes, usually the first one, in its biosynthetic pathway.*

As will be more fully presented in a later section, the pattern of inhibition of one or more enzymes in a biosynthetic pathway by its end product serves to create an unexpected situation: the operation of these pathways is governed not so much by the availability of their starting materials as by the *rate of utilization of their end products in polymerization reactions!* This fact is highly significant for growth metabolism, for it leads to the counterintuitive result that *cells do not grow as fast as their building blocks can be made—they make their building blocks as fast as they can grow!* We shall return to this curious situation in a later chapter on regulation (see Chapter 12).

These characteristics of biosynthetic pathways (branching patterns, use of NADPH and ATP, and control by allosteric enzymes) are summarized in Fig. 7.2.

One fact should be clear from this discussion. Biosynthetic pathways are real biological entities, not intellectual constructs designed to help one organize the network of biosynthetic reactions into bite-size pieces. The microbial cell, not the microbiologist, has defined the pathway as a metabolic unit. Evidence continues to be found that the enzymes of a pathway are physically associated in the cell. This organizational point is of some significance because it speaks to the issue of whether the cell is merely a bag of enzymes (as was once thought by some to be the case) or a highly ordered set of enzyme complexes. Physical association of metabolically related enzymes can accelerate processes by minimizing the diffusion time of metabolites from one catalyst to the next and by lowering the concentration of intracellular pools of metabolites.

None of the 13 precursor metabolites contains nitrogen or sulfur, yet many building blocks and other cellular constituents contain one or both of these elements. At this point, it would be well to examine how nitrogen and sulfur become incorporated into metabolites (the fancy word for such processes is **assimilation**).

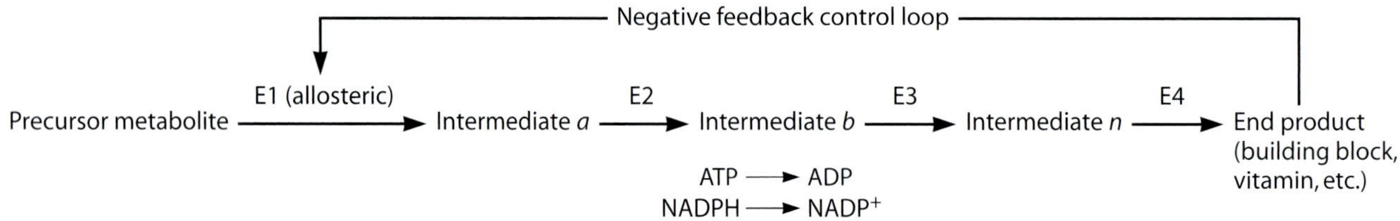

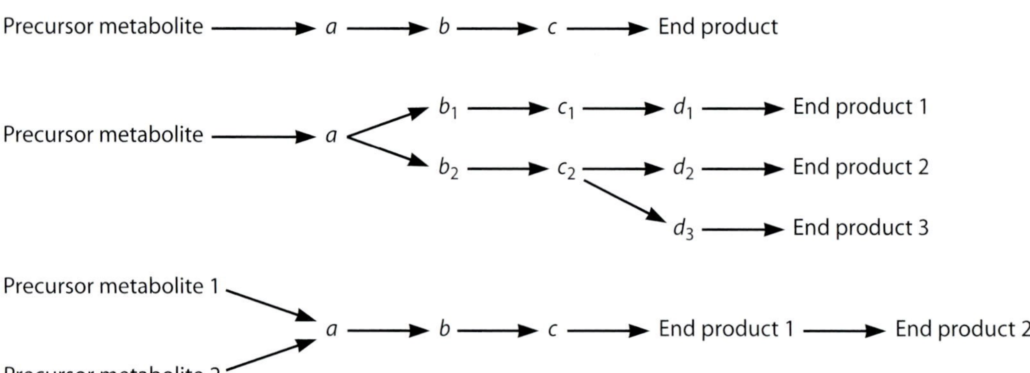

Figure 7.2 Characteristics of microbial biosynthetic pathways. (A) Generalized features of a biosynthetic pathway. E1 to E4, enzymes in the pathway. **(B)** Patterns of different pathways, illustrating multiple precursors and pathway branching.

ASSIMILATION OF NITROGEN

Nitrogen is a significant component of protoplasm: it is the main element by weight after carbon. It is found in all the amino acids and nucleotides, as well as other compounds, and comprises about 15% of the dry weight of most cells. Assimilation of nitrogen does not occur in fueling, but rather later on in biosynthesis, at various steps along the pathways leading from the precursor metabolites to the building blocks of the cell's macromolecules. Nitrogen enters these pathways in the reduced state (in the -3 oxidation state as an amino or amido group) and remains reduced in almost all cellular constituents (as amino, amido, or imino groups or as part of heterocyclic organic molecules). (There are, as is almost always the case for biological generalizations, exceptions, but they are few. A notable one is the natural antibiotic, chloramphenicol, in which nitrogen occurs as a nitro [$—NO_2$] group.)

There are an enormous number of environmental sources of nitrogen: nitrogen gas (the major component of Earth's atmosphere), various inorganic nitrogenous oxides, and of course, all the nitrogenous components of protoplasm. As we survey how these sources are utilized, it will be helpful to follow the discussion by consulting Fig. 7.3, which presents a broad overview of nitrogen assimilation.

Central actors: glutamate and glutamine

In startling contrast to the uniformity of the valence state (-3) of nitrogen in cellular constituents, microorganisms can use nitrogen sources in the environment that run the gamut of oxidation states from -3 to $+7$. Thus, nitrogen metabolism of microorganisms can be viewed as a pyra-

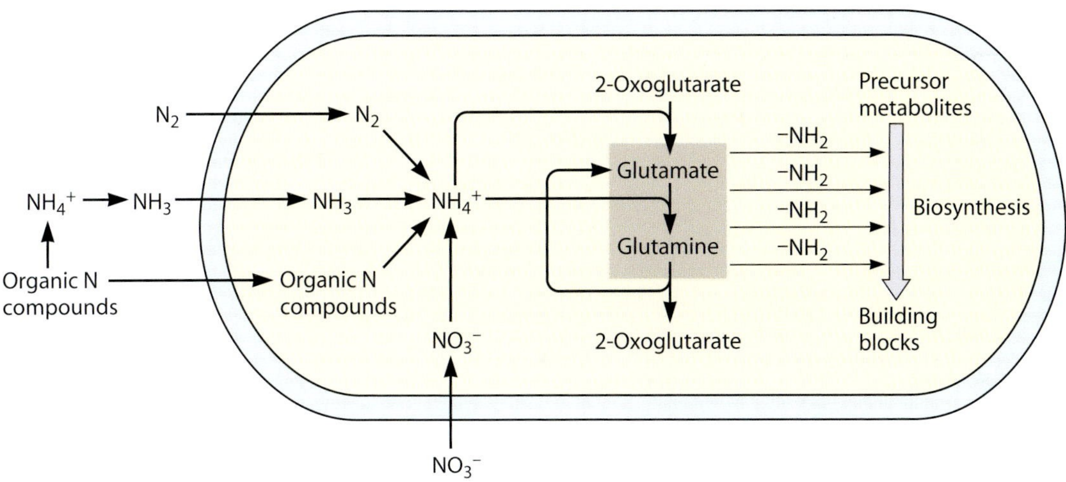

Figure 7.3 Biosynthetic assimilation of nitrogen. The details are described in the text.

mid with a broad base of highly varied starting materials converging to relative uniformity when it enters cellular metabolism. The pathways from nutrients progressively converge, and quite remarkably, at one point *almost all flow through two compounds, the amino acids **glutamate** and **glutamine**.* Glutamate and, to a lesser extent, glutamine donate their nitrogen to the various pathways of biosynthesis of building blocks. Quantitatively, glutamate is by far the more important: about 90 percent of the cell's nitrogen flows through it.

Glutamate donates its amino group by **transamination** in the biosynthesis of amino acids, such as phenylalanine:

Glutamate + Phenylpyruvate $\xrightarrow{\text{Aromatic amino acid transaminase}}$ 2-Oxoglutarate + Phenylalanine

COOH–CH_2–CH_2–HC(NH_2)–COO^- C_6H_5–CH_2–C=O–COO^- COOH–CH_2–CH_2–C=O–COO^- C_6H_5–CH_2–HC(NH_2)–COOH

Glutamine donates its amido group to various other biosyntheses, including that of the nucleotide cytidine triphosphate (CTP):

Glutamine + UTP + ATP + H_2O $\xrightarrow{\text{CTP synthase}}$ Glutamate + CTP + ADP + P_i

O=C(NH_2)–CH_2–CH_2–HC(NH_2)–COOH   UTP: HN–C(=O)–CH=CH–N(Ribose–5′PPP)–C(=O) ring   COOH–CH_2–CH_2–HC(NH_2)–COO^- CTP: N=C(NH_2)–CH=CH–N(Ribose–5′PPP)–C(=O) ring

Synthesis of glutamate and glutamine

How are glutamate and glutamine synthesized? Perhaps not surprisingly, their pathways of synthesis are interconnected. Some bacteria can synthesize glutamate in two ways. Which one they choose is not arbitrary but depends on the nutritional conditions, a neat example of how ecology prescribes physiology.

1. Glutamate can be synthesized by incorporating ammonia into a precursor metabolite of the tricarboxylic acid cycle, 2-oxoglutarate, in an ATP-independent reaction catalyzed by glutamate dehydrogenase:

$$\text{2-Oxoglutarate} + NH_3 + H^+ + \text{NADPH} \xrightarrow{\text{Glutamate dehydrogenase}} \text{Glutamate} + H_2O + \text{NADP}^+$$

$$\begin{array}{l} \text{COOH} \\ | \\ CH_2 \\ | \\ CH_2 \\ | \\ C{=}O \\ | \\ COO^- \end{array} \qquad\qquad \begin{array}{l} \text{COOH} \\ | \\ CH_2 \\ | \\ CH_2 \\ | \\ HC{-}NH_2 \\ | \\ COO^- \end{array}$$

Glutamine can be synthesized from glutamate by incorporating a second molecule of ammonia in an ATP-driven reaction catalyzed by glutamine synthetase (GS):

$$\text{Glutamate} + NH_3 + \text{ATP} \xrightarrow{\text{Glutamine synthetase (GS)}} \text{Glutamine} + P_i + \text{ADP}$$

$$\begin{array}{l} \text{COOH} \\ | \\ CH_2 \\ | \\ CH_2 \\ | \\ HC{-}NH_2 \\ | \\ COO^- \end{array} \qquad\qquad \begin{array}{l} O \\ \| \\ C{-}NH_2 \\ | \\ CH_2 \\ | \\ CH_2 \\ | \\ HC{-}NH_2 \\ | \\ \text{COOH} \end{array}$$

2. In some microbes, the amido nitrogen of glutamine, rather than ammonia, can also be used to make glutamate from 2-oxoglutarate in a reaction catalyzed by glutamine synthase (also called GOGAT, for glutamate-oxoglutarate amido transferase):

$$\text{Glutamine} + \text{2-Oxoglutarate} + H^+ + \text{NADPH} \xrightarrow{\text{GOGAT}} \text{2 Glutamate} + \text{NADP}^+$$

$$\begin{array}{l} O \\ \| \\ C{-}NH_2 \\ | \\ CH_2 \\ | \\ CH_2 \\ | \\ HC{-}NH_2 \\ | \\ \text{COOH} \end{array} \qquad\qquad \begin{array}{l} \text{COOH} \\ | \\ CH_2 \\ | \\ CH_2 \\ | \\ C{=}O \\ | \\ COO^- \end{array} \qquad\qquad \begin{array}{l} \text{COOH} \\ | \\ CH_2 \\ | \\ CH_2 \\ | \\ HC{-}NH_2 \\ | \\ COO^- \end{array}$$

Thus, microbes that have genes for all three enzymes can use two alternative routes to make glutamate, the central actor in the flow of nitrogen metabolism: (i) via glutamate dehydrogenase or (ii) via GS plus GOGAT.

Why might it be selectively advantageous to have two routes for synthesizing glutamate? The answer seems to lie in a trade-off between being able to scavenge low concentrations of environmental ammonia and conserving ATP. The glutamate dehydrogenase route requires no expenditure of ATP but functions only in the presence of substantial concentrations of ammonia. The second route (GS plus GOGAT) has the opposite set of properties: it requires a greater expenditure of ATP (one molecule per glutamate), but it functions well at low concentrations of ammonia. *If ammonia is abundant in the environment, the cell saves ATP by using the first route; if ammonia is scarce, it pays the energy price to use the second route.*

Getting ammonia

You will note from the above discussion that the nitrogen atoms that glutamate and glutamine donate to biosynthesis are derived exclusively from ammonia. We should then ask how microbes obtain ammonia.

Nitrogen in synthetic culture media is generally supplied in the form of ammonium ion, but this form is relatively rare in natural environments. When anhydrous ammonia fertilizer is added to soil, for example, it disappears rapidly, as some microbes (heterotrophs) assimilate it and others (nitrifying autotrophs) oxidize it to nitrate (see Chapter 6). Needless to say, the ability to scavenge all possible sources of nitrogen in the environment has been a potent force in evolution, leading to the selection of microbes with a great diversity of talents in this endeavor. As a result, among present-day microbes, nitrogen can be acquired as part of almost any molecule, organic or inorganic, or it can be drawn directly from the great reservoir of inorganic dinitrogen (N_2) present as 80% of earth's atmosphere (Fig. 7.3). Recall that in Chapter 5 we introduced the concept of biogeochemical cycles of matter. Earth's supply of nitrogen exists in various chemical forms and ecological locations. The interconversion of these forms and their transfer between physical locations is to a great extent the result of microbial activity. The nitrogen cycle is described in more detail in Chapter 18.

When ammonia is available, it diffuses into most bacteria through transmembrane channels as dissolved gaseous ammonia (NH_3) rather than ionic ammonium (NH_4^+). Curiously, the protein that forms these channels is closely related to the vertebrate Rh protein, which forms CO_2 channels in red blood cells. When ammonia is not available, most microbes obtain it by processing other nitrogenous compounds, generally intracellularly. Some microbes secrete deaminases into their environment or periplasm. *Helicobacter pylori*, for example, secretes an active extracellular urease (which hydrolyzes urea to ammonia and CO_2), and *Klebsiella pneumoniae* produces a periplasmic asparaginase (which hydrolyzes asparagine to ammonia and aspartate), a cytidine deaminase, and a cytosine deaminase. In each case, the resulting ammonia is then taken up by the cells. Once inside, organic sources of nitrogen are metabolized by a variety of distinct pathways, with deaminases often responsible for the actual release of ammonia. Many microorganisms are able to obtain ammonia from nitrate ion in a two-step pathway catalyzed by assimilatory nitrate reductase (nitrate to nitrite) and assimilatory nitrite reductase (nitrite to ammonia).

Ammonia from dinitrogen

The biological reduction of atmospheric dinitrogen (N_2) to ammonia or nitrate is called **nitrogen fixation**. This is a uniquely prokaryotic process, mediated by both Bacteria and Archaea, and is crucial to life on Earth. Fixed nitrogen is being continuously depleted through prokaryote-mediated **denitrification** (the cascade of **anaerobic** respirations that converts nitrate to dinitrogen) and the **anammox reaction**, which converts ammonia and nitrite into dinitrogen (see Chapter 18). Nitrogen fixation is the only means of replenishing the nitrogen sources used by most microbes and by all plants and animals. In geological terms, such cycling of Earth's nitrogen supply through atmospheric dinitrogen is rapid. Its half-life is only about 20 million years, which is very fast, considering the immensity of the nitrogen gas reservoir (80% of Earth's atmosphere). More startling is the estimate that were the microbes of Earth to cease fixing nitrogen, the soil would be depleted of this nutrient in 1 week.

Until the 20th century, nitrogen fixation was the near-exclusive province of prokaryotes. (Small amounts are products of electrical storms and volcanic activity.) Then, with the discovery by the German chemist Fritz Haber of a method of converting dinitrogen to ammonia, we humans became major participants. Now, about half the world's supply of fixed nitrogen for fertilizers is produced industrially using chemical methods.

Biological nitrogen fixation is mediated by the **nitrogenase enzyme complex**, which consists of two components: **dinitrogenase** and **dinitrogenase reductase**. Their names imply their functions: dinitrogenase mediates the actual eight-electron reduction of N_2. It is a tetramer of two α and two β subunits that works in conjunction with an iron-molybdenum cofactor (FeMoCo). This process yields two molecules of ammonia, with nitrogenase reductase (an α_2 dimer) supplying the electrons. The biological invention of the nitrogenase reductase complex may have occurred only once, because it is so highly conserved. Its primary sequence is 70% identical among all the species of Bacteria and Archaea that possess it. Mixtures of dinitrogenase and dinitrogenase reductase derived from different species, even from species in different domains, actively reduce dinitrogen. This reaction is dramatically energy intensive—not because of the catalyzed reaction itself, which is slightly **exergonic** (yielding heat energy), but because of the high activation energy of breaking the very strong triple bond that joins the two nitrogen atoms in N_2. Somewhere between 20 and 24 molecules of ATP are needed to reduce 1 molecule of dinitrogen to ammonia.

Another remarkable property of the nitrogenase enzyme complex is its extreme sensitivity to oxygen. It is among the most highly oxygen-labile proteins known. Prokaryotes that mediate nitrogen fixation have evolved a variety of strategies for protecting their nitrogenase enzyme complexes from oxygen (see Chapter 18).

ASSIMILATION OF SULFUR

Sulfur, one of the 10 most abundant elements on Earth, exists in a large number of oxidation states, some more oxidized than elemental sulfur (S^0) and some more reduced. As we saw in Chapter 6 (see Table 6.1), prokaryotes exploit these multiple states in fueling reactions by using sulfur compounds as both donors and recipients in energy-generating electron transport, including both respiration and photosynthesis. As a result, prokaryotes are major agents in the sulfur biogeochemical cycle (see Chapter 18).

How do prokaryotes incorporate sulfur into the sulfur-containing compounds of the cell? Just as nitrogen always enters biosynthetic pathways as the inorganic ammonium ion, so sulfur *always* enters as an inorganic form, sulfide (as H_2S or S^{2-}). An overview of the process of sulfur assimilation is shown in Fig. 7.4. H_2S is incorporated through synthesis of the amino acid L-cysteine. (L-Cysteine is made by the enzyme O-acetylserine sulfohydrylase, which catalyzes the reaction of H_2S with O-acetyl-L-serine to produce L-cysteine, acetate, and water.) L-Cysteine then serves directly or indirectly as the source for most other sulfur-containing compounds in the cell (for example, L-methionine, biotin, thiamine, and coenzyme A [CoA]).

Having enough H_2S available for this important reaction can be a critical issue, because H_2S is spontaneously oxidized by O_2 and exists in nature primarily in anaerobic environments. Strict anaerobes in such places face no problem, and some use exclusively H_2S for biosynthesis, as well as energy generation. Most other bacteria, and all aerobes, however, must reduce oxidized sulfur compounds to H_2S for assimilation. The major form of sulfur in aerobic soil is organic (organic sulfates, amino acids, and other C-bonded sulfur compounds). Sulfur-containing amino acids can be assimilated as such, but acquiring sulfur from most organic sulfur compounds requires oxidizing them to sulfate. What happens next is *energetically expensive*, because the reduction of sulfate through sulfite to sulfide occurs in a series of four reactions requiring the consumption of three high-energy phosphate bonds, three NADPH, and a molecule of reduced thioredoxin for every sulfate molecule reduced to sulfide. Organisms lacking this expensive pathway are, as we noted, confined to the use of sulfur in sulfide (H_2S) and therefore are chiefly anaerobes. There is, of course, one other way around the expense of sulfate reduction: import all the preformed sulfur-containing building blocks. Not surprisingly, some bacterial pathogens and obligatory parasites do just that.

Figure 7.4 Biosynthetic assimilation of sulfur. The details are described in the text.

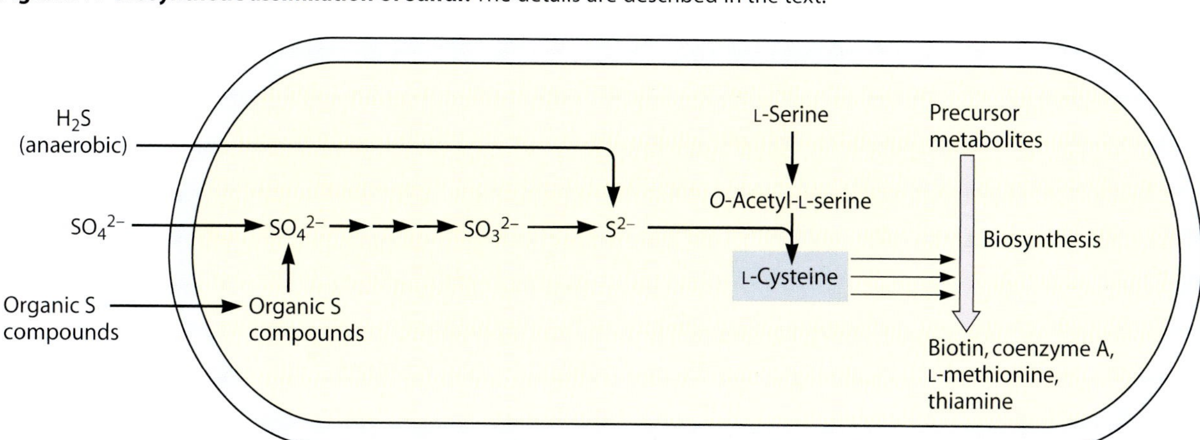

ASSIMILATION OF PHOSPHORUS

The large cellular content of phosphorus—a major component of nucleic acids, phospholipids, and many coenzymes, as well as phosphorylated sugars and proteins—speaks for a hearty appetite of cells for sources of this element. As gardeners and ecologically minded individuals know, phosphate is limiting in many environments, and it is often its availability which determines the crop of desirable or undesirable plant life. We can thus correctly expect evolutionary pressures to have favored the development and persistence of effective means for microbes to assimilate this vital component in competition with other organisms.

Phosphorus is always assimilated in metabolic reactions as inorganic phosphate, though organic phosphate compounds also serve as environmental sources. Most cells are impermeable to organic phosphates, and in only a few cases can such compounds enter bacteria. Commonly, bacteria hydrolyze phosphate esters outside the cell, or in the periplasm of gram-negative bacteria. The entry of phosphate then occurs via either general or specific transport systems. In well-studied cases, the enzymes responsible for hydrolysis of organic phosphates are repressed when adequate phosphate is present in the environment. In Chapter 13, we deal with how phosphate is sensed and how the appropriate decision is made about how to acquire it.

Unlike the assimilation of nitrogen or sulfur, phosphorus is neither oxidized nor reduced in the assimilation process, and—a major difference—its assimilation occurs during fueling rather than in biosynthetic reactions. This fact is related to the central role of phosphate in energy transduction. The assimilatory reactions are numerous; some prominent and familiar ones occur in the central fueling pathways, and several are described in detail in Chapter 6. The ATP produced by these reactions then serves as the major phosphate donor in the cell.

PATHWAYS TO BUILDING BLOCKS

The overview of biosynthesis shown in Fig. 7.1 provides a highly simplified summary of the hundreds of enzyme-catalyzed reactions that convert the carbon skeletons of the precursor metabolites into the cellular building blocks. The pathways are known in great detail and can readily be examined at the EcoCyc website. Here, we shall simply clarify the abbreviated overview in Fig. 7.1 and indicate some general features of the pathways leading to the main classes of building blocks.

Amino acids

There are 21 amino acids found in proteins, including microbial proteins. Actually, counting pyrrolysine, norleucine, and other modifications of the standard amino acids, there are more, but only 21 are incorporated into growing polypeptide chains at the direction of a specific transfer RNA (tRNA)-related triplet nucleotide codon; selenocysteine's tRNA recognizes the nonsense codon UGA. Some bacteria synthesize all the amino acids, others only some, and still others none of them. We humans, incidentally, can synthesize 12 of the 21, having lost the ability somewhere among our evolutionary ancestors to make the other 9. In our anthropomorphism, we call these nine "essential" in a dietary sense. (Clever us; we have given up making the most difficult ones, i.e., the ones with the longest pathways of synthesis.)

There are six so-called *families* of amino acids, defined simply by the fact that the amino acids in each share a common precursor metabolite or combination of precursor metabolites (Fig. 7.1 and Table 7.3).

Nucleotides

Not readily apparent in Fig. 7.1 is the fact that the building blocks for nucleic acids are manufactured in an activated state. That is, the end products of the purine and pyrimidine biosynthetic pathways are ribonucleoside and deoxyribonucleoside triphosphates, ready for polymerization. The purine nucleotides, both the ribose and deoxyribose forms, are made from the precursor metabolite, **ribose-5-phosphate,** in a multistep, branching network catalyzed by 21 enzymes. The pyrimidine nucleotides are made from two precursor metabolites, ribose-5-phosphate and **oxaloacetate,** in a branching network catalyzed by 24 enzymes. Needless to say, in addition to the precursor metabolites, these pathways consume much energy (as high-energy phosphate), reducing power (from both NADH and NADPH), and assimilated nitrogen (both as amino groups

Table 7.3 Families of protein amino acids and their precursor metabolites

Amino acid family	Precursor metabolite(s)
Serine family	
Serine	3-Phosphoglycerate
Glycine	3-Phosphoglycerate
Cysteine (and selenocysteine)	3-Phosphoglycerate
Aspartate family	
Aspartate	Oxaloacetate
Asparagine	Oxaloacetate
Threonine	Oxaloacetate
Methionine	Oxaloacetate
Isoleucine	Oxaloacetate, pyruvate
Lysine	Oxaloacetate
Glutamate family	
Glutamate	2-Oxoglutarate
Glutamine	2-Oxoglutarate
Arginine	2-Oxoglutarate
Proline	2-Oxoglutarate
Pyruvate family	
Alanine	Pyruvate
Valine	Pyruvate
Isoleucine	Pyruvate, oxaloacetate
Leucine	Pyruvate, acetyl-CoA
Aromatic family	
Tyrosine	Erythrose-4-PO_4, phosphoenolpyruvate
Phenylalanine	Erythrose-4-PO_4, phosphoenolpyruvate
Tryptophan	Erythrose-4-PO_4, phosphoenolpyruvate, pentose-5-PO_4
Histidine family	
Histidine	Pentose-5-PO_4

donated by transamination from glutamine and as ammonium ion in the case of a single reaction in the pyrimidine network).

Given the fact that most fast-growing bacteria are rich in nucleic acids (up to 25% of the cell's total dry weight), the demands on biosynthesis for these building blocks are high. As we saw in Chapter 4 ("How Is the Physiology of the Cells Affected by the Growth Rate?"), a major aspect of the economy and efficiency of growth consists of conserving the resources needed for RNA synthesis by modulating the amounts of RNA made at different growth rates.

Sugars and sugar-like derivatives

Many of the functions and much of the specificity of each variety of microbial species depend on the synthesis of sugars and sugar-like derivatives that are components of the capsules, wall, and outer membrane of these cells. The fate of bacteria invading humans, for example, can be greatly affected by whether or not the microbial surface components are recognized by the host immune system. In Table 7.1, a small number of the carbohydrate-like molecules found as envelope components in one kind of bacterium are listed; selective pressures have led to the evolution of dozens of different sugar derivatives in various bacteria. The complete list of known building blocks is very large. Few, if any, of these components can be assimilated ready-made from the environment; almost all must be made from the precursor metabolites (four-, five-, six-, and seven-carbon sugars) shown in Fig. 7.1.

A different need for production of sugars comes from their storage function. Many prokaryotes store carbon reserves as glycogen, much as humans do. In some instances, these polymers are stored in organelle-like structures sometimes called inclusion bodies (see Chapter 3).

Fatty acids and lipids

Life is a phenomenon of surfaces, it has frequently been remarked, and membranes composed of **phospholipid bilayers** are the biological surfaces that help define living organisms on Earth. (The membrane structure of the Archaea is a variation on this theme [see Chapter 2, "Cell envelopes of the Archaea"]). As you recall, phospholipids are diglycerides that contain fatty acids in ester linkage with glycerol. Some fatty acids are also found in a few other compounds, such as **lipoprotein** (the protein that anchors the murein wall to the cell membrane in gram-negative bacteria), but the amounts in these compounds are minor compared with those in the phospholipids of bacterial membranes and in lipid A of the outer membrane in gram-negative bacteria. In spite of the vast diversity of fatty acids found in bacteria, the syntheses of all of them follow the same modular scheme, rather like making many different structures by stacking the same kind of bricks. The brick for making fatty acids is a **two-carbon unit** derived from acetyl-CoA, and units are added over and over until a fatty acid of the appropriate length has been made.

Fatty acids differ in the number of carbon atoms, the number of double bonds and their placement, and whether they are branched. The diversity of phospholipids reflects in part differences among species and in part specific responses to the physical environment of the cells (see Chapter 13). More double bonds, shorter chains, and less branching increase the fluidity of lipids; evolution has led to the development of mechanisms to modulate these characteristics in newly synthesized fatty acids as a function of environmental temperature and pressure. Branched-chain and unsaturated fatty acids are synthesized from pathways that diverge at various points, depending on the species, from the common stem of biosynthesis. The fatty acid end products of these pathways do not exist free in bacteria but exist attached to CoA until they are incorporated. They become incorporated into one of the several membrane phospholipids, lipid A, or a lipoprotein.

SUMMARY

Two microbial themes in biosynthesis should be emphasized. First, what we know of the ways all living creatures synthesize the building blocks of protoplasm has been learned largely from studies of bacteria. The reasons are simple: (i) many bacteria can make all the building blocks, including those we cannot (which are therefore called "essential" in the terminology of human nutrition), and (ii) in the middle of the 20th century, bacteria provided the best practical opportunity for a concerted genetic and biochemical exploration of biosynthesis. As a result, the very concept of a biochemical pathway grew out of microbial studies.

Second, one of the important take-home messages from the discovery of biosynthetic pathways in microbes was that the theme enunciated by biochemists early in the 20th century as "unity of biochemistry" was indeed correct. This theme is the concept that all life on this planet is related, i.e., has a common ancestry, and that this underlying relatedness is reflected in similar, if not identical, biochemistry.

Biosynthesis, then, more than fueling, reflects the essentially close relation of microbes to plants and animals. While the fueling metabolism of microbes expresses the potential for life forms to scavenge for food and thereby recycle organic material, *the biosynthetic metabolism of microbes reminds us of what we have lost in evolving into beings higher in the food chain.*

STUDY QUESTIONS

1. Bacterial species that inhabit nutritionally rich environments tend to lack the ability to make many building blocks. Explain how this matching of genetic characteristics to ecological niches may have come about.

2. Using the pathway information displayed in the website EcoCyc, determine the longest biosynthetic pathway leading to an amino acid and the shortest.

3. Choose any biosynthetic pathway from EcoCyc that illustrates all the typical features of these pathways.

4. What selective pressure might have led to the evolution of two coenzyme carriers of hydrogen atoms (NAD and NADP) rather than a single one?

5. Explain in what way the biosynthesis of amino acids and other nitrogen-containing components of human cells depends ultimately on the biochemical activities of Earth's prokaryotes.

6. In what sense are glutamate and cysteine unique in the biosynthesis phase of growth metabolism?

7. Cite a major way in which assimilation of phosphorus differs from that of sulfur or nitrogen.

8. Explain the costliness of the biosynthesis of nucleic acid building blocks (compared to making amino acids and sugars, for example).

chapter 8 building macromolecules

INTRODUCTION

Fueling and biosynthesis provide the energy and building blocks needed for the next stage in growth metabolism: building the macromolecules that form the structure of the cell. Every course in biochemistry describes the essentials of this process, which (largely, though not exclusively) involves polymerizing like molecules (the building blocks) together into long chains, modifying and folding the products, and placing them in their appropriate cellular locations. At least, this description applies to proteins, nucleic acids, and polysaccharides. Lipids and their conjugates, murein, and certain complex carbohydrates do not fit this description because they are composed of disparate moieties—sugars, fatty acids, amino acid derivatives of metabolites, and unusual sugars—linked together by a variety of chemical bonds.

Each macromolecule must be placed in its proper cellular location and juxtaposed accurately with partners for cooperative functions. This assembly of functional structures is no mean feat. Certain proteins that for the most part present a hydrophilic surface must be translocated across distinctly hydrophobic membranes, and building the outer structures of the envelope, such as the murein wall or the gram-negative outer membrane, involves reactions that require energy but take place in the periplasm, an environment that lacks ATP. As one might expect, multiple assembly tactics have evolved. In some instances, topological arrangement is accomplished during the very act of polymerization, as in the formation of murein and certain other envelope structures. In other instances, proper placement occurs after completion of the molecule, as for many **secreted proteins.** In the case of ribosomes, several RNA molecules and dozens of proteins must be assembled correctly and in proper order to construct this most dominant machinery of the cell. The other stable

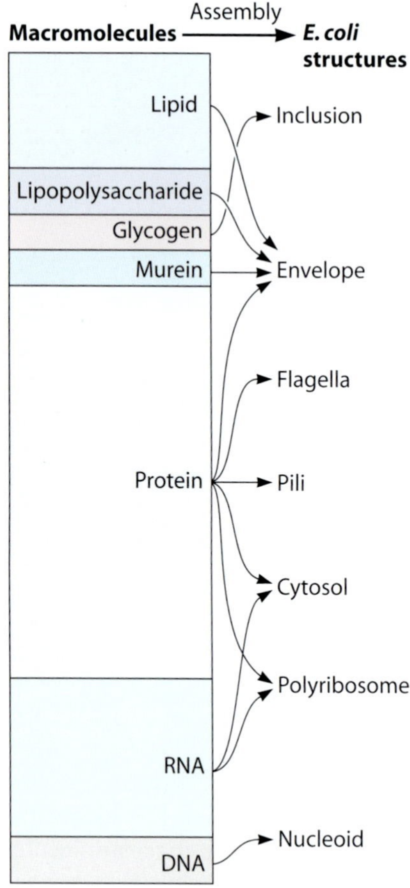

Figure 8.1 Macromolecules that form the structures of a bacterial cell. The size of each box indicates the relative amount of each macromolecule in the structures of the common bacterial cell, *E. coli.*

RNA species, **transfer RNA (tRNA)**, must be extensively and specifically modified after completion of their polynucleotide chains.

So far, we have been dealing largely with the biochemistry of molecules; now, we get to see a cell take shape from its individual building blocks. In order to focus on specifically prokaryotic features of macromolecule synthesis, we shall once again review only briefly topics normally covered in biochemistry courses. Figure 8.1 provides an overview of the focus of this chapter.

DNA

Overview of replication

Prokaryotic synthesis of DNA conforms to a general rule: each chain acts as a template for forming its companion, **complementary** chain. The program for **replication** follows a common pattern (with at least one interesting exception called rolling-circle replication [see Chapter 10]). At the site where replication is occurring, the double-stranded molecule separates, allowing enzymes and other proteins to bind and form a **replication fork** that allows deoxyribonucleoside triphosphates (deoxyadenosine triphosphate [dATP], deoxyguanosine triphosphate [dGTP], deoxycytidine triphosphate [dCTP], and deoxyribosylthymine triphosphate [TTP]) to pair with their complementary bases on each of the two exposed single strands. These nucleotides can then be polymerized into two new strands with the splitting off of pyrophosphate. Such replication is called **semiconservative:** each new double strand consists of an old strand and a newly synthesized one.

What follows is an account of replication in *Escherichia coli*, where the process has been most thoroughly studied. One should keep in mind that this description does not apply universally, because even in this most fundamental of cellular processes, microbes exhibit unexpected diversity (one indication: not all prokaryotic chromosomes are circular). Replication among archaeal species follows the general bacterial outline except that many of the proteins involved are much more similar to those found in eukaryotes than to the bacterial ones.

Replication does not begin randomly, but rather at a specific location, called ***oriC*** (for origin of chromosome replication). Initiation of the process takes place at a specifically determined frequency. With some exceptions, the chromosomes of bacteria are circular, not linear, and replication is **bidirectional:** it proceeds at a uniform pace and in both directions until the two replication forks have reached the **terminus *(terC)*.** The pace of replication differs greatly among different species, and the known range is a surprising 10-fold. Except at very low cell growth rates, the speed of polymerization in each species is almost constant, as far as is known (see Chapter 9). Thus, DNA replication is regulated, not by the intrinsic velocity of the replication apparatus, but by *how frequently the process is initiated at the origin* (see below and Chapter 9).

DNA synthesis takes place in the $5' \rightarrow 3'$ direction (relative to the phosphate linkages between adjacent deoxyribose residues). Consequently, the synthesis of each strand must proceed by a somewhat different mechanism, because the two template strands of the parent double helix are **antiparallel** (one strand runs in the $5' \rightarrow 3'$ direction; its complement runs

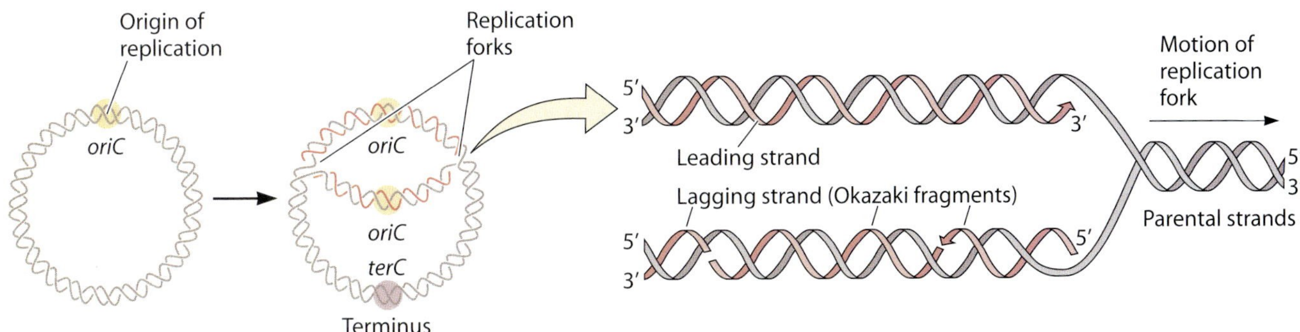

Figure 8.2 Overview of DNA replication. See the text for a description.

in the 3′→5′ direction). The strand synthesized in the same direction the replication fork is moving is conventionally called the **leading strand**, and the other is called the **lagging strand**. Figure 8.2 depicts these overall features of replication.

Initiation of replication

Getting replication started is quite complicated—but one would hardly expect a process that regulates synthesis of the cell's genetic blueprint to be simple. The details are best known in *E. coli*, but even there the story is incomplete.

Three questions arise immediately.

1. What biochemical steps are required for initiation?
2. How is the frequency of initiation regulated, i.e., how does the cell know it needs more DNA?
3. How does the cell prevent premature initiation?

To answer these questions, contemplate the structure of *oriC* (Fig. 8.3) and its interactions with various key proteins.

1. No biochemical transactions can normally take place on DNA as long as it is wound tightly in a double helix. Replication, as well as **transcription** and **recombination**, requires separation of the two strands. In the case of replication, the separation must be more than local; eventually, the two strands are totally separated, which must involve unwinding of the duplex. In the case of initiation of replication, DNA unwinding begins with the binding of a positive regulator of initiation, the protein **DnaA**, to special sites on *oriC* called **DnaA boxes** (five 9-base-pair [bp] segments, R1 to R4 and M [Fig. 8.3]). The DnaA-*oriC* complex is next acted upon by **DnaB**, a **helicase** (or unwinding enzyme), which is helped to bind to the origin by a **loading factor (DnaC)**. Unwinding begins nearby at three repeated 13-bp sequences (L, M, and R [Fig. 8.3]) that are rich in AT pairs, which makes them easier to separate than GC-rich regions (because they have one fewer hydrogen bond per pair). Molecules of **single-stranded binding (SSB) protein** then coat the separated strands and prevent them from **reannealing** (re-forming the hydrogen bonds that hold the strand together). In this opened region of the DNA duplex, sometimes called a **bubble**, the machinery for replication, or **replisome** (Fig. 8.4), can be assembled.

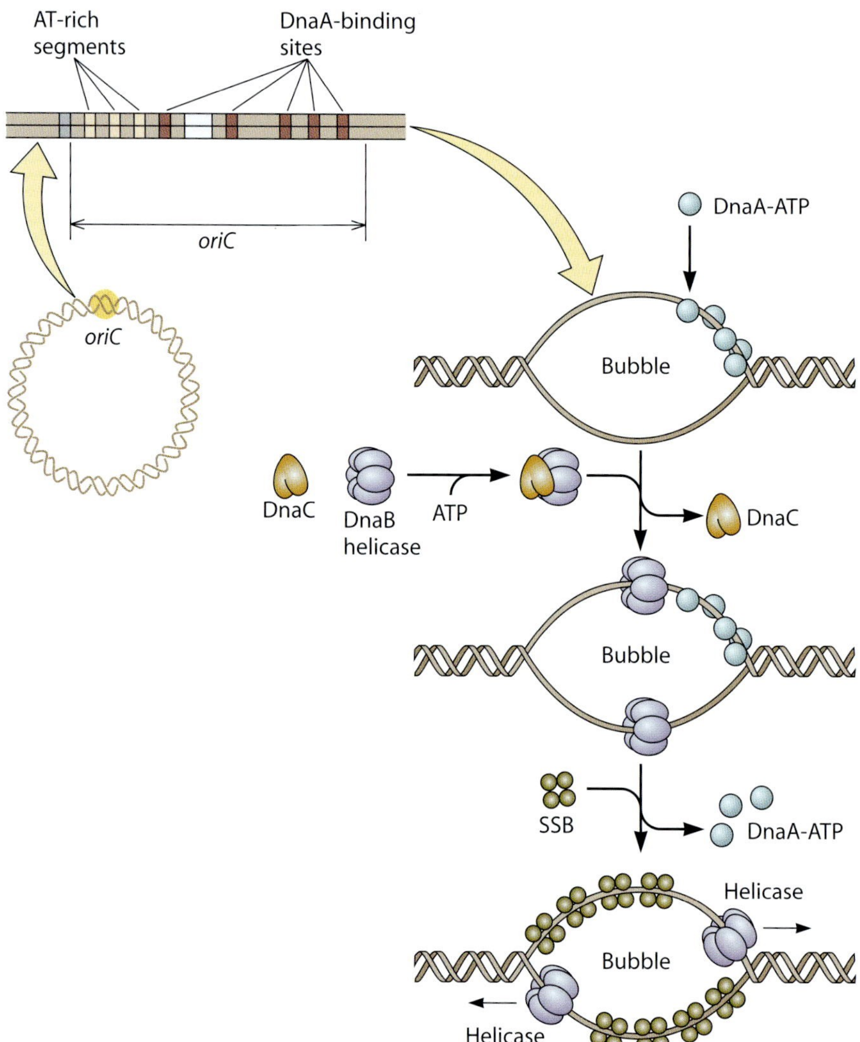

Figure 8.3 Initiation of DNA replication in *E. coli*. The replicative origin, *oriC*, has three AT-rich 13-mers and five DnaA boxes (DNA-binding sites). Once DnaA binds to these boxes, the helicase DnaB binds (with the help of DnaC) and begins to unravel the DNA. Single strandedness is maintained by the binding of an SSB protein. Initiation can then proceed.

2. When an appropriate cellular concentration of DnaA is achieved, two replisomes assemble at each strand of *oriC* and begin to move, one clockwise and the other counterclockwise. Because the concentration of DnaA or any cell constituent changes as the mass of the cell changes, replication initiation and cell mass are connected. Thus, initiation is tied to cell growth and, generally speaking, does not take place unless cells are growing. In addition, in order to be active, DnaA must bind ATP (the product is represented as **DnaA-ATP**). As shown in Fig. 8.3, for DnaA-ATP to be effective, many molecules must bind to *oriC* at the same time. Their coordinate binding is facilitated by bending the DNA into a special structure, which brings all the DnaA boxes into close proximity. Bending is carried out by the action of DNA-bending proteins.

3. The prevention of premature reinitiation matters, because if *oriC* permits initiation too often, the chromosome is likely to become tangled and thus to malfunction. The final answer to how this pre-

vention happens is not available, but much is known about the process. The origin is prevented from participating in initiation for some time, a phenomenon called sequestration. The mechanism of sequestration relies on a way bacteria have to distinguish between the newly made and the old strands. This distinction is made on the basis of their respective states of methylation. Scattered throughout *oriC* are 11 GATC **palindromic sequences** (sequences of bases that are the same in reverse order in the complementary strand). The A residues of GATC can be methylated by an enzyme named DNA adenosine methyltransferase (puckishly called **Dam methylase**). Dam methylase acts on these residues, but only after several minutes have elapsed. During this time, the parental strand of DNA carries its previously methylated adenine residues, but the newly synthesized strand will temporarily be lacking methylated adenines. The result is a temporary **hemimethylation** of newly replicated DNA. What is the relevance of this? A protein called **SeqA** inhibits initiation by binding to *oriC*, covering its DnaA boxes, which prevents their methylation. SeqA binds preferentially to hemimethylated DNA and thus acts selectively at *oriC*, which is rich in GATCs. As long as SeqA is bound, new initiations are prevented. A further safeguard against inappropriate repetitive firing of *oriC* is provided by conversion of DnaA-ATP to DnaA-ADP during initiation, which lowers the concentration of this positive regulator. At the opened duplex, the machinery for replication (Fig. 8.4) can now assemble.

DNA strand elongation

As judged by measurements of *E. coli*, the chromosome replicates at a rate approaching 1,000 nucleotides per second (a number that is greatly variable among different species) and does so with an error frequency of only 1 in 10 billion bp replicated. This is incredible, for it is the equivalent of 8 million pages of text without a single typographical error. Such a feat becomes all the more remarkable when one tries to picture what is happening in the process.

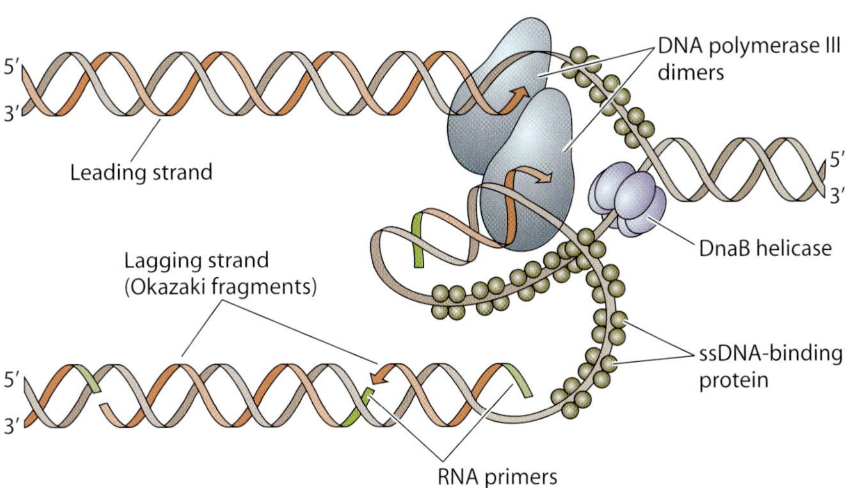

Figure 8.4 Elongation of DNA. Both strands are made by the DNA polymerase Pol III. The lagging strand is made *opposite to the general direction* in small segments called **Okazaki fragments** using RNA primers. Pol III acts as a dimer and, because the lagging strand is bent around, faces in the same direction for both strands. ssDNA, single-stranded DNA.

The chromosome, which here is a DNA molecule a thousand times the length of the cell and consisting of two intimately entwined, helically coiled strands, is replicated in about 40 minutes at 37°C. The process has to include separation of the strands, which cannot be done without unwinding the helix, which in turn means that the duplex immediately ahead of the replication fork will be twisted rapidly into positive supercoils. The polymerase that makes DNA can grow chains only in the 5′→3′ direction, so one new strand must be made (backward) in short 1,000-bp 5′→3′ segments and then spliced together.

Clearly, the process of chain elongation must require the cooperative involvement of many proteins. The DnaB helicase, using energy from ATP hydrolysis, moves along the lagging strand unwinding the duplex with the assistance of the SSB proteins. As the replisomes progress, the DNA ahead of the fork rotates due to the unwinding of the helix. The positive supercoils in the double-stranded DNA ahead of the fork are countered by introducing negative coils by the action of **DNA gyrase** (also called **topoisomerase II**), which cleaves, twists, and then reseals both strands.

Polymerization does not proceed the same way on each of the strands. On the leading strand, it proceeds via DNA polymerase III (Pol III), but for the lagging strand, it additionally requires RNA polymerase (RNA-P). Why the latter? The answer is that Pol III can *extend* a polynucleotide chain but cannot initiate one. That is, it needs a **primer** segment on which it can add nucleotides. RNA-Ps, on the other hand, do not need primers. Pol III can grow chains only in the 5′→3′ direction, which can proceed continuously on the leading strand in the direction of the replicating fork. However, the lagging strand is made backwards (in the opposite orientation) in short 1,000-bp 5′→3′ segments called **Okazaki fragments** (named for their discoverer). Thus, *lagging-strand synthesis is discontinuous.* As the replicating fork moves, the Okazaki fragments are spliced together by an enzyme called **DNA ligase** (Fig. 8.5). The function of the RNA-P (or **DnaG primase**) is to initiate synthesis of an RNA segment about 10 nucleotides in length. Pol III can then use this short RNA segment as a primer for Okazaki fragments by adding onto the 3′-OH end of the RNA. As shown in Fig. 8.4, Pol III sits on the fork as two multisubunit molecules, each complex synthesizing one of the two growing DNA strands. The active catalyst of DNA polymerization in Pol III (the α-subunit) needs the assistance of another subunit (the β-clamp) that forms a sliding bracelet on each parental strand ahead of the processing Pol III. Placing the β-clamp in position requires another complex of polypeptides (called γ, δ, χ, and ψ) that function as a **clamp loader.**

Topologically curious readers may wonder whether the two replisomes course their way around the chromosome much as model slot car racers on a track or whether they are fixed more or less solidly to some structure while the duplex DNA is fed through them. A definitive answer cannot yet be given. As yet, we do not know for sure.

Figure 8.5 Ligation of nicked DNA. In order to make a continuous lagging-strand molecule, Okazaki fragments are ligated by DNA ligase, using NAD in the process.

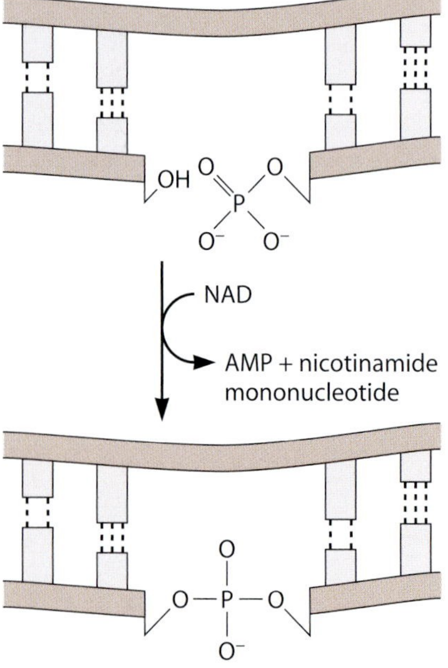

Repair of errors in replication

Spontaneous changes in DNA—**mutations**—are due mainly to **mismatches** (matches other than A with T and G with C) between bases during DNA replication. If the wrong base is inserted through a mismatch, pairing will not occur with its complement on the other DNA strand. As we have seen above, spontaneous mutations in *E. coli* take place approximately only once for every 10^{10} bases synthesized. This miraculous-sounding faithful copying of DNA is due to several factors. Base pairing during DNA synthesis is itself a very precise process. Accuracy is augmented by a special property of Pol I and III called $3' \rightarrow 5'$ exonuclease activity. These polymerases cannot progress unless there is a properly matched nucleotide pair behind them. If not, *they pause and move backward*, remove the mismatched base using their exonuclease activity, and then proceed. This **proofreading** activity is in all probability the reason that Pol I and III require a 3′-OH end on a DNA or an RNA primer (that is, they must have a properly matched base pair behind them, so they cannot start from scratch). That a primer-independent polymerase has not evolved testifies to the importance of this proofreading function.

In addition, all cells, not just bacteria, have mechanisms for detecting and repairing mismatches that may have escaped proofreading. Mutant cells that lack a **mismatch repair (MMR) system** mutate spontaneously about 1,000 times more often than normal cells. The MMR pathway is not just a prokaryotic property but is widespread among living organisms. The human MMR system has received a great deal of attention because families with certain hereditary cancers have mutations in MMR genes.

Much of what we know about MMR stems from work performed on the system in *E. coli*, called **methyl-directed MMR.** In this system (as well as others), MMR involves three steps: **recognition** of the mismatch, **excision** of the misincorporated base and its surrounding DNA, and **repair synthesis** to replace the excised DNA. This takes several steps (Fig. 8.6). First, a protein, **MutS,** recognizes the mismatch. Another protein, **MutL,** stimulates a third protein, **MutH,** to cut the DNA on the newly synthesized DNA strand in order to remove the incorrect base. How do these proteins know which strand is the old one (with the faithful sequence) and which is the new one (with the wrong base inserted)? Some bacteria, notably *E. coli* and its relatives, possess a mechanism for *distinguishing between newly made DNA and that made earlier*, and we have already encountered it in another context when we discussed the initiation of replication: the activity of Dam methylase. This enzyme adds a methyl group to the adenine in the sequence GATC, a sequence found about once every 256 stretches of tetranucleotides. A newly synthesized segment of double-stranded DNA is hemimethylated because, as we noted, it takes some time for the Dam enzyme to methylate the newly synthesized strands; the old one (the original template strand), of course, is fully methylated. The steps, then, that result in repair are the following. The protein MutS binds to the mismatched pair. The protein MutL binds to MutS. The MutL-MutS complex bound to the mismatch site translocates DNA through it until the nearest hemimethylated GATC site is reached and recognized by MutL. At that point, the MutH endonuclease joins the game and cuts the unmethylated strand at this GATC site; helicase unwinds the DNA from the point of the cut, and an exonuclease chews this strand back, finally removing the mismatched base. The missing segment is then replaced, and the repaired strand is made intact by DNA ligase.

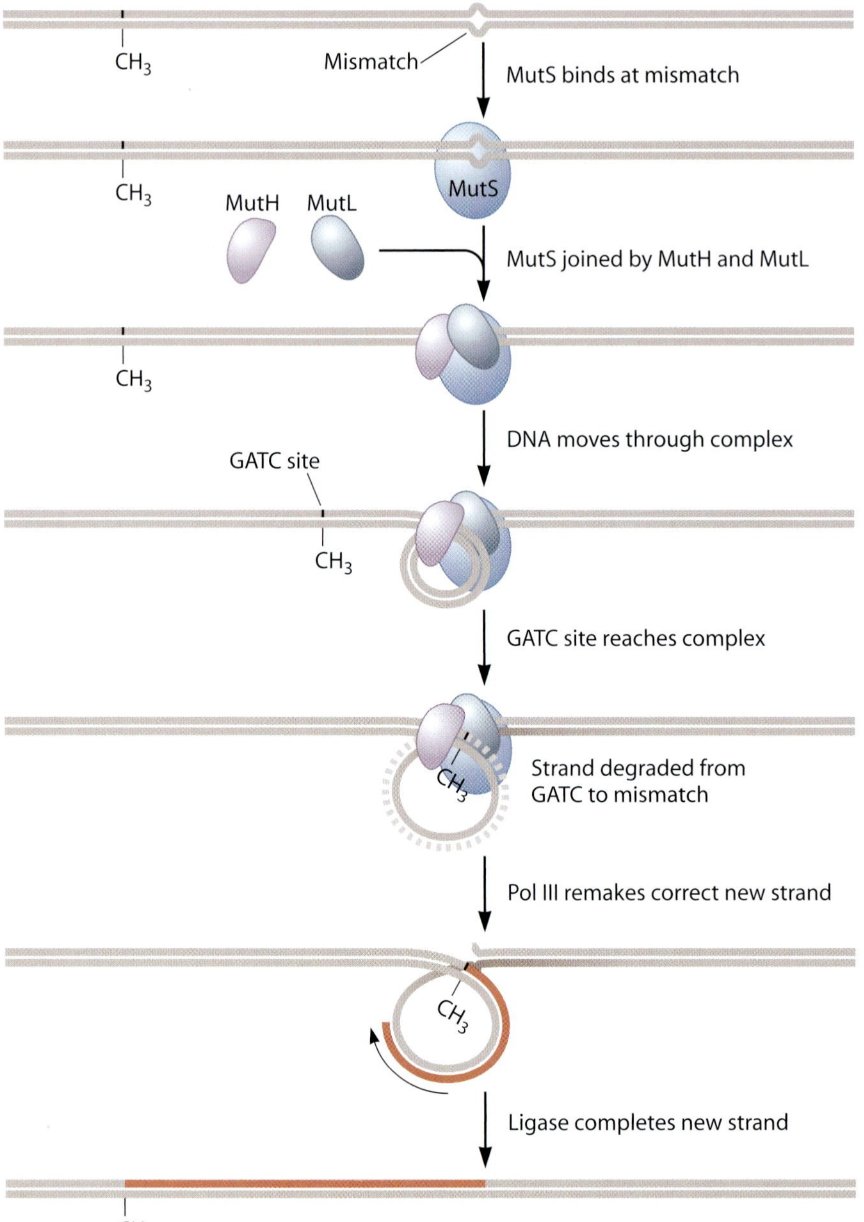

Figure 8.6 Mismatch repair. A mismatch is recognized by the protein MutS. MutH and MutL then bind to the MutS-DNA complex. DNA moves through this complex until it reaches a GATC site (the site for methylation by the Dam methylase). There the as-yet-unmethylated new strand with the mismatch is degraded, and the gap is filled by Pol III and DNA ligase.

Termination of replication

As they near completion of their task, the two replisomes converge upon the region halfway (180°) around the chromosome from *oriC*. A collision is avoided by means of a well-regulated termination process. In this region, there are several special sequences, called ***ter***, arranged in separate **terminator** groups, **T1** and **T2**, of opposite polarity. T1 blocks the replisome moving counterclockwise; T2 blocks the one moving clockwise. The terminator sites are aided in their function by a protein, **Tus (terminator utilization substance)**, which binds to *ter* sequences and halts the replisome's progress by inhibiting its DnaB helicase.

Because most bacterial chromosomes are circular, the final step in replication requires some further maneuvers. At the close of termination, the completed daughter chromosomes are topologically locked together (intertwined) into a structure called a **concatenate** (i.e., two links in a chain), which is the inevitable result of unwinding a helical double-stranded closed circle (Fig. 8.7). Decatenation is carried out by enzymes called topoisomerases that insert a double-strand break and permit the other DNA molecule to pass through the gap. The same enzymes later reseal the gapped strand.

Before resolution into separate circular chromosomes, the catenate is susceptible to **homologous recombination,** i.e., crossover events between homologous sequences of the two chromosomes (see Chapter 10). Since a single-crossover event would lead to the formation of a dimer consisting of two chromosomes (a lethal event if not reversed), the action of special **recombinases** (enzymes that cut and splice DNA) becomes especially important.

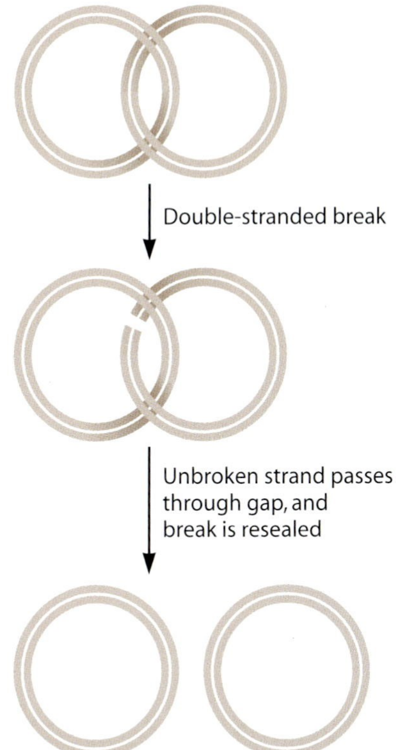
Double-stranded break
Unbroken strand passes through gap, and break is resealed

Figure 8.7 Decatenation of sister chromosomes. See the text for details.

Protecting the DNA

During the synthesis of DNA, many bacterial and other microbial cells mark their DNA in a distinctive manner (a process akin to branding cattle) so as to identify it as their very own. To what purpose? Microbial cells are frequently invaded by "foreign" DNA introduced by plasmids or viruses. The result of such invasion is seldom beneficial to the cell (in the case of viral invasion, death and lysis is often the result). Bacteria have evolved powerful DNA-hydrolyzing endonucleases that destroy unwanted invading DNA. Many hundreds of these **restriction endonucleases** (so named because they were discovered through their property of restricting the growth of bacterial viruses) have been discovered among bacteria. Each recognizes a different specific sequence of bases (**recognition site**) in DNA and produces double-strand breaks, which lead subsequently to the complete degradation of the DNA because other nucleases can attack the molecule at these exposed ends (Table 8.1). The cell *protects its own chromosomal DNA by methylating an adenine or a cytosine residue in the recognition site.* This **modification** by methylation is the "branding" that marks the DNA as self and protects it from its own restriction endonuclease.

Many microbes, especially bacteria, utilize this **restriction-modification** strategy. In all likelihood, you have encountered restriction enzymes in the context of their role in recombinant DNA technology, where their many uses can make us forget their important biological role. No laboratory dealing with genomic analysis and in vitro DNA manipulation can function without a ready source of purified restriction enzymes that can bring about highly specific and reproducible cleavage of DNA at specified sequences. DNA **cloning,** a hallmark of biotechnology, depends heavily on restriction enzymes. The many hundreds of prokaryotic restriction-modification systems fall into a few broad classes. Type I systems combine the modification (methylating) and **restriction** (cleaving) activities into a single multifunctional protein; type II systems, the most useful in research laboratories, consist of separate endonucleases and methylases. In either case, it is essential that during replication of the cellular DNA the methylase (or restriction-methylase) act on the recognition site just made in the new strand and modify it by methylation.

Table 8.1 Actions of certain restriction endonucleases

Class	Enzyme	Producing microorganism	Recognized DNA sequence[a]
Six base pairs recognized; complementary single-stranded ends produced	EcoRI	*Escherichia coli* (R)[b]	G▼A A T T C C T T A A▲G
	HindIII	*Haemophilus influenzae*	A▼A G C T T T T C G A▲A
Six base pairs recognized; blunt ends produced	HpaI	*Haemophilus parainfluenzae*	G T T▼A A C C A A▲T T G
	HindII	*Haemophilus influenzae* Rd	C ▼A G T (T) (G) A C C A (A) (T) T G G ▲C
Four base pairs recognized; complementary single-stranded ends produced	HhaI	*Haemophilus haemolyticus*	G C G▼C C▲G C G
	MboI	*Moraxella bovis*	▼G A T C C T A G▲
Four base pairs recognized; blunt ends produced	HaeIII	*Haemophilus aegypticus*	G G▼C C C C▲G G

[a]Arrow-head indicates site of single-strand cleavage. Upper sequence of bases is written in the 5′→3′ direction.
[b]Encoded by genes that are plasmid borne.

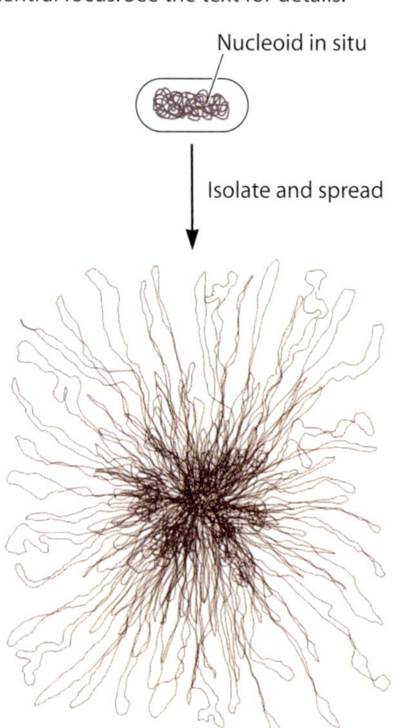

Figure 8.8 Diagram illustrating the compact nature of the nucleoid. After isolation and spreading, the DNA is seen to consist of loops emanating from a central focus. See the text for details.

Assembly of the nucleoid

We noted in Chapter 3 that the chromosome of prokaryotic cells is extraordinarily tightly packed—as is the DNA of all cells. In eukaryotic cells, the chromosomal DNA is tightly wound around a complex of histone protein molecules to form **nucleosomes** and then further compacted by tertiary coiling. In bacteria, the story is somewhat different, though perhaps analogous. Instead of nucleosomes, *E. coli*'s chromosomal DNA possesses physical domains (the number is variously estimated between 30 and 200) that consist of supercoiled, compacted loops. The loops are sufficiently distinct from each other that single-strand breaks in any one of them can relax the supercoiling without affecting that of the neighboring loops. The loops can even be visualized in the electron microscope; gentle lysis of *E. coli* cells and spreading of the released DNA reveals the "spaghetto" not as a formless mess but as a flowerlike structure with loops emanating from a central focus (Fig. 8.8).

How the nucleoid is structured is largely unknown, but it is likely that several factors come into play. Proteins called SMC (for structural maintenance of chromosome) act to condense it into a tight structure. SMC proteins consist of antiparallel strands that bind to DNA at both ends. In the middle, these molecules have a hinge, which allows the two arms to come together and bring the attached DNA into close proximity, reminding one of tweezers with sticky ends. Note that the two DNA segments may be far removed on the chromosome, and thus distant regions may be brought together by this process. ATP is consumed in the process

(SMC proteins are ATPases). Some SMCs, including one called Muk (Japanese for "nucleus"), are essential for proper chromosome segregation. SMC proteins have been found in almost all organisms and are quite conserved, suggesting that they originated early in evolution and that their role is indeed important.

Other elements believed to be involved include several basic proteins associated with the nucleoid (**HU**, **H-NS**, and **Fis**). Also, there is interest in the possibility that **REP** sequences (short for repetitive extragenic palindrome) play some role. REP elements are 38-bp palindromic sequences that occur (with some variations in individual bases) throughout prokaryotic genomes. There are many hundreds of these in a genome, and they always occur between genes, never within a gene. It is speculated that they participate in domain definition.

The whole point of replicating the genome is to prepare for the birth of two daughter cells. What little we know about the mechanism of this segregation of chromosomes at the event of cell division is discussed in Chapter 9.

RNA

Overview of transcription

RNA-P and its function

Transcription means the synthesis of an RNA molecule, called a **transcript,** with a base sequence complementary to a segment of one strand of the DNA. Eukaryotic cells have several RNA-polymerizing enzymes, but prokaryotic cells have only one, called **DNA-dependent RNA-P** (abbreviated as **RNA-P**). This enzyme synthesizes all the cellular species of RNA, stable (**ribosomal RNA [rRNA]**, tRNA, and regulatory RNA) and unstable (**messenger RNA [mRNA]**). In spite of there being only one RNA-P, it can be modified to transcribe selectively certain sets of genes by associating with one of several **sigma (σ) factors** (replaceable subunits that confer DNA-recognizing specificity [see below]).

The function of RNA-P as a catalyst of transcription is easily stated: it links ribonucleoside triphosphates (ATP, GTP, CTP, and UTP) in the $5' \rightarrow 3'$ direction by sugar phosphate bonds, in an order dictated by one strand of DNA. A pyrophosphate (two terminal phosphoryl groups) is split off in the process. The core RNA-P has relatively low affinity for DNA and binds indiscriminately. Achieving selective binding is the role of σ factor. Without σ factor, the core has been likened to a closed clamp (picture your index finger touching your thumb). Binding of σ factor opens the clamp (the thumb and finger separate), allowing more contact of the core with the DNA. Later (see below), when σ factor is released, the clamp shuts again, holding RNA-P to the DNA.

Not surprisingly, RNA-P molecules are rather complicated (though with fewer components than the DNA replisome). Possibly the RNA-Ps of all bacteria share the same core subunit structure, designated $\alpha_2\beta\beta'$ because it is composed of four subunits, one of which (α), is present in two copies. Transcription among the Archaea as well is performed by a single RNA-P, but one quite different from the bacterial one. Archaeal

RNA-P has as many as 14 subunits rather than the 4 typical of bacteria, and in this respect it more closely resembles that found in eukaryotes. One consequence is that antibiotics (such as rifampin) that inhibit the bacterial enzyme do not affect the archaeal polymerase.

Products of transcription

The product made by RNA-P is called a **transcript.** Transcripts of protein-encoding **structural genes** (more precisely called **cistrons,** after the ***cis-trans*** genetic test for functionality) are called mRNA. Almost all bacterial mRNA transcripts are **polycistronic**—they encode several polypeptides. This uniquely prokaryotic arrangement comes about because most protein-encoding genes occur on the chromosome in clusters, usually of related function, called **operons,** which are **transcriptional units** because they start with a promoter and end with a transcription termination signal. Near the promoter (in some cases between the promoter and the first structural gene of the operon) is a region called the **operator,** a site at which regulatory proteins that control transcription initiation act. (Actually, the control of some operons is quite complex and includes regions far upstream from the promoter.) Transcription of the protein-encoding genes of an average bacterium is unusual in that the mRNA molecules have a half-life of only a few minutes (but this property is highly variable; for some mRNAs, it is shorter than a minute, for others much longer). Proper regulation of mRNA synthesis involves second-by-second response to the physiological state of the cell (see Chapter 12).

RNA-P also transcribes the nearly 50 genes that encode tRNA and the half dozen redundant genes encoding each of the three major rRNAs. Though involving relatively few genes, this activity can at any one time tie up most of these RNA-P molecules. It must therefore be carefully and economically modulated in response to the growth-promoting potential of each environment. The genes encoding stable RNA are generally clustered. Their transcripts must be processed by cleavage as an initial step.

As with all macromolecular syntheses, transcription consists of three distinct phases: initiation, elongation, and termination (Fig. 8.9), and there is much to say about each.

Initiation of transcription

The title of this section may seem to herald a bland tale about RNA-P and its work. Quite the contrary; the story is fascinating, and the three words "initiation of transcription" encapsulate much of the excitement of 20th-century biology, *including the birth of molecular biology.* What researchers working with microbes gave to biological thought in the last century was nothing less than the discovery of the nature of genes and how their expression is regulated. It is a story now largely summarized by recounting the features of the gene *lacZ*, which encodes **β-galactosidase,** the enzyme that breaks down lactose, and the regulation of the *lac* operon by both the Lac repressor and the catabolite repressor protein.

Much gene regulation occurs by varying the frequency of transcription initiation. Once started, the RNA chain elongation rate is nearly constant at a given temperature. Many steps of gene expression—not just initiation—are points for exerting regulation in one gene or another. The

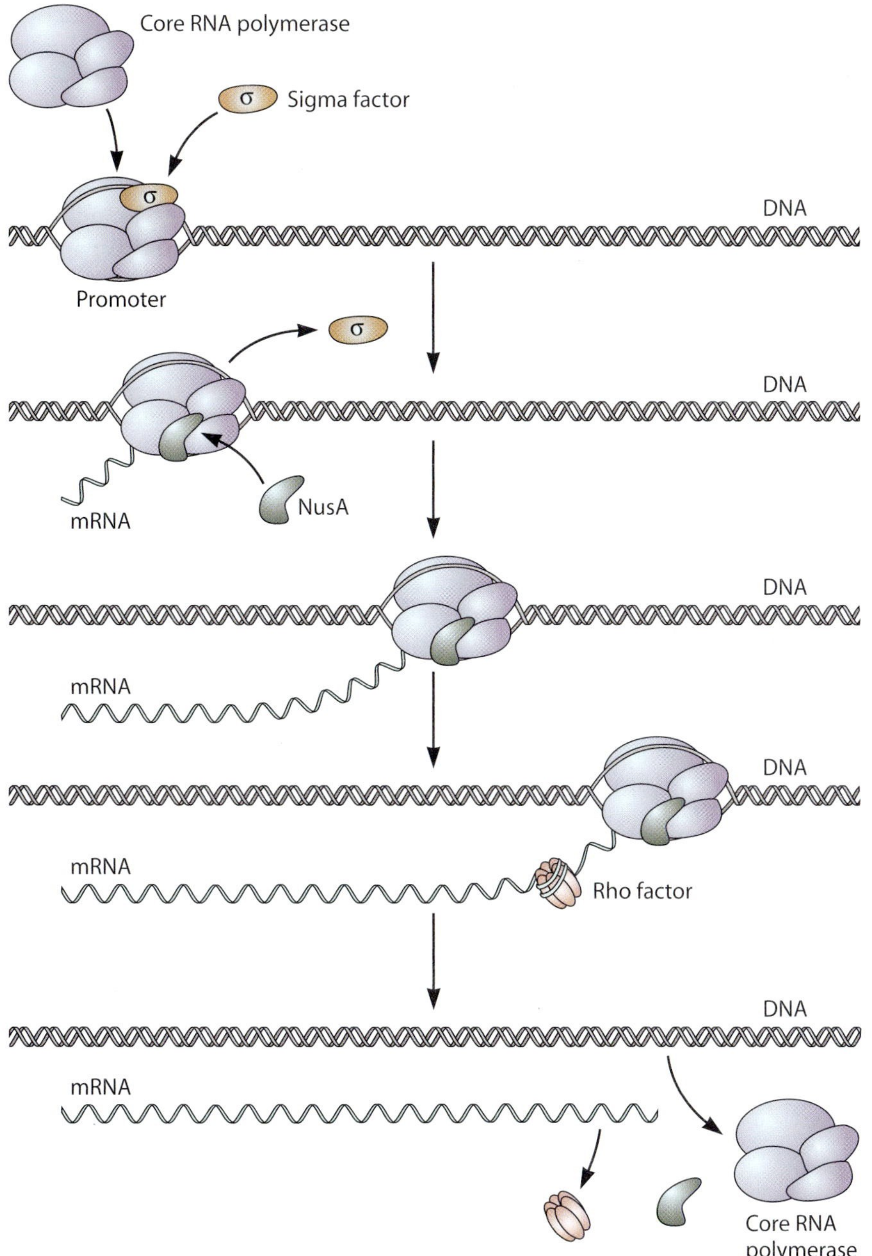

Figure 8.9 Overview of transcription. See the text for details.

dominant regulatory devices, however, involve control of initiation in some way, perhaps because blocking gene expression at its very first step is the most economical. More about that later (see Chapter 12).

The promoter

Initiation of transcription occurs when the α and σ subunits of RNA-P locate specific DNA sequences called **promoters,** which precede all transcriptional units (operons). Given their essential role in defining where RNA-P should start, promoters might be expected to share a distinctive, effective, and easily recognized sequence, and one that would be strongly conserved in evolution. Not so. It took considerable time and much experimental work before a promoter within a DNA segment could be recognized with any confidence. A moment's reflection can make sense of this ambiguity in promoter structure. Some genes are transcribed less than once per generation, others once every second. The magnitude of gene expression varies over a wide range—as much as 10,000-fold. This variation reflects the functions of various regulatory devices operating on promoters of vastly different inherent strengths. Thus, there is no single promoter sequence, either within one organism or among different species. Instead there are *general patterns of sequences* rather than unique nucleotide sequences that spell "promoter" to RNA-P.

What determines a promoter's strength? Obviously, the promoter's nucleotide sequence must be important, and to see how sequence comes into play, we should examine a bacterial promoter. Figure 8.10 depicts the major features of one—in this case, the promoter recognized by *E. coli*'s RNA-P when guided by its major sigma subunit, σ^{70} (named after its molecular weight of approximately 70,000). The DNA segment shown in the figure is labeled to show three principal regions: a **promoter core** flanked on the left (in the 5′ direction) by an **upstream region (UP element)** and on the right by a **downstream region.** For the present, we shall have little to say about the up- and downstream regions other than to note that (i) for some promoters, the upstream area is the site of action for several regulatory proteins, and (ii) the downstream area comprises the initial portion of the **coding region** of the gene in question and also has regulatory significance.

The bacterial promoter core has three parts: a **−35 hexamer** and a **−10 hexamer** (the minus signs indicate distance upstream from the codon initiating translation of the gene) and a **17-bp spacer** that separates the hexamers. What you may find disappointing is that *no definitive nucleotide sequence is indicated for the promoter core*; this is not an omission, simply a reflection of the fact that no universal sequence exists. What are shown (in the top row of the diagram) are two **consensus sequences,** TTGACA and TATAAT, for the −35 and −10 regions, respectively. (The numerical names of the two regions indicate the approximate nucleotide residue near the center of the hexamer, counting backward from the first base

Figure 8.10 Structure of the consensus promoter recognized by σ^{70} in *E. coli*. The core promoter includes the −35 (TTGACA) and −10 (TATAAT) hexamers and the spacer between them. The numbers printed in red beneath the nucleotide bases and the spacer indicate their percentage occurrence in known σ^{70} promoters in this organism. (Note: all bases indicated are those in the nontemplate strand of DNA.) The positions of bases in the DNA are numbered relative to the first base transcribed into mRNA.

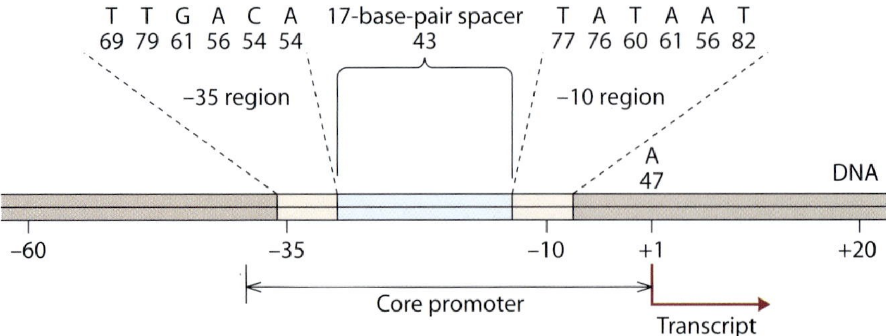

pair of the downstream, or protein-coding, region.) They are called consensus sequences because they depict a sequence made up of the most statistically probable base to be found at each nucleotide residue. (Watch out for a confusing convention employed in discussing transcription: the strand of DNA that is copied to make RNA is called the **antisense** or **template strand;** the *other* strand (the complement to the template strand) is called the **coding strand** for the simple fact that its sequence of bases is identical to the sequence in the RNA product with U substituted for T. *The inherent strength of a promoter is determined by its nucleotide sequence,* including both the −35 and −10 regions, as well as the sequence around the +1 position.

Starting the transcript

For simplicity, let us confine our attention to initiation by σ^{70}-directed RNA-P. The first step is a reversible binding of RNA-P to a promoter to form a **closed complex.** The promoter not only signals where transcription begins, but also determines which DNA strand will be transcribed, because the sequence of bases on the template strand determines the orientation of binding of the RNA-P molecule. In the second step, a relatively slow isomerization takes place in which an **open complex** is formed. This structure consists of a **transcription bubble** of locally melted DNA duplex, about 12 bp in length, within which polymerase is free to act on the strand to be transcribed. The third step is the migration of the polymerase and the bubble (now grown to its mature length of 18 bp) from the promoter area as the RNA transcript is formed. In Chapter 12, we shall see that promoter activity is greatly changed by some of the many protein regulators of initiation.

Transcription initiation in the Archaea differs considerably from the bacterial mode. The archaeal RNA-P initiates at a **TATA box** (a nucleotide sequence in DNA) rather than at a bacterial-type promoter—a further similarity to the Eukarya.

RNA chain elongation

When RNA-P has formed a transcript about a dozen nucleotides in length, the σ factor is released and is free to associate with another RNA-P core. Loss of σ factor clamps the RNA-P to the DNA and moves it down the template DNA strand; the bubble moves with it (base pairs of DNA being opened in front and closed behind), and the nascent RNA transcript trails behind it, paired to the template DNA strand. Each step of elongation involves nucleoside triphosphate binding, bond formation with the nascent chain releasing pyrophosphate, and movement of polymerase along the DNA strand. The rate of chain growth in *E. coli* at 37°C is about 45 nucleotides added per second for mRNA and close to double that for rRNA, tRNA, and other stable RNA species. We mention these numbers not simply to display how much has been learned about transcription, but because they have significance in the overall program of protein synthesis (see below).

Although we have emphasized transcription initiation as a crucial event in regulating gene expression, chain elongation is not a constant, uneventful process. It has its own share of drama. After σ factor leaves the polymerase, other protein factors exert their influence on the course of transcription. Proteins in *E. coli* named **NusA** and **NusG** (named for roles in the synthesis of bacterial viruses) associate transitorily with RNA-P. They affect the behavior of the polymerase when it encounters regions of the DNA called **pausing sites** (GC-rich regions of dyad symmetry that form hairpin loops in the growing RNA strand). Pausing increases the chance that transcription will be interrupted; NusA and NusG modulate the probability of interruption. Therefore, keep in mind that regulation of gene expression involves steps beyond initiation; we return to this aspect of transcription in Chapter 12.

Termination of transcription

Our interest in termination goes beyond the fact that it is an obligatory step. Unlike the case in eukaryotes, termination in prokaryotes participates in regulation as well. The important case of regulatory termination called **attenuation** occurs before the operon's structural genes are transcribed and is discussed in Chapter 11. Usually, termination at the end of an operon is brought about either by a **simple** process, mediated solely by a special sequence of bases in the template DNA, or by a more **complex** process that is also signaled by a sequence of bases but requires the participation of an accessory protein called **rho** (ρ). Complex termination is referred to as **rho-dependent termination.**

Simple termination, which is rho independent, occurs after the polymerase has transcribed a GC-rich region of DNA arranged in an **inverted repeat** that enables it to form a **stem-and-loop structure** (lollipop shaped, with a double-stranded stem and a single-stranded head), followed by a region that contains a series of A residues. The stem-and-loop structure is thought to form within the transcription bubble, thereby causing the process to stall. The subsequent AU pairs between transcript and template are easily broken, releasing the transcript.

Rho-dependent termination occurs whenever RNA-P pauses for at least 10 seconds (as is caused by a similar stem-and-loop structure). When this happens, rho interacts with the stalled RNA-P complex and releases the transcript. Rho is blocked, however, from reaching the polymerase by ribosomes translating the mRNA.

Fate of transcripts

Transcripts are of three general sorts. They may be (i) mRNA molecules that encode proteins (polypeptides, really, since a protein consists of one or more polypeptide subunits, frequently modified), (ii) the stable RNA components of the protein-making machinery of the cell, or (iii) small regulatory RNAs, the function of which we shall discuss in Chapter 12.

We have noted that there are hundreds of species of mRNA, which differ greatly in abundance in each cell. A hallmark of prokaryotic life is that these instructions for making proteins are constantly being turned over, with an average half-life of under a minute but, as we have said, with a huge variation. rRNA and tRNA molecules are stable (at least under nonstarvation conditions) and function repetitively in the job of synthesizing polypeptide chains.

There is more of interest to discuss about the cellular fates of these two disparate classes of RNA.

Life and death of mRNA

A hallmark of prokaryotic organisms is their turnover of mRNA. The average **half-life** (the time for 50% of the molecules to be destroyed) of mRNA in *E. coli*, for example, is just under 1 minute at 37°C. Some mRNA species exist longer, others for a shorter time. Details of which nucleases are involved in the destruction of mRNA are still being explored, but the main point is that the instructions that dictate which proteins should be made are constantly changing as the cell senses both its physiological health and the nature of its current environment. By continually erasing its instructions for making proteins, the bacterial cell is in a state of constant readiness to respond to its environment. Though one might wonder about the cost of hydrolyzing mRNA so liberally, it is more economical and advantageous than making useless protein during tough times.

Modification and assembly of stable RNA

In the bacteria that have been most thoroughly studied, all of the stable RNA species (largely rRNA and tRNA) are transcribed into products that must then be further processed. In *E. coli*, the genes encoding the three rRNA molecules (5S, 16S, and 23S) contained in each ribosome are clustered and transcribed as a single unit, i.e., they are arranged in operons. This scheme helps ensure synthesis of rRNAs in the stoichiometry in which they appear in ribosomes. There are seven clusters of these ***rrn*** genes in *E. coli*, and the general structure of each operon is:

16S gene—spacer tRNA gene—23S gene—5S gene—distal tRNA

Bacteria are not noted for having redundant genes, so seven copies of rRNA genes is noteworthy. The extraordinarily high demand for the products of these genes is undoubtedly one basis for retention of this redundancy: during rapid growth, half of the cell's transcripts come from *rrn* genes. Not all bacterial species have this many *rrn* genes; the number is generally correlated with the cellular growth rate of the species, with the range being from one to a dozen per genome. Multiple copies by themselves cannot account for this extraordinary rate of rRNA synthesis; multiple highly active promoters are largely responsible. The inclusion of select tRNA genes within these *rrn* operons is curious and may simply be related to the fact that they encode the most abundant tRNAs.

Simple transcription of any of these *rrn* operons would produce a useless giant molecule—but this does not happen. As transcription proceeds, a group of **ribonucleases** cut and trim the growing product into usable rRNA and tRNA precursors (Fig. 8.11). The ribosomal precursors are then acted upon by dozens of enzymes that methylate or otherwise modify selected bases—10 such enzymes for 16S rRNA and 13 for 23S rRNA. Almost all of these modifications reside in the regions of the rRNA molecules that comprise their active sites within the ribosome. The implication is that evolution has sharpened the effectiveness of the protein-synthesizing machinery by these structural adjustments. They must be important, because *the cell invests far more genetic information encoding the modifying enzymes than encoding the RNA itself.*

The two subunits of the bacterial ribosome, 30S and 50S, assemble by a sequence of reactions whereby various ribosomal proteins (**r-proteins**)

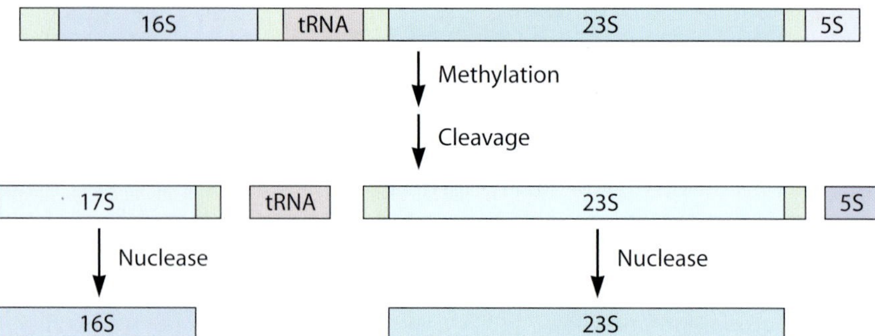

Figure 8.11 Processing of rRNA transcripts. The three kinds of rRNA arise from a common precursor molecule.

bind in a certain definite order to the rRNA core of the particle. To a large extent, the process seems to occur spontaneously, that is, without guidance or facilitation by other molecules. Ribosome assembly is a splendid example of how some biological macromolecules have built into them the ability for morphogenesis, the assembly of complex structures.

With the exception of the cases mentioned above, the genes for the approximately 50 tRNA species of the cell are in operons separate from those encoding rRNA. The need for processing these tRNAs to produce an active product is just as great. Dozens of enzymes act on the raw tRNA transcripts to convert them into functional molecules. Producing most tRNA molecules involves cleaving large polycistronic transcripts into a monocistronic precursor tRNA, removing any extra nucleotides at each end, adding terminal CCA ends (to which amino acids become attached) if these are not already in place, and finally, extensive modification of many base or ribose residues to produce methylated bases, as well as distinctive moieties practically unheard of in nucleic acids, such as inosine and pseudouridine. An astounding number of genes and their enzymatic products are involved in tRNA maturation. In all, there are more than 30 different modified nucleosides in the tRNAs of *E. coli* (more than 80 in the various microbes that have been studied), produced by more than 45 different enzymes. These modified nucleosides in mature tRNA have subtle functions. Individually, many can be mutationally eliminated without destroying the cell's ability to grow, but there is little doubt that they contribute to the efficiency of tRNAs' role in translation.

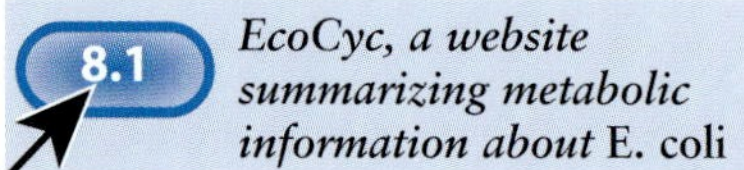

EcoCyc, a website summarizing metabolic information about E. coli

Polymerase collisions and genome organization

Picture a replication fork with its dual DNA polymerase replisome moving along the chromosome, replicating both strands of DNA at 1,000 nucleotides per second (see "DNA strand elongation" above). The process resembles a train moving along its track, but the track is not clear! Far from it. Ahead of the train are RNA-P molecules—perhaps thousands of them—engaged in transcribing genes. Some of these RNA-P molecules are transcribing one strand, some the other, depending on the orientation of the operon (or what is the same thing, on which strand is the **sense strand** for the particular operon). Recall (i) that RNA-P progresses only in the $5' \rightarrow 3'$ direction and (ii) that the two strands of DNA run in opposite directions. From these facts, it is clear that some of the RNA-P molecules ahead of the replisome are moving toward it, and some are running away from it. *In either case, collisions with the replisome are* *inevitable*, because RNA-P moves at only 45 nucleotides per second (see "RNA chain elongation" above), some 20 times slower than the DNA-P of the replisome, and therefore, even those traveling in the same direction as the replisome will quickly be overtaken.

One question occurs immediately: is the collision catastrophic? As one might surmise, the event is not lethal—but it does have consequences. The outcome turns out to depend on the direction of transcription, though in all cases the RNA-P is the loser. Head-on collisions are the worst; they halt replication, if only briefly, and abort transcription. Same-direction collisions have less impact; they only slow replication, and they permit continued transcription. Also, in any actively transcribed region there will be slightly fewer collisions if the RNA-P molecules are traveling in the same direction as the replication fork.

For both of these reasons (fewer collisions and milder consequences), there is an advantage in having genes oriented in the same direction as replication. Analysis of approximately 100 completely sequenced bacterial genomes confirms that this advantage has led to a strand bias in the locations of various genes. The results are complex, but it appears that in many species there is a strong bias toward having genes on the strand that orients them in the direction of replication (the *leading* as opposed to the *lagging* strand, as can be deduced from Fig. 8.4). In some cases, the bias applies chiefly to genes that are highly expressed; in other cases, it applies more to those with essential functions, whether they are highly expressed or not.

Collisions between RNA-P and the replisome are not the only factor influencing the organization of the genome. For example, genes near the origin of replication are in greater **copy numbers** than those near the terminus (described in "DNA Replication during the Cell Cycle" in Chapter 9). This gene dosage effect is another (of several) gradients, or nonsymmetrical features, of the chromosome that influence the evolution of the genome's organization.

PROTEIN

Special tempo and mode of protein synthesis

It is on the ribosome that so much of life depends, for the efficient and faithful production of proteins is how the cell's genetic information becomes operational.

Rapidly growing bacterial cells are bustling protein-making factories. Over half the prokaryotic cell's dry weight is protein, and depending on the growth rate, from 30 to 60% of the cell consists of machinery for making protein. This machinery, called the **protein-synthesizing system (PSS)**, consists of ribosomes; tRNA; amino acid-activating enzymes; proteins for initiation, elongation, and termination; enzymes that modify the completed product; and proteins that aid the folding and translocation of newly made proteins. Electron micrographs of thin sections of the cell (see Fig. 3.1) confirm visually the number of machines in this factory, because the overwhelming appearance of the cytoplasm is that of a compartment packed with ribosomes. Bacterial ribosome machines are not only numerous, they are smaller and faster than eukaryotic ribosomes. Also, in contrast to the 10 protein **initiation factors** used by eukaryotic cells, bacteria need only 3. The entire PSS is more streamlined in prokaryotes.

While the number of new protein chains being made and the speed of their growth are impressive, even more distinctive is the *mode* of prokaryotic protein synthesis. In the absence of a nuclear membrane, prokaryotic chromosomes (those of Archaea included), with few exceptions, are exposed directly to the cytoplasm, offering the possibility of an extraordinarily compact arrangement of their PSS. Since mRNA transcripts do not have to be translocated from a nucleus to a cytoplasm before functioning, *they can begin their work directing the synthesis of polypeptides long before they themselves have been fully transcribed.* As soon as a ribosome-binding site is formed at the 5′ end of the nascent mRNA transcript, protein synthesis starts. This strictly prokaryotic situation is called **coupled transcription-translation** (Fig. 8.12). Finally, another unique prokaryotic feature, already noted, is the fact that most of their mRNA is polycistronic, an arrangement that further facilitates efficiency and speed in making a variety of proteins.

Several aspects of this overall process are similar in eukaryotic translation, despite the striking difference brought about by the feature of cotranslation-transcription in prokaryotes. Nevertheless, the molecular structure and mode of action of the bacterial PSS is so different from the analogous components of the eukaryotic PSS that the differences form a strong basis for medical therapy of bacterial infections. Components of the bacterial PSS, especially the ribosome itself, are favored targets of many antibiotics.

What about protein synthesis in the Archaea? The story is interesting, because it is a blend of bacterial and eukaryotic features, along with some unique archaeal components. Archaeal ribosomes resemble those of bacteria in size, but not in detailed structure. As a result, the array of antibiotics (such as streptomycin, erythromycin, chloramphenicol, and **tetracycline**) that target bacterial ribosomes are ineffective against archaeal ribosomes. We shall note other differences in translation along the way.

On a weight-for-weight basis, few eukaryotic cells can match fast-growing prokaryotes in the rate of protein synthesis.

Figure 8.12 Coupling of transcription and translation. Translation (protein synthesis) takes place as the mRNA molecule is being made (transcription). This simultaneous synthesis can take place only in prokaryotes, where there is no nuclear membrane separating the two processes.

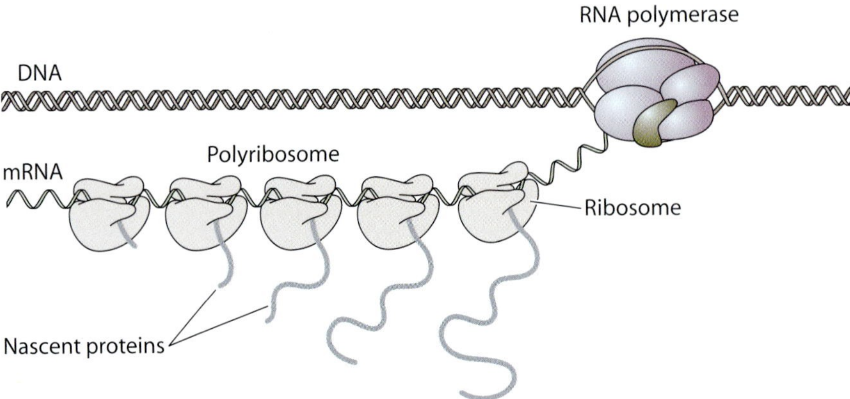

Translation

Protein synthesis involves **translation**—a word, like transcription, used by analogy with the processing of language. Whereas *transcription* in linguistics refers to rewriting information in the same language (copying a voice recording into a written document, for example), *translation* involves expressing information in a different language (Swahili to Japanese, for example). In biology, *transcription* occurs when the information inherent in a nucleotide sequence (the language of nucleic acids) in DNA is copied into a nucleotide sequence in mRNA; *translation* is required to convert this information into a sequence of amino acids (the language of proteins) in a polypeptide chain.

Translation, whether in linguistics or biology, depends on a translator. The biological translator that interprets a nucleotide sequence and converts it into an amino acid sequence consists, not of a single molecule, but of a large set of *pairs* of molecules: **aminoacyl-tRNA synthetases** and **tRNA** molecules. There are 20 aminoacyl-tRNA synthetases; each recognizes a particular amino acid and attaches it to specific tRNA molecules. These recognize the nucleotide **codons** (nucleotides read in groups of three, or **triplets**) in mRNA. By attaching the proper amino acid to each of the nearly 50 different tRNA molecules (the number varies in different species), the aminoacyl-tRNA synthetases prepare the tRNAs to match amino acids with the codons in mRNA (Fig. 8.13), thereby accomplishing the feat of translation. In a few cases, there are fewer tRNA molecular species than there are codons, and in these cases, the cell depends on wobble (the ambiguous reading of the third nucleotide) to utilize certain codons.

Figure 8.14 illustrates general features of tRNA molecules, including the critical three nucleotides called the **anticodon triplet** (within loop II), which pair with the codon in mRNA. Amino acid attachment to tRNA at the CCA end produces the activated form of the protein building block. Attachment by the aminoacyl-tRNA synthetase takes place in two stages. First, the amino acid reacts with ATP to form an enzyme-bound molecule of aminoacyl-adenylate:

$$\text{ATP} + \text{amino acid} \rightarrow \text{aminoacyl-AMP} + \text{inorganic phosphate } (P_i\text{-}P_i)$$

Second, aminoacyl-tRNA is formed by a transfer reaction:

$$\text{Aminoacyl-AMP} + \text{tRNA} \leftrightarrow \text{aminoacyl-tRNA} + \text{AMP}$$

Importantly, the second reaction is reversible, and if an incorrect amino acid is mistakenly attached to a given tRNA, the synthetase can remove it. Why is this **proofreading** important? Because whatever amino acid is attached to the tRNA is the amino acid that will be incorporated into protein at the position called for by the codon corresponding to the tRNA, not the amino acid. An incorrect amino acid on the tRNA results in an incorrect amino acid in the protein, and possibly an inactive protein.

Initiation of protein synthesis

Initiation begins at a **start site** on the mRNA with the formation of an **initiation complex** consisting of the two ribosomal subunits (50S and 30S)

First letter	Second letter: U	C	A	G	Third letter
U	UUU Phe	UCU Ser	UAU Tyr	UGU Cys	U
	UUC Phe	UCC Ser	UAC Tyr	UGC Cys	C
	UUA Leu	UCA Ser	UAA Stop	UGA Stop	A
	UUG Leu	UCG Ser	UAG Stop	UGG Trp	G
C	CUU Leu	CCU Pro	CAU His	CGU Arg	U
	CUC Leu	CCC Pro	CAC His	CGC Arg	C
	CUA Leu	CCA Pro	CAA Gln	CGA Arg	A
	CUG Leu	CCG Pro	CAG Gln	CGG Arg	G
A	AUU Ile	ACU Thr	AAU Asn	AGU Ser	U
	AUC Ile	ACC Thr	AAC Asn	AGC Ser	C
	AUA Ile	ACA Thr	AAA Lys	AGA Arg	A
	AUG Met	ACG Thr	AAG Lys	AGG Arg	G
G	GUU Val	GCU Ala	GAU Asp	GGU Gly	U
	GUC Val	GCC Ala	GAC Asp	GGC Gly	C
	GUA Val	GCA Ala	GAA Glu	GGA Gly	A
	GUG Val	GCG Ala	GAG Glu	GGG Gly	G

Figure 8.13 The genetic code. The possible triplet codons of mRNA are listed with the amino acids they encode. Nonsense codons, called ocher (UAA), amber (UGA), and opal (UGA), which cause termination of translation, are shaded and labeled Stop.

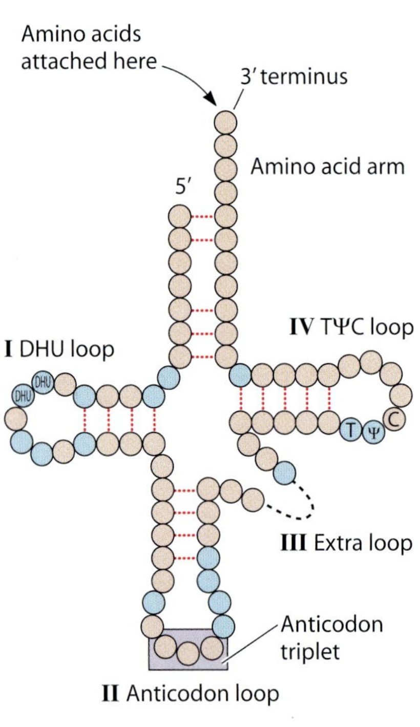

Figure 8.14 Generalized structure of tRNA. Ψ stands for pseudouridine; DHU stands for dihydroxyuridine.

plus a special initiator tRNA (usually methionine tRNA, recognizing AUG) to which a derivative of that amino acid, **formyl-methionine (fMet),** is attached in bacteria, though not other microbes. In a few cases, fMet is attached to a valine (GUG) or leucine (UUG) tRNA rather than the much more common methionine (AUG) tRNA. In any case, since these codons appear frequently within genes, how can they function as **start codons?** What prevents initiation from occurring wherever there is an AUG within a gene? The answer is that an authentic start codon is preceded approximately 10 nucleotides upstream by a sequence of 4 to 6 bases that is complementary to the 3′ end of the 16S rRNA of the 30S ribosomal subunit. These bases, called the **Shine-Dalgarno sequence,** are believed to help position the 30S ribosomal subunit at the proper site by hydrogen bonding with the 16S rRNA. Among the Archaea, start signals are based on the same principle of affinity with 16S rRNA, but the initial amino acid is methionine rather than fMet.

The steps leading to the formation of the initiation complex are shown in Fig. 8.15. The 70S ribosome released from a previous mRNA dissociates into its 30S and 50S components by the intervention of GTP and three accessory protein initiation factors (**IF1, IF2,** and **IF3**), which bind to the 30S subunit. These factors promote the association of fMet-tRNA and the binding of this complex to an initiation site on mRNA. This **30S initiation complex** is joined by a partner 50S subunit, and GTP is hydrolyzed, which helps eject the initiation factors and in some way stabilizes the complex. By now, the three factors have left, and the mature 70S initiation complex is ready to go.

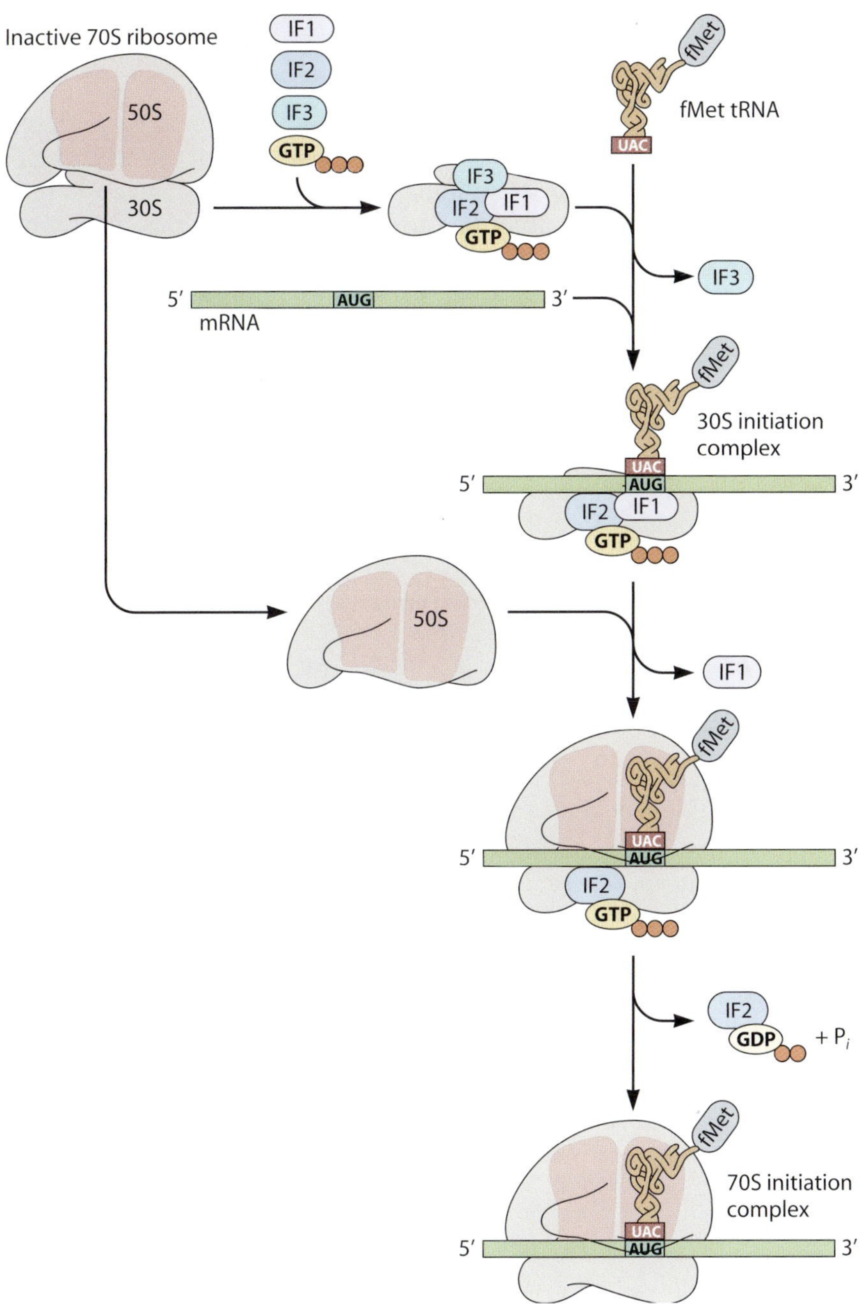

Figure 8.15 Initiation of bacterial protein synthesis. See the text for details.

Polypeptide chain elongation

Once the initiation complex has formed, a nascent polypeptide grows by an **elongation cycle,** which adds an amino acid at each round as the 70S ribosome advances along the mRNA. The cycle requires three protein **elongation factors** and the expenditure of two high-energy bonds (supplied by hydrolysis of two molecules of GTP). The growing polypeptide is bound to the ribosome through the most recent aminoacyl-tRNA to join the action. Soon after a ribosome has cleared the start site, another initiation complex forms. Repeated initiations result in the formation of a **polyribosome** consisting of a still-growing mRNA chain to which 70S ribosomes are continually being added at the translation start site near its **5′ terminus** (Fig. 8.12). The number of rounds of the elongation cycle corresponds to the number of amino acids in the protein.

One of the major triumphs of molecular biology, along with the breaking of the genetic code, has been the elucidation of the structure of the ribosome and the chemical and physical means by which it sews together the amino acids according to inherited instructions. We should savor some of this knowledge.

The nascent peptide begins to grow by the start of the elongation cycle (Fig. 8.16). Each round of the cycle results in the addition of one amino acid through the formation of one peptide bond (an amide linkage between the amino group of one amino acid and the carboxyl group of the next) in a carefully orchestrated process perfected over the billions of years of bacterial evolution. To visualize the beginning of the cycle, think of a 70S ribosome bearing a partially completed peptide chain, attached at a certain point to an mRNA. This ribosome has three sites that bind

Figure 8.16 Elongation cycle in bacterial protein synthesis. See the text for details.

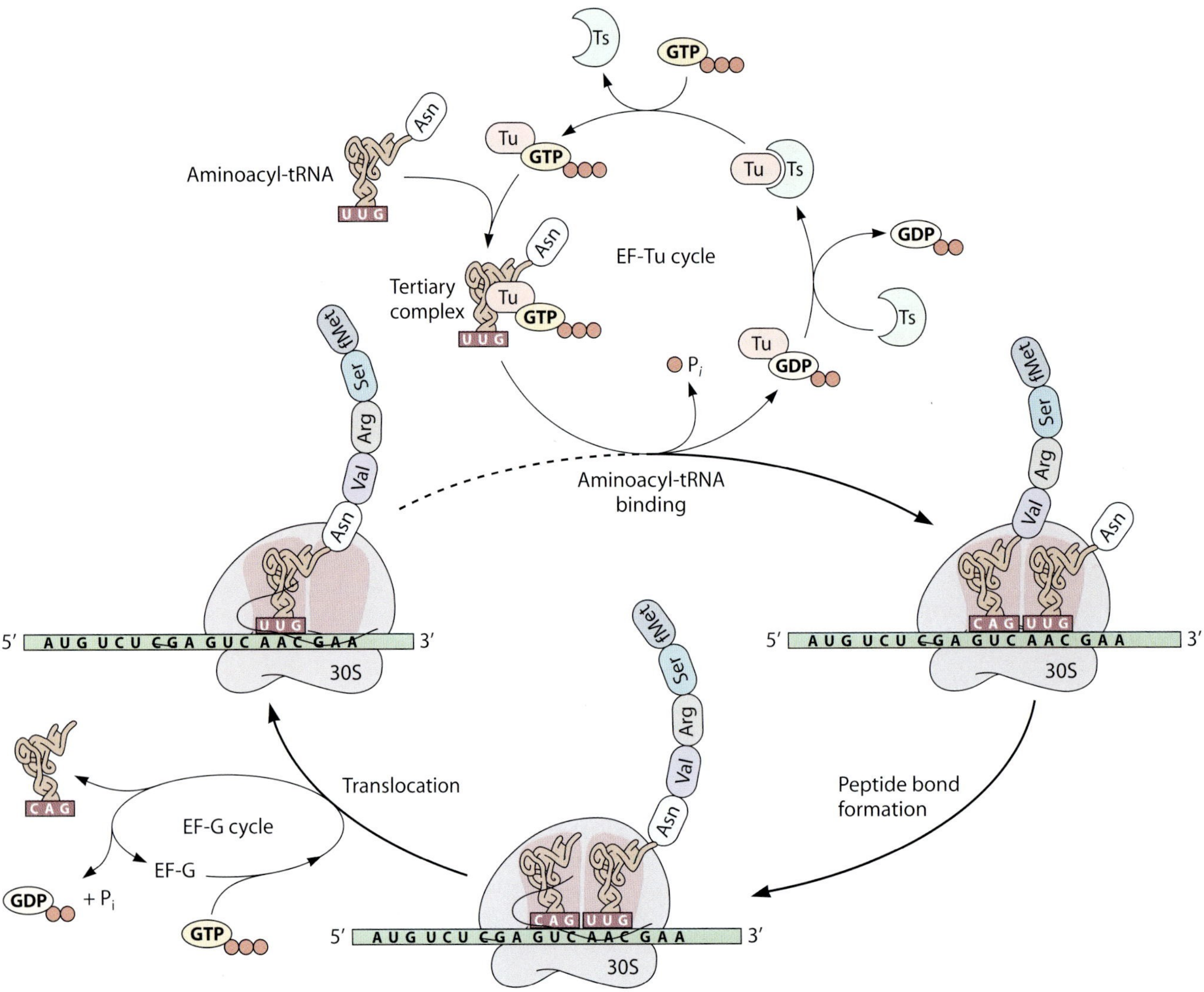

tRNA. The **A site** accepts the molecule of aminoacyl-tRNA, and the **P site** is occupied by a molecule of tRNA to which a partially completed peptide chain is attached. The **E site** (not shown in the figure) is occupied by an uncharged tRNA, which has just transferred its peptide chain to the new charged tRNA and is about to exit the ribosome. The anticodons of the tRNA molecules at all three sites are paired with codons of the mRNA.

In the first reaction of the elongation cycle—**aminoacyl-tRNA binding**—an aminoacylated tRNA with an appropriate anticodon binds to the exposed codon of the A site. This reaction requires two accessory proteins (elongation factor Tu [**EF-Tu**] and **EF-Ts**) and expenditure of energy in the form of the hydrolysis of the terminal phosphate of GTP. Aminoacyl-tRNA does not enter the A site alone but as a **ternary complex**, aminoacyl-tRNA–EF-Tu–GTP. Following hydrolysis of the GTP, the protein EF-Ts removes GDP from the complex so that EF-Tu is free to form a new ternary complex (this is called the EF-Tu cycle).

The tRNAs in the A and P sites are properly positioned so that the amino group of the amino acid attached to the tRNA in the A site lies next to the terminal acyl group of the peptide in the P site. The second step, **peptide bond formation,** breaks the acyl bond and forms the peptide bond—*a reaction catalyzed not by a protein but by a segment of the 23S RNA of the 50S ribosomal subunit* (which thus is an example of a **ribozyme,** an RNA molecule with catalytic activity). This action results in transfer of the peptide (now one amino acid longer) to the tRNA in the A site. The uncharged tRNA is then moved from the P to the E site during the third step of the cycle, **translocation,** whereby the peptide-bearing tRNA is transferred to the P site and the ribosome is moved one codon down the mRNA. Translocation requires the participation of the third accessory protein, EF-G, and hydrolysis of a second GTP (the EF-G cycle). This completes the elongation cycle.

For all its complexity, the cycle is speedy, adding approximately 15 amino acid residues to the nascent chain per second (in *E. coli* growing at 37°C). However, speed is not the only feature. You may recall that the rate of synthesis of mRNA is approximately 45 nucleotides per second. Since ribosomes advance along the mRNA at a speed of 15 codons per second, and since a codon consists of three nucleotides, *the ribosomes move along the message at the same rate that the message is being made.* Thus, each mRNA molecule is coated with a parade of ribosomes busily translating the growing mRNA, and the ribosome leading the parade is keeping pace with RNA-P at the transcription bubble—a very elegant process (Fig. 8.12). One consequence of this synchrony is that, as long as translation is proceeding, there is never a large segment of mRNA exposed to nucleases.

Elongation of nascent peptides proceeds similarly in the Archaea, though once again, the components of the system differ structurally from their bacterial counterparts. The archaeal EF-G is more like the eukaryotic than the bacterial factor.

Termination of translation

Termination is not spontaneous. It requires two events (hydrolysis of the peptidyl-tRNA and release of the completed peptide) and is triggered when the ribosome encounters one of three **termination codons** (UAA, UAG, or UGA; also called **nonsense codons,** because ordinarily no tRNA can read them, and hence they convey no sense). One fascinating exception, evolved only in bacteria, is the use of UGA for the tRNA for selenocysteine (cysteine with a selenium atom in place of sulfur). Apparently, the context in which UGA appears (that is, the kinds of codons surrounding it) determines whether it will function as a **stop codon** or as a sense codon calling for selenocysteinyl tRNA.

What happens next depends on the actions of two accessory protein factors. One, **release factor 1** (**RF-1**; or **RF-2**, depending on the termination codon encountered), binds to the A site and activates a **peptidyl transferase** to hydrolyze the polypeptide from the tRNA in the P site. The second, **RF-3,** removes the other release factor from the ribosome. The 70S ribosome then dissociates into its subunits, which are recycled to translate another mRNA molecule.

Processing of proteins

To be precise, translation of an mRNA produces a *polypeptide*—a specific linear array of amino acids in peptide linkage. A polypeptide becomes a *protein* by folding into its mature three-dimensional shape, sometimes after one or more chemical modifications of some residues. Molding a polypeptide into a protein, however, does not wait until the termination of translation; some folding and modification occur while the peptide chain is growing.

Three processes convert a polypeptide chain into a working protein, whether it has a catalytic, a regulatory, or a structural role. Covalent **modification** cleaves extra residues from either end of the polypeptide chain and may covalently add other substituents to the primary peptide structure. Before or after these modifications, the chain must undergo the process of **folding** to change it from a random linear molecule into its mature, active, three-dimensional shape. And third, the protein must be **translocated** to its proper location in, on, or outside the cell.

Covalent modification

Cleavage of amino acid residues from the N termini of proteins is fairly common and may involve simply the initial f-Met residue or removal of sizeable leader peptides that have served various functions, including **signal peptides** that function in the translocation (export and secretion—to be dealt with later) of many proteins. Formation of S-S bonds is another example of covalent modification. In *E. coli*, and probably most other bacteria, S-S bonds do not form within the cytoplasm because it is highly reduced. They form within certain proteins after they have been exported to the periplasm, which is less reducing. (One exception, and there could well be more, is that S-S bonds do form within the more oxidizing cytoplasm of cyanobacteria.)

Finally, addition of phosphate, methyl groups, nucleotides, fatty acids, sugars, or in a few cases, cyanide groups is quite common (Table 8.2). These modifications fall into two functional classes. The first class simply completes protein molecules that need special chemical groups for their functions (such as lipoproteins or glycoproteins). We refer to these as **assembly modifications.** The second class is **modulation modifications,**

Table 8.2 Examples of bacterial protein modifications

Modification type	Protein[a]
Chiefly assembly reactions	
Signal peptide cleavage	Secreted proteins
Formation of S-S bonds	Many proteins
Addition of lipid moiety	Murein lipoprotein
Addition of sugar moiety	Membrane glycoproteins
Attachment of prosthetic group	Many enzymes
Chiefly modulation reactions	
Phosphorylation	Ribosomal protein S6; isocitric dehydrogenase; many proteins with regulatory functions
Methylation	Chemotactic signal transducers
Acetylation	Ribosomal protein L7
Adenylylation	Glutamine synthetase

[a]Examples are drawn from *Escherichia coli*.

which regulate (or modulate) the activity of the mature protein. They provide flexible adjustments in protein structure and activity, and most are reversible. More will be said about them in Chapter 12.

Protein folding

A general assumption has been that polypeptides have information in their primary amino acid sequence structure to direct their folding into a mature, biologically active protein. This notion is not entirely correct. Most polypeptides can fold into any of several final three-dimensional forms, only one of which may be biologically functional. The path from the nearly one-dimensional polypeptide chain to the final protein has not been thoroughly mapped, but much has been learned in the past decade. In bacteria, the impetus and occasion for most of the discoveries about protein folding have come from research on the **heat shock response** (see Chapter 12). The story has many fascinating details, but here we shall simply summarize the main aspects of the bacterial protein-folding pathway quite apart from its modulation by environmental stress.

The protein-folding pathway shown in Fig. 8.17 operates from two principles. First, left to its own devices, a polypeptide will likely misfold, and therefore it should not be left alone for long. Second, misfolding must be corrected quickly. Bacteria contain large quantities of a dozen or so proteins appropriately called **chaperones.** They associate temporarily with polypeptides as they are folding in order to modulate the speed and path of the process. (In fact, there are several families of these molecules, and they are critical in the life of all cells, prokaryotic and eukaryotic alike. The chaperone families have been strongly conserved in all living organisms.) In *E. coli*, the first chaperone encountered by a polypeptide is the ribosome-associated **trigger factor.** This multidomain protein has a **peptidyl proline isomerase** activity and mediates *cis-trans* conversions of proline peptidyl bonds in the growing polypeptide. Proline residues introduce bends in the secondary structure of protein chains, and trigger factor helps ensure that the proline bonds are appropriately rotated. Once off the ribosome, a polypeptide still needing assistance is joined by

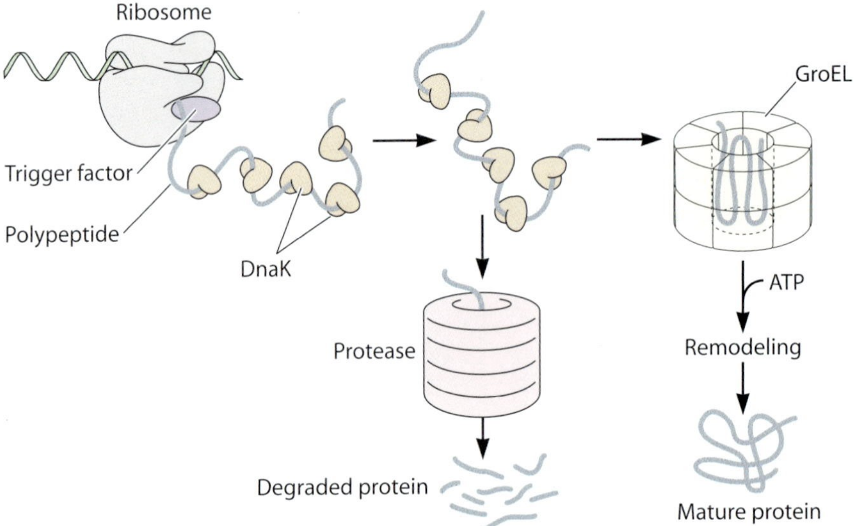

Figure 8.17 Protein-folding pathway. See the text for details.

another chaperone, **DnaK,** which, with the aid of companion chaperones, has the ability to remove small misfolded hydrophobic regions. Some maturing polypeptides need further assistance, and this is provided by a folding machine consisting of GroEL and GroES. GroEL molecules associate to make very large molecular chambers within which polypeptides can mold themselves, protected from proteases; GroES controls the entrance to the GroEL chamber.

It is important that the cell have an alternative to the path of chaperone-assisted folding. Some polypeptides accumulate in a useless misfolded state, either through errors of synthesis or as a result of damage caused by environmental conditions (Fig. 8.17). These misfits are speedily destroyed by proteases, and their amino acids are reused.

Translocation of proteins

Properly folded protein molecules in aqueous environments, such as the cytoplasm or the periplasm, have a hydrophilic surface; most hydrophobic amino acid residues are buried in the protein interior. Were it otherwise, such proteins would aggregate and precipitate; hydrophobic proteins, such as those imbedded in the cell's lipid membranes, are notoriously insoluble in aqueous solutions unless coated with a detergent. With this hydrophilic surface, protein molecules cannot readily enter or pass through the hydrophobic phospholipid membranes, yet this must be done. *The bacterial cell membrane contains over 300 proteins* with a large variety of metabolic functions (see Chapters 2 and 6), whereas *the periplasm and outer membrane of gram-negative bacteria have another hundred or more proteins not found in the cytoplasm.* Furthermore, gram-positive and, to a lesser extent, gram-negative bacteria secrete proteins (enzymes and toxins) into their environments. Some bacteria even inject proteins directly into eukaryotic host cells (see Chapter 20). How is this accomplished?

One's first thought may be that proteins entering membranes are exceptionally hydrophobic. This notion has a kernel of truth. Many cell membrane proteins are strongly hydrophobic, projecting only hydrophilic tails into the cytoplasm or the exterior; the hydrophobic loops or tails that may be in the membrane can be small or quite extensive. On the other hand, many periplasmic and secreted proteins are as hydrophilic as common cytoplasmic ones. There must be another answer. In fact, bacteria have evolved at least a half dozen means for moving proteins into and through membranes. **Translocation** (movement of a molecule from one location to another) has become a fertile research topic, and novel mechanisms are being continually reported for both **protein export** (translocation out of the cytoplasm) and **protein secretion** (translocation through all membranes into the external environment). A word of caution is indicated here: nomenclature in the subject of translocation is not yet consistent, and the terms "export" and "secretion" (and even "excretion," used by some authors) are not employed uniformly. We shall use "translocation" to mean the transfer of a protein from one compartment to any other destination, "export" to indicate passage out of the cytoplasm, and "secretion" to indicate passage to the external environment. Proteins destined for final placement in the cell membrane, periplasm, or outer membrane are deemed to be *exported*; those being sent into the environment are said to be *secreted*.

Two questions arise immediately. How does the cell recognize the appropriate destination of a protein? And, once a protein needing translocation is recognized, how is movement accomplished? Recognition is an enormous puzzle, because it entails not just knowing that a protein must be moved but also what its destination should be. Is the protein meant to reside in the cell membrane, the periplasm, or the outer membrane, or is it to be secreted into the environment? The answers are beginning to be learned. As usual, most information comes from studies of *E. coli*.

Insertion into the cell membrane is the simplest case. As we have said, proteins or the portions of them that reside within membranes are hydrophobic. This property by itself is sufficient for these proteins to become incorporated into the cell membrane; they do not need the help of other proteins. However, export of proteins across the cell membrane is considerably more complex. First, let us consider how this occurs via the widely dispersed **Sec** system, which is responsible for the export of most periplasmic and outer membrane proteins of gram-negative bacteria and the secreted proteins of gram-positive bacteria. Recognition by the Sec system occurs early, usually while the peptide is still being made on the ribosome. A recognition site, called a **signal sequence,** is at the protein's N terminus and therefore is the first part to be synthesized; it consists of 15 to 30 amino acids in a distinctive but variable sequence (Fig. 8.18). One of the proteins of the Sec system (SecA) recognizes this sequence, attaches to it, and leads the protein to a cell membrane-spanning channel composed of three other Sec proteins (SecY, -E, and -G) (Fig. 8.19). Because this channel has a hydrophilic inner surface, hydrophilic proteins can pass through it. In transit to the membrane, the protein is coated with yet another Sec protein (SecB), which is a specific chaperone for exported proteins that prevents them from folding, thereby keeping them in an extended form so they can thread their way through the channel. During this passage, a "signal peptidase" (Lep) clips off the protein's signal sequence; it has now completed its job. Passing a protein through the

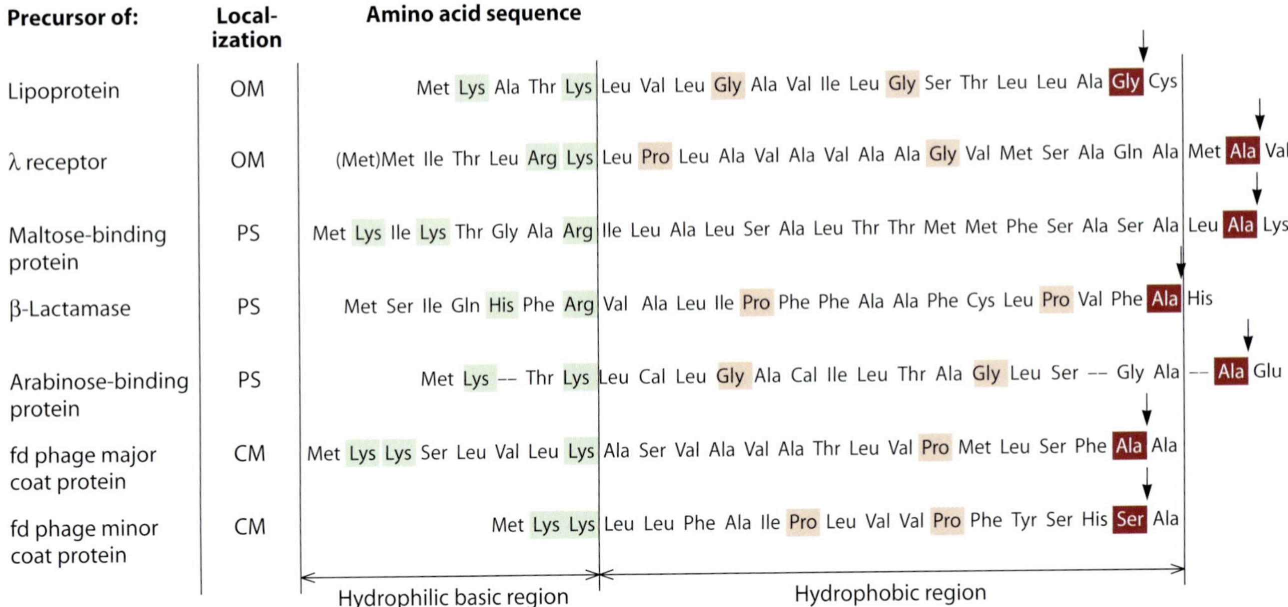

Figure 8.18 Signal sequences. The amino-terminal ends of the precursor forms of seven proteins that enter the envelope of *E. coli* are shown. Basic amino acids of the hydrophilic region are in grey squares, glycine and proline residues of the hydrophobic region are in grey circles, and the residues at the cleavage sites (arrows) are shown in red. OM, outer membrane; PS, periplasmic space; CM, cell membrane.

SecYEG channel is not free; it requires energy in the form of proton motive force, ATP, or both. SecG and SecA hydrolyze a molecule of ATP, providing sufficient energy to force an approximately 20-amino-acid length of protein through the channel. The protein now folds in the periplasm (or the environment in the case of gram-positive bacteria). All this is shown in Fig. 8.19A.

Proteins that are to be inserted into the cell membrane rather than secreted from the cell commonly follow other paths (Fig. 8.19B). The N termini of these proteins display a signal anchor sequence that is recognized by a signal recognition particle consisting of a protein (Ffh) and a small (4.5S) RNA molecule (ffs). The signal recognition particle escorts the nascent protein, while it is still being translated, to a protein receptor in the cell membrane, FtsY. Translation of the future membrane protein continues at FtsY, during which some proteins are inserted into the membrane with the aid of SecYE (leftward path in Fig 8.19B) while others are inserted without SecYE mediation (rightward path).

In gram-positive bacteria, proteins are secreted directly, but proteins secreted from gram-negative bacteria must traverse the outer membrane as well. In certain cases, the two processes are sequential: the Sec system exports them to the periplasm, and then a different system takes them across the outer membrane. Such outer membrane transporters can be quite complex; for example, 14 different proteins are needed to transport an amylase across the outer membrane of *Klebsiella*. Other proteins (e.g., toxins, bacteriocins, and proteins that build flagella and pili) are transported across both membranes by systems specific for each protein.

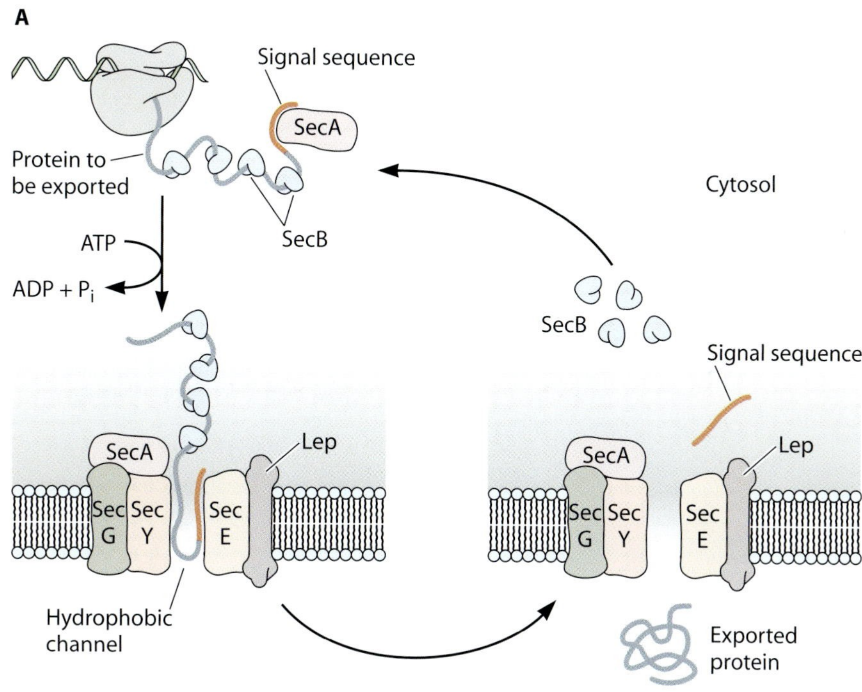

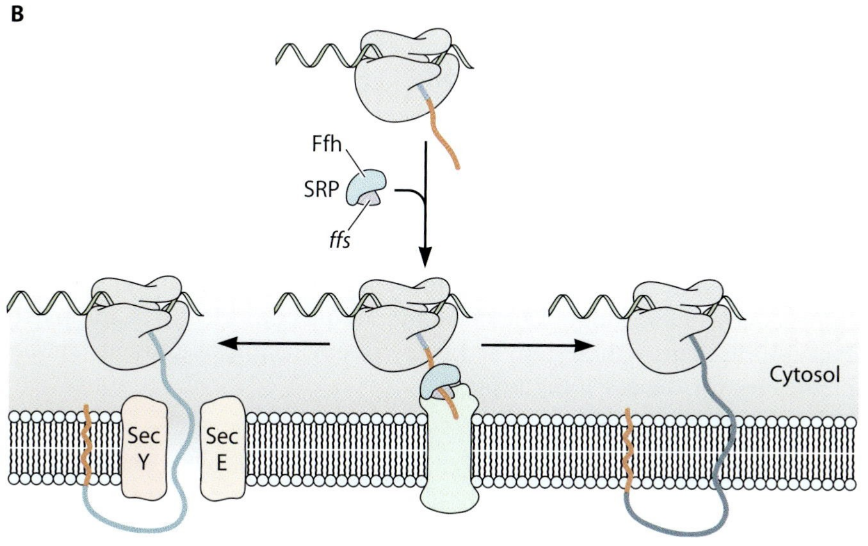

Figure 8.19 Two modes of Sec-mediated protein translocation. **(A)** Export across the cell membrane. **(B)** Insertion into the cell membrane. SRP, signal recognition particle. See the text for further details.

Not all the mechanisms fall neatly into one of these categories. A fascinating instance of translocation that is neither cotranslational nor Sec dependent occurs in many pathogenic bacteria (*Yersinia*, *Salmonella*, *Shigella*, and *Pseudomonas* species). It is called **contact-dependent secretion.** Secretion is triggered upon contact of the bacterium with a host cell and results in the direct injection of the protein into the host cell, where it participates in the infection. This manner of secretion, known also as **type III secretion,** is mediated by a large complex of approximately 20 proteins (including a chaperone for the protein to be secreted and ATP-binding protein to energize the system) (see Chapter 20).

Table 8.3 Some protein secretion systems in gram-negative bacteria

System	Name	Sec dependency	Description
Type I	ABC exporter	Sec independent	Consists of three proteins, one of which is an ATP-binding cassette; sometimes called the ABC transporter; operates on proteins lacking a signal sequence
Type II	Two-step system	Sec dependent	Moves protein into the periplasm by the Sec system in one step; then, 14 accessory proteins move it across the outer membrane in a second step
Type III	Contact-dependent system	Sec independent	Involves 20 or more proteins (including an ATPase) that span the envelope from cytoplasm to surface; activated by contact with a host cell, and then injects a toxin protein into the host cell directly
Type IV	Conjugal transfer system	Unknown	Uses same system that transfers DNA from some bacteria to eukaryotic cells
Type V	Autotransport	Sec dependent	Two steps involved, as in type II, but no helper proteins needed for transfer through the outer membrane

The repertoire of ways to export proteins seems to have no end; new ones continue to be discovered. A partial list is given in Table 8.3 and illustrated in Fig. 8.20.

ENVELOPE FORMATION

Challenges of envelope formation

The bacterial envelope carries out a large number of critical functions—far more than the relatively simple surface of eukaryotic cells. Appropriately, its structure is extremely complex, in either its gram-positive or -negative form (see Chapter 2). Assembly of the bacterial envelope has a typically bacterial flavor, partly because its various components—cell membrane, outer membrane, wall, periplasm, capsule, and appendages—have unique chemical and architectural features and partly because rapid growth in a great variety of environments presents special challenges for the assembly process.

Consider these challenges.

- Topologically, all layers of the envelope are closed surfaces and must be physically continuous for cell integrity and viability, yet they must always expand during cell growth by the addition of new material.
- All constituents of the envelope must grow coordinately and with special regard for their final destinations.
- Proteins, lipids, and complex polysaccharides must be correctly incorporated into their final destinations, which might be, in the case of gram-negative bacteria, the cell membrane, the periplasm, the outer membrane, or the exterior.
- Newly assembled envelopes must facilitate cell division.
- The envelope, as the first sensor of environmental change, must participate in the cellular response to these changes, in many cases by modifying its own structure or composition.

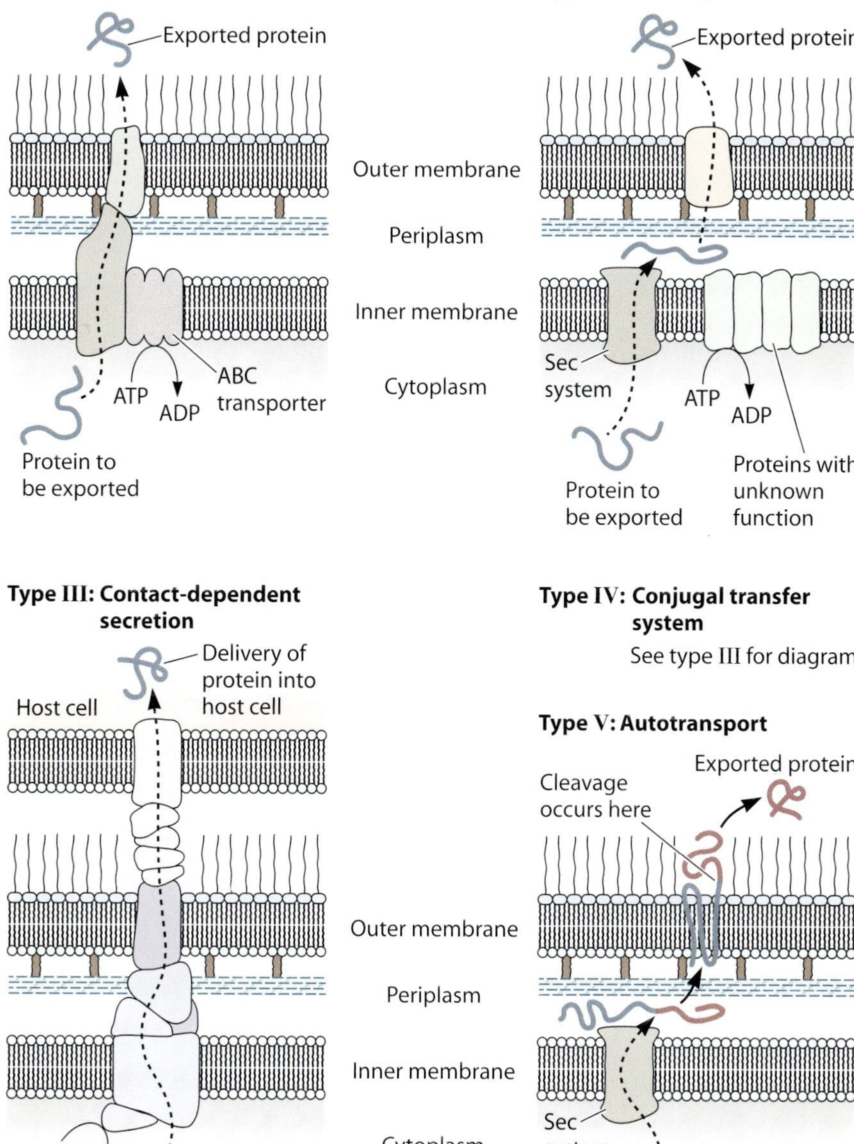

Figure 8.20 Several mechanisms for protein export. See Table 8.3 for details.

All this brings us to the question of how the envelope components are assembled with speed, accuracy, and adaptive flexibility. Microbiologists have only a general picture of this process. Many specific questions are unanswered.

Here, we present a general overview of what is entailed in constructing this complex part of the bacterial cell.

The cell membrane

It is not known with certainty whether the membrane is assembled at a few or many sites. Newly formed phospholipid molecules appear first in the inner leaflet of the cell membrane, and some must be translocated to become part of the external leaflet, though exactly how this is accomplished

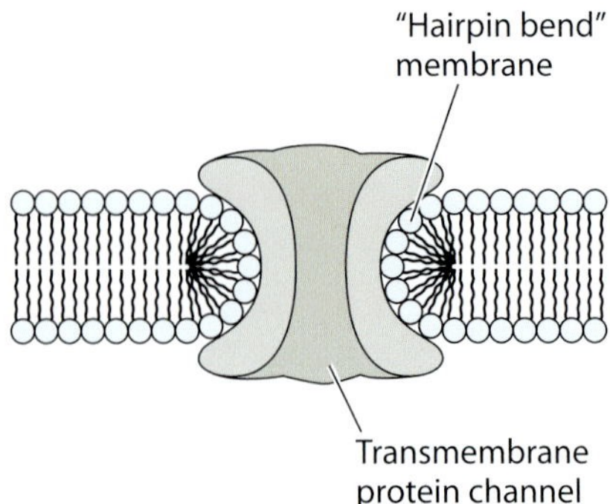

Figure 8.21 Incorporation of phospholipids into the cell membrane bilayer. See the text for details.

is not known. A hypothetical mechanism is shown in Fig. 8.21. Interestingly, different phospholipids are distributed asymmetrically among the leaflets of bacterial membrane bilayers. Little is known about what ensures such asymmetry.

Insertion of membrane proteins occurs by the mechanisms discussed above.

The cell wall

Murein is assembled in a series of defined steps: synthesis of precursors in the cytoplasm, binding to a lipid carrier (a 55-carbon-long alkyl chain with a phosphate at the end, called undecaprenylphosphate) for transfer across the cell membrane, and polymerization by addition to the preexisting molecular scaffold (Fig. 8.22). In some bacterial species, wall assembly takes place at many sites along the cell surface, while in others,

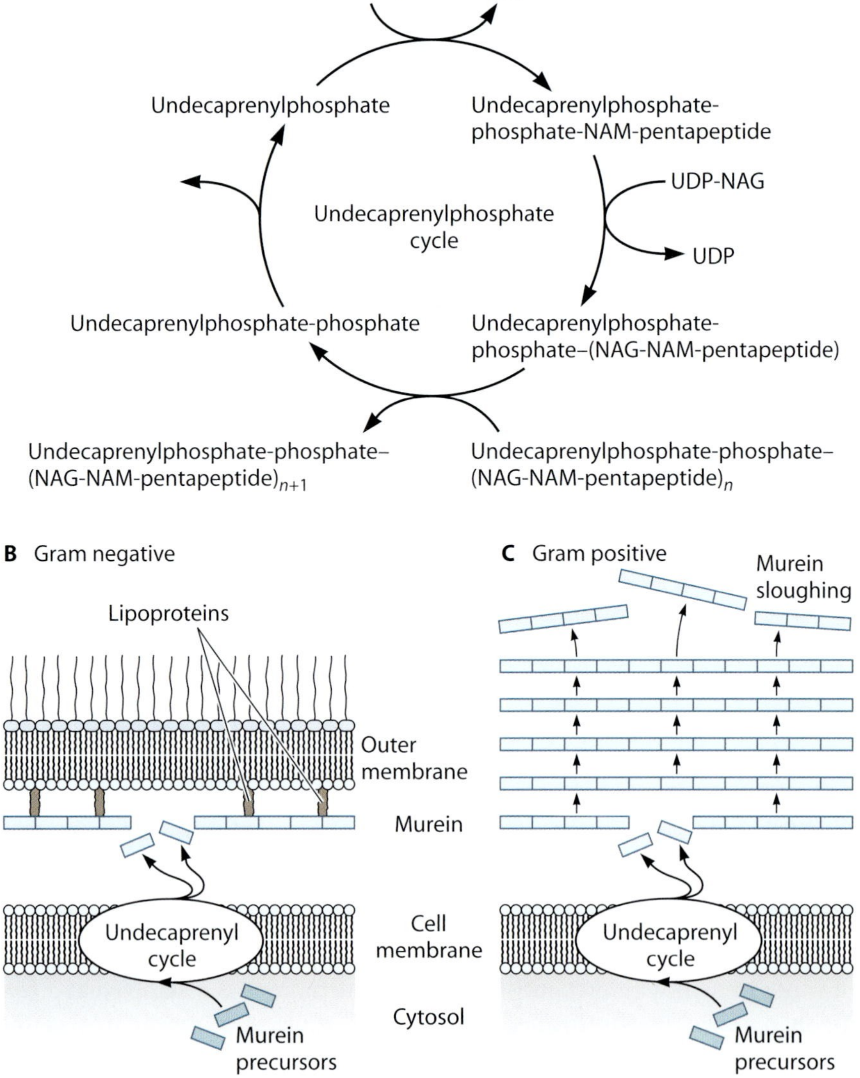

Figure 8.22 Synthesis and assembly of the murein wall. (A) Undecaprenylphosphate cycle. Synthesis of the murein precursor undecaprenyl-P-P-[NAG-NAM-pentapeptide] and its polymerization (shown in red) into glycan is shown. P, phosphate; NAM, *N*-acetylmuramic acid; NAG, *N*-acetylglucososamine; UDP, uridine diphosphate; UMP, uridine monophosphate. **(B)** Wall assembly in gram-negative bacteria. The diagram shows a region where the murein precursor units (NAG-NAM-pentapeptide) are shuttled through the cell membrane on undecaprenylphosphate, polymerized into glycan in the periplasm, and covalently attached to the existing murein sacculus. **(C)** Wall assembly and turnover in gram-positive bacteria. As in gram-negative bacteria, the undecaprenyl cycle operates in the cell membrane, turning out murein precursor units that are polymerized and attached to the preexisting murein sacculus. In many species, the newly formed murein progresses from its site of synthesis to the periphery of the cell, where it is eventually hydrolyzed and sloughed off.

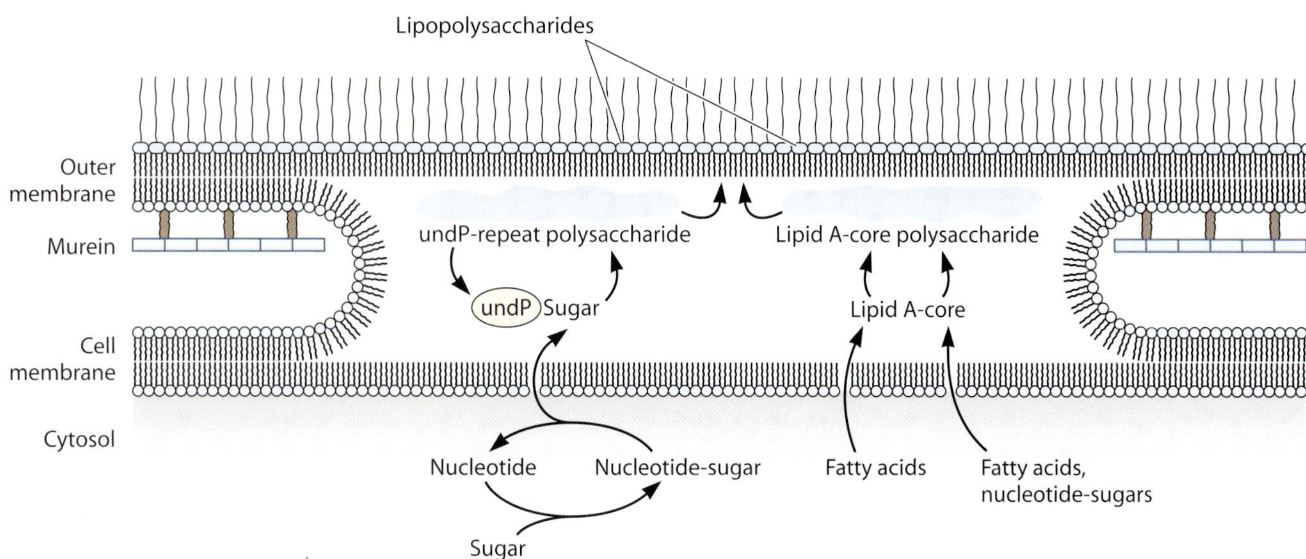

Figure 8.23 Lipopolysaccharide synthesis and assembly. Lipopolysaccharide subunits are synthesized in the cell membrane by two parallel processes. One path makes the repeating polysaccharide side chain, built up on undecaprenyl phosphate (undP), the same carrier that functions in murein polymerization. The other makes the core polysaccharide built up on lipid A, which functions as both a primer and the carrier that transports the core across the cell membrane. These two sets of precursors are made on the inner surface of the cell membrane, translocated by proton motive force to the outer surface, and covalently joined there.

such as some gram-positive cocci, this process is restricted to the equator of the cells.

The lipoproteins of gram-negative envelopes are modified after their synthesis by the addition of three fatty acids. This hydrophobic anchor within the outer membrane facilitates the covalent bonding of one-third of the lipoprotein molecules to the underlying murein, thereby providing attachment of the wall to the outer membrane (Fig. 8.22).

The outer membrane

The outer membrane carries to an extreme the asymmetry of the cell membrane, as shown by the lipopolysaccharide being present solely in the outer leaflet. The components of this hallmark gram-negative molecule are made by separate pathways that independently synthesize the lipid A core portion and the side chain polysaccharide. These two constituents are then joined at the outer membrane. Assembly of phospholipids and proteins into the outer membrane requires energy from the proton motive force of the cell membrane. Lipopolysaccharide is thought also to assist in the incorporation of proteins. As shown in Fig. 8.23, if there were junctions between the outer and cell membrane, it would greatly assist us in visualizing the growth of these two structures. Unfortunately, firm evidence for such areas of fusion or adhesion has proven hard to come by.

Appendages

As mentioned in Chapter 2, flagella are assembled from the interior outward, component by component (basal body, hook, and filament). The long filament is made by the extrusion of flagellin molecules through the core of the structure and their polymerization at the growing tip. Pili, on the other hand, grow by addition at the basal end.

Capsules

Bacterial capsules are secreted as preformed polymers or are polymerized by extracellular enzymes.

CONCLUDING REMARKS

We have reviewed in the past four chapters the metabolic processes that enable a newly formed bacterial cell to double its components. The large cell is ready now for the biggest event in its relatively short life: becoming two living cells. In a sense, the mother cell we have been following is about to cease existence and be represented by its two daughters. The intricate events that bring about this feat are the topic of the next chapter in our story.

STUDY QUESTIONS

1. It is said that prokaryotic cells are "streamlined" for rapid growth. What characteristics of macromolecule synthesis illustrate this feature of the prokaryotic cell?

2. Some aspects of DNA replication are different in prokaryotes and eukaryotes; others are the same. List the similarities and differences.

3. Methylation of DNA plays major roles in prokaryotic cells. Name these roles, and briefly indicate how each functions.

4. Explain how prokaryotic cells can divide faster than their chromosomes can replicate.

5. The average lifetime of mRNA molecules in prokaryotes is a few minutes. This is clearly costly in terms of energy. What might be the selective pressure for rapid turnover?

6. It took many years for microbiologists to understand how RNA-P in prokaryotes recognizes where to start transcription (i.e., what a promoter looks like). Why should this have been a difficult problem? What is the structure of prokaryotic promoters?

7. The prokaryotic chromosome is the site of much activity in a growing cell, including replication and transcription. Are these processes ever in conflict? If so, what is the outcome of the conflict?

8. Prokaryotic transcripts are almost always polycistronic. What does this mean? What is its significance?

9. What prevents ribosomes from initiating protein synthesis at AUG sequences within cistrons?

10. If one were to collect all the polyribosomes that were engaged at a given instant in producing the same protein in a bacterial cell, would they be the same size (i.e., each have the same number of ribosomes)? Explain your answer.

11. Ribosomes protect prokaryotic mRNA from degradation during protein synthesis. Explain how this protection depends on the relative speeds with which RNA-P and ribosomes work. Predict the result if a mutant RNA-P were to arise that worked twice as fast as normal.

12. Rather than a single efficient mechanism for exporting proteins from the prokaryotic cytoplasm, evolution has produced a great number of mechanisms, falling into at least five different classes (Table 8.3). Why must biology be so complicated?

chapter 9 the cell division cycle

INTRODUCTION

A bacterial cell leads an ephemeral existence. Its life span begins on the division of its mother cell and ends shortly thereafter with its own division. This period of time is the **cell cycle** of the individual bacterium. What happens to a bacterial cell as it matures from inception to duplication? Do biochemical events take place in a sequential series or concurrently? The answer depends on which event we are considering. Two major events, **DNA replication** and **cell division**, dominate the cycle. Each produces a large structural unit, a progeny chromosome or a progeny cell, and each takes up a considerable portion of the cell cycle. By contrast, those components that are present in great numbers (e.g., individual enzymes or ribosomes) are made in hundreds or thousands of short overlapping bursts throughout the cell cycle.

STRATEGIES FOR STUDYING THE BACTERIAL CELL CYCLE

In spite of its fundamental importance, the bacterial cell cycle has been studied less intensely than the growth of populations. In a bacterial culture, the individual cells take about the same amount of time to grow and divide, but under normal circumstances, they are not in synchrony. Because they are dividing **asynchronously**, a sample of a growing culture contains cells in all stages of the cell cycle and cannot provide information about any particular stage.

How can one study the cell cycle? A single bacterium is obviously too small for conventional chemical measurements. However, specific constituents of an individual cell, such as certain DNA sequences or specific proteins, can be visualized and located within a cell by tagging them with **fluorescent dyes**. It is also possible to label a given protein by fusing its gene to the gene encoding **green fluorescent protein** (GFP). The locations and movements of fluorescence-tagged constituents can be followed in living bacteria.

However ingenious such studies with single cells may be, it is sometimes desirable to carry out measurements on a larger scale, which requires the use of synchronously dividing populations. Because bacteria do not divide synchronously in normal culture, a synchronously growing culture must be either induced or selected by artificial means. Care must be taken to avoid altering the cells' physiology in the process. For that reason, methods that induce or force synchrony (such as temperature shifts and inhibitors of cell division or macromolecule synthesis) are inappropriate for most purposes. Selecting cells that are of the same age (and size) avoids many problems. It is possible to retrieve a sample of the youngest cells without disturbing their physiology. Such a culture will divide almost synchronously for a few doublings before the natural variation in the length of the cell cycle among individual cells makes the culture again become asynchronous. The device used to obtain young cells is called the **baby machine**. This technique involves filtering a culture through a thin membrane filter that retains the bacterial cells, followed by inversion of the filter over the holder and the slow pouring of fresh medium through it (Fig. 9.1). Most cells stick to the filter by electrostatic attraction; those that do not are simply washed away and discarded. A variation on this theme is to make the bacteria stick to glass beads via their flagella and then to stop flagellin synthesis.

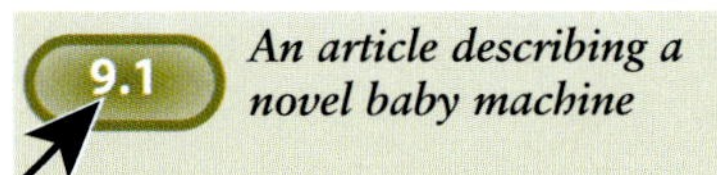

An article describing a novel baby machine

When a stuck cell divides, one of its progeny cells is released. The other cell remains on the filter and continues to grow and divide, eventually releasing another progeny cell (a "baby") into the medium. This method sorts cells by *age*, not by *size*: the first cells to detach come from the oldest cells placed on the membrane, the next ones to detach result from the division of cells that were middle aged, and the last ones to be collected are from the youngest cells in the original culture. Collecting the effluent from a baby machine for a short time yields a culture that divides synchronously for a few doublings. Certain bacteria, such as *Caulobacter* (see Chapter 14), act as natural baby machines, because they stick to surfaces and release babies upon division.

Another technique that is useful in the study of the cell cycle is **flow cytometry**. Flow cytometers employ lasers to determine the distribution of cells in a population according to their content of a fluorescent tag, usually a fluorescent dye, sometimes linked to antibodies that bind to a particular cell constituent. Flow cytometers are used extensively with eukaryotic cells but can also be used to measure the contents of individual bacteria. This allows one to estimate the distribution of cell size and of any particular constituent, for example, DNA per cell (Fig. 9.2).

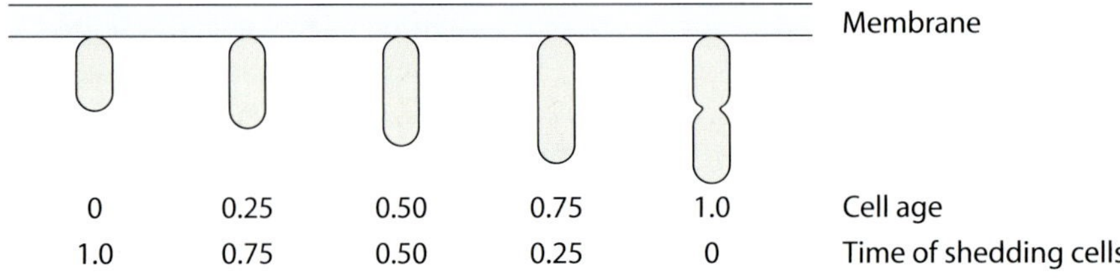

Figure 9.1 Kinetics of detachment of cells from a baby machine. The products of division of old cells are released first, followed by those of progressively younger cells. The numbers for cell age refer to fractions of a generation.

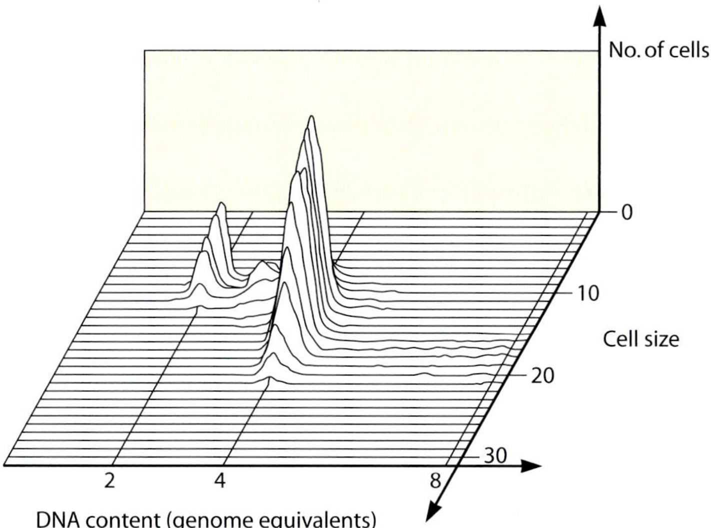

Figure 9.2 Analysis of a bacterial population by flow cytometry. *x* axis, content of DNA per cell; *y* axis, distribution of cell sizes; *z* axis, number of cells. The culture was treated with rifampin for 90 minutes before a DNA-specific fluorescent **probe** was added. The drug treatment allowed all chromosomes in the process of replication to terminate their cycles but not initiate new rounds. There are very few cells that have other than two or four genome equivalents. This suggests that within each cell, the chromosomes initiated their replication in synchrony.

However, the baby machine remains the method of choice for studying certain events that take place during the bacterial cell cycle, such as the synthesis of cell envelopes.

DNA REPLICATION DURING THE CELL CYCLE

In fast-growing bacterial cultures, DNA replication takes place over the greater part of the cell cycle. Only in slow-growing cells is there a period with no DNA synthesis. In *Escherichia coli* growing at 37°C, the replication period occupies about 40 minutes. Surprisingly, this time period is nearly the same in both fast- and relatively slow-growing cultures.

As described in Chapter 8, DNA replication in bacteria begins at a unique **origin** and ends at another unique region, the **terminus**, 40 minutes later. This introduces a conundrum. Many bacteria, including *Bacillus subtilis* and *E. coli*, can double more quickly than every 40 minutes. How can a cell grow and divide faster than the time it takes to make a genome's worth of DNA? The answer is that in such cultures *more than one round of replication goes on simultaneously*. In other words, several replication forks are traversing the chromosome at the same time (Fig. 9.3). When the replication fork going in one direction has just started out, another one preceding it along the same half of the DNA molecule will not yet have reached the terminus. Each chromosome will therefore contain more than one replication fork, a phenomenon known as **multifork replication**. Consequently, after cell division, *each newborn cell receives a chromosome that is undergoing replication in addition to a completed one*. Being born with more than one genome's worth of DNA is a uniquely bacterial characteristic: eukaryotic cells provide their

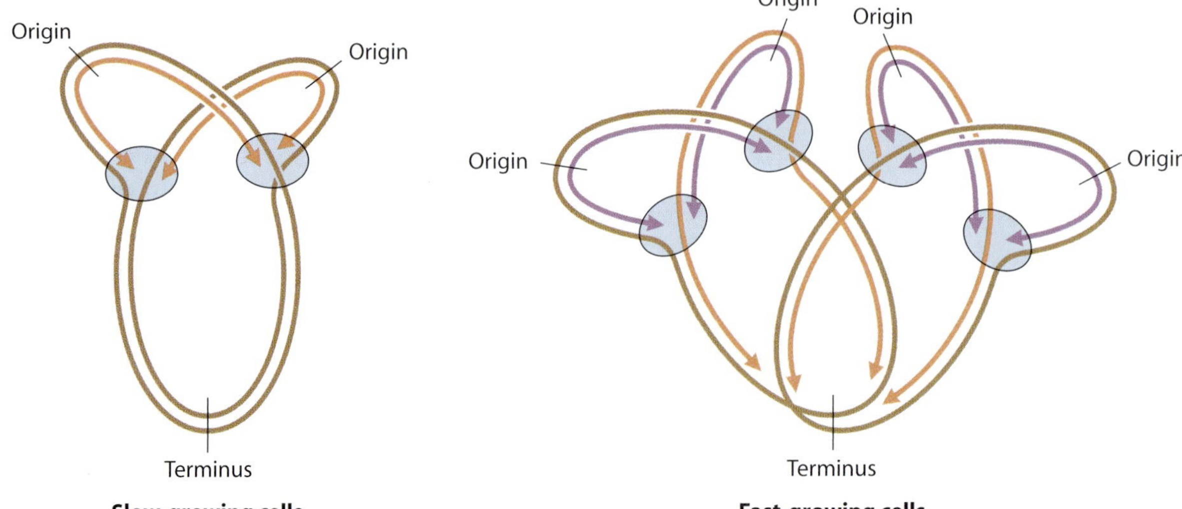

Figure 9.3 DNA replication in slow- and fast-growing *E. coli*. Replication starts at the origin and proceeds bidirectionally toward the terminus. The process takes about 40 minutes at 37°C. In a culture doubling every 20 minutes, initiation must take place every 20 minutes, that is, before the previous round of replication has terminated. This is known as multifork replication.

progeny cells with finished chromosomes only. This difference illustrates that the cell cycle has different meanings in the two domains. Eukaryotic cells tend to follow a precise clock, undergoing a complex and well-timed series of episodes, with **mitosis** and cytokinesis dominating the morphological picture. In the prokaryotes, although DNA replication and growth are linked, they do not follow a unique schedule. For example, fast-growing bacteria contain more *nucleoids per cell* than slow-growing ones. This means that, unlike in eukaryotes, their number of genomes per cell is readily adjustable and dependent on how fast they grow (see Chapter 3).

It is worth noting that as a consequence of multifork replication, the genes that replicate first, those located near the origin, will be present in greater number than those toward the terminus. Accordingly, the *ratio* of the genes near the origin to those near the terminus can be as high as 4, 8, or more in fast-growing cells and between 1 and 2 in cells growing slowly (Fig. 9.4). The position of genes along the chromosome, therefore, is not inconsequential and affects the level of their expression. The closer to the origin, the greater the amounts of gene products that can be made. Genes with products that must be produced in large amounts tend to be located near the origin. These include genes for ribosomal proteins and others involved in making the protein synthesis machinery. This effect, known as **gene dosage**, suggests that the position of genes on a chromosome is determined by natural selection.

HOW IS DNA REPLICATION REGULATED?

Bacteria growing rapidly must synthesize a larger amount of DNA per unit time than those that grow slowly. But, as we have said, the rate of travel of each replication fork does not vary greatly with the growth rate. How can this happen? The amount of DNA made in a cell cycle must

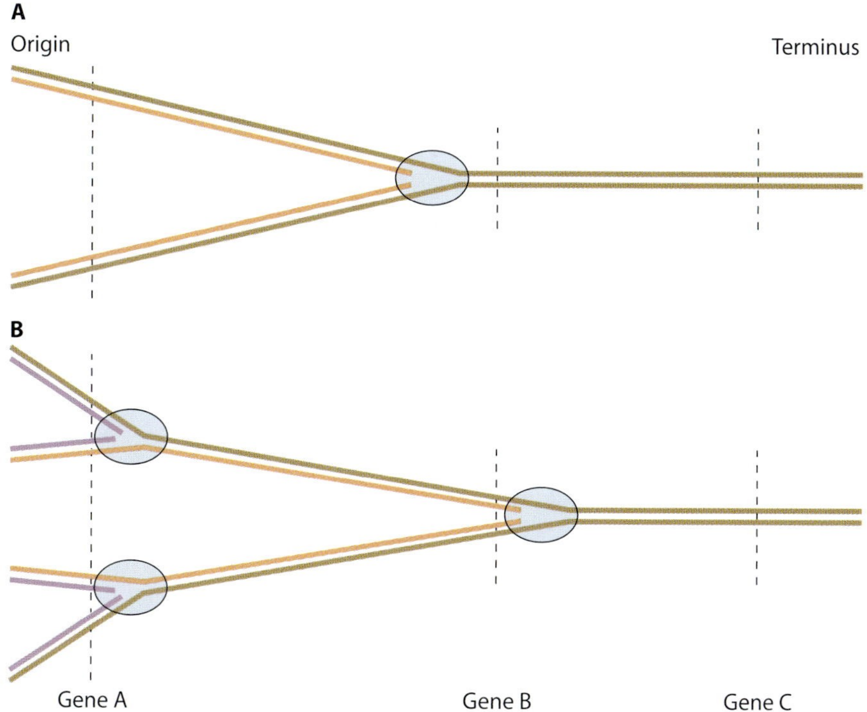

Figure 9.4 Multifork replication and gene dosage. (A) In slowly growing cells with a single replication fork, the ratio of genes close to the origin (gene A) is twice that of downstream genes (genes B and C). **(B)** In a cell with several replication forks, the ratio of genes near the origin is higher (2 for gene A/gene B, 4 for gene A/gene C). For simplicity, the chromosomes are drawn as linear rather than circular structures.

depend on the **frequency of initiation** of rounds of replication. The faster the cells grow, the more often they start a new round of replication, indicating that the *frequency of initiation of rounds of replication is a regulated process*. How is this regulation accomplished? We don't have a final answer, but an examination of the available facts and ideas is illuminating.

Experimental facts, not just theory, bolster the belief that initiation of DNA replication is a regulated process. For example, in bacteria containing more than one nucleoid, replication of the nucleoids initiates with an astounding degree of synchrony. This was demonstrated by flow cytometry measurements, which showed that a population of fast-growing *E. coli* consists of cells with mainly two, four, or eight nucleoids (Fig. 9.2). Very few cells had the odd numbers that would be expected from nonsynchronous replication. The initiation of replication of each nucleoid must be a well-regulated process.

Another experiment suggests what this regulation may entail. The antibiotic rifampin inhibits RNA synthesis, and thereby protein synthesis. Rifampin inhibits the initiation of replication but allows any that has already started to continue. Thus, the effect of the drug on replication suggests that *initiation of a round of replication requires the synthesis of new proteins*. As discussed in some detail in Chapter 8, one protein, called **DnaA**, has indeed been found to play a distinctive regulatory role in the initiation process. We know this from both **in vitro** and **in vivo** experiments. The addition of DnaA is essential for initiation to take place in the test tube. Also, DnaA binds to the replicative origin at characteristic sequences, where it helps open the double helix, which allows other proteins to enter that site and replication forks to form (see Chapter 8). In vivo studies revealed that mutants that don't make DnaA cannot initiate replication. There is a quantitative relationship between the amount of DnaA a cell contains and its capacity to carry out the initiation step. The presence of a large amount of DnaA, for instance, when the encoding gene is present on a multicopy plasmid, leads to extra rounds of replication. These could be deleterious, but they abort a short time after initiation because cells have a mechanism to avoid making too much DNA. Of course, realizing the importance of the DnaA concentration only pushes the question back one step. What regulates the amount of DnaA made? Although there are several hints, a fully satisfying mechanistic answer is not yet available.

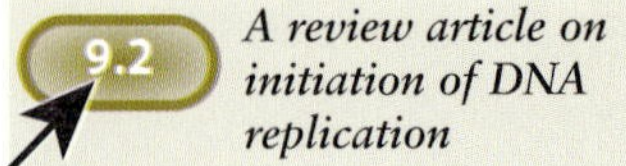

A review article on initiation of DNA replication

CELL DIVISION

Most "ordinary" bacteria divide by binary fission into two equal progeny cells. Still others have a more intricate division pattern. For example, the bacterium *Caulobacter* divides into two cells that are unequal with respect to size, shape, flagella, and a stalk-like structure (see Chapter 14). Some bacteria divide not by fission but rather by budding (resembling yeasts in this regard). In many bacteria, progeny cells separate from one another upon division and move away without forming characteristic arrays of progeny cells. Others stick together, forming pairs or chains of cocci, chains of rods, grapelike bunches, square sheets, or cubes, depending on the planes of division (Fig. 9.5). Even more intricate patterns are seen in prokaryotes that undergo complex developmental cycles, such as the myxobacteria, the cyanobacteria, and the actinomycetes (see Chapter 11). Some of these bacteria resemble fungi and algae in the morphological complexity and arrangements of their progeny cells.

Morphological considerations

Cell division involves invagination of the cell membrane, the cell wall, and, in the case of gram-negative bacteria, the outer membrane as well. Only when these processes are completed can progeny cells separate. Cell division, therefore, is a multistep process, and it differs in detail among species. Generally speaking, most gram-negative bacteria divide by mak-

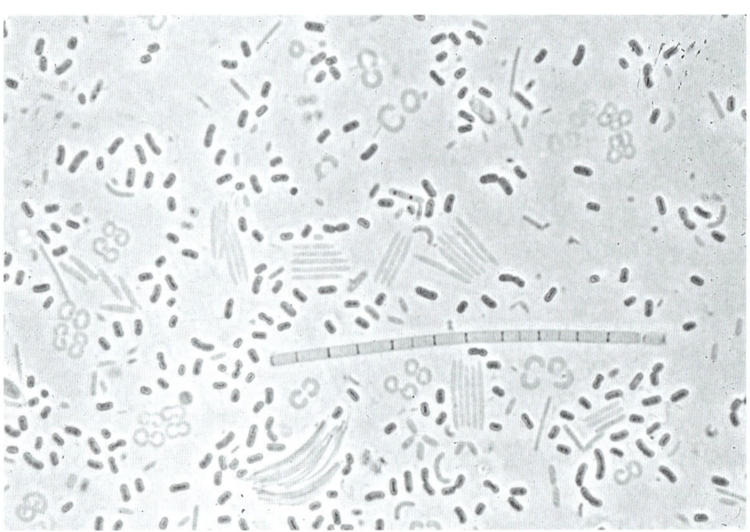

Figure 9.5 The many shapes of bacteria and archaea. The microbial community shown was found in the sediment-water interface of Burke Lake, near East Lansing, Michigan. The rich variety of shapes brings to mind foods, such as bananas, string beans, sausages, a stalk of sugarcane, doughnuts, dates, and others.

ing a constriction of their envelope layers at midcell (Fig. 9.6). This furrow curves progressively inward until two hemispherical caps are formed. In some gram-positive cocci and rods, cell division proceeds without apparent constriction of their girth. Here, a division septum is formed by the envelopes invaginating at 90° to the surface (Fig. 9.7). The poles of such cells are usually more blunt and less rounded than those of rods that divide with constrictions, giving these cells a squared-off cylindrical rather than a sausage-like shape.

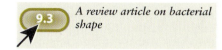

9.3 *A review article on bacterial shape*

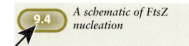

How is the septum formed?

What induces the cell envelopes of a growing cell to change the direction of their extension and start making a septum? A surprising finding made in 1991 proved to be key to our understanding. The investigators found that a protein called **FtsZ** has the unexpected ability to make a **constricting ring** where the cell septum will eventually form. This structure, called the **Z-ring**, closes gradually (like the iris of the eye) as the septum forms.

9.4 *A schematic of FtsZ nucleation*

The Z-ring may well be the functional equivalent of the contractile ring that mediates cytokinesis in eukaryotic cells.

Several properties of the FtsZ protein are illuminating; but first, how was the constricting ring discovered? The investigators used antibodies specific to FtsZ labeled with gold particles that could be seen under the electron microscope in ultrathin sections of bacteria. They found that the FtsZ protein appeared to be localized only in the region of the septum (Fig. 9.8). However, this was true for large cells only, not the small younger ones, which led to the proposal that the Z-ring is formed only in the late portion of the cell cycle. This suggestion was confirmed by looking under a fluorescence microscope at living cells in which FtsZ was fused to GFP. Thus, in response to yet-undiscovered stimuli, at the proper time, FtsZ protein monomers polymerize into filaments that form the Z-ring. Once the ring has done its job and the cells divide, the FtsZ proteins

Figure 9.6 Cell division in a gram-negative bacterium. This thin section of *Escherichia coli* shows that the cell constricts before it divides.

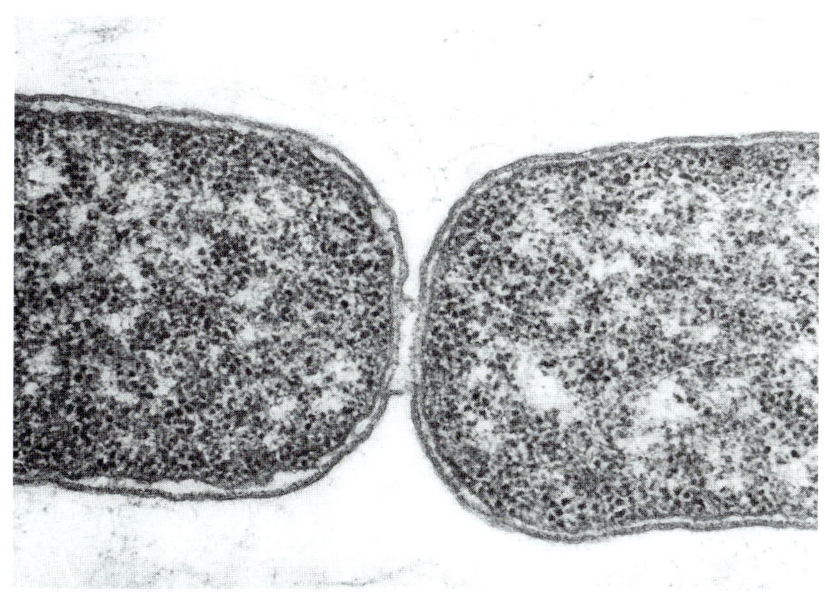

Figure 9.7 Cell division in a gram-positive bacterium. This thin section of *Bacillus subtilis* shows no cell constriction but a completed septum made before the progeny cells begin to detach.

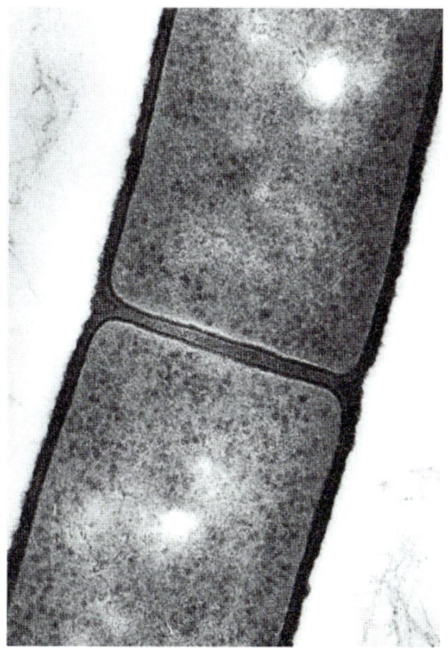

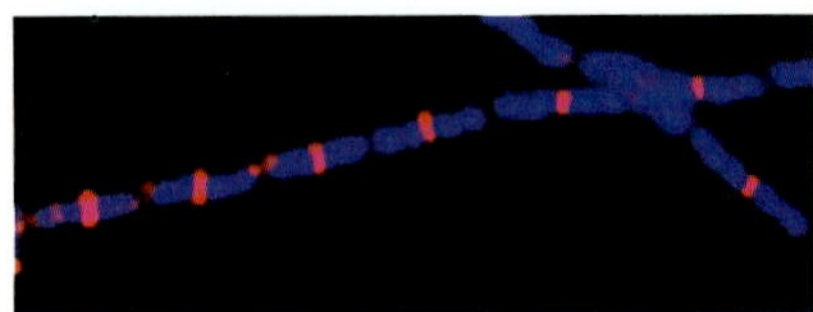

Figure 9.8 Z-rings in growing bacterial cells. The location of FtsZ, the protein that makes Z-rings, at the division site is revealed by fluorescence microscopy of *B. subtilis*, using an antibody directed against FtsZ (red). The DNA is stained blue with a dye called DAPI (4′,6′-diamidino-2-phenylindole).

disperse in the cytoplasm of the daughter cells to reaggregate again during the next cell cycle.

The ability of FtsZ to aggregate into filaments reminds one of cytoskeletal proteins in eukaryotic cells. Indeed, FtsZ does resemble **tubulin**, a protein of eukaryotic microtubules. FtsZ and tubulin are structurally quite similar, and their three-dimensional arrangements align quite well, even though their amino acid sequences have only 20% **homology.** Both proteins are **GTPases** and require GTP (guanosine triphosphate) hydrolysis to be organized into filaments. The two proteins share a glycine-rich GTP-binding sequence motif (for aficionados, the sequence is GGGTG[T/C/G]G). One may conclude that FtsZ is the bacterial homolog of tubulin and that the two probably share an evolutionary ancestor.

FtsZ is not the only bacterial protein that is similar to those of the eukaryotic cytoskeleton (see Table 2.3). A protein called MreB (for murein regulation gene cluster e) has a high degree of structural similarity to the eukaryotic protein actin and appears to be involved in the segregation of chromosomes and plasmids.

What causes the Z-ring to close? Because we know of no contractile mechanism in bacteria, one plausible notion is that FtsZ polymerizes at the leading edge of the ring and depolymerizes in step at the trailing edge. But how does FtsZ locate the middle of the cell and polymerize only there? We do not know the answer, although there are hints that selection of the site for assembling the Z-ring requires that the machinery for synthesizing the murein of the septum be properly positioned.

FtsZ was first identified in *fts* mutants of *E. coli*, which grow normally at 30°C but cannot divide at higher temperatures. (Note that by convention, names of genes are italicized and those of proteins are capitalized but not italicized). The term *fts* stands for "filamentous temperature sensitive" and denotes a family of genes each of which, when mutated, exhibits the same filament-forming phenotype. *ftsZ* is a highly conserved gene found in nearly all bacteria and many archaea. In addition to bacteria and archaea, *ftsZ* is found in several eukaryotic organelles, such as chloroplasts and some mitochondria. FtsZ also plays a role in the division of these structures, confirming the ancient bacterial ancestry of the eukaryotic organelles.

We will deal below in more detail with the question of how the cell knows where to divide. We know more about what happens thereafter, *after* the Z-ring begins to form. The Z-ring is associated with the cell membrane, as are a number of other Fts proteins. The Z-ring is thought to form a scaffold for the recruitment of other key cell division proteins. These additional proteins bind sequentially, and all of them depend on the presence of FtsZ at the site of division. The assemblage of these proteins at the site of the Z-ring has been called the **divisome.** Interestingly, *ftsZ* is also found in mycoplasmas, the cell wall-less bacteria that lack murein and many other bacterial division proteins. This fact alone suggests that its product, FtsZ, may be capable of driving cell division without a functional divisome.

Some divisome proteins may regulate the division process, others may be involved in the synthesis of the murein of the septum, and still others may have a stabilizing or structural role. One of these proteins, called PBP 3, is specifically involved in making the murein at the septum. PBP, which stands for **penicillin-binding protein**, designates a group of proteins that covalently bind to penicillin and other **β-lactam antibiotics** and are involved in murein metabolism. How was the role of PBP 3 determined? A β-lactam antibiotic, cephalexin, which inhibits septation without impairing cell elongation or net murein synthesis, binds with special affinity to PBP 3, thus suggesting that this protein is responsible for the synthesis of septal murein.

How does a bacterium find its middle?

Rod-shaped bacteria, such as *E. coli* or *B. subtilis*, tend to divide quite precisely in the middle. What sort of ruler do they use to locate this position along the cell? There are several hypotheses, none of which has proven conclusive so far. However, exciting and surprising facts have been discovered. The story begins in 1969 (prehistory by current standards of science), when researchers found that certain mutants of *E. coli* produce small anucleated cells by dividing abnormally near their ends (Fig. 9.9). They called these mutants "min," for minicell formers. An imaginative interpretation of this finding is that a rod-shaped bacterium has three sites where division can potentially take place: one in the middle and one near each end (the poles). The last two may be residues of previous septum sites that were shut off when the previous division was completed. However, min mutants altered in one of three genes, *minC*, *minD*, or *minE*, lack this shutting-off mechanism and divide abnormally at a polar division site. The protein products of these genes have roles in regulating normal division. How they work is a bit complicated, but understanding it may be rewarding. MinC and -D work together to inhibit polymerization of FtsZ into a Z-ring. MinE counteracts this inhibition, *but only at the normal septum site*. In line with this model, MinE is localized at the middle of the cell. Normally, the inhibition of Z-ring formation is overturned at the center of the cell only, which allows normal division to take place. Mutants with defective MinC or MinD will form minicells, supposedly because they cannot inhibit the abnormal divisions at polar sites.

The Min system holds a further surprise: its proteins move rapidly from one pole of the cell to another, an unexpected event in a bacterial cell. This movement can be seen in living cells by time-lapse microscopy using GFP fusions of these proteins.

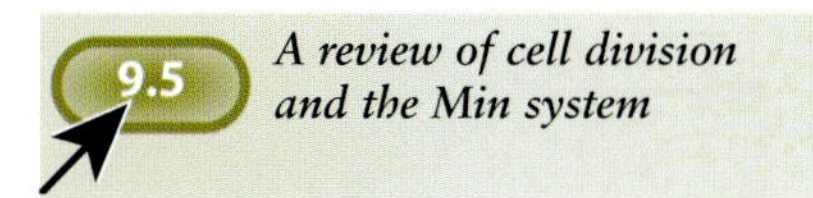

A review of cell division and the Min system

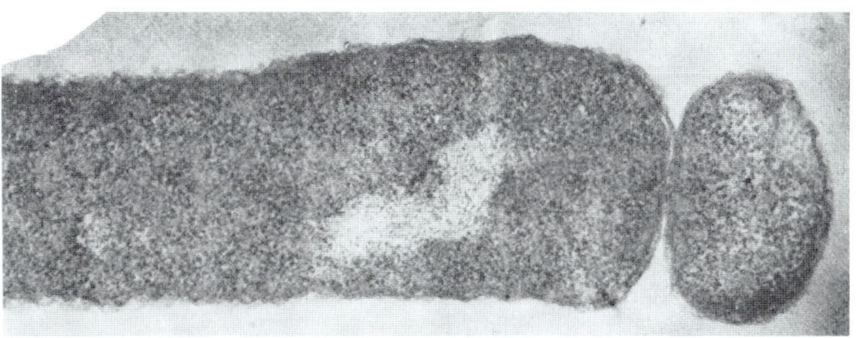

Figure 9.9 Minicell production in *E. coli*. A thin section shows an abnormal division at a pole, leading to minicell formation.

At any one time, the Min proteins are localized at one pole of the cell, only to reappear soon thereafter at the other. The movement is remarkably rapid, taking about 20 seconds for MinD to go from pole to pole. The proteins do not move through the cytoplasm; instead, they slither along the cell membrane, making coiled structures that wind around the cell from pole to pole. For what function do Min proteins play this kind of microscopic game? It has been proposed that this oscillation may serve as some kind of a kinetic measuring device, helping to determine the middle of the cell, but more work is needed for a fully satisfying explanation.

Min proteins are not the only molecules that travel within a microbial cell. Other proteins tour the inside of the cell, and so does DNA, as will be discussed below. As an aside, the development of fluorescent-tagging techniques to locate macromolecules in bacteria has spawned a new field now known as bacterial cell biology. We've learned that bacteria are not just bags containing static arrays of macromolecules. These findings suggest that much work is needed before we have a good sense of what their interior is like (see Chapter 3).

THE CONNECTION BETWEEN CELL DIVISION AND DNA REPLICATION

In principle, a cell should not be able to divide until each progeny cell has its own genome. In normal bacterial growth, cell division and DNA replication are indeed coupled, yet the two processes can be uncoupled. For example, certain mutants are able to divide without making DNA, thus producing **anucleated cells**. But one cannot tamper with DNA without soon stopping cell division. This is especially true when DNA is damaged, e.g., by ultraviolet or ionizing radiation, which introduce single- and double-strand breaks, or by the action of drugs, such as mitomycin C, that cross-link the two DNA strands. If the insult is not overly massive, cells so treated will continue to grow without dividing, thus forming long **filaments** (Fig. 9.10). The damaged DNA is repaired quite soon, and DNA replication can then resume, resulting in filaments with evenly spaced nucleoids along their lengths. Later, cell division resumes and the filaments break up to make proper-length cells.

How is inhibition of cell division connected to DNA damage? The best-studied mechanism is appropriately called the **SOS response** (as in the marine distress signal). SOS is one of the global response systems exhibited when bacteria are subjected to chemical and physical insults (see Chapter 13). Single-strand breaks in the DNA double chain activate a number of genes that code for normally dormant proteins involved in **DNA repair** and cell division. These genes have one property in common: their expression is held in check by the same repressor, a protein called **LexA**. As long as LexA is present, these genes cannot be transcribed. However, when DNA is damaged, the resulting single strands bind to an inactive protein called **RecA**, making it capable of cleaving the LexA repressor (Fig. 9.11). Actually, LexA has the ability to self-cleave, but only when stimulated by the activated RecA. Thus, RecA is not actually a protease but rather a cofactor necessary for proteolysis. RecA is a multifunctional protein that also plays a major role in homologous recombination, which makes it especially interesting.

Figure 9.10 Inhibition of cell division. A culture of *E. coli* was subjected to a sublethal dosage of ultraviolet light, which inhibited cell division without interfering with growth and nucleoid segregation. The result is a long filament with normally spaced nucleoids.

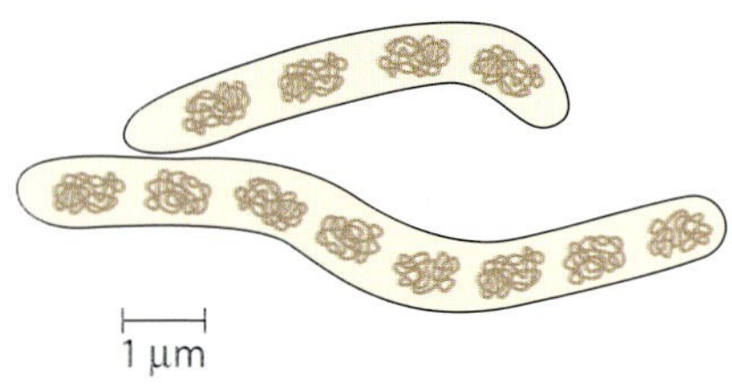

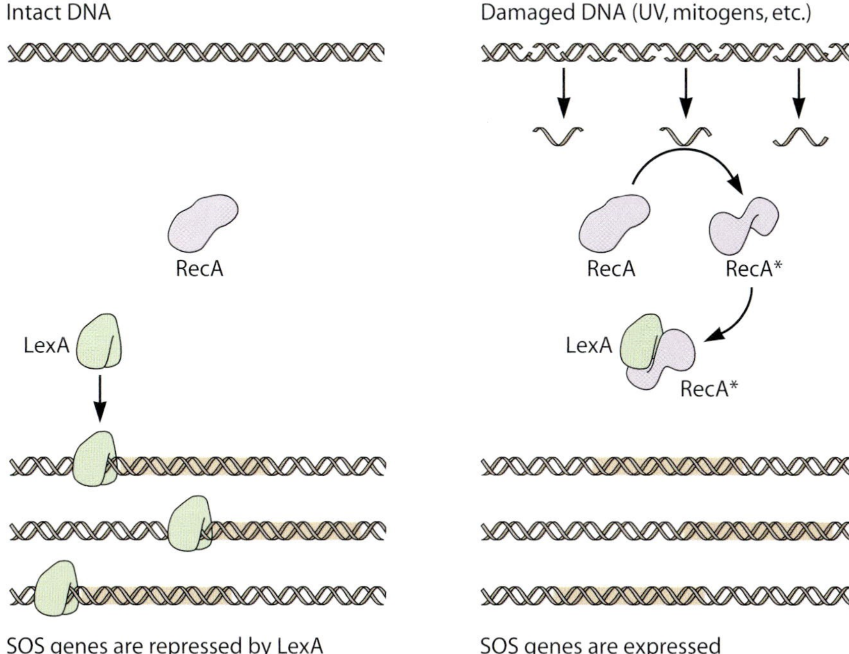

Figure 9.11 SOS response. Damage to DNA that causes double-strand breaks (right) results in the formation of small DNA segments. These segments activate RecA to become a coprotease, RecA*. Now RecA* cleaves a repressor of several operons called LexA, permitting the genes under its control (the SOS genes) to be expressed. The normal, nonstressed situation is shown on the left, where LexA is intact and represses these genes.

Among the many genes under the control of LexA is *sulA*, which encodes an inhibitor of cell division that is normally repressed. The **SulA** protein binds to the septal-ring-forming protein, FtsZ, thereby inactivating it. SulA is negatively controlled by LexA. Thus, when DNA damage induces RecA-mediated inactivation of LexA, SulA can be synthesized, and it blocks cell division. Other genes under LexA control are part of a complex machinery involved in DNA repair and are also activated when LexA is inactivated. Thus, the SOS response is quite complex and involves several systems, including DNA repair, as well as cell division. When DNA repair is completed, the stimulus for the SOS response is gone and the cells resume division. This resumption is accelerated by the action of a specific protease called **Lon**, which destroys SulA.

The portion of the SOS mechanism that deals with division inhibition comes into play only when DNA is damaged. Under normal conditions, the SOS mechanism does not have a role in regulating division, which proceeds normally in mutants lacking SulA. Thus, the SOS response has evolved as an *anticipatory mechanism,* one that is called upon only when DNA is damaged. SOS is not the only mechanism that couples DNA damage to cell division. There are others, also complex. Why do bacteria go to such lengths to ensure that they do not divide when their DNA is damaged? Several notions seem plausible. Stopping cell division ensures that cells have the time to repair their damaged DNA before dividing, thus avoiding the production of progeny cells with damaged DNA. In addition, in the filaments that form, scarce cell factors that may be required for recovery from damage to DNA can be shared among several cell equivalents.

CELL DIVISION AND PLASMID REPLICATION

Mechanisms for watchful guard over cell division are not limited to sensing the integrity of the cell's own DNA. Plasmids (extrachromosomal elements further discussed in Chapter 10) also encode mechanisms that ensure their retention by their "parent" or host cell. For example, an *E. coli* plasmid called P1 (which in a different guise is also a bacteriophage) has evolved a way to *prevent the production of plasmid-free cells*. How does P1 ensure its persistence in the population? The mechanism has been dissected genetically and shown to be quite intricate. P1 plasmids carry the genes for two proteins; one is a **toxin** called **Doc**, and the other is an **antitoxin** called **Phd** (Fig. 9.12). (Yes, humor crops up in microbial terminology.) Interestingly, Doc is resistant to proteolytic degradation, while Phd is not. Under normal circumstance, Phd, though unstable, is made in sufficient amounts to neutralize the deleterious effects of Doc. What happens when plasmid replication is halted? The cells continue to divide and produce plasmid-free offspring. The cytoplasm of these cells initially contains appreciable amounts of both Doc and Phd, but because neither of these proteins can be renewed and Phd antitoxin is unstable, the toxic activity of Doc can no longer be neutralized and the cells will die. These and other plasmids have evolved a rather demanding relationship with their hosts: "Leave me behind, and you die!" The two-gene array of toxin and antitoxin has been called an **addiction module**. A similar arrangement exists in chromosomal genes, which has led to the proposal that bacterial cells, like eukaryotic cells, possess a mechanism for programmed cell death. Such mechanisms come into play under conditions of starvation and are thought to be responsible for some cells dying in order to provide food for the remaining ones. How altruistic!

Figure 9.12 Toxin-antitoxin pathway in a plasmid. In this plasmid, the *doc* gene encodes a toxin, Doc; the *phd* gene encodes an antitoxin, Phd. Doc is a stable protein; Phd is unstable and is cleaved by a protease (called ClpPX). When the plasmid carrying these genes is lost, the Doc toxin remains active and is no longer neutralized by Phd, resulting in the death of the cell.

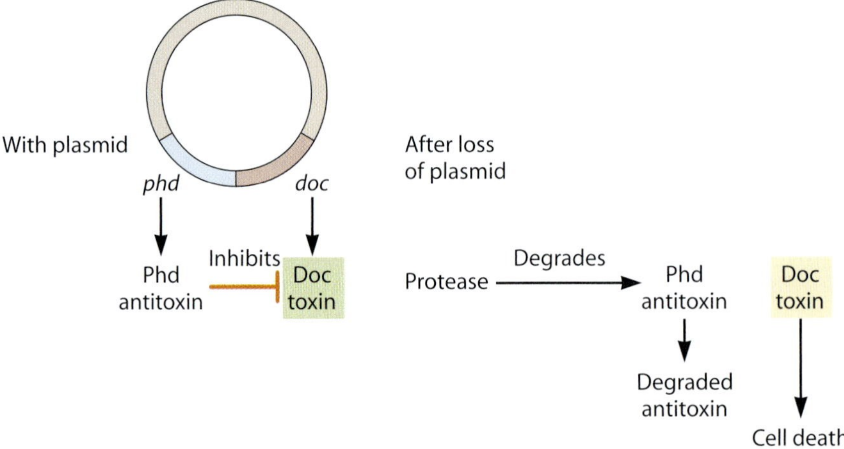

THE PROKARYOTIC EQUIVALENT OF MITOSIS

We know a great deal about mitosis, the elaborate process by which eukaryotes ensure the distribution of replicated chromosomes to progeny cells. We know very little about the equivalent process in prokaryotes. For eukaryotes, we know that several structures are needed: the region of the chromosome (the centromere) to which the segregation apparatus attaches, the mechanical apparatus itself (the mitotic spindle), and the machinery that separates the segregated chromosomes (microtubular gliding). For prokaryotes, we still have little insight into equivalent aspects of chromosome segregation. We do know, however, that the *bacterial chromosome segregates with great fidelity.* Anucleated bacteria, which would arise if segregation were unregulated, are very rare in normal cultures.

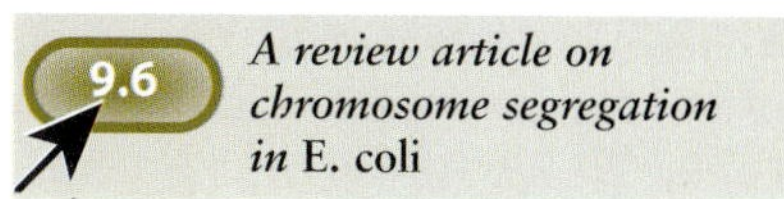

A review article on chromosome segregation in E. coli

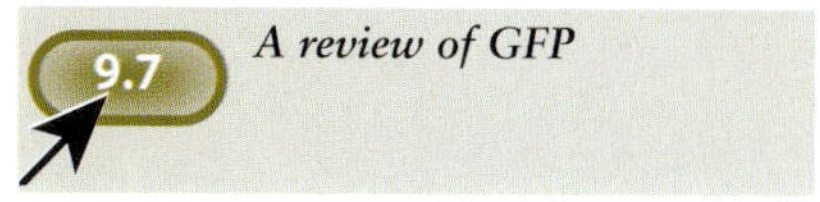

A review of GFP

In the best-studied bacteria, chiefly *B. subtilis* and *E. coli,* the processes of replication and chromosome segregation are not sequential, as they are in eukaryotic cells. Instead, as we discussed above, they take place concurrently, at least in fast-growing cells. The replication machinery begins its work at the origin and proceeds bidirectionally until it reaches the terminus region. Recently developed techniques have permitted us to visualize the locations of various chromosome regions within living cells. This may sound surprising, because offhand, one may think that it would be difficult to localize a relatively small target, such as a part of a chromosome, in cells as tiny as bacteria. Nonetheless, such techniques do exist. They make use of ultrasensitive **fluorescence microscopy** techniques in combination with some clever genetic constructs. Thus, specific regions of the chromosome can be tagged by making the following genetic construct. A DNA sequence that binds a specific repressor (e.g., that of the lactose operon), which has been fused to GFP, is inserted near the origin. Because a single molecule of fluorescent repressor may be hard to see, investigators often insert more than one operator, usually several dozen of them in a row, thus amplifying the fluorescent signal.

In young cells that are just starting their cycle of growth and DNA replication, the origins are located at the one-quarter and three-quarter positions along the cell length, and the terminus is at the cell center. As the cell grows and the origin replicates, the two new origins move apart, while the terminus remains in the cell center for some time (Fig. 9.13).

Figure 9.13 Cellular locations of the origin and terminus of DNA replication. The origin was labeled with GFP; the terminus was labeled with a red fluorescent protein (see the text). Before the onset of replication, the replicative origin is seen in the one-quarter and three-quarter positions in the cell, and the terminus is in the cell middle. The two are shown in the same cells by an optical overlay (right). The origins progressively move toward the ends of the cell, while the terminus remains in the center. Eventually, the terminus doubles and moves to the one-quarter and three-quarter positions along the cell's length, which become the cell centers upon division. Bar, 2 μm.

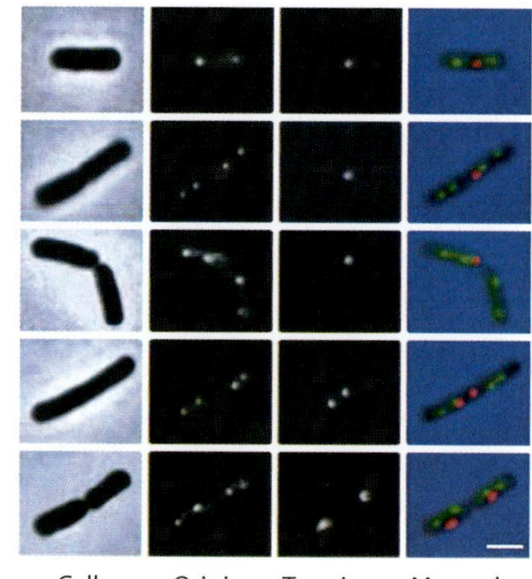

As discussed in Chapter 8, the final step in replication of a circular chromosome requires the resolution of a **catenate**. Mutants defective in this process continue to make DNA but cannot separate the progeny chromosomes. Such cells continue to grow as filaments and accumulate a large amount of DNA in their cell centers (Fig. 9.14). There is yet another impediment to chromosome segregation: every so often, about once in six generations, sister chromosomes still in the same cell undergo **homologous genetic recombination** (between homologous sequences) (see Chapter 10). A single recombinational event, which is more common than multiple ones, produces a chromosome dimer. Proper segregation could occur only if the dimer is converted to two monomers by another recombination event, but two such events are relatively rare. Instead, bacteria use an efficient dimer resolution system that kicks in with great frequency. Near the chromosome terminus, there is a special recombination site called *dif*. Specific enzymes carry out recombination at this site, leading

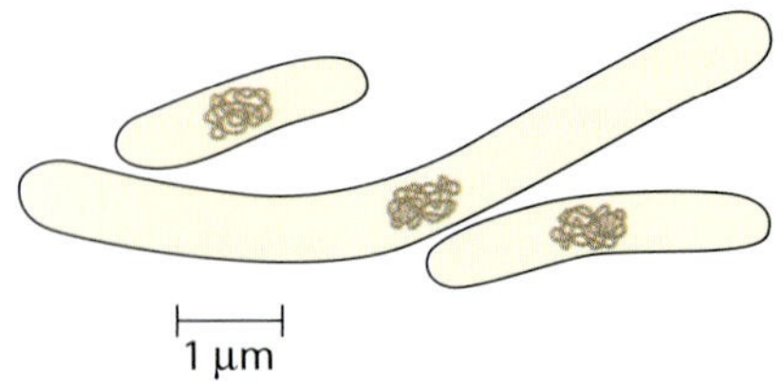

Figure 9.14 Effects of inhibiting decatenation of sister chromosomes. A mutant defective in the decatenation of chromosomes after replication continues to grow but cannot divide. The DNA does not segregate into individual nucleoids but remains as a large mass in the center of the cell.

to the conversion of the dimer into two sister chromosomes. Why would these cells worry about an event that occurs relatively rarely? Something that happens once every six generations affects only 1 cell in 64. But, in the context of competition in the environment, a population with a defect in 1 to 2% of its cells would be at a distinct competitive disadvantage.

In summary, the cell cycle of bacteria differs in several significant ways from that of eukaryotic cells, reflecting the differences in cellular organization. Although structurally simpler, bacteria have intricate regulatory mechanisms to ensure that their genome is suitably apportioned between progeny cells and that cell division is a well-regulated process. In the next chapter, we explain how genetic changes take place and influence present-day populations and how they may help us think through their evolutionary origins.

STUDY QUESTIONS

1. What techniques allow the study of the cell cycle of bacteria?

2. What happens to DNA replication during the bacterial cell cycle? In fast-growing cells, how can this process take longer than the cell cycle?

3. How does cell division in prokaryotes differ from that of eukaryotes?

4. How is the cell septum formed during bacterial division?

5. How are cell division and DNA replication connected in bacteria?

6. How do bacteria find their middle for the purpose of dividing?

7. Discuss the bacterial equivalent of mitosis.

section IV inheritance

chapter 10 genetics

INTRODUCTION

Genetics is remarkably unified. The genetic information of all cellular organisms is encoded the same way, in sequences of bases in **deoxyribonucleic acid (DNA)**; the **genetic code** is universal; and the mechanisms of replication and expression of genetic material are, in essence, the same among all organisms. Still, there are intriguing differences between prokaryotes and eukaryotes in the ways that genetic information is exchanged between individuals, in the patterns of its storage, and in its susceptibility to change.

Because genetics is so unified and because prokaryotes are such excellent experimental tools, much of the fundamentals of modern molecular genetics, including its most fundamental discovery—that DNA is the molecule in which the genetic information of all cellular organisms is encoded—was learned from studies of bacteria. This breakthrough and its confirmation came from studies of the ways prokaryotes pass DNA between cells. It is one of the spectacular examples of how much we have learned about ourselves by studying prokaryotes.

EXCHANGE OF DNA AMONG PROKARYOTES

The ways that prokaryotes and eukaryotes exchange DNA and the consequences are fundamentally different in at least five important respects.

1. Unlike most eukaryotes, prokaryotes do not exchange DNA as an obligatory step in their reproduction. Of course, neither do some eukaryotes: some are capable of indefinite asexual reproduction, and some are capable of parthenogenesis, thereby circumventing the lockstep connection between reproduction and sexual exchange of DNA. However, prokaryotic reproduction *never* depends on genetic exchange.

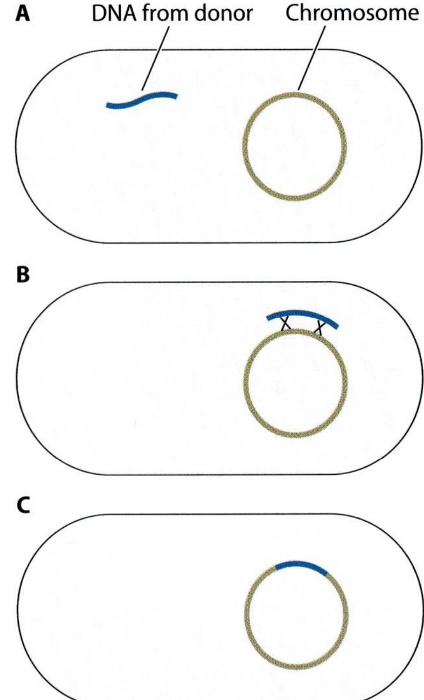

Figure 10.1 Genetic exchange between prokaryotes. The sketch shows how a fragment of double-stranded DNA can become a part of the genome of the recipient cell. **(A)** During most genetic exchanges between prokaryotes, a small piece of double-stranded DNA from a cell enters a recipient cell. **(B)** As a consequence of two recombinations (crossovers, indicated by X's), a portion of the donor fragment can be incorporated into the chromosome of the recipient, replacing a portion of it. **(C)** The consequence is a merozygote. Unincorporated portions of the donor fragment and the displaced portion of the chromosome are destroyed by nucleases in the cytoplasm of the recipient.

2. Rather than a mixing of their complete complements of DNA, as occurs when eukaryotic gametes fuse, usually only a small bit of DNA is transferred from one prokaryotic cell (the **donor cell**) to its partner (the **recipient cell**). Such transfer has a high probability of being genetically futile, because the donated DNA usually cannot replicate within the recipient cell, and it is likely to be destroyed. The donated bit of DNA can replicate only if it becomes incorporated (by a pair of recombination events) into a preexisting **replicon** (see Chapter 8) within the recipient or if the donated DNA itself happens to be a replicon (Fig. 10.1).
3. In nature, genetic exchange among prokaryotes is a random event, but when conditions are made optimal in the laboratory, it can be mediated with near certainty.
4. Most higher eukaryotes are diploid. They carry more than one set of **chromosomes**, so recessive alleles are not expressed. In contrast, prokaryotes are haploid; their recessive alleles are expressed. The difference between the consequences of expressing and not expressing recessive genes is profound. Put in human terms, only about 1 in 3,500 white Americans suffers from the recessive genetic disease cystic fibrosis; if recessive genes were expressed, 1 in 30 would experience it.
5. Although genetic exchange between eukaryotes occurs by one mechanism—fusion of gametes—it can occur between prokaryotes by three totally different mechanisms, called **transformation, transduction,** and **conjugation.** The exchange of chromosomal genes by two of these mechanisms, transduction and conjugation, is apparently an accidental byproduct of some other event—infection by and replication of a bacteriophage in the case of transduction and movement of certain kinds of **plasmids** (small DNA replicons that are present in many prokaryotes) from one cell to another in the case of conjugation. However, the third mechanism, transformation, might actually have evolved to mediate genetic exchange.

Although genetic exchange among prokaryotes seems accidental, even trivial, its impact on prokaryotic evolution has been profound and its consequences for scientists who study the topic have been, at times, exasperating. Modern molecular studies of prokaryotes have revealed that their evolutionary history has been marked by numerous examples of **horizontal gene transfer** (from one cell to another; also called **lateral gene transfer**), not exclusively by **vertical gene transfer** (from mother to progeny cells), as one might expect for organisms that reproduce overwhelmingly by asexual means. Sometimes, such horizontal gene transfer must have occurred between distantly related individuals. Horizontal gene transfer confounds molecular studies of evolution because it introduces the possibility that organisms sharing similar genes might not necessarily have evolved from a common ancestor that contained the progenitor of those genes; the similarity might be a consequence of horizontal gene transfer between two distantly related cells.

Transformation

The word **transformation** (change in appearance or nature) has two distinct meanings in biology: (i) the change of a normal healthy eukaryotic cell into a cancerous one and (ii) genetic change of a cell as a consequence of taking soluble DNA from its environment and incorporating by recombination a bit of this DNA into its genome. (In the 19th and early 20th centuries, transformation was a [usually whispered] euphemism for a lady's wig: "Did you know she wears a transformation?")

Cells capable of taking up DNA from their environment and thereby becoming transformed by it are said to be **competent. Genetic competence** is a transient state. It develops naturally among some species of bacteria (cells of these species acquire or lose competence depending on their physiological state). Cells of most species that lack the natural capacity to develop competence can nevertheless be induced to become competent by artificial means.

Artificial transformation

Most prokaryotes (probably all if the proper conditions were discovered) can be transformed artificially by treatments that transiently alter the permeability of a cell's membrane, thereby making it competent. These treatments include such diverse manipulations as exposure to high concentrations of calcium ions on ice followed by a heat shock, and **electroporation** (making transient holes in the cytoplasmic membrane by exposing cells to millisecond jolts of tens of thousands of volts). Such **artificial transformation** is usually restricted to the uptake of circular plasmids; because they have no exposed ends, they are resistant to attack by the intracellular exonucleases that most prokaryotes contain, so they can survive within the transformed recipient. Mutant strains of *Escherichia coli* lacking two exonucleases can be successfully transformed artificially with linear DNA.

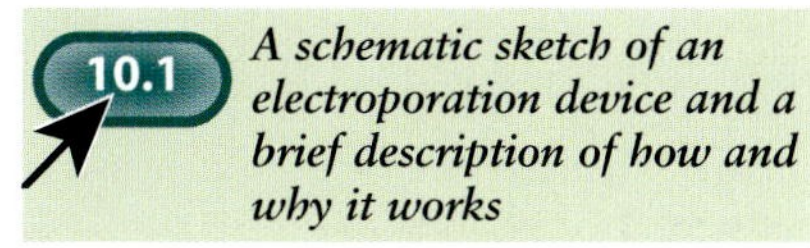

A schematic sketch of an electroporation device and a brief description of how and why it works

Artificial transformation is a critical step in gene cloning, which is fundamental to **recombinant DNA** technology.

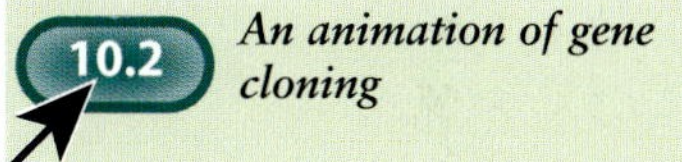

An animation of gene cloning

Natural transformation

Some species of bacteria undergo transformation naturally during growth and **development** (Table 10.1). Under certain conditions of

Table 10.1 Some species of bacteria known to be capable of natural transformation

Bacterium	Remarks
Gram positive	
Streptococcus pneumoniae	A pathogen
Bacillus subtilis	A soil-dwelling, mesophilic sporeformer
Bacillus stearothermophilus	A soil-dwelling, thermophilic sporeformer
Enterococcus faecalis	A gut bacterium
Gram negative	
Acinetobacter calcoaceticus	A soil dweller
Moraxella urethralis	An opportunistic pathogen
Psychrobacter spp.	Psychrophiles
Azotobacter agilis	A free-living, soil-dwelling nitrogen fixer
Pseudomonas stutzeri	A soil-dwelling denitrifier
Haemophilus influenzae	A pathogen
Neisseria gonorrhoeae	A sexually transmitted pathogen
Campylobacter jejuni	An intestinal pathogen
Helicobacter pylori	A stomach pathogen

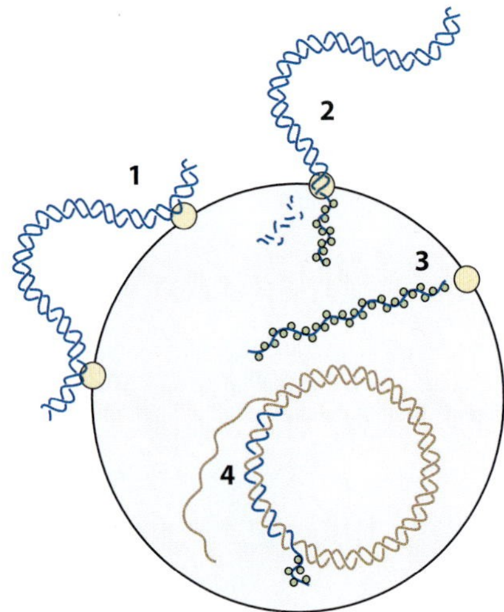

Figure 10.2 Principal features of transformation of *S. pneumoniae*. **(1)** A fragment of double-stranded DNA binds to the cell surface at several sites. **(2)** Bound DNA is nicked and cut, and one strand is degraded by a competence-specific nuclease. **(3)** A single DNA strand enters the cell and becomes coated by a specific protein. **(4)** The entering DNA displaces an existing strand, forming a heteroduplex.

growth, they become competent. Then, they can become genetically altered by taking up soluble DNA that might be present in their environment and incorporating it into their genomes. Usually, competence develops in dense cultures, perhaps reflecting natural selection for a higher probability of successful gene transfer between closely associated cells, such as those in biofilms (see "Coping with Stress by Community Effort" in Chapter 13). One might expect that transformation within biofilms would be rampant, but more studies are needed to confirm this prediction. Natural transformation systems are relatively rare among prokaryotes. The most thoroughly studied, those of *Streptococcus pneumoniae*, *Bacillus subtilis*, and *Haemophilus influenzae*, are distinct in some details, but they have many common properties (Fig. 10.2). These systems are complex, being encoded by 20 to 30 genes, but they all process the transforming DNA in similar ways. DNA in the cell's environment binds to the cell surface. It is fragmented there into smaller pieces, only one strand of which enters the cell; the other is destroyed by a transformation-specific nuclease. What usually takes place next is that the entering strand becomes coated with a transformation-specific DNA-binding protein, which helps it survive its hazardous trip through the exonuclease-laden cytoplasm. If it is homologous to the cell's resident DNA, the strand becomes incorporated by **single-strand displacement**, forming a stretch of hybrid DNA consisting of one strand of resident DNA and one strand of transforming DNA. On subsequent division, only one daughter cell bears transformed DNA—the one that inherited the duplicated transforming strand of DNA.

The existence of natural transformation systems raises several intriguing questions. First, we might ask about the source of transforming DNA in a natural environment. Let us keep in mind that only transforming DNA that is closely related to the recipient's DNA is capable of transforming because it must be able to hybridize with DNA of the recipient's genome. It is commonly proposed that the source of transforming DNA is random lysis of cells in the environment, but if we assume that transformation evolved as a means of genetic exchange, such an explanation would seem implausible. DNA in solution is short-lived in natural environments because **deoxyribonucleases (DNases)** are widespread. It seems improbable that a complex, multigene-encoded process for taking up DNA would have evolved to be dependent on random release of DNA from other cells. Indeed, there is some evidence that release of DNA from donor cells is also genetically programmed to occur when the probability of its being taken up is favorable. For example, natural transformation of *Pseudomonas stutzeri* occurs when donor and recipient cells touch. Free DNA can be detected in *Streptococcus pneumoniae* cultures only when populations are dense, governed by a population-dependent phenomenon called quorum sensing (see Chapter 13).

Second, we might ask about the biological function of natural transformation. Some have proposed that it is nutritional—that it evolved as a means of using DNA as a nutrient. It seems, however, that natural transformation is far too complex to be explained so simply. Possibly the most telling counterargument is that some naturally transformable bacteria, including *Haemophilus* and *Neisseria*, have evolved mechanisms to recognize and thereby take up DNA only from closely related stains. The system found in *H. influenzae* is particularly specific and elaborate. Its chromosome contains 600 copies of an 11-base-pair (bp) (5′-AAGTGCGGTCA-3′) **recognition site.** Competent cells of *H. influenzae* take up only DNA that contains one of these sites. However, other naturally transformable species, including *Acinetobacter calcoaceticus* and *P. stutzeri*, do take up DNA indiscriminately.

Some speculate that natural transformation evolved as a mechanism of DNA repair. However, it does not selectively target damaged DNA. It seems likely that natural transformation evolved as a means of exchanging DNA among strains of a particular bacterial species, a substitute for sexual exchange among eukaryotes. However, that raises another critical but still unanswered question. Why is natural transformation so relatively rare among prokaryotes? At the moment, there are no good answers.

Natural transformation holds a special position in the history of genetics because its study led to the discovery that genes are made of DNA. In 1928, Frederick Griffith did a well-known, simple, but elegant experiment that pointed the way. Using the pneumococcus (*S. pneumoniae*), he showed that capsule-bearing (called S) strains but not capsule-free (called R) strains killed mice. Because heat-killed S cells did not kill, he concluded that the capsule itself was innocuous. But might the capsule itself render R cells lethal? Indeed, he found that a mixture of heat-killed S cells and live R cells was deadly, but the two components did not act synergistically, as he had anticipated. The dead mice contained only live S cells, which Griffith then showed had other properties characteristic of the injected R cells. These cells were recombinants, possessing certain properties of S cells and others of R cells. Griffith concluded (rightly) that something, which he called the **transforming principle**, from the dead S cells had "transformed" the R cells, rendering them smooth (Fig. 10.3). Sixteen years later (1944), O. Avery, C. MacLeod, and M. McCarty purified this transforming principle and showed that it was DNA.

10.3 *Images of Avery, MacLeod, and McCarty and information about them and their discovery*

Transduction

Although a good case can be made for natural transformation's having evolved as a mechanism for genetic exchange among prokaryotes, a similar case seems harder to make for transduction. Transduction is apparently a consequence of mistakes that occur regularly as the **virions** (viral particles) of certain **bacteriophages** (**phages,** or bacterial viruses) are assembled,

Figure 10.3 Frederick Griffith's experiments that led initially to the discovery of the transforming principle and subsequently to the discovery that it is composed of DNA. (A) Mice injected with live R cells. **(B)** Mice injected with live S cells. **(C)** Mice injected with heat-killed S cells. **(D)** Mice injected with live R cells and heat-killed S cells. R cells lack capsules; S cells have them.

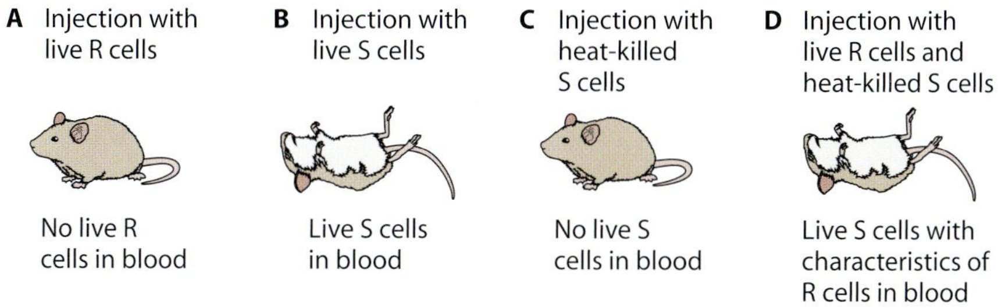

A Injection with live R cells — No live R cells in blood

B Injection with live S cells — Live S cells in blood

C Injection with heat-killed S cells — No live S cells in blood

D Injection with live R cells and heat-killed S cells

Live S cells with characteristics of R cells in blood

although we have to admit that such mistakes might have persisted because they offered the selective advantage of mediating genetic exchange. The heads of these imperfect virions, called **transducing particles**, contain host cell DNA that partially or totally replaces the phage's own DNA. When such a transducing particle subsequently attaches to and donates its DNA to another host cell, it mediates a genetic exchange between the cell in which the transducing particle was assembled and the cell that it subsequently infected.

Two different kinds of mistakes can occur, and each is characteristic of a particular type of phage. One kind results in the formation of generalized transducing particles, which mediate **generalized transduction,** and the other results in the formation of specialized transducing particles, which mediate specialized transduction. The names generalized and specialized refer to how many host cell genes are transduced (moved from one cell to another). Generalized transducing phages can move any of the host's genes; specialized transducing phages can move only certain ones.

Generalized transduction

Among the many phages capable of generalized transduction, P22, which infects *Salmonella*, is probably the most thoroughly studied and is a good example for all generalized transducing phages.

Within cells infected by P22, phage DNA is synthesized (probably by a rolling-circle mechanism [see Fig. 10.6B]) as long concatemers consisting of about 10 phage genomes joined end to end (Fig. 10.4). These are then packed into the phage head by what is called a "head-full" mechanism. Two phage-encoded proteins recognize a frequently occurring 17-bp sequence (pac) in each genome's length of the concatemer, cut there, and lead the cut end of the double-stranded phage DNA into an empty phage head. When the head is filled, a second nuclease cuts the concatemer again, leaving a neatly filled phage head and a remaining concatemer

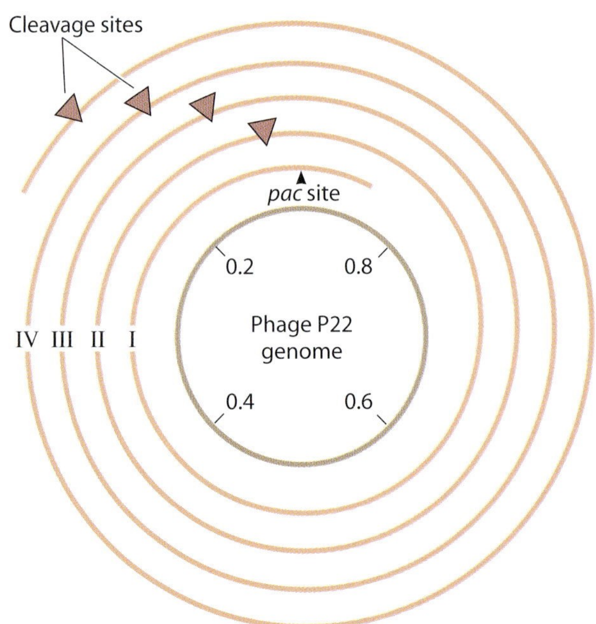

Figure 10.4 Packaging DNA by phages that mediate generalized transduction. The inner circle represents the genome of the phage. The outer spiral represents a newly synthesized concatemer of phage DNA. Packaging starts at the *pac* site and continues until the phage head is filled. Cleavage occurs at the triangles. Successive lengths of DNA (I, II, III, and IV) are each packaged into a separate phage head. Note that the packaged lengths are longer than a phage genome, thus ensuring that each phage particle gets a complete set of genes, no matter where packaging begins.

ready to enter another empty phage head. The head-full mechanism of phage assembly makes generalized transduction possible: pac-like sites also occur in the host cell's chromosome. The host chromosome can be cut at such sites, yielding an end to be packaged into phage heads. When this happens, it generates a set of generalized transducing particles starting from that pac-like site. Such generalized transducing particles contain only host cell DNA. As we said, any host cell gene can be incorporated into a generalized transducing particle, but, as you probably suspect, not with equal frequency. A gene's probability of being incorporated depends on its proximity to a pac-like site. Indeed, wild-type P22 transduces some host cell genes 2,000-fold more frequently than others. However, certain mutant phages, designated HT, for high frequency of transduction, transduce all host cell genes at nearly equal frequencies.

Abortive transduction

Mistakes leading to the formation of generalized transducing particles are surprisingly common: the resulting defective phage particles account for about 2% of the progeny in a P22 lysate. However, when used to infect another culture, DNA from only about 10% of these particles becomes incorporated into the genomes of host cells. Most of the rest of it remains free within the host. Amazingly, the linear transducing DNA from P22 is either rapidly integrated in the host cell's chromosome or it never is. It survives indefinitely in the cytoplasm because a phage-encoded protein (Tdx) joins the ends together, forming a circle. Cells that contain this stable DNA that can be expressed but not replicated are called abortive transductants. Thus, a colony that develops from an abortive transductant contains only a single cell that expresses the transduced genes. If growth depends on their expression, the colony will be quite small, because expression of the genes on that single piece must satisfy the entire colony. If such a tiny colony is picked and plated again, only one of its 10^5 or so cells will develop into a colony, and that colony will also be a tiny one. Note that the abortive transductant continues to divide, producing cells that cannot make the essential gene product. However, their cytoplasm contains half the amount of that protein that is in their mother cells, so they can produce a few more cells.

The phenomenon of **abortive transduction** was discovered in a series of simple but beautifully insightful experiments done in connection with studies of bacterial motility. These experiments are a tribute to careful observation and deserve a revisit now, nearly 50 years later. To score the motility of progeny from P22-mediated crosses, the researchers transferred colonies to semisolid agar, on which nonmotile colonies develop as a compact mass and motile colonies as a spreading swarm of growth. However, the experimenters noticed a third type of colony, which consisted of a "trail" of nonmotile colonies extending from a compact mass. They reasoned that one motile cell as it moved across the plate left, at each division, a series of nonmotile cells, each of which developed into a compact colony (Fig. 10.5). From this observation, they hypothesized, and then proved, the phenomenon of abortive transduction. The one motile cell was an abortive transductant.

Abortive transduction proved to be a valuable tool for the then developing field of microbial genetics because it provided a way to test for the

Figure 10.5 Colonies formed from a nonmotile cell (A), a motile cell (B), and a motile abortive transductant (C).

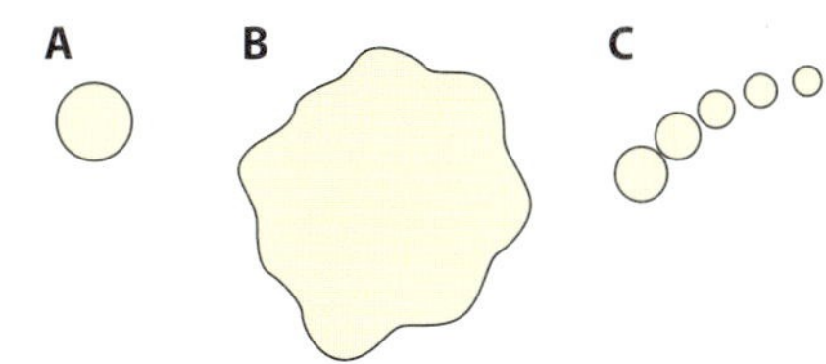

dominance of mutant alleles, since abortive transductants are stably diploid for a small portion of their genome. They are called merodiploids (cells with a complete copy of one genome and a partial copy of another).

Specialized transduction

Specialized transduction is mechanistically distinct from generalized transduction (Table 10.2). Rather than being a consequence of packaging errors, it is a consequence of errors that occur earlier in the life cycles of certain temperate phages (discussed in Chapter 17). The genomes of these phages become integrated into a host cell's chromosome, a phenomenon called lysogeny, and are later excised to undergo vegetative reproduction. Usually, excision is accurate, but rarely, the register of the cuts is shifted to include some host cell genes on one side. As a consequence, some host cell DNA becomes attached to the phage genome and enters the phage head with it. (A corresponding piece of phage DNA is left behind in the bacterial chromosome.)

Most of our understanding of specialized transduction comes from experiments on the well-studied phage λ. Lysogeny, the fundamental characteristic of phage λ, is discussed in Chapter 17. We shall talk here only about how it can result in specialized transduction.

Phage λ lysogenizes *E. coli* by inserting itself at a specific location (*attB*) on the bacterial chromosome. There, it remains under the control of a repressor, called the λ repressor, until some natural or imposed event (for example, exposure to ultraviolet [UV] light) lowers the concentration of repressor, thereby inducing virulent reproduction leading to cell lysis and the release of phage virions (see "Viral Replication" in Chapter 17).

The first step of induction is the excision of the prophage, which usually occurs precisely at the two ends of the genome. Rarely, however—extremely rarely, in fact (at a frequency of about one in a million virions made)—the register of excision is shifted, resulting in some virions (called specialized transducing particles) that contain a bit of host DNA attached to their phage DNA. Because the register might shift in either direction, specialized transducing particles might contain host genes that flank *attB* on either side. In *E. coli*, genes encoding biotin synthesis, *bio*, lie on one side of *attB*, and those encoding galactose degradation (*gal*) lie on the other.

When such a specialized transducing particle infects another bacterial cell and its DNA becomes integrated into that host cell's genome, the host DNA contained in the transducing particle also becomes integrated, creating a short diploid region. If the shift in register that occurred during formation of the transducing phage particle is relatively slight, the phage will not have lost any essential phage genes as a consequence of the incorrect excision. In this case, the specialized transducing particle retains the

Table 10.2 Differences between generalized and specialized transduction

Property	Generalized transduction	Specialized transduction
Mediator	Virulent or temperate phages	Only temperate pages
DNA in transducing particle	Only host cell DNA	Phage and host cell DNA
Genes transduced	Any host gene	Only genes close to insertion site of prophage

capacity for independent growth. If a larger piece of phage DNA is lost, the transducing particle still might be able to reproduce in a host cell if it is complemented (that is, if it is provided with some essential functions) by simultaneous infection with an intact **helper phage.** Such particles are called defective phages and designated λ dgal or λ dbio, depending on which adjacent genes they carry.

Because incorrect excision of the genome of a wild-type phage is a rare event, only a small fraction of phages made are low-frequency transducing particles. However, because precise excision of the genome of a defective lysogen is the overwhelmingly most common event, most of the phages produced from them are transducing particles and become high-frequency transducing particles.

Specialized transduction has been a powerful tool of microbial genetics, made even more so when techniques were developed to engineer bacterial genes near *attB*, thereby making possible the formation of high-frequency transducing lysates that carry any host gene.

Conjugation

Conjugation, the third mechanism of genetic exchange, depends on the properties of certain plasmids. These particular plasmids have evolved the remarkable property of being able to transfer themselves from one prokaryotic cell to another. Some of them, called broad spectrum or promiscuous, have developed this skill to a remarkable degree. They can transfer themselves between species that are only distantly related. For example, some can transfer between almost any pair of gram-negative bacteria, and some such plasmids are able to transfer genes to eukaryotes. The Ti plasmid of *Agrobacterium* can transfer genes to plants, and certain plasmids (R751 and F) can transfer genes to the yeast *Saccharomyces cerevisiae*. The claim has also been made that *E. coli* plasmids can transfer genes by conjugating with animal cells in culture. Such plasmids may be responsible for much of the horizontal gene transfer that torments those who study the evolution of prokaryotes. In addition to genes that encode transfer, some plasmids, designated **R plasmids** or **R factors,** carry additional genes that encode resistance to antibiotics. The impact of these plasmids on public health is profound, because they are largely responsible for the rapid spread of antibiotic resistance within populations of bacteria, with consequences that now present serious problems for clinical medicine.

Conjugation among gram-negative bacteria

The most thoroughly studied **conjugative plasmid** is the F (for fertility) plasmid, or **F factor,** which transfers itself between cells belonging to the same or different strains and species of enteric bacteria. An operon containing 13 *tra* genes on this relatively large (94.5-**kilobase** [**kb**]) circular plasmid encodes its capacity for self-transfer. One of these genes encodes the protein subunits of several pili, called sex pili. Cells bearing such pili are called **F$^+$**, or male (Fig. 10.6A). The ends of these sex pili can attach to cells (designated F$^-$, or female) that lack an F plasmid. Then, the pilus retracts by depolymerizing its subunits, thereby drawing the mating pair together and rendering their attachment more stable. By a rolling-circle mechanism, a single strand of DNA is synthesized from a specific site (*oriT*) on the F plasmid and donated to the female cell (Fig. 10.6B).

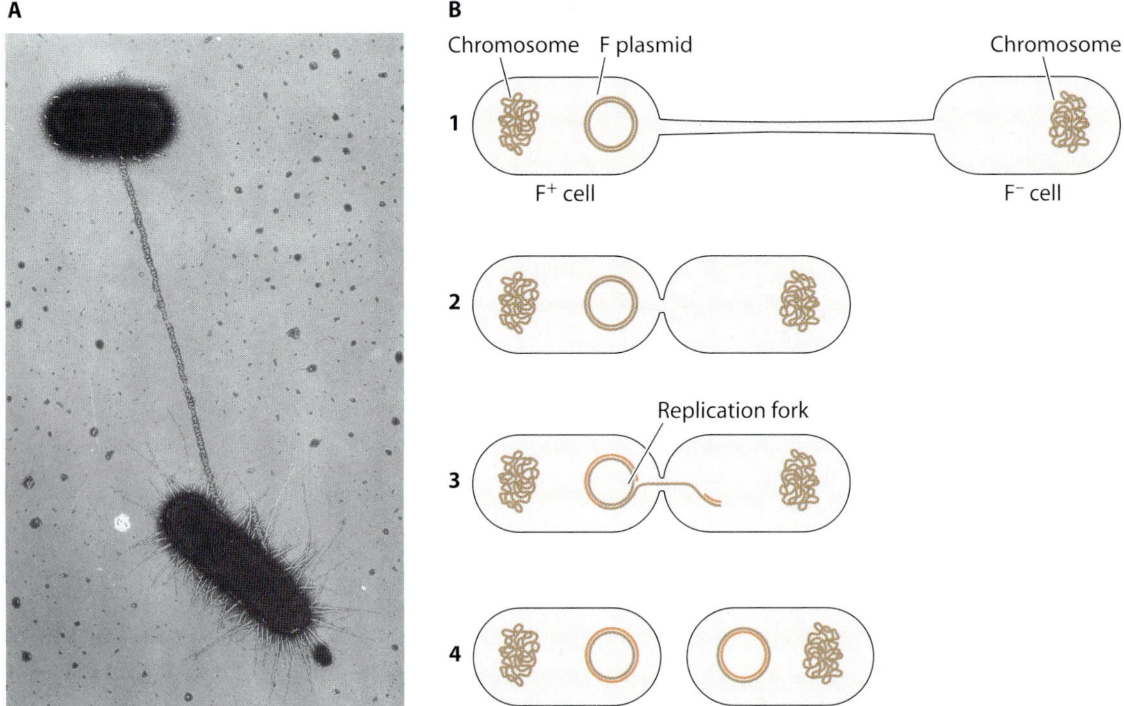

Figure 10.6 F-plasmid-mediated conjugation. (A) Electron micrograph of two *E. coli* cells held together by a sex pilus (step 1 in panel B). **(B)** Steps in the process of conjugation. **(1)** Conjugation begins when a sex pilus on an F^+ cell attaches to an F^- cell. **(2)** The sex pilus retracts, drawing the cells together. **(3)** By a rolling-circle mechanism, the F plasmid is replicated. A nick is made in the double-stranded DNA, forming a replication fork at which DNA replication occurs as it does for the leading strand in chromosomal replication (see Chapter 8). As a result, the single strand of F-plasmid DNA that enters the F^- cell is synthesized. **(4)** In the F^- cell, the single strand is duplicated and circularized, converting the F^- cell into an F^+ cell.

There, it is duplicated and circularized. As a consequence of the mating process, the F^- cell becomes F^+. When cultures of F^+ and F^- cells are mixed, the culture rapidly becomes completely F^+. "Maleness," in this case, is infectious.

Rarely (at a frequency of about 10^{-7}) chromosomal genes enter the F^- cell along with the F plasmid itself, presumably because the F plasmid becomes attached to the chromosome and drags it along into the recipient cell. Indeed, the F plasmid can become integrated into the host cell chromosome by a recombination event between homologous regions, usually matching **insertion elements (insertion sequences)** on plasmid and chromosome. Clones that carry an integrated F plasmid are designated **Hfr** (for high frequency of recombination). They retain properties of F^+ strains and acquire a new one: the F plasmid still expresses its ability to transfer itself but now always brings chromosomal genes with it.

Mating is a rather slow process, requiring about 100 minutes at 37°C for a complete copy of the *E. coli* chromosome to enter an F^+ cell. The DNA enters at a constant rate; therefore, the time delay between the initiation of mating and the entry of a particular gene can be used to locate that gene with respect to the position of *oriT* (the point on F that first enters the recipient cell). Mapping of the relative positions of genes on the *E. coli* chromosome was first accomplished by experiments based on this

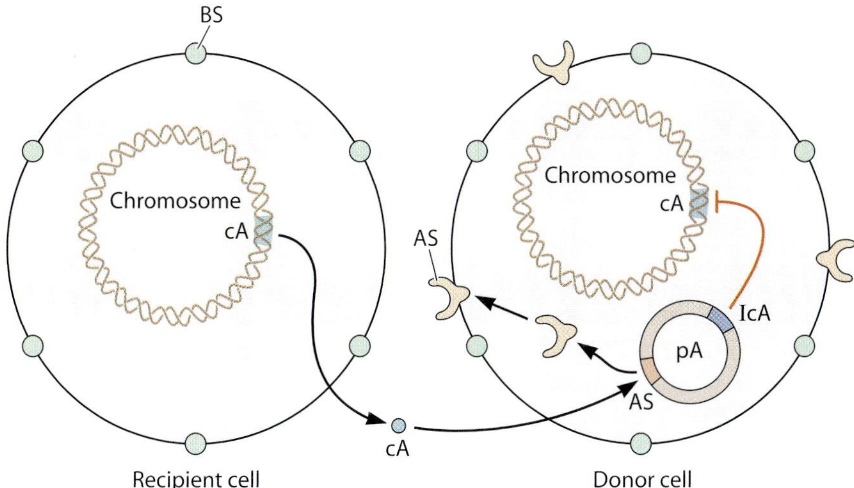

Figure 10.7 Conjugation of *E. faecalis*. The recipient cell produces a chromosomally encoded pheromone (cA) that interacts with a plasmid (pA) in the donor cell, causing it to produce aggregation substance (AS), which migrates to the cell surface. At the surface, AS binds to binding substance (BS) on the surfaces of other cells, forming a clump of cells within which genetic exchange occurs. Although donor cells also carry a gene encoding cA, its expression is repressed by IcA, encoded on pA. Thus, aggregation occurs only among groups of donor and recipient cells.

principle. These interrupted mating experiments involved mixing an Hfr and an F^- strain, agitating them vigorously at intervals to break apart mating pairs, and scoring which genes had entered by the time the mating pair was disrupted.

Conjugation among gram-positive bacteria

Pilus-mediated conjugation is well represented among gram-negative bacteria but seemingly absent among gram-positive bacteria. Instead of mating pairs being attached by a pilus, mating gram-positive cells clump together, forming a mass of cells within which DNA is transferred from donor to recipient by an unknown mechanism. As is the case with gram-negative bacteria, plasmids play critical roles in the conjugation of gram-positive bacteria. The gram-positive gut bacterium *Enterococcus faecalis* illustrates the process (Fig. 10.7). Recipient cells excrete chromosomally encoded pheromones (secreted compounds that signal other members of the same species) that invite donor cells to clump with them. Although donor cells carry the same genes, they do not produce pheromones, because the plasmids in donor cells contain genes that repress their expression. The pheromone enters donor cells, signaling them to produce aggregation substance (also chromosomally encoded), which causes them to clump with recipient cells. Then, genetic exchange occurs.

Curiously the recipient produces pheromones all the time. This constant invitation to aggregate must reflect the importance of conjugation to these species.

MUTATION AND SOURCES OF GENETIC VARIATION AMONG PROKARYOTES

Mutation (heritable change in DNA) is the engine that drives biological variation and, thereby, evolution. Experiments using mutants probably constitute the most valuable set of methods for analyzing cellular functions. Mutant technology depends on comparing mutant strains (those carrying specific mutations) to the strains from which they were derived.

The mutant technique in its broader context requires inducing, inserting, or discovering particular mutations and analyzing their consequences.

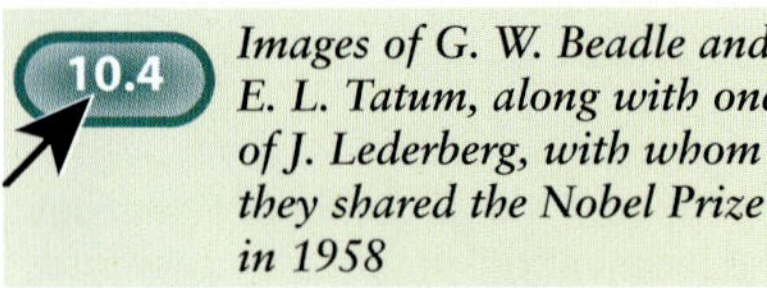

Images of G. W. Beadle and E. L. Tatum, along with one of J. Lederberg, with whom they shared the Nobel Prize in 1958

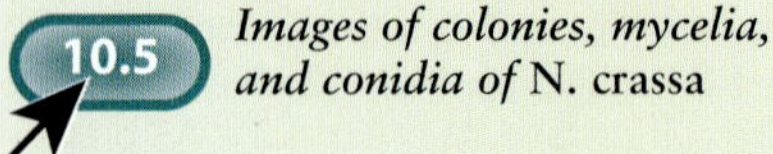

Images of colonies, mycelia, and conidia of N. crassa

This has been and continues to be such a productive and ubiquitous tool of modern experimental biology as to appear almost self-evident, but of course, it had to be discovered. The use of mutants for elucidating biochemical pathways was discovered in 1941 (before genetic information was known with certainty to be encoded in DNA!) from experiments by G. W. Beadle and E. L. Tatum. They UV irradiated (a treatment known to cause mutations) spores of *Neurospora crassa* (the red bread mold) and allowed them to develop into filaments. Because such filaments are haploid, recessive as well as dominant phenotypes are expressed. The UV irradiation induced a rich diversity of mutant strains—including auxotrophic mutants. These are mutants that, unlike their unirradiated parents, were unable to grow on a minimal medium unless it was supplemented with a particular small molecule—an amino acid, a nucleotide base, or a vitamin, for example. By standard genetic crosses and biochemical studies, the sites of the mutations were assigned to particular genes. The consequences of these mutations, for the most part, proved to be the loss of function of a particular biosynthetic enzyme. These results unfortunately became summarized by the appellation "the one gene, one enzyme theory." Later, it was found that the implications of this term are not entirely correct. Several genes encode the various heteropolymeric subunits of certain enzymes, and some proteins encoded by a single gene have more than one enzymatic activity. However, the Beadle-Tatum experiments established a fundamental principle of biology: most genes encode a protein, and that protein's physiological role can be surmised from the phenotypes of strains that have a mutation in it. The Beadle-Tatum experiments laid the foundation of the mutant technique, which has been applied successfully to such diverse topics as pathways of metabolism and complex cascades of signal transduction.

Kinds of mutations

Because the term mutation (any change in a cell's DNA) is so broad, it is useful to subdivide it. Mutations can be classified in two ways: by their **phenotype** (the observable change in appearance or function of a strain carrying the mutation) or by their **genotype** (the chemical change that occurred in the cell's DNA). Some of the more frequently used terms to describe the phenotypes of mutant strains are shown in Table 10.3.

Table 10.3 Types of mutant strains

Designation	Phenotype
Auxotroph	Requires an exogenous growth factor, e.g., an amino acid or vitamin
Carbon source	Unable to use a particular compound as a source of carbon
Nitrogen source	Unable to use a particular compound as a source of nitrogen
Phosphorus source	Unable to use a particular compound as a source of phosphorus
Sulfur source	Unable to use a particular compound as a source of sulfur
Temperature sensitive	Loses a particular function at a high or low temperature
Heat sensitive	Loses a particular function at a high temperature
Cold sensitive	Loses a particular function at a low temperature
Osmotic sensitive	Loses a particular function at high or low osmolarity
Conditional lethal	Unable to grow in a particular environment (e.g., high temperature) in any medium

Table 10.4 Small mutational changes: point mutations and microlesions

Kind	Description	Consequences
Base pair	Change in a base pair	Varies from no effect to complete loss of function
Transition	Substitution of one pyrimidine for another or one purine for another in a base pair, e.g., change of AT to GC	Varies from no effect to complete loss of function
Transversion	Substitution of a purine for a pyrimidine and a pyrimidine for a purine in a base pair, e.g., AT to CG	Varies from no effect to complete loss of function
Silent	Change in a base pair leading to a redundant codon (one that encodes the same amino acid as the original codon encodes)	Usually none
Missense	Change in a base pair leading to a codon that encodes a different amino acid	Varies from little effect to complete loss of function
Nonsense	Change in a base pair leading to a stop (nonsense) codon (UAG is called amber, UAA is called ochre, and UGA is called opal)	Usually loss of function
Frameshift	Gain or loss of one or several base pairs, designated +1, −1, +2, −2, etc.	Usually loss of function

The genotypes of mutations reflect all manner of changes in DNA—some small (called variously **point mutations** or microlesions, small changes) (Table 10.4) and some large (called variously rearrangements or macrolesions, large changes) (Table 10.5).

Sources of mutations

Mutations occur as DNA is being replicated or when it suffers chemical damage. Because mutations are so frequently associated with replication, the **mutation rate** is generally calculated as the number of mutants formed per cell doubling, not per hour or day.

Replication is extraordinarily precise (see "Repair of errors in replication" in Chapter 8). The principal DNA polymerases are abetted by a highly sensitive proofreading capacity: they sense mismatches resulting from incorrect bases that might have been incorporated and eliminate them using their built-in $3' \rightarrow 5'$ nuclease activity. The result is about one mistake in a million. Additional surveillance mechanisms, including a DNA repair system, reduce the error rate to an impressive 1 in 10 billion.

Mutations also result from chemical damage, such as the spontaneous deamination of cytosine, the effects of irradiation, the action of internally generated active oxygen species (including hydrogen peroxide and superoxide), and the results of the insertion of **transposons.** Spontaneous deamination, perhaps the most frequent chemical damage, converts cytosine to uracil, which during the next round of replication pairs with adenine, thereby causing a CG-to-TA transition (uracil is "read" as thymine in DNA replication).

Table 10.5 Large mutational changes: rearrangements and macrolesions

Kind	Description	Change	Consequences
Duplication	Formation of a supplemental copy of DNA, usually in tandem	ABCDEFGHI → ABCDEF-DEFGHI[a]	No loss of function, unless duplication lies within a single gene; increases gene dose
Deletion	Loss of a segment of DNA	ABCDEFGHI → ABC-GHI	Always causes a loss of function
Translocation or insertion	Movement of a fragment of DNA from one location to another	ABCDEFGHI→UVW-DEF-XYZ	Usually causes a loss of function at the site of insertion
Inversion	Reversal of the order of a segment of DNA	ABCDEFGHI → ABC-FED-GHI	Frequently causes a loss of function

[a]Hyphen represents the location of an improper or novel junction. Underlining indicates the changed region of DNA.

Genetics used to be confined to studying genes that are stably sitting on chromosomes. No longer. Our view has changed to include so-called **mobile elements,** such as transposons and retroposons. Transposons, also called jumping genes, are genetic elements that move from one location in DNA to another. All organisms—prokaryotic and eukaryotic—carry transposons in their DNA, and they are abundant. Most prokaryotes carry several different transposons, and it appears that about half of our own DNA might be transposons. There are different kinds of transposons, and they move (or hop) by different mechanisms (Table 10.6), but they all encode a **transposase** that mediates the movement.

When inserted into a new location on the genome, a transposon will, with high probability, interrupt the protein-encoding sequence, causing complete loss of function, just as a **deletion** mutation does. In contrast to deletion mutations, however, certain transposon-caused mutations can revert because some transposons can be excised precisely. Indeed, insertion sequence-caused mutations were first identified as something new because they appeared to be revertible deletions. They have no other recognizable phenotype. In contrast, mutations caused by the insertion of composite transposons (Table 10.6) are usually quite readily identified because of the expressible genes they carry. Transposon Tn*5*, for example, carries a gene encoding an enzyme that catalyzes **phosphorylation** of the antibiotics kanamycin and neomycin, thereby inactivating them. Thus, when Tn*5* has been inserted in the genome of a bacterial strain, that strain becomes resistant to these antibiotics. Likewise, Tn*10* carries genes that encode enzymes that pump the antibiotic tetracycline out of cells, rendering that antibiotic harmless.

Mutagens

A number of chemical and physical agents (**mutagens**) induce mutations (Table 10.7). Some of these mutagens are alarmingly potent. Nitrosoguanidine, for example, can cause a mutation in almost every surviving cell of a treated culture, and many cells will have multiple mutations

Treating a microbial culture with a physical and chemical mutagen can cause a mutation in any of its genes with nearly equal probability. For

Table 10.6 Properties of transposons

Designation	Characteristics
Kinds of transposons	
Insertion sequences	Relatively short pieces of DNA, 750 to 2,000 bp long, that encode only a transposase; designated IS followed by an italicized number, e.g., IS*1*, IS*2*, IS*3*
Composite transposons	One or more genes flanked by matching insertion sequences; designated Tn followed by an italicized number, e.g., Tn*5*, Tn*10*
Mechanism of transposition	
Cut and paste	The transposon is cut out of the DNA where it resides and is inserted in a new location.
Replicative	The transposon is replicated; one copy remains at its original location, and the other is located at a new one.

Table 10.7 Some physical and chemical mutagens

Agent	Mutagenic action
Physical agents	
X rays	Cause double-strand breaks in DNA, the repair of which leads to macrolesions
UV light	Cause adjacent pyrimidines in DNA to join at positions 4 and 5, forming dimers, which in the process of their repair result mostly in transversions, but also in frameshifts and transitions
Chemical agents	
Base analogs	Become incorporated in DNA and then, owing to their ambiguous pairing on subsequent replication, cause transitions
2-Aminopurine	Can pair with either thymine or cytosine
5-Bromouracil	Can pair with either adenine or guanine
DNA modifiers	
Nitrous acid	Deaminates bases; deamination of cytosine produces uracil and then a CG-to-TA transition
Hydroxylamine	Hydroxylates 6 amino group of cytosine, causing CG-to-TA transition
Alkylating agents (e.g., nitrosoguanidine and ethyl methane sulfonate)	Alkylate DNA bases, distorting DNA structure and resulting in a variety of types of mutations
Intercalating agents (e.g., acridine orange and ethidium bromide)	Intercalate between stacked bases in DNA; replication results in frameshift mutations

this reason, if an experimenter is seeking one particular type or location of mutation, it is necessary to select or enrich for it and then, if necessary, screen for it. The ways of doing this are limited only by the investigator's imagination. Two methods can be readily adapted to many schemes.

One, an enrichment method, exploits the fact that penicillin kills only those cells that are growing. Therefore, to enrich for a particular mutation, it is necessary only to put the mutagenized culture in an environment in which wild-type cells can grow but the desired mutant cells cannot—for example, when seeking a particular amino acid **auxotroph,** grow the culture in a medium lacking that amino acid. Then add penicillin. Wild-type cells will start to grow and be killed. The desired mutant (and some others too, of course), which cannot grow because its required amino acid is missing, will not be killed, and thus, its abundance in the surviving culture will be enriched. Once the antibiotic is diluted away, the surviving cells grow normally.

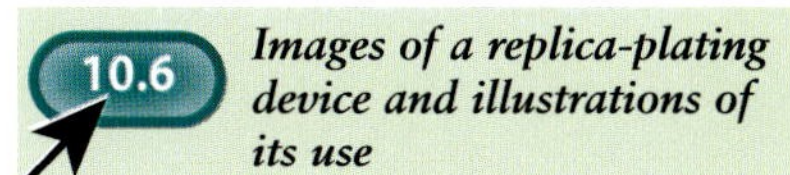

Images of a replica-plating device and illustrations of its use

A second screening method is **replica plating,** which is as powerful as it is simple. Here is how it is used to search for an auxotrophic mutant, one that requires a particular amino acid, for example. A petri-dish-size metal or wooden block covered with a fuzzy fabric, such as velveteen (filter paper also works), is pressed down on a plate with both wild-type and mutant colonies. Two plates, one containing the amino acid and the other lacking it, are then pressed down in sequence on the fabric. Each plate then receives an inoculum from each colony—a replica of the original. By comparing these plates after a period of incubation, colonies that do not grow under certain conditions of nutrition or incubation are apparent, thereby identifying the desired mutant clone. The more modern (and expensive) version of this technique depends on robots that "pick" large numbers of colonies from the master plate and deposit each of them on the two test plates. Replica plating can be extended to search for other kinds of mutants, for example, **temperature-sensitive mutants**. Here, the two replicated dishes contain the same medium, but they are incubated at different temperatures.

Site-directed mutagenesis

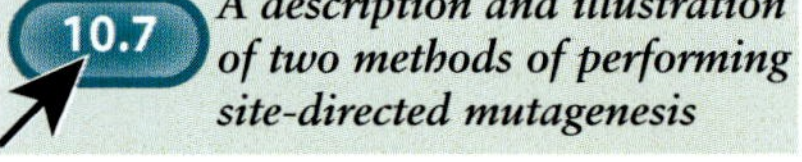

A description and illustration of two methods of performing site-directed mutagenesis

A set of modern methods make it possible (if the sequence of the genes is known) to make specific alterations of DNA in specific genes. An altered bit of DNA is synthesized and inserted in place of the wild-type gene. This technology, called **site-directed mutagenesis,** has become a powerful new tool of experimental biology.

GENOMICS

The traditional pathway of genetic discovery leads from phenotypes and their patterns of inheritance to the genome. This well-traveled route of discovery is as old as genetics itself. Gradually, however, as more was learned about the genome itself, traffic on the pathway began to flow in the opposite direction, the direction that genetic information actually flows, from genome to phenotype. That was the beginning of **genomics,** nowadays the comparative study of whole DNA sequences of organisms. Studying the genome led to glimpses of understanding the cell and learning how it works.

At first, such studies were directed largely toward revealing the relatedness of microbial species and groups. For example, determining the moles percent G+C of a group of organisms' genomic DNA (by measuring its density or melting point) provided some such information. If the values for two organisms were similar, they might not necessarily be closely related, but if the values were significantly different, they most certainly were not closely related. Determining the similarity of genomic DNA from two organisms by their ability to hybridize proved to be an excellent index of relatedness, one that is still used today as a defining property of prokaryotic species. Relative genome size (see "The Nucleoid" in Chapter 3) seemed to offer great promise as a possible index of a cell's complexity or array of talents. Although such information provided interesting correlations, it also revealed perplexing inconsistencies. For example, the two cyanobacteria *Nostoc* and *Synechocystis* are quite similar in most respects, but the genome of the former (9.06 megabases [Mb]) is 2.5-fold larger than that of the latter (3.6 Mb). The genome of the newt (an amphibian) (15,000 Mb) is five times larger than ours (3,000 Mb), and the lily's (100,000 Mb) is over 30 times as large.

Then, a major change occurred. Genomics expanded dramatically when methods were developed to determine the sequence of bases in DNA. The advent of such **sequencing,** as it is called, made possible totally new ways to study microbes (and, of course, other organisms as well). All the information needed for an organism to reproduce and carry out its myriad activities is encoded in the sequence of bases in its DNA. That sequence must also contain all the information we need to understand an organism completely. But how do we extract it?

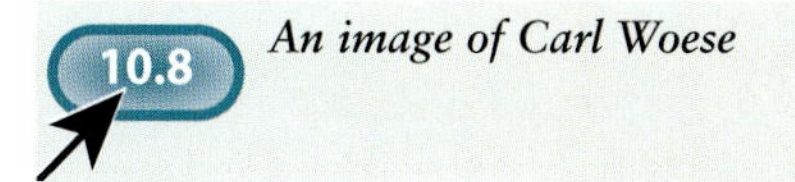

An image of Carl Woese

The first startling information to come from sequencing (actually, RNA sequencing) was the discovery of the Archaea in 1977 by Carl Woese. The sequence of bases in their ribosomal RNA was so different from those of other prokaryotes and eukaryotes that he concluded (rightly, as a wealth of subsequent data now supports) that they constitute a distinct third branch of the biological world.

In the beginning, sequencing DNA "by hand" was so slow, laborious, and boring that it seemed to offer only marginal promise of attaining sufficient data to sequence any organism's DNA completely, much less to acquire enough data for enough organisms to make meaningful comparisons. However, with technological improvements, the situation changed. In 1995, the genome of the pathogenic bacterium *H. influenzae*, which has a medium-size genome, was completely sequenced. More complete sequences were soon announced. Then, in the late 1990s, high-speed sequencing machines entered the scene. Whereas sequencing 20,000 to 50,000 bp by hand was 1 year's work for an individual, the machines could accomplish the same task in about an hour. But even machines can determine only about a 300-bp sequence in a single run. These 300-bp segments must then be arranged in the order in which they occur in the genome, which could be millions or billions of base pairs long. Computers provided the answer. If the random bits of DNA are sequenced at a level of about sevenfold redundancy (repeating the procedure seven times), enough information about overlaps and similar data is gained for a computer to line the bits up to form a contiguous sequence. Still, some hand sequencing (finishing) is required to close seemingly inevitable gaps in the sequences. Now, the sequences of the genomes of hundreds of organisms (including our own species) are available. The problem is no longer acquiring information, it is how to deal with it.

Annotation

A completely sequenced genome is just a list of A's, G's, T's, and C's, albeit a very long one. The process of converting this raw data into meaningful information is called **annotation.** The first step is to ask a computer to search for presumed genes, called **open reading frames (ORFs)**—that is, stretches of DNA each of which is presumed to encode a protein—by scanning the sequence for stop and start sequences separated by gene-like distances. Then, the computer is asked to compare sequences of the discovered ORFs with a database (e.g., a big one called GenBank) of the sequences of genes with "known" functions (we shall discuss these quotation marks below). By such annotation, it is usually possible to assign functions to about half the genes of a newly sequenced organism. Using additional computer programs, these individual functions can be arranged into pathways, those of biosynthesis, for example.

You will note that annotation is not based on experimentation but is based on making comparisons. Subsequent conclusions are only as good as the previous assignments. This powerful approach to understanding biology is not without hazards, and it is comforting to note that we humans still have a critical role to play—both in making biological judgments and in catching mistakes.

Even if there is no match in existing databases for a deduced protein sequence derived from a newly discovered ORF, often it is still possible to learn a lot about it by examining it for sequences, called motifs, with known functions. For example, certain motifs indicate the product protein's ability to bind to particular compounds, such as ATP, **nicotinamide adenine dinucleotide (NAD)**, or DNA. Others might be characteristic of hydrophobic regions likely to be associated with a membrane or to pass through one. Still others might be likely to encode signal sequences (see "Processing of proteins" in Chapter 8), indicating that the product protein is probably secreted into the periplasm or the external environment. Thus, an ORF with an ATP-binding motif may well be a kinase, and one with an NAD-binding motif is probably a dehydrogenase.

As we have seen, annotation is based on comparisons, and even comparisons to "known" functions can be somewhat risky. Such functions were determined by experience in the laboratory. Many of a cell's proteins carry out more than one function, some of which may not be obvious under artificial conditions. It is likely, then, that the function attributed to a gene is the one that is most obvious to the researcher, not necessarily the one that plays a major role for the organism under natural conditions. Annotation of such genes will turn out to be faulty. At best, annotation, as valuable as it is, should be considered tentative.

Relatedness

Genomics has already told us much about the relatedness of various microorganisms, and it promises more. Just as exciting, it tells us about the relatedness and evolution of proteins. On the basis of sequence similarity of genes, their product proteins from various sources across the total spectrum of organisms can be grouped into families, and families of proteins on the basis of lesser similarity can be grouped into superfamilies. These groupings have evolutionary significance. Members of the same family or superfamily, no matter in which organism they are presently found, presumably evolved from a common ancestral protein, regardless of whether they were spread through vertical or horizontal transmission. Deciphering the relatedness of protein sequence similarity is decidedly more straightforward than deciphering the relatedness of organisms.

We have only touched on the wealth of scientific and application benefits that are likely to derive from genomics. Give your imagination free rein.

CONCLUSIONS

There is a common thread to the genetics of all organisms, but the details of genetic exchange differ profoundly between prokaryotes and eukaryotes. Because of the ease and precision with which prokaryotes can be studied, most of our knowledge of basic molecular biology has come from studies of prokaryotes and viruses. Now, with the advent of genomics, studies of genetics have entered a new phase, and the wealth of knowledge accumulated about prokaryotes will be invaluable.

STUDY QUESTIONS

1. A bacterial strain that is defective in the recA gene is unable to undergo homologous recombination.

a. Would such a strain be able to serve as a donor or a recipient strain for generalized transduction? Why?

b. Would it be able to serve as a donor (if it were F^+) or a recipient (if it were F^-) of chromosomal genes? Why?

2. What would be three possible phenotypes and genotypes of a bacterial strain that, as a consequence of mutation, lost the ability to grow on a minimal medium that supported the growth of its wild-type parent?

3. After generalized transducing phages that were grown on a wild-type bacterium were added to a suspension of 10^8 cells of an auxotrophic mutant and the mixture was spread on a petri dish containing a minimal medium (lacking the nutrient that the mutant requires), several dozen colonies developed. How would you determine whether any of these colonies developed from an abortive transductant?

4. Could a **frameshift mutation** generate a **nonsense mutation?** How?

5. How would you use penicillin counter selection to enrich for mutant strains unable to utilize ribose as a carbon source?

chapter 11 evolution

OVERVIEW

Evolution is biology's unifying concept. It provides the framework that supports and makes order out of biology's voluminous and otherwise disparate facts. It provides the guidance for rational research and comparisons. Finally, it carries its own intrinsic fascination by addressing our most fundamental questions. How did the rich diversity of life on our planet arise? Where did all these species we see about us, including our own, come from? We talk about evolution throughout this book. We focus on its microbiological aspects in this chapter.

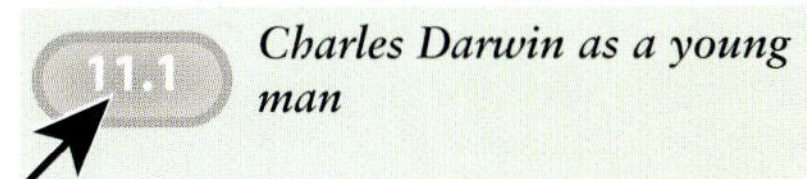

Charles Darwin as a young man

The concept of evolution is a relatively recent human achievement, not yet 150 years old. Charles Darwin's gentle and carefully reasoned masterpiece, *On the Origin of Species by Means of Natural Selection*, which first synthesized the concept into coherent form and convincingly laid out its rational basis—natural selection—was published in 1859. Quickly thereafter, the general outlines of the evolutionary history of the various plant and animal phyla were discovered. The major source of guiding information was the fossil record. By knowing which fossils were present in various geological strata, which at first could be sequentially ordered and later could be precisely dated by radioisotope decay, biologists could follow the rise in complexity of members of certain groups and the disappearance of others almost as though they were looking at a series of time-lapse photographs. Some of these fossils revealed remarkable details about ancient organisms. The feathers of *Archaeopteryx lithographica*, a dinosaur-like bird, are clearly preserved in a famous 150-million-year-old fossil.

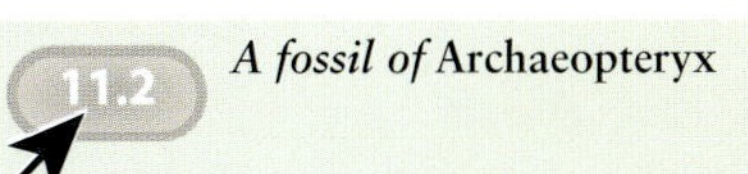

A fossil of Archaeopteryx

Remarkable microscopic details, such as evidence of fungal associations with the root tissues of the earliest land plants, are clearly visible in the fossilized remains. Even fungal spores are sometimes preserved.

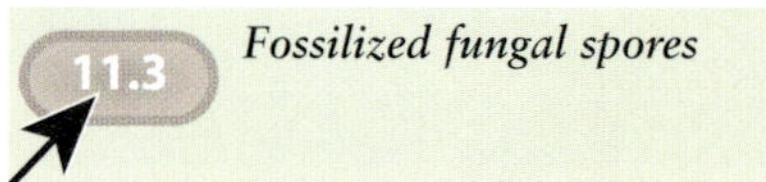

However, this rich, fossil-derived record of evolution is largely limited to the time since the beginning of the Cambrian period, a mere 500 million years ago. For the 3.5 billion years before that, Earth was populated only by microbes, and they did not leave a useful fossil record, except for a few notable exceptions. Thin sections of carbonaceous **chert** (a hard sedimentary rock) from Australia have revealed fossilized remnants of filamentous prokaryotes (Fig. 11.1) that closely resemble modern cyanobacteria. The cherts in which these fossils are imbedded were formed during the early Precambrian, 3.5 billion years ago. Thus, we know that microbes were present, apparently fully developed, only a little over a billion years after Earth was formed (4.55 billion years ago by most estimates) and only about 300 million years after it assumed its relatively smooth spherical shape (3.8 billion years ago).

Therefore, the major portion of evolutionary history occurred when only microbes were present on Earth. They did the evolutionary experiments to develop the genetic and metabolic systems that we inherited and use. We stand at the end of the long evolutionary trail they began to blaze. Before information about the sequence of monomers in cellular macromolecules (protein, RNA, and DNA) became available, there seemed to be no good way to study this long, exclusively microbial era. Moreover, the morphology of modern prokaryotes is too simple to reveal their evolutionary origins.

SEQUENCE OF BASES IN MACROMOLECULES

Soon after the structure of DNA was deciphered and the **central dogma** of gene expression (DNA to RNA to protein) was established, it became clear that the sequence of monomers in an organism's macromolecules held a wealth of information about its evolutionary history. The identity and position of each monomer in the macromolecule was recognized as being a distinct character, capable of individual variation. Therefore, determining the sequence of monomers comprising homologous macromolecules from a set of organisms should reveal relationships among those organisms and their evolutionary connections. Moreover, if the changes in sequence were selectively **neutral** (neither beneficial nor harmful) and occurred randomly, they should constitute a useful **molecular clock**. The number of differences between homologous macromolecules in two organisms should be a measure of the time since they shared a common ancestor. The rate of the clock's ticking could be estimated by comparing the probable appearance of certain microbes with some outside event, such as the time in geological history when Earth's atmosphere became sufficiently enriched in oxygen to produce strata containing iron oxide. That must have been the time when oxygenic photoautotrophs became dominant. But which macromolecule should be chosen for these studies?

Figure 11.1 Fossils of filamentous cyanobacteria found in a rock in the Bitter Springs Formation of central Australia. The formation, and therefore the fossils, is approximately 850 million years old.

Small-subunit ribosomal RNA

In the early 1970s, Carl Woese began to accumulate partial sequences of the small subunit of ribosomal RNA (ssRNA) from a number of prokaryotes by an innovative method: he hydrolyzed this RNA into oligomers and sequenced them. His results were the basis of the discovery that the

Archaea are a distinct domain, and they also revealed unexpected relationships among various groups of bacteria. For example, the wall-less mycoplasmas do not constitute a distinct third major group of bacteria, as was commonly believed; rather, they are so closely related to gram-positive bacteria that the two constitute a single group. ssRNA, which Woese chose to examine, had the dual advantages of being present in all cellular organisms and having a single evolutionary origin. ssRNA also proved to offer another significant advantage. Some regions within the molecule (primarily the double-stranded regions) evolve slowly, thereby preserving sufficient similarity over long stretches of evolution for relationships between distantly related organisms to be discerned. Other regions change rapidly, generating sufficient differences between closely related organisms to reveal their evolutionary connections.

Soon, rapid methods of DNA sequencing became available, and large numbers of ssRNA sequences were determined by the much easier method of sequencing their genes. Computer programs were developed to process this huge volume of data and to present it in a graphic summary called a **tree**. The diagram that links representatives of all groups of organisms is sometimes called the **tree of life** (Fig. 11.2). It shows all the evolutionary branches leading to all extant groups of organisms. Phylogenetic trees can be presented in the form of a **radial tree** or as a **dendrogram** (Fig. 11.3). Sometimes the branches of trees are drawn as wedges of various breadths to indicate how many species are represented (Fig. 11.4). In any of these forms, the trees contain more information than just which organisms share a common ancestor. The lengths of the lines

Figure 11.2 The "tree of life," a representation of the evolutionary branches leading to the extant groups of organisms. Note that most of the groups in the three domains of life correspond to microorganisms.

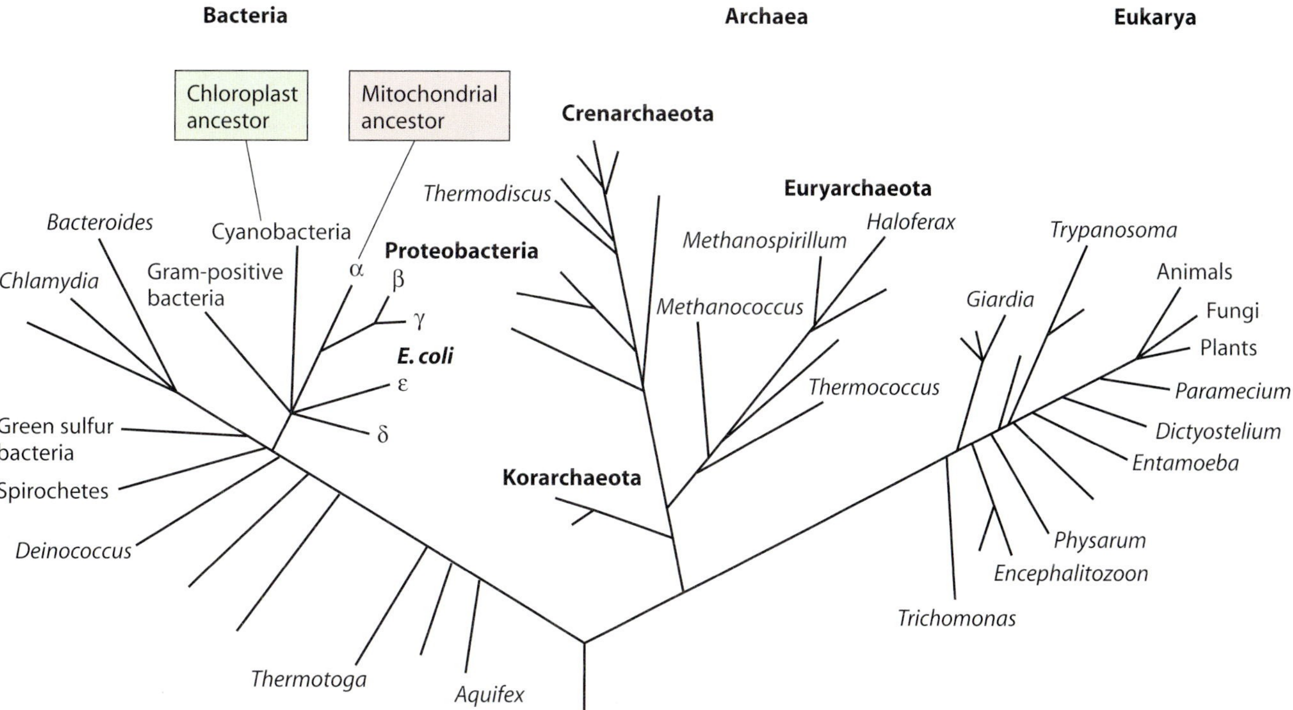

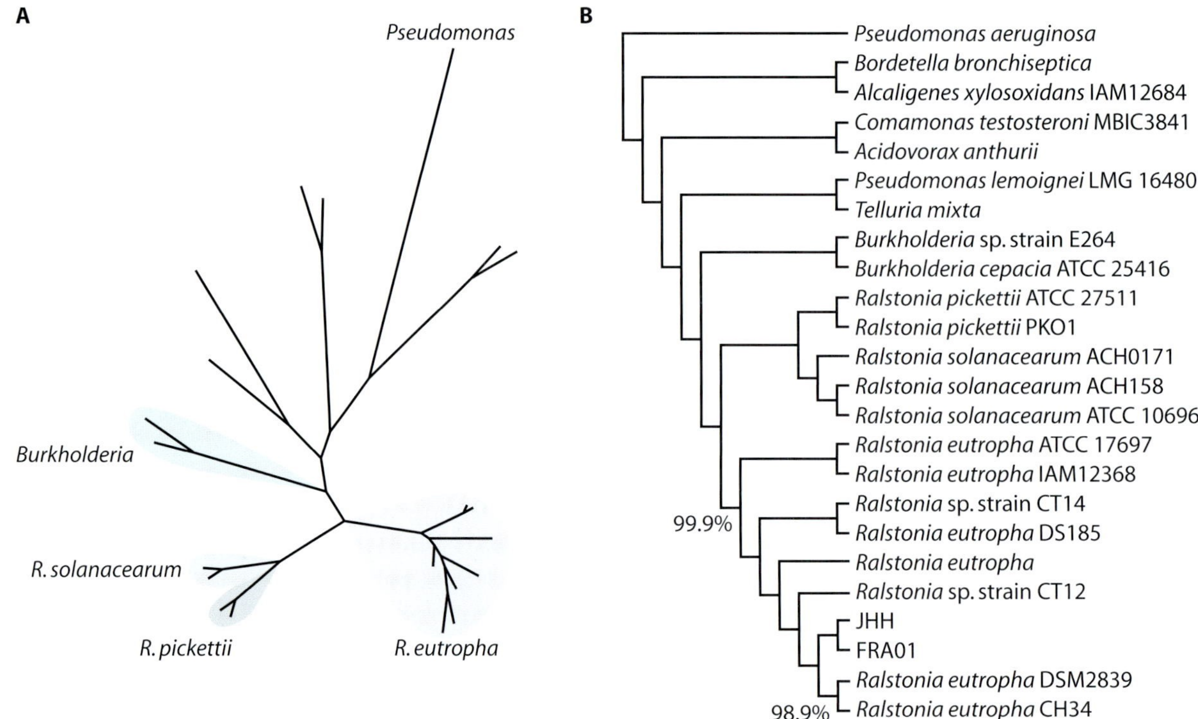

Figure 11.3 Relationships among Bacteria. Relationships among certain Bacteria are shown in the form of a radial tree **(A)** and as a dendrogram **(B)**.

between **nodes** (branches or end points) are proportional to phylogenetic distance; therefore, the tree offers information about the phylogenetic distances separating all the organisms that are included.

The information gained from ssRNA sequences has led to a completely new picture of microbial evolution. Microbiologists once speculated that evolution must have been linear. That is, the most ancient organisms, represented by present-day Archaea, gave rise to Bacteria, which in turn gave rise to eukaryotes. The information gained from ssRNA sequences suggests a different story. It says that early in evolution, these three major lineages branched from a common stem, and each lineage underwent independent evolution (Fig. 11.2).

Of course, significant questions remained unanswered. For example, what were the properties of the ancient, probably now extinct, ancestors of the three major lineages of present-day microbes? In particular, what were the properties of the last common ancestor of all of them? We'll return to these questions below.

Proteins as markers of evolution

Of course, the sequence of nucleotides in ssRNA isn't the only repository of evolutionary information. The sequences of amino acids in proteins also hold such information. As more complete genome sequences became available, the amino acid sequences of various proteins from widely distributed microbes were examined in the same way as the nucleotides in ssRNA had been studied. The results were unsettling. Sometimes trees constructed from protein sequences differed from those based on ssRNA sequences. In fact, trees based on one particular protein differed from

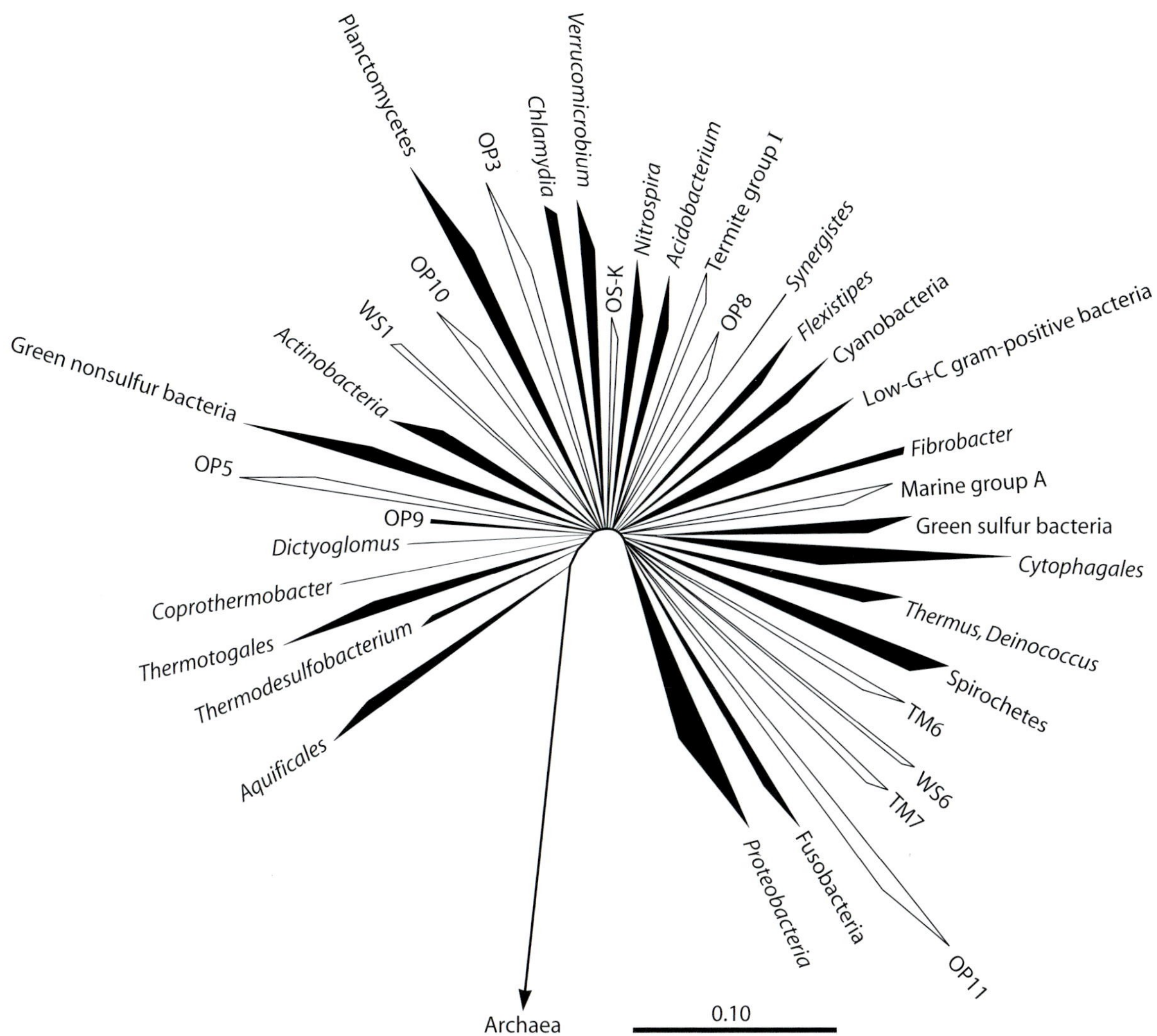

Figure 11.4 Relationships among Bacteria presented as wedge diagrams. The depth of each wedge reflects the branching depth of the group. Wedges shown in outline are for groups that contain no species that has been cultivated.

those based on another protein. Quite clearly, all the macromolecules in a particular microbial cell could not have evolved together; their encoding genes could not have been passed solely by **vertical gene transfer** from one generation to the next. Some of these genes must have been passed by **horizontal gene transfer** between lineages. We know that horizontal gene transfer among microbial lineages is possible because some of it occurs today, for example, when a broad-host-range plasmid is passed from one bacterial species to a distantly related one.

Horizontal gene transfer confounds tracing **phylogeny** deduced from comparing sequences (Fig. 11.5). If a transfer occurred between lineages, the evolutionary history of that lineage is obscured. The evidence that this has happened is powerful. The aminoacyl-transfer RNA (tRNA) synthetases are an example. These enzymes all catalyze similar reactions—attaching an amino acid to its tRNA molecule—but there are about 20 of them, each specific for a particular amino acid. It is common to see archaeal versions of some, but not all, of them scattered throughout the Bacteria.

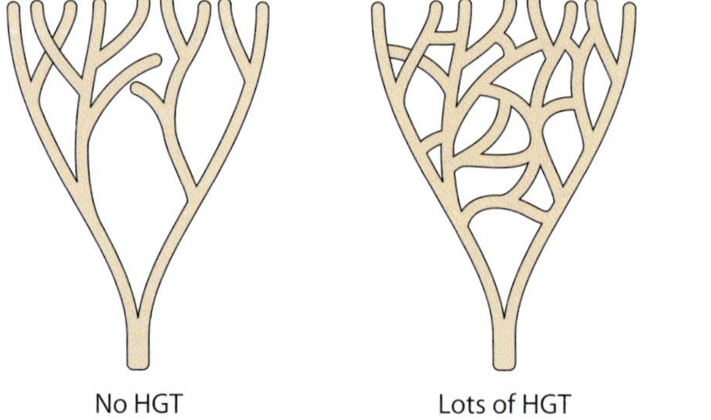

Figure 11.5 Horizontal gene transfer (HGT). Three possibilities are shown. **(A)** There has been no HGT. **(B)** HGT was frequent at all times. **(C)** Some HGT took place, mainly early in the history of evolution.

Did horizontal gene transfer occur so often that it erased traces of the evolutionary trail? Most microbial evolutionists believe (or at least hope) that is not the case. There are reasons to be optimistic. In higher organisms, where the path of evolution is readily mapped by consulting the fossil record, molecular methods based on ssRNA indicate the same path. One might argue that although successful horizontal gene exchange might have been most frequent during the early stages of microbial evolution, when microbial functions were primitive and readily improved, such transfers would be increasingly rare as cells became more complex and less readily improved. Highly interdependent properties are more resistant to evolutionary change.

There are inevitable difficulties in using sequences to reconstruct evolution. One is the problem of multiple changes in sequence. Differences in the sequences of two extant organisms have not always occurred since they evolved from a common ancestor. Intermediary changes may have occurred, for example, a difference between "a" in one organism and "b" in another does not preclude the possibility that the change was from "a" to "c" to "b." Even identity at corresponding positions in a sequence may be the result of forward and backward changes.

Most biologists hold the view that life on Earth began only once. The universality of the genetic code, the common themes of metabolism among organisms, and the many other similarities of living things speak powerfully for that view. It is striking, for example, that lactic acid bacteria, human muscle, and pea seeds make lactic acid from glucose by an identical series of enzyme-catalyzed reactions and that the enzymes that catalyze homologous reactions in these various organisms are themselves similar. Moreover, it seems probable that the molecules from which the most primitive life form was assembled could accumulate only in an environment without the presence of organisms that would certainly consume them as nutrients. As pointed out in Chapter 1, this precludes new life forms from emerging in a buggy world.

THE UNIVERSAL ANCESTOR

The conviction that life began only once leads to its own quandaries. The best data now available indicate that the three major lineages of extant life—Bacteria, Archaea, and Eukarya—separated early in their evolutionary development. At that branch point, they must have shared a common ancestor, the **universal ancestor.** What was the universal ancestor like? By any logical analysis of evolution theory, the universal ancestor had to possess all the characters shared by the three domains, and that is a long list. It includes not only the genetic code but also a large package of other properties—fueling pathways, the mechanism of protein synthesis, and cellular structure, to name just a few. This logic leads immediately to a major quandary. Such a universal ancestor must have been highly sophisticated (at least metabolically) to possess all these properties. In addition, it must have existed in this full-blown state only around a half billion years after life appeared. How could such a complex organism have evolved so quickly? This quandary causes problems for what many, but not all, scientists believe about evolution, namely, that it occurs through slow, gradual changes. The universal-ancestor dilemma implies a sudden evolution producing a complex organism and then, over the next 3 billion years, further diversification. Explosive evolution of this sort has been postulated to occur periodically at other, more recent, stages in evolution as well, when new or unoccupied ecological niches become available. This theory has been called **punctuated equilibrium.**

ORIGIN OF LIFE

The puzzle of a complex universal ancestor pushes us back further toward considering the origin of life and how the first self-replicating entity developed metabolic complexity. The first living entity must have been a very simple one, possibly even too simple to be called an organism. The term **progenote** is sometimes used. Many scenarios have been advanced for how the progenote might have come into being—compounds accumulating on some adsorptive surface, such as a clay particle, coming from another planet (although this proposal is not an explanation for how life began.), etc. All proposals share the supposition that molecular components of living things, purines, pyrimidines, amino acids, etc., must have accumulated in the prebiotic world. Chemical evidence supports the feasibility of such accumulation. The first evidence was a celebrated experiment carried out by Stanley Miller and Harold Urey in the early 1950s. By passing an electrical discharge through a mixture of gases (hydrogen, water, ammonia, and methane) that they presumed to be the constituents of Earth's prebiotic atmosphere, they generated many of the constituents (building blocks) of living things, including amino acid and nucleic acid bases. Although earth scientists now believe Miller and Urey guessed wrong about the composition of Earth's prebiotic atmosphere, the experiment convincingly established that life's building blocks form readily and could have accumulated in a prebiotic world.

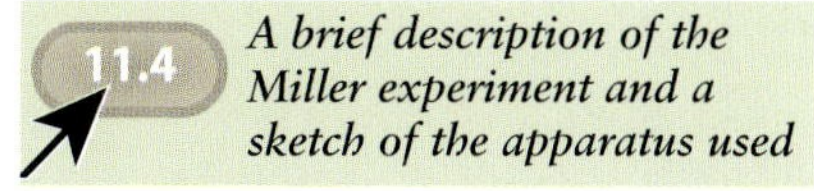

A brief description of the Miller experiment and a sketch of the apparatus used

Most such proposals of how life might have begun on Earth are variations of the one called the "warm little pond" that Charles Darwin offered in a letter to a friend in 1871.

> My dear Hooker,
>
> . . . But if (and oh, what a big if) we could conceive in some warm little pond, with all sorts of ammonia and phosphoric salts, light, heat, electricity, etc., present that a protein compound was chemically formed,

ready to undergo still more complex changes, at the present day such matter would be instantly devoured or absorbed, which would not have been the case before living creatures were formed. . . .

Yours affect'
C. DARWIN

The consensus today is that the earliest forms of life were based on RNA, i.e., that there first was an RNA era. This idea seems plausible, because we know now that RNA molecules can do it all. They can serve as a repository for accumulated genetic information, as they do for certain viruses; they can catalyze chemical reactions as ribozymes (RNA molecules with enzymatic activity). Indeed, one of the most fundamental metabolic reactions, the formation of the peptide bonds that link amino acids to form proteins, is catalyzed by a ribozyme in the large ribosomal subunit. Is this a remnant of the progenote's first experiments with making proteins? One can imagine the RNA era evolving into an RNA-protein era and finally the present DNA-RNA-protein era.

Progenotes probably did not do anything very well by present standards, but being the only life form, they were spared having to compete with more competent organisms. Certain aspects of their weaknesses even offered advantages. Their genetic information probably had a high rate of mutation, which in turn led to rapid evolution.

Figure 11.6 The Luria-Delbrück experiment. Cultures started from small inocula that were unlikely to contain phage-resistant cells were grown to full density and plated on phage-containing plates. **(A)** Multiple samples from one culture all produced about the same number of phage-resistant colonies. **(B)** Numbers of phage-resistant colonies resulting from samples taken from individual cultures showed considerable variation.

A

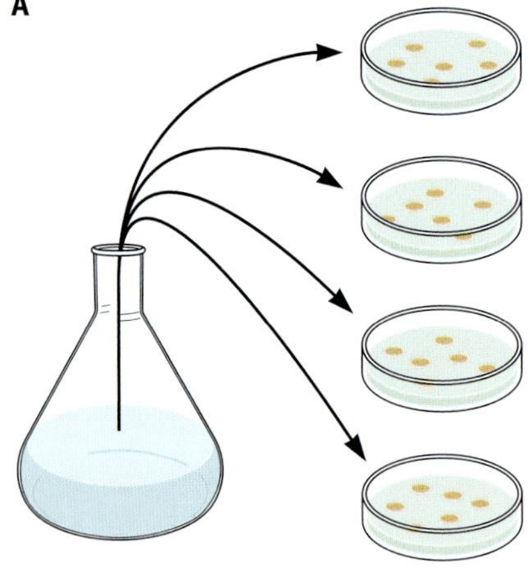

B

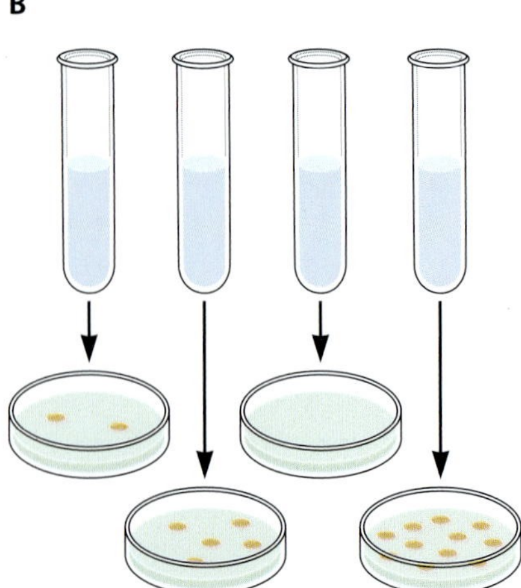

MECHANISMS OF BACTERIAL EVOLUTION

In his seminal book, *The Origin of Species,* Darwin proposed that evolution proceeds by natural selection, which, by acting on naturally occurring variation, favors advantageous characteristics. The essence of his theory is the existence of characteristics prior to the intervention of an environmental challenge. The challenge merely selects a favorable characteristic; it does not cause one, as his predecessor Jean-Baptiste Lamarck suspected.

Darwinism

In 1943, Salvatore Luria and Max Delbrück provided elegant support for Darwin's proposal for natural selection of preexisting variations. As was well known, if large numbers of susceptible bacteria are mixed with virulent phages and plated, a small number of phage-resistant colonies always develop. Luria and Delbrück designed an experiment to determine whether exposure to phages induces some bacterial cells to become phage resistant (as Lamarck would have argued) or merely reveals the preexistence of phage-resistant cells by killing all the phage-sensitive ones (as Darwin would have predicted). Luria and Delbrück showed that the latter was the case (Fig. 11.6). They grew a large number of separate cultures of susceptible bacteria from small inocula, each too small to contain, with reasonable probability, phage-resistant cells. When the cultures were fully grown, a sample from each culture was mixed with phage and plated. The number of colonies that developed was highly variable—much more variable than multiple samples taken from a single culture. These results convincingly fit a mathematical model that attributes the high degree of variation among the individual cultures to the point during the culture's growth at which a mutation may have occurred. The cultures in which mutations occurred early would have large numbers of resistant progeny cells, whereas mutations occurring later would produce fewer resistant progeny (Fig. 11.7). If exposure to phage induced phage resistance, variation among samples from different cultures and from the same culture would be invariant.

Nine years later, Joshua Lederberg and Esther Lederberg, using their newly devised technique of replica plating (see Chapter 10), did a much simpler but possibly even more convincing experiment to prove the same point. They replicated a plate of microcolonies to fresh plates and infected one of them with phage. Picking cells on the phage-free plate from the region corresponding to where phage-resistant colonies had developed on the phage-infected plate, they grew another plate of microcolonies and repeated the procedure. Eventually they isolated a clone of phage-resistant cells that had never been exposed to phage. Without doubt, its resistance was *not* phage induced.

Neo-Lamarckism

The elegant experiments of Luria and Delbrück and those of the Lederbergs convincingly established that some mutations arise independently of selection, but of course, that does not prove that all mutations do. In 1988, John Cairns and associates published experiments suggesting that stress might induce specific mutations to overcome the restrictions it imposed. Specifically, they found that when cells unable to utilize lactose because of a mutation in a gene encoding β-galactosidase were challenged to utilize that substrate in order to grow, they apparently underwent corrective mutation, thereby permitting growth. These mutations occurred at a rate about 100-fold higher than if the cells were unchallenged. It seemed as though stress had stimulated favorable mutations to occur. This implication was controversial, because it suggested a Lamarckian mechanism rather than an adaptive (Darwinian) mechanism as the cause.

Figure 11.7 Variation of numbers of mutant progeny in a clone. The number of mutant progeny in a clone depends on when the mutation occurs. As shown in culture 1, if a single mutation occurs during the first generation, there will be eight mutant cells after the fourth generation. However, as shown in culture 2, two mutations result in only six mutant cells if they occur during the second and third generations.

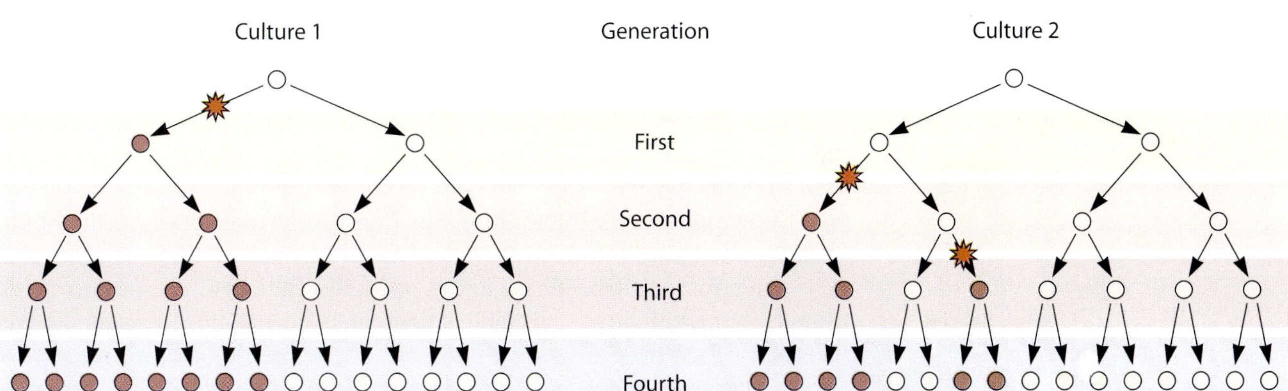

In 1998, John Roth and associates showed that Darwin remained unchallenged: the imposed stress was not selectively or even generally mutagenic. They demonstrated that the mutant strains were slightly leaky, that is, they expressed a small amount of β-galactosidase activity. In these mutants, the gene in question was duplicated, and up to 10 copies were present in each cell. In such a leaky mutant, this was sufficient over several days to produce a microcolony containing about 10^5 cells. In a population of this size, it is quite likely that a favorable mutation will occur. The likelihood is increased because **duplications** are corrected. The process of correction transiently produces single-strand stretches of DNA, which in turn induce a stress response known as SOS (see "The Connection between Cell Division and DNA Replication" in Chapter 9). Part of this response is due to the induction of a DNA polymerase that is particularly prone to making errors. This increases the mutation rate.

In other words, these researchers showed that stress did not induce specific favorable (beneficial) mutations. Rather, the imposed conditions *selected* for preexisting duplications, which, by permitting some growth, generated large numbers of target genes, thereby increasing the probability of mutations, some of which could correct the defect in β-galactosidase.

This mechanism, called **amplification-mutagenesis**, depends, of course, on tandem duplications forming frequently. This is indeed the case. Duplications form frequently by recombination between sister chromosomes and are lost frequently by recombination between regions of their own redundancy (Fig. 11.8). As a consequence, cells with duplications of any particular gene are present in a bacterial culture at a frequency of about 10^{-4} to 10^{-5}.

Duplications probably occur at comparably high frequencies in other organisms as well. Thus, in humans, amplification mutagenesis may also account for other mysterious hard-to-explain biological phenomena, such as the occurrence of groups of linked mutations that are associated with certain cancers and the evolution of new genes.

EARLY EUKARYOTES AND ENDOSYMBIOSIS

Many aspects of early microbial evolution are just speculation, but the origins of mitochondria and chloroplasts in eukaryotes are clearly established. Both of these organelles are significantly changed descendents of bacteria that were engulfed by a primitive eukaryotic cell. Comparing DNA sequences establishes that mitochondria are derived from a captured bacterium in the group *Proteobacteria* and that chloroplasts are derived from a captured cyanobacterium (these groups of bacteria are described in Chapter 15).

Capturing bacteria and keeping them to perform specific metabolic tasks is not limited to mitochondria and chloroplasts. For example, cells of aphids maintain bacterial endosymbionts that produce amino acids, thereby supplementing the aphid's protein-deficient diet of plant sap (see Chapter 19). These endosymbionts are undoubtedly of more recent origin than mitochondria and chloroplasts. Although adapted to being endosymbionts long enough to no longer be capable of independent growth,

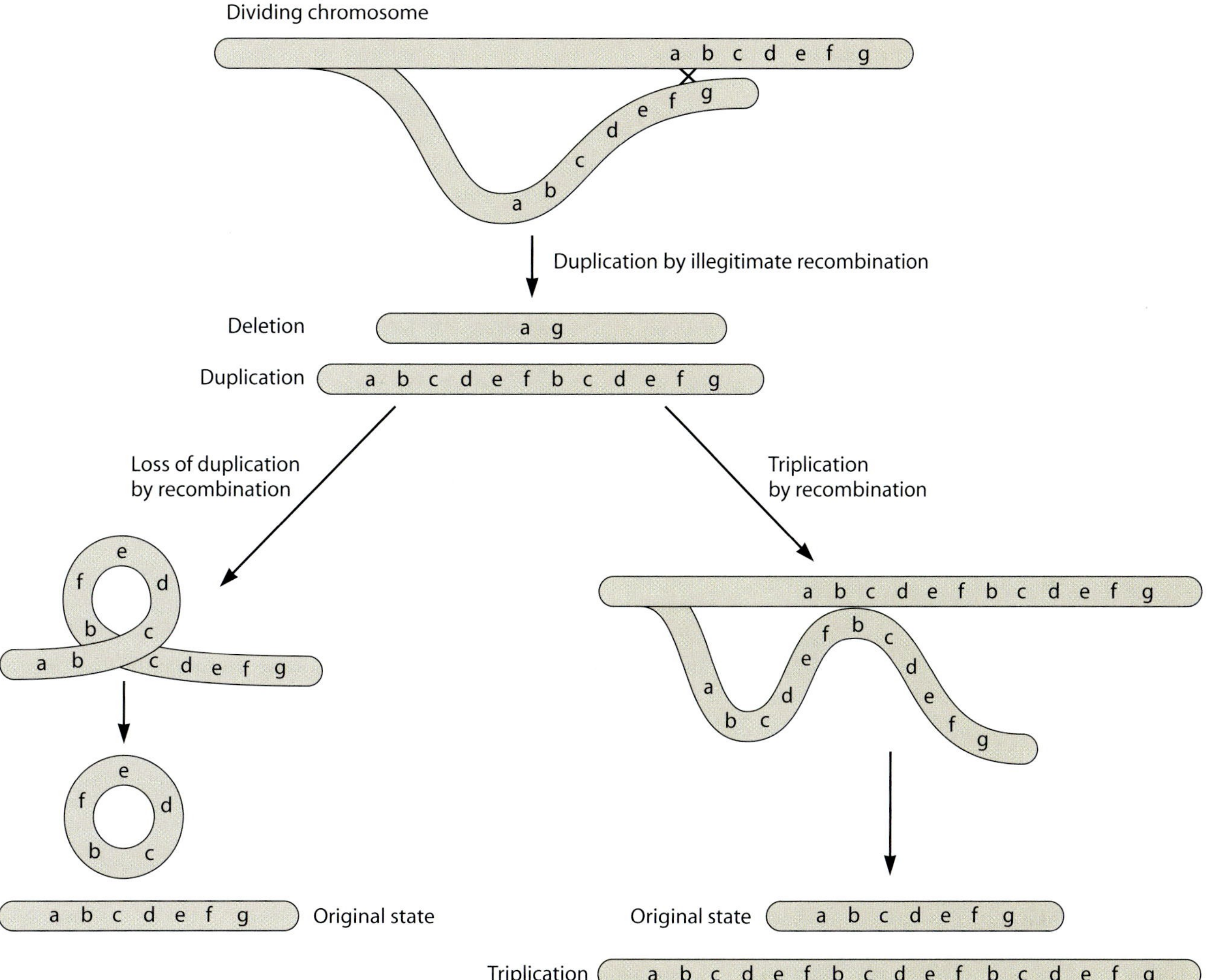

Figure 11.8 Possible mechanism for the formation, loss, and amplification of duplications.

they are still readily identifiable as bacteria *(Buchnera)* related to *Escherichia coli.* The eukaryotic world is filled with other examples of bacterial endosymbionts, although the selective advantages offered by most of them remain unexplained.

Certain primitive eukaryotic microbes (those located on deep branches of the tree of life, such as the groups that include the waterborne pathogens *Cryptosporidium* and *Giardia*) lack mitochondria, but there is evidence that they once had them. Some bacterial-type genes that might have come from a previously lost endosymbiont exist in the nuclear DNA of these organisms. The organisms belong to a group of eukaryotes (as judged by the tree of life) that branched from the main line of evolution at an ancient time, and gain followed by loss or genomic reduction is a characteristic more easily associated with more recently branching **clades** (groups of species derived from single ancestors). Recent evidence indicates that *Giardia* may contain **mitosomes,** remnant mitochondrial organelles bounded by double membranes, indicating that they are a consequence of reductive evolution.

EVOLUTION OF MOLECULES

The progenote's genome must have been small, and its expansion as evolution progressed probably occurred by gene duplication. This process is extremely common today (see Chapter 10), as it must have been throughout evolutionary history. A pattern of duplication and subsequent improvement and specialization implies a different form of evolution than the evolution of species, namely, the evolution of genes, and it introduces the concept of protein evolution. Protein evolution is possibly the most fundamental aspect of the overall process, and it might be the easiest to reconstruct. Proteins are, after all, the primary agents of biological processes. Their evolution can be studied independently of their distribution among organisms.

With the wealth of data now available from the sequences of so many organisms, it has become obvious that proteins, like organisms, are related. On this basis, proteins are assigned to **protein families**, members of which presumably share a common evolutionary history. Groups of protein families that exhibit distinct similarities are collected into **protein superfamilies**. About 7,500 protein families have been identified so far, a number that may be plateauing.

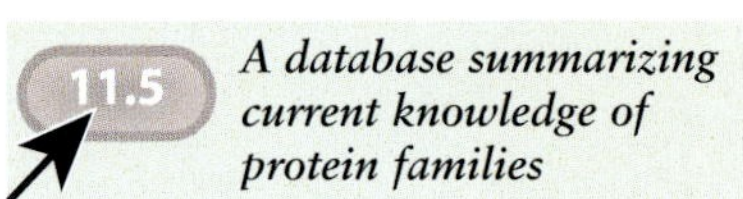

A database summarizing current knowledge of protein families

Remarkably, members of the same protein family are spread through the entire world of organisms. Even more remarkably, enzymes belonging to the same protein family do not necessarily catalyze the same reaction or, at first glance, even closely similar reactions. This new insight into protein evolution offers a plausible explanation of how it could have occurred as rapidly as it did. Each protein need not have evolved independently. Rather, it might merely be the product of a side branch of a complex evolutionary tree.

Even more insightful revelations have come from the study of **protein domains** (regions of a protein with distinct morphologies or functions). The existence of domains with specific functions, such as binding ATP or NAD, in many proteins has profound evolutionary implications. Presumably, proteins evolved, at least in part, by combining domains from different sources into a single protein and thereby generating a new enzymatic activity. For example, combining into a single protein an NAD-binding domain with one that binds a specific substrate might generate a new dehydrogenase. It seems probable that many enzymes might have evolved by this modular mechanism. Indeed, some scientists have proposed that **introns** (intervening noncoding regions within genes), which are common in the genes of eukaryotes and also occur in some prokaryotic genes, may have evolved to facilitate recombination leading to new enzymes composed of new sets of domains.

In addition to recombining domains, protein evolution must have proceeded by two mechanisms.

1. A gradual change of an existing activity by mutation and selection, along with vertical transmission; proteins belonging to such a line of descent are **orthologous**.

2. Duplication of a protein's encoding gene and independent mutation and selection of the two resulting lines; proteins that are related in this way are **paralogous**.

Progress over the past three decades toward understanding microbial evolution has been astoundingly rapid, but many questions remain. Possibly the most profound change that has occurred during this period is an increased sense of optimism for those involved in the research and those who follow it. Before information about sequences of macromolecules was available, most microbiologists thought the details of microbial evolution might forever remain a mystery. Now a broad consensus believes that in time microbial evolution will be understood.

CONCLUSIONS

The concept of evolution is the guiding principle of biology that organizes and makes sense of its voluminous diversity and disparate details. Most evolutionary accomplishments—metabolic, genetic, and cellular—occurred among prokaryotes before eukaryotes appeared. Evolutionary relationships among prokaryotes remained unknown, owing to a sparse fossil record, until methods for determining the sequences of macromolecules became available. Because it is universal and some of its parts evolve slowly, ribosomal RNA became the molecule of choice for molecular studies of microbial evolution. Studies of proteins have largely supported these conclusions but have also highlighted the complexities introduced by horizontal gene transfer. Studies of bacteria have confirmed theories relating to certain aspects of evolution. Overwhelming evidence suggests that all life on Earth stems from a single origin and that mitochondria and chloroplasts are the modern evolutionary products of anciently captured bacteria. Sufficient molecular data have accumulated to begin to study the evolution of proteins.

STUDY QUESTIONS

1. How would you define biological evolution?

2. Both the fossil record and the sequence of bases in the DNA of extant organisms provide information about evolution.

a. Why is the latter more useful for the study of the evolutionary relationships among microbes?

b. In using the fossil record to study the evolutionary connections between pairs of organisms, the term "missing link" is sometimes used to suggest that one more fossil is needed to establish a direct connection. Is there an analogous concept when base sequences are used to establish connections between organisms? If so, what would it be?

c. The fossil record of microbes, although thin, has contributed information that studies of base sequences could not supply. What was this information, and why could base sequence data not supply that information?

3. What properties must a gene (or set of genes) possess for its sequence to be useful in assessing the evolutionary relationships among organisms?

4. Explain how studies of the sequences of two different genes might suggest different evolutionary relationships among a group of organisms.

5. What arguments support the existence of an "RNA era" during evolution?

6. Support the statement that bacteria are responsible for all biologically mediated oxygen production and utilization.

7. How might protein domains speed the rate of protein evolution?

section V physiology

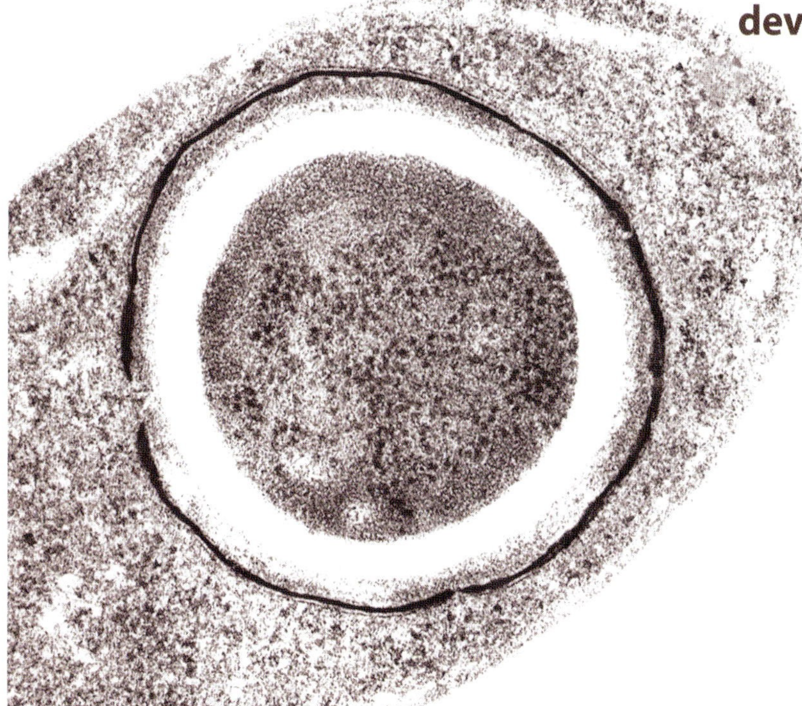

chapter 12 coordination and regulation

INTRODUCTION

Our survey of bacterial metabolism and growth (see Section III) noted that a bacterium is built—and performs its life functions—by the activities of perhaps 2,000 chemical reactions. We introduced the notion of phases of metabolism (fueling, biosynthesis, polymerization, and assembly), we discussed the organization of metabolic pathways, and we described the complex overall sequence of biochemical events leading to the synthesis of a living cell from simple organic or even completely inorganic materials.

We would be greatly deluded, however, to believe that this is all it takes to make a living cell. There is no way that each of 2,000 *independent* chemical reactions could automatically cooperate with all the others to produce the orderliness and efficiency of growth. *What we covered in Section III would produce not a cell, but chaos.* In this chapter, we will examine how order is made out of chaos and how independent chemical reactions are welded into a single, coordinated network—an *integrated circuit* of enormous complexity and elegance, surpassing the highest achievements of human engineers.

EVIDENCE FOR COORDINATION OF METABOLIC REACTIONS

A few simple experiments can provide concrete evidence of the cell's ability to coordinate its working parts and become a single integrated system. Coordination can be demonstrated in each of the major stages of metabolism. For simplicity, we shall begin with biosynthesis.

Coordination in biosynthesis

Prepare a medium consisting of a mixture of inorganic salts and just enough radioactive ^{14}C-labeled glycerol as the sole source of carbon and energy (substrate) to permit the growth of a bacterial species such as

Escherichia coli to a density of 10^8 cells per ml (well below what the medium would permit if more glycerol were present). Inoculate the medium with a few cells, and incubate it aerobically.

Samples taken from this culture during its growth would reveal several simple but informative facts. First, analysis of the culture medium from which the cells had been removed by filtration or centrifugation would indicate that throughout growth, the chief radioactive materials were unused glycerol and the CO_2 formed from oxidative metabolism. Then, when growth ceased due to the exhaustion of the glycerol, almost all the radioactivity was dissolved CO_2. Only traces of labeled amino acids, nucleotides, or other metabolites would be found in the medium; the growing cells did not spill out metabolic intermediates. These data suggest that *the rate of formation of each building block matches its rate of utilization for macromolecule formation* and that *the flow of carbon through each metabolic pathway is coordinated with the flow through each of the others.* Rather neat and simple.

Repeat this experiment, but now include an amino acid, for example, histidine (unlabeled), in the medium. Analysis of this culture at the end of growth would reveal a remarkable situation. Analyzing the total protein of the cells would show that the histidine residues, alone among all the amino acids, contained almost no traces of radioactivity; they must therefore have been derived almost exclusively from the nonradioactive histidine supplied in the medium rather than from the radioactive glycerol. As before, the radioactive substance in the medium would be almost exclusively CO_2; in particular, there would be no radioactive histidine or intermediates of the histidine biosynthetic pathway. *Thus, almost no histidine was made from glycerol.* The cells not only used the histidine supplied in the medium, they completely turned off the entire pathway leading to the synthesis of histidine. This experiment could be repeated with virtually any of the building blocks, vitamins, or cofactors (that can be readily taken up by the cells) with the same result. *The presence of any one of compound in the medium stops endogenous synthesis of that compound.*

This evidence for coordination can easily be made more extensive. The same outcome is observed if the experiment is performed under different growth conditions—different temperature, pH, O_2 tension, carbon source, mixture of nutrients, and so on. Also, the outcome is independent of the growth rate and the chemical composition of the cells. This result shows that *the coordination of one pathway with the others is adjustable, not fixed.* The rate at which each pathway functions relative to all other pathways can be adjusted over a wide range.

Coordination in fueling

Does this adjustable coordination exist in fueling reactions as well? Many bacterial species can grow on a variety of substrates that supply different amounts of precursor metabolites, energy, and reducing power. Because the cellular needs for these products of fueling are virtually identical no matter what the carbon and energy source, there must be mechanisms to adjust the ratios of these resources yielded from the various fueling reactions. Another demonstration of flexibility occurs when a rich supply of nutrients (containing ready-made building blocks) is added to a medium with glycerol as the carbon and energy source. Under these conditions, the needs for products of fueling are altered. The precursor metabolites are no longer utilized for biosynthesis and the need for reducing power is decreased, but the cell still has a demand for ATP to drive polymerization and assembly. Almost as though it had a brain to think through the problem, the cell metabolizes just sufficient glycerol through the fueling pathways to generate an adequate supply of ATP; it metabolizes precursor metabolites to CO_2 and reoxidizes reduced coenzymes. Thus, both unneeded precursor metabolites and reducing power are converted to ATP.

Coordination in polymerization

Recall from "How Is the Physiology of the Cells Affected by the Growth Rate?" in Chapter 4 that the macromolecular composition of a bacterial cell is not greatly affected by the nature of the growth medium. Cells growing on a mixture of amino acids are not richer in protein; those growing on fatty acids are not richer in lipid; those growing on nucleosides are not richer in nucleic acids; those growing on sugars are not richer in carbohydrates. Quite the contrary; media of very different chemical natures may produce cells of nearly identical macromolecular compositions. Some changes in medium composition do affect cellular composition, but *only to the extent that they affect growth rate*. Thus, cells growing with a supply of amino acids are richer in RNA than the same cells growing more slowly in a minimal medium. Therefore, there must be regulatory devices that sense the potential for the cell's attaining a particular growth rate and appropriately control the synthesis of each macromolecule.

In summary, all the major fueling, biosynthetic, and polymerization pathways are subject to powerful controls that bring order out of the potential chaos of a system with thousands of individual working parts. These examples demonstrate the coordination. The question now becomes, how is it brought about?

TWO MODES OF REGULATION

If you wanted to design a cell that would work in this organized fashion, you would probably consider the possible ways by which the rates of metabolic reactions *might* be regulated. In general, the rate of a metabolic reaction such as the following (where S_1 and S_2 are substrates and P_1 and P_2 are products):

$$\longrightarrow S_1 + S_2 \xrightarrow{\text{Enzyme}} P_1 + P_2 \longrightarrow$$

might be controlled by any of three means: (i) varying the activity of the enzyme, (ii) changing the amount of the enzyme, or (iii) varying the intracellular concentration of one or both substrates. Of the three, which would you choose?

Bacterial cells use all these means, but we shall focus on the first two. The third mechanism, varying the concentration of substrate(s) for a reaction, is brought about by one of the other mechanisms acting on some reaction that produces these compounds. Thus, the primary basis of metabolic coordination is varying either the *activities* or the *amounts* of enzymes.

Controlling enzyme activity

Controls on *protein activity* are of two sorts. In some cases, a protein enzyme is reversibly inactivated (or activated) by **covalent modification** (phosphorylation is common, but addition of adenylyl, acetyl, methyl, and other residues also occurs). In other cases, activity is modulated by reversible association with another molecule—called a **ligand** if it is a small molecule and a **modulator** if it is a protein or nucleic acid. This modulation of protein activity is called **allostery**. Modification and allostery are both important, though, as we shall see, the latter is far more common and pervasive in microbial regulation.

Controlling enzyme amounts

Controls on enzyme levels affect either the rates of synthesis or, more rarely, the rates of degradation of individual proteins.

As a general rule, differentiated cells of higher plants and animals display more controls on enzyme activity than on enzyme amount, at least compared to bacteria. This is not to say that bacteria do not control enzyme activity (in the next section, we shall see how prevalent and important such controls are), but rather that bacteria modulate the *levels* of their individual enzymes to a much larger extent.

So, why do bacteria and cells of higher organisms differ? At least four reasons can be given. First, bacteria are often substrate limited. It is uneconomical to make an enzyme and not use it (recall the energy it takes to make a protein [see Chapter 8]). Bacteria cannot afford to make useless or redundant enzymes, nor can they survive without producing essential enzymes; both situations call for rapid adjustment of enzyme synthesis.

Second, bacteria are admirably suited to adjust the levels of enzymes upward or downward by controlling their synthesis. Up regulation can start within seconds and can quickly increase an enzyme level by a thousandfold or more, as we shall see below. Down regulation also starts right away. By turning off the synthesis of an enzyme, a bacterial cell will halve the level of that protein with each doubling of cell mass. If the generation time is 20 minutes, the level can be reduced eightfold in an hour.

Third, because bacterial mRNAs turn over rapidly, the blueprints for protein synthesis are completely renewed every few minutes, affording a marvelous opportunity to control enzyme synthesis rapidly at the transcriptional level.

Fourth, the multicistronic nature of bacterial mRNA facilitates unitary control of whole pathways or other groups of functionally related enzymes. By a single adjustment of the synthesis of one mRNA, the cell can change the rate of synthesis of an entire pathway.

Why two modes of regulation?

If reactions can be regulated equally well by controlling the amounts or the activities of enzymes, why do bacteria employ *both* mechanisms? To examine this key question of cell physiology, we need more information about each mode of regulation. We shall then be in a more informed position to evaluate the rationale for microbes having both modes of regulation.

The overview of regulatory devices presented in Fig. 12.1 shows the various microbial means of regulating protein activity and amount. We shall examine regulation of enzyme activity first and then regulation of enzyme level.

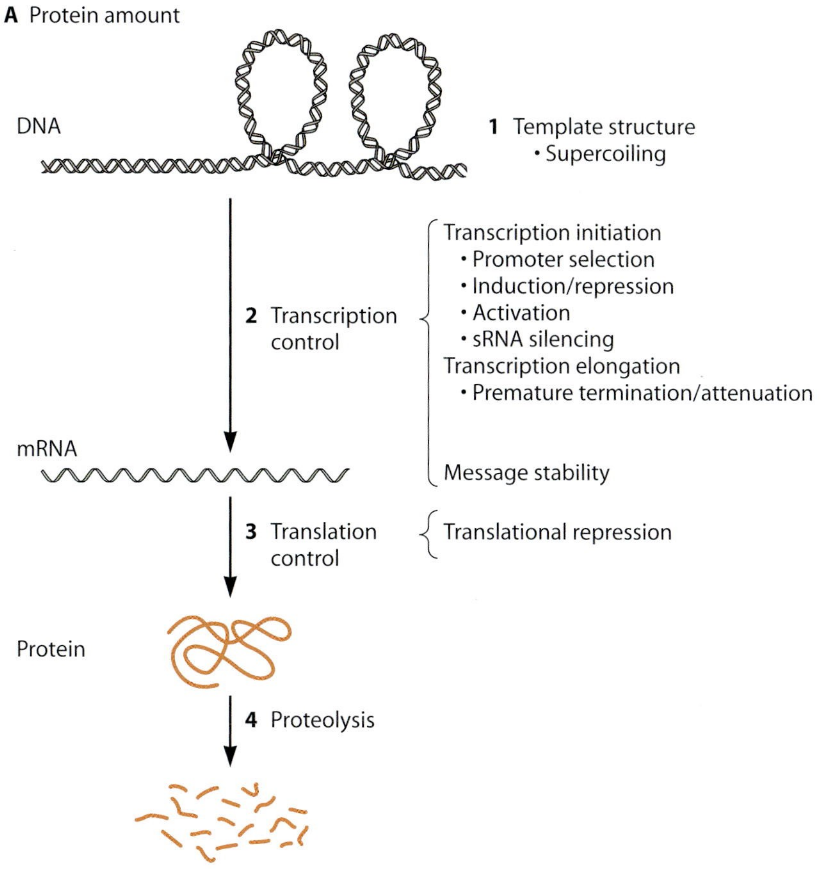

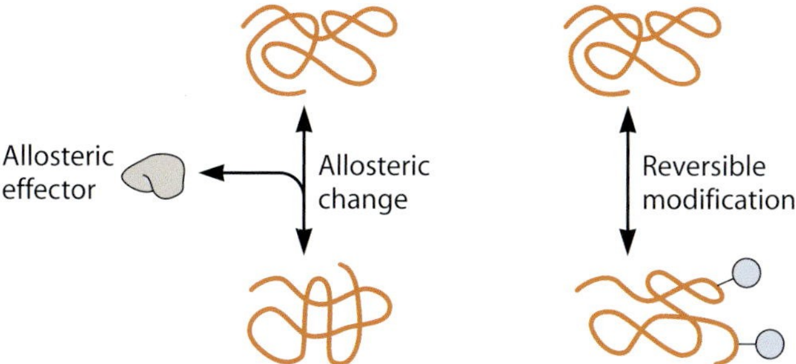

Figure 12.1 Overview of metabolic regulatory devices. The figure shows a synopsis of the two forms of control that operate to coordinate metabolism: regulation of the amount of a protein in the cell **(A)** and regulation of its activity **(B).**

MODULATION OF PROTEIN ACTIVITY

Modulation of enzyme activity occurs rapidly, almost instantaneously, bringing about adjustments of metabolic flow in the microbial cell in a fraction of a second.

Allosteric interactions

The most prevalent mode of controlling the activities of enzymes (and other proteins) is **allosteric interactions** (**allostery** means different shape). Allostery involves a change in the conformation and hence the activity of a protein after it binds a ligand called an **allosteric effector** (Fig. 12.1B). In the case of an enzyme, the change in shape alters catalytic activity. Allosteric changes have two attributes: (i) the **regulatory site** at which the allosteric effector binds is separate from the enzyme's catalytic site and

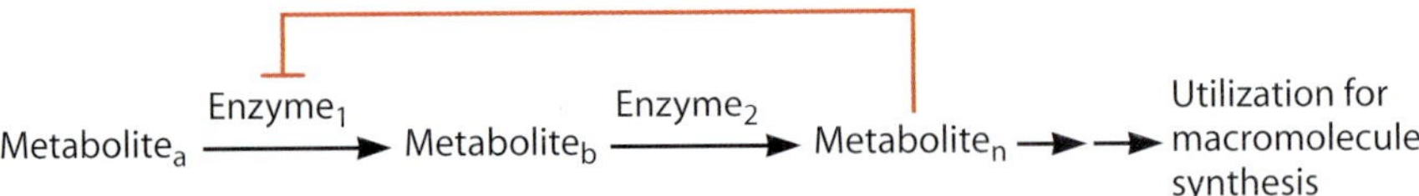

Figure 12.2 Feedback inhibition. The final product of a series of enzymatic reactions, metabolite$_n$, has the property of binding to the regulatory site of the allosteric protein, enzyme$_1$, and thereby inhibiting it. A scheme of this sort can guarantee that metabolite$_n$ is produced only as rapidly as it is used in some subsequent process, such as macromolecule synthesis.

Figure 12.3 Patterns of feedback inhibition found in bacterial biosynthetic pathways. (A) Isofunctional enzymes for the regulated reaction allow differential feedback effects by the two pathway end products. **(B)** Cumulative feedback inhibition involves multiple allosteric sites on the regulated enzyme, ensuring that there will be some activity unless all the end products are in excess. **(C)** Sequential feedback inhibition involves different end products operating separately on the various branches of the pathway. **(D)** Inhibition plus activation uses both positive (+) and negative (−) allosteric effectors to coordinate complex pathways.

A Isofunctional enzymes

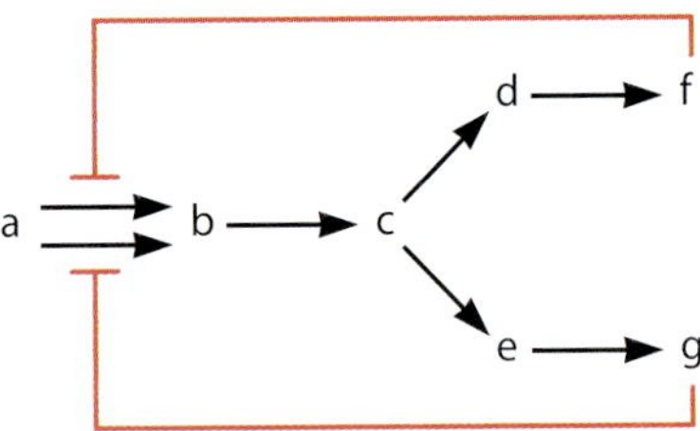

B Cumulative feedback inhibition

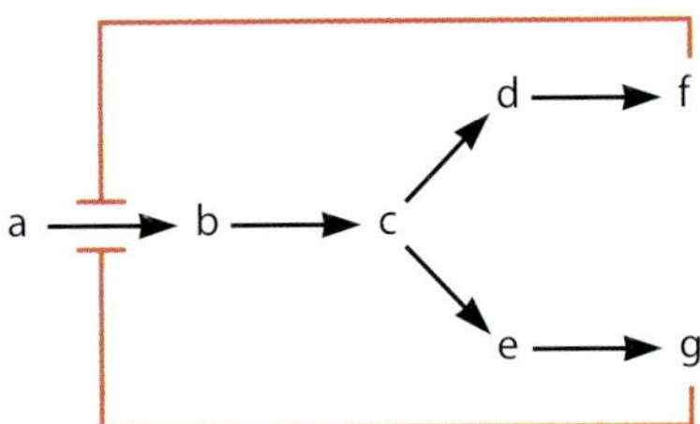

C Sequential feedback inhibition

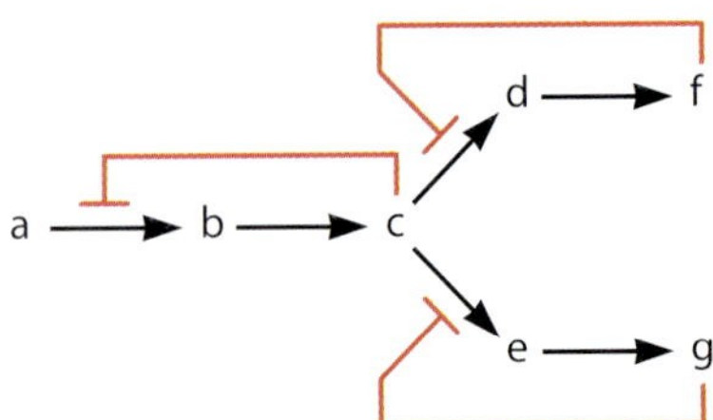

D Inhibition plus activation

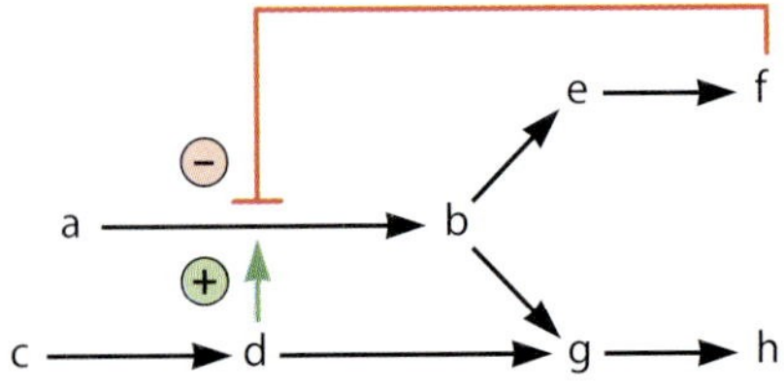

(ii) the effector need bear no steric resemblance to the substrates (or products) of the enzyme. This is worth emphasizing: *allostery provides a means for modifying the activity of an enzyme by substances not even remotely resembling the substrates or products*. Allostery differs fundamentally from competitive inhibition, in which a substance resembling the substrate competes for binding to the active site of the enzymes. How does allostery work? Allosteric enzymes can exist in at least two conformations, one with high and another with low activity. The binding of the effector favors one conformation over the other. In some cases, what is changed is the velocity (V_{max}, the maximum rate of reaction) of the enzyme; in others, it is affinity for the substrate (K_m, the Michaelis constant). Allosteric effectors may increase or decrease the activity of the enzyme, depending on the specific example; some enzymes have both **positive** and **negative** allosteric effectors.

Allostery in biosynthesis

We opened this chapter by noting that a bacterium will utilize the building block, histidine, supplied in the medium rather than make it. We can press this point further by examining the immediate consequences of adding histidine to a culture already growing in a medium composed of radioactively labeled glycerol and inorganic salts. Samples taken very rapidly reveal that *within seconds after the addition of histidine, almost all flow through the histidine biosynthetic pathway ceases*. This result shows that the coordination is due to a rapid mechanism. The suspicion that it is brought about by allostery is confirmed by the finding that the first enzyme in the histidine biosynthetic pathway is indeed an allosteric enzyme and that histidine is its negative allosteric effector. Thus, the addition of histidine immediately shuts off the entire pathway at its very first step.

This immediate response is simply a more dramatic version of a process that takes place continuously in the cell to regulate histidine biosynthesis: the internal concentration of histidine, whether made by the cell or collected from the medium, determines the rate of flow from precursor metabolites to end product. Histidine is an example, not an exception—*virtually all building blocks control their own synthesis by acting as negative allosteric effectors of the first enzyme in their biosynthetic pathway*. This control process is called **feedback inhibition** (or **end product inhibition**). Since each building block is being drained off by its utilization in macromolecule synthesis, biosynthetic pathways work by *demand feeding*: they produce their end products at the rate they are being consumed by polymerization. Figure 12.2 summarizes in general terms the elements of feedback inhibition in biosynthetic pathways. Feedback inhibition can be adapted to accommodate the branching of biosynthetic pathways or their other complexities (Fig. 12.3).

Allostery in fueling

Allosteric inhibition and activation also regulate the flow of metabolites through fueling pathways. Here, the simple device of end product control of the first or an early step in a pathway cannot apply. The pathways of formation of the 13 precursor metabolites are so interrelated (some pathways are cyclic) that *there are no unique early steps*. An early step in growth on glucose, for example, is a late step in growth on malate (see Fig. 6.16). Thus, controls work internally in each of the main fueling pathways, and both positive and negative allosteric interactions abound (Fig. 12.4). Note that one metabolite (acetyl coenzyme A, for example) may act as a positive effector for one enzyme and a negative one for another.

The rates of formation of ATP and reduced nicotinamide adenine dinucleotide (phosphate) (NAD[P]H) must also be regulated appropriately. Because ATP synthesis and utilization involves a cyclic flow through

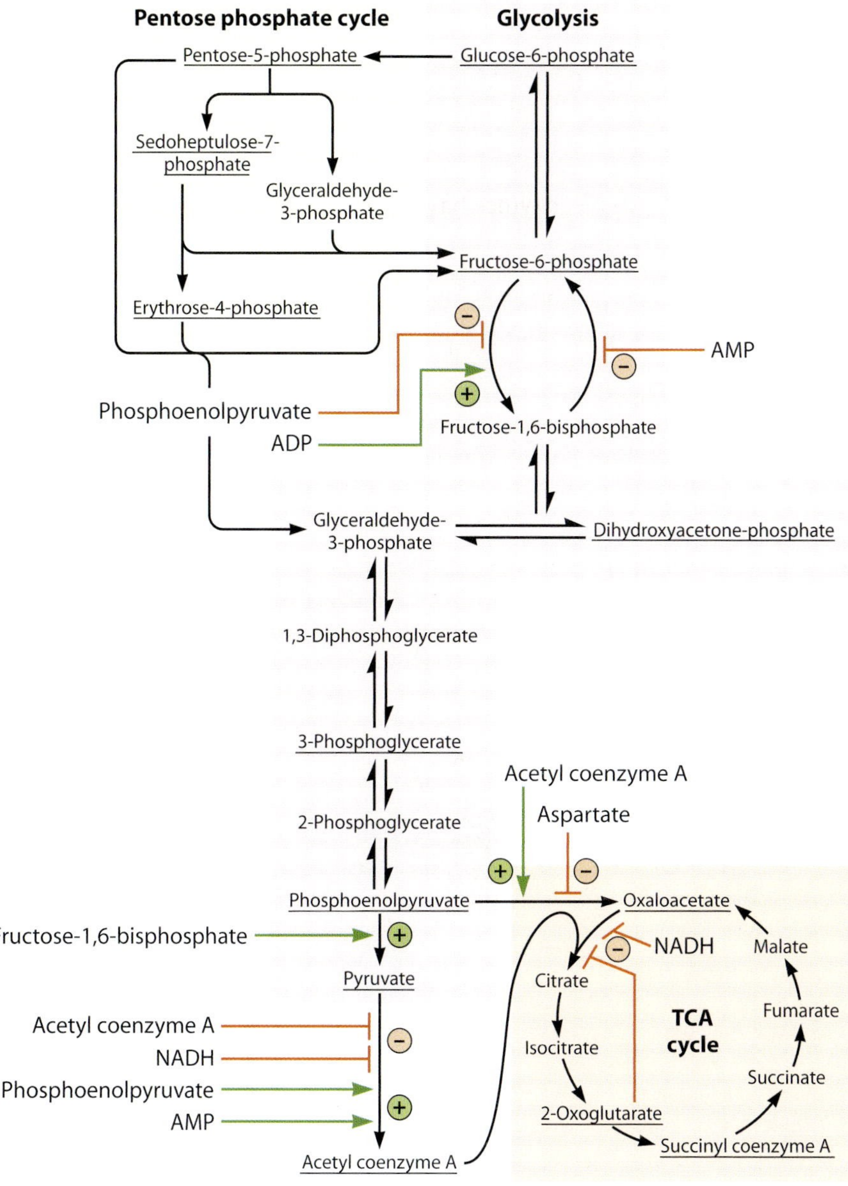

Figure 12.4 Central pathways of fueling reactions showing some of the allosterically controlled steps. −, negative allosteric effector; +, positive allosteric effector.

ADP and AMP, it makes sense that all three adenylates play regulatory roles in fueling reactions (as well as in biosynthetic pathways). A useful index of a cell's energy status is the **energy charge** of the cell, defined as follows:

$$\text{Energy charge} = \frac{([\text{ATP}] + [\text{ADP}]/2)}{([\text{ATP}] + [\text{ADP}] + [\text{AMP}])}$$

Mathematically, the energy charge of the cell mathematically could vary from 0 (all AMP) to 1 (all ATP), but in fact, the energy charge of bacteria under normal conditions is held narrowly between 0.87 and 0.95. Starving a cell for a very long time reduces its energy charge slowly; when it reaches approximately 0.5, the cell is dead. In general, ATP-replenishing pathways (fueling) are inhibited by high levels of energy charge, and ATP-utilizing pathways (biosynthesis and polymerization) are stimulated.

Analogous allosteric interactions maintain appropriate working levels of the coenzymes NADH and NADPH.

Covalent modification

Control of enzyme activity by covalent modification occurs in both eukaryotes and prokaryotes. Bacterial examples are given in Table 12.1. Some of these (phosphorylation, in particular) are extremely prevalent in bacterial physiology; we shall encounter them in detail in our discussions of motility and chemotaxis and sensory signal transduction (see Chapter 13). Others may not be involved in *many* cellular processes, but the processes in which they do participate are major ones (for example, adenylylation in the regulation of glutamine synthetase, a key enzyme in nitrogen assimilation).

MODULATION OF PROTEIN AMOUNTS

Even when all the processes described above are functioning optimally, unneeded proteins could still be made. However, such waste does not occur; other regulatory mechanisms intervene. Learning about them brings us to the topic that generated great excitement throughout biology during the past half century and gave rise to the field of microbial molecular genetics: modulation of enzyme levels by the regulation of **gene expression**.

Table 12.1 Examples of covalent modification of bacterial enzymes and other proteins

Enzyme	Organism(s)	Modification
Glutamine synthetase	*E. coli* and others	Adenylylation
Isocitrate lyase	*E. coli* and others	Phosphorylation
Isocitrate dehydrogenase	*E. coli* and others	Phosphorylation
Chemotaxis proteins	*E. coli* and others	Methylation
P_{II}[a]	*E. coli* and others	Uridylylation
Ribosomal protein L7	*E. coli* and others	Acetylation
Citrate lyase	*Rhodopseudomonas gelatinosa*	Acetylation
Histidine protein kinases	Many bacteria	Phosphorylation
Phosphorylated response regulators	Many bacteria	Phosphorylation

[a]P_{II} is a regulatory protein in nitrogen metabolism.

In the mid-20th century, intensive efforts were made to discover how bacteria change the rates of synthesis of their enzymes. The initial studies were done with *E. coli* and other closely related enteric bacteria, *Salmonella* and *Klebsiella*, but all bacteria are extremely adept at turning genes on and off, thereby raising and lowering the levels of enzymes.

Evidence that flexibility of gene expression is widespread among bacteria can be quickly obtained by either of two tools in wide use today: **proteomic monitoring** and **genomic** (or **transcriptional**) **monitoring.** Proteomic monitoring allows one to survey the levels and rates of synthesis of almost all the proteins of the cell (traditionally by using two-dimensional polyacrylamide gels to separate individual cellular proteins); transcriptional monitoring (using DNA microchips containing all the cell's individual genes to hybridize the cellular mRNA) allows one to assess the profile of mRNA molecules in the cell. (These techniques are more fully described in "Coping with stress by escaping" in Chapter 13.) Examination of many different species of bacteria has shown that their protein and messenger RNA (mRNA) profiles change radically during growth under different conditions. Most proteins in the cell are subject to dramatic changes in levels and rates of synthesis—some by many orders of magnitude. Similar changes in levels of mRNA indicate that in many (but not all) instances the modulation of protein levels occurs by changes in the synthesis of mRNA.

Gene regulation is an epic tale, best told in two parts. In the first, we ask how individual operons are controlled, temporarily ignoring the fact that many (probably most) operons are members of larger regulatory units. In the second part, we examine how sets of operons are coordinately regulated and tied into a cellular regulatory network.

Regulation of operon expression

The operon is a hallmark of the prokaryotic cell. As we noted in Chapter 8, an operon is a unit of transcription, consisting of a DNA segment containing a promoter, two or more (typically more) cistrons (genes encoding polypeptides), and a terminator. Transcription of an operon yields a single mRNA molecule that encodes one or more polypeptides. A quick review of the operon concept can help us recall its main features.

A historic picture of the regulation of an operon, as first proposed by F. Jacob and J. Monod in 1961 for the genes encoding the enzymes metabolizing lactose, designated *lac*, is shown in Fig. 12.5. Figure 12.5A depicts the condition of the *lac* operon in *E. coli* cells growing in the *absence* of lactose. The promoter, which is the binding site for RNA polymerase (RNA-P), is available for action, but immediately adjacent, and intervening between the promoter and the three protein structural genes (*lacZ, lacY,* and *lacA*) of the operon, lies the operator. To this site is bound an allosteric protein **repressor,** LacI, the product of the nearby gene *lacI*. The repressor prevents the RNA-P from transcribing the operon (either by preventing polymerase from binding or by blocking its movement), and thus, none of the three enzymes encoded by the operon is made. These products are LacZ (β-galactosidase, the enzyme that hydrolyzes lactose), LacY (a membrane protein, or **permease,** that transports lactose into the cell), and LacA (a transacetylase of uncertain function). Occasionally, the repressor dissociates from the operator and a few

Authors' note The reader is referred to the chapter by J. Beckwith entitled "The operon: an historical account" (Chapter 78, p. 1227–1231, *in* F. C. Neidhardt, R. Curtiss III, J. L. Ingraham, E. C. C. Lin, K. B. Low, B. Magasanik, W. S. Reznikoff, M. Riley, M. Schaechter, and H. E. Umbarger (ed.), Escherichia coli *and* Salmonella: *Cellular and Molecular Biology,* 2nd ed., vol. 1 (ASM Press, Washington, D.C., 1996) (available at most college and university libraries).

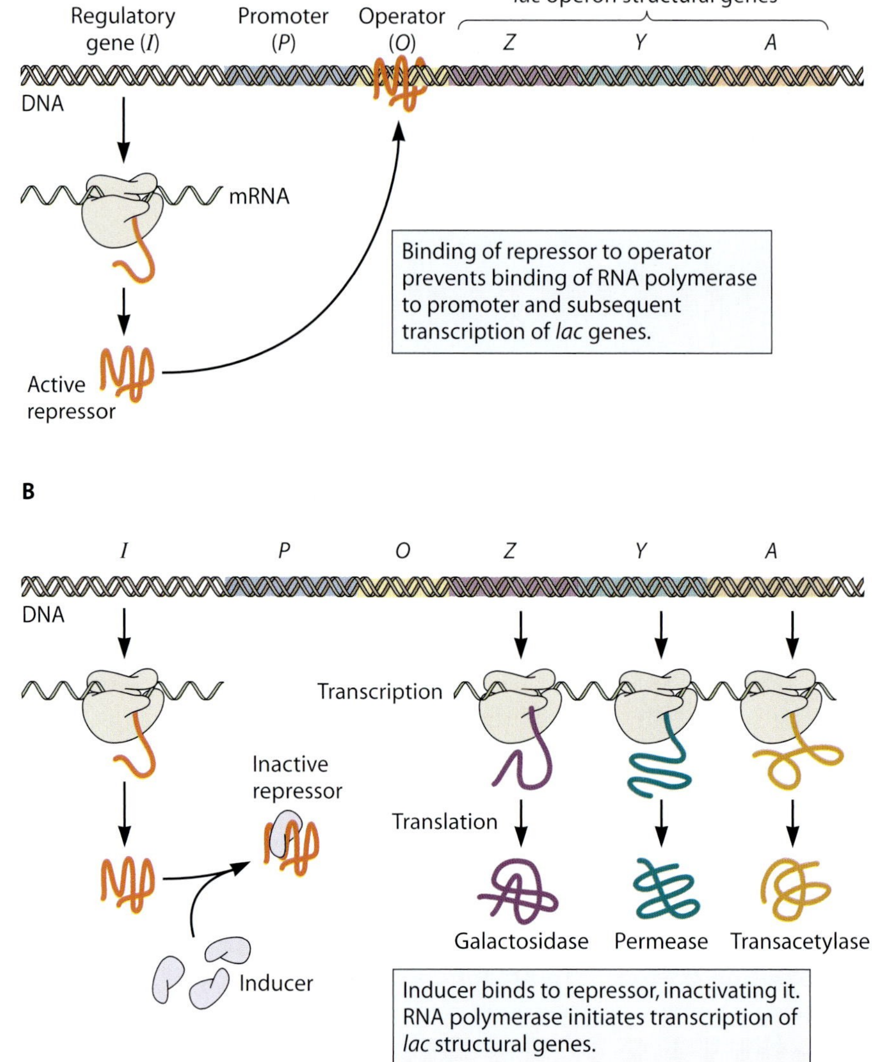

Figure 12.5 Original operon model of Jacob and Monod, proposed in 1961 for the regulation of the lac *genes* of *E. coli.* (A) Uninduced state of the *lac* operon in cells growing on substrates other than lactose; **(B)** induced state during growth on lactose. See the text for an explanation.

transcript molecules are made, ensuring the cell will have a low **basal level** of the three products of the operon.

Figure 12.5B shows the conditions in cells growing in the *presence* of lactose. Lactose is brought into the cells by LacY permease and acted upon by β-galactosidase to form a metabolite called allolactose. Allolactose molecules bind to the LacI repressor, triggering a change in this allosteric protein to an inactive conformation, that is, a form that cannot bind at the operator. With nothing to prevent its function, RNA-P is free to make a transcript. The operon products are produced, and the cells grow on lactose. You might ask how the cells make the transition from the **uninduced state** depicted in Fig. 12.5A to the **induced state** shown in Fig. 12.5B when lactose becomes available in the growth medium. How does lactose enter the cells when the permease to transport it is not induced? The answer is that initially a small amount of lactose is transported into the cell by the basal level of permease, and then it is converted to allolactose by the basal level of β-galactosidase. (Readers versed in molecular biology will likely spot the omission of a major feature added later in history: the **CAP** [for **catabolite gene activator protein,** the product of the *crp* gene] site within the promoter, at which catabolite repression works. We will discuss that below, in the context of regulatory networks.)

In summary, the *lac* operon is expressed (induced) when its allosteric protein repressor is made inactive by the binding of an **inducer** made from the lactose substrate. Induction, therefore, consists of removing an inhibition.

Spectacular advances in research on the *lac* operon of enteric bacteria led many to the optimistic view that understanding its mechanism would explain all gene regulation. Gradually, it became clear that the *lac* operon is a perfect paradigm, but only for the *lac* operon! The central notions of the original operon model are fairly universal, *but only if expressed in conditional terms.*

1. Initiation of transcription at the promoter of an operon *can* be a site of regulation.
2. Initiation of transcription *can* be controlled by regulatory proteins with allosteric properties affected by the binding of specific ligands.
3. Increase in the expression of an operon *can* be caused by relief of a negative control.

The reason why each of these statements is stated conditionally is that the *lac* operon displays only one example of how microbial operons are regulated. Here, as with so many topics in microbiology, we encounter the fact that microbes exhibit a dazzling diversity of solutions to metabolic and physiologic challenges. The modes of gene regulation that have been produced by microbial evolution are so numerous that one can almost say that every operon is uniquely controlled. Operons differ as to (i) the site at which control is exerted, (ii) the mode of control (positive or negative), and (iii) the particular molecular device used to bring about the regulation. Some operons even employ more than one promoter—some within the operon provide a second regulatory site for the downstream portion of the operon—to permit greater flexibility and sensitivity of regulation. As shown in Fig. 12.6, *every conceivable step that leads from a gene to its finished protein product has been singled out for control in some operon or some bacterium.* In addition, some of the 10 regulated steps or conditions shown in the figure can be controlled by a multitude of molecular mechanisms. Let us examine some tricks that have evolved, using a few well-studied examples. We shall consider briefly the regulatory mechanisms that occur at each of the 10 steps, referring to them by the numbers shown in Fig. 12.6.

Initiation is the most commonly controlled step in transcription and is controlled in numerous ways at many steps (processes numbered 1, 2, 3, 4, and 5). This is not surprising, because no effort is wasted to make a transcript if the final gene product is destined not to be made. Since initiation involves the binding of RNA-P to a promoter and the subsequent opening of the DNA to allow transcription, some of the possibilities for control are obvious. Recall from "The promoter" in Chapter 8 that promoters differ in their inherent or basal strengths. Thus, a strong promoter can be dampened by some inhibitory agent, and a weak promoter can

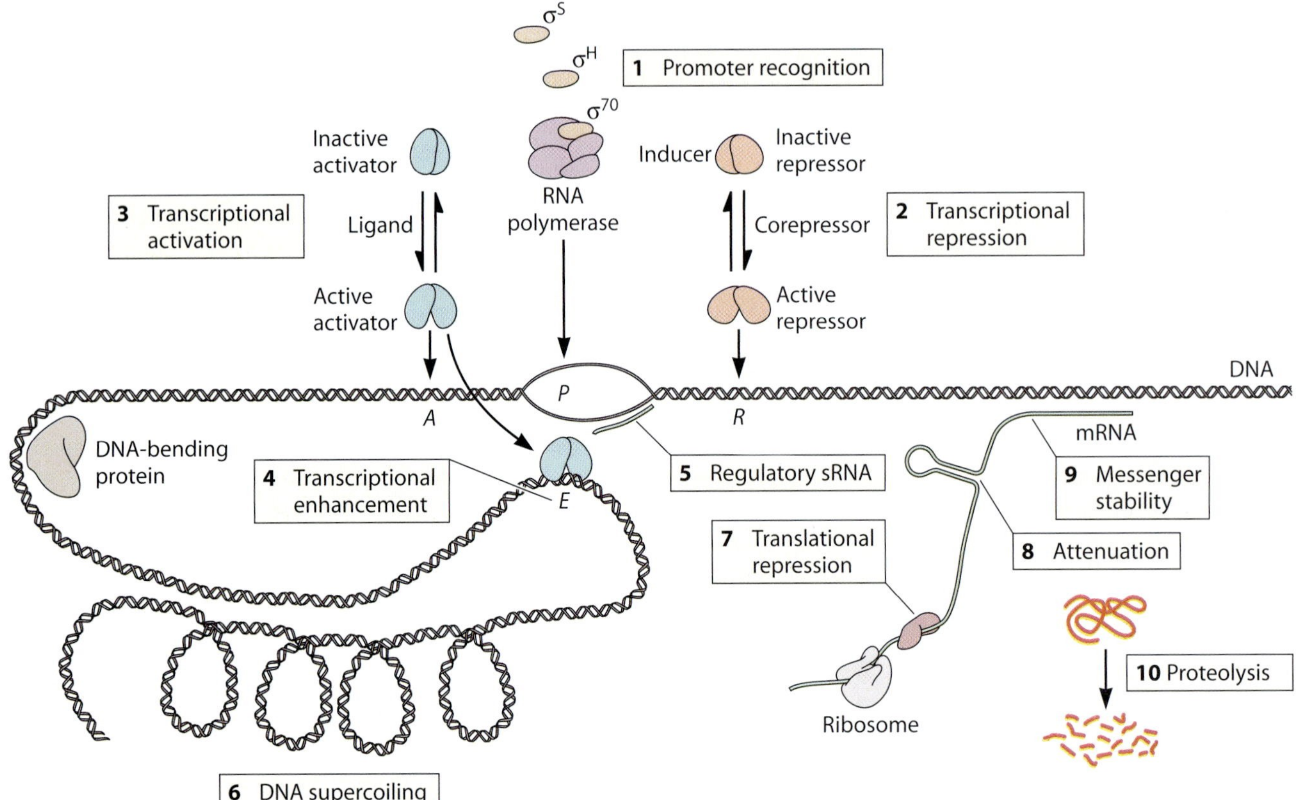

Figure 12.6 Diagram depicting some regulatory processes that can affect the cellular level of a protein. See the text for an explanation.

be strengthened by some enhancing agent. How these inhibitions and enhancements are brought about is fascinating.

Mechanisms at site 1: promoter recognition

The mechanisms at site 1 operate not through a control region on the DNA, but on the structure of RNA-P itself. Recognition of promoter sequences is conferred on RNA-P by the addition of a sigma (σ) subunit, which binds to the core RNA-P, thus forming the holoenzyme (see "Initiation of transcription" in Chapter 8). Bacterial cells have multiple sigma subunits, each recognizing a different promoter structure. As a result, transcription is initiated only at operons having the ability to bind a particular sigma subunit. Which sigma subunit binds to the core determines which operons will be transcribed. Regulation comes about by any of several means that determine which sigma subunit is joined to core RNA-P. This mechanism is important in higher-level regulatory mechanisms (see "Regulation beyond the operon" below), including those leading to stationary phase (see Chapter 13), spore formation (see Chapter 14), or the heat shock response (see Chapter 13).

Mechanisms at sites 2, 3, and 4: transcriptional repression, activation, and enhancement

DNA sequences called **control regions** are located at various places relative to the specific operons they control. Some are adjacent to or overlapping the −35 and −10 regions of the promoter, but others are many dozens of nucleotides upstream of the promoter or even downstream of the operon terminator. In most cases, these control regions are binding sites for allosteric **regulatory proteins** and are called **boxes**. Many are named after the protein they bind (such as the **CAP box**) or the process they regulate (such as the **nitrogen box**). Those boxes near the promoter were initially called **operators**, and you will still see that term used, but it is less useful now that we know that control regions can function in ways other than by binding regulatory proteins. For example, some control regions are promoters for genes transcribed from the opposing DNA strand. Others direct the folding of their mRNA transcripts. Control regions located distant from the operon they control are called **enhancers**, because they usually stimulate initiation. How can a DNA sequence influence a promoter a great distance away? The answer lies in the **bending of DNA** to bring the bound regulatory protein to the promoter region, where it can influence initiation. Bending does not happen spontaneously but is facilitated by **DNA-bending proteins**.

Some regulatory proteins (called **activators** or **positive regulators**) bind to DNA control regions and *increase* the frequency of initiation; others (called **repressors** or **negative regulators**) *decrease* the frequency. Is the choice of one or the other mechanism a matter of chance? One suggestion that has experimental support is that operons encoding products in high demand tend to be positively regulated, while those with intermittent or low demand are negatively controlled.

Enzymes involved in utilizing carbon and energy sources (feeder pathways) are encoded in operons that are usually turned off and are controlled by **induction**; that is, the enzymes for utilization of a particular substrate are produced only when that substrate is presented by the environment. In many cases, the substrate, or a substance closely related to it, acts as an **inducer**, a ligand that binds to a regulatory protein and either releases it from inhibiting the relevant operon (if it is negatively regulated) or enables it to bind to an enhancer or activator site (if it is positively regulated).

Operons encoding biosynthetic enzymes are generally turned on unless their building block end product is present in the environment, in which case the end product acts as a **corepressor ligand**. The end product can perform double duty, acting both as a ligand that controls an allosteric enzyme, usually the first, in the biosynthetic pathway and as a ligand that binds to the relevant **repressor** protein and turns off the biosynthetic operon. **Repression** (together with a control on transcription termination, discussed below, and feedback inhibition) was operative in those cells we experimented with at the beginning of this chapter—the ones growing in glycerol salts medium to which we added end products.

Mechanisms at site 5: targeting by sRNAs

Not all regulatory molecules are proteins. Some are **small RNA (sRNA) molecules** that bind near the 5′ end of an mRNA, preventing ribosomes from translating that message (mechanism 7 [see below]). A few of these act by base pairing directly with a strand of DNA near the promoter.

Mechanisms at site 6: DNA supercoiling

Most promoters are sensitive to the degree to which the DNA is supercoiled. One of the effects of DNA supercoiling is that it can change the tertiary structure of the molecule from its common B form to one with a left-handed twist, called Z-DNA. The activities of some promoters are increased by a transition from B- to Z-DNA, and those of others are decreased. The extent to which DNA structural transitions are actually employed as a general regulatory mechanism for controlling transcription is under study.

We shall turn now to mechanisms (7 to 10) that act after transcription has been initiated.

Mechanisms at site 7: translational repression

Some bacteria have evolved **posttranscriptional mechanisms.** In several well-studied cases, the controlled step is the initiation of translation. Perhaps the most noteworthy examples are the operons that encode ribosomal proteins (r-proteins). Their mRNAs contain nucleotide sequences similar to some in rRNA that serve as the sites where certain r-proteins (not yet assembled into ribosomes) can bind (see Chapter 8). This binding blocks the initiation of translation of the mRNA, resulting in a control called **translational repression.** In this way, the pool of free (that is, not assembled into ribosomes) r-proteins can balance the rate of r-protein synthesis with the rate of assembly of ribosomes, as well as coordinating the rates of synthesis of individual r-proteins with each other. Not all r-proteins serve as **translational repressors**—just those that bind directly to ribosomal RNA (rRNA) during assembly—but at least one of them exists in each of the principle r-protein-encoding operons.

The regulatory sRNA molecules described for mechanism 5 have base sequences complementary to the mRNAs they target for control. Binding of the sRNA to the mRNA (near its beginning, the 5′ end) blocks ribosomes from initiating translation, and hence, this constitutes another version of translational repression (though in this case, it is an RNA molecule rather than a protein doing the job). As simple as this may seem, the manner in which sRNA brings about useful modulation of gene expression is far from simple. Something has to control the action of the sRNA. That *something* is not yet understood, but in many cases, the story involves a set of proteins, called **RNA chaperones**, that have the ability to bind the sRNA molecules, escorting them to their targets.

Mechanisms at site 8: attenuation

One common means for regulating operons, particularly biosynthetic operons, relies solely on the structure of the mRNA; there are no special regulatory molecules, either protein or RNA. In this mechanism, called **attenuation**, up to 90% of transcripts are prematurely terminated shortly after initiation, but the extent of this termination can be modulated. (Since no regulatory protein is involved, imagine the challenge to researchers who, guided by the *lac* operon model, failed over and over again to discover regulatory proteins and binding boxes for these operons.) The best-studied example of attenuation (and the system in which it was first discovered) is the *trp* operon of *E. coli*, encoding the enzymes responsible for biosynthesis of tryptophan. It deserves a close look.

The promoter of this operon is located far (160 base pairs) upstream from the five structural (enzyme-encoding) genes. The transcript thus includes a long sequence, called the **leader sequence** (*trpL* [Fig. 12.7]),

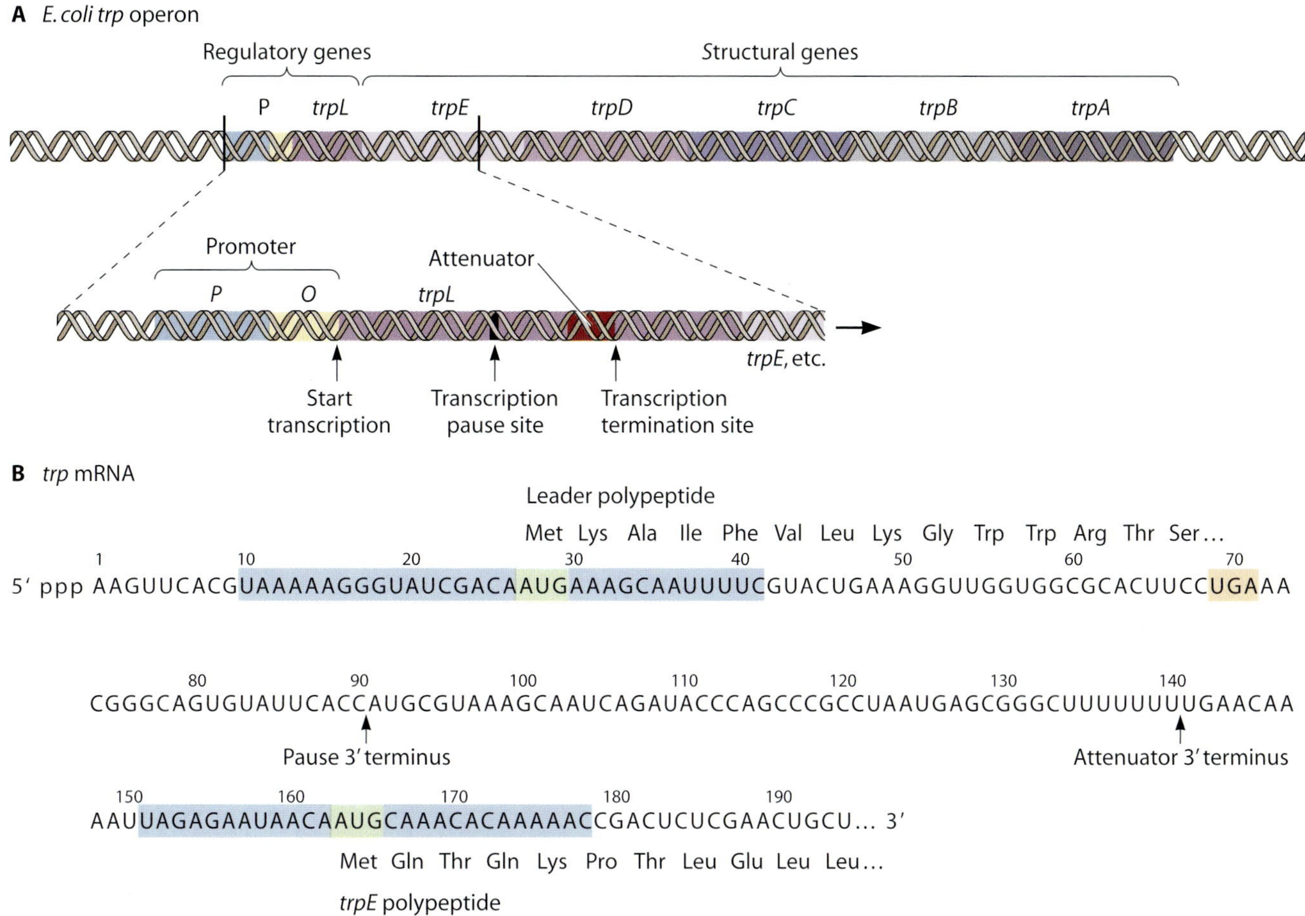

Figure 12.7 Regulatory and structural regions of the *trp* operon of *E. coli.* (A) Overall view of the structure and major functional features of the operon. Transcription initiation is controlled at a promoter-operator, and transcription termination is regulated at an attenuator in the transcribed 162-base-pair leader region, *trpL*. All RNA-P molecules transcribing the operon pause at the transcription pause site before proceeding further. **(B)** Nucleotide sequence of the 5′ end of the *trp* mRNA. The nonterminated transcript is shown; upon termination at the attenuator, a 140-nucleotide transcript is produced (the **3′ terminus** is indicated by an arrow). The 3′ terminus of the pause transcript (nucleotide 90) is also indicated by an arrow. The two AUG-centered ribosome binding sites in this transcript segment are shaded. The boxed AUGs are where translation starts, and the shaded UGA is where it stops. The predicted amino acid sequence of the leader peptide is shown.

between the promoter and the first gene. *This sequence is the only player in attenuation—the other components are simply ordinary parts of the translation apparatus.* The RNA transcribed from the leader sequence, the **leader transcript**, has two properties: (i) it can exist in two configurations determined by the speed with which the leader is translated, one causing termination of transcription and the other allowing it to proceed, and (ii) it has a **ribosome binding site** (Fig. 12.7B), which permits translation of the leader transcript into a **leader peptide** of 14 amino acids, two of which are sequential tryptophans. The contiguous codons for tryptophan in the leader mean that the cellular abundance of tryptophan (actually tryptophanyl-transfer RNA [tRNA]) affects the speed of translation of the leader and thus the frequency of transcription termination. An adequate supply of tryptophan lets translation of the leader proceed, and termination occurs. A deficiency of tryptophan causes ribosome stalling at the tryptophan codons and formation of the transcript configuration that attenuates termination; thus, the enzymes for making tryptophan can be produced abundantly. Figures 12.8 and 12.9 diagram the action, explain the role of RNA-P pausing, and show how transcription can be stopped by a structure behind the transcribing polymerase (a process not easily visualized).

In different operons and different bacteria, attenuation displays different wrinkles. In some *Bacillus* species, for example, attenuation is

Figure 12.8 Alternative secondary structures of the *trp* leader region of *E. coli*. The numbers 1, 2, 3, and 4 indicate the RNA segments that form the secondary structures pictured. **(A)** Termination configuration. The arrowhead indicates the site of transcription pausing. **(B)** Antitermination configuration.

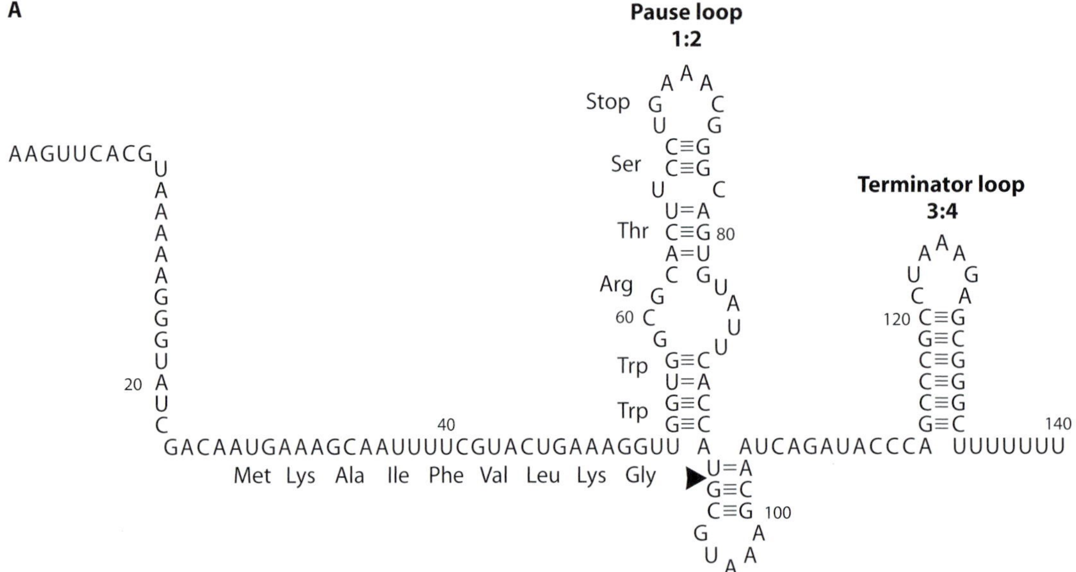

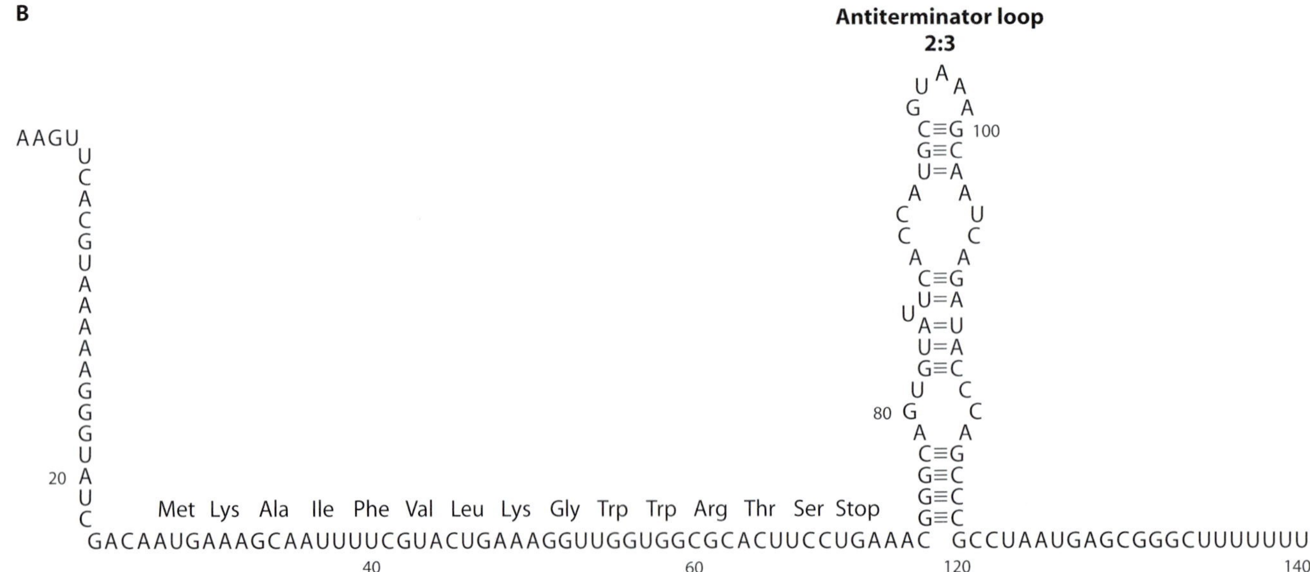

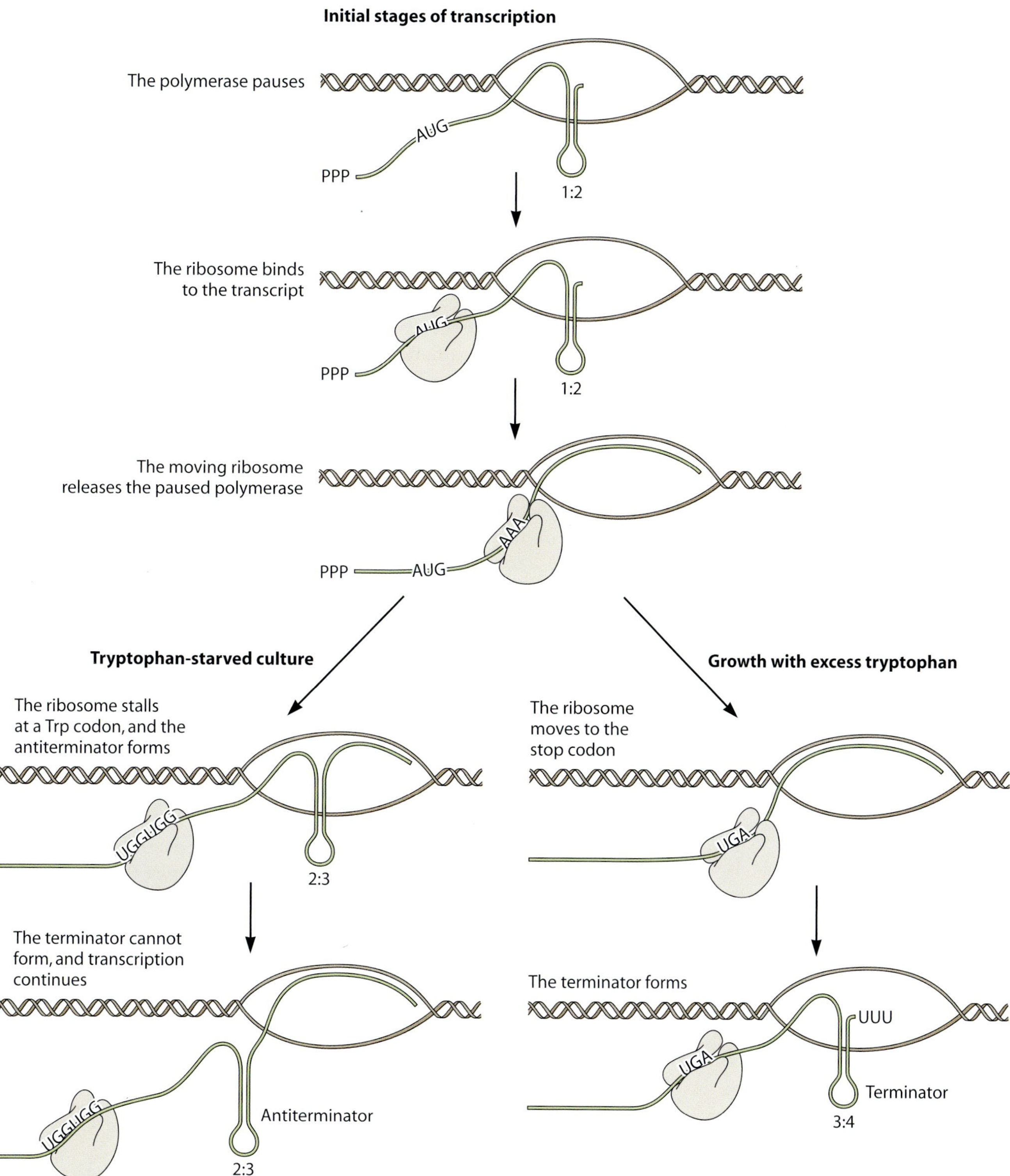
Initial stages of transcription
The polymerase pauses
AUG
PPP
1:2
The ribosome binds
to the transcript
AUG
PPP
1:2
The moving ribosome
releases the paused polymerase
AAA
PPP
AUG
Tryptophan-starved culture
The ribosome stalls
at a Trp codon, and the
antiterminator forms
UGGUGG
2:3
The terminator cannot
form, and transcription
continues
UGGUGG
Antiterminator
2:3
Growth with excess tryptophan
The ribosome
moves to the
stop codon
UGA
The terminator forms
UUU
UGA
Terminator
3:4

Figure 12.9 Representation of the roles of RNA-P and the first translating ribosome in attenuation control. See the text for an explanation.

brought about, not by ribosome stalling at tryptophan codons, but by a specific protein that, when bound to free tryptophan, prevents formation of an **antiterminator** structure in the transcript and therefore leaves a rho-dependent terminator free to abort further transcripts. In *E. coli*, the operon that codes for aspartate carbamylase (an essential enzyme in pyrimidine biosynthesis) is controlled by attenuation brought about by the stalling of RNA-P at a U-rich site of the leader sequence. When uridine triphosphate (UTP) levels are low, the leader-translating ribosome catches up to the polymerase and prevents formation of a terminator hairpin in the transcript.

Mechanisms at site 9: mRNA stability

The transcripts of bacteria are on average short-lived (see Chapter 8). Obviously, the longer a particular mRNA exists and functions, the more of its protein products will be made. Changing the stability of mRNA, therefore, can be a means of regulating the cellular amounts of its protein products. This possibility is actually employed, and the principal agents in this regulation turn out to be the same class of molecules we met in transcriptional and translational repression: sRNAs. Instead of blocking ribosomes, some sRNAs are known to trigger the degradation of their target mRNA molecules. Just as for the other regulatory processes in which sRNAs are implicated, the details of how they can increase and decrease their effects on mRNA stability remain to be discovered.

Mechanisms at site 10: proteolysis

Though not high on most bacterial cells' lists of means by which to coordinate metabolic pathways and regulate responses to the environment, proteolysis definitely plays a role. In a few instances, enzyme degradation provides a means to control a metabolic reaction. More commonly, however, proteolysis is employed to remove a protein that regulates other proteins. For example, alternative sigma factors (see "RNA-P and its function" in Chapter 8), which program core RNA-P to recognize special sets of promoters during an emergency, are constantly being made and degraded. Proteolysis keeps their cellular level lower than that of the major sigma factor (σ^{70} in *E. coli*). When an environmental change calls for a particular sigma factor to come into play, its degradation is stopped, leading to a rapid rise in its concentration. Protection of a sigma factor from proteolysis is brought about in part by its increased binding to core RNA-P. This process is significant in controlling the sigma factor that programs RNA-P to transcribe heat shock genes upon a rise in temperature (see Chapter 13). Variations on this theme are also common, such as modulation of the actions of chaperones that present a particular sigma factor to a special protease (ClpP in *E. coli*) for degradation. Sporulation (see Chapter 14) entails programmed degradation of regulatory proteins at key steps.

Regulation beyond the operon

The operon is the prokaryotic unit of transcription, and from what we have just seen about its control, one might expect it to be the unit of regulation as well. This mode of organization—whereby the genes of an entire pathway can be united as a single transcriptional unit—achieves a simple solution to the problem of regulating genes with related functions. Why go further? Why would natural selection move beyond this obviously successful strategy? There are at least two answers to this question. The first is that some processes involve too many genes to be accommodated in a single workable operon. The translation machinery, for example, comprises at least 150 gene products, all of which must be coordinately regulated if growth is to be efficient. Therefore, the coordination of multiple operons is a necessity.

The second answer is that some processes involve genes that must be independently regulated *and* subject to an overriding, coordinating control. A large number of genes encode enzymes that enable the utilization of sugars, amino acids, and other compounds for fueling. If an environment contains a mixture of such compounds, economy would demand that only a premium substrate, one best able to satisfy the carbon and energy needs of the cell, be metabolized. (For enteric bacteria, glucose is such a substrate; for pseudomonads, succinate is.) Therefore, operons encoding enzymes for less valuable substrates should be repressed. And yet, each operon must be induced individually if its cognate substrate is present and a premium substrate is no longer available. This double requirement calls for a level of organization above the operon.

Regulatory units above the operon

Estimates vary, but the general thought is that each bacterium contains 100 or so multioperon systems. Their discovery and analysis is still under way (in some respects, it has just begun). Some of the better-known examples are shown in Table 12.2. The entries are arranged in broad functional networks: response to nutrient limitation, managing the **oxidation-reduction reactions** and electron transport, responding to damage by chemical and physical agents in the environment, and changing cell physiology and morphology. This table contains a *very* small sample of the systems under intensive research. Even so, the variety is readily apparent. The complex and overlapping nature of these systems leads into realms of study familiar to chemical and electronics engineers but new to most microbiologists (systems analysis, control circuit design, redundant controls, system stability, and the like).

Bacteria have devised multiple ways to weave individual operons into coordinated units. In some cases, an allosteric protein regulator has simply been borrowed from operon regulation: a protein repressor or activator recognizes a particular sequence common to the controlling regions of the member operons. This device is used in the **SOS, oxidation damage,** and **anaerobic electron transport** systems of enteric bacteria. In other prominent cases, an alternative sigma factor reprograms RNA-P to recognize the promoters of member operons. The **heat shock** and **sporulation** systems of various species illustrate this situation. Other cases involve a combination of protein regulators *and* sigma factors, as found in the **nitrogen utilization** systems of many bacteria, as well as the universal process of entry into stationary phase. One of the largest networks, the stringent-control system, has no protein modulator at all; the member operons are regulated by the nucleotide guanosine tetraphosphate (ppGpp), which in some way affects RNA-P to bring about a multitude of effects on the expression of many genes.

Table 12.2 Some regulons and modulons of bacteria

Stimulus/condition	System	Organism(s)	Regulatory genes (and their products)	Regulated genes (and their products)	Type of regulation
Nutrient utilization					
Carbon limitation	Catabolite repression	Enteric bacteria	*crp* (transcription activator, CAP); *cya* (adenylate cyclase)	Genes encoding catabolic enzymes, (*lac*, *mal*, *gal*, *ara*, *tna*, *dsd*, *hut*, etc.)	Activation by CAP protein complexed with cAMP as a signal of carbon source limitation (see text)
Amino acid or energy limitation	Stringent response	Enteric bacteria and many others	*relA* and *spoT* [enzymes of (p)ppGpp metabolism]	Genes (>200) for ribosomes, other proteins involved in translation, and biosynthetic enzymes	(p)ppGpp thought to modify promoter recognition by RNA polymerase (see text)
Ammonia limitation	Ntr system (enhances ability to acquire nitrogen from organic sources and from low ammonia concentrations)	Some enteric bacteria	*glnB*, -*D*, -*G*, -*L* (transcriptional regulators and enzyme modifiers)	*glnA* (glutamine synthetase), *hut*, others encoding deaminases	Complex
Ammonia limitation	Nif system (nitrogen fixation)	*Klebsiella aerogenes*, many others	Multiple genes, including those controlling ammonia assimilation	Multiple genes encoding nitrogenase (for nitrogen fixation)	Complex
Phosphate limitation	Pho system (acquisition of inorganic phosphate)	Enteric bacteria	*phoB* (PhoB, response regulator); *phoR* (PhoR, sensor kinase); *phoU* and *pstA*, -*B*, -*C*, -*S* (facilitate PhoR function)	*phoA* (alkaline phosphatase), and approx 40 other genes involved in utilizing organophosphates	Two-component regulation: transcriptional activation by PhoB upon signal of low phosphate from the sensor kinase, PhoR
Energy metabolism					
Presence of oxygen	Arc system (aerobic respiration)	*E. coli*	*arcA* (ArcA, repressor); *arcB* (ArcB, modulator)	*Many genes* (>30) for aerobic enzymes	Repression of genes of aerobic enzymes by ArcA upon signal from ArcB of low oxygen
Presence of electron acceptors other than oxygen	Anaerobic respiration	*E. coli*	*fnr* (Fnr)	Genes for nitrate reductase and other enzymes of anaerobic respiration	Transcriptional activation by Fnr
Absence of usable electron acceptors	Fermentation	*E. coli* and other facultative bacteria	Unknown	Genes (>20) for enzymes of fermentation pathways	Unknown

Response to damage					
UV and other DNA damagers	SOS response	*E. coli* and other bacteria	*lexA* (LexA, repressor); *recA* (RecA, modulator of LexA)	Approx 20 genes for repair of DNA damaged by UV and other agents	Transcriptional repression by LexA, relieved upon cleavage of LexA by RecA
DNA alkylation	Ada system (alkylation response)	*E. coli* and other bacteria	*ada* (Ada, an activator)	Four genes involved in removal of alkylated bases from DNA	Transcriptional activation by Ada
Presence of H_2O_2 or similar oxidants	Oxidation response	Enteric bacteria	*oxyR* (OxyR, a repressor)	Approx 12 genes involved in protection from H_2O_2 and other oxidants	Transcriptional repression by OxyR
High temperature	Heat shock	All bacteria	*rpoH* (sigma-32)	Dozens of genes involved in protein synthesis, processing, and degradation	Programming RNA polymerase by sigma-32
Low temperature	Cold shock	*E. coli* and other bacteria	*cspA* (CspA)	Several (12) genes for macromolecule synthesis	Unknown
High osmolarity	Outer membrane porin response	*E. coli* and other bacteria	*ompR* (OmpR, response regulator); *envZ* (EnvZ, sensor kinase for osmolarity)	*ompF* (OmpF porin); *ompC* (OmpC porin)	Phosphorylated OmpR, when phosphorylated by EnvZ, is a negative regulator of *ompF* and a positive regulator of *ompC*
Miscellaneous global systems					
Growth-supporting property of environment	Growth rate control	All bacteria	*fis* (Fis); *hns* (H-NS); *relA* (RelA); *spoT* (SpoT)	Hundreds of genes, many involved in macromolecule synthesis	Complex; involves availability of RNA polymerase/sigma-70 holoenzyme influenced by passive control; see discussion in text of Chapter 13
Starvation or inhibition	Stationary phase	All bacteria	*rpoS* (sigma-S); *lrp* (Lrp); *crp* (CAP); *dsrA*, *rprA*, and *oxyS* (regulatory sRNA molecules); many other regulatory genes	Hundreds of genes affecting structure and metabolism	See discussion in text of Chapter 13
Starvation	Sporulation	*Bacillus subtilis* and other sporeformers	*spo0A* (activator); *spo0F* (modulator); many other regulatory genes	Many (>100) genes for spore formation	See discussion in text of Chapter 14

What do we call the units that are above the operons? Two terms are in common use: **regulons** and **modulons.** Before defining them, we want to qualify their usage. Bacteria are so adept at designing regulatory networks that assigning just *two* names to distinguish among all the varieties is not realistic. Nevertheless, at the moment, the terms *regulon* and *modulon* are in common use, so let us proceed. It will be helpful to refer to Fig. 12.10 in this discussion.

Start with the term **regulon**, illustrated in Fig. 12.10A. It refers to *a group of independent operons governed by the same regulator, usually a protein repressor or activator.* This can be illustrated by dozens of examples; the one for which the term was first used is the *arg* regulon of *E. coli*, consisting of a group of operons scattered around the chromosome encoding enzymes of the arginine biosynthetic pathway. (In contrast, its tryptophan and histidine pathways are encoded by single operons.) Regulons provide a means by which sets of operons can be coordinately regulated. Above regulons are modulons. The term **modulon**, illustrated in Fig. 12.10B, refers to *a group of independent operons subject to a common regulator, even though they are members of different regulons.* Modulons provide a means by which regulons can be coordinated and may consist of individual operons, as well as regulons. An example of a modulon is the **catabolite repression system**, which in enteric bacteria unites virtually all the operons and regulons related to substrate utilization (**catabolism**), repressing them even in the presence of their inducing substrates if the cellular level of the nucleotide **cyclic AMP (cAMP)** is low (Table 12.2). This modulon ensures that cells growing in a mixture of potential carbon and energy sources can select the premium substrate from among them and thereby avoid making unneeded enzymes.

Examples of global regulatory systems

Some modulons are so pervasive they are referred to as **global regulatory** or **global control systems.** Two modulons that fit this category are the catabolite repression system (described above) and the stringent-response system. Together, these two systems directly or indirectly control *probably three-fourths of the protein-synthesizing capacity of the cell.* For this reason alone they deserve a closer inspection.

Catabolite repression has extraordinary importance in the physiologies of many bacteria. The **inducible** catabolic operons in this network are capable of very high levels of expression when induced by the presence of their substrates. Preventing redundant expression of these operons is so critical that their control in a modulon is very general in the bacterial world, even though the molecular mechanisms of the control differ widely. The ability of enteric bacteria to ensure that a premium substrate, glucose, is utilized to the exclusion of lesser substrates depends on at least three different devices, only one of which is the catabolite repression module (the other two are the constant synthesis of high levels of glucose-metabolizing enzymes in any medium and the inhibition by glucose of the entry of other substrates). How does the catabolite repression module work? Each operon has a regulatory region upstream of the transcriptional start site, at which a protein regulator, CAP, can bind. As its name implies, CAP stimulates transcription initiation from the promoters of these operons, all of which have low inherent strengths. CAP is an allosteric protein, and it binds to its sites only if it is bound by a small nucleotide ligand, cAMP. The

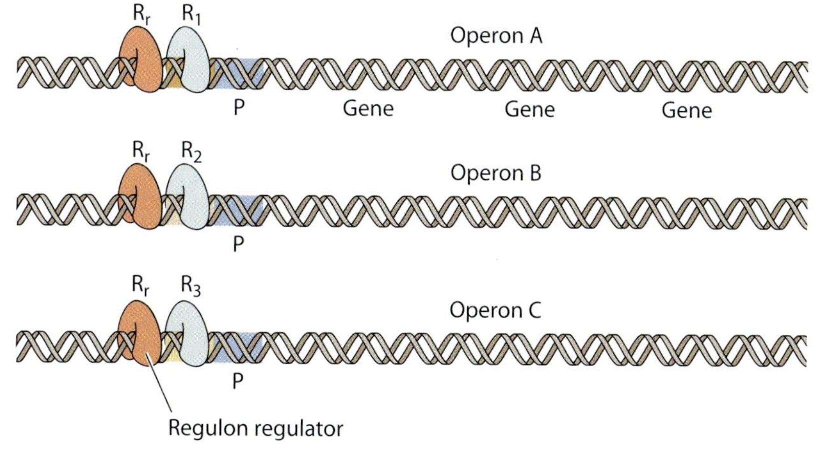

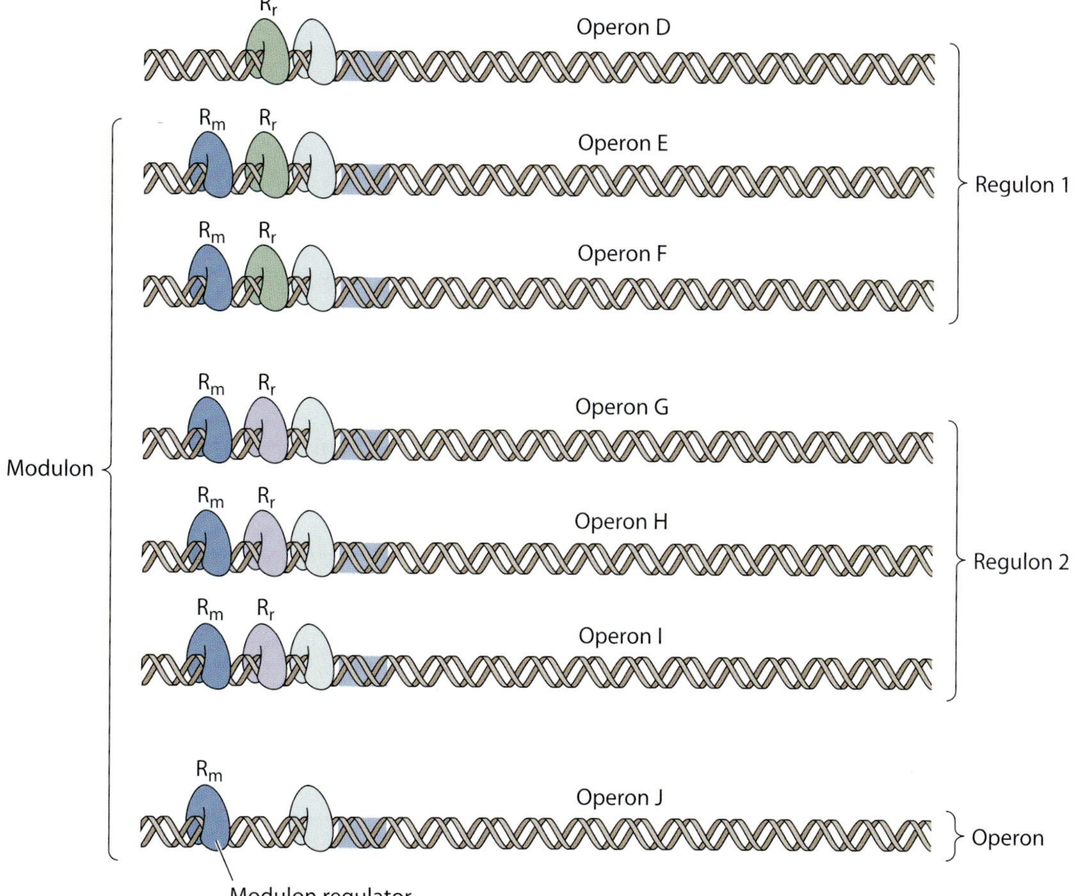

Figure 12.10 Patterns of operon organization into higher (global) regulatory units. (A) Regulon organization. Each operon (A, B, or C) is depicted as a segment of DNA consisting of a control site at which a common protein regulon regulator (R_r) and operon-specific protein regulators (R_1, R_2, and R_3) act to control transcription of the genes of the operon. The promoter (P) of each operon is also indicated. **(B)** Modulon organization. A modulon consisting of six operons (E to J), controlled by a single modulon regulator (R_m) or global regulator, is depicted. Operons E and F are part of one regulon (regulon 1), along with operon D (which is not in the modulon). All three operons (G, H, and I) of regulon 2 are included in the modulon, along with the independent operon J. The two individual regulon regulators are shown, as are the individual regulators of each operon.

binding of cAMP-CAP then increases transcription if the operon has been properly induced. This is important: *this activation of transcription by cAMP-CAP works only on operons within the modulon that have already been induced by their cognate substrate in the medium.* Glucose lowers the cellular level of cAMP, decreasing the amount of cAMP-CAP, and as a result preventing high-level expression of even those operons that have inducing substrates present. This inhibition of the synthesis of inducible catabolic enzymes is called, in the case of enteric bacteria, **glucose-mediated catabolite repression.** This repression and the exclusion by glucose of other substrates from the cell are collectively responsible for what has been called the **glucose effect.** Glucose affects the cAMP level of the cell through inhibition of the enzyme that makes cAMP, adenylate cyclase. This is a membrane-bound enzyme that becomes inhibited when glucose enters the cell by the phosphotransferase system (see Chapter 6). Different bacterial species achieve the same result without employing cAMP; gram-positive bacteria, such as *Bacillus*, lack this molecule yet possess an effective catabolite repression system.

The **stringent-response modulon** is a gigantic network geared to respond to starvation, for either energy or amino acids. The response is extensive. It affects a large number of operons and regulons with diverse functions (Fig. 12.11). Many details of the mechanism remain to be learned. What is known is that limitation of energy or amino acids restricts the formation of one or more species of aminoacyl-tRNA. Some subset of the cell's ribosomes, when they are stalled in the act of making a protein by lack of an aminoacyl-tRNA, have the ability to make a GTP derivative, ppGpp. What happens next is a bit fuzzy, but somehow ppGpp alters RNA-P's ability to recognize many regulatory sites on DNA, thereby altering the global pattern of transcription. The result (Fig. 12.11) makes physiological sense: inhibitions and stimulations of various metabolic processes serve to increase the capacity of the cell to cope with the nutrient restric-

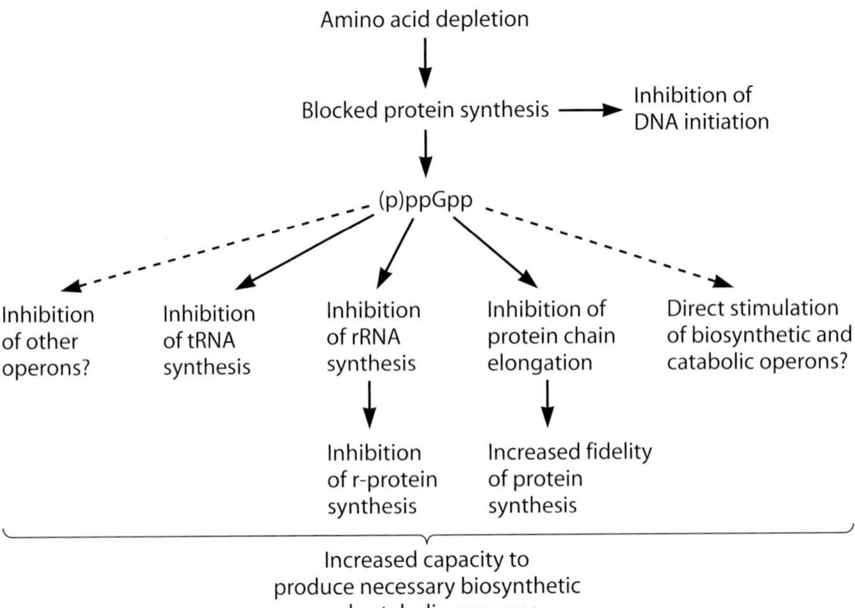

Figure 12.11 Stringent response. The series of events ensuing after amino acid restriction of bacteria are shown. The solid arrows indicate processes known on good evidence to result from ppGpp accumulation. The dashed arrows depict more speculative relations.

tion. The complete story of the stringent response is more involved than this brief account; you can guess that there are elaborate biochemical paths for making and degrading ppGpp and interesting properties of mutants lacking part or all of the components needed for the response.

COOPERATIVE INTERACTION OF REGULATORY DEVICES

In "Why two modes of regulation?" above, we posed the question, why should a cell regulate its metabolism by changing both protein activity and protein amount? We are now in a better position now to tackle this question.

Control of allosteric enzyme activity is both swifter and more precise than control of enzyme level in adjusting the flow of material through the many pathways of metabolism. The results we saw in the experiments described in "Evidence for Coordination of Metabolic Reactions" above could only be achieved by end product inhibition operating by means of allosteric enzymes. Regulation of the histidine biosynthetic pathway, for example, by modulating the expression of the genes encoding the enzymes of the pathway certainly occurs, but it could not provide the near-instantaneous shifts in histidine biosynthesis observed when that amino acid is added to or removed from the medium. The major job of coordinating metabolic flow so as to produce the components for a new cell in an orderly fashion could be accomplished by allosteric enzymes.

However, we have just seen the enormous lengths to which bacteria go in order to regulate the synthesis of proteins. Our cursory review of 10 vastly different ways to govern gene expression did not even include all the varieties of each mechanism. Evolution has forced bacteria to be highly sensitive about what proteins they make in a given situation and how much they make of each. A reasonable guess about the force behind this evolution is the need for economy and efficiency. Over half the cell's dry mass is protein, and making proteins is expensive. Optimizing the rate of growth in each environment requires that cells not make redundant or irrelevant proteins.

By this argument, metabolic coordination—making order out of chaos, as was said in "Evidence for Coordination of Metabolic Reactions" above—would seem to be the domain of allosteric control of enzyme activity. Competing successfully in a world filled with competitors for food and space would seem, on the other hand, to demand the elegant and effective controls on transcription and translation.

From these considerations, and because allosteric proteins play major roles as regulators of gene expression, one might speculate that allostery developed at an earlier stage of evolution than did controls on gene expression.

SUMMARY AND CONCLUSIONS: NETWORKS FOR COORDINATION AND RESPONSE

Our discussion has gradually led from a consideration of how bacterial cells coordinate their metabolism to the issue of how they cope with their environment.

Growth and maintenance of a bacterial cell require integration of its 2,000 individual chemical reactions into a single complex system, with controls on each reaction so that chaos is avoided. Specific control devices have evolved to govern enzyme activity and enzyme synthesis, and we have had a glimpse into the world of feedback control and complex circuitry that integrates the activity of all the cell's parts and processes. Even in an ideal and constant environment (probably nonexistent in nature and only approached in the growth of cultures in laboratory chemostats), all this coordination *would still be required for the life and growth of the cell.*

In real life, nothing is constant. All niches in the biosphere of Earth are subject to changes in temperature, chemical composition, and all the other parameters important for life. The environment of most microbial cells is subject to catastrophic changes of a sort not usually met by the individual cells of higher plants and animals.

Therefore, an extraordinary burden must be added to the regulatory devices that coordinate the internal processes of the bacterial cell. This cell must cope with an environment that continually presents new challenges: changes in the nutritive value of the environment; changes in temperature, pH, redox potential, barometric pressure, and the other physical parameters of the environment; and changes in the presence of toxic factors. One cannot discuss coordination of metabolism for long without encountering the necessity of dealing with a changing and at times hostile environment.

The story of how metabolic reactions are coordinated cannot be separated from the story of how microbial cells cope with environmental stresses. Therefore, the next chapter is a continuation of this one.

STUDY QUESTIONS

1. Bacteria rely more on adjusting the levels of their enzymes than do cells of plants and animals. Cite four reasons for this difference.

2. Which type of control is generally faster: modulation of protein activity by allosteric interactions or adjustment of enzyme levels? What advantage would be sacrificed if all metabolic coordination were achieved solely by modulating protein activity?

3. Modulating protein activity by covalent modification, such as by phosphorylation, would seem to be more costly than doing it by simple allosteric binding of a ligand. What advantage(s) might be offered by covalent modification?

4. Describe how the ratio of the concentration of ATP to those of ADP and AMP is maintained almost constant in bacterial cells over a wide range of energy supply and demand.

5. All proteins are made by ribosomes, but in some cases, ribosomes also play a role in regulating the rates of synthesis of particular proteins. Of the dozen or so different mechanisms for regulating the synthesis of specific proteins, name two that involve direct participation of ribosomes.

6. In terms of energy, what are particularly expensive ways for a bacterial cell to adjust the cellular level of a given protein? What ways of adjusting the protein level are best at avoiding energy waste?

7. Operons encoding biosynthetic enzymes are regulated in a number of ways, two of which are transcriptional repression and attenuation. Compare these two mechanisms with respect to the number of regulatory genes involved.

8. Explain how control regions located hundreds of nucleotides away from a promoter can control gene expression from that promoter. What name is given to such control regions?

9. Allosteric interactions modulate the activities of enzymes. Explain how allostery also plays a role in modulating the amounts of enzymes by controlling gene expression.

10. Give two reasons for the evolution of controls that govern sets of operons together as regulatory units.

11. What is the difference between regulons and modulons?

12. What is the sole reason for designating some modulons global regulatory systems? Give an example of such a modulon.

chapter 13

succeeding in the environment

MICROBES IN THEIR HABITAT

A thumbnail summary of the prototype bacterium we have described so far might run as follows: *a one-celled organism, invisible to the unaided human eye, rich with structural detail at the molecular level, and capable of rapid rates of self-synthesis and asexual reproduction by binary fission in a variety of biochemical environments.*

So far, so good. But this picture is akin to describing a human as *a multicellular bipedal omnivore supported by a vertebral column, of medium size, reproducing over years by a complex process involving sexually differentiated individuals.*

In each case, it is not just that some major attributes of the organism have been omitted—rather, the essential life and being of the organism have been neglected. In this chapter, we shall demonstrate that the bacterial thumbnail misses the boat as fully as the human sketch fails to do justice to, let us say, *you*. What has been missing up to now is an account of the bacterial cell in its natural environment.

We shall present two aspects of microbial life in this chapter. First, we shall examine how bacteria sense their environment and respond as individual cells to specific environmental challenges; largely, this portion of our narrative will describe well-delineated responses to various stresses and directed motility. Second, we shall indicate how bacteria act cooperatively, displaying such communal activities as cell-cell communication, group mobilization, and community formation. Later chapters will deal with other activities of microbes in their natural habitat, including their ability to interact in subtle and not so subtle ways with humans.

COPING WITH STRESS AS INDIVIDUAL CELLS

In the preceding chapter, we saw how bacteria regulate the flow of material through metabolic pathways. We noted how they optimize their synthesis of macromolecules to achieve rapid growth. The sensing molecules and integrated circuits that achieve these flexible internal adjustments are effective and elegant, but they alone cannot account for the remarkable ability of bacteria to cope with catastrophic events. And cope they must, for the same feature that contributes to their success in colonizing planet Earth—microscopic size—leaves microbes exceptionally exposed to environmental forces.

Picture a bacterial cell suddenly expelled from the intestinal tract of some large mammal. From a constantly warm, dark, nutritionally complex, anaerobic environment where it may have been living for weeks, this cell is thrust without warning into what—depending on circumstance—could be a world that is cold, nutritionally lean, aerobic, of different pH, and bathed with light that includes ultraviolet wave lengths. The stress on this small cell is enormous, and yet this event is neither uncommon nor exaggerated. Many microbial cells dwell in environments in which the temperature can change quickly from moderate to very hot or very cold; where the water available to them can change from isosmotic to something approaching distilled water (rain) or, perhaps, to heavily salted brine; where the pH can depart from neutrality and become either very acidic or very alkaline; where the oxidation potential can rise from that of a completely anaerobic environment to one that is highly oxidizing; and where the threat of radiation damage is ever present. Differential gene expression and allosterically regulated enzymes may suffice to handle a new diet, but the kind of food on the table is of secondary import to a cell faced with sudden and inescapable environmental stress.

Table 13.1 Environmental properties that can produce stress for microbes

Physical factors
Temperature
Hydrostatic pressure
Osmotic pressure
Radiation
Ionizing
Nonionizing
Chemical factors
Presence of deleterious agents
Inorganic compounds and ions
Heavy metals
Oxygen and its derivatives
Protons
Hydroxyl ions
Organic compounds
Antibiotics
Acids, phenols, alcohols, etc.
Nutrient depletion or restriction
Carbon and energy sources
Nitrogen, sulfur, and phosphorus sources
Metal ions (Fe, Zn, Mg, Mo, Se, etc.)
Oxygen or other electron acceptors
Endogenously produced compounds
Protons
Organic acids, alcohols, etc.

Nature of stress

First, we should have a clear notion of what is meant by environmental stress. The defining feature is *change*. A thermophilic bacterium living constantly at 100°C is not particularly stressed, nor are halophilic microbes living in salt brine or acidophilic species living in acid mine water. To such organisms, known as **extremophiles**, living under such conditions is just another day at the office. What *is* stressful for any microbe is a radical change in the environment to which it has become adapted by evolution and its recent history.

Each of the environment's physical factors and chemical components (Table 13.1) can vary over a wide range, and thus, each can be the source of significant stress on microbial cells. Indeed, each can potentially become lethal. When populations of microbes become large, they even create their own problems, and not just by depleting the local environment of nutrients. Microbial cells, like all living things, pollute. Compounds produced during metabolism, particularly by the fueling reactions, are significant constituents of the environment and can be quite toxic. Changes in the pH of the medium and the production of short-chain organic acids and related by-products, for example, are prominent among the stress factors produced by metabolism. The ability to live in one's own metabolic waste has clear advantages, especially if one's competitors cannot.

For many microbial species, the ranges of temperature, pressure, pH, and the like to which the organism can successfully adapt are remarkably broad, thanks to stress responses that have evolved over the past couple of billion years. We shall examine some general features and specific examples of these responses now.

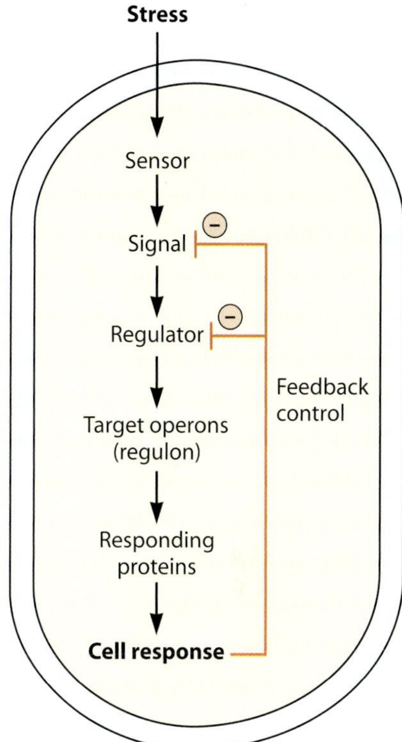

Figure 13.1 A simplified stress response circuit.

Overview of stress responses

Change triggers change. A potentially dangerous shift in the environment invokes a modified behavior. The cell's response usually has some cost associated with it and must be reversed when conditions improve. The speed, simplicity, and effectiveness of its response almost obscure the fact that the response is made by a nonconscious creature incapable of purposeful decision and action. The organism's initial response, and the management of that response, must be entirely automatic. How this is accomplished deserves a close look. We shall concentrate on bacteria because their behavior has been particularly well studied.

A simplified diagram of a stress response system, as shown in Fig. 13.1, is a good place to start in understanding responses to stress. To begin, a stress brought about by some change in the environment is sensed by the bacterium. Commonly, though not universally, the **sensor** is a protein within the cell membrane. The environmental change causes an allosteric alteration of the sensor, which leads, commonly, to its autophosphorylation, in which case it is termed a **sensor kinase**. Next, the modified sensor generates a **signal** that is transmitted—commonly through a train of phosphorylated intermediates—to a **response regulator**. This passage of information is called **signal transduction**. The response regulator frequently is a protein that **modulates** (adjusts) transcription by binding to the DNA that governs the expression of a set of target operons tied together in a regulon (the regulatory unit we met in the previous chapter). This regulon produces the responding proteins that bring about the cellular response. Finally, feedback controls match the cellular response to the problem, i.e., to the magnitude and length of the period of stress.

More needs to be said about this process, both about how cells sense their environment and about how they manage to mount very complex cellular responses.

Sensing the environment

Responding to changes in the environment requires the ability to monitor its key nature: pH, temperature, redox potential, light, osmolarity, concentrations of individual organic and inorganic nutrients, and presence of toxic substances. Evolution has been active in selecting many sensory mechanisms. A large number of them consist of a sensor kinase and a response regulator and thus are called **two-component regulatory systems**.

Many other response systems, however, contain more than two components. These systems include **signaling proteins**, which transmit the phosphate signal between sensor and regulator, and are thus **multicomponent regulatory systems**. Components of bacterial response systems have been highly conserved throughout the microbial world.

As illustrated in Fig. 13.2, sensor kinases are usually transmembrane proteins capable of autophosphorylation. Specific environmental changes

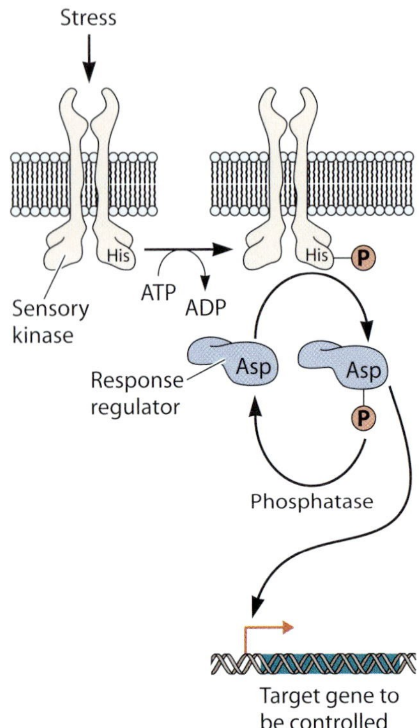

Figure 13.2 Function of sensor kinases. An external stress affects the N-terminal portion of a sensor kinase, an integral membrane protein, activating its autokinase activity, which phosphorylates a conserved histidine residue. The phosphoryl group is transferred to a conserved aspartate residue of the response regulator protein, inducing a conformational change that activates its DNA-binding ability to regulate a target gene(s).

cause phosphorylation of a histidine residue in the sensor kinase. In most cases, the sensor kinase spans the cell membrane; in others, it resides entirely within the cytoplasm. The phosphate donor is ATP. In either case, the phosphorylated kinase next transfers the phosphate group to an aspartate residue near the amino-terminal end of a second protein, usually the response regulator protein, which can then activate the appropriate cellular response, either by activating a set of genes or interacting with other proteins.

There are a great variety of multicomponent regulatory systems in the microbial world. Table 13.2 lists a sampling.

Complex circuitry for complex responses

The diagram in Fig. 13.3 conveys more realistically than Fig. 13.1 the true complexity of many stress response circuits. Note the added features: a stress is shown triggering *more than one* signal and activating *more than one* regulatory cascade. Multiple regulons thus become involved in the response, some of them being activated secondarily as consequences of the activation of more primary regulons. Some of the effector proteins involved in the primary response may themselves be regulatory proteins involved in the subsequent activation of other regulons. (By now you may have realized that some of the responding networks fit the definition of modulons, described in Chapter 12, rather than regulons.)

We need a name for the large ensemble of responding genes in stress responses of this complexity, and that name is **stimulon.** A stimulon is the set of all genes encoding proteins that have their rates of synthesis significantly changed (positively or negatively) as a result of a cell's response to a stimulus (Fig. 13.3). It is relatively easy to determine the constituents of a stimulon. One simply tallies the proteins that have been induced or repressed. The more difficult task is sorting the members of each stimulon into their cognate regulons and modulons (regulatory units) and learning the regulatory devices that achieve the overall response (Table 13.3). These tasks, which have hardly begun, require all the approaches of traditional molecular genetics and physiology, along with more recently developed techniques for monitoring the cell's total transcription and translation pattern.

Table 13.2 Multigene networks controlled by homologous phosphotransfer proteins[a]

System	Protein kinase	Response regulator	Organism
Nitrogen regulation (Ntr)	NRII	NRI	*Escherichia coli, Salmonella, Klebsiella* spp., *Rhizobium* spp.
Phosphate regulation (Pho)	PhoR	PhoB	*E. coli, Bacillus subtilis*
Porin regulation (Omp)	EnvZ	OmpR	*E. coli*
Symbiotic nitrogen fixation (Fix)	FixL	FixJ	*Rhizobium meliloti*
Aerobic respiration (Arc)	ArcB	ArcA	*E. coli*
Virulence (Vir)	VirA	VirA/VirG	*Agrobacterium tumefaciens*
Virulence (Vir)	BvgC	BvgA	*Bordetella pertussis*
Sporulation (Spo)	SpoIIJ	Spo0A/Spo0F	*B. subtilis*

[a] In some cases, the primary information is not the demonstration of phosphorylation but sequence homology.

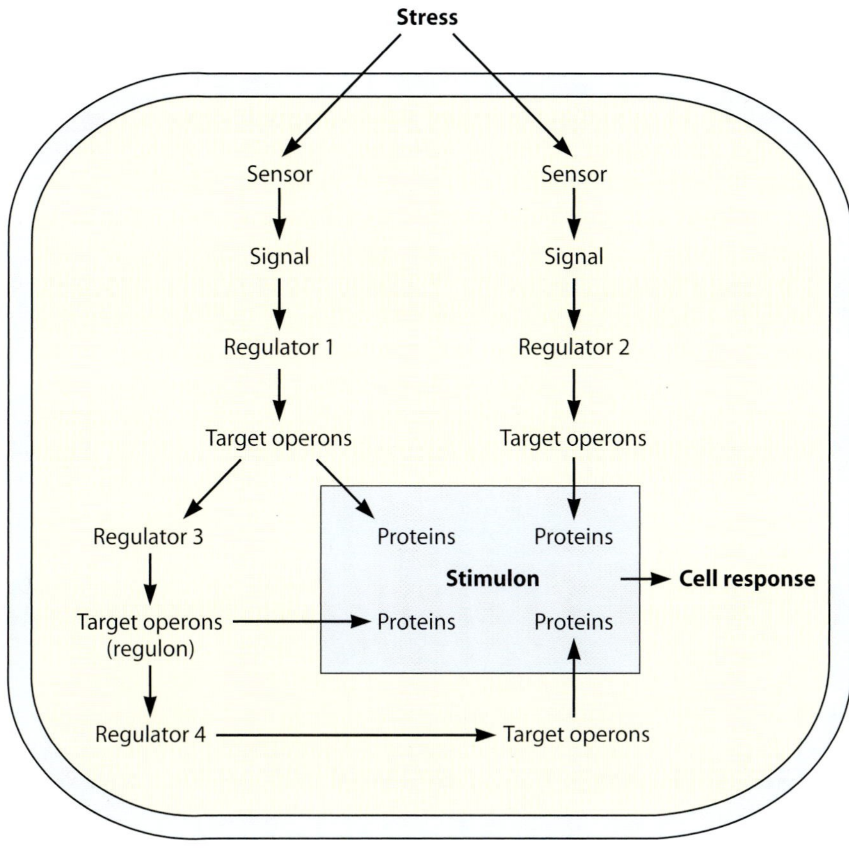

Figure 13.3 A generalized stress response circuit. The imposition of a stress is shown triggering the response of its complex stimulon.

Monitoring stimulons

In the mid-1970s, techniques were developed that permitted the monitoring and measurement of the synthesis of nearly the entire complement of individual proteins made by a cell (the cell's **proteome**). The analysis consists of feeding radioactively labeled amino acid precursors to the cells during a stress response, followed by two-dimensional polyacrylamide **gel electrophoresis** to separate the individual cellular proteins. Each protein appears as an individual spot, and its radioactivity can be visualized by autoradiography (exposing the gel to an X-ray film, which is developed after a suitable time). This approach, called **proteomic monitoring**, proved a watershed in bacterial physiology, particularly with respect to the study of stress responses. It permitted discovery of the full extent of the bacterial cell's response to a particular stress, free of the investigator's preformed notions. Instead of asking about the synthesis of this or that protein of personal interest, the investigator could learn what proteins *the cell* considered important for a given stress response.

The power of proteomic monitoring was significantly augmented in the 1990s by the application of mass spectrometry to identify protein spots on polyacrylamide gels. Knowing the sequence of the complete genome of an organism makes it possible to calculate the precise molecular mass, as well as the **isoelectric point** (the pH at which the acidic [negative] and basic [positive] charges are exactly balanced), of each of the proteins it can make. By mass spectrometry, the mass of a protein can be measured with great

Table 13.3 Prominent global response regulatory systems

System	Function	Regulatory components	No. and kinds of genes regulated	Type of regulation
Catabolite repression (CRP regulon)	Ensure priority use of premium substrates	In gram-negative bacteria, regulatory protein is CRP, nucleotide is cyclic AMP (cAMP); in gram-positive bacteria, regulatory protein is CcpA (no cAMP involvement)	Large numbers of genes encoding catabolic enzymes	CRP complexed with cAMP stimulates transcription initiation from many operons; low cAMP level during metabolism of premier substrate causes low expression of genes for catabolic enzymes; gram-positive bacteria use CcpA protein complexed with Hpr of the phosphotransfer system.
Stringent response (RelA regulon)	Divert metabolism to fueling and biosynthesis	RelA and SpoT enzymes make nucleotide, ppGpp; ribosomal RelA acts when ribosome is "hungry" for aminoacyl-tRNA	Perhaps hundreds of genes in fueling and biosynthesis, as well as negative effects on genes encoding polymerization enzymes and RNA	Not known completely; evidence suggests an effect of ppGpp on RNA polymerase; high ppGpp stimulates some fueling and biosynthetic operons; high ppGpp shuts off expression of genes for r proteins, rRNA, and translation factors.
Growth rate control	Adjust transcription and translation to match the growth-promoting ability of the medium	Components of RelA regulon; Fis protein; H-NS protein (see text and Table 13.4)	Hundreds of genes encoding ribosomal proteins, rRNA, and translation factors are adjusted to match the rate of protein synthesis possible in a given medium	A major unsolved problem; stringent response functions transiently during shift to leaner medium, as does a protein, Fis, that controls rRNA synthesis; steady-state condition may be due to "passive control" (repression of biosynthetic and fueling operons in rich medium may free RNA polymerase or other resources for making translation machinery).
Stationary phase (RpoS regulon)	Change cell structure and convert metabolism to new mode to optimize survival during nongrowth	RpoS (σ^{38} or σ^{s}); cooperative with other global regulators, such as CRP, Lrp, H-NS	Direct or indirect effects on most genes of the cell	Multiple modes of regulation in a complex network; involves several global regulatory systems, in addition to selection of genes with promoters recognized by σ^{s}.

Heat shock response (RpoH regulon)	Assist protein folding at high temperature; repair and/or degrade damaged proteins	RpoH (σ^{32} or σ^{H})	Dozens of genes encoding proteases, protein chaperones, and other genes of uncertain function	Primary mode of control is selection of genes with promoters recognized by RpoH; cooperative with stringent response.
Envelope stress response (RpoE regulon)	Assist protein folding or degrade unfolded membrane proteins; monitor outer membrane proteins, particularly porins	RpoE (σ^{E}, σ^{24}), RseA (anti-σ^{E}), and RseB and the proteases DegS and YaeL	Many genes encoding proteases, periplasmic chaperones, and biosynthetic enzymes for the lipid A component of lipopolysaccharide in gram-negative bacteria	In nonstress conditions, RseA, RseB, and RpoE form a complex that prevents RpoE from binding to RNA-P. Envelope stress, as by heat, denatures periplasmic proteins, which are bound by RseB, allowing protease DegS to cleave RseA, thereby freeing RpoE to bind to RNA-P, directing it to promoters of the regulated genes.
Envelope stress response (CpxAR regulon)	Assist protein folding or degrade unfolded membrane proteins; monitor envelope surface proteins, such as pili	CpxA (sensor kinase) and CpxR (response regulator); CpxP, an inhibitor of CpxA	Many genes involved in protein folding and trafficking within the envelope; partially overlapping with the RpoE regulon	In nonstress conditions, CpxP is bound to CpxA, preventing it from phosphorylating itself. When envelope stress leads to misfolded proteins, they bind CpxP, freeing CpxA to act as a kinase, phosporylating itself and the cytosolic response regulator CpxR. The activated CpxR then serves as an activator of the genes of the CpxAR regulon.
Leucine response (Lrp regulon)	Assist cells to adjust to major changes in nutrient availability, particularly amino acids	Lrp (leucine response protein)	Many genes encoding proteins involved in amino acid biosynthesis, degradation, and transport; also genes encoding pilin proteins	Lrp is a dual regulator, serving as a transcriptional activator of some operons (especially those encoding bioysynthetic enzymes) and a transcriptional repressor of others (especially those encoding degradative enzymes). In some, but not all, cases, leucine modifies Lrp activity, either positively or negatively.

precision; the location of a protein in the polyacrylamide gel provides an indication of its isoelectric point. Together, these two kinds of data make it possible to ascertain the identities of spots of interest.

Another development, **transcriptional monitoring** (also called **genomic monitoring**), which came on the heels of microbial-genome sequencing in the 1990s, provides an important complement to proteomic monitoring. Such monitoring enables one to examine the cell's **transcriptome**, that is, the individual gene transcripts (messenger RNAs [mRNAs]) being made at any given time. Transcriptional monitoring is accomplished through the use of DNA **microchips**, which are small glass or plastic supports (perhaps a square inch) on which small quantities of DNA corresponding to every cellular gene are firmly affixed in a grid pattern. The total RNA from a culture being studied is rapidly extracted, and **complementary DNA (cDNA)** is made from the RNA sample by use of the enzyme RNA-dependent DNA polymerase (also called **reverse transcriptase**). During the synthesis of the cDNA, a fluorescent nucleotide is incorporated. The fluorescence-tagged cDNA is then allowed to hybridize to the gene-loaded chip (called a **microarray**), and subsequent scanning with appropriate detection devices displays which genes in the culture were actively producing mRNA—they fluoresce. Comparison of two cultures can be made by the use of fluorescent tags of contrasting colors, one for the cDNA of each sample. One can easily imagine the usefulness of this technique for studying the total cell response to its environment.

Proteomic monitoring and transcriptional monitoring are far from redundant. Rather, they are overlapping. Each provides a measurement of gene activity, but one measures a short-lived intermediate (mRNA) and the other measures a more stable product (protein), although one that is subject to degradation or modification. Thus, differences in results can be expected. Also, the numbers of protein molecules produced by a single mRNA molecule differ from gene to gene, and for a given gene, this number can vary with physiological conditions. Variation in the yield of protein from each mRNA molecule may be caused by changes in the rate of degradation of the mRNA or by changes in the regulation of translation initiation (see Chapter 9). Finally, the two approaches differ in their sources of error. As a result, microbial research utilizes both techniques.

Proteomic monitoring and transcriptional monitoring *do*, however, yield one consistent and surprising picture: *even simple stress or minor alterations in conditions of growth lead to changes in the rates of synthesis of hundreds of cellular proteins.* The responding proteins turn out to consist of several functional kinds. Some you might expect, such as proteins that deal directly with the given stress, including hydrogen peroxide-destroying catalase when that toxic agent is the cause of stress or RecA, the key DNA-splicing enzyme, when the cause is DNA damage (see Chapter 9). However, there are also responder proteins, such as chaperones and proteases, that assist in coping with many different stresses, including heat shock. The number of responder proteins also varies; as many as 75 are reduced in synthesis when cells grow more slowly. These include various transcription and translation factors, as well as the ribosomal proteins and ribosomal RNA (rRNA) itself. Finally, a bevy of proteins are part of the regulatory cascade involved in preparing the cell for the nongrowth state called **stationary phase** (see below); their synthesis is triggered under severe stress conditions.

Major stress response networks

Considering the complexity of each response, it is a wonder that individual bacterial cells can marshal defenses against such varied types of challenges. Several facts help explain this difficult feat. First, a cell does not have to mount a totally different response for each stress. Many toxic substances target the same cellular structure, which can be protected or repaired by common mechanisms. Simple modification of the cell envelope, for example, may suffice to diminish the deleterious effects of different chemical agents. Second, several stress responses are effective against hundreds of different specific problems. These major networks, widely distributed among bacterial species, deserve special attention.

Global response networks

As few as a dozen networks collectively equip a cell to cope with most conditions. These networks affect many cell functions, are active in response to many environmental stresses, and are widely distributed throughout the bacterial world. For these reasons, the networks are appropriately designated **global response networks** (or **global regulatory systems**), and we examined two of them (catabolite repression and the stringent response) in Chapter 12. Other prominent networks are briefly presented in Table 13.3, which supplements the list of multigene systems in Table 12.2. The widespread distribution of these global response networks among bacterial species speaks not only to their effectiveness, but also to their early appearance in evolution.

Three of their collective features of global responses should be highlighted. First, these systems employ a variety of molecular mechanisms: unique sigma factors; regulatory proteins that act as repressors, activators, or both; and nucleotides that seem to instruct RNA polymerase (RNA-P) in ways still unknown. Second, these systems are not independent of each other: starvation for an essential amino acid, for example, invokes at least three global response networks, catabolite repression, the stringent response, and the Lrp (leucine response) regulon.

Finally, global response networks are all best understood in the context of the entire intact cell; that is, their quantitative details depend on the integral physiological state of the cell. The best example of this dependence is found in the growth rate-related regulation of the protein synthesizing system (PSS), which we should examine now.

Regulation of the PSS. We noted in Chapter 4 (see "How Is the Physiology of the Cells Affected by the Growth Rate?") that the faster cells grow, the greater the proportion of their resources they devote to making ribosomes and the many other components of translation. Synthesis of the PSS can be a massive metabolic task. In *Escherichia coli*, for example, during fast growth in rich media, rRNA transcripts alone constitute over half the RNA made by the cell, even though the encoding genes make up less than 0.5% of the genome. In addition, a hundred or so proteins constitute the PSS, including initiation and elongation factors, aminoacyl-transfer

RNA (tRNA) synthetases, and naturally, the ribosomal proteins. When the cells are growing slowly in nutritionally lean medium and require less of the translation apparatus, efficiency demands that cells curtail this massive synthesis of the PSS. How cells achieve this efficiency—and all microbial cells *do* finely tune protein-synthesizing capacity to the growth rate—has occupied investigators for half a century. The complete answer, even for the most studied microbe *(E. coli),* remains elusive. What is known, however, is both surprising and instructive.

We shall examine only the control of synthesis of ribosomes, which, by weight, are the major constituents of the PSS. Ribosomal-protein synthesis, it may be recalled, is controlled by translational repression (see "Regulation of operon expression" in Chapter 12), through which their rate of synthesis is governed by the availability of rRNA to bind them and relieve the repression they cause. Thus, synthesis of rRNA is the rate-limiting (governing) step in ribosome synthesis, and we should focus our attention there.

In brief, the seven rRNA operons of *E. coli* possess strong promoters, and several of them have special upstream sequences (**UP elements**) that provide especially strong binding for RNA-P. The activities of rRNA promoters can be affected by many factors (Table 13.4), including two proteins: **Fis** and **H-NS**. The protein called Fis (for factor for inversion stimulation) binds to sites near the UP sequences and attracts RNA-P to the promoter; it is a *positive* controlling factor. The protein H-NS (for histone-like nucleoid structuring protein) affects **DNA bending** and is a *negative* controlling factor because it inhibits rRNA promoter activity. The cellular levels of both Fis and H-NS vary in a manner consistent with their presumed roles. In addition to these regulatory proteins, the guanosine triphosphate derivative ppGpp (guanosine tetraphosphate) has a powerful effect on rRNA synthesis. This nucleotide is the effector of the stringent-response modulon (see "Examples of global regulatory systems" in

Table 13.4 Factors contributing to regulation of synthesis of the PSS

Factor	Nature of factor	Role
UP elements	DNA sequences, rich in A-T residues, found immediately upstream of the promoters of rRNA operons (*rrn*)	UP elements bind the α subunits of RNA-P and thereby increase the activity of *rrn* promoters. Transcription initiation may be increased as much as 50-fold.
Fis	Protein with 3 to 5 binding sites upstream of the UP elements	Fis aids binding of RNA-P to *rrn* promoters and facilitates early steps in transcription initiation. Transcription initiation may be increased as much as eightfold. Cellular levels of Fis are high when rRNA synthesis is rapid, and vice versa.
H-NS	Protein that binds to DNA at regions where bending occurs, though no specific binding sequence is involved	H-NS inhibits transcription initiation at *rrn* promoters. The mechanism is unknown but may involve trapping RNA-P at the promoter by bending the DNA. Cellular levels of H-NS are low when rRNA synthesis is rapid, and vice versa.
ppGpp	Guanosine tetraphosphate; derivative of guanosine triphosphate (GTP) produced by the ribosome-associated RelA protein (Table 13.3)	ppGpp is a potent inhibitor of rRNA (and tRNA) synthesis, both directly and possibly indirectly by favoring the association with RNA-P of sigma factors other than σ^{70}, the required version for *rrn* promoter recognition.
RNA-P(σ^{70})	RNA polymerase with σ^{70}	RNA-P(σ^{70}) transcribes most of the operons of *E. coli*. Recruitment of polymerase to other promoters diminishes the number available to transcribe *rrn*. Also, binding of other sigma factors has the same effect (Fig. 13.4).

Chapter 12), and one of its properties is the ability to totally inhibit the synthesis of rRNA and tRNA. When a nutritional restriction slows growth, ppGpp accumulates quickly and changes RNA-P in such a way that it cannot function at promoters of rRNA (or tRNA, either). Evidence indicates that ppGpp favors the association of RNA-P with sigma (σ) factors other than σ^{70}, the one that recognizes rRNA promoters. With rRNA synthesis cut off, continued slow growth lowers the amount of the PSS. This stringent response accounts for the immediate halt in rRNA and tRNA synthesis upon starvation for even a single amino acid. Conditions that limit the cells' ability to synthesize nucleoside triphosphates, the substrates of RNA-P, will of course also restrict RNA synthesis.

One might imagine that this array of controls would account for growth-related regulation of PSS formation. *The surprising result from mutant studies is that knocking out FIS, H-NS, and the stringent response results in a cell that still adjusts PSS formation in a growth rate-dependent manner.* The elusive missing mechanism is thought to be the availability of RNA-P itself, and in particular, RNA-P containing the σ^{70} subunit (the common one), which recognizes not only rRNA promoters but also most other promoters of the cell, including genes with fueling and biosynthetic functions. How could RNA-P(σ^{70}) availability affect initiation of transcripts from rRNA promoters more than from the hundreds of other operons using this form of the polymerase? One possible explanation is that rRNA promoters have higher maximal activities and therefore are more difficult to saturate with RNA-P than all the other σ^{70}-dependent promoters. The poorer the nutritional environment, the more RNA-P will be engaged in transcribing the fueling and biosynthetic operons instead of rRNA operons. This explanation, known as **passive control**, is diagrammed in Fig. 13.4.

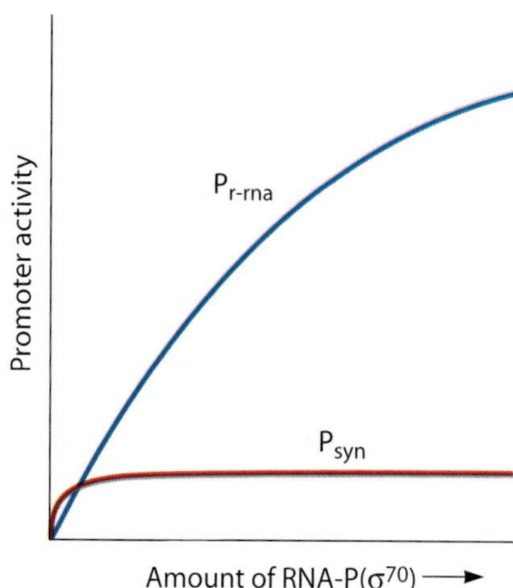

Figure 13.4 A model of passive control of rRNA synthesis. The responses of two hypothetical promoters to varying availability of RNA-P(σ^{70}): one promoter ($P_{r\text{-}rna}$) governs an rRNA operon; the other (P_{syn}) governs an operon encoding biosynthetic enzymes. $P_{r\text{-}rna}$ has a high maximal rate of activity and is difficult to saturate; P_{syn} has lower activity but is more easily saturated. Synthesis of rRNA can be curtailed by reduction in the available RNA-P(σ^{70}) brought on by increased activity of other σ^{70}-dependent operons or by ppGpp-stimulated association of other sigma factors to core RNA-P. In the same fashion, massive repression of other σ^{70}-dependent operons can make RNA-P(σ^{70}) available for rRNA promoters.

In summary, two aspects of PSS regulation command attention. (i) Multiple, partially redundant mechanisms involving small molecules (ppGpp and nucleoside triphosphate substrates), as well as proteins (Fis, H-NS, and σ subunits), cooperate to achieve rapid adjustment of rRNA transcription initiation at rRNA promoters. These agents come into play whenever there is a significant change, up or down, in the nutritional richness of the medium. (ii) During steady-state growth, these mechanisms may operate, but they are not needed to match rRNA synthesis to the growth rate; *growth-dependent regulation can occur solely as a consequence of the activities of all the other genes of the cell.*

The ultimate stress response: stationary phase

What happens when the best responses are not sufficient to enable growth under particularly restricting environmental circumstances? No single regulon or modulon can deal with the situation in which growth is no longer possible; rather, an emergency program initiates preparations for the cell to enter the broadly resistant nongrowth state known as **stationary phase** (Fig. 13.5). Forming endospores or various other differentiated resistant cells is an alternative strategy practiced by some bacteria (see Chapter 14). Some such emergency program is responsible for the survival of each species.

Bacteriologists have long known that nongrowing cells are different from growing ones. For one thing, they look different; cells in stationary

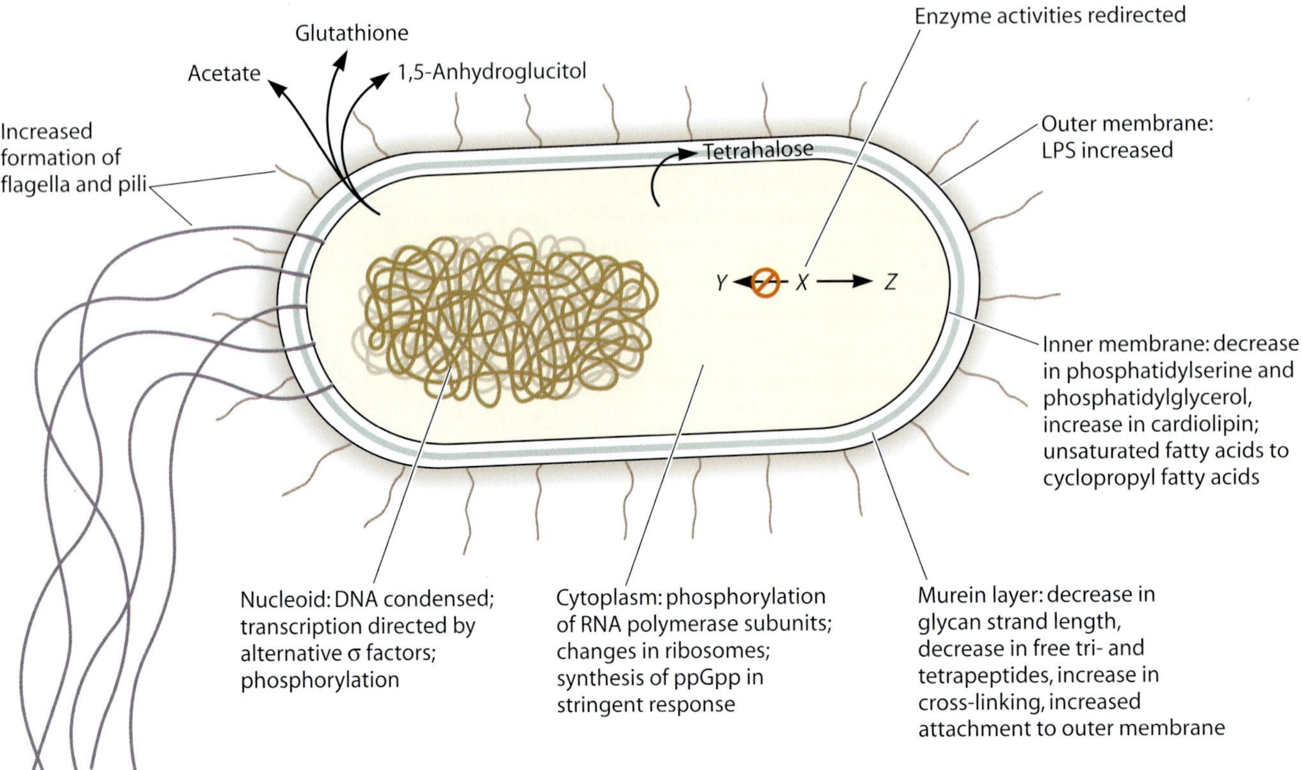

Figure 13.5 A stationary-phase cell. The outcome of the differentiation process governed by σ^s is shown by indicating some of the many differences between the stationary-phase cell and one in active growth.

phase are small. They are also tougher than growing cells; customary ways of lysing bacteria (such as sonic disruption, grinding with alumina, or subjecting them to cycles of high and low pressure) are not as effective with stationary-phase cells as with growing cells. In the stationary phase, metabolism is altered in ways that promote survival rather than synthesis for growth. These readily observable changes are brought about by alterations in the overall chemical composition of stationary-phase cells and in the changes in the chemical and/or physical nature of each cellular component (appendages, outer membrane and periplasm of gram-negative bacteria, and murein wall, cell membrane, nucleoid, and ribosomes) (Fig. 13.5). All of these changes contribute to the cell's toughness and ability to survive in a harsh environment.

One further feature of the stationary phase is noteworthy: to a first approximation, it matters little what caused the cells to stop growing; the end result seems almost the same. Cells may become unable to grow for innumerable reasons, such as depletion of the carbon and energy source or of any required nutrients; accumulation of toxic by-products of their own metabolism; environmental changes in temperature, osmotic pressure, or pH; or encounter with deleterious chemical agents. This approximation, however, is far from the whole story; a cell in stationary phase as a result of depletion of glucose will *not* be identical to one that has been poisoned by hydrogen peroxide or a heavy metal, but the differences will be reflected largely in the profile of enzymes made under the two conditions, not in the appearance and overall chemical toughening of the cell.

How does a bacterium change itself so radically and in such a predictable way? And how does it enter this program of toughening up no matter what the cause of its inability to grow? You may anticipate two answers right away: the process is complex, and not all of it has been elucidated. But what is known is both intriguing and satisfying.

A regulatory circuit of enormous complexity guides the process of turning a growing cell into a nongrowing stationary-phase cell. Part of the process involves **small RNA (sRNA)** molecules regulating translation initiation (see Chapter 12). In Fig. 13.6, a greatly simplified overview of the entry into stationary phase is diagrammed. At the left are represented some of the many circumstances that may halt cell growth. At the right are represented some of the hundreds of target operons whose products must be increased or decreased to transform the cell into stationary phase. Between the stimuli and the target genes are two columns of controllers. The left column lists some sRNA molecules (DsrA, RprA, and OxyS). In essence, these sRNAs are **integrators**; they *process signals from different stresses and integrate them into a single, coherent action*, in this case, stimulating the synthesis of σ^s, a sigma factor directing RNA-P to operons having a distinctive promoter. These operons constitute the **stationary-phase regulon**; their products are responsible for producing the distinctive structures and activities of the stationary-phase cell. Thus, σ^s is the major player in directing entry into the stationary phase. Association of σ^s with core RNA-P diminishes the amount associated with σ^{70}, and in this way, it diverts the polymerase from transcribing operons related to growth to transcribing those related to survival in the nongrowth state.

σ^s is not the only regulator of stationary phase. Other global regulators (H-NS, CRP, Lrp, and FhlA, some of which have been discussed above) play important roles as well by modulating and fine-tuning the expression of individual operons of the σ^s-dependent stationary-phase regulon. A network of this sort, in which regulators govern regulators, which in turn may govern other regulators, is called a **regulatory cascade**.

Figure 13.6 The regulatory cascade governing entry into stationary phase. See the text for an explanation.

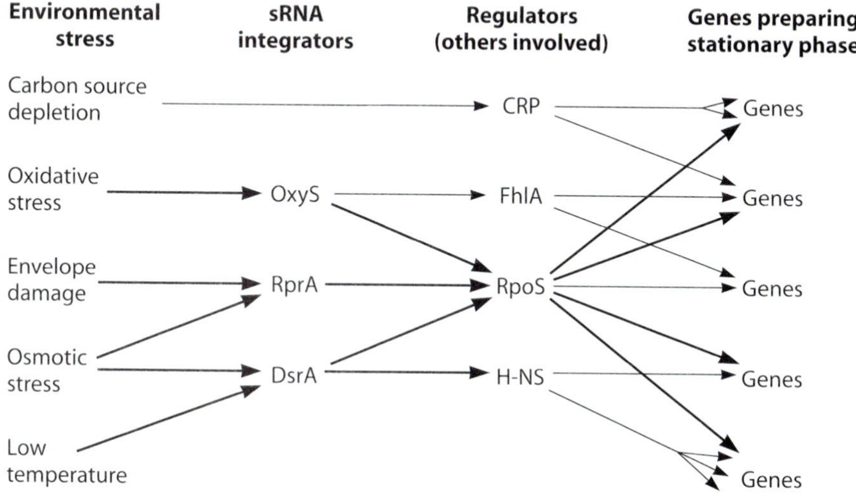

A few words should be said about the sRNAs and their manner of action, not least because this aspect of stress response will likely be intensely investigated in the next several years. There are approximately 50 different sRNA species in *E. coli*, most of which have unknown functions and mechanisms. Not restricted to *E. coli*, regulatory sRNAs are widespread among prokaryotes. The ones known to be involved in stationary-phase transition, however, appear to have a single target, the control of translation of the mRNA for σ^s, the product of the *rpoS* gene (Fig. 13.7). This mRNA has a very long leader region that is thought to form a looped secondary structure that hides the ribosome binding site and thereby prevents ribosomes from initiating translation of RpoS (Fig. 13.7A). The postulated role of the sRNAs is to bind to this loop, freeing the ribosome site from its inhibitory pairing (Fig. 13.7B).

Stress responses and microbial diversity

Not all bacteria are equally skilled at withstanding physical and chemical challenges. Of particular interest are the challenges that the medically important bacteria encounter in the body of their host. These challenges are unique and complex. They include facing the formidable innate and adaptive immune defenses of animals against infection (see Chapter 20). Whether or not they cause disease, bacteria that have solved the problem of living intimately with animal hosts possess special systems to respond to the stresses encountered upon entering and traveling through a potential host (see Chapter 21). As one might suspect, while selecting for the mechanisms for coping within a host, evolution gradually deprives such microbes of many that are necessary in the wider world.

Figure 13.7 Regulation of synthesis of σ^s via translational control by sRNA. The mRNA made from the *rpoS* gene encodes σ^s. Translation of this message is controlled by sRNA. **(A)** Absence of DsrA sRNA; **(B)** presence of DsrA sRNA. See the text for an explanation.

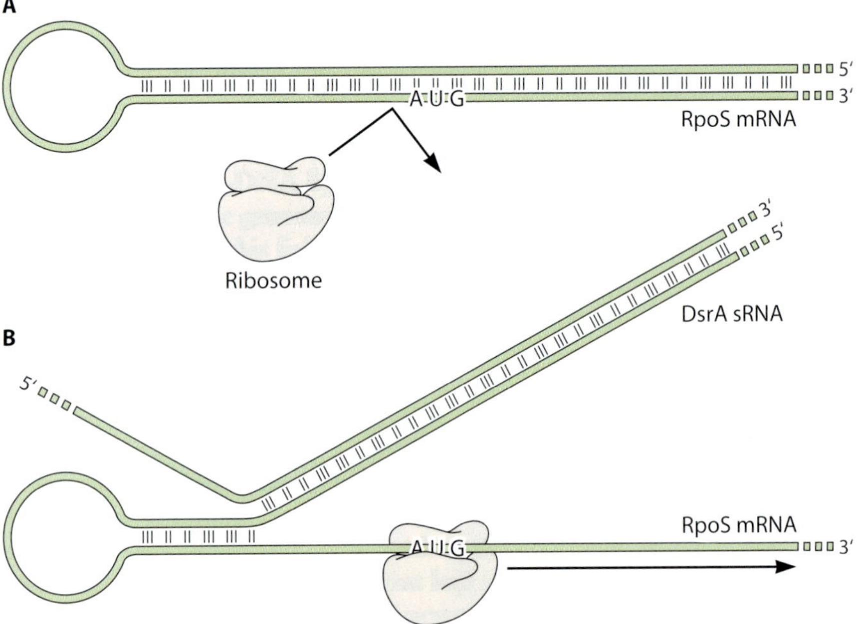

Stress responses are expensive; they require special genes, enzymes, and sensing and regulatory proteins and the expenditure of energy. If they are not used, selective pressures tend to eliminate them. In broad strokes, all microbes, and especially those living in a narrow ecological niche, *retain only those stress responses that benefit inhabitants of their environmental niche.* Analysis of sequenced microbial genomes bears out this rule: a general relationship has been discovered between the lifestyle of an organism and the nature of its genome. Genome size generally reflects the extent to which an organism faces environmental stresses. The more stresses an organism faces, the larger its genome *and* the greater the fraction of its genome devoted to regulatory genes. *Pseudomonas aeruginosa*, noted for both its metabolic versatility and its ubiquity in the varying environments of soil, waters, and diverse hosts, has a genome of approximately 5,570 open reading frames (ORFs), of which nearly 470 (8.4%) appear to encode regulatory proteins. *E. coli,* not as versatile as *P. aeruginosa*, spends most of its life in the guts of animals but must still survive transit from one host to another. It manages this with 4,289 ORFs, of which nearly 250 are putative regulators (5.8%). *Helicobacter pylori*, an organism largely restricted to the stomach as a habitat, has 1,553 ORFs with only 17 putative regulatory genes (1.1%). The smallest known free-living cell, *Mycoplasma genitalium,* possesses only 480 genes, and far fewer than 1% exhibit the motifs of regulatory genes. Not all the data from sequenced genomes support this corollary; genome size seems to depend on many still-obscure factors.

A website listing genome sizes

Stress responses and safety in numbers

Usually, not all bacteria in a given population actually survive a particular shift in living conditions. After a precipitous change in some environmental parameter, a large (99%) or even *very* large (99.99%) fraction of a microbial population may be killed. Still, the survivors from, say, 10^8 cells will be of sufficient number (here, 10^6 and 10^4, respectively) to maintain a viable local population, ready to multiply and quickly restore the original number when conditions ameliorate. For example, such a mild treatment as rapid cooling of a fast-growing culture of *E. coli* from 37 to 5°C can kill over 90% of the cells, and yet this seeming catastrophe goes almost unnoticed by microbiologists who perform this manipulation routinely in the course of experiments.

Further, there is another aspect to the safety-in-numbers method of microbial survival. Coping with environmental change is a feat accomplished by the stress response systems of individual free-floating (**planktonic**) microbial cells, but this is not the whole story. Microbiologists are discovering that bacteria have evolved ways to work together in responding to stress. We shall turn our attention to these "united-front" methods of coping with the environment, but first, we have to learn some of the basics of how bacteria move in response to their environment.

Coping with stress by escaping

Microbes often cope with environmental change by migrating to a more hospitable location. The movement of many, but not all, motile bacterial species is achieved by swimming with the aid of flagella. The complex structure of a flagellum—its filament, hook, and basal body—was presented in

Chapter 2. Recall that bacterial flagella serve the same function as eukaryotic flagella but are totally different in structure and means of propulsion. The helical filament functions as a propeller, the hook as a universal joint, and the basal body with its rod and rings as bushings and bearings in the envelope. The driving motor lies just internal to the basal body. In other bacteria, motion is imparted by very different means. We shall first concentrate on the flagellar mode and then turn to the others.

Flagellar motility

How do flagella confer on the bacterial cell the ability not just to move, but to move *seemingly with purpose*? We shall answer this question with information gained from studies of *E. coli*.

First, how do the cells move? The flagellar motors turn the filaments by using energy directly via a flow of protons—about 1,000 per turn—from the electrochemical gradient (proton motive force) of the cell membrane rather than from ATP. The filament can be rotated either clockwise or counterclockwise. The flagella of each cell are synchronized to rotate simultaneously in the same direction. Counterclockwise rotation results in productive vectorial motion (a **run**), because this rotation causes the flagella to entwine into stiff bundles, producing a propeller-like force. Clockwise rotation of the flagella dissociates the bundles and causes the cell to **tumble** in place (Fig. 13.8). The flagella alternate between periods of clockwise and counterclockwise rotation according to an endogenous schedule (the basis of which we shall see directly). As a result, motile bacteria move in brief runs interrupted by periods of tumbling. Each new run occurs in whatever random direction the cell happens to be facing when the clockwise mode changes to counterclockwise.

Figure 13.8 Flagellar behavior and motility. (A) A cell with a single polar flagellum moves from right to left when the flagellum turns counterclockwise and tumbles when the flagellum turns clockwise. **(B)** In a cell with many flagella all over its surface, counterclockwise rotation produces a coherent flagellar bundle and smooth movement. Clockwise rotation causes the bundle to fly apart; tumbling results.

A Polar flagellum

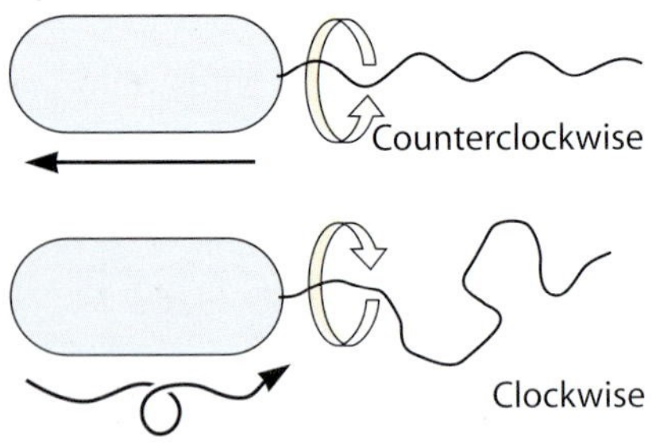

B Many flagella (peritrichous)

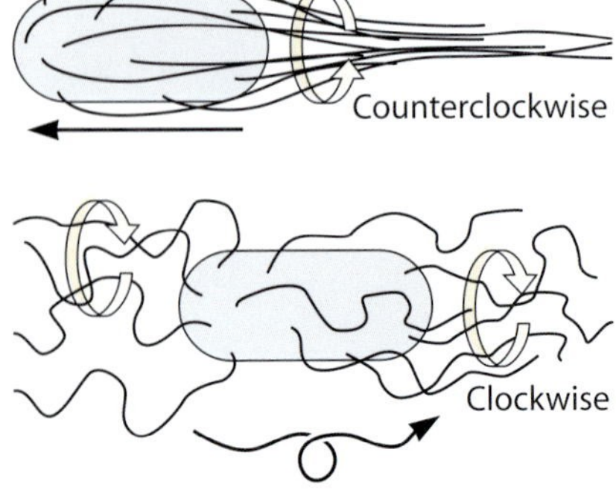

How could alternate running and tumbling produce other than random movement of the cell? There is no doubt that something looking like purposeful migration does occur. Take, for example, a population of bacterial cells in a nutritionally lean area that is not far from a source of sugar diffusing away from, say, a drop of maple sap. Moving toward this windfall would clearly be advantageous. Indeed, this is exactly what occurs; most of the population will in a short time congregate near the source of the food. This migratory process driven by a chemical concentration gradient is called **chemotaxis**. Our question can thus be phrased, how can chemotaxis be accomplished by a random process of running and tumbling?

At one level, the answer is simple and satisfying. Individual cells in this population can be seen to get nearer the source by a sequence of runs and tumbles, illustrated in Fig. 13.9. The productive runs (i.e., those leading to progress toward the food source) are prolonged, while the neutral or counterproductive runs are shortened. As a result, the cell inexorably approaches the goal through what are called **biased random walks**.

These observations sharpen the problem for us. How the cells migrate purposefully has now become the question, how do these cells regulate the frequency of tumbling? An increased frequency when the cells are headed in the wrong direction, or a suppressed frequency when they are on the right course, would produce the observed result through biased random walks.

Now comes the big question. How can "right direction" be distinguished from "wrong direction" by these brainless cells? If you guessed that the bacterial cell might be measuring the difference in sugar concentration between its "head" end and its "tail" end and using these data to terminate runs in the wrong direction (higher sugar at the tail) and prolong them in the correct direction (higher sugar at the head), you would get an "A" for effort, but you would be wrong. What you did not reckon is that the distance from head to tail of a bacterial cell (a couple of micrometers at most) is so short that it would take an extraordinarily steep concentration gradient to produce a perceptible difference in concentration from one end to the other. The concentration gradients to which these cells can respond are much shallower.

The answer is almost in the realm of science fiction. Contrary to what one might expect, bacteria do not sense a gradient from their front to their back. *These cells distinguish up-the-gradient from down-the-gradient by memory.* That is, they continually compare the concentration of sugar (in the example we have been using) *now* with what their memory tells them it was *then*, a short time ago. Binding of an attractant—the molecules of sugar from the drop of maple sap—at a receptor near the cell surface alters the endogenous routine schedule of runs and tumbles. It does so by interrupting a **phosphorylation cascade** that governs the direction of rotation of the flagellar motors, which has the effect of prolonging the run. **Accommodation** by a **methylation system** restores the endogenous schedule and resets the cell's sensitivity to the attractant to require a higher concentration to prolong the run. (Accommodation accounts for many properties of human perception, such as our gradual failure to smell an odor after being in its presence for an extended time.) In effect, accommodation constitutes a **molecular memory** that enables change in attractant concentration to be sensed and ensures progress toward it. This remarkable molecular sensory system thus possesses many of the characteristics that would be expected of behavioral systems in higher animals, including adaptation to a prolonged stimulus.

Besides the genes encoding flagellar proteins (called *fla*, for flagella), more than 30 genes (called *mot*, for motility, and *che*, for chemotaxis) encode the proteins that make this system work: receptors, signalers, transducers, tumble regulators, and motors. You can follow the detailed action in Fig. 13.10.

Flagellar motility extends to many species, some of which exhibit variations on the *E. coli* theme, but many display other modes of achieving random walks to a source of nutrients. As an example, some bacteria have single flagella, in which formation and dissolution of a bundle obviously cannot occur. *Rhodobacter sphaeroides*, which has a single flagellum inserted medially, swims by rotating it in a clockwise direction. This organism does not change the direction of flagellar rotation; rather, it reorients its swimming direction by ceasing rotation. When it does, the flagellum coils tightly back on itself, and the cell is reoriented by the forces of Brownian motion. Resumption of rotation in the clockwise direction re-forms the productive, helical shape of the flagellum, and another run ensues.

Chemotaxis is both a survival device (for avoiding toxic substances) and a growth-promoting device (for finding food). It can also be a virulence

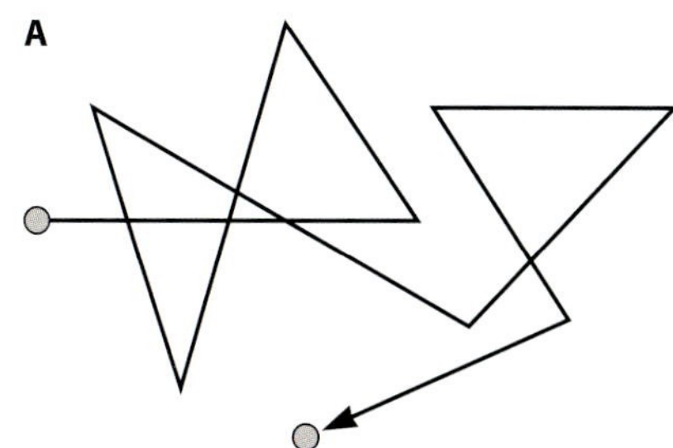

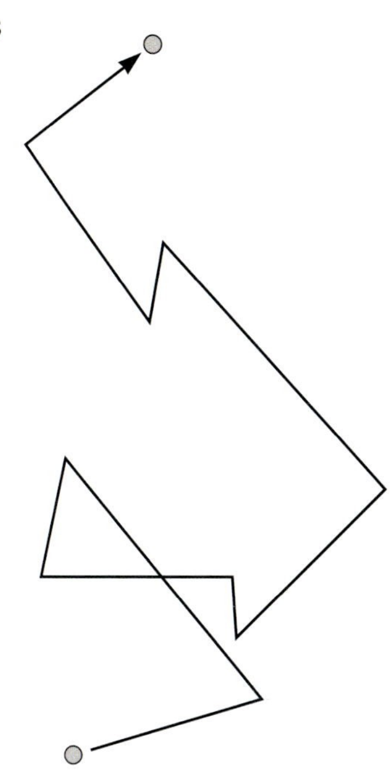

Figure 13.9 A biased random walk. **(A)** Typical random path of a motile bacterium such as *E. coli* in the absence of a chemotactic stimulus. Periods of runs and tumbling alternate by an endogenous schedule. **(B)** Path taken by this cell when an attractant is present toward the top of the figure. Runs last longer when they are in the direction of the attracting chemical.

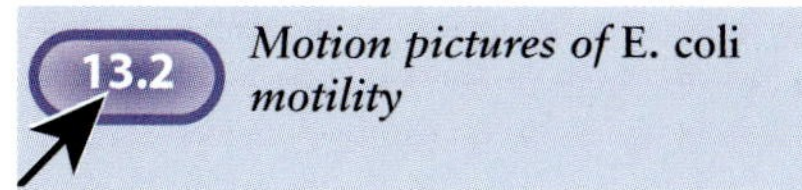

Motion pictures of E. coli *motility*

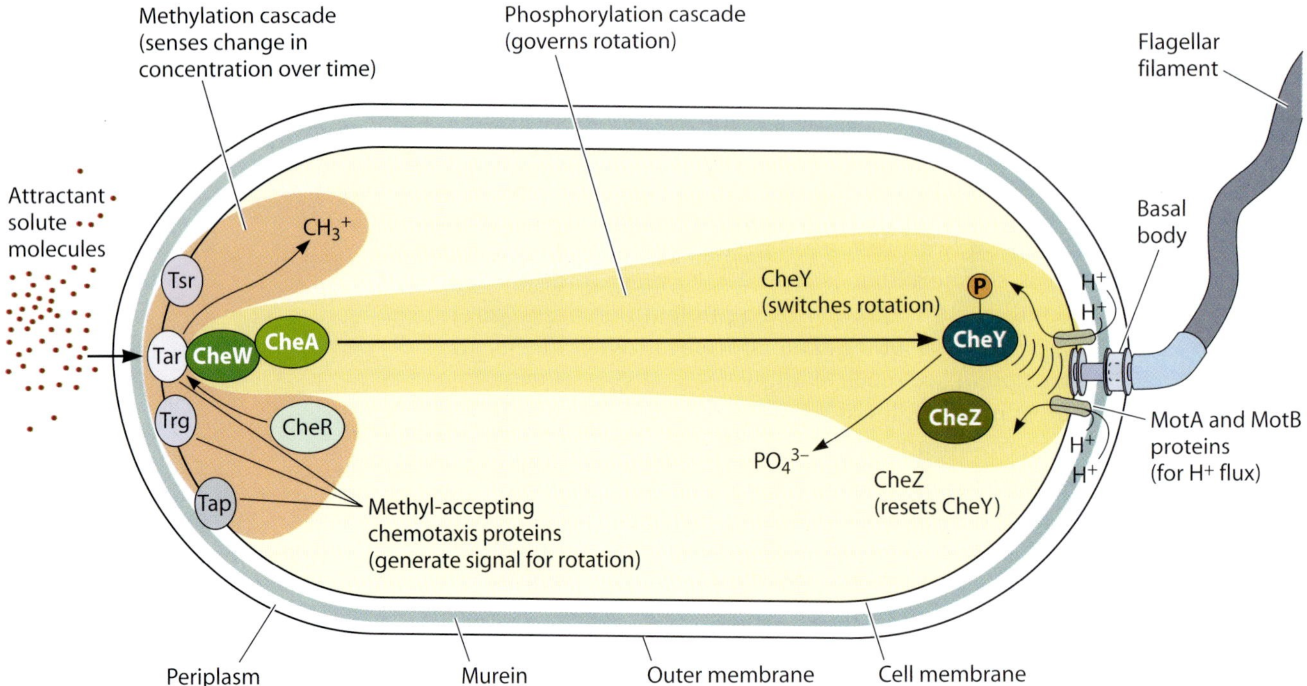

Figure 13.10 The *E. coli* chemotaxis circuit. A diffusible chemoattractant is shown entering the periplasm and complexing with a periplasmic binding protein. In this case, it interacts specifically with Tar, one of the membrane methyl-accepting proteins, triggering a phosphorylation cascade that activates counterclockwise rotation of the flagellum, leading to a run. Methylation of the membrane Tar soon leads to a reversal of this activation, and clockwise rotation causes the cell to tumble in place. The timing of this cycle depends on the new concentration of the attractant that is presented to Tar.

factor facilitating colonization of the host. Also, as we indicated at the end of our discussion of stress responses, chemotaxis has been recently shown to play a major role in the communal response to stress. A signal directing cells to assemble into aggregates is an important selective advantage of chemotaxis in many circumstances, as we shall see below (see "Quorum-sensing, motility, and biofilm formation"), and may be a key factor in the evolution of this ability.

Can properties of the environment other than a food source lead to directed motility? Absolutely. Many bacterial cells can sense gradients of pH, temperature, light, salinity, and oxidation potential and move in a direction appropriate for optimizing these conditions for the particular bacterial species. Some aquatic bacteria exhibit a particularly arcane version of directed motion called **magnetotaxis**—motility guided by sensing the Earth's magnetic field. This talent has nothing to do with north-south migration but with the ability to know which way is up; that is, the magnetic force lines in most northern latitudes actually point mainly *downward* toward north and thus provide a guide for cells to move deeper into the mud, which they prefer, since they are **microaerophilic.** (Does this mean that magnetotaxic bacteria in the northern and southern hemispheres follow magnetic lines in opposite directions? Yes!) Table 13.5 lists some examples of bacterial tactic responses.

The importance of cell translocation in the natural world is underscored by the great variety of movement devices bacteria have evolved. A

Table 13.5 Bacterial tactic responses

Tactic response	Description
Chemotaxis	Directed movement in response to chemicals (chemoeffectors), which can be either attractants (positive chemotaxis) or repellents (negative chemotaxis); widespread property of prokaryotes; well studied in *E. coli.*
Aerotaxis	Directed movement in response to oxygen; widespread property.
pH taxis	Movement either toward or away from acid or alkaline conditions; *E. coli* moves from either acid or alkali to reach a neutral pH.
Magnetotaxis	Directed movement along geomagnetic lines of force; believed to function in guidance up or down rather than north or south and useful in moving toward microaerophilic environments.
Thermotaxis	Directed movement toward a range of temperatures usually optimal for the bacterium's growth; widespread property.
Phototaxis	Directed movement toward wavelengths of light, usually related to photopigments and their function in the metabolism of the bacterium; property associated with virtually all photosynthetic bacteria.

brief look at some of these will supplement the *E. coli* story and demonstrate the lengths to which cells have gone to achieve motility.

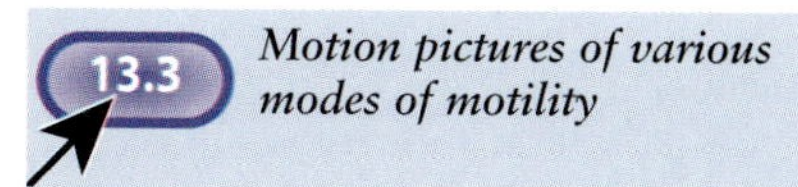

Motion pictures of various modes of motility

Swarming motility

Some bacteria use flagella for a different kind of motility—not that commonly seen in liquid but one that takes place on surfaces, such as on agar plates. This phenomenon is called **swarming.** This is a group process; individual cells do not swarm, but "rafts" of cells arranged side by side do. The reason for this cooperative need is not clear. It is puzzling, because individual cells otherwise capable of swarming sometimes differentiate into long cells with many new laterally located flagella. Swarming requires a slime layer secreted by the swarmer cells; one component of this slime is **surfactin**, a lipopeptide with exceptional surface tension-lowering power, synthesized and secreted by several swarming species. Several species (e.g., *Proteus vulgaris* and *Vibrio parahaemolyticus*) are well known among microbiologists for their ability to swarm, but the swarmers include *E. coli*, *Salmonella*, and *Bacillus subtilis*.

Gliding motility

A dozen or more genera of bacteria display a form of motility on solid surfaces called "gliding." This movement occurs without the use of flagella. The term may include a number of different mechanisms that share only the fact that they produce movement across a solid surface without flagella. Gliding has been observed in such diverse groups as cyanobacteria, myxobacteria, and mycoplasmas. In one well-studied example, that of the myxobacterium *Myxococcus xanthus*, there appear to be two independently operated systems, one (called **social motility**) that is responsible for the migration of cells as a raft and one (called **adventurous motility**) that enables individual cells to move out from such a group. Social motility involves a kind of pili called type IV; adventurous motility, at least by mutant analysis, seems to depend on the

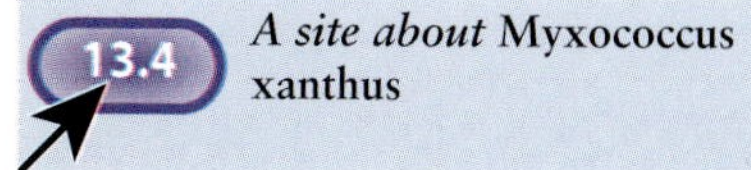

A site about Myxococcus xanthus

lipopolysaccharide of the gram-negative cell envelope by an unknown mechanism (see Chapter 14).

Here is a story that illustrates how tactic responses operate with surprising results in the natural environment. This example happens to involve gliding. Two kinds of gliding bacteria cooperate to produce a diurnal tactic response. One genus, *Beggiatoa*, possesses three negative (phobic) tactic responses, to H_2S, to O_2, and to light. These are sulfur-oxidizing bacteria that prosper at interfaces between **aerobic** and **anaerobic** environments, where two of their nutrients—O_2 and H_2S—are both available at the right concentrations. *Beggiatoa* is found only in this zone, which comes about as a consequence of its phobic (negative) tactic response to each nutrient. In this narrow zone, the concentrations of both nutrients are decreased by the metabolic activity of the *Beggiatoa* cells. The *Beggiatoa* bacterium is trapped here by the higher concentrations of O_2 above the zone and H_2S below it. *Beggiatoa* lives in microbial mats together with the second genus, the cyanobacterium *Oscillatoria*. The latter is photosynthetic and phototropic; using gliding motility, the cyanobacteria congregate on the surfaces of the mats, producing large amounts of O_2, which could be disastrous for the oxygen-sensitive *Beggiatoa*. To avoid this, at night, when the cyanobacteria are not producing O_2, the *Beggiatoa* organisms migrate to the surface of the mat, following the rising aerobic-anaerobic interface. If *Beggiatoa* responded phobically only to O_2 and H_2S, they would be trapped away from H_2S at the mat surface in the morning when oxygenic photosynthesis begins. However, *Beggiatoa* is also phobically tactic to light, so the bacteria are able to migrate back into the mat to reach the aerobic-anaerobic, O_2-H_2S interface.

Twitching motility

Several species of gram-negative bacteria, including *Pseudomonas aeruginosa*, *Neisseria gonorrhoeae*, and *E. coli*, employ yet another kind of movement. Whether they have flagella or not, cells of these species can move in a jerky fashion across solid surfaces (as, for example, on an agar medium). This type of motility, called **twitching motility,** depends on the presence of type IV pili and takes place by a "grappling hook" mechanism of extension of the pilus, its attachment, and subsequent retraction back into the cell (see Chapter 2). Twitching motility contributes to biofilm formation (discussed below) and to a kind of aggregation that causes differentiation in the myxobacteria (see Chapter 14).

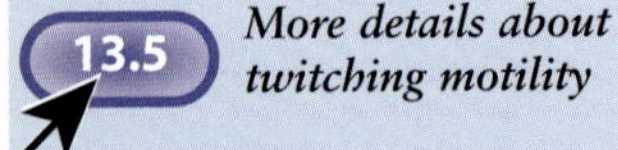

More details about twitching motility

COPING WITH STRESS BY COMMUNITY EFFORT

Sensing the population

Biology is rich in examples of major new insights arising from the study of seemingly minor, esoteric subjects. Such is the case with the realization that bacteria are not the lone players once thought but rather are social individuals that communicate with each other and exhibit cooperative behavior. This revolutionary concept, which is still being developed through research, came to be appreciated only in the 1990s but had its origin many decades earlier in studies of the light-emitting bacterium called *Vibrio fischeri*. These bioluminescent cells are free-living marine organisms that, although capable of emitting light, do so only when present at high density—which they are when acting as normal symbionts inhabiting the light organs of squid and many fish. (The value of this symbiosis for the bacterium is the provision of shelter and food; for the squid, it has been speculated that the light emitted may protect the animal from predators beneath it by avoiding the casting of a moon shadow by the squid; this camouflage hypothesis is little more than a guess at this point.) The cell density dependence of light emission depends on the following facts. The proteins responsible for converting the energy of ATP into light (**luciferase** is one) are the products of a seven-gene operon (called *lux*). This operon is not expressed unless it is activated by a protein, LuxR, when it is complexed with a small ligand with a big name: *N*-(3-oxohexanoyl)-L-homoserine. This molecule is an **autoinducer.** When produced by the bacteria, it diffuses into the surrounding environment and is lost to the cells. In the confined space of the squid's light organ, however, the bacterial density becomes very high, and so does the extracellular concentration of the autoinducer. The autoinducer molecules diffuse back into the cells and bind to the LuxR protein. Then, active LuxR triggers the expression of the *lux* operon, and light is produced. The biochemical reaction that leads to light is catalyzed here, as well as in fireflies, by the enzyme luciferase.

Light emission from a squid may not be a very compelling event for some of us, but it explained how this action depends on the size of the bacterial population. This phenomenon is called **quorum sensing.** The term is apt, because it evokes the idea that it takes a quorum of individuals, that is, a certain number, to take some kinds of actions. Quorum sensing is a form of cell-to-cell communication that enables cells to exhibit certain behaviors only when the population exceeds some threshold value. It turns out that this is an extremely common feature in host-parasite interactions (although far from confined to these situations). An example is the lung infection of patients with cystic fibrosis caused by *Pseudomonas*. Here, the organisms survive in the lungs of patients in part by becoming organized in complex communities called biofilms (see below), which arise as a result of quorum sensing.

The *V. fischeri* version of quorum sensing is unusually simple. Far more common is the involvement of a multicomponent response regulator system. Here, the secreted autoinducer does not simply diffuse back into the cell, but it is sensed by a typical sensor kinase on the cell surface, triggering the transmission of a phosphorylation signal to some response regulator protein. We might add that in one well-studied instance, the sexual transmission of genetic material in some gram-positive species, population density is signaled by the secretion of an autoinducer called, for obvious reasons, a **pheromone** (see Chapter 10). We wonder if this exchange of molecular signals is not the bacterial forerunner of the chemical communication so universal among insects.

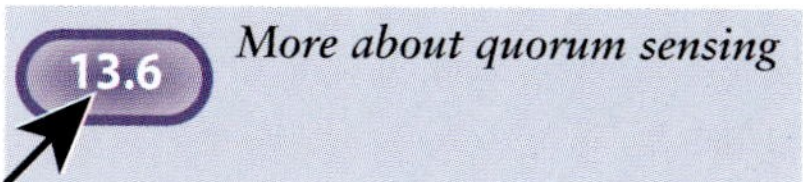

The widespread existence of quorum sensing is now appreciated, even though many details remain to be worked out. Most of the gram-negative species use a variety of *N*-acyl-homoserine lactones (relatives of the *Vibrio* variety) as autoinducers, whereas among gram-positive bacteria, it is more common for the signaling molecules to be peptides. Table 13.6 lists a sampling of known quorum-sensing systems.

Table 13.6 Small sampling of bacterial quorum-sensing systems

Genus	Autoinducer signal molecules	Physiological function(s)
Vibrio	Homoserine lactones	Bioluminescence
Pseudomonas	Homoserine lactones	Pathogenesis
Agrobacterium	Homoserine lactones	Conjugation
Bacillus	Peptides	Competence, development
Enterococcus	Peptides	Conjugation, plasmid maintenance, pathogenesis
Myxococcus	Peptides	Development
Streptococcus	Peptides	Transduction
Staphylococcus	Peptides	Pathogenesis

Formation of organized communities

A natural corollary of the bacterial ability to communicate with each other and to sense their own numbers is their ability to aggregate into multicellular formations. Such formations offer protection and other advantages in the natural environment (see Chapter 18).

We call bacterial cells associated with each other in natural environments a **biofilm.** Biofilms are found on the surfaces of rocks in a high mountain stream, on a prosthetic medical device implanted in a human patient, or inside a corroded water pipe. They are found everywhere there is a solid surface bathed by water. A large proportion of all microbes are in fact represented in biofilms. In every case, the microbes are **sessile,** they stick to surfaces (the word *sessile* is derived from the Latin "to sit"). Most research on the physiology of bacteria has been conducted on **planktonic** cells, which leads us to expect many discoveries about additional properties and behaviors of sessile cells.

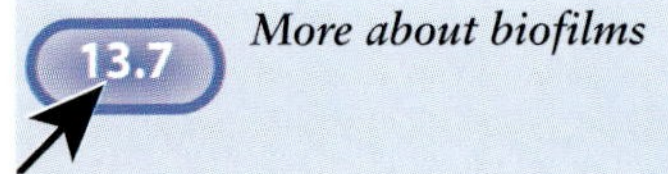

More about biofilms

Surfaces have physicochemical properties distinct from those governing free solutions. For our purposes, the differences have to do with the adherence of dilute solutions of organic and inorganic substances to surfaces, attracted largely by electrostatic charge. These properties and the natural stickiness of bacterial cells lead to the formation of biofilms. Once adhered to each other and to a surface, bacteria develop large communities, often visible with the naked eye. Such biofilm communities are architecturally complex, often made up of pedestals and mushroom-like structures surrounded by enclosed channels. Among the singular properties of biofilm is a greatly enhanced resistance to some antibiotics and other toxic substances; in some cases, the inhibitory concentration is increased as much as 1,000-fold as a result of the biofilm forming a protective shell of cells around a central space that excludes the antibiotic or in which it can be destroyed. Further, as biofilms form, enzymes are induced that render individual cells within the films intrinsically more resistant to antibiotics.

In nature, biofilms and other communities often consist of several different microbial species in cooperative arrangements. For example, the formation of dental plaque (which begins to occur just as you leave the dental hygienist's office) requires that certain bacterial species, notably *Streptococcus mutans*, stick to the tooth surface first. Bacteria of different species, such as some belonging to a large group called the actinomycetes (see Chapter 15), stick to these pioneers. Eventually, others stick to the actinomycetes, leading in time to the development of a complex scaffold of different kinds of organisms, a multispecies biofilm. Not all such biofilms exist solely for architectural purposes. Biofilm communities may also serve a cross-metabolic purpose. Several examples of these will be encountered later in this book, but one is worth describing now. Certain marine Archaea are capable of oxidizing methane, an abundant gas in their environment. This process, however, is thermodynamically unfavorable (see Chapter 18). In order to drive the reaction, these Archaea associate with Bacteria that metabolically remove the hydrogen resulting from methane oxidation. Together in biofilms, they make methane oxidation happen. It is likely that such cooperative enterprises are common in the microbial world.

Quorum sensing, motility, and biofilm formation

A fitting conclusion to this chapter about the life of bacterial cells out in the world is to consider how these cells can put it all together. Work with *E. coli* has demonstrated the combined operation of chemotaxis, quorum sensing, and biofilm formation. These cells secrete small amounts of amino acids, one of which is glycine. These same cells possess a chemotaxis receptor protein, **Tsr**, which binds glycine (or L-serine, L-alanine, or L-cysteine). In a large liquid culture, nothing unusual occurs, but in a small culture vessel containing a tiny central enclosure (250 by 250 μm) accessed by very narrow (40-μm) channels, the cells migrate into the enclosure and accumulate there in sufficient numbers to establish a significant concentration of attractant. Through the use of appropriate mutants, it can be shown that it is their own secretion of glycine that attracts the cells. Once they have packed themselves into the chamber, the cells form dense granular aggregates, a biofilm precursor. If the same experiment is done with the bioluminescent *Vibrio harveyi*, light emission—a quorum-dependent response—from the central chamber occurs.

These results indicate the ability of bacteria to form multicellular communities spontaneously. Genetic exchange (see Chapter 10), communal degradation of antibiotics, biofilm formation, and other quorum-dependent behaviors are thereby enabled.

CONCLUSIONS

What remains for microbiologists to learn about how bacteria cope in their environments? Far more than has already been discovered. Microbiologists have merely scratched the surface in seeking to understand the domination of earth by microbes. If the truth were known, not a single microbial stress response can now be accurately and quantitatively modeled. That is, none can be described with sufficient mathematical precision to permit predictions of its behavior under various different circumstances. It is not just that we do not have enough quantitative information about the cell (though that is part of the issue); it is that we do not sufficiently understand the design of the cell's regulatory circuitry. Understanding the cell's response to its environment will require the use of tools more traditionally associated with engineering than with chemistry or biology. Learning how genes act in context with each other in the intact cell will require analytical techniques of systems analysis not yet available to most microbiologists, and indeed, many techniques of analysis that are still to be developed. To these challenges we must add the relatively new subject of communal response to stress, an area of investigation just getting under way.

Let us summarize. As individual cells, and as organized communities of cells, microbes have evolved a dazzling array of coping strategies for life in the environment. To their potent reproductive capability, bacteria have added a large set of molecular responses to stress. These systems overcome the inherent vulnerability of being small and subject to inconstant environments. In this chapter, we have provided a mere taste of what is known about microbial strategies. In the next chapter, we extend our analysis by considering microbial participation in two highly evolved biological processes: differentiation and development.

STUDY QUESTIONS

1. What is the defining characteristic of the term *stress* for microbes (i.e., what brings about stress)?

2. Explain how inoculating bacterial cells of an unknown species into a medium at 37°C can lead to either a heat shock or a cold shock response.

3. What are the usual components of a bacterial sensory or response-regulatory system? For each component, describe its chemical nature, its role in the system, and where it is located in the cell.

4. The large group of genes that encode proteins in a bacterial cell that become elevated or depressed in amount in response to a given stress is called a stimulon. How can the constituents of a stimulon be discovered?

5. The faster bacteria grow, the more PSS (including ribosomes and related factors) they make. Describe the postulated role of each of the following molecules in regulating the amount of the bacterial PSS as a function of growth rate: Fis, H-NS, UP elements, ppGpp, nucleoside triphosphates, and RNA-P(σ^{70}). What makes it reasonable to expect that synthesis of the PSS would be under such complex and redundant controls?

6. In what ways does a bacterial cell in stationary phase differ from the same cell in exponential growth? For a given bacterial species, are stationary-phase cells identical no matter what caused the cessation of growth?

7. The differentiation of a growing bacterial cell into a stationary-phase cell involves many significant changes. Name a half dozen major changes. In what respects are all nongrowing cells of a given species similar no matter what caused their transition into stationary phase? How might they differ?

8. Describe the roles of sRNAs and of σ^s in bringing about the transition of bacteria from growth to stationary phase.

9. Describe how simply knowing the size of its genome can tell something of the physiology of a newly discovered bacterium.

10. Evaluate the usefulness of a claim that a given agent will "kill 99% of the bacteria" on a kitchen table or similar surface.

11. Bacterial chemotaxis based on flagellar motility needs both a phosphorylation cascade and a methylation system in order to function. Describe in general terms the distinctive role of each in chemotaxis by flagellar motility.

12. Explain how an individual bacterial cell can learn how many of its kind are present in the near neighborhood. Of what value is this information?

13. Define a biofilm, and describe its general properties. Cite three advantages that a biofilm offers over the planktonic mode of microbial life.

chapter 14

differentiation and development

OVERVIEW

In the past two chapters, we considered how microbes adjust their activities in order to survive and prosper, with respect both to coordinating the processes of growth and to coping with environmentally imposed stress. We saw how individual cells **differentiate** (become structurally and functionally specialized) to become much tougher stationary-phase cells when, for one reason or another, continued growth is no longer possible. Finally, we noted that groups of cells of many bacteria undergo **development** (progressive change in form and function in ways that play prominent roles in their life cycles) leading to their becoming biofilms.

"Differentiate" and "develop" are words that fit comfortably with the biology of plants and animals. The early progeny cells of their zygotes *differentiate* into specialized cells that form specialized tissues, which undergo *development* to become a mature organism. When applied to microbes, the terms and their usages tend to merge. We shall do our best to distinguish them as we proceed.

Differentiation in plants and animals is for the most part a one-way street: specialized cells only very rarely go back (dedifferentiate) to become **stem cells** (undifferentiated cells that are capable of differentiation). The development of plants and animals is genetically programmed and is largely undirected. In contrast, prokaryotic cells that differentiate are usually capable of returning to their original cellular form—but there are exceptions. One is the formation of **heterocysts** by cyanobacteria. Heterocysts are thick-walled, nondividing, specialized cells that are dedicated to fixing nitrogen and, notably, cannot dedifferentiate to become vegetative cells again. They serve the specialized role of supplying fixed nitrogen to neighboring vegetative cells, which in turn supply nutrients derived from the carbon dioxide that they have fixed. Another example of terminal differentiation is discussed later in this chapter in connection with endospore formation. The spore mother cells are dedicated to nurturing the developing endospore. When their task is completed, they lyse.

Another prominent feature of prokaryotic development is that it is almost always environmentally triggered. In this respect, it resembles the stress responses discussed in Chapter 13. Heterocyst development in cyanobacteria, for example, occurs when the organism is deprived of fixed nitrogen, and resistant spores form when a culture capable of producing them runs out of nutrients.

Because prokaryotic differentiation is generally reversible, the edges of what should be included under the term are blurred. For example, at one time enzyme induction was included. Now the term is generally reserved for distinct morphological changes.

Almost all the cells that comprise a plant or an animal carry identical genes. Understanding how they become differentiated into distinctly different types and how they arrange themselves to develop into a complete organism has long been a central challenge of experimental biology. It is, indeed, a very tough one. In light of the great success of revealing the fundamentals of genetics and biochemistry through studies of the more amenable prokaryotes, the same approach was taken for cracking the problems of differentiation and development in eukaryotes. Here, the value of prokaryotic models turned out to be somewhat disappointing. The molecular mechanisms mediating development in prokaryotes differ from those in plants and animals and even among prokaryotes themselves.

In spite of the vast diversity of prokaryotic differentiation and development, certain unifying principles are becoming apparent, and one generalization is possible—within the limits of our present state of knowledge: no Archaea undergo morphological development, although archaeal cells differentiate. For example, under the direction of appropriate environmental signals, *Halobacterium* spp. alter the structure of their membranes, forming purple patches capable of mediating a simple form of photosynthesis. However, no archaeon undergoes readily visible morphological changes as it goes through life: none form endospores; none form heterocysts. Still, they get by and have done so for eons.

We shall survey bacterial differentiation and development, as well as multicellular life, by looking at examples and noting the underlying themes. Certainly, common approaches have been used to unravel the molecular mechanisms of differentiation and development. The mutant technique (see Chapter 10) proved invaluable in this enterprise. More recently, high-throughput methods, such as microarray technology, have been employed to advantage. Genomics has become a major aid because the genomes of most model organisms have been sequenced. The eventual goal and proof of our understanding of these processes is to construct a mathematical model of them. That is still a long way off, but there is a feeling that it will happen.

Within the past decade, defining the limits of bacterial differentiation has been further complicated. Laboratory bacterial strains grow in culture as dispersed **planktonic** (floating) cells and are, therefore, readily amenable to manipulation in biochemical, physiological, and genetic studies. Microbiologists have come to realize that these strains are domesticated variants of strains that exist in nature. Most laboratory strains have been selected, overtly and as an inevitable consequence of long-term laboratory culture, for their convenient but atypical growth habits. In nature, however, most bacteria are not planktonic. Instead, as was discussed in Chapter 13, they accumulate on solid surfaces, forming **biofilms,** which on liquid surfaces are sometimes called pellicles. Biofilms and pellicles are not merely random associations of cells sticking together. Many develop into distinctive shapes that are products of some of the same mechanisms that lead to the more distinctive examples of development that we discuss in this chapter. For example, cells of myxococci move, aided by twitching motility, to form elaborate fruiting bodies; similarly, by twitching motility, cells of *Pseudomonas aeruginosa* climb up primordial stalks to aggregate into the globular heads of the tiny mushroom-shaped structures that develop on their biofilms.

In the following sections, we will discuss some specific examples of prokaryotic development.

ENDOSPORES

Properties

Many microbes produce cells called spores, which are agents of dispersal. Most such cells, also called **exospores,** are formed by a pinching off of the tips of the filamentous cells of fungi and actinomycetes that stick out in the air. Exospores disperse in the air to initiate new filamentous growth someplace else. However, some species of bacteria form totally different cells that are also referred to as spores. These are **endospores,** so called because they develop inside another cell. *Endospores are agents of survival, not dispersal.* In this sense, they are functionally analogous to the stationary-phase cells discussed in Chapter 13, but endospores are extraordinarily resistant. They can survive in a metabolically inert state for extended periods, certainly hundreds of years, perhaps thousands. When appropriate nutrients again become available, endospores rapidly germinate to become vegetative cells. In addition to being long-lived, endospores are highly resistant to extremes of temperature, radiation, reactive oxygen, acid, and alkali. Indeed, they are the most resistant of all known biological structures. For this reason, their resistance determines the parameters of sterilizing treatments. For example, the heat treatment applied in canning food is designed to kill bacterial endospores. If the treatment is sufficient to kill endospores, certainly other microbial life forms that are likely to be present will also be eliminated. In 2003, however, hyperthermophilic species of Archaea that thrive at temperatures lethal to endospores were described. However, their discovery need not alter the sterilization regimes in use for most purposes because these hyperthermophiles do not grow at moderate temperatures.

Phylogenetic distribution

In spite of its prominent status in basic and applied microbiology, endospore formation is restricted to a handful of bacterial genera in the gram-positive phylum Firmicutes (Table 14.1). This suggests that the ability to make endospores has evolved only once, after this phylum branched off.

Table 14.1 Genera of endospore-forming bacteria

Genus	Properties
Bacillus	Aerobes and facultative anaerobes; the most thoroughly studied endospores; includes important pathogenic species, e.g., the agent of anthrax
Clostridium	Anaerobes; includes important pathogenic species, e.g., the agents of tetanus, botulism, and gas gangrene
Thermoactinomyces	Thermophilic aerobes; closely related to *Bacillus*
Sporolactobacillus	Endospore-forming lactic acid bacteria
Sporosarcina	Obligate aerobic cocci (the only endospore-forming cocci), with cells arranged in packets of four or eight
Sporotomaculum	Anaerobes that carry out anaerobic respiration, using sulfate as a terminal electron acceptor
Sporomusa	Anaerobic, acetate-forming (acetogenic) bacteria
Sporohalobacter	Anaerobic, salt-resistant bacteria from the Dead Sea

Formation

The complex sequence of events leading a vegetative cell to differentiate into an endospore has been most intensively studied using *Bacillus subtilis* as a model organism. This process, **sporulation,** is triggered by near depletion of any of several nutrients (carbon, nitrogen, or phosphorus), provided that the culture is relatively dense. The setting of this trigger is extremely critical. Were a cell to redirect its metabolic activities toward forming an endospore when adequate concentrations of nutrients were still available, it would waste precious nutrients that otherwise could have been used for growth and proliferation. However, an even greater disaster would ensue if the trigger were set at a level of nutrients too low for the energy-requiring process of spore formation to be completed before the nutrients were completely exhausted. For this reason, cells facing near-complete starvation are unable to form endospores, thereby avoiding being trapped in a vulnerable state where they are committed to sporulation but not yet resistant to environmental challenges.

Although there are significant minor variations by species, endospore formation occurs in all of them through essentially the same series of events that occur in *B. subtilis*, the model organism used for such studies (Fig. 14.1). First, the nucleoid lengthens, becoming a structure called an **axial filament,** which extends the length of the cell. Then, the cell begins to divide by forming a septum, but rather than at the center of the cell (as is the case in all vegetative-cell divisions), this septum forms at a point about a quarter of a cell length from one end of the cell. For this reason, it is called the **polar septum.** The larger cellular product of this division becomes the **mother cell,** which nourishes the developing spore, and the smaller one, the **forespore,** becomes the actual spore. As the septum begins to form, only about 30% of the chromosome is on the forespore side; the rest of it enters the forespore by a mechanism similar to the transfer of DNA that occurs during conjugation (see Chapter 10). The mother cell then behaves as a phagocyte, sending out pseudopod-like extensions that surround the forespore. When the tips of these extensions meet at the other side of the forespore, their cytoplasmic membranes fuse, engulfing

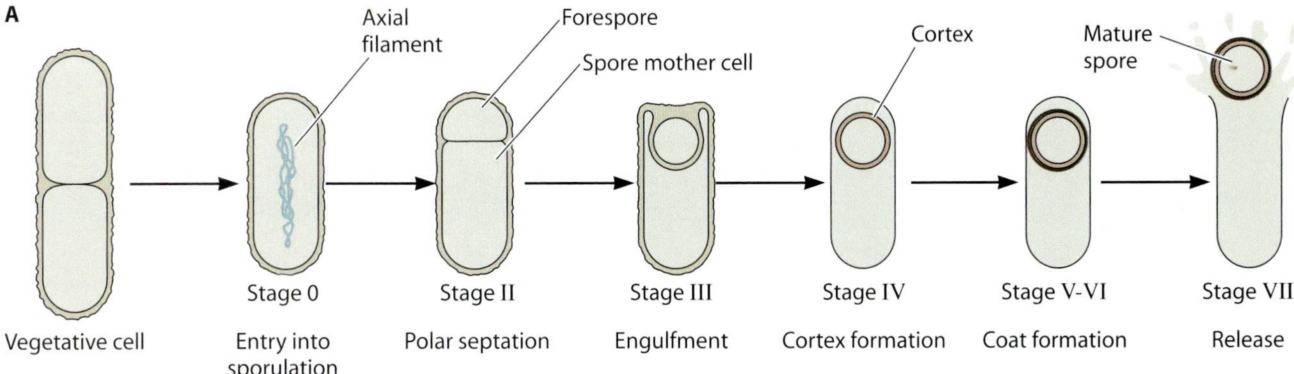

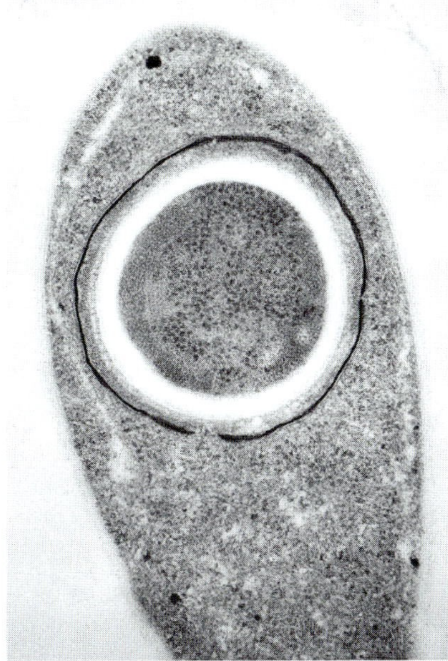

Figure 14.1 **(A)** Stages in the process of endospore formation. **(B)** An electron micrograph of a thin section of an almost completely developed spore of *Clostridium tetani* (stages V and VI). The core is almost completely dehydrated and packed with ribosomes. The cortex and spore coat are clearly seen.

the forespore completely. The forespore (as is the case with phagocytosed cells) is then surrounded by two cytoplasmic membranes: its own and the mother cell's. The outer surfaces of the two membranes face one other. A thick layer, called the **cortex,** made up of murein, is laid down between the two membranes. The murein of the cortex differs in several chemical properties from that of the cell wall of vegetative cells. Outside these membranes (from the viewpoint of the forespore), a protein **spore coat** forms, and it might be surrounded by a membranous layer called the **exosporium.** During this process of becoming surrounded by layers of protective coverings, the forespore changes internally. For example, it synthesizes large quantities of the spore-specific compound dipicolinic acid and becomes extremely dry. Finally, the mother cell lyses, releasing the mature spore.

The process of sporulation we have just described is divided by convention into **stages,** designated 0, before any indication of sporulation is detectable, through VI, when the endospore is mature (Fig. 14.1). Mutant alleles affecting the process are designated by the particular stage at which the process is arrested. For example, mutations in *spo0A* arrest spore formation at stage 0.

A mature endospore appears to be completely inert. It has no detectable metabolism. It contains no ATP or reduced pyridine nucleotides. Its interior (called the **core**) is extremely dry and takes up water only upon germination, accounting, perhaps, for its remarkable resistance to the moist heat of pressurized steam at temperatures exceeding the boiling point.

Programming and regulation

The genetic programming and regulation of endospore formation has been investigated intensely for the past 50 years. It is, without doubt, the most thoroughly studied example of prokaryotic differentiation.

As we have seen, limiting a culture's source of carbon, nitrogen, or phosphorus initiates sporulation if the population is sufficiently dense. Proper density is determined by a form of **quorum sensing** (see Chapter 13), whereby accumulation of certain peptides that each cell secretes must rise to a critical threshold before efficient sporulation can occur. All the signals triggering initiation of sporulation are perceived by various components of a series of reactions that lead to the phosphorylation, and thereby activation, of Spo0A, which is a DNA-binding protein. This cascade of events is called the **Spo0A phosphorelay** (Fig. 14.2). Activated Spo0A (Spo0A~P) plays a central role in sporulation; its intracellular concentration determines which sporulation-associated processes occur. Indeed, the genetic control of sporulation can be conveniently divided into two parts: (i) how Spo0A is activated and (ii) when activated, what Spo0A~P does.

Activation of Spo0A

The Spo0A phosphorelay is a relatively simple phosphorylation cascade, much like the ones described in Chapter 13: a **kinase** autophosphorylates and then transfers its phosphate group to another protein (Spo0F), which through the mediation of a phosphotransferase (Spo0B) donates its phosphate group to Spo0A. Although the Spo0A phosphorelay system consists of a single linear pathway, it is capable of sensing several different sporulation signals. It is precisely regulated by physiological and environmental factors. Its first component, the kinase, is made up of not one but three proteins, each of which is activated by different signals, including near depletion of any of several nutrients. The overall system is further regulated and fine-tuned by three phosphatases that remove

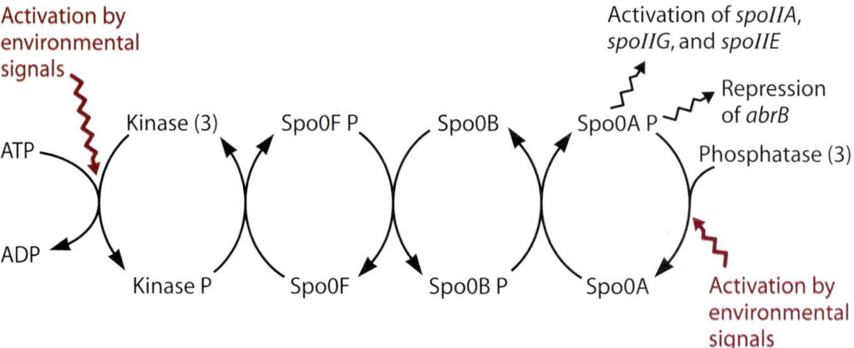

Figure 14.2 The phosphorelay system that regulates sporulation and sporulation-associated events in *B. subtilis*. Sites of regulation are shown in dark red.

phosphate groups from the end product (Spo0A~P) or from one of the intermediates. Thus, both phosphorylation and dephosphorylation act to set the optimal internal concentration of Spo0A~P. Both kinases and phosphatases are activated by a particular signal; thus, these enzymes respond to a large repertoire of environmental cues. Sporulation, then, can be set in motion by many different changes in environmental circumstances.

At low levels, Spo0A~P is involved not only in sporulation, but also in other processes associated with starvation. When a cell faces incipient starvation, it might still be able to obtain additional nutrients and thereby postpone the need to sporulate. These starvation-associated food-seeking processes include formation of antibiotics and toxins (chemically mediated predation?), induction of competence for transformation (to use foreign DNA for food?), turning on of motility and chemotaxis (to seek nutrients?), and expression of transport systems and catabolic pathways (to utilize alternative sources of nutrients that might be present?).

Triggering of reactions by low Spo0A~P concentrations occurs indirectly. During growth when nutrients are abundant, these reactions are *repressed* by AbrB (one of several proteins that control these genes). Later, when sporulation signals cause the concentration of Spo0A~P to increase, it represses the synthesis of AbrB, thus allowing the enzymes that catalyze sporulation-associated reactions to be made.

Activity of Spo0A~P

Perhaps not surprisingly, the regulatory pathways leading to sporulation are much more complicated than those controlling the expression of the starvation-associated reactions. Sporulation depends on genes being expressed at the proper time—some immediately, to start the process, and others later, at particular stages of spore development. In addition, genes must be expressed in the proper location, some within the mother cell, some within the forespore. This differential expression occurs in spite of the fact that the mother cell and forespore are genetically identical.

Although our understanding of all the genes that regulate sporulation and how they interact is far from complete, the general outlines of the process are taking shape. The central actors are **sigma factors** (proteins that bind to RNA polymerase and thereby confer specificity on certain promoters [see Chapters 8 and 12]). During vegetative growth, a sigma factor designated sigma A (σ^A) predominates. Then, during sporulation, five other sigma factors are formed that cause various spore genes to be expressed at various times in specific locations (Table 14.2). This prominent role of sigma factors is reminiscent of the major role played by σ^s in preparing growing cells to enter stationary phase (see Chapter 13).

Four of these sigma factors are synthesized as inactive proteins that become activated by proteolytic cleavage at the proper time and location. Spo0A~P initiates four independent cascades that lead to expression of these sigma factors.

Sporulation: a group activity

We have discussed sporulation as though it were the activity of a single bacterial cell. Indeed, that is the way it has been studied. We have already seen that sporulation depends on quorum sensing, a group activity—but

Table 14.2 Sigma factors that direct expression of sporulation genes

Sigma factor	Encoding gene	Location	Stage at which active
σ^H	*sigH*	Predivision sporangium	0
σ^F	*sigF*	Forespore	II
σ^E	*sigE*	Spore mother cell	II
σ^G	*sigG*	Forespore	III
σ^K	*sigK*	Spore mother cell	IV

there is more. The laboratory strains of *B. subtilis* used in these studies have undergone long-term microbial domestication, having been selected for rapid growth in liquid media. Cultures of such laboratory strains grow as dispersed single cells. In contrast, almost all the cells in standing cultures of strains freshly isolated from nature migrate to the surface, forming a thick, sculptured **pellicle.** Finger-like projections extend from the pellicle, and spores form at their tips. Such structures resemble the fruiting bodies of fungi, and possibly they offer a similar selective advantage: they might aid in dispersal.

Although studies of laboratory strains revealed a wealth of information about sporulation, they missed for a long time the fascinating fact that spore-forming bacteria form communities that behave as multicellular organisms.

DEVELOPMENT OF *CAULOBACTER CRESCENTUS*

One group of **prosthecate** bacteria (a prostheca is an appendage) is **dimorphic** (meaning it has two shapes). When one cell of this sort of organism divides, the two products of division differ in both appearance and behavior. These bacteria undergo a developmental cycle that is essential to their reproduction.

In the case of *Caulobacter crescentus*, the most thoroughly studied of the prosthecate dimorphic bacteria (Fig. 14.3), one product of division, called a **swarmer,** is motile by means of a polar flagellum, and the other is nonmotile, or sessile. The nonmotile cell bears a polar prostheca, called a **stalk** because it usually anchors the cell to a solid surface by means of a terminal organ of attachment called a **holdfast.** Both stalk and flagellum are located at the cell's "old" pole (in distinction to the "new" pole formed by division). For *Caulobacter* to attach, almost any surface will do, e.g., a rock in a pond or the faucet of a sink.

A stalked cell continues to divide, each time producing a new swarmer, which actively swims away by means of its polar flagellum. But that is about all a newly formed swarmer can do. It cannot replicate its DNA or divide—not until it differentiates to become a stalked cell, which, like its mother cell, is capable of dividing and producing more swarmer cells. Swarmers, which are also chemotactic, are clearly agents of dispersal: they leave their tethered, stalked mother cell and can colonize areas with new sources of nutrients.

The cell cycle

Each swarmer cell goes through a highly structured developmental progression before it can reproduce; altogether, the process takes about 150

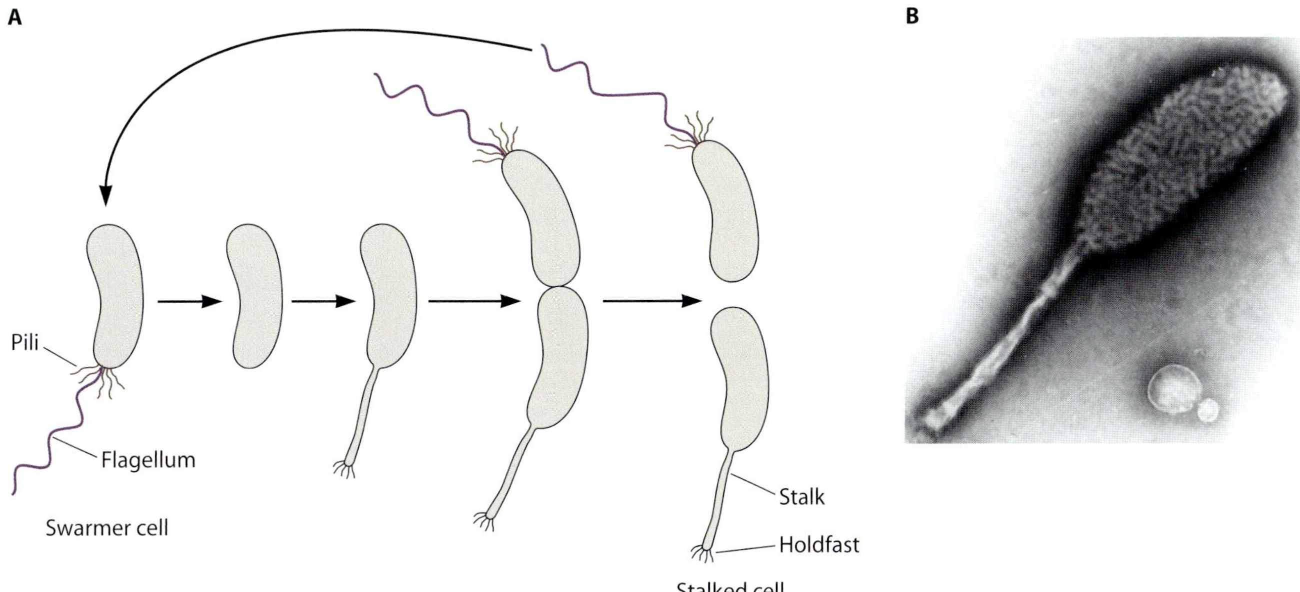

Figure 14.3 ***Caulobacter crescentus.*** **(A)** Life cycle; **(B)** photomicrograph.

minutes at 30°C (Fig. 14.3). For about the first 15 minutes, a swarmer swims chemotactically. Then, it dismantles its distinctive swarmer cell features: it sheds its flagellum, along with its nearby pili, and degrades its chemotaxis apparatus (which is located at the same pole). Next, it replicates its DNA and synthesizes a stalk and a holdfast. At 110 minutes, it starts to divide to produce a new swarmer.

The swarmer's developmental progression is morphologically and biochemically elaborate. Throughout the process, genes are turned on and off and cellular components are degraded. About a fifth of *Caulobacter*'s genes are activated or repressed at discrete times during the cell cycle.

Genetic control of development

The cell cycle of *Caulobacter* has two particularly distinctive features. One is the powerful influence of cellular polarity. As has been pointed out by students of bacterial differentiation and development, bacterial cells are inherently asymmetric, some more obviously so than others. *Caulobacter* certainly belongs to the extremely asymmetric group. Most of the significant events in *Caulobacter* development—formation of flagella, pili, chemotactic apparatus, and holdfast-bearing stalk—occur at the cell's old pole. The cell seems to be compartmentalized, although it has no internal membranes to define separate compartments. However, protein-labeling methods (such as green fluorescent protein tagging) show that certain constituents, including regulatory proteins, accumulate at particular intracellular locations. Indeed, localization of regulatory molecules is essential for such spatially differentiated development. The detailed basis for such localization remains obscure, but it is known to operate by a "diffuse and capture" method: the molecules move rapidly through the cell by diffusion and are captured by some cellular component at the pole.

The other distinctive feature is the precisely programmed turning on and off of the biosynthesis of cell components—in fact, it is startlingly precise. *Caulobacter* appears to follow a "just-in-time" policy of gene expression. Genes encoding components of particular cellular structures and functions—for example, flagellar biosynthesis and associated chemotaxis systems—are expressed just before their products are used. Likewise, when no longer needed, these systems and their components are quickly destroyed by specific proteases that are also synthesized just in time.

Several phosphotransferase cascades, of the sort we encountered in spore formation, seem to be dominant controllers of the *Caulobacter* cell cycle as well. Each appears to control a modular set of proteins that work together and interact to carry out a particular cellular function. Regulatory linkages "talk" to the various modules and act to synchronize them. One particular response regulator protein, CtrA, is the master regulator. In *Caulobacter*'s development, it plays a role parallel to that of Spo0A in sporulation. Directly or indirectly, it controls about a quarter of the genes that are regulated in the cell cycle.

DEVELOPMENT OF MYXOBACTERIA

The myxobacteria are a group of proteobacteria in the class *Deltaproteobacteria*. Myxobacteria undergo an elaborate process of differentiation as an integral part of their life cycle (Fig. 14.4).

Individual cells of these largely terrestrial aerobic organisms gather into swarms that move by **gliding motility** over solid surfaces, killing and consuming other microbial cells in their path. No wonder they have been called wolf packs. The swarms are not selective: they devour filamentous fungi, yeasts, and protozoa, as well as other bacteria. Being prokaryotes, myxobacteria are incapable of ingesting their prey. Instead, they secrete antibiotics and enzymes that kill and lyse cells, along with other enzymes that break down the released macromolecules. They then take up the resulting small molecules; their nutrition is primarily based on amino acids. This mode of killing and consuming microbial cells depends on group action: only by pooling their resources can myxobacterial cells create an environment sufficiently rich in lytic enzymes to mediate the process. Indeed, dilute cultures of myxobacteria are incapable even of metabolizing the protein casein, although denser cultures can readily break it down and utilize the amino acids that are released.

When prey becomes scarce (as signaled by near exhaustion of amino acids in their environment), the developmental cycle begins. Individual cells switch from "adventurous" (A) to "social" (S) motility, which has a component of pilus-driven twitching motility (see Chapters 2 and 13). Cells become more closely packed in "aggregation centers." Then, they

Figure 14.4 Life cycle of *M. xanthus*.

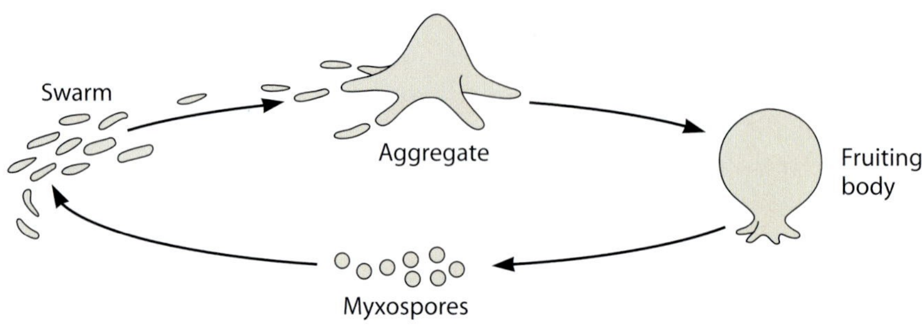

pile on top of one another, and in so doing, construct a **fruiting body**, which in some species can be nearly a millimeter high. Fruiting bodies are not haphazard piles of cells; they have a definite form and size, characteristic of the particular species that produces them (Fig. 14.5). At the tips of these fruiting bodies, **sporangioles** (globular spore-containing structures) form. As they do, about half the cells within them lyse, and materials released from them are used by survivors, which become resistant resting cells called **myxospores.** Although not nearly as hardy as endospores, they survive considerably more heat, UV irradiation, and desiccation than vegetative cells. They are known to survive for 15 years in soil stored at room temperature. When conditions become favorable, myxospores germinate, initiating a new growth cycle.

The fruiting bodies of various myxobacteria are quite diverse in shape (Fig. 14.5). Those of *Myxococcus xanthus*, the most thoroughly studied myxobacterium, are quite simple: they consist of a single dome-shaped sporangiole. In contrast, those of *Stigmatella* spp., which are usually found on the bark of rotting wood, are elaborately shaped and richly colored. These bright-orange structures consist of several sporangioles, each of which sits at the tip of a branched stalk.

The function of myxospores is clear. They enable myxobacteria to survive periods of starvation and harsh conditions, but what is the selective advantage of their being grouped together in a raised sporangiole? Most students of myxobacteria speculate that there are two advantages. First, the sporangiole is probably an agent of dispersal. Worms or insects moving toward a more nutrient-rich environment might carry the sporangiole along to a new location that has not been depleted of microbial cells. Also, the myxospores being collected into a group within the sporangiole ensures that germination will give rise to a community of cells; this is important because, as we have seen, myxococcal feeding is a group effort.

Regulation of development

Because the developmental cycle of myxobacteria depends more heavily on interaction between groups of cells than on differentiation of individ-

Figure 14.5 Fruiting bodies of myxobacteria vary in complexity from the relatively simple one of *Myxococcus stipitatus* (A) to the elaborate structure produced by *Stigmatella aurantiaca* (B). Bars, 50 μm.

A

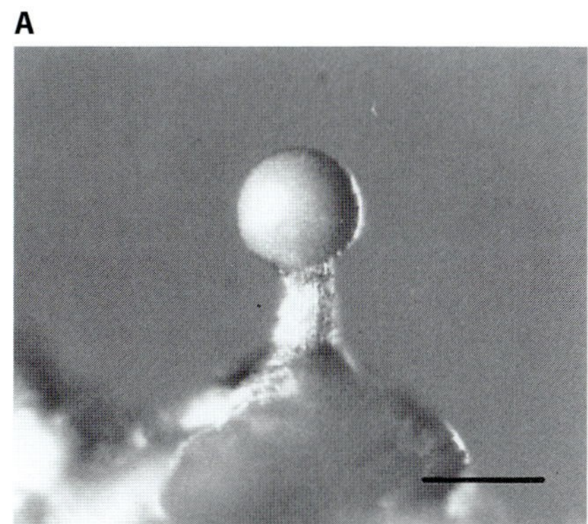

B

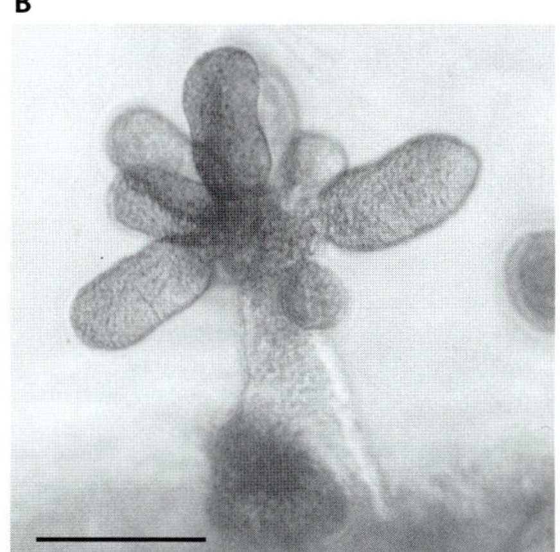

uals, we probably should not be surprised that cell-to-cell signaling directs the process.

At appropriate times during development, cells of *M. xanthus* produce at least five molecular signals (designated A, B, C, D, and E) that direct and coordinate the process. Some, but not all, of these intercellular signals have been identified chemically. For example, A signal, which is produced when cells sense starvation, is a mixture of amino acids and peptides. C signal is a mixture of two proteins. The identity of B signal remains unknown. The discovery of these molecular signals and of the existence of those that have not been characterized chemically depends on genetic evidence. A strain that by itself is unable to initiate or complete the developmental cycle can do so when mixed with wild-type cells (or cells of certain other developmentally defective mutants).

OTHER BACTERIA THAT UNDERGO DIFFERENTIATION AND DEVELOPMENT

Many other bacteria differentiate and undergo development. Some of the most thoroughly studied are listed in Table 14.3.

Possibly, *Nostoc punctiforme* and related cyanobacteria win the differentiation record for prokaryotes (Fig. 14.6). Here are the kinds of differentiation it undergoes.

1. Upon the signal of nitrogen limitation, cells within a filament differentiate to become nitrogen-fixing specialists called **heterocysts.** They are induced to synthesize nitrogenase, which they protect from destructive oxygen by degrading their oxygen-generating pho-

Table 14.3 Some well-studied examples of bacterial differentiation and development

Group	Example	Differentiation and development
Actinobacteria	*Streptomyces coelicolor*	Forms both aerial and substrate mycelia. Forms exospores at tips of aerial mycelium.
Cyanobacteria	*Nostoc punctiforme*	Forms nitrogen-fixing heterocysts. Forms resting cells called akinetes. Produces migrating groups of cells called hormogonia. Forms nitrogen-fixation symbioses with certain plants, ferns, and bryophytes.
Firmicutes	*Bacillus subtilis*	Forms endospores largely in the tips of protuberances that extend from pellicles on liquid surfaces.
Myxobacteria	*Myxococcus xanthus*	Individual motile, feeding cells aggregate and develop into fruiting bodies that form myxospores within themselves that can germinate to form feeding cells.
Stalked bacteria	*Caulobacter crescentus*	Stalked cells divide, producing a stalked cell and a swarmer, which develops to become a stalked cell.
Many kinds of bacteria	*Proteus mirabilis*	Groups of cells perform swarming migration when individual cells differentiate into elongated (20- to 80-μm-long) multiflagellate (up to 50) cells that move together; then, they dedifferentiate into the original cell form.
Chlamydiae	*Chlamydia trachomatis*	Undergoes a mandatory developmental cycle in which an infectious, metabolically inactive elementary body enters a host cell by phagocytosis; within the phagocytic vacuole, it differentiates into a noninfectious but metabolically active reticulate body, which multiplies to fill the cell. The reticulate bodies dedifferentiate, becoming elementary bodies that are released as the host cell lyses.
Endosymbiotic nitrogen fixers	*Sinorhizobium meliloti*	Free-living bacterial cells enter the root hairs of leguminous plants through an infection tube to differentiate into nitrogen-fixing bacteroids within a nodule that the plant forms; bacteroids are capable of dedifferentiating to become free-living bacterial cells.

tosystem II and by forming a thick, oxygen-impenetrable wall around themselves.

2. Upon nutritional limitation, some cells in the filament differentiate to become **akinetes**, resistant resting cells. Whereas vegetative cells remain viable for about 2 weeks in a dark, dry state, akinetes can survive for 5 years.
3. The cyanobacterial filament can also differentiate to form agents of dispersal called **hormogonia**—short lengths of smaller cells that become capable of gliding motility; terminal cells in the hormogonium develop characteristic pointed ends.
4. Upon receiving an appropriate chemical signal from certain plants, ferns, or bryophytes, homogonia migrate toward them, enter pockets that the host forms, and differentiate to form a mass of nitrogen-fixing heterocysts.

Certainly this elaborate system of differentiation will challenge microbiologists for some time.

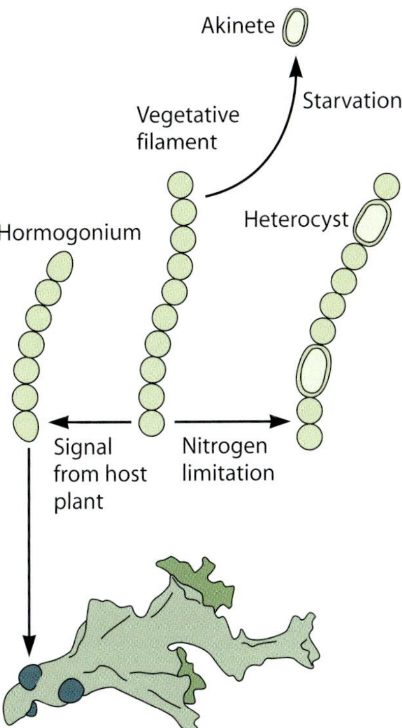

Figure 14.6 ***N. punctiforme*** **patterns of differentiation.**

SUMMARY

Certainly bacteria are not just single-celled creatures that enlarge and divide. Many have remarkable capacities to differentiate and undergo development, and the overwhelming majority of them live together in multicellular groups as biofilms or pellicles. The heterogeneous prokaryote domain is rich with biological properties beyond those described here; we shall survey this remarkably diverse group of organisms in the next chapter.

STUDY QUESTIONS

1. What characteristic do heterocysts (in cyanobacteria) and spore mother cells (in *Bacillus*) share?

2. Distinguish between endospores and exospores with respect to formation and function.

3. How many membranes surround the forespore? What is their origin?

4. How is it possible for several different environmental signals to regulate the single pathway leading to the formation of Spo0A~P?

5. How do the swarmer and stalked cells of *Caulobacter* differ morphologically and functionally?

6. Is there evidence for functional compartmentalization within a stalked *Caulobacter* cell? What is it?

7. What selective advantage is conferred on swarming myxobacteria by their ability to secrete antibiotics as well as lethal and lytic enzymes?

8. What selective advantages are conferred by the ability of myxobacteria to produce myxospores?

section VI diversity

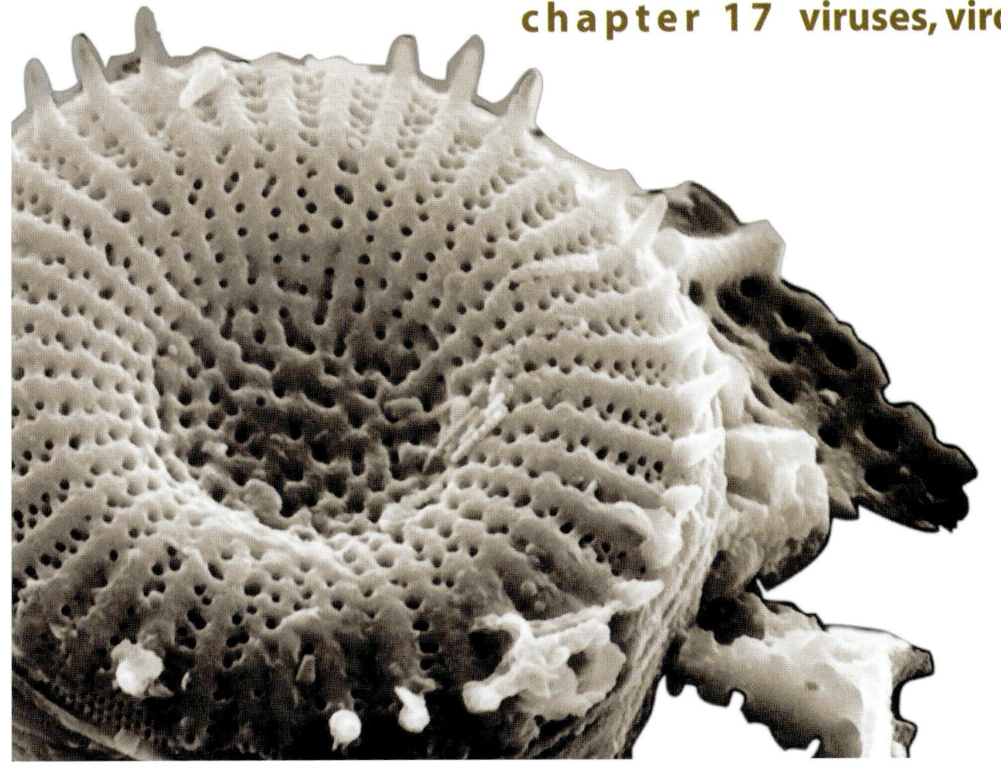

chapter 15 prokaryotic microbes

INTRODUCTION

In part I, we dealt with microbial physiology, a study that, by its nature, seeks generalizations that apply across the microbial world; its goal is unity. Now, we shall shift dramatically and focus on microbial diversity. We shall begin in this chapter with the diversity of prokaryotes and then go on in subsequent chapters to eukaryotic microbes and viruses. As we shall see, there is no doubt that the greatest diversity, even though not yet fully cataloged, is found among prokaryotes.

MAKING ORDER OF PROKARYOTIC DIVERSITY

How does one extract coherence and a sense of order out of a sea of information as vast as even our present knowledge of the myriad kinds of extant prokaryotes? Certainly since Aristotle, and probably well before, humans have taken the same general approach to making sense of seemingly bewildering diversity: identify groups of individuals that resemble one another, and collect them into larger assemblages by seeking similarity among the various groups. In this way, we simplify by focusing on a more manageable number of groups rather than on overwhelming numbers of individuals. We have long since applied this approach to biological diversity, which is a special case because it has been guided by prevailing paradigms: initially by the concept of a species being a special act of creation, and later by the principles of evolution.

Prokaryotes, fungi, and protists are grouped under the umbrella term microbes because they are small (although, as we have seen, there are exceptions). However, sharing the property of smallness does not indicate **phylogenetic** (evolutionary) relatedness, nor is smallness unusual. Microorganisms—usually defined as creatures invisible to the unaided eye (i.e., smaller than about 100 micrometers [μm])—occur in all three biological domains: Eukarya, Bacteria, and Archaea. The two prokaryotic domains, Bacteria and Archaea, are composed chiefly of organisms that fit the definition of microbes.

Frustration and frequent changes of direction have characterized the history of classifying microorganisms (Fig. 15.1). Complex morphology and an abundant supply of fossils—so critical for the traditional classification of plants and animals—are sparsely available for microbes, especially for the prokaryotic ones. In the earliest stages of dealing with prokaryotic diversity, during the 17th and 18th centuries, even recognizing individual microbes as being distinct proved difficult and controversial. Because the relative abundance of various microbes in natural popula-

Figure 15.1 Evolution of the major schemes for classifying organisms.

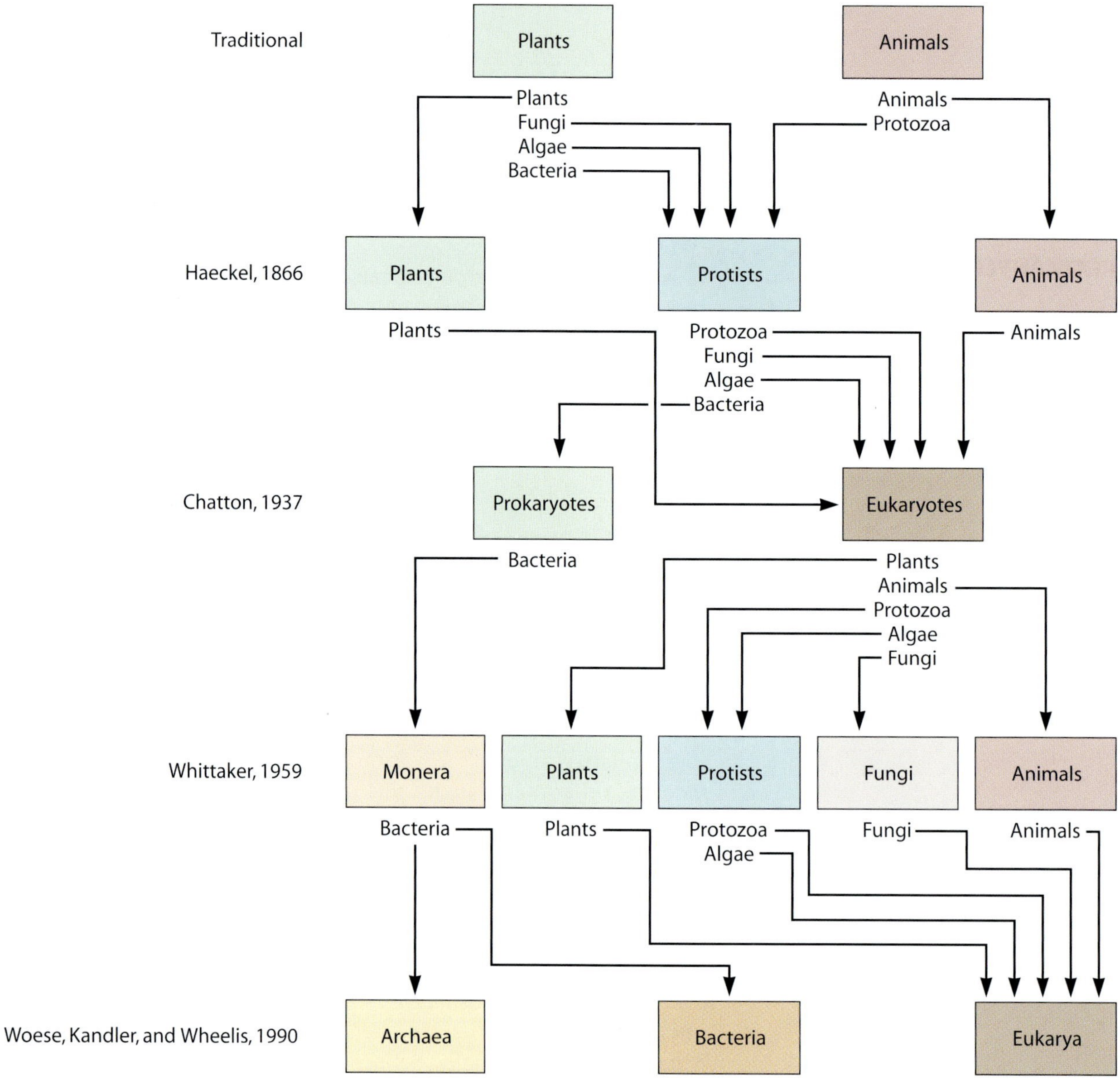

Traditional
Plants
Animals
Plants
Fungi
Algae
Bacteria
Animals
Protozoa
Haeckel, 1866
Plants
Protists
Animals
Plants
Protozoa
Fungi
Algae
Bacteria
Animals
Chatton, 1937
Prokaryotes
Eukaryotes
Bacteria
Plants
Animals
Protozoa
Algae
Fungi
Whittaker, 1959
Monera
Plants
Protists
Fungi
Animals
Bacteria
Plants
Protozoa
Algae
Fungi
Animals
Woese, Kandler, and Wheelis, 1990
Archaea
Bacteria
Eukarya

tions shifts rapidly as they alter their microenvironment by their own metabolic activities, microscopists misinterpreted the progression of forms they saw while observing a particular environment as being stages of an individual microbe's life cycle, a phenomenon they called **polymorphism**. At the time, some biologists concluded that all prokaryotes belonged to a single species.

The polymorphism misconception was challenged in the 19th century, when microbiologists began to work with **pure cultures** (clones derived from a single cell), and it died abruptly when Robert Koch (a German physician turned microbiologist) developed simple means of studying them, such as growing them on agar plates. Then, it became apparent that there are vast numbers of different kinds of prokaryotes, each with a characteristic morphology that (with a few exceptions, such as those we considered in Chapter 14) does not change much during growth and development. The burgeoning studies of pure cultures led to new dogmas: particular prokaryotes are distinct in morphology, physiology, and (if pathogens) the disease they cause. They do not change. Although the new dogmas exaggerated the genetic stability of microorganisms, they encouraged research into distinguishing various types of prokaryotes, and hence, the need to name and classify them. The Linnean binomial system of Latinized genus and species names that had long been used to name plants and animals was adopted.

PROKARYOTIC SPECIES

Because the species is the fundamental unit of biological classification, an agreed-upon meaning is crucially important. All **higher taxa**—genus, family, order, etc.—are readily recognized as being human constructs, but *species* maintains a special conceptual status. To Carolus Linnaeus (1707–1778), the founder of modern taxonomy, who dealt only with plants and animals, the concept was crystal clear: species are distinct groups of "individuals that breed true, are immutable, and were created as such." The discovery of evolution in the mid-19th century and the principles of genetics in the early 20th called for new definitions. Possibly the most influential of these came in 1942 with Ernst Mayr's **biological species concept,** namely, "a group of actually or potentially interbreeding populations." In Mayr's and most other 20th-century definitions of eukaryotic species, the idea of a breeding group dominates. It accounts for the similarity of the members of a particular species (because they interbreed) and their drifting apart from members of others (because they cannot interbreed with them).

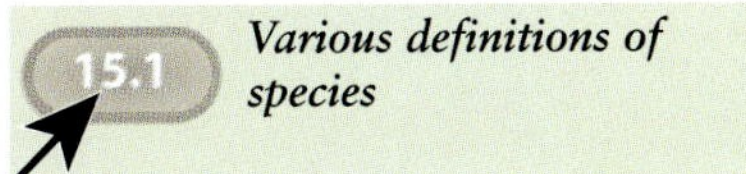

Various definitions of species

Such breeding-group-based definitions work well for most animal species and less well for plant or protist species, but they have no relevance whatsoever to prokaryotes. Prokaryotes do not have breeding groups because prokaryotes reproduce asexually. They do exchange genes by a variety of mechanisms, and usually such interchange is restricted to related groups, but reproduction never depends on such exchange (see Chapter 10). Free of the restraints imposed by dependence of reproduction on genetic exchange, prokaryotes might be expected, through mutation, to have evolved into a vast continuum of individuals, each slightly different from the next. However, experience shows that in fact, clusters of similar individuals do exist among prokaryotes. Microbiologists call these clusters **species** and name them using the binomial Linnean system. The resulting names, such as *Escherichia coli* or *Pseudomonas aeruginosa,* for example, are indeed useful descriptors of prokaryotic clusters. They quickly tell us what properties a group of **strains** (individual isolates or genetic variants) are likely to share. We might ask, however, why there are species-like clusters of prokaryotes. Mayr addressed this question with respect to all organisms by asking "why the total genetic variability of nature [is] organized in the form of discrete packages, called species." He suggested that they share a "gene complex that is specifically adapted to a particular ecological situation."

But how should one define prokaryotic species in order to identify their members? The earliest approach was to identify a small number of key *defining* characteristics (for example, the ability to ferment lactose while producing acid and gas, which was one of the definitions of *E. coli*) that are shared by members of the group. However, reliance on such characters, particularly small numbers of them, carries obvious hazards: they change readily through mutation. Membership in a species should not hinge on a single mutational event.

Later an opposite approach, called **numerical taxonomy,** was favored. It involved determining **similarity indices** among various strains, calculated from large numbers of plus-or-minus phenotypic characteristics, e.g., produces or does not produce spores. Strains sharing a similarity coefficient greater than 70% were usually considered to constitute a species.

At present, the generally accepted definition of a prokaryotic species is based largely on the similarity of the DNAs of its members, although a clear phenotypic distinction from other species is also required. Boundaries of prokaryotic species that most microbial taxonomists now accept are 70% or greater DNA-DNA relatedness (as determined by forming hybrid DNA with a change in melting temperature [ΔT_m] of less than 5°C) and less than 5% difference in G+C moles percent among members of a species. Although admittedly highly arbitrary, the definition has proven to be practical and useful.

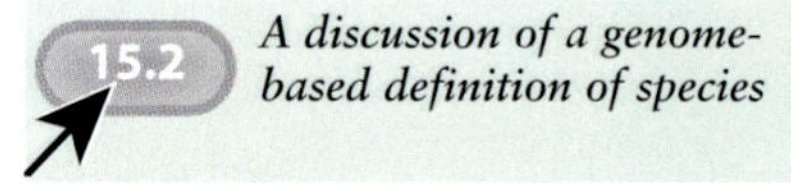

A discussion of a genome-based definition of species

Recently, with the advent of methods for surveying uncultured microbial populations by isolating and sequencing DNA from the environment, the definition of microbial species has become further complicated. In such samples, DNA-DNA hybridization is not possible, because usually only a portion of the genome is retrieved. Reflecting these complications, some researchers in this field use the term **phylotype,** or **genomic species,** instead of species and usually define it as a group with 94% sequence identity among the genes of its members.

EXTENT OF PROKARYOTIC DIVERSITY

In spite of the powerful array of methods now available to study prokaryotes, the full extent of their diversity remains largely unknown. There are prokaryotes everywhere, and we know little or nothing about most of them. Only about 5,000 prokaryotic species have been described (contrasted with, for example, about half a million insect species). Many times this number of kinds of prokaryotes must exist. Some microbiologists estimate that 10 million species would be a more accurate estimate of the diversity of prokaryotes. Certainly, the potential extent of prokaryotic diversity is enormous, because there are so many of them. They have been subjected to a vast variety of selective pressures because they thrive in all life-supporting environments on Earth, even those far too hostile to support other forms of life. Reliable estimates assess the total amount of carbon in prokaryotes to be nearly as much as the total in plants (and individual plants are, of course, much larger than a microbe), the total content of phosphorus to be 10-fold greater than the collective amount in plants, and the total number of their cells to be a mind-boggling 4×10^{30} to 6×10^{30}. Altogether, they weigh more than 50 quadrillion metric tons! This huge population, coupled with a capacity for rapid multiplication and their ubiquity, constitutes a reservoir for immense genetic variation among prokaryotes. That means that a vast number of prokaryotic species must exist.

Why, then, has only such a small fraction of the prokaryotes been recognized and studied? There are many reasons. At a technical level, we could cite the enormous genetic variation among individual strains normally grouped into a signal prokaryotic species—much greater than that found within the taxon "family" of animals. Indeed, it would take many eukaryotic species, genera, or families to match the diversity within what we designate a single species of prokaryote. By the criteria used with prokaryotes, humans and worms would fall within the same species! However, other factors are more significant and important. Just viewing a prokaryote, with very few exceptions, is not sufficient to identify it. Cultivating most prokaryotes, the sine qua non of characterizing and, by convention, naming them (uncultivated species can be assigned only tentative names preceded by "*Candidatus*"), can be quite difficult. From many environments, only 1% or so of the number of apparently viable cells visible by microscopy can now be cultivated. However, this is an active area of research and much progress is being reported. Many previously uncultivable bacteria and archaea have been coaxed into making colonies on agar plates or growing in liquid media. The secret seems to be patience—it takes weeks or months for growth to appear—and attention to the nutritional conditions. Thus, many marine bacteria do not grow on nutritionally rich media but require a "thin soup" similar to that found in seawater.

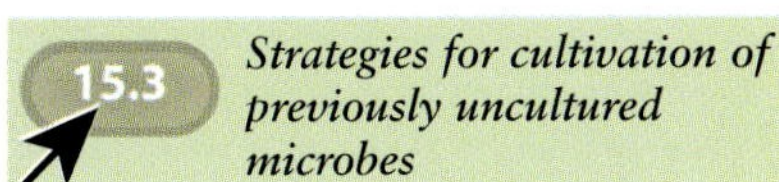

15.3
Strategies for cultivation of previously uncultured microbes

For now, molecular probes tell a similar story. For example, using probes to analyze genes encoding RNA from masses of cells harvested in the open ocean reveals the presence of large numbers of Archaea that no one has yet been able to cultivate. Curiously, most pathogens that infect humans, animals, and plants (although there are dramatic exceptions, for example, *Mycobacterium leprae*) can be cultivated with relative ease. One explanation for pathogens' being cultivatable may be that many of them must have the capacity to grow in near-pure culture; they do so while infecting their hosts.

We might then ask why it has proven so difficult to cultivate so many of the prokaryotes that flourish in nature. To do so, it is only necessary to reproduce in the laboratory the chemical and physical environment that supports their growth, but that is not always easy, for numerous reasons. For example, in nature, interdependent nutritional **consortia** (interdepen-

Table 15.1 Some metabolic functions mediated only by prokaryotes

Name	Conversion
Nitrogen fixation	N_2 to ammonia
Denitrification	Nitrate to N_2
Anammox	Anaerobic conversion of ammonia and nitrite to N_2
Nitrification	Ammonia to nitrite and nitrite to nitrate
TMAO reduction	Trimethyl amine *N*-oxide (TMAO) to trimethyl amine
Sulfate reduction	Sulfate to sulfide
Arsenate reduction	Arsenate to arsenite

dent groups of individuals) of prokaryotes are common. One microbe in such a consortium might depend on a continuous supply, at very low concentrations, of end products produced by an intimate neighbor. In turn, the producer might be dependent on its neighbor's ability to constantly use its end products. Such a situation occurs in the consortium "*Methanobacillus omelianskii,*" once thought to be a single prokaryote that could grow in pure culture by converting ethanol and CO_2 to methane. In fact "*M. omelianskii*" proved to be a mixture of two species of prokaryotes, neither of which can grow singly on a mixture of ethanol and CO_2. One species can grow by producing H_2 from ethanol, but the equilibrium of the reaction is growth supporting only if hydrogen is maintained at an extremely low concentration, which the second species accomplishes by utilizing it as fast as it is produced. The second species scavenges the H_2 produced by the first, using it to reduce CO_2 to methane.

In other cases, growth as consortia is probably necessary to establish growth-sustaining concentrations of O_2 or to supply constant levels of required, but unstable, metabolites, for example, H_2S in the presence of O_2. Species of the bacterium *Beggiatoa*, for example, derive their metabolic energy by metabolizing these incompatible gases. In nature, they locate themselves at an interface where the two gases meet, where microbially produced H_2S diffuses toward them from one direction and oxygen from the other (see Chapter 13). Such conditions could be established in the laboratory, but it is not easy. Of course, there must be reasons that we still do not understand for the existence of certain consortia. Those microbiologists who have been most successful in culturing recalcitrant prokaryotes attribute most of their achievements to methods that supply substrates at extremely low concentrations and to the realization that some microbes grow only very slowly in nature. These microbiologists are prepared to wait weeks or months for their cultures to produce a colony that is visible under a dissecting microscope. Such approaches have markedly increased the number of microbes from certain environments that can be cultured, from less than 1% to roughly 7%, for example. Still, culturing the vast majority of the microbes that exist in nature remains an important goal and a major challenge.

Metabolic diversity—particularly with respect to their fueling reactions—is the hallmark of prokaryotes (see Chapter 6). These metabolic innovators are uniquely capable of an impressive list of chemical conversions, some of which are essential for the continued existence of all other forms of life (Table 15.1). For example, until the 20th century, when we humans learned how to do it industrially, only prokaryotes were capable of converting atmospheric dinitrogen to ammonium ion, the fixed form of nitrogen that flows directly or indirectly into the metabolisms of all organisms. Also, only prokaryotes are capable of the reverse processes, **denitrification** and **anaerobic ammonia oxidation (anammox)**, by which fixed nitrogen is reconverted to dinitrogen (see Chapter 18). Such reconversion is also an essential life-sustaining process, because otherwise, all fixed nitrogen would accumulate in the ocean. It accumulates there because, as we will discuss in Chapter 18, highly soluble nitrogen-containing salts enter river systems and flow to the sea. Only by returning it to the atmosphere can nitrogen fixation resupply this essential nutri-

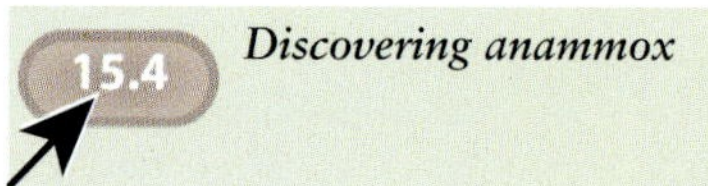

Discovering anammox

ent to Earth's landmasses. The biochemical details of these uniquely prokaryotic processes were discussed in Chapter 6; their environmental impact is considered further in Chapter 18.

Now let us turn to the second step of assessing and making sense of prokaryotic diversity: grouping species into a hierarchical set of related groups.

HIGHER TAXA OF PROKARYOTES

Until the 1970s, when significant amounts of information about DNA and protein sequences became available, there was no rational basis for assigning prokaryotes to phylogeny-based taxa above the level of genus. Earlier editions of ***Bergey's Manual of Systematic Bacteriology***—the internationally recognized authority on prokaryotic identification and classification—conceded the issue forthrightly by disregarding completely the phylogenetic relatedness of prokaryotic species. Instead, they distributed all known prokaryotic species among four "divisions" and 31 "sections" (terms without taxonomic standing) on the basis of phenotypic similarity alone. Sections contained phenotypically related, not necessarily phylogenetically related, species.

Now, with the wealth of material contained in the sequences of macromolecules, prokaryotes can be grouped phylogenetically. The sequence of DNA encoding 16S ribosomal RNA has been used most extensively. Such sequences offer the advantages of being represented in all free-living organisms and containing regions that are highly conserved, thus allowing comparison of distantly related groups. The latest edition of *Bergey's Manual* (the second edition, published in 2001) uses this information to arrange prokaryotic species into the traditional Linnean hierarchy: species, genus, family, order, class, and phylum, adding **domain** to accommodate the largest groups, Archaea and Bacteria. Now let us look at these domains and some of their members.

ARCHAEA

The realization that the group of prokaryotes now called Archaea differs profoundly from all other prokaryotes (now called Bacteria) occurred in the mid-1970s, when Carl Woese of the University of Illinois began to sequence 16S ribosomal RNA from various prokaryotes. The sequences of nucleotides in 16S RNA from some were so different from the rest that Woese and his colleagues concluded they must constitute a distinct group of prokaryotes, which he called **Archaebacteria** to suggest their suspected ancient origins. Now a preponderance of biologists agree that they are no more closely related to Bacteria than they are to eukaryotes, and to reflect this fact, they are called **Archaea.** In addition to Bacteria and **Eukarya** (eukaryotes), Archaea constitute a third major division (a domain) of living things. Some biologists contest this tripartite division of the biological world. Their reasons include the unequal sizes of the groups and inconsistent distribution of **signature sequences** (short sequences of nucleotides shared by a group of organisms but not present in others). For the moment, however, it has strong support.

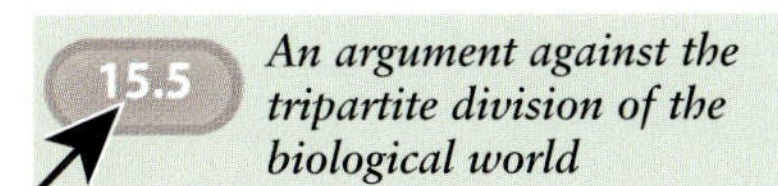

An argument against the tripartite division of the biological world

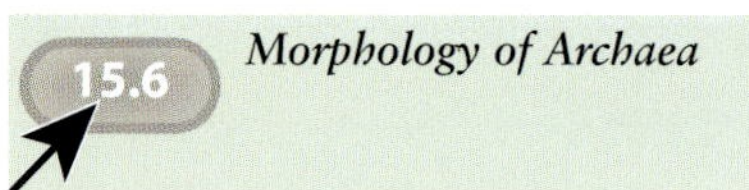

Acceptance of the three-domain classification of organisms relegates the term prokaryote to a mere description of the particular cell type that two domains (Bacteria and Archaea) happen to share. As the term prokaryote ("before a nucleus") implies, these cells (with the rare exception of the planctomycetes [see Chapter 3]) lack the well-defined, double-membrane-enclosed nucleus of eukaryotic cells. Instead, their DNA exists as an ill-defined structure (called a **nucleoid** [see Chapter 3]). Prokaryotes share other properties as well. Most are only about 1/10 the size of eukaryotic cells (a few micrometers compared with 10 μm or so). They contain 70S ribosomes rather than the 80S ribosomes of eukaryotes. They lack the complex intracellular, membrane-bound organelles (e.g., the Golgi apparatus and mitochondria) typical of eukaryotes. It is worth noting that most of the defining properties of prokaryotic cells (other than their ribosomes) are negative—what they lack rather than what they have—and hence do not imply relatedness. Still, the term prokaryote is useful if for no other reason than that it links small-celled organisms, which as a consequence of their size share many similarities of cell physiology.

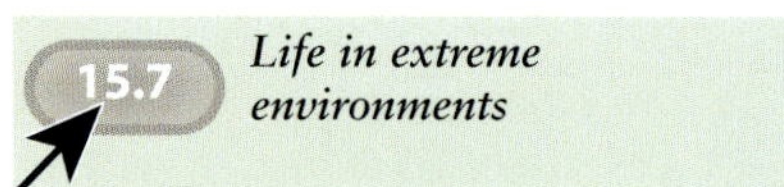

In many respects, the archaea are startlingly different from all other organisms. Notable among these distinctions is the ability of many of them to thrive in extremely hostile environments. Some grow well at temperatures higher than the boiling point of water. Astoundingly, one *(Pyrolobus fumarii)* can grow at 113°C, and even more astoundingly, one has been found growing at 121°C, the temperature of an autoclave. Other archaea set biological records for tolerating high concentrations of salt and high acidity. Because such environments were probably more common on Earth during its earliest development, the ability of archaea to exploit them leads to the suspicion that some of our earliest ancestors might have been members of the Archaea. Some bacteria do thrive in extreme environments, and some archaea do grow under quite ordinary conditions, but growth in extreme environments is a general characteristic of the Archaea.

In certain biochemical respects, archaea resemble eukaryotes. The structures of their DNA and RNA polymerases are like those of eukaryotes. Like eukaryotes, their newly made peptides do not begin with an *N*-formylmethionine residue, as those of bacteria do. Further biochemical differences from bacteria are reflected in the fact that archaea are resistant to antibacterial antibiotics. In other biochemical respects, archaea are unique, differing from both eukaryotes and bacteria. Most remarkably, their cytoplasmic membranes are composed of **glycerol ethers** instead of the phospholipids (glycerol diesters) found in the cytoplasmic membranes of all other organisms (see Fig. 2.12). Rather than being composed of fatty acids attached by ester bonds to a molecule of glycerol, archaea have isoprenoid groups attached to glycerol by ether bonds. Some archaeal membranes are made up of bilayers of glycerol diethers, much as phospholipid bilayers are constructed: their hydrophobic tails (in the archaeal case, the isoprenoid moiety) interact at the middle of the membrane; their hydrophilic heads are exposed to the aqueous exterior. The other form of archaeal membrane, composed of diglycerol tetraethers, is not even a bilayer, although it has similar properties. Each constituent tetraether fulfills the role of a pair of diethers in the membranes of other archaea or phospholipids in the membranes of other organisms (see Fig. 2.12). It seems likely that their unique biochemical properties represent adaptations to the extreme environments in which many Archaea are found. In addition, some Archaea (those that make methane) produce coenzymes found nowhere else in nature. We will discuss them later in this chapter.

Other biochemical properties of the Archaea distinguish them clearly from their fellow prokaryotes, the Bacteria. The walls of almost all bacteria are composed principally of the peptidoglycan called **murein.** No archaea contain murein in their walls. Most have walls made of protein. A few have walls of a different peptidoglycan called **pseudomurein,** which lacks the *N*-acetylmuramic acid and D-amino acids found in the murein of Bacteria. Archaea are also distinguished from bacteria by the composition of their RNA polymerases. The RNA polymerases found in all Bacteria share the same subunit structure ($\alpha\beta\beta'\sigma$), but not those in Archaea. Theirs resemble the subunit structure of eukaryotic RNA polymerases.

Another curious distinction of the known Archaea is the absence of pathogens for either plants or animals (even though some are found as commensals in animals). One can only speculate on the reasons for this absence. Possibly the major groups of Archaea evolved and became largely fixed in their lifestyles before eukaryotic hosts became available for them to attack, or perhaps they evolved in environments that were devoid of metazoans.

The latest edition of *Bergey's Manual* distributes the Archaea into two phyla, the Crenarchaeota and the Euryarchaeota. A third group, the **Korarchaeota,** consists of organisms from hyperthermophilic environment that have not been cultured. Their properties and relation to other Archaea are inferred from sequencing their genomes.

Crenarchaeota

The Crenarchaeota, sometimes called **thermoacidophiles,** consist of extreme thermophiles (growing over the temperature range of 70 to 113°C) and acidophiles (the optimal pH of some is as low as 2.0). Most of these organisms are found in highly acidic hot springs. This is the group to which the extreme thermophile *P. fumarii* belongs. Most Crenarchaeota in some way metabolize elemental sulfur as part of their fueling reactions. Some are aerobic chemolithotrophs that oxidize sulfur. Others are chemolithotrophs that reduce it to H_2S while oxidizing H_2.

Surprisingly, large numbers of Crenarchaeota have been detected (by molecular methods) in the open ocean, although none of these has yet been cultured. Because of their huge numbers, they must play an important, but still unknown, ecological role in the marine environment. Their metabolism must differ markedly from those of other Crenarchaeota: they are not thermophiles or acidophiles and are unlikely to metabolize sulfur because only minuscule amounts of it exist in the open ocean.

Euryarchaeota

The Euryarchaeota, which make up a larger group of Archaea than do the Crenarchaeota, comprise several physiological types, including the

methanogens (methane producers), extreme **halophiles** (those that grow in the presence of high concentrations of salt), and those that lack a cell wall.

Methanogens

Methanogens are metabolically specialized to a remarkable degree: their only fueling reactions produce methane as an end product. However, they are ecologically diverse: they prosper in all carbon-rich anaerobic environments, from those in the psychrophilic (cold) to those in the hyperthermophilic (very hot) temperature ranges of growth.

Authors' note To test for methane, disturb the bottom of the pond with a stick. Collect the released gas in a water-filled inverted funnel closed with a short piece of pinched-off tubing. Open the tubing, and light the escaping gas.

Bubbles that rise from most quiet ponds contain methane produced by methanogens in the organically rich, anaerobic mud at the bottom, a fact that can be easily verified by collecting the gas and noting that it burns with a blue flame. Gas released when cattle and other ruminants belch also contains methane produced by methanogens in the animals' rumens. Collectively, ruminants produce huge amounts of methane, which account for 10 to 20% of all the methane in Earth's atmosphere. About a third of us humans also produce methane as an intestinal gas as a consequence of our being colonized by methanogens. Those of us who do produce methane do so for life. The rest of us produce hydrogen, one of the substrates for methane production in methanogen-colonized humans. Although they were once incriminated as being carcinogenic, there is no evidence that methanogens harm their human hosts in any way.

In many other environments, including landfills and anaerobic sewage digesters, methanogens produce large quantities of methane. Methane (called natural gas by the petroleum industry) is, of course, a valuable fuel, but it is also a powerful **greenhouse gas,** thereby constituting an ecological hazard by contributing to global warming. This concern is exacerbated by the fact that the total global production of methane is continuously increasing, currently at the rate of about 1% per year.

Methane formation can be viewed as a form of anaerobic respiration (see Chapter 6). In the simplest case (mediated by many methanogens), H_2 is the electron donor and CO_2 is the **terminal electron acceptor:** as eight electrons are transferred from four molecules of H_2 to a molecule of CO_2, one molecule of methane (CH_4) is made, and a transmembrane proton gradient that can be used to generate ATP or to drive other energy-utilizing cellular processes is created.

Among the various species of methanogens, there are three variations on this general theme.

1. An organic compound (formate or one of several alcohols, for example, ethanol, 2-propanol, 2-butanol, or *c*-pentanol) serves in place of H_2 as an electron donor.
2. A methyl-group-containing compound (e.g., methanol, trimethylamine, or dimethyl sulfide) serves in place of CO_2 as an electron acceptor.
3. Acetate serves as both a donor and an acceptor. The acetate molecule is cleaved. Its methyl group, which serves as an electron donor, is oxidized to CO_2, and its carboxyl group, which serves as an electron acceptor, is reduced to methane.

Many of the various substrates for methanogens (CO_2, H_2, formate, acetate, and short-chain alcohols) are abundantly formed from higher-molecular-weight organic compounds by fermentative bacteria that are present in anaerobic environments. By converting these products of fermentation to a gaseous product (methane), methanogens provide a route by which organic materials escape from anaerobic environments. As methane rises into an aerobic environment, some of it is oxidized by **methylotrophs** (methane-oxidizing bacteria) to CO_2. However, as mentioned earlier, much of the methane produced by methanogens escapes into the atmosphere, where it acts as a greenhouse gas.

Methanogens are biochemically unique in that they produce certain cofactors not found elsewhere in nature. The stepwise reduction of CO_2 occurs while attached to these cofactors.

Extreme halophiles

The extremely halophilic Archaea (**haloarchaea**) are another remarkable group. They grow abundantly in near-saturated brine (4 M) and are dependent on it. If their environment is diluted even to 1 M saline (still a high concentration of salt), their protein walls weaken and the cells lyse. Haloarchaea flourish in salt evaporation ponds, which they turn bright red because of the carotenoid pigments they produce. From an airplane flying over San Francisco Bay, evaporation ponds appear as bright-red checkerboard squares (Fig. 15.2). Humans have used the appearance of this red color since biblical times as an index of when to drain off the "**bitterns**" (brine high in magnesium and sulfate) from the crystallized salt that sinks to the bottom of the pond, thereby producing a better-tasting product.

Figure 15.2 Aerial photograph of evaporation ponds in San Francisco Bay. As the salt concentration rises, halophilic archaea proliferate, turning the ponds red.

We might ask how haloarchaea are able to tolerate such high concentrations of salt. Unlike moderately halophilic bacteria that exclude salt from their intracellular environment, haloarchaea avoid plasmolysis by concentrating KCl intracellularly—up to 5 M when growing in near-saturated brine. Necessarily, their proteins are resistant to high concentrations of salt that would denature the proteins in most organisms.

Being strict aerobes, haloarchaea are precariously dependent on a supply of oxygen, a gas with low solubility in their high-salt environment. Some haloarchaea, including the well-studied *Halobacterium halobium*, possess an emergency alternative to aerobic respiration, namely, a primitive system of photosynthesis that generates sufficient ATP, when oxygen becomes limiting, for them to move by aerotaxis closer to the oxygen-richer surface of a pond. Low concentrations of ambient oxygen signal the cell to synthesis specialized regions in its cytoplasmic membrane called **purple membranes.** These regions of the membrane house light-driven pumps that extrude protons, creating a transmembrane proton gradient capable of generating ATP and energizing certain transport mechanisms. These light-driven pumps, consisting of only a single protein (**bacteriorhodopsin**) arranged in a crystalline lattice, are a marvel of photosynthetic simplicity. Bacteriorhodopsin, a transmembrane protein, consists of a protein moiety, **bacterioopsin,** which is covalently bonded to a chromophore, **retinal.** A similar protein, **rhodopsin** (which is also associated with retinal), located in the retinas of animals, is the light receptor in the chain of events leading to vision.

In the purple membrane, light energy deprotonates retinal on the inner surface of the membrane; the proton is then transferred to the bacterioopsin and released outside the cell. Each such cycle pumps one proton out of the cell, thereby generating a proton motive force capable of forming ATP.

BACTERIA

The domain Bacteria is much the larger and more complex of the two prokaryotic domains. However, the perceived difference might be greater than the actual one: new species and groups of Archaea continue to be discovered. As we saw, at present, there seem to be only three major lines of descent in the Archaea, but 23 (called phyla B1 through B23) are recognized in the Bacteria (Table 15.2). The sizes of the various phyla of Bacteria vary enormously. The smallest, Dictyoglomi, contains only a single recognized species. The largest, **Proteobacteria** (which is divided into five classes, alpha through epsilon) is huge (by the usual standards of prokaryotic taxonomy), with 1,300 species. In spite of this enormous variation in size, knowing which phylum a bacterium belongs to says a lot about it. Phylum designation has become the accepted shorthand method—in publications, as well as conversation—for microbiologists to communicate something about an unfamiliar bacterium: "It belongs to this or that phylum." The phylum Proteobacteria is somewhat of an exception because it is so large. Its members are usually described by their "class," i.e., Alphaproteobacteria, Betaproteobacteria, etc.

We will not discuss all 23 phyla, although members of some of them are discussed elsewhere in this book (Table 15.2). Rather, we shall pick and choose, commenting briefly on the general properties of notable representatives of some of the phyla that might otherwise be neglected.

Phylum B4: Deinococcus-Thermus

Phylum B4 contains a species, *Deinococcus radiodurans*, that is remarkable for its resistance to intense gamma radiation and ultraviolet light, as well as to prolonged desiccation. We think of archaea as being the record-setting tolerators of harsh environments, but no other prokaryote—archaeon or bacterium—comes close to matching *D. radiodurans*' ability to survive radiation. A radiation dose of 500 to 1,000 rads will kill the average human; *D. radiodurans* is not harmed by 5 million rads. It can grow while being subjected to constant radiation of 6,000 rads per hour. *E. coli* cannot survive two or three double-strand breaks in its DNA; *D. radiodurans* survives hundreds of them by constantly repairing them. Indeed, *D. radiodurans*' exceptional ability to repair double-strand breaks is probably the basis for its hardiness, including its resistance to prolonged desiccation, a condition that also causes such lethal breaks.

D. radiodurans does produce sufficient carotenoid pigments (compounds known to offer screening protection against radiation) to give its colonies a distinctive bright-red color, and it does have an especially thick cell wall. But its remarkable toughness depends on its ability to repair DNA, not on protecting it from damage. Two mechanisms act to repair double-strand breaks in *D. radiodurans*: single-strand **annealing** and RecA-mediated homologous recombination. The former, as its name implies, repairs double-strand breaks by reattaching DNA fragments with protruding single-stranded ends. RecA identifies an intact piece of DNA that spans a break and uses that piece as a patch to repair it. Neither of these repair mechanisms is unique to *D. radiodurans*, although they might be somewhat more active in it. How, then, can *D. radiodurans* repair double-strand breaks so effectively? Of course, RecA-mediated repair depends on the availability of an intact piece of DNA that spans the break. Its availability might be the basis of *D. radiodurans*' success. Four to 10 copies of the genome are present in each of its cells, larger than the usual number in other bacteria but hardly enough to explain *D. radiodurans*' extraordinary resistance. However, proximity might be the answer. The several genome copies are held together in a highly unusual, possibly unique form: as a ring-shaped object in which homologous regions of DNA are presumed to be adjacent to one another. Also, there appears to be yet another backup for supply of repair DNA to the RecA system. *D. radiodurans* cells occur in tetrads (Fig. 15.3), and after DNA damage, the DNA unfolds and migrates through passages to an adjoining cell in the tetrad.

Figure 15.3 Electron micrograph of a thin section of *Deinococcus radiodurans*.

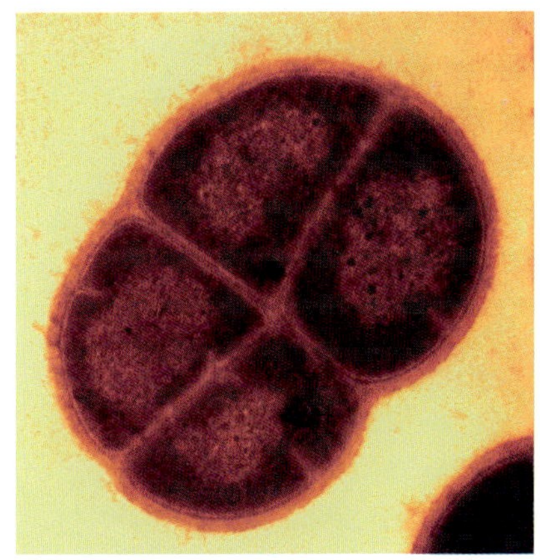

D. radiodurans raises a raft of ecological questions. It is resistant to radiation exceeding that encountered in any known natural environment on Earth. What selective pressures could account for its having evolved such remarkable resistance to radiation? Possibly it was not radiation at all. Prolonged desiccation also causes double-strand breaks in DNA, and as possible support for the notion that this was the selection pressure, *D.*

Table 15.2 Some representatives of various phyla of Bacteria

Phylum			
No.[a]	**Name**	**Properties**	**Representative species or genus**
B1	Aquificae	Gram-negative, nonsporulating, thermophilic rods or filaments	*Aquifex pyrophilus*
B2	Thermotogae	Gram-negative, nonsporulating rods with a sheathlike outer layer, or "toga"	*Thermotoga maritima*
B3	Thermodesulfobacteria	Gram-negative, rod-shaped cells; outer membrane forms protrusions; thermophilic sulfate reducers	*Thermosulfobacterium commune*
B4	Deinococcus-Thermus	Includes gram-positive radiation-resistant cocci and rods, as well as gram-negative thermophiles	*Deinococcus radiodurans, Thermus aquaticus*
B5	Chrysiogenetes	Represented by a single species	*Chrysiogenes arsenatis*
B6	Chloroflexi	Gram-negative, filamentous bacteria with gliding motility. Some are anoxygenic phototrophs; others are chemoheterotrophs.	*Chloroflexus aurantiacus*
B7	Thermomicrobia	Represented by a single species	*Thermobacterium roseum*
B8	Nitrospirae	A metabolically diverse group containing nitrifiers, sulfate reducers, and magnetotactic forms	*Nitrospira marina*, "*Candidatus*[b] Magnetobacterium bavaricum"
B9	Deferribacteres	Heterotrophs that respire anaerobically. Terminal electron acceptors include Fe^{3+}, Mn^{4+}, S, Co^{3+}.	*Deferribacter thermophilus*
B10	Cyanobacteria	Gram-negative unicellular, colonial, or filamentous oxygenic photosynthetic bacteria	*Nostoc punctiforme*
B11	Chlorobi	Gram-negative, anoxygenic photoheterotrophs ("green sulfur bacteria")	*Chlorobium limicola*
B12	Proteobacteria	Largest bacterial phylum, containing 384 genera and 1,300 species. Contains five classes (listed below).	
		Alphaproteobacteria	*Rickettsia rickettsii, Caulobacter* spp., *Rhizobium* spp., *Nitrobacter* spp.
		Betaproteobacteria	*Bordetella pertussis, Neisseria meningitidis, Nitrosomonas* spp., *Zoogloea ramigera*
		Gammaproteobacteria	*Beggiatoa* spp., *Francisella tularensis, Legionella pneumophila, Pseudomonas aeruginosa, Azotobacter* spp., *Vibrio cholerae, Escherichia coli, Salmonella enterica* serovar Typhi
		Deltaproteobacteria	*Desulfovibrio* spp., *Bdellovibrio bacteriovorus*
		Epsilonproteobacteria	*Campylobacter jejuni, Helicobacter pylori*

B13	Firmicutes	Contains low-G+C-content gram-positive bacteria and mycoplasmas	*Clostridium tetani, Clostridium botulinum, Mycoplasma pneumoniae, Bacillus anthracis, Listeria monocytogenes, Staphylococcus aureus, Lactobacillus* spp., *Pediococcus* spp., *Oenococcus oeni, Streptococcus pneumoniae*
B14	Actinobacteria	Contains actinomycetes and mycobacteria	*Corynebacterium diphtheriae, Mycobacterium tuberculosis, Nocardia* spp., *Propionibacterium acnes, Streptomyces griseus, Frankia* spp., *Bifidobacterium* spp., *Gardnerella vaginalis*
B15	Planctomycetes	Gram-negative bacteria. Some reproduce by budding, some have appendages.	*Planctomyces bekefii*
B16	Chlamydiae	Nonmotile, obligately parasitic coccoid bacteria that reside within vacuoles in the cytoplasm of host cells	*Chlamydia trachomatis*
B17	Spirochaetes	Gram-negative, spiral-shaped, flexible bacteria; motile by periplasmic flagella	*Borrelia burgdorferi, Treponema pallidum, Leptospira interrogans*
B18	Fibrobacteres	Gram-negative anaerobes associated with the digestive tracts of herbivores	*Fibrobacter*
B19	Acidobacteria	Gram-negative, aerobic, acid-tolerant heterotrophs and gram-negative anaerobes	*Geothrix*
B20	Bacteroidetes	A phenotypically diverse group of gram-negative bacteria containing aerobic rods, anaerobic rods, and gliding, sheathed, and curved bacteria	*Bacteroides gingivalis, Cytophaga* spp.
B21	Fusobacteria	Anaerobic, gram-negative rods with heterotrophic metabolism	*Fusobacterium* spp.
B22	Verrucomicrobia	Gram-negative, mesophilic heterotrophs. Some produce prosthecae; some reproduce by budding.	*Prosthecobacter fusiformis*
B23	Dictyoglomi	A single genus of gram-negative, rod-shaped, extremely thermophilic, obligately anaerobic, heterotrophic bacteria	*Dictyoglomus*

[a]Numbers are those assigned to phyla in the second edition (2001) of *Bergey's Manual of Systematic Bacteriology.*

[b]Species that have not been cultivated in pure culture are designated "*Candidatus*."

radiodurans has been discovered to be present in the extremely dry valleys of Antarctica, as well as more expected places, such as irradiated canned meats and irradiated medical equipment. Alternatively, perhaps *D. radiodurans* evolved on the primitive Earth when it was subjected to more intense radiation. RNA-based trees of bacteria show *D. radiodurans* as being on a deeply diverging branch on the tree of life, suggesting an ancient origin of this group of microbes.

Phylum B10: Cyanobacteria

Prochlorococcus marinus, a component of the marine phytoplankton, is arguably the most abundant organism on our planet and certainly one of the most ecologically important. It was discovered only recently, in 1986, and probably only then because it is unlike most other cyanobacteria. Its oval cells are small, around 0.6 μm in diameter (which makes it the smallest known oxygen-evolving photoautotroph). A slightly larger (about 0.9 μm in diameter) cyanobacterium than *P. marinum*, *Synechococcus*, is only a bit less abundant.

Cyanobacteria are a highly diverse group, but the cells of most of them are large for members of the Bacteria—several micrometers in diameter—and many are highly differentiated (Fig. 15.4). *Nostoc punctiforme,* for example, displays elaborate differentiation. It forms filaments composed of cells that differentiate to become various forms called heterocysts, akinetes, or hormogonia. In addition, it can form intimate nitrogen-fixing symbioses with certain vascular and nonvascular plants (see chapter 14). Only some cyanobacteria are nitrogen fixers (*Prochlorococcus* is not), but all are oxygen-evolving photoautotrophs. Indeed, one can make a case for cyanobacteria being directly or indirectly the producers of all the oxygen in our atmosphere. Plants and algae (also photoautotrophs) are, of course, major producers of oxygen, but their oxygen-producing capacity resides in their **chloroplasts** (intracellular photosynthetic organelles),

Figure 15.4 Cyanobacteria. (A) A chain of cells of the cyanobacterium *Nostoc paludosum*. Het indicates heterocysts, differentiated cells that carry out nitrogen fixation Bar, 20 μm. **(B)** Electron micrograph of a thin section of a *Synechococcus* sp. This marine cyanobacterium contains abundant membranous structures, typical of many photosynthetic and chemosynthetic organisms. Bar, 100 nm.

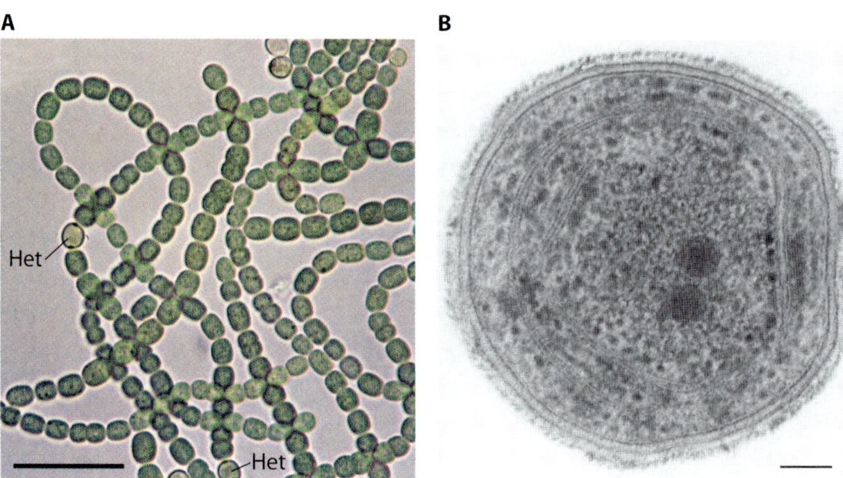

which are modern remnants of anciently captured cyanobacteria. Cyanobacteria have existed and have produced the oxygen of Earth's atmosphere for a very long time. The oldest known fossil, from over 3 billion years ago, closely resembles modern filamentous cyanobacteria (see chapter 11).

In spite of its abundance in the upper layers of the open ocean (where it attains densities of 10^5 cells/ml), discovering *P. marinus* was a challenge. That these waters contained some unknown phototroph was known from spectroscopic analyses, but its identity was elusive. *P. marinus* was finally discovered in the North Atlantic by using a **flow cytometer** (a laser-based laboratory instrument normally used to sort out various kinds of cells).

About half of Earth's photosynthesis (CO_2 fixation and O_2 production) is carried out by marine phytoplankton, which include eukaryotic microbes. *P. marium* stands out not only for its abundance and size. Its considerable metabolic capacities, reflected by its requirements only for light, CO_2, and mineral salts in order to grow, are encoded in a genome of only 1.7 million base pairs containing only 1,716 genes. It might represent the minimal genome for a photoautotroph. Certainly the unchanging nature of *P. marinus'* environment is one explanation for its genomic simplicity. The ocean just does not change rapidly. As a consequence, *P. marinus* has little need for, and very few genes dedicated to, regulatory functions. Specialization might also contribute: there are two groups of strains of *P. marinus*, one adapted to grow in bright light near the surface and the other adapted to grow at very low light intensity at a greater depth. The growth rates of both of these sets of strains are limited by the availability of iron. Their growth rates increase from 1.1 to 1.8 doublings per day when their habitat is fertilized with iron. Such widespread fertilization is a controversial option that some have proposed to moderate the rise of the greenhouse gas CO_2 in Earth's atmosphere.

Synechococcus, Prochlorococcus' close relative and intimate neighbor, has evolved a surprising route to conserve scarce iron by using nickel and cobalt, in place of iron, as cofactors for certain enzymes. *Synechococcus* is highly unusual, possibly unique, in another respect: it swims actively, while lacking flagella, by some as yet unexplained mechanism.

Phylum B12: Proteobacteria

Alphaproteobacteria

Agrobacterium tumefaciens is a genetic invader of broadleaf (dicotyledonous) plants. This ordinary-sized (about 1 by 3 μm), gram-negative, peritrichously flagellated bacterium does something quite extraordinary: it shares some of its genes with its plant host. *A. tumefaciens* attaches to a plant at the site of a wound and inserts some of its genes into the genomes of surrounding plant cells, thereby subverting them to become better hosts. The hosts do this by differentiating to form a protective habitat stocked with nutrients that only *A. tumefaciens* can utilize. The consequence is an ugly tumorlike growth on the plant, usually near the **crown** (the connection of stem and root, where medieval theologians presumed the plant's soul to be located), called **crown gall** (Fig. 15.5). Crown gall does not kill plants (most commonly affected are young plants that are

Figure 15.5 A gall on an oak tree caused by *Agrobacterium tumefaciens*.

members of the rose family, such as apple, peach, cherry, raspberry, and roses themselves), but nevertheless, considerable effort has been expended to control the disease because it reduces the value of nursery stock.

The disease-causing capacity of *A. tumefaciens* resides in a large plasmid, called **Ti** (tumor inducing), that it carries. If *A. tumefaciens* loses Ti, it becomes nonpathogenic and is then commonly called *Agrobacterium radiobacter*. If Ti is transferred to *Agrobacterium*'s close relative, the symbiotic nitrogen fixer *Rhizobium*, it too can attack plants and form galls.

The relationship between *A. tumefaciens* and the plant it infects is surprisingly intimate, even cooperative. The interaction begins when the wounded plant releases (among other compounds) a phenolic compound (acetosyringone) to which *A. tumefaciens* responds chemotactically at concentrations as low as 10^{-7} M. As *A. tumefaciens* moves closer to the wound, where the concentration of acetosyringone is as high as about 10^{-5} M, it induces the expression of certain of *A. tumefaciens*' **virulence factors.** One of these is an **endonuclease** (an enzyme that cuts DNA at locations within the molecule) that excises a piece, called transfer DNA **(T-DNA),** of the Ti plasmid. This excised T-DNA then leaves *A. tumefaciens*, enters a plant cell, integrates into a chromosome, and directs its metabolism for the benefit of *A. tumefaciens*. It causes the plant to produce proliferation-inducing plant hormones, which form the gall, and to synthesize a set of compounds called **opines** (unusual amino acids) and unusual phosphorylated sugar derivatives (agrocinopines), which *A. tumefaciens* uniquely can use as nutrients. In other words, by means of a genetic invasion, *A. tumefaciens* creates a near-ideal habitat for itself at the plant's expense.

Incorporating foreign DNA into the T-DNA of the Ti plasmid of *A. tumefaciens* has become the favored route for genetically engineering plants. The *A. tumefaciens* Ti plasmid has been used to make transgenic plants resistant to insects and herbicides.

Crown gall can be controlled with antibiotics, but such treatments are not economically feasible for control of plant diseases. Fortunately, a biological agent, *A. radiobacter* strain K84, provides an effective preventative treatment. It is only necessary before planting to dip seeds, seedling, or cuttings in a suspension of these bacterial cells. *A. radiobacter* acts by an unusual and highly specific mechanism. It produces agrocin 84 (a toxic analogue of adenine), which is selectively toxic to *A. tumefaciens* because only it can take the analogue into its cells via one of its unique and specific agrocinopin permeases.

Figure 15.6 *Magnetospirillum magnetotacticum.* The dark bodies are magnetosomes.

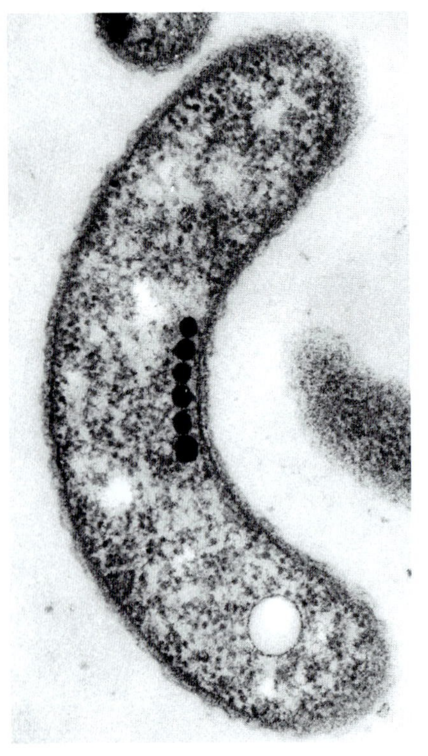

The Alphaproteobacteria also include an aquatic bacterium, *Magnetospirillum magnetotacticum*, that has the peculiar capacity of being able to sense and follow magnetic lines of force (see "Flagellar motility" in Chapter 13). The sensing organelles are **magnetosomes,** which are angular inclusions of **magnetite** (a magnetically sensitive iron oxide) aligned centrally in the cell (Fig. 15.6). By **magnetotaxis,** *M. magnetotacticum* can follow magnetic lines of force, which lead it toward the bottom of a body of water. The usually mud-covered floor contains only low concentrations of oxygen and therefore offers an optimal environment for this bacterium because it is **microaerophilic** (it requires oxygen, but only at low concentrations). *M. magnetotacticum* is not *pulled* to the bottom by the force that Earth's magnetic field exerts on the magnetosome. That attraction is too slight to move even such a small object as an *M. magnetotacticum* cell. Instead, the magnetosome is the cell's compass needle, which aligns the cell with magnetic lines of force. The cell's flagella provide its locomotive power.

Betaproteobacteria and Gammaproteobacteria

We should say something about the Beta- and Gammaproteobacteria because they are huge groups with members that have great impact on human affairs. We have encountered some of them repeatedly throughout this book.

The group Betaproteobacteria contains two important genera of pathogens, *Bordetella* and *Neisseria. Bordetella pertussis* causes whooping cough, which is largely a childhood disease. *Neisseria gonorrhoeae* causes the sexually transmissible disease gonorrhea, and *Neisseria meningitidis* causes meningococcal meningitis.

The group Gammaproteobacteria is even larger. It contains 13 orders, including the *Enterobacteriales, Vibrionales, Pasteurellales,* and *Pseudomonadales.* The *Enterobacteriales,* also commonly called the enteric bacteria, include human pathogens, most of which infect the digestive system. *Salmonella enterica* serovar Typhi causes typhoid fever; *Shigella* spp. cause shigellosis, a form of dysentery; *Yersinia pestis* causes bubonic plague; and the well-studied *E. coli* has a few pathogenic strains. The *Vibrionales* include *Vibrio cholerae,* the species that causes cholera. The *Pasteurellales* include two genera of devastating pathogens, *Pasteurella* and *Haemophilus.* The *Pseudomonadales* include a species, *P. aeruginosa,* that commonly causes infections in weakened hosts, especially burn victims and cystic fibrosis patients.

Phylum B14: Actinobacteria

Streptomyces spp. are important ecologically, industrially, medically, and scientifically. They belong to a group of spore-forming Actinobacteria and are known as **actinomycetes.** They are abundant in soil, accounting for much of the breakdown of organic matter that occurs there. In fact, the pleasant odor of freshly turned soil comes from the volatile compounds (**geosmins**) that these bacteria produce. When a sample of soil is streaked on a simple inorganic medium, many of the colonies that develop are species of *Streptomyces.* They are readily identified by their striking pastel colors: various shades of violet, blue, orange, yellow, and red. You can be almost certain that a colony is a species of *Streptomyces* by touching it with an inoculating needle. The colony sticks together: you can pick up all of it but not just part of it. You can also smell the characteristic soil odor of the plate.

The color and integrity of a *Streptomyces* colony reflect its structure (Fig. 15.7). *Streptomyces* is filamentous. It grows into a fungus-like mycelium, although in cross section, a filament looks like a perfectly ordinary gram-negative bacterium. Part of the mycelium (the **substrate mycelium**) grows into the agar, anchoring it there. On top, the **aerial mycelium** extends upward, and at the tips of its filaments, long chains of spores form. They give the colony it distinctive pastel color. The intertwining of the filaments is what makes the colony stick together.

Figure 15.7 ***Streptomyces.*** **(A)** Pastel-colored orange colonies on an agar plate; **(B)** drawing of a cross section of a colony.

A

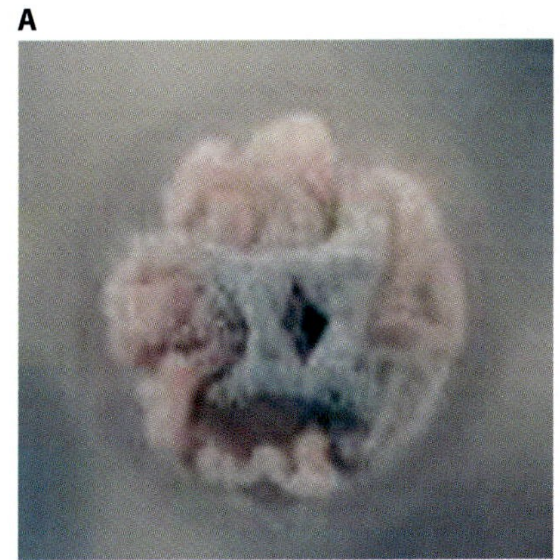

B

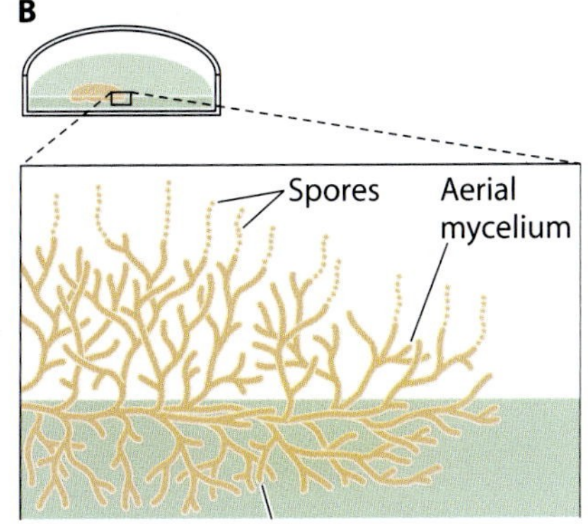

Table 15.3 Some *Streptomyces* species and the antibiotics they produce

Species	Antibiotic
S. aureofaciens	Tetracycline
S. erythreus	Erythromycin
S. fradiae	Neomycin
S. griseus	Streptomycin
S. lincolnensis	Clindamycin
S. noursei	Nystatin

Possibly the greatest impact that *Streptomyces* spp. have on our lives comes from the ability of some of them to produce antibiotics. The majority of all antibiotics in use today (with the exception of the penicillins, which are made by fungi, and some antibiotics made by species of *Bacillus*) are made by various species of *Streptomyces*. Each commercially relevant antibiotic is made by a particular species, perhaps reflecting patent protection more than biological distinction (Table 15.3; also see Table 19.2). Species of *Streptomyces* are also used to produce antiparasitic agents, herbicides, immunosuppressants, and several enzymes used in the food and other industries.

Antibiotics are products of the **secondary metabolism** typical of *Streptomyces*, as well as of fungi and plants. They and other **secondary metabolites** are formed when nutrients become growth limiting. One rationale for production of antibiotics by *Streptomyces* spp. is a desperate attempt to eliminate competition and thereby have the remaining scarce nutrients to themselves. Other microbiologists argue that antibiotics do not eliminate competition because in a natural setting they are ineffective. Some—streptomycin, for example—are tightly bound to clay sols, inactivating them. Also, there is the intriguing coincidence that the major antibiotic producers—*Streptomyces, Bacillus,* and fungi—all produce spores of one sort or another. Curiously, in spite of the huge impact antibiotics have had on human and animal health, the selective advantage to the producer remains unexplained.

The genome of one species, *Streptomyces coelicolor,* which has been the object of intense genetic studies, has recently been sequenced. It is one of the largest bacterial genomes yet sequenced, with 8.5 million base pairs and 7,825 predicted genes, probably not a surprising result in view of *Streptomyces*' complex developmental cycle and abundant capacity for producing secondary metabolites.

CONCLUSIONS

As we admitted at the outset, the diversity of prokaryotes is so vast that we had to pick and choose examples. We've drawn the fundamental distinctions between the Archaea and the Bacteria and discussed representatives of each group. Because the Archaea are a smaller group, we were able to say something about each of the major subdivisions. Not so for the Bacteria. We commented in some detail on a few especially intriguing examples, and we relied heavily on Table 15.2 to give a bare-bones summary of bacterial diversity. Many of the species listed in the table are discussed in other chapters. You can use the table in two ways: use the index to find where in the book species listed there are discussed further, and use the table to determine the relatives of species you might encounter elsewhere in the book.

Now, we shall continue our survey of microbial diversity by examining in Chapter 16 the eukaryotic microbes—the protists and fungi.

STUDY QUESTIONS

1. What are the similarities and differences between a prokaryote and a eukaryote?

2. What facts support the contention that the number of prokaryotic species that exist must be enormous?

3. What are the meaning of and logic behind the taxonomic term *"Candidatus"*?

4. It has been argued that Bacteria and Archaea are prokaryotes but are otherwise unrelated. What is the basis of this argument?

5. How are the Archaea ecologically and biochemically unique?

6. If you saw bubbles rising from a pond, what might you suspect would be at the bottom?

7. How does *D. radiodurans* survive intense radiation?

8. How does the Ti plasmid of A. *tumefaciens* benefit its host?

chapter 16 eukaryotic microbes

INTRODUCTION

Stepping across the great cellular divide of the biological world, we shall now consider the other large group of microscopic organisms, the eukaryotic microbes (Table 16.1). They constitute a very large portion of the living world, challenging the prokaryotes (Bacteria and Archaea) in total mass and diversity. Eukaryotic microbes are larger than most prokaryotes and vary enormously in size and shape. They can be found in a wide variety of environments: in bodies of water and in soil, as well as in and on plants and animals, including humans, both healthy and diseased.

The first eukaryotic cells arose from prokaryotes by acquiring greater structural complexity. They acquired a nuclear membrane, and hence a "true" nucleus, and various organelles. The earliest eukaryotes were undoubtedly microscopic in size. The details of how some prokaryotes became eukaryotes are not known with certainty, especially with regard to the invention of the nucleus, but it is widely believed that major organelles, such as mitochondria and chloroplasts, were acquired by the ingestion of prokaryotic microbes. This topic is discussed in detail in Chapter 19.

Prokaryotes were the sole form of living organisms from about 3.8 billion to 2 billion years ago, when the eukaryotes (Eukarya) emerged. Together, microbes of all three domains (Bacteria, Archaea, and Eukarya) had the planet to themselves until nearly 1 billion years ago, when the first multicellular organisms arose.

The taxonomy of eukaryotic microbes is particularly complex, befitting the large number of organisms in this group. We will not discuss this taxonomy here in detail but will consider them as belonging to two large groups, the fungi and the protists. **Protozoa,** perhaps a more familiar

Table 16.1 Main groups of eukaryotic microbes

Group	Examples of types	Characteristic cell envelope and constituents
Fungi[a]	Yeast, molds	Cell wall containing glycoproteins (e.g., mannoproteins), polysaccharides (e.g., chitin, glucans), other
Protozoa	*Paramecium*, amoebas, *Giardia*, *Plasmodium* (agents of malaria), *Tetrahymena*	Pellicle, interior to plasma membrane; consists of interconnected protein molecules; gives shape to the cell
Algae[a]	*Euglena*, *Chlorella*, diatoms, dinoflagellates	Chrorophyll, cell wall containing cellulose, silica in some, calcium in some

[a]Not all are microscopic, e.g., mushrooms(fungi) or seaweed (algae).

term, are a diverse group of some 50,000 nonphotosynthetic, mostly motile unicellular species. They and the microscopic algae (photosynthetic) are often lumped into a group called the **protists.** The terms protists, protozoa, and algae are well entrenched and still useful, but they have little taxonomic meaning, because the organisms have not evolved as separate, distinct groups. For example, certain algal groups (e.g., *Euglena*) are more closely related to certain protozoa (e.g., trypanosomes) than they are to other algae. Most, but not all, fungi and protists have a typical eukaryotic organization—nuclei, mitochondria, and in those that photosynthesize, chloroplasts. Some, like *Giardia*, the agent of hiker's diarrhea, and its relatives have no mitochondria. It may be tempting to think that these organisms represent descendants of a primitive line of eukaryotic cells that had not yet acquired the endosymbiotic bacterium destined to become a mitochondrion. Perhaps not: the genomes of these organisms contain what are probably bacterial genes. This evolutionary clue suggests that *Giardia* had acquired a premitochondrial endosymbiont and then shed it, though retaining some of its genes. To date, convincing candidates for the progeny of ancestral premitochondrial eukaryotic cells have not been found.

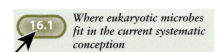

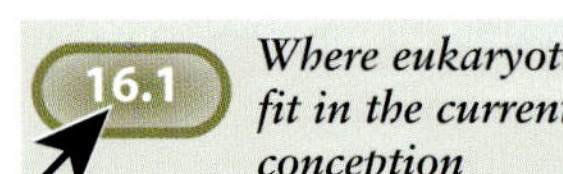

Where eukaryotic microbes fit in the current systematic conception

Some protozoa, such as the sexually transmitted pathogen *Trichomonas vaginalis*, reveal yet another variation on the theme of evolution through endosymbiosis. These organisms contain **hydrogenosomes,** membrane-bound organelles that function differently than mitochondria. Hydrogenosomes convert pyruvate to acetate, CO_2, and H_2. Unlike mitochondria, typical hydrogenosomes contain no DNA. However, the nuclear genes that encode their reactions have bacterial counterparts, suggesting that these organelles may have surrendered *all* their genes to the host cell's nucleus. There is an exception that seems to confirm this inference: a protozoan (*Nyctotherus ovalis*) found in the hindgut of termites (a teeming microbiological garden made up of a large number of bacterial and protozoan species) has hydrogenosomes that do contain DNA, and apparently ribosomes as well. Below, we concentrate on the lifestyles of fungi and two protozoa, *Paramecium*, a well-studied protist, and *Plasmodium*, the parasite that causes malaria, which exemplifies how intricate a protist's lifestyle can be.

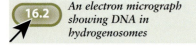

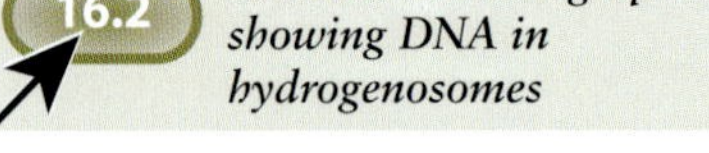

An electron micrograph showing DNA in hydrogenosomes

FUNGI

The term "fungi" evokes delectable dishes made with wild or cultivated mushrooms; foods made with the help of **yeasts,** such as beer, wine, and bread; athlete's foot and other infections; moldy food forgotten in the refrigerator; or mildew on old leather shoes. The fungi are, in fact, as diverse in size and shape as these images suggest. They range from simple unicellular organisms to gigantic bracket fungi in old forests. (One, the largest specimen known, over 1 m across, is *Bridgeoporus nobilissimus*, the "most noble polypore." It is a large shaggy shelf fungus growing from old tree stumps with "the upper surface reminiscent of a green pizza with a crew cut," according to someone. It is not edible.) For all their diversity, all fungi share several properties.

- They are not photosynthetic and depend on preformed organic nutrients for growth. Like many bacteria, their nutritional requirements are usually quite simple.
- They take up nutrients by absorption only, not by phagocytosis.
- They possess cell walls unlike those of plants or bacteria, consisting of **chitin** (a polymer of *N*-acetylglucosamine also found in the shells of crustaceans) and other polymers.
- They can grow vegetatively by extension of filaments without undergoing a sexual cycle.
- They can grow without free water under conditions of high humidity, as witnessed by the mildew that covers walls and leather shoes in damp environments. Such fungi make spores that stick out from the surface and thus can readily disperse in the air, much to the discomfort of those who are allergic to them.
- They have a great propensity to interact with other organisms both in beneficial symbioses and as disease-causing agents.

Most fungi are filamentous (yeasts are the exception). A fungal filament is called a **hypha,** and a collection of hyphae is called a **mycelium.** Commonly, the diameters of both fungal filaments and yeast cells are similar to that of typical vertebrate cells, about 4 to 10 micrometers (μm).

A major activity of fungi on earth is recycling vegetable matter. Fungi are indeed the great decomposers. They can break down many complex organic compounds, including cellulose and lignin, the main components of wood. Termites and other wood-eating insects depend on fungi and bacteria in their guts to breakdown these otherwise indigestible compounds. Were it not for fungal activity, dead plants and trees would accumulate to great depths and become a colossal accumulation of carbon. Without fungal activity, there would not be enough carbon dioxide to sustain plant and microbial photosynthesis. Thus, all animals, which ultimately depend on plant life, would not survive. Without fungi, we would be denied not only porcini risotto, but our very existence.

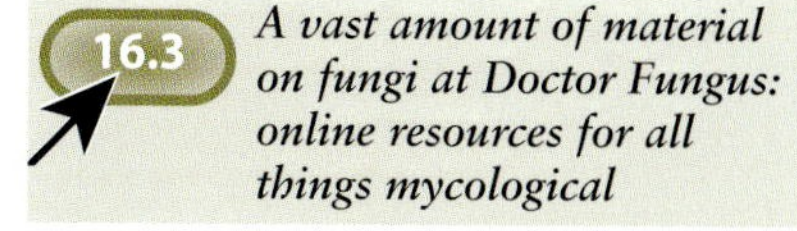

A vast amount of material on fungi at Doctor Fungus: online resources for all things mycological

The yeasts

The best-known fungi are yeasts, which in recent years have become the most intensely studied model for eukaryotic cells. Studies carried out with yeasts have provided us with a detailed understanding of the regulation of gene expression under diverse circumstances and the control of the cell cycle and cell division.

By "yeast," most people mean the common baking or brewing yeast, *Saccharomyces cerevisiae*, which, like most yeasts, replicates by budding (a bud is the bubble-like extrusion of a progeny cell). This is the species

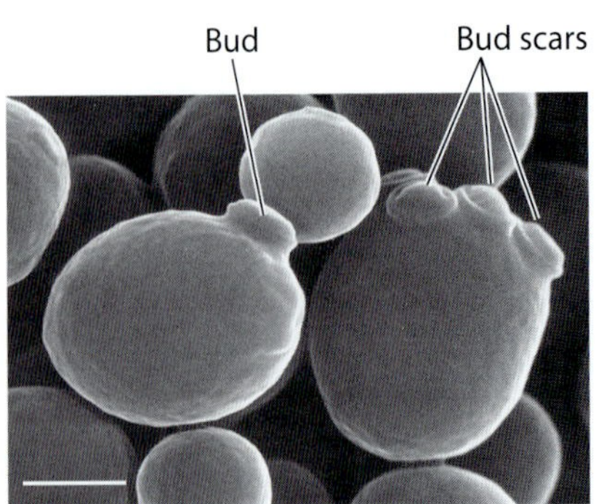

Figure 16.1 Scanning electron micrograph of a yeast cell showing a bud and several bud scars. Bar, 1 μm.

favored by most fungal researchers. There are many other kinds of yeast, including some, such as *Schizosaccharomyces*, that replicate the way bacteria do, by binary fission (splitting in half). Many fungi grow either in the yeast form (single cells) or as filaments, depending on the circumstances. The reasons for this **dimorphism** are not fully understood, but this property most likely extends the organisms' biological repertoire, enabling them to thrive in different environments.

S. cerevisiae grows in simple media containing only a single carbon source, such as glucose or acetate, plus mineral salts. In rich media, it can double every 2 hours. It produces visible colonies on agar plates within 2 days, which makes it highly desirable for experimental work. *S. cerevisiae* grows **aerobically** or **anaerobically.** In the presence of oxygen, it respires, i.e., oxidizes carbon sources to carbon dioxide and water. Anaerobically, it ferments sugars to carbon dioxide and ethanol. Because of the difference in the energy yield between respiration and fermentation, would aerobic or anaerobic conditions be chosen for making, on one hand, beer or wine and, on the other, lots of yeast cells? Brewers, being interested in boosting alcohol production, use anaerobic conditions; yeast suppliers, who need to maximize cell yield, use aerobic ones.

The fungal lifestyle

In budding yeasts, division is **asymmetrical.** The bud starts out as a protrusion from the "mother cell" and enlarges until it and the mother cell are nearly equal in size. At this point, the newly formed cell separates from the mother. Unlike binary fission, budding has a traumatic consequence: at the site where the bud separates from the mother cell, a differentiated little plate, called a **bud scar,** is formed (Fig. 16.1). (The term bud scar is also used to describe the mark left when a leaf of a plant detaches from its stem.) No new bud can form at the site of a bud scar. Notice that the number of bud scars is an indication of the age of a yeast cell. Unlike prototypical "higher" organisms that alternate at prescribed points in their life cycles between **haploid** and **diploid** states, many yeasts can grow in either state for prolonged periods. Some yeasts, such as *Candida albicans* (the agent of **thrush, vaginal infections,** and other types of infections), are always diploid; other species are forever haploid. A variation on the theme is found in most of the mushrooms. Their filaments, which arise by the joining of two gametes, have two nuclei sharing the same cytoplasm. In such cells, called **dikaryons,** the two nuclei coexist without fusing, as if they were detached roommates rather than a committed couple. The cells of a mushroom remain in the dikaryon stage until they are ready to make spores. At that time, the two nuclei fuse and undergo meiosis right away, leading to the production of haploid spores (Fig. 16.2). Note how remarkably short the diploid stage is in these organisms.

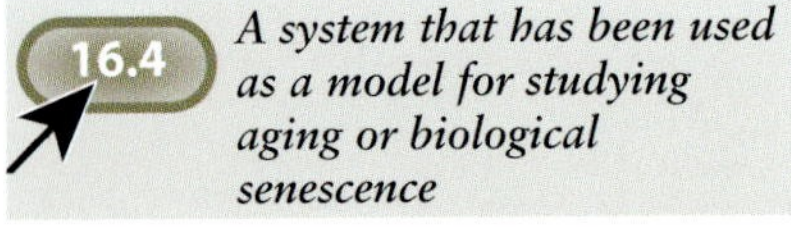

A system that has been used as a model for studying aging or biological senescence

Fungi have no obligatory germ line—no specialized sperm cells and/or eggs. As in higher organisms, fungal **gametes** can mate to make **zygotes** only with cells of the opposite sex. Cells of one sex (also called mating type) go about their metabolic business and do not become involved in sex with cells of the same mating type. They differentiate into gametes only when cells of the opposite mating type are present. What is the mating call? The signals are diffusible molecules called **pheromones**

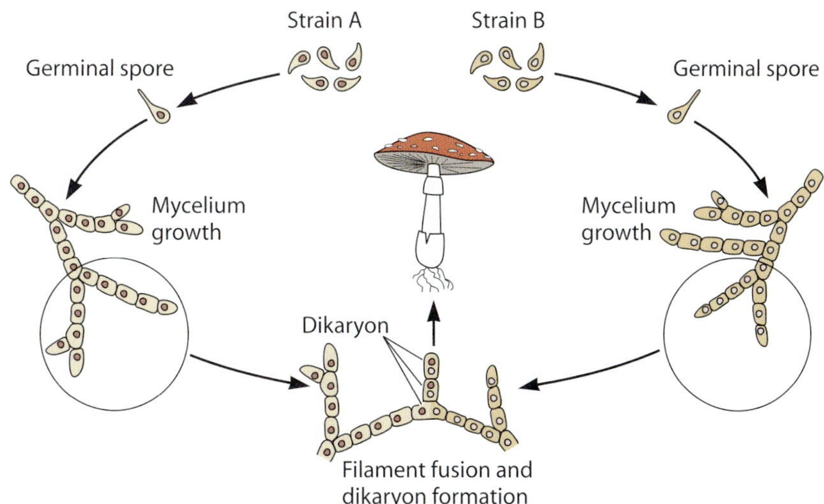

Figure 16.2 How to make a mushroom. Spores of different mating types (shown as having red and tan nuclei) germinate and form filaments (hyphae). The hyphae fuse to make a new filament with *both* kinds of nuclei, called a dikaryon. The dikaryon eventually differentiates to make a fruiting body, which we call a mushroom.

(secreted chemicals that affect behavior). Each mating type secretes its own kind of pheromone into the medium. When a signal from the opposite mating type is perceived, a cell becomes a gamete, i.e., it becomes competent to mate. This is a reciprocal situation: cells of each mating type are equally alerted to the possibility that mating is at hand. When yeast cells become gametes, they stop dividing and elongate to become pear-shaped cells (Fig. 16.3). They then fuse to become a structure that resembles two pears joined at their small ends. The two haploid nuclei then fuse into a diploid nucleus, becoming a zygote.

Yeast mating pheromones are small peptide molecules that diffuse through the environment. When cells of one mating type are grown on agar, a high concentration of the pheromone accumulates around their zone of growth. If cells of the other mating type are placed on this zone, they rapidly transform into pear-shaped gametes. This relatively simple morphologic test for pheromone activity can be used to purify and identify these compounds.

After mating, cells may proliferate in the diploid state for a long time (unlike mushrooms). Only when nutritional conditions become unfavorable do diploids respond by turning into haploid **spores.** Unlike bacterial endospores, which are extraordinarily resistant to heat and chemicals, fungal spores are only partly able to withstand such harsh conditions. However, they survive in nutritionally fallow environments, an ability that helps them endure starvation.

How are spores made? A diploid culture must undergo **meiosis** to make haploid spores. The signal for the onset of meiosis is, as just mentioned, nutritionally poor conditions. Meiosis leads to the formation of four cells, which in yeast are encased in a common envelope, looking something like potatoes in a bag. In other ascomycetes, these cells undergo another division, resulting in a sac with eight spores (Fig. 16.4). This structure, called an **ascus** (plural, **asci**), gives this group of fungi their taxonomic name, the **Ascomycetes.** The Ascomycetes include most yeast and

Figure 16.3 Yeast cells mating. Three hourglass-shaped mating yeast cells are shown together with a round diploid cell.

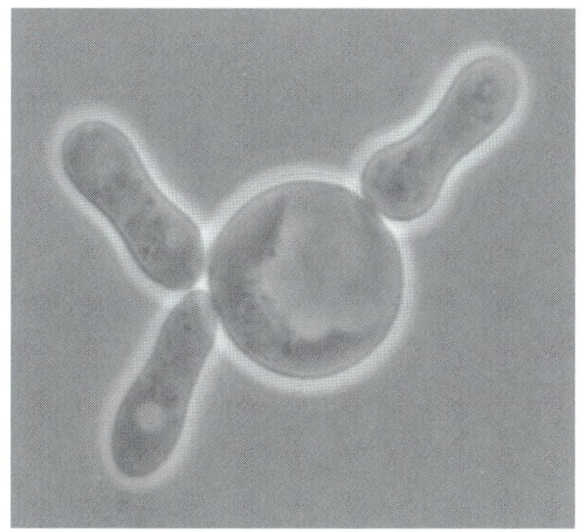

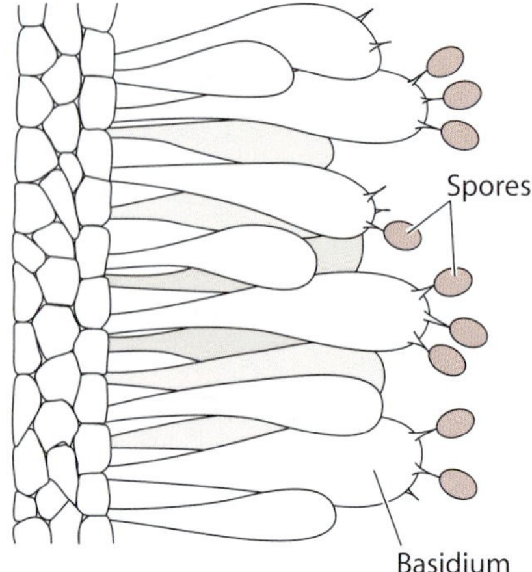

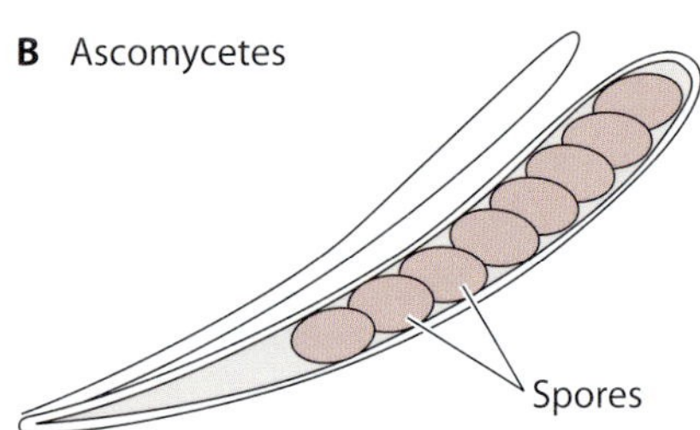

Figure 16.4 Two ways that fungi produce sexual spores. (A) In the Basidiomycetes, which include most mushrooms, the spores bud from a cell called a basidium. **(B)** In the Ascomycetes, which include most yeasts, molds, and a few mushrooms, the spores are encased in a sac called an ascus.

molds, many human and animal pathogens, truffles, and morel mushrooms. Other fungi, including most mushrooms, make a cell called the **basidium,** from whose surface haploid spores emerge by budding. These fungi are called the **Basidiomycetes** (Fig. 16.4).

In a nutritionally suitable environment, ascospores germinate and undergo vegetative growth to become haploid cells. Following Mendelian rules, two of the cells in each ascus will be of one mating type, and the other two will be of the opposite mating type. If a single germinated spore were separated from the others and kept separate, it would be seem to be deprived of sex. Fate is not so unforgiving, because cells of each mating type regularly convert into the other mating type. The process requires a rearrangement of genes in a manner that broadly resembles the generation of antibody diversity. Here is how this works. Each yeast cell has two **silent** genes, one encoding one mating type and other encoding the other (Fig. 16.5). These silent loci are packaged in chromatin structures, which keep them silent by preventing them from being transcribed. However, each of the silent genes can serve as the "cassette tape" that can be inserted into a site (the "cassette player") where the gene can be expressed. DNA recombination replaces the expressed mating-type cassette with a version of the opposite mating type, copied from its silent locus. This mating-type switching during growth results in a colony containing cells of both mating types, even if the population has arisen from a haploid spore of only one type.

Why is yeast such a popular genetic tool?

The lifestyle of yeasts makes them relatively easy to work with, which is one reason for their popularity among researchers. Mating can be induced at will by mixing cultures of different mating types. Each of the four products of a meiosis, which are enclosed in a single ascus, can be separated by using a micromanipulator. Few other eukaryotes can be genetically

Figure 16.5 Switching of mating types in yeast. There are two mating types, called α and **a,** only one of which is expressed at a time. The genes for both are silent, unless a copy of one of them is transferred to a site, *MAT,* where transcription can take place. This is formally similar to inserting an audiocassette in a tape player, which is the reason why this is known as the **cassette model.** With some frequency, the gene that is not being expressed exchanges with the resident one, which results in a switch of mating types.

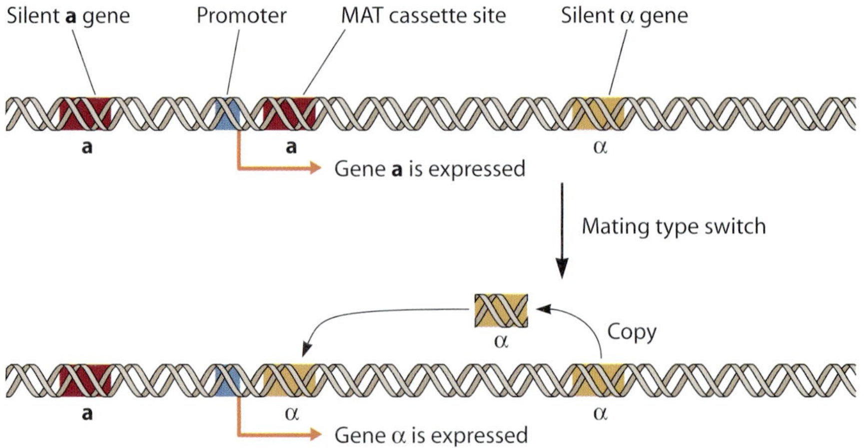

analyzed so rapidly, conveniently, and precisely. There are other advantages to using yeast for genetic research:

- Yeast has a relatively small genome (14 million bases), only about three times the size of *Escherichia coli*'s. The genome is divided into 16 tiny chromosomes averaging 800 kilobases in size that behave in most regards like typical eukaryotic chromosomes. Their small size makes yeast chromosomes attractive models for the study of eukaryotic DNA replication and chromosome behavior. Yeast can be thought of as a stripped down eukaryotic cell that does all of a cell's essential business without the ability to do the fancy stuff (such as differentiate into a neuron or produce antibodies).
- Having stable haploid and diploid phases allows lethal mutations to be expressed (in haploid cells) and maintained (in the diploids). Mutations can then be examined immediately, and lethal ones can be saved for further analysis.
- Foreign yeast DNA can be readily introduced into a yeast cell by transformation. Although the cell wall of yeast is very tough, it can be breached by electric shock (electroporation) or by treatment with certain salts, such as lithium acetate. Once inside, introduced DNA undergoes homologous recombination with high efficiency, which makes it easy to change the structure of any gene at will.
- Constructing plasmids that replicate in yeast is relatively easy. Some even reproduce in *E. coli* as well as in yeast. Such **shuttle vectors** take advantage of both genetic systems, those of yeast and *E. coli*. Plasmids containing remarkably large amounts of foreign DNA can also be engineered. They are called **yeast artificial chromosomes (YACs)**. These properties make yeasts convenient vehicles for the study of DNAs from many sources, including humans. Characterization of eukaryotic genomes usually depends on cloning large chromosomal fragments. It is possible to construct YACs with fragments as large as 800 kilobases. In addition, making proteins by genetic engineering in yeast has several advantages. Being eukaryotic, yeasts allow most eukaryotic proteins to become properly folded and posttranslationally modified, e.g., by glycosylation. The first genetically engineered human vaccine, hepatitis B core antigen, and the first genetically engineered enzyme used in food production, rennin, were produced in yeast.

PROTISTS

Some of the major groups of protists are listed in Table 16.2.

Paramecium

To stir up interest in science, high school teachers often ask their students to look at a drop of pond water under the microscope. Slithering by are all sorts of creatures, going to and fro in apparently haphazard motion. Among the larger ones are particularly active creatures that look like hairy slippers, the paramecia. These creatures fascinate not only teenagers, but

Table 16.2 Some of the major groups of protists

Group	Examples	Means of locomotion	Main mode of nutrition
Flagellates	*Trypanosoma* (African: sleeping sickness; American: Chagas' disease), *Giardia* (hiker's diarrhea), *Trichomonas* (sexually transmitted infection)	Flagella	Absorption (uptake of soluble food)
Amoeboids	*Entamoeba histolytica* (dysentery, abscesses), *Naegleria fowleri* (encephalitis)	Pseudopodia	Phagocytosis (uptake of particulate food)
Ciliates	*Paramecium, Balantidium* (human intestinal infection)	Cilia	Ingestion (of particulate food via a mouth-like organ)
Apicomplexa (so called because they produce a structure called an apicoplast)	*Plasmodium* (malaria), *Toxoplasma* (toxoplasmosis)	Nonmotile (except for some stages in the life cycle)	Absorption (uptake of soluble food)

professional biologists as well, because they are among the largest and most complex single-celled organisms and they carry out their physiological and genetic transactions in intriguing ways. One can argue about what constitutes large size in cells. After all, an ostrich egg is a single cell. However, it does not move about like a *Paramecium* and it is programmed for a single mission. Paramecia and their allies, other unicellular giants called the **ciliates,** are well adapted to a free life. They are found not only in pond water, but also in the oceans, lakes, rivers, and soils, where they live by eating bacteria and other smaller organisms. Paramecia are grazers and live by ingesting particles such as bacteria. But even paramecia can be ingested and become part of the food chain (Fig. 16.6). *Paramecium caudatum* will serve here as an example of the world of protists.

For unicellular organisms, paramecia are enormously complex. They have specialized structures that resemble, at least vaguely, a mouth, a gullet, and an anus. Because they may accumulate too much water, they have kidney-bladder combinations called **contractile vacuoles** that expel excess liquid. The fact that they consist of single cells is puzzling, because other organisms of about the same size are composed of many cells (Fig. 16.7).

Nobody has yet given a convincing argument for this refusal to adopt the more popular solution and partition biological functions among different cells. Paramecia are so big (although not especially so for a ciliate) that one could expect them, hippopotamus-like, to go about their business in a deliberate and unhurried manner. (Not so. One paramecium can divide every 10 hours, and some of its relatives can divide every 2 hours.) For these large and complex cells, rapid growth represents special challenges. In an environment that is replete with ingestible and digestible bacteria, nutrition may not be a problem, as food is plentiful. Even so, rapid growth requires that a lot of things fall into place. The cell must

Figure 16.6 Predation among ciliates. Scanning electron micrographs of a ciliate protozoan, *Didinium*, swallowing another ciliate, *Paramecium*, are shown. **(Left)** Side view of the initial stages; **(right)** top view showing the prey being almost completely ingested.

double every one of its organelle-like components in the time it takes to divide. This requires that the machinery involved in duplicating components work at high efficiency. In turn, this translates into requiring a lot of enzymes to make nucleic acids, proteins, carbohydrates, and lipids. Incidentally, **RNA enzymes,** or **ribozymes**—the exception to enzymes being proteins—were discovered in a relative of *Paramecium* called *Tetrahymena*. So was RNA splicing.

The problem then becomes how a paramecium makes a lot of enzymes efficiently. This requires a lot of mRNA, and there is a limit to how rapidly mRNA can be made from DNA. Ciliates solve this problem in a way unique in biology: they make a large extra nucleus, called a **macronucleus.** This is a bag of repeatedly copied selected genes that are excised from the "real" nucleus (the **micronucleus**). The macronucleus contains many copies of these genes, between 40 and 1,000, depending on the ciliate. The micronucleus, then, serves as the information repository of the cell, a safe where genes can be securely stored. The macronucleus, on the other hand, is a busy place, with molecular copying machines buzzing. The macronucleus contains the genes needed for growth. In some species, they constitute no more than 15% of the total present in the micronucleus; in others, they include up to 90%. Forming a macronucleus from a micronucleus is a complex process involving a selective reduction in the cell's genetic inventory: because the genes to be included in the macronucleus are not contiguous in the chromosomes of the micronucleus, they must be reshuffled. As the keeper of information for future generations, the micronucleus must duplicate with extreme precision by mitosis. Not so with the macronucleus. It has so many copies of each gene that they need not be partitioned precisely. The macronucleus does not divide by mitosis, as the micronucleus does, but by simply dividing in half, like a piece of chewing gum being pulled apart.

Paramecia possess other interesting features. They swallow any particle in their vicinity, which includes other paramecia. In some instances, the swallowed ones defend themselves by carrying a bacterial endosymbiont

Figure 16.7 *Paramecium caudatum*. This complex single-celled organism has structures involved in food uptake (gullet), food digestion (food vacuoles), secretion (contractile vacuoles), locomotion (cilia), safeguarding of the genome (micronucleus), and using genetic information (macronucleus).

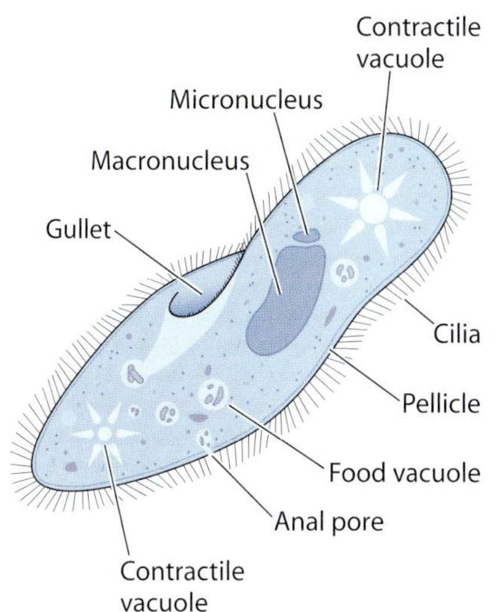

in their micronuclei. When these paramecia are ingested, the bacteria release a toxin that kills the assailant. The killer strains that carry such endosymbionts are immune to the toxin. Interestingly, several levels of biological interactions operate here: the toxin of the bacterial endosymbiont is encoded by a defective prophage (see Chapter 17).

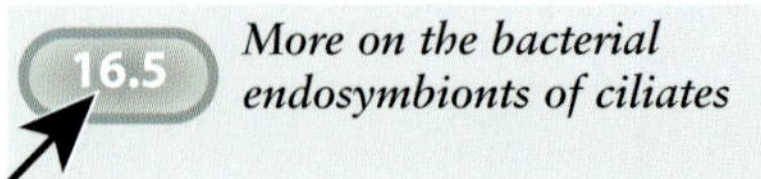

More on the bacterial endosymbionts of ciliates

Paramecia and certain other protists do not divide indefinitely: they die after a number of cell divisions. This phenomenon of **senescence** is reminiscent of animal cells. In the case of *Paramecium*, senescence is not caused by shortening of telomeres (the ends of linear chromosomes), as occurs in cells of vertebrates. The cause of senescence in protists is not known. How come paramecia do not die out? It turns out the clock is reset when the organisms undergo sexual reproduction; thus, these species can endure as long as they occasionally engage in sex. How this works is a mystery.

Yet another contribution of ciliates to our understanding of genetics and evolution comes from the phenomenon called **cortical inheritance.** This refers to the inheritance of surface properties *without the participation of genes*. Mating in paramecia involves cell fusion, and later, separation of the conjugating cells. The separation is not always perfect: occasionally one of the partners will pick up a patch of the cortex ("skin") from the other. The cortex carries cilia that are oriented in a certain way. The ones in the new patch will be oriented in the opposite sense. This is important, because cilia allow a paramecium to swim. If enough cilia are in the wrong orientation, swimming will be aberrant. This natural incident can be reproduced experimentally by surgically reversing part of the cortex. The point here is that the new pattern of cilia, and therefore the altered swimming behavior, are inherited. This kind of inheritance does not involve any change in genes; it is dependent entirely on the local geometry of the cortical patch involved. We call such a phenomenon **epigenetics** (see Chapter 17 for a similar situation involving prions).

Plasmodium, the parasite that causes malaria

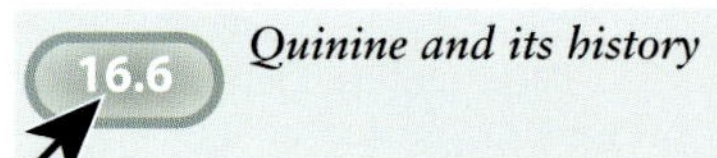

Quinine and its history

Malaria has been and continues to be one of the great scourges of humankind. It affects some 300 million people, especially in tropical areas of the world, and causes between 1 million and 1.5 million deaths per year; it ranks among the major deadly infectious diseases. It is transmitted by biting mosquitoes found mainly, but not only, in tropical regions. These mosquitoes fly not much further than 2 miles, an important consideration in attempting their control. To a large extent, the disease can be controlled by sanitary measures, such as the use of insecticides, drainage of pools of water where mosquitoes develop, and mosquito netting. Such measures are expensive on a large scale and beyond the means of some countries in the developing world. Several drugs can prevent and treat malaria. Some, such as quinine, have been used for centuries. However, some of the most virulent species of the parasite have become resistant to many antimalarial drugs, and insecticides are now ineffective because mosquitoes develop resistance to them.

Human malaria is caused by four species of the genus *Plasmodium*, the most virulent one being *Plasmodium falciparum*. *Plasmodium* parasites alternate between two obligatory life stages, one in the mosquito—their only vector—and the other in their vertebrate host. *Plasmodium* goes through an extraordinarily intricate choreography, including a mandatory sexual phase, to complete its life cycle. The parasite has a dozen or so distinctive stages, each of which has been given a name. We will discuss only a few of these.

As with most bloodsucking insects, only female mosquitoes feast on vertebrate hosts. Unlike the males, females need a rich source of protein for making their eggs. When feeding on a malaria-infected host, the mosquito ingests parasitized red blood cells. The parasites proliferate in the mosquito's mouth, so that as it bites another host, it delivers an infective load of parasites. However, matters are not so simple. The parasites must undergo a complex series of developmental changes before they reach an infective load to cause disease (Fig. 16.8).

The blood meal that the mosquito acquires from the infected host contains parasites in several stages of development, but only the so-called **gametocytes** differentiate into **gametes** and invade the insect. Once in the mosquito's gut, the gametes exit the red blood cells, and the male gametes

Figure 16.8 Life cycle of the malaria parasite. Parasites **(sporozoites)** released from the salivary gland of a female mosquito are injected into a human **(1).** They travel through the bloodstream and enter the liver **(2).** There, they mature into cells called tissue **schizonts (3).** They are later released back into the bloodstream **(merozoites [4]),** where they invade red blood cells **(5).** However, some parasites remain dormant in the liver **(2).** In some forms of malaria (due to *Plasmodium vivax* and *Plasmodium ovale*), these cause relapsing fever and chills. In the red blood cells, the parasites mature into ring-shaped forms **(6)** and other asexual stages (trophozoites and schizonts **[7** and **8]).** When fully mature, the parasites lyse the red blood cells and are released to invade uninfected red blood cells **(9).** Within the red blood cells, some parasites differentiate into sexual forms (gametocytes) **(10).** When taken up by a mosquito, the gametes further differentiate **(11)** and mate to produce a zygote **(12),** which penetrates the mosquito's gut **(13)** and develops into a structure called an **oocyte.** In time, oocytes produce cells **(14)** that migrate to the mosquito's salivary glands **(1)** and repeat the cycle.

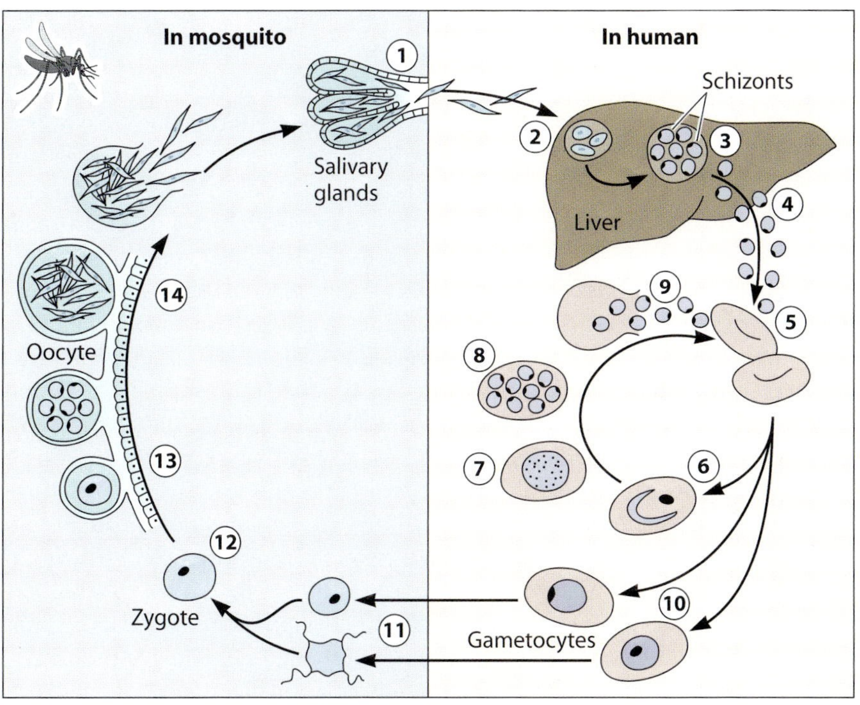

give rise to eight wildly motile, sperm-like cells that seek out female gametes. Within 30 minutes of entering the mosquito, mating is complete and diploid **zygotes** are formed. The zygotes then differentiate into motile cells that invade the mosquito gut. During this process, meiosis takes place to yield haploid cells. These divide to produce tens of thousands of offspring that are released into the hemolymph (blood) of the mosquito. In this form, the parasites invade the salivary glands and are ready to be injected when the mosquito bites the next person or animal.

In humans, *Plasmodium* rapidly travels through the bloodstream to the liver. Depending on a number of factors, including the immune state of the host, it either multiplies rapidly within liver cells or remains dormant but able to reactivate to cause relapsing bouts of malaria at a later time. When parasites are released from the liver, they enter the circulation and infect red blood cells. The detour through the liver may have evolved so the parasites can persist in the body.

After the parasites multiply, the infected red blood cells become stiff. The spleen recognizes them as being "old" and destroys them. However, the parasites have evolved a protective mechanism to avoid being destroyed along with the cell. They induce knob-like structures on the red blood cells that make them adhere to the surface of the blood vessels and keep them from ending up in the spleen. Despite this, many red blood cells are lysed, not all of which are infected. Many uninfected ones are also lysed, which suggests that infection by *Plasmodium* may trigger some autoimmune mechanism that causes the symptoms of malaria. (The subject of immunity in malaria and the possibility of creating a vaccine is intricate and will be left to other sources.)

Coinciding with the lysis of blood cells is the typical symptom of malaria, namely, chills that can be so severe as to lead to violent shaking and loud chattering of teeth, even in hot weather. This is usually followed by a feeling of great weakness, throbbing headaches, vomiting, and high fever. The patient then breaks out in a drenching sweat until the fever subsides. It is thought that red blood cell lysis releases parasite-produced molecules that induce the production of fever-inducing cytokines. This clinical picture is repeated, sometimes quite regularly every few days (depending on the species of *Plasmodium*), because the multiplication and release of the parasites takes place with a certain degree of synchrony. The destruction of red blood cells can be so great that vast amounts of hemoglobin are excreted, giving the urine a deep blackish-red color (hence the name of this condition, "blackwater fever").

One of the best-studied aspects of the *Plasmodium* life cycle is its penetration into the red blood cells. These cells are not phagocytic; they must be coaxed into allowing *Plasmodium* to penetrate them. The initial interaction between parasite and red blood cell consists of binding mediated by ligand-receptor interactions. This is followed by a reorientation of the parasite, so that its "apical," or pointed, end is in contact with the host cell surface. *Plasmodium* species belong to a group of protists, the Apicomplexa, that have a special DNA-containing organelle called the **apicoplast** (see also Chapter 19). The red blood cells have a two-dimensional submembrane **cytoskeleton** that impedes phagocytosis, and it must be disrupted in order for the parasite to enter. The parasites exploit these cytoskeletal proteins: they glide along their surfaces and, in a sense, "crawl" inside the red blood cells. Some of the *Plasmodium* parasites that infect new red blood cells differentiate into gametes that begin the life cycle anew.

What forces selected for such an intricate lifestyle? We do not really know. Unlike most other human protozoan parasites, they have a sex life. Sex is complicated and requires differentiation into gametes, mating, and meiosis. These parasites invade two kinds of human cells, those of the liver and the red blood cells, and this aspect of their life cycle must also contribute to their survival and transmission.

Other parasites, including relatives of *Plasmodium*, do not follow such a complicated design. Trypanosomes, which cause sleeping sickness in Africa, for example, do not undergo marked morphological changes in the insect or in the mammalian host. Other trypanosomes that cause Chagas' disease in South America and the more distantly related *Leishmania* species (the agents of leishmaniasis, a disease that affects both deep tissues and the skin) have only two morphologically distinct forms, one in the insect and one in the human. Why such disparate lifestyles have evolved evades a simple explanation. However, only a few malarial plasmodia enter the bloodstream after the insect bites, whereas trypanosomes and *Leishmania* cause a local lesion at the site of the insect bite, proliferate, and enter the circulation in larger numbers.

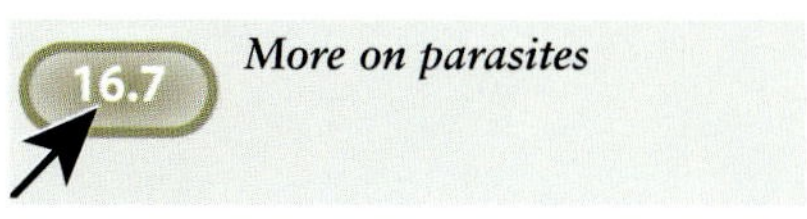

More on parasites

One may think that the multifaceted life of *Plasmodium* would give great opportunities for interfering with one or another stage with drugs or vaccines. Although much progress has been made, it has not as yet been sufficient to control the disease in many parts of the world.

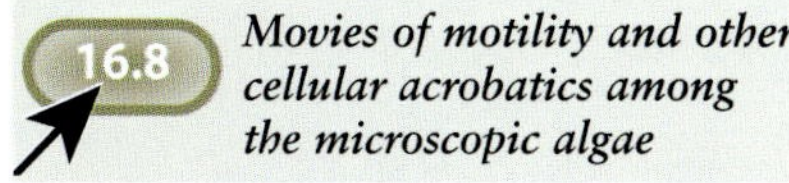

Movies of motility and other cellular acrobatics among the microscopic algae

Diatoms and others

In the futuristic writings of H. G. Wells or an episode of *Star Trek*, we encounter alternative forms of life based on silicon rather than carbon. Why not? Silicon and carbon share chemical properties, including the ability to form polymers. But there are problems with silicon, such as the need for a greater amount of energy to break an Si—O bond than a C—O bond. In addition, oxidized silicon makes an insoluble substance (silicon dioxide) rather than a gas (carbon dioxide). Thus, a silicon-based organism would either have to be anaerobic or make frequent deposits of silicon dioxide "bricks." We will not go that far but will describe organisms on Earth that make good use of silicon. Silicon is one of the most abundant elements on Earth. In the form of silicon dioxide or silica, it is the main component of glass. It is also used as structural protective material by certain protozoa, sponges, and plants, and, most conspicuously, by all diatoms.

Silicon is also found in the bones and connective tissue of animals. An adult human contains about 10 g of silicon, but we focus here on the diatoms, organisms that play a vital role in aquatic environments. The oceans are full of exciting forms of life that cannot be seen with the naked eye. Collect the small particles that float in the sea, and you would observe, under the microscope, in addition to many prokaryotes, an array of eukaryotic microbes of great diversity and endless beauty. Many display a body structure remarkable for its eye-pleasing geometric arrangement. Some of the organisms are round, others oblong, and yet others are joined in chains. These organisms make up the **plankton**, an essential

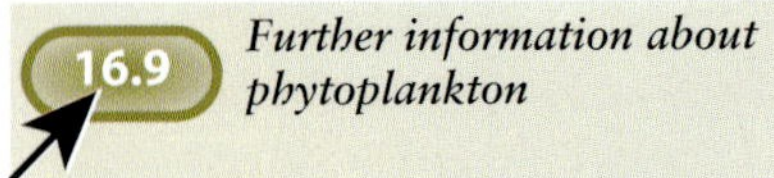

Further information about phytoplankton

element of the oceans' food chain. Examining a drop of pond water is a different but equally rewarding experience.

Diatoms are one-celled algae. They are extraordinarily abundant both in the plankton and in the sediments in marine and freshwater ecosystems. Diatoms either lead a solitary existence or are joined together in chains of various lengths. Some species are capable of active movement on surfaces. Individual diatoms range in size from 2 μm to several millimeters, although very few species are larger than 200 μm. They are widely varied: the species extant today number over 50,000. A striking feature of the diatoms is their cell walls, which are made of silicon dioxide (in other words, glass) enveloped by organic material. These shells have complex geometric patterns and are some of the loveliest forms in nature (Fig. 16.9). People who have observed them tend to wax eloquent and call them by such names as "nature's gems" or "plants with a touch of glass."

Why silica? The design and material of the diatom cell wall makes for a very tough structure that is resistant to mechanical breakage. However, the silica in diatom shells is far from inert. It is thought to speed up photosynthesis, possibly by acting as a buffer to keep the pH within an optimal range. The ornate designs in the shells may also facilitate photosynthesis because their pores and indentations increase the amount of surface exposed to water and carbon dioxide. Diatoms make further use of silica. Silicon activates a variety of genes, including the one coding for the diatoms' DNA polymerase.

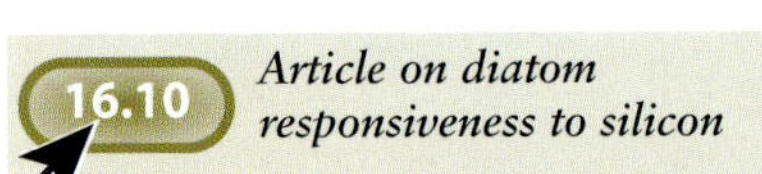

Article on diatom responsiveness to silicon

Diatoms are photosynthetic and are thought to be responsible for 20 to 25% of all organic carbon fixation on the planet. Because they are so plentiful, they are both an important food source for marine organisms and a major producer of oxygen for our atmosphere. Not all diatoms float freely, though; many cling to surfaces, such as those on aquatic plants, mollusks, crustaceans, and even turtles. Some whales carry dense growths of diatoms on their skin. Diatoms are an important element of the food chain. Many protists and small plankton consume the smaller diatoms whole, but some invade large diatoms and eat them out of their shells. Even though diatoms get eaten, their shells are almost indestructible and accumulate over geological time, going back at least to the Cretaceous period, to make enormous deposits. White chalky rocks consisting almost entirely of fossil diatom shells are known as **diatomite** or **diatomaceous earth.** These deposits are mined commercially to make abrasives, cleansers, and paints and for filtering agents for various liquids,

Figure 16.9 Examples of diatoms. Scanning electron micrographs of differently shaped diatoms are shown.

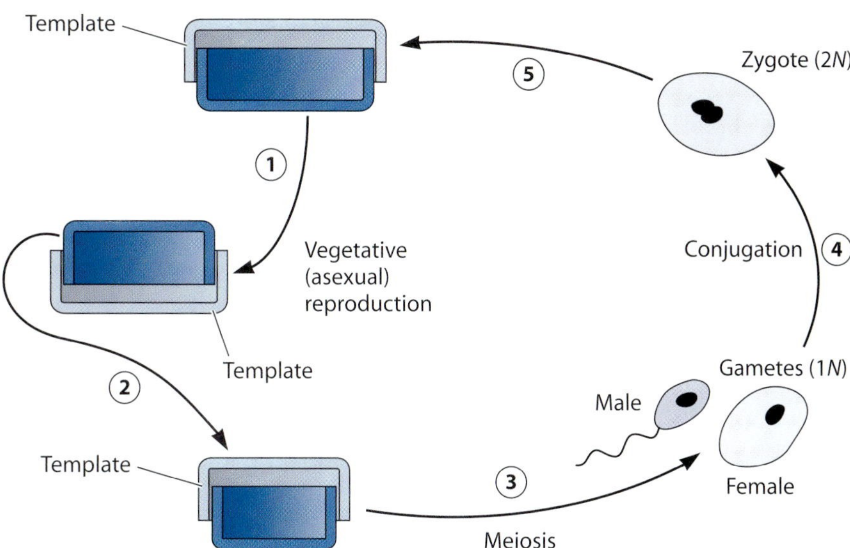

Figure 16.10 Life cycle of petri dish-shaped diatoms. Upon asexual reproduction **(1),** the "top lid" shell of a cell serves as a template for the "bottom." In turn **(2),** this bottom becomes a top and serves as the template for another bottom. With each division, the cells become encased in a smaller shell (blue). When the shell size becomes too small, the cells exit, undergo meiosis, and become gametes **(3).** These can fuse to make a zygote **(4),** which can grow a shell to start the process again **(5).** For convenience, the alternating half shells are drawn as light and dark, and only a few of the divisions are shown.

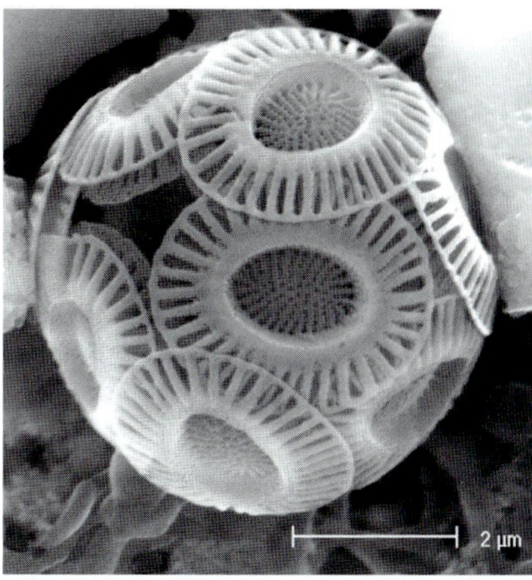

Figure 16.11 A coccolithophore. Shown is "Ehux," short for *Emiliania huxleyi,* the most abundant of all coccolithophores. Under favorable conditions, it makes marine blooms the size of England, outnumbering all other members of the phytoplankton by as much as 10:1. The source of their beauty is the **coccoliths,** platelets of calcium carbonate that cover the cells and give the organisms their name. The blooms are highly reflective, causing more light and heat to be reflected into space instead of heating the ocean. The construction of huge numbers of coccoliths and their sinking to the ocean floor make a difference in how CO_2 can be stored in the atmosphere to contribute to the greenhouse effect. However, the organisms not only influence climate, they eventually form chalk and limestone rocks, for instance, the white cliffs of Dover in England.

including wine. Brushing one's teeth puts one in contact with diatoms because some types of toothpaste contain processed diatomaceous earth. Because individual diatom species thrive under different climatic conditions, the analysis of fossil diatoms provides information on past environments.

The shells of some diatoms are made up of two unequal parts that fit into one another, an arrangement that in round species resembles a pillbox or a petri dish. This results in an unusual mode of replication, where each half serves as a template for a new shell (Fig. 16.10).

Diatoms are distantly related to other photosynthetic members of the plankton. One large group, the **coccolithophores** (round stone bearers), is so named for their calcium carbonate cell walls. They rival the diatoms in their fanciful shapes (Fig. 16.11). They make huge "blooms" that can reach enormous sizes, covering as much as 100,000 km^2 of ocean surface (almost the size of England, or about the size of the state of Ohio). Coccolithophore blooms have been described as giant chemical factories, responsible for a large proportion of all photosynthesis. Coccolithophores periodically shed their tiny scales, called coccoliths, the world's greatest source of carbonates. Coccoliths turn the normally dark water a bright, milky blue, making coccolithophore blooms easy to see in satellite imagery. Despite their enormous importance in the cycle of matter, coccolithophores are little known and have not gotten their "day in the sun."

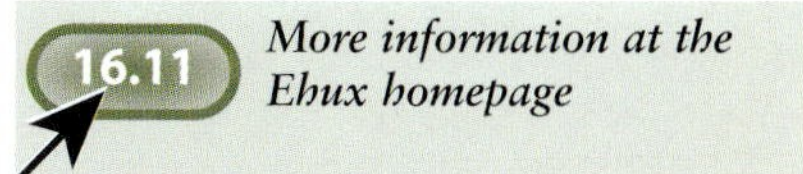

16.11
More information at the Ehux homepage

Other cousins of the diatoms make a rigid cell wall made up of cellulose or other organic polymers rather than minerals. These are the **dinoflagellates,** some of which cause algal blooms, called "red tides" (Fig. 16.12). Dinoflagellates make an interesting spectacle at night: bioluminescence is

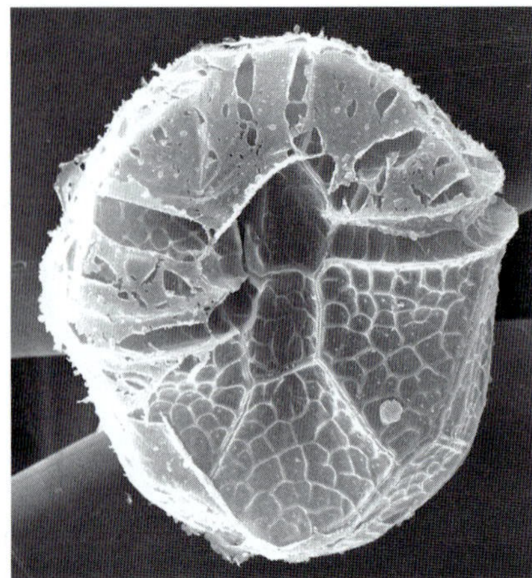

Figure 16.12 A freshwater dinoflagellate, *Peridinium willei*, seen under a scanning electron microscope. Note the armor-plated appearance due to a cellulose-containing cell wall and a characteristic groove known as the cingulum. Most dinoflagellates are marine, but some are also found in freshwater lakes and rivers. About half of them are photosynthesizers; the others are heterotrophs. When present in large numbers, they produce red tides that are poisonous to vertebrates.

oxygen dependent, and the presence of the organisms is revealed by flashes of light on the crest of the breaking surf or the wake of a ship. Some dinoflagellates produce potent neurotoxins that can be transmitted through the food chain to affect and sometimes kill shellfish, fish, birds, marine mammals, and even humans.

There are many questions peculiar to the diatoms, regarding their unusual use of silicates, their varied shapes, their peculiar mode of reproduction, and others. Perhaps readers of this book may someday become motivated to answer some of these questions. Keep in mind that diatoms can be cultured in the laboratory.

CONCLUSIONS

Eukaryotic microbes comprise a huge number of life forms that vary greatly in morphology and size. Many play essential roles in the cycles of nature; some cause disease in humans, animals, and plants, and some are the subjects of intensive research. Yeasts serve as a model for all eukaryotes and rival *E. coli* for the position of the best understood of all organisms. Others, especially certain ciliates, have greatly amplified our understanding of molecular biology, permitting the discovery of such unexpected phenomena as gene splicing and some RNAs acting as enzymes. We can expect more from the study of these fascinating organisms. We are only beginning to understand their roles in the major biogeochemical cycles and the recycling of carbon. In the next chapter, we continue with the theme of biological diversity and consider another central ingredient of the biological world, the viruses.

STUDY QUESTIONS

1. Do the main groups of eukaryotic microbes fall into readily distinguishable categories? Name some examples.

2. What are the central characteristics of the fungi?

3. On what basis do researchers favor yeast as a model system for eukaryotic cells?

4. In what ways do the micronucleus and the macronucleus of *Paramecium* differ?

5. By what means do malaria parasites harm their host?

6. What are the most readily distinguishing characteristics of diatoms and coccolithophores? What role do these organisms play in the cycles of nature?

chapter 17 viruses, viroids, and prions

INTRODUCTION

If viruses did not exist, could anyone have conjured them up? They are inert particles that commandeer cells for their own advantage and in doing so cause profound changes in the cell's metabolic and genetic activities. Could one imagine small infective particles that multiply by losing their bodily integrity, by coming apart inside a host cell? Could anyone have predicted that they would influence all forms of life, that they would be spectacularly varied in shape and size, and that they would be present in very large numbers in the environment? Viruses play a central role in biology: they have shaped evolution, are constantly involved in ecological relationships between living organisms, and of course, cause disease. Yet the viruses do not appear in the conventional depictions of the "tree of life" (e.g., see Fig. 1.2). The reason is that even though they possess some of the qualities of living things, such as the ability to replicate, mutate, and recombine, they are not organisms.

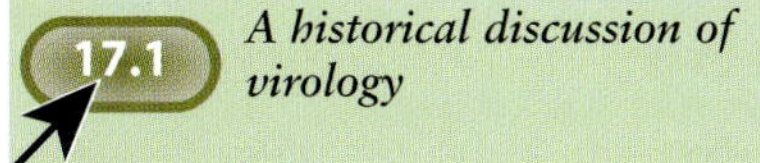

A historical discussion of virology

SIZE AND SHAPE

Viruses infect a great variety of hosts and vary greatly in size, shape, and chemical composition. (Fig. 17.1) The largest known virus is as large as the smallest bacterium and has several thousand times the volume of the smallest virus. Some viruses look like sticks, some like geodesic domes, and others like lunar landers, but all these shapes have a common structural basis, namely, the arrangements of components in the protein shell that all viruses possess.

A virus particle consists of nucleic acid, either DNA or RNA (never both), surrounded by a protein shell. The particle itself is called a **virion,** the shell is called a **capsid,** and the capsid-nucleic acid complex is called

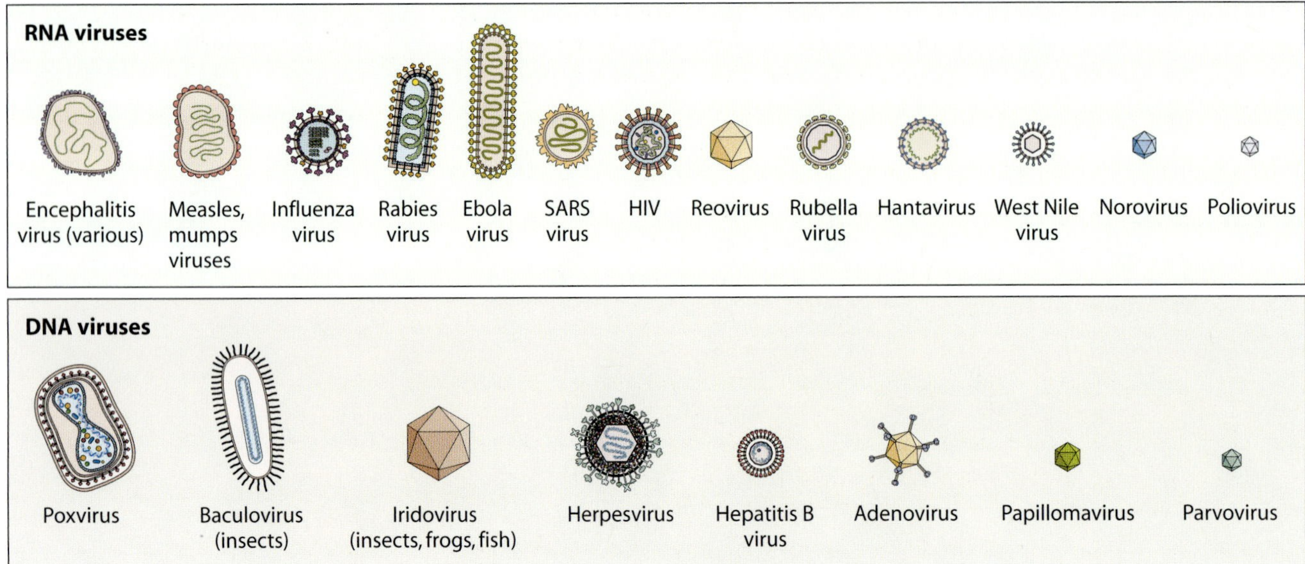

Figure 17.1 Examples of shapes and sizes of viruses. Note the wide range of shapes and sizes. All viruses are drawn to scale.

a **nucleocapsid.** The structural subunits of the capsid are known as **capsomers,** made up of a single or several types of proteins. In the simplest and smallest viruses, capsomers are arranged to form one of two structures: an **icosahedron,** made up of 20 equilateral triangular faces, or a **helical filament,** made up of capsomers arranged as a spiral with a hollow core (Fig. 17.2). Both shapes are determined by built-in properties of the capsomers: isolated capsomers can self-assemble (crystallize) spontaneously into the shape of the virion even in the absence of nucleic acids. The sizes of icosahedral viruses are determined by their self-assembly property alone. The lengths of filamentous viruses, on the

Figure 17.2 Basic viral forms. (A) Icosahedral, nonenveloped; **(B)** icosahedral, enveloped; **(C)** helical, nonenveloped; **(D)** helical, enveloped.

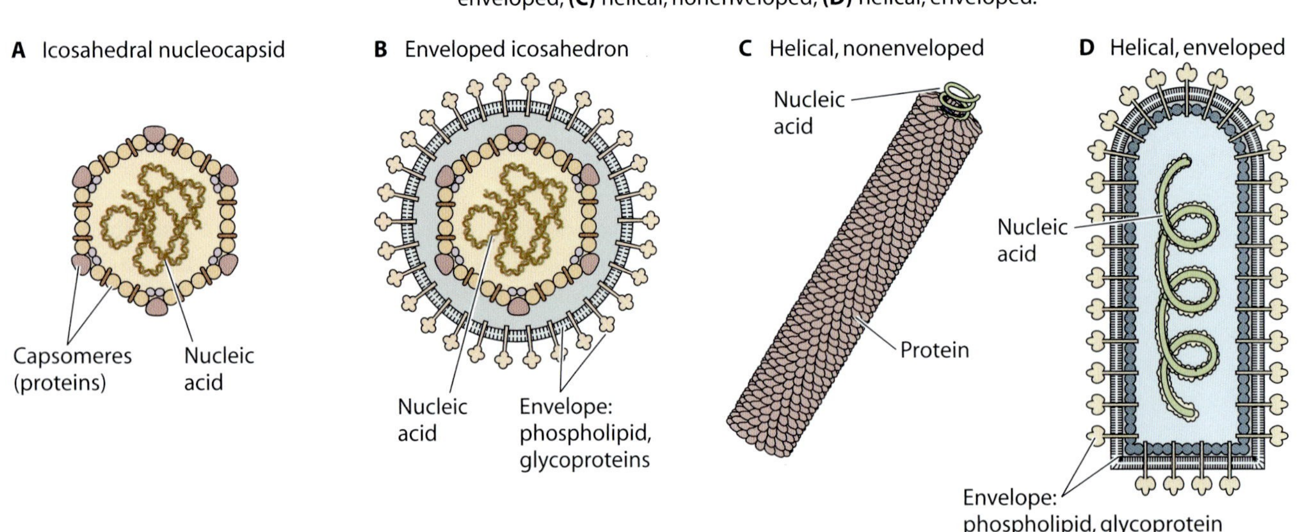

other hand, are determined by the lengths of the nucleic acids they contain. If a filamentous virus acquires foreign nucleic acid, the virions simply become longer (a useful property if you want to make proteins by genetic engineering).

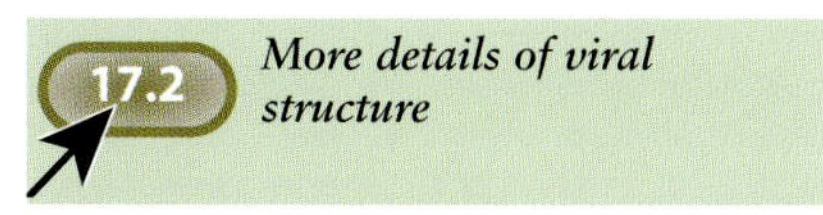

A simple icosahedral virion is made of 60 capsomers, 3 in each of its 20 faces. Larger icosahedral viruses are made of many capsomers (as many as 1,500), but because capsomers are asymmetric, the number of them that can assemble to form a symmetric equilateral triangular face follows strict rules. Thus, icosahedral virions do not contain just *any* number of capsomers. The next step after 60 is 180, then 240, 540, 960, and 1,500.

Some viruses, such as human immunodeficiency virus (HIV), have a **mixed morphology.** HIV has an icosahedral capsid with a filamentous nucleic acid core. Some viruses are surrounded by a lipid-carbohydrate-protein **envelope,** or membrane, that surrounds the capsid. The proteins of this structure are encoded by the viral genome, but the lipids and carbohydrates are derived from one of the host cellular membranes (e.g., the cytoplasmic, nuclear, or Golgi membrane or the endoplasmic reticulum). These cell components are picked up as the virus extrudes (buds) through a host cell membrane in its process of maturation (Fig. 17.3). In addition, some viruses have specialized structures that are involved in attachment to the host cell, such as the spokes of the influenza viruses or the lunar lander aspect of some bacterial viruses.

Although virions are metabolically inert, they are not biochemically helpless. Many **virally encoded enzymes** are used for attachment to host cells, or replication and even modification of their nucleic acid (this will be discussed in detail below). Some of the larger viruses, such as the one that causes smallpox (or a huge virus found in some amoebas called mimivirus), are very complex and invite the question: what would it take for them to become cellular entities able to replicate independently?

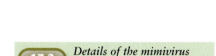

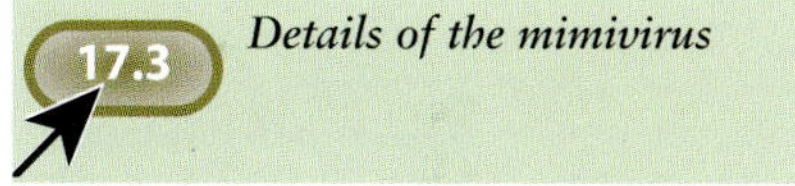

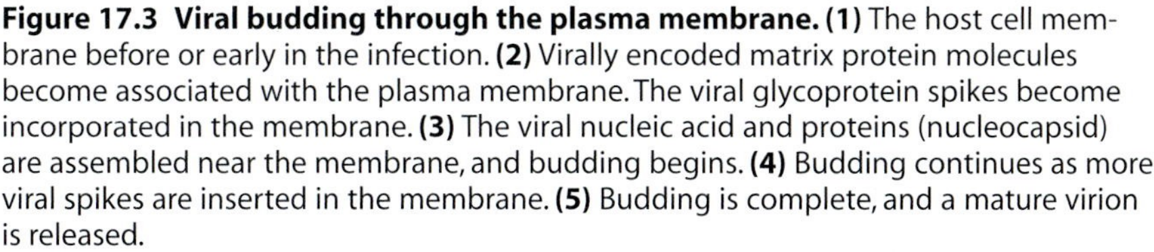
Figure 17.3 Viral budding through the plasma membrane. (1) The host cell membrane before or early in the infection. **(2)** Virally encoded matrix protein molecules become associated with the plasma membrane. The viral glycoprotein spikes become incorporated in the membrane. **(3)** The viral nucleic acid and proteins (nucleocapsid) are assembled near the membrane, and budding begins. **(4)** Budding continues as more viral spikes are inserted in the membrane. **(5)** Budding is complete, and a mature virion is released.

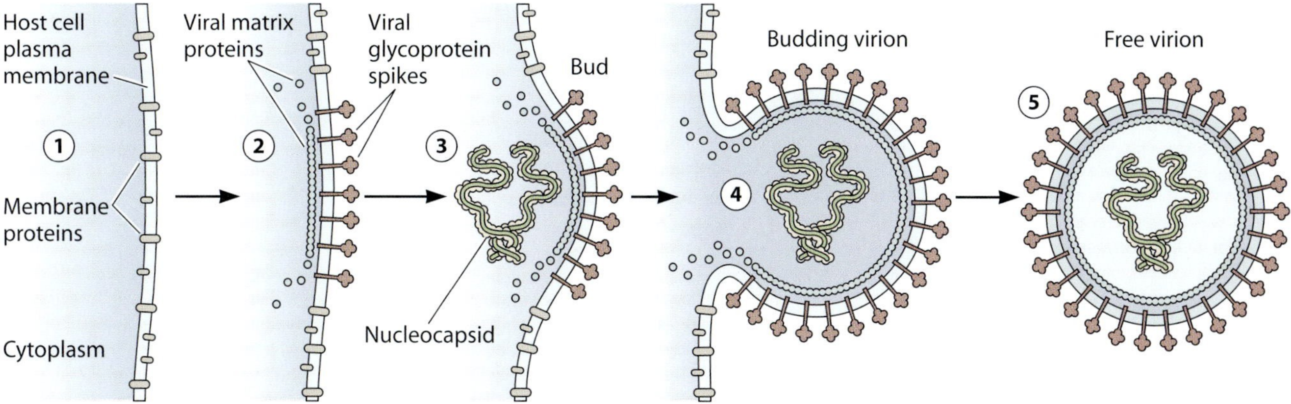

ECOLOGY AND CLASSIFICATION

Few, if any, forms of life escape being infected by viruses, so we can expect that there are a very large number of different kinds. There are hundreds of viruses that cause human diseases (Table 17.1). Even such "simple" organisms as *Escherichia coli* can be infected by dozens of different viruses. Each kind of virus consists of different **strains** that differ in virulence and antigenic properties (serotypes). Viruses tend to be quite host specific, but exceptions abound, as for instance the swine flu virus, which affects humans as well as pigs. At present, thousands of viruses are known, and we know this to be a substantial underestimate.

Are there cellular forms of life that are not infected by viruses? Probably not. Moreover, we are aware of the existence of many viruses without knowing which hosts they infect. For example, the oceans contain a huge number of viruses, upward of 10 million presumably **bacteriophage** (a virus that infects bacteria) particles per ml of seawater, as determined by counting them under the electron microscope. Because most of the bacteria in seawater cannot yet be cultivated, we can only guess which ones are hosts to these viruses. This ignorance obscures exceedingly important ecological and evolutionary relationships.

Table 17.1 One host, many viruses: major human diseases caused by viruses[a]

RNA viruses
Influenza
Common cold (caused by over 100 kinds of viruses)
SARS[b]
West Nile encephalitis
Hantavirus pulmonary syndrome
Rabies
Mumps
Measles
Rubella
Polio
Gastrointestinal disorders (several different kinds of viruses, e.g., Norwalk agent, rotavirus)
Ebola hemorrhagic fever
HIV infection, AIDS
DNA viruses
Fever blisters (herpes simplex)
Genital herpes
Hepatitis B
Smallpox
Chickenpox, shingles (varicella, zoster)
Infectious mononucleosis (Epstein-Barr virus)
Cytomegalovirus infection
Adenovirus infection
Papillomatosis (warts)

[a]Several hundred viruses are known to infect humans. Some of the diseases are specific for humans (e.g., smallpox); others (e.g., influenza) are not. Some viruses cross kingdom barriers and can affect both animals and plants.

[b]SARS, severe acute respiratory syndrome.

Viruses can be classified in many ways: by their host (bacterial, plant, animal, etc.), their size and shape, the presence or absence of an envelope, and the nucleic acid they contain—DNA or RNA—and whether it is single or double stranded, linear, or circular (Table 17.2 and Fig. 17.4).

VIRAL REPLICATION

For a virus to replicate, it *must penetrate a susceptible host cell.* The virion attaches to the cell, its nucleic acid enters and directs viral components to be made, and these are then assembled into progeny particles and released from the cell. Each of these steps is convoluted and idiosyncratic for each kind of virus.

Attachment and penetration

Viruses bump into their host cells at random, and once in a while, in every 10^3 to 10^4 collisions, they stick. This step, called **adsorption,** does not require energy, but it does require specific ionic and pH conditions. Next, **ligands** on the adsorbed viruses bind specifically and tightly to **receptors** on the host cell. The capsomers of the simplest viruses may do double duty, acting both as ligands and as structural components, but the more complex viruses (such as influenza virus and other enveloped viruses) have dedicated binding structures, such as spokes that jut out from the surface. The host's receptors are glycolipids or proteins (often glycoproteins), the

Table 17.2 Some animal viruses

Group	Nucleic acid polarity[a]	Example(s) of disease[b]
RNA viruses		
Single stranded		
Poliovirus	Positive	Polio
Coronavirus	Positive	Colds, SARS
Hepatitis A virus	Positive	Hepatitis A
HIV	Positive	AIDS
Rabies virus	Negative	Rabies
Measles virus	Negative	Measles
Influenza virus	Negative	Influenza
Double stranded		
Rotavirus		Gastroenteritis
DNA viruses		
Single stranded		
Parvovirus		Rash (humans), GI disease (dogs)
Double stranded		
Adenovirus		Colds, eye infections
Herpesvirus		Herpes, encephalitis, infectious mononucleosis, chickenpox
Poxvirus		Smallpox, vaccinia
Hepatitis B virus[c]		Hepatitis B

[a]Positive indicates that the virion's RNA can serve directly as mRNA. Negative indicates that the virion's RNA must first be copied into a complementary strand, which then serves as mRNA.

[b]SARS, severe acute respiratory syndrome; GI, gastrointestinal.

[c]Hepatitis B virus DNA has a single-stranded stretch.

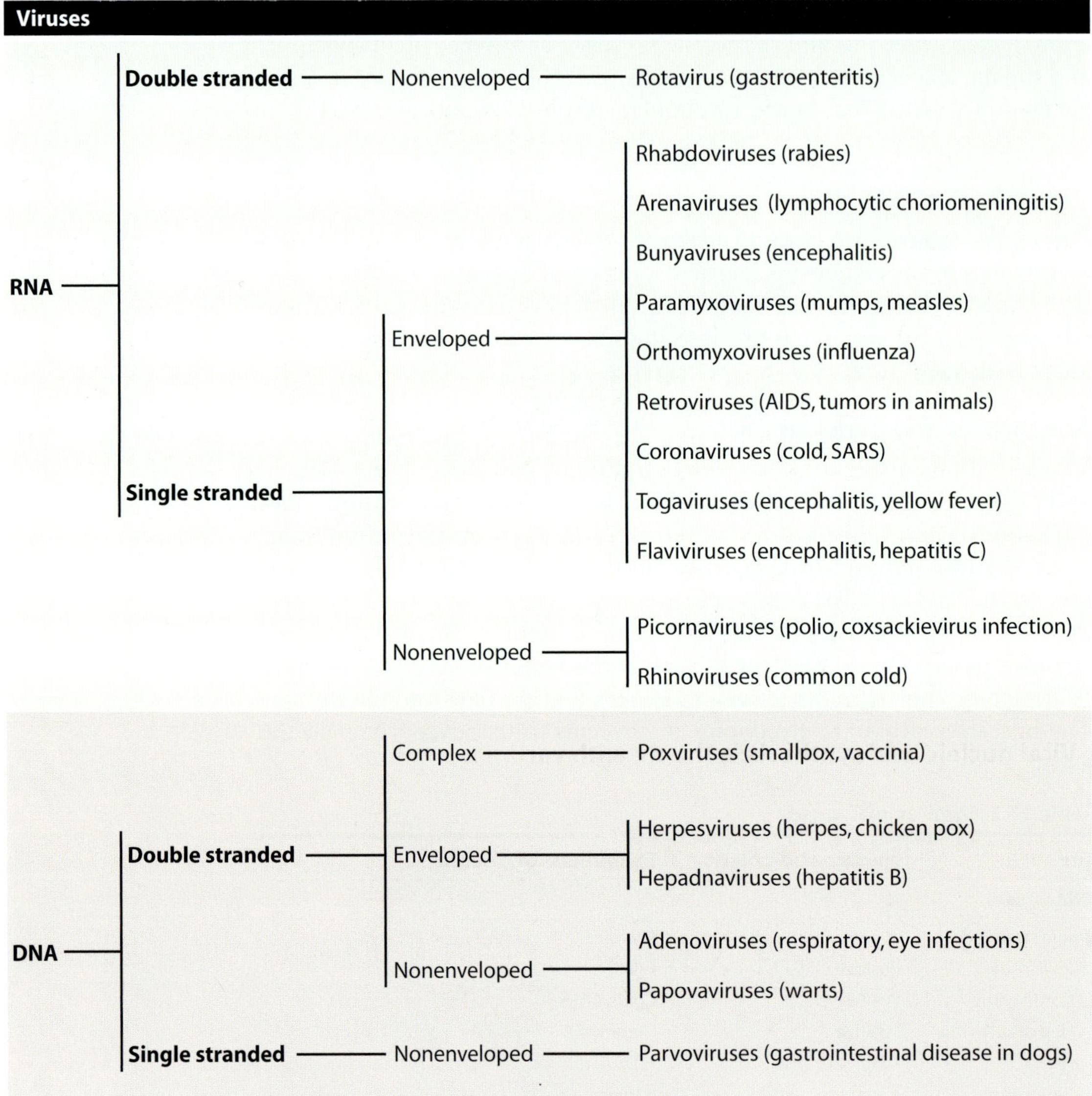

Figure 17.4 Main groups of human viruses. This is not a representation of viral phylogeny and is shown for practical purposes only.

structures of which have been determined in only a few instances. In the case of the influenza virus, for example, there is a nice complementary fit between the ligand protein on the virus and the cell receptor.

Some viruses enter the host cell completely or virtually intact and then release their nucleic acid by a process called **decoating.** Certain enveloped viruses, such as herpesviruses and HIV, enter by fusing into the cell membrane. Other enveloped viruses, such as influenza virus, enter by endocytosis, thereby trapping the virion in an endocytic vesicle. Decoating occurs as this vesicle becomes acidified, which induces conformational changes in proteins on the virion's surface. When the pH drops to 5, the amino-terminal end (called the **fusion peptide**) of one of these proteins (hemagglutinin HA2 in influenza virus), which is normally embedded in the capsid, flips outward and becomes exposed to the aqueous environment. Because the fusion peptide is highly hydrophobic, it fuses the viral envelope to the vesicle membrane. The released viral genome then enters the cell's cytoplasm, where it undergoes replication.

There are many variations on this theme. Some viruses uncoat as they enter the cell. Bacterial viruses (bacteriophages, or phages), for instance, penetrate the tough envelopes surrounding most bacteria. One way to do this is via a syringe-like apparatus that penetrates the cell wall and cell membrane (Fig. 17.5). This structure has a tail with a base plate at its far end and tail fibers that stick out from the plate. After the collision of virus and bacterium, the tail fibers bind to the receptor and the base plate attaches firmly to the bacterial surface. The tail is a hollow tube through which the nucleic acid can pass to infect the host cell. Penetration is carried out by the contraction of the tail proteins (conformational change again!), which eventually leads to the ejection of the nucleic acid into the host cell cytoplasm. The actual process of ejection is poorly understood. The viral "head" does not change shape in the process and thus does not function like a bulb that is squeezed to deliver its contents.

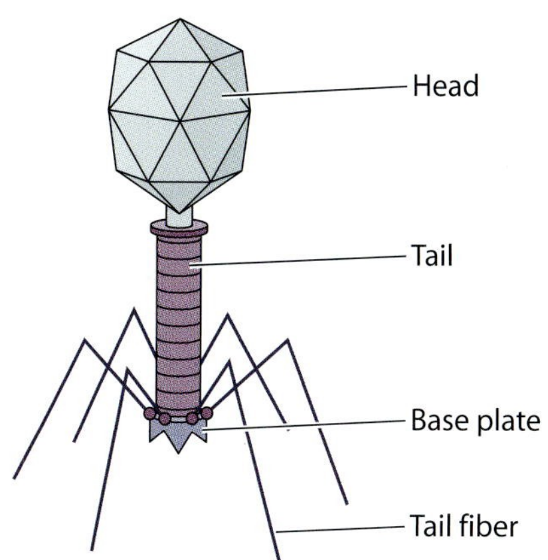

Figure 17.5 Structure of a phage (the *E. coli* phage T4). The virion attaches to the host cell by interactions of the tail fibers and base plate with receptors on the bacterial surface. Then, the DNA contained in the phage head is introduced into the host cell.

Modes of penetration may be quite varied, but they all share one central aspect: they inevitably lead to the *dissociation of the viral nucleic acid from its capsid*. It is this process of *disassembling in order to replicate* that most clearly distinguishes viruses from cellular forms of life.

Viral nucleic acid synthesis: a theme with variations

Just when it was thought that the synthesis of macromolecules could be described by the mantra "DNA makes RNA makes proteins" (the "central dogma"), along came the realization that many viruses deviate from this theme. Some viruses, e.g., the papillomaviruses, which cause warts, follow the scheme explicitly. Their genome consists of **double-stranded DNA,** which can be replicated, transcribed, and translated. However, other viruses have more original ways of doing business. For example, some viruses contain **single-stranded DNA.** Host DNA polymerases replicate their genomes by these steps:

Single-stranded DNA (infecting virus) → double-stranded DNA (replicating form)
→ single-stranded DNA (progeny)

Transcription into messenger RNA (mRNA) takes place while the viral DNA is double stranded and makes use of a host RNA polymerase (a DNA-dependent RNA polymerase). This might seem like a small variation on the usual theme, but consider the **single-stranded RNA viruses.** They must make RNA from an RNA template, which is a novel biochemical reaction because host cells do not have enzymes that make RNA from RNA. What's the solution? The answer is that RNA viruses encode a special enzyme, an **RNA replicase** (an RNA-dependent RNA polymerase), that can make RNA from RNA. How is this enzyme made? In its single strandedness, the viral RNA is actually an mRNA molecule that can be used directly to synthesize the protein.

Matters are not so simple for the single-stranded RNA viruses whose RNA is *not* mRNA but its *complementary strand*. Seemingly, such a virus cannot replicate even if its genome encodes an RNA replicase (how is it going to make its mRNA?). The solution, elegantly enough, is for the virus particle to carry its own fully formed RNA replicase in the virion. Immediately upon entering the cell, the replicase copies the viral RNA into an mRNA complement. Such viruses are called **negative-polarity RNA viruses** (negative-stranded RNA viruses). This group includes the viruses that cause influenza, mumps, and many plant diseases. In contrast, the viruses whose RNA can serve directly as mRNA are called **positive-polarity RNA viruses** (or positive-stranded RNA viruses). They include such familiar viruses as the agents of polio, hepatitis C, and the common cold. One more complication: the RNA of negative-polarity RNA viruses must be fully accessible to the enzyme that copies it, and thus, it must be devoid of structural impediments, such as double-stranded regions. To ensure ready access, the RNA's sugar-phosphate backbone is sometimes stuck to a protein scaffold; thus, its nucleotide bases are exposed to the replicase.

Yet another alternative is seen in **retroviruses,** such as HIV, the virus that causes **acquired immunodeficiency syndrome (AIDS).** Its genome is a single-stranded positive-polarity RNA, but it replicates via a DNA intermediate by the following steps:

Single-stranded RNA (infecting virus) → single-stranded DNA → double-stranded DNA (integrated into host genome) → single-stranded RNA (progeny)

The virus carries an enzyme called **reverse transcriptase** that converts the viral RNA into DNA. (This enzyme is unique and therefore a good target for anti-HIV drugs, as is a protease carried in the virion that is needed for the maturation of viral proteins [see "HIV Infection and AIDS" in Chapter 21].) In the double-stranded form, the DNA is integrated into the host genome by the activity of integrase, another enzyme carried in the virion. The host RNA polymerase then transcribes this DNA into viral RNA. Is HIV an RNA or a DNA virus? The answer is both, but not at the same stage in its life cycle. The selective advantage of this complex replication cycle is derived from its DNA intermediate: it can integrate into the genome of the host and thereby be carried with it indefinitely.

If you feel compelled to include every possible known replication strategy, add to the list the double-stranded RNA viruses and a few others with mixed strategies (Fig. 17.6). In summary, there is considerable latitude in how the information is copied and how it flows from the genome to proteins, but it never flows backward from proteins to nucleic acids. What brought about so many variations on the central theme? No one knows.

Making viral proteins

The viral cycle begins with infection of the host and ends with the assembly and release of progeny virions. In between, many events take place. Right after infection, there are no intact virions in the cell, because as we said, in order to replicate, the virion must come apart. The period from infection to the assembly of whole virions is called the **eclipse period** (Fig. 17.7). However, this is a busy time: a lot is going on within the infected cell.

Making viral nucleic acids and proteins is not a haphazard business, but rather, it is a symphony of well-timed and interrelated events. Except in the case of a positive-stranded RNA virus, the first thing that needs to

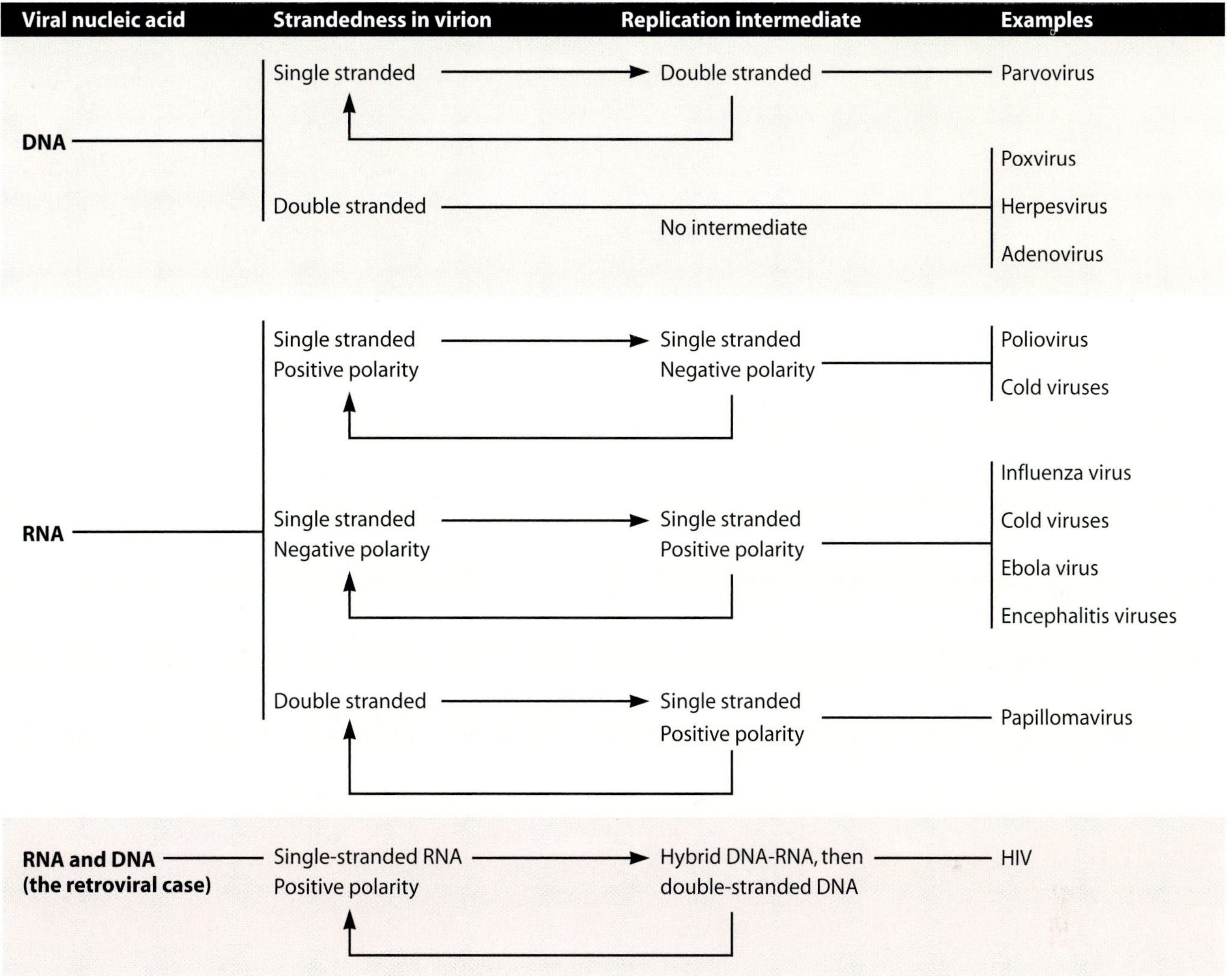

Figure 17.6 Main viral replication strategies. The arrows indicate replication resulting in the nucleic acid of the virion. Positive polarity means that an RNA molecule can serve directly as mRNA. Negative polarity means that it cannot serve as mRNA but must first be copied into its complementary nucleic acid.

be synthesized in order to make viral proteins is their mRNA. Once this is available, viral proteins can be synthesized by the host cell's own machinery. Offhand, it may seem that the only proteins that need be made are the constituents of progeny virions. Many viruses do more than that: they subvert the host cell's metabolism and induce it to make a lot of new virions. Some virulent bacteriophages shut down many host functions, making their machinery available to viral components. In certain bacteriophages, such subversion reaches a pinnacle of thoroughness. Soon after infection, the bacterial DNA is broken down, cannibalized to support viral synthesis. Any small amount of the host cell DNA that survives cannot be transcribed because a virus-encoded **sigma factor** takes over, restricting the host's RNA polymerase to transcribing the viral DNA. Certain virus-encoded proteins prevent translation of the bacterial mRNA into proteins. Thus, by clobbering almost every aspect of the

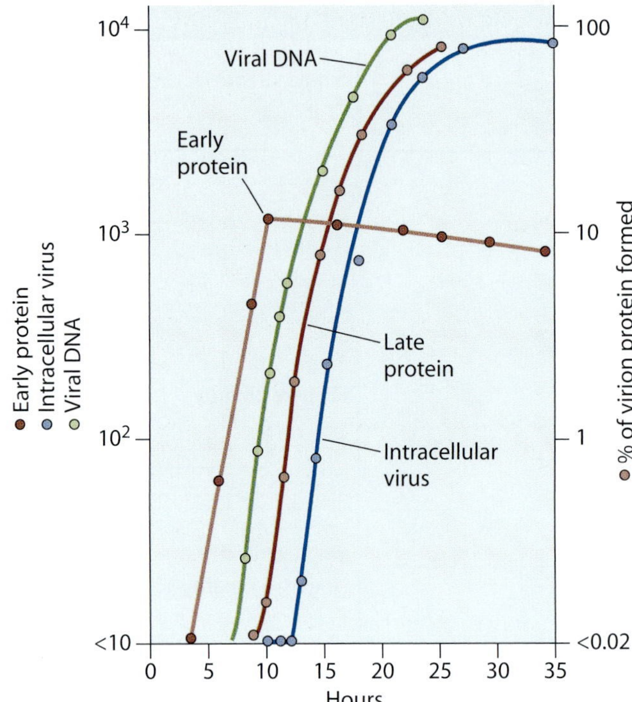

Figure 17.7 A typical "one-step growth curve" of an animal virus. Adenovirus replication in cell culture is shown. Various steps in viral replication proceed according to an orderly scheme. First, viral early proteins are made; they direct the host cell to make viral DNA and late proteins (including virion structural proteins). After some time, complete infective virions are formed within the host cell.

host machinery, these viruses become enormously efficient biosynthetic parasites. Not surprisingly, such bacteriophages reproduce very quickly; in many cases, they make hundreds of progeny virions in 20 minutes or less. Note that for this scheme to work, redirecting host functions must happen quickly, mediated by **early proteins.** The structural proteins of the virions are made later and are appropriately called **late proteins.** Thus, viral replication is a carefully planned and executed affair (Fig. 17.7).

Some viruses employ additional strategies to induce host cells to make more viruses. Cytomegalovirus, a cousin of the common herpesvirus, fools its host cell into replicating only viral DNA. Some viruses that persist for long periods within their host cells practice another viral subversion strategy. When such viruses enter epithelial cells that normally have short life spans, infection leads to an inhibition of **apoptosis** (programmed cell death). Infected cells become "immortalized" and produce virus for much longer periods. Viruses that use this strategy include the one that causes infectious mononucleosis (Epstein-Barr virus; see "Infectious Mononucleosis: the 'Kissing Disease'" in Chapter 21). Immortalized cells can sometimes produce cancers.

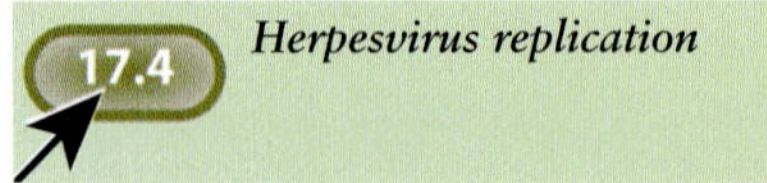

Virion assembly and release from the host cell

Once enough viral nucleic acid and proteins are made, the virions can be assembled. The capsid is made first, by the self-assembly of viral capsomers into crystalline-appearing arrays in the general shape of the virion. The capsids are then filled with the viral nucleic acid, which in the nonenveloped viruses completes the process of producing viable virions. The virions are then released as the host cells lyse, a process induced by viral proteins that disturb the membrane or the cytoskeleton. **Virions of enveloped viruses** have a more complicated birth. They are released from the host cell by **budding,** becoming surrounded by a piece of the host cell's membrane, a process that does not necessarily damage the cell. Note that this strategy benefits the virus by extending the life of its host so it can produce progeny viruses for a longer period. Before budding begins, the surface viral proteins encoded are incorporated into discrete patches of the host cell's membrane. Viral capsids bind to these patches and are extruded, coated by an envelope (Fig. 17.3).

17.5 *A short film on how bacteria lyse during phage infection*

Visualizing and quantitating viral growth

An animation of viral infection

Imagine mixing a small number of bacterial or animal virions with a large number of sensitive bacteria or animal cells in culture and placing (plating) them on an agar plate. After a suitable period of incubation (a few hours for bacterial viruses; a day or more for animal viruses), the plate will be overgrown with uninfected host cells. However, here and there, there will be circular areas that look like holes in the "lawn" of host cells. These areas are called **plaques** and are the results of a microepidemic. Each plaque started from a single infected host cell that eventually lysed. The released virus particles attached to adjacent uninfected host cells, and the process repeated itself, leading to a centrifugal spreading of the infection. The process stops when the bacteria stop growing and enter the stationary phase.

It is a simple matter to use plaque counts to quantitate the number of virus particles or infected cells in a preparation. As in the case of colony counts, one multiplies the number of plaques on a plate by the dilution factor of the preparation. This will enumerate only virions capable of proliferation, and if the total number of virus particles is to be determined, the sample must be examined under an electron microscope and the number of particles in a given volume must be counted. A simple way of doing this is to mix the preparation with a known concentration of latex beads and determine the ratio of virus particles to beads.

17.7 *More information about animal viruses: "All the Virology on the WWW" (many links)*

There are variations on this theme. In the case of plant viruses, plaque counts can also be carried out by spreading viruses on an abraded surface of a leaf of a susceptible host. The number of lesions that develop is proportional to the number of virus particles in the preparation.

LYSOGENY AND INTEGRATION INTO THE HOST GENOME

Introduction

So far, we have viewed viruses as destructive agents that damage their host cells and cause disease. This is an imperfect view of what viruses really do. We know that many viruses coexist with their host cells for long periods. Such associations often have momentous consequences because they can lead to genetic changes within the host cell and the exchange of genetic material between cells. As a consequence, viruses play a key role in evolution. There are, then, two kinds of viruses: **virulent** viruses, which usually kill their host cells by inducing lysis, and **temperate** viruses, which sometimes live in harmony with their hosts. Viruses of both eukaryotes and prokaryotes (phages) can be virulent or temperate.

The genome of a temperate virus can become *integrated into that of its host cell.* In a real sense, the genes of such a virus become part of the host's genetic makeup. In the integrated state, the viral genome is carried along with that of the host and replicated at each cell division. It follows that only DNA viruses can be integrated (although, as in the case of HIV, the virus may contain RNA in its virion and make DNA during part of its replicative cycle). The concept of temperate virus includes one other key aspect: integration into the host genome is reversible; that is, under some conditions, the *genome of a temperate virus is released from that of the host*. This reversibility allows the virus to become virulent and multiply, usually leading to the death of the host.

A few definitions, used mainly for phages, are useful: **temperate phages** that can integrate into the host genome are said to be **lysogens;** the integrated form is called a **prophage,** the bacteria that carry prophages are said to be **lysogenized,** and the phenomenon is referred to as **lysogeny.** A simplified version of the **lysogenic cycle** is shown in Fig. 17.8.

There are variations on these themes, too. In an alternative mode of lysogeny, the prophage is not integrated into the host genome; rather, it exists as a plasmid. Like other plasmids, it replicates in synchrony with the bacterial chromosome and is transmitted to the progeny. An example of a phage that can become a plasmid is P1. Thus, integration is not an

Figure 17.8 Two lifestyles of a temperate phage: lysogenic and lytic. Phage attachment to a sensitive host bacterium **(1)** is followed by the injection of viral DNA **(2)** into the host. In typical cases, the viral DNA circularizes and either replicates **(3),** leading to the production of progeny phage and cell lysis **(4),** or integrates into the host chromosome **(5)** to become a prophage. In this state, the host cell has become a lysogen that can replicate for many generations **(6).** On rare occasions, the prophage excises spontaneously from the bacterial chromosome **(7),** a process that can be augmented by induction (e.g., treatment with UV light or mutagens). Upon excision, the phage DNA can initiate a lytic cycle **(3** and **4).**

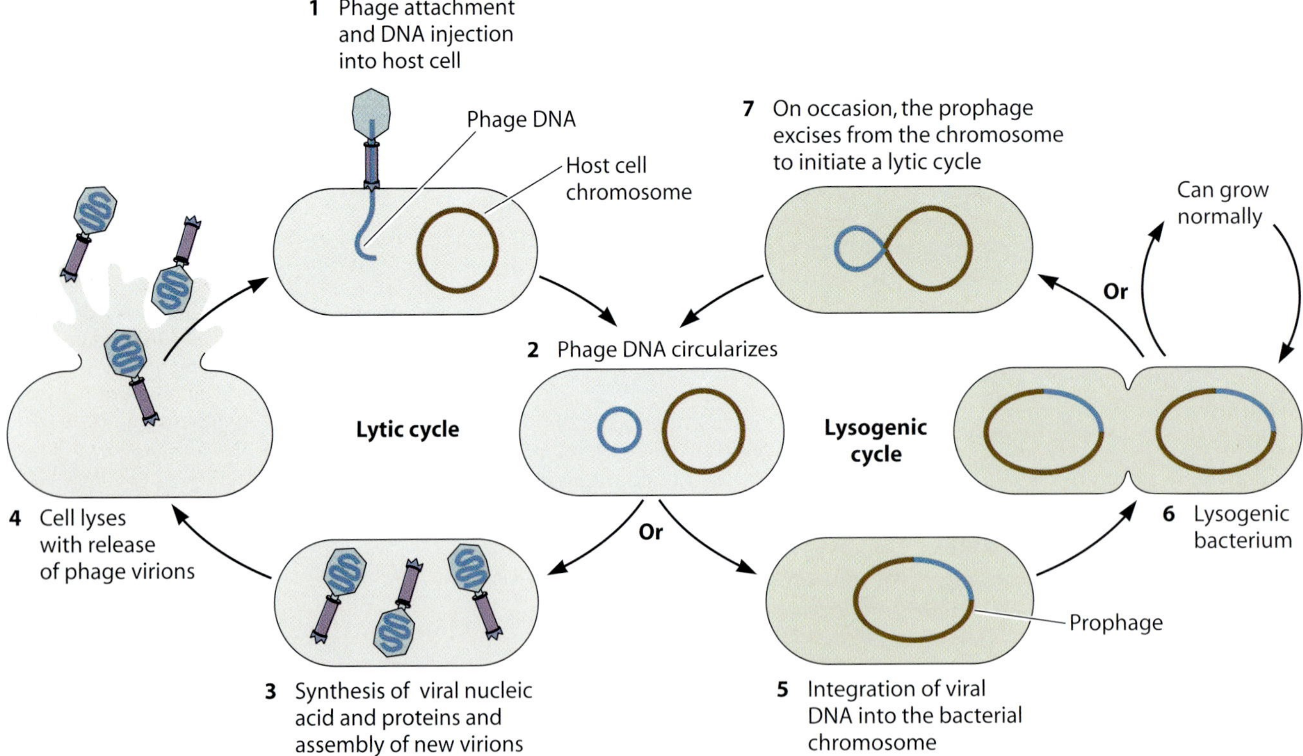
1 Phage attachment and DNA injection into host cell
Phage DNA
Host cell chromosome
7 On occasion, the prophage excises from the chromosome to initiate a lytic cycle
Can grow normally
Or
2 Phage DNA circularizes
Lytic cycle
Lysogenic cycle
4 Cell lyses with release of phage virions
Or
6 Lysogenic bacterium
Prophage
3 Synthesis of viral nucleic acid and proteins and assembly of new virions
5 Integration of viral DNA into the bacterial chromosome

essential feature of lysogeny; what is important is that the phage does not actively replicate in its host cell when it pursues the temperate lifestyle.

The association between a lysogenic phage and its host bacterium can be interrupted, generally by conditions that threaten the life of the host. For example, if a lysogenized *E. coli* cell is irradiated with ultraviolet (UV) light, its prophage is excised and replicates freely. This phenomenon is called **viral induction.** It suggests a "bailing out" mechanism used by temperate viruses to survive even if their host is killed.

Which would better foster the survival of the virus over time, being virulent and making a large number of progeny immediately or being temperate and preserving its genome within the host? Unbridled replication of virulent viruses could lead to the destruction of a large proportion of host cells and thus to a dead end. On the other hand, being continuously temperate would mean perpetual confinement. A good solution would be to alternate between virulence and temperance. The virus would make particles on occasion but keep its genome in a safe repository. This is what lysogenic phages do.

How do lysogenic phages manage to balance the virulent and temperate lifestyles? The most studied temperate phage is an *E. coli* phage called **lambda,** which has become a model for our understanding of the phenomenon of lysogeny. (So much is known about lambda that a course dealing with it alone would cover the most fundamental principles of molecular biology.)

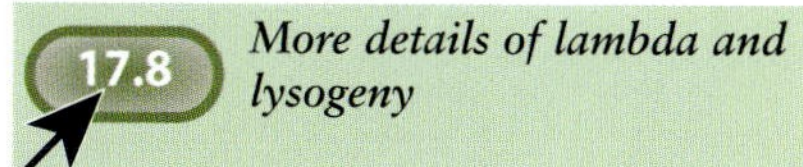

More details of lambda and lysogeny

How does the genome of a temperate phage become integrated into that of a host cell?

In the lambda virion, the DNA is a double-stranded *linear* molecule. Shortly after being injected into an *E. coli* cell, it becomes circular. Circularization is possible because the ends of the molecule have single-stranded complementary protruding sequences. These can base pair to form a circle. The resulting circle has two single-strand nicks that are closed by DNA ligase to form a **covalently closed circle** that integrates into the host DNA *by a single crossover event* (Fig. 17.9). This process has unique features: if the phage DNA were linear, a single crossover would

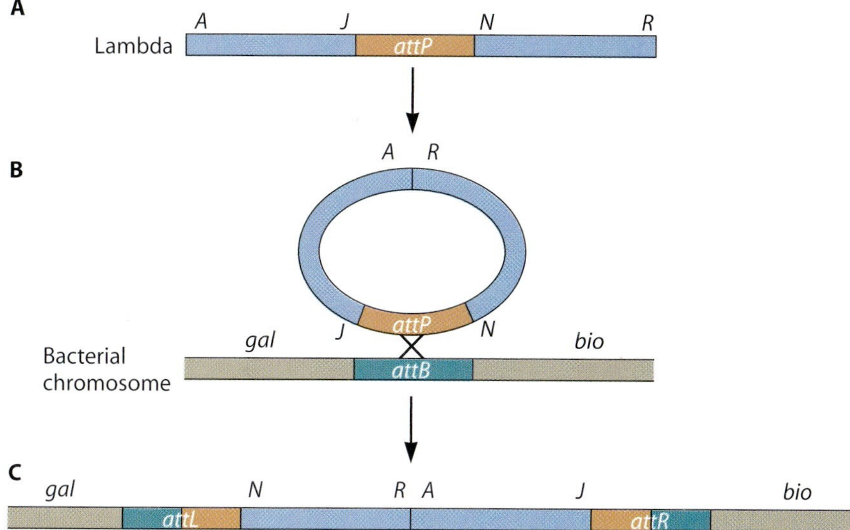

Figure 17.9 Mechanism of integration of a temperate-phage genome into the host cell chromosome. The ends of the linear lambda phage genome **(A)** come together to form a circle **(B)**. The phage attachment site, *attP*, recombines with a site on the bacterial genome, *attB*. As the result of a crossover between these two sites, the phage genome is integrated into that of the host **(C)**. In the process, the attachment sites have been changed to *attL* and *attR*.

make the host chromosome linear; it would be degraded by intracellular nucleases and, in any case, could not be replicated (*E. coli* can replicate only circular DNA). A double crossover is required to maintain circularity. Recombination between lambda and the chromosome takes place only at a *specific site on each molecule*. Why do most, but not all, bacteria put such a premium on circularization? Bacterial cells have powerful exonucleases that degrade linear DNA unless the ends are protected. A circular chromosome is simply an easy way of doing so.

Viruses that integrate into their host's genomes risk inactivating genes essential for the life of their host. To avoid this danger, most such viruses integrate into specific locations only. Such a recombination event requires the activity of a special protein called **integrase,** which recognizes *two specific but dissimilar sites,* one on the bacterial chromosome and one on the viral DNA flanking a short region of **homology.** This region is too short to allow a crossover to take place without the integrase that recognizes the adjacent DNA. Many other integrases effect integration of other viruses, but that of lambda is the best studied.

How does the integrated viral genome remain quiescent?

A prophage remains a prophage and does not direct the formation of virus particles due to the fact that *most of its genes are silenced.* Silencing is mediated by a special protein, the **lambda repressor,** which keeps all viral genes, *except for the one that encodes the repressor itself,* from being expressed. The lambda repressor acts as a typical repressor and binds to the operator in the promoter region (see Chapter 12). As long as active repressor is present, the prophage genes encoding excision and replication remain silent, but when active repressor is no longer present, the prophage is excised and new phage particles are made. The eventual result is lysis of the host cell with the release of many progeny phages. You will agree that it should be possible to convert lambda from a temperate virus into a virulent one *by mutation.* Two kinds of mutants do in fact accomplish this: in one kind, the mutation inactivates the repressor protein; in the other, it renders an operator insensitive to the repressor. These mutants are analogous to constitutive mutants with mutations in the *lac* operon (see Chapter 12). In fact, the two systems were discovered almost at the same time by two interactive groups of scientists working in an attic at the Pasteur Institute in Paris during the 1950s.

The crux of how a prophage is activated to replicate as a phage is the manner in which the repressor is made and inactivated. It is advantageous for the virus to have an *optimal level of repressor.* Too little, and phage replication would kill the cell. Too much, and viral induction would be difficult to achieve. The "just so" concentration of repressor can be maintained because the repressor protein is capable of regulating the expression of its own gene. Such **autoregulation** is a feedback loop, or if the analogy works better, a device such as a thermostat (see Chapter 12). Too much heat, and the circuit is off; too little, and it is on.

The importance of the repressor is illustrated by two experiments. (i) If an *E. coli* culture is infected with a small number of virus particles, only a few enter each cell, little repressor is made, the phages multiply, and most cells lyse. (ii) If the cultures are infected with a large number of virus particles, many enter each cell, quickly synthesizing a high level of repressor, and most cells become lysogenized. We discuss the lytic-versus-lysogenic decision further below.

What causes viral induction?

Environmental conditions deleterious to bacteria can inactivate the repressor, leading to induction of the prophage and the initiation of a **lytic cycle.** An example of an inducing agent is UV light, which damages the host cell's DNA and provokes a cascade of molecular events. In the process of repairing damaged DNA, cells accumulate short stretches of single-stranded DNA. These segments set off the **SOS response** (see Chapter 13), which leads to the activation of a normally dormant coprotease called RecA. Activated RecA protein cleaves the lambda repressor, allowing phage genes to be expressed. One of these genes encodes an enzyme called **excisionase,** which helps the integrase carry out the crossover that excises the prophage. Integrase needs help to excise because excision is not simply the reverse of integration. The two crossovers take place at somewhat different sequences.

If no induction took place, how could one know that a prophage is lurking in a bacterial genome? Usually, one cannot simply look at a culture or a colony and tell whether the bacteria are lysogenic, but a simple test can. If after infection with the same phage, lysis occurs, the culture is not lysogenic; if it doesn't lyse, the culture is lysogenic. Lysogenic bacteria do not permit phage to develop because the repressor present in them prevents it. However, an alert reader will notice a problem with this experiment. How does lambda act as a **lytic phage** without being induced? Read on.

Deciding between lysogeny and lysis

Let us go back to the beginning, infecting a virgin, nonlysogenic culture. We painted the picture of how the viral genome becomes integrated and the cells become lysogenic. This is far from the inevitable outcome; in fact, most of the time, the opposite happens. Lambda replicates, many progeny particles are made, and the cell lyses. The outcome depends on the environment, on the number of infecting particles per cell, and on the physiological state of the host. For example, when *E. coli* cells are growing in a rich nutritional environment (e.g., in the presence of high glucose concentrations), they tend to be lysed upon infection with lambda. In a poor medium, with low glucose concentrations, they tend to become lysogenized. How is the decision made to "go lytic" instead of lysogenic?

The properties of the lambda repressor ultimately govern the outcome. The repressor protein is unstable because it can be cleaved by proteases in the cell. We have already seen that one such protein, the RecA protein, comes into play when the cell's DNA is damaged. However, another protease that affects the repressor, called Hfl, is regulated by environmental conditions, e.g., the growth rate of the culture. The activity of this protein is influenced by the cellular level of cyclic AMP (cAMP). From Chapter 12, you may remember that the level of cAMP in *E. coli* varies with the presence of glucose in the medium. If the medium has lots of glucose, levels of cAMP are low, and the reverse is true for glucose-deficient media. It turns out that the level of the Hfl protease is inversely proportional to the concentration of cAMP. Thus, more protease is made in glucose-containing

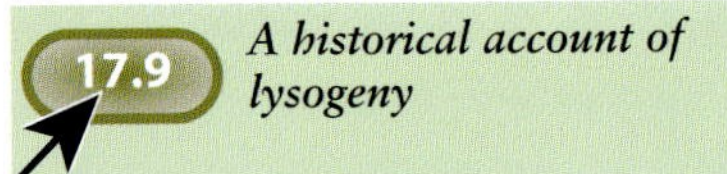

medium than in a glucose-deficient medium; the lambda repressor will be cleaved to a greater extent in a glucose-rich medium, and the lytic cycle will be set in motion. Proteolysis of the repressor is not the only factor in the decision. Both the rate of synthesis of the repressor and its activity are important, but other proteins are involved. Some of these protect the repressor protein from proteolysis; others compete with it for operator sites and thus affect repressor activity. This is a complex choreography.

Let us return to the test to determine if an *E. coli* culture is lysogenic. The experiment is usually done in a high-glucose medium. Here, the level of cAMP in the cells is low, and hence, that of the Hfl protease is high and that of the lambda repressor is low. Consequently, the lytic cycle will be favored over the lysogenic one. If, on the other hand, the recipient cells were lysogenic, enough repressor would be present to inhibit the lytic cycle.

The switch between a lytic and a lysogenic cycle is not just a laboratory phenomenon and probably has selective value to virus and host. *E. coli* alternates between life in a nutrition-rich environment (the colons of animals) and in nutritionally poor ones (waters, soils, and sediments). In fact, several natural isolates of *E. coli* are lysogenic for phage lambda or its close relatives. It seems possible that switching between two lifestyles helps both phage and host survive, although this has not been demonstrated experimentally.

What are the genetic consequences of lysogeny?

Being a lysogen means carrying extra genes. The prophage may contain not only typical phage genes, but also some that are distinctly bacterial. Where do these genes come from? It is surmised that they originated during infections with earlier bacterial hosts. One can witness today how prophage-borne bacterial genes are acquired. When a prophage is excised from the host genome, recombination does not always take place *precisely* at the ends of the prophage. Rarely, excision takes place a little to the right or to the left. In this case, the resulting phage genome includes neighboring *bacterial* genes. A subsequent temperate infection introduces these genes into another cell. The resulting phenomenon, **specialized transduction,** is discussed in Chapter 10.

The acquisition of genes by lysogeny has profound consequences for the physiology of host bacteria and their ability to survive and cause disease. A surprising number of temperate phages carry genes that are involved in bacterial virulence. These genes render otherwise-placid bacteria pathogenic. Bacteria converted from nonvirulent to virulent by lysogenization include the agents of diphtheria, cholera, scarlet fever, and staphylococcal food poisoning. In many of these cases, prophage genes code for powerful toxins; in others, they code for new antigenic types. In addition, because these genes are carried on a mobile element, they may be transferred from one bacterium to another.

Lysogens can also be genetically altered *even if the prophage carries no bacterial genes*. A lysogenic bacterium will not allow a second infection by the same phage, a seemingly minor but potentially life-saving consequence for the bacterium. Lysogenized bacteria are **immune** to infections that could kill them. Lysogeny may have yet other consequences. Lambda and many other temperate phages integrate at precise sites on the host chromosome, but others are less meticulous. Some phages (e.g., Mu) can integrate at random sites on the chromosome. Such integration inactivates the gene at such a site, thus producing an **insertion mutation.** Note that this is equivalent to the insertion of a transposon (see Chapter 10). Indeed, Mu has been called "a transposon masquerading as a phage."

In animals, a phenomenon akin to lysogeny causes some forms of cancer. This process is called **cell transformation,** a term otherwise used with bacteria to describe the genetic changes effected by the uptake of naked DNA (see Chapter 10). Cell transformation is caused by genes called **oncogenes** that subvert mechanisms that control cell replication. Oncogenes are altered forms of normal regulatory genes that disrupt normal signal transduction pathways responsible for altering cellular gene expression. Oncogenes may either be carried by the virus or be host genes whose structure or expression is altered by insertion of the viral genome or by products of viral genes. Cell transformation by viruses usually requires the persistence of all or a part of the viral genome and the continual expression of oncogenes. However, certain viruses are capable of a "hit-and-run" kind of oncogenesis. Among viruses capable of transformation are retroviruses, such as Rous sarcoma virus and human T-cell leukemia virus type 1; papillomaviruses; and herpesviruses.

What is the effect of lysogeny in evolution?

It is widely believed that phages strongly affected bacterial evolution. Acting as predators, **virulent phages** select for phage-resistant survivors, which favors the emergence of new bacterial types. However, it is likely that temperate phages play an even more important role in shaping their hosts' evolution. By effecting gene transfers, they may lead to genetic combinations more pervasive than single gene mutations (see Chapter 10). Surely, lysogeny was common in the past and is frequently taking place at present. Many bacterial genomes contain sequences characteristic of phages, which suggests that a lysogenic phage was integrated in the past. Such prophage remnants persist because they do not have all the genes needed for excision and replication. In addition, as we've said, lysogeny appears to be a favorable strategy for survival of the virus. More than 80% of the known phages are temperate.

VIROIDS AND PRIONS

We probably agree that the world of virology is far from monotonous. Viruses vary greatly in shape, structure, and mode of replication. They encompass every version of nucleic acid: single or double strands, linear or circular, RNA or DNA. Yet there is more. There are other agents, **viroids** and **prions,** that do not conform to the definition of viruses, yet they too are very small and infectious. The kinship of viruses with viroids is closer; viroids are **naked nucleic acid molecules** made up entirely of RNA circles. Prions, on the other hand, are entities apart; they consist only of **single protein molecules.** If there were a more appropriate place in this book for prions, we would gladly place them there, but we have not found it.

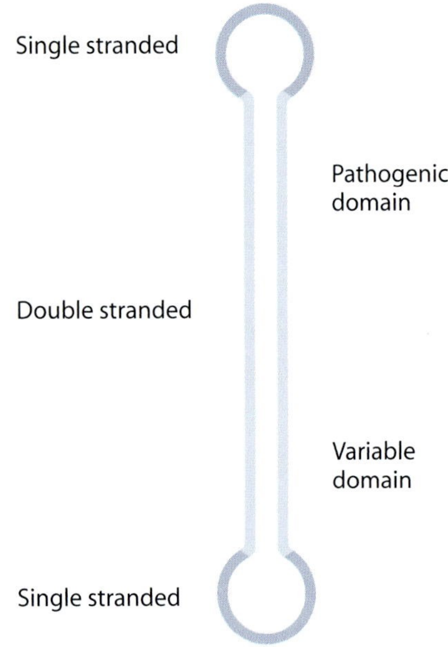

Figure 17.10 Structure of a typical viroid.

Viroids

Viroids, which cause diseases in plants, consist of RNA molecules not enclosed in a protein shell. That is all. They do not encode even a single protein. A viroid is just a circular RNA molecule curled upon itself to create an extensive double-stranded segment (Fig. 17.10). Viroids cause such diseases as avocado sunblotch, peach latent mosaic, and coconut cadang-cadang and represent serious economic problems. In fact, coconut cultivation in the Philippines and chrysanthemum growing in the United States have been seriously threatened by viroid diseases. How do viroids cause disease? By multiplying, they divert cell resources, but that alone does not matter much to the host. They are thought to cause pathogenic effects by a direct interaction between the viroid RNA and one or more cellular targets. The detailed mechanism is not yet known. One of the striking characteristics of viroids is their small size. The spindle tuber viroid that infects potatoes contains only 359 nucleotides!

Viroids are not protected by a capsid and seem not to need one. They have extensive regions of double strandedness that are resistant to destruction by ribonucleases. How do viroids reproduce? Since they do not face the problem of making capsid proteins, they are not burdened by genes to encode them. However, they do face the same problem as the RNA viruses do, namely, how to make RNA from an RNA template, plus, in addition, how to make it circular. The way viroids accomplish this is quite intricate. The RNA of infecting viroids is replicated by a **host RNA polymerase** that can make RNA using an RNA template. Such enzymes are common in plants but not in uninfected animals or bacteria. Replication is carried out using the RNA circle as a template and making a complementary copy of it by going around and around the circle. This so-called **rolling-circle replication** mechanism is also used by certain DNA viruses and plasmids. The result is not a single copy of the template but a long string of copies repeated in tandem, like an unrolled roll of toilet paper. To make viroids, this long molecule must be cleaved into segments of the proper size. This can be done in one of two ways. In some viroids, the RNA itself has enzymatic activity; it is a **ribozyme,** capable of cutting the long string into single copies. In other viroids, cutting is done by a host endonuclease. In both cases, the resulting copies must be ligated into circles that then assume the viroid's molecular structure.

The only viroid-like human agent is called (wrongly) **hepatitis D virus;** it is actually something between a virus and a viroid (hence, another name for it is "**virusoid**"). It also consists of naked RNA but differs from plant viroids in that it does encode proteins. However, unlike true viruses, the hepatitis D agent does not encode its own capsid. Instead, the proteins it encodes appear to play a role in packaging. The hepatitis D agent uses the capsid of an authentic virus, the hepatitis B virus, to become packaged. The disease hepatitis D therefore occurs only if the host is simultaneously infected with hepatitis B virus and the hepatitis D agent. This "two-for-the-price-of-one" infection results in a more severe disease than hepatitis B virus infection alone.

The hepatitis D agent may cause disease by inactivating an essential cell component. Its RNA has extensive sequence homology with a cytoplasmic 7S RNA particle that is involved in signal recognition and translocation of secretory and membrane-associated proteins. It is likely that the virusoid RNA sequesters or cleaves the 7S particle. This would kill the cell.

Prions

Degeneration of the central nervous system is a serious matter, often with lethal consequences for humans and animals. Recently, the most notorious of these conditions has been mad cow disease, but there are others with similar manifestations, such as Creutzfeldt-Jakob disease in humans and scrapie in sheep. Collectively, these diseases are called **spongiform encephalopathies** because holes form in the brain, making it appear spongelike. The **infectious nature** of one of these human diseases, kuru, was suspected when it was realized that the affected people had consumed human brains as part of their rituals. Indeed, some of these diseases can be transmitted to laboratory animals. Researchers suspected that the agent might be a virus, but they were unable to find viruses, bacteria, or any other agents from known groups. In time, it was established that the agent, then named a "prion," was highly unusual in that it *did not contain nucleic acids; it was composed entirely of protein.* This finding was startling because all other infectious agents, indeed all other self-replicating entities, possess nucleic acid-based genomes.

The mystery was partially solved when it was discovered that prion proteins have unusual properties. Like other proteins, they can fold into a number of different three-dimensional structures. In one configuration, they are *normal constituents of the cells* of the central nervous system and cause no harm. In another configuration, they become prions and gain the unusual ability to act as templates that can convert molecules of the normal proteins into prions. Prions and their normal precursors have the *same amino acid sequence* and are encoded by the same genes: they differ only in how they are folded. The precursor proteins are rich in α-helices and have little in the way of β-sheets; prions are the converse. In contrast to their normal counterparts, prions are highly resistant to proteases, harsh chemicals, and high temperatures. Disinfecting prion-containing material requires extreme measures. For example, contaminated objects must be soaked in 1 normal sodium hydroxide for at least 1 hour to destroy prion proteins.

How do prions impose their folding on normal molecules? It is thought that prions *induce* normal proteins to fold or refold into prion form. It is not known, however, whether they act on proteins that are still in the process of folding or on those that have already folded into their natural configuration. The process has not been reproduced in vitro. The process is thought to be irreversible—prions do not revert to become normal proteins.

Whatever the mechanism, the accumulation of enough prions and their spread to adjacent cells impairs normal brain function. In cows, "mad" refers to insanity, not to anger. In sheep, the disease is called scrapie because sick animals rub themselves intensely against any surface, scraping off their wool and skin. As clusters of cells die, the brain resembles a Swiss cheese full of holes (Fig. 17.11). Prion proteins aggregate into waxy translucent fibrils called **amyloid** that induce programmed cell death. The term "spongiform encephalopathy" refers to the spongelike appearance of the brain tissue due to extensive cell necrosis. Prion diseases

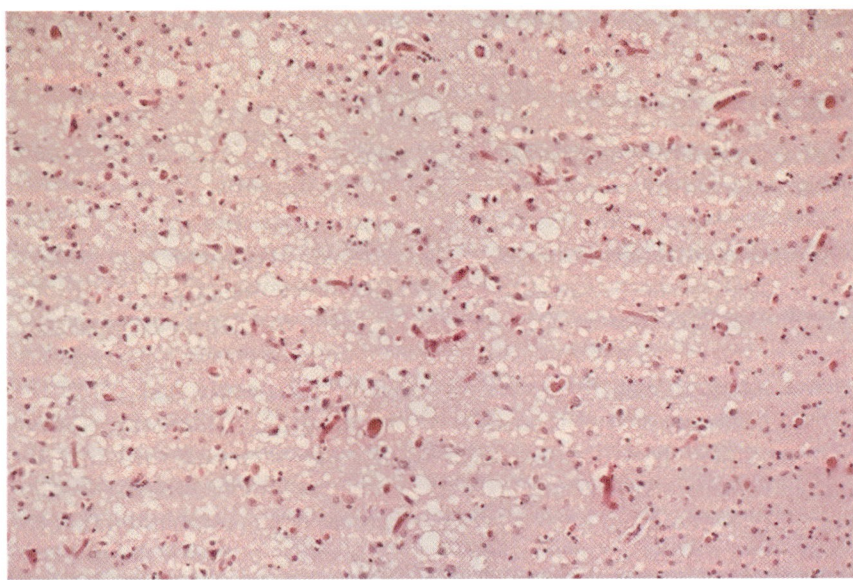

Figure 17.11 Pathology of prion disease. A section of a brain with spongiform encephalopathy revealing numerous small cavities in the tissue (light areas) is shown.

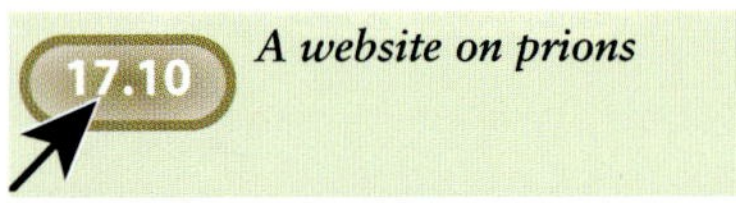

are always slow in developing. Interestingly, the function of the normal form of prions is unknown. Mutant mice that are defective in this protein do not display symptoms. One can speculate that if prions affected an essential protein, the manifestations of disease would be more immediate.

In general terms, the activity of prions falls into the realm of **epigenetics,** inheritable changes that do not result from altered genome sequences. Examples in complex organisms abound—think of the ability of liver cells, for example, to reproduce their own kind even though they have the same genome as, say, muscle cells. In some instances, modifications of DNA by methylation are blamed for the change in the pattern of gene expression that characterizes a given cell type. In unicellular organisms, the phenomenon appears to be more limited. However, yeasts have proteins that act like prions. They induce inheritable modifications when introduced into other cells by mating. Like animal prions, yeast prions aggregate into fibrils and inactivate their natural forms. It remains to be seen how widespread the prion mechanism is in other organisms. Prions may be involved in other epigenetic phenomena, such as long-term memory or even somatic-cell differentiation.

CONCLUSIONS

Viruses, viroids, and prions reveal the great plasticity of the biological world, extending our basic concepts from the cell as the basis for life to chemical messengers of genetic information. We learn from these entities that certain phenomena central to our concept of what is living, such as replication and mutation, are not the province of cells alone. These processes can be unlinked from the principal characteristics of cells, to divide and form progeny cells. Such intracellular entities do not operate in a vacuum: they need cells to provide them with energy and the machinery for synthesizing proteins. If viruses that are already capable of making their own nucleic acid were endowed with a mechanism to obtain energy and to make proteins, would they become capable of unaided replication? In other words, would they be cells? This seems particularly relevant to some of the large enveloped viruses, such as those that cause smallpox, which are endowed with a significant complement of nucleic-acid-synthesizing enzymes (see Chapter 22).

This wonderland invites further speculation. How did viruses arise? Are they "reduced cells," the result of loss of properties required for independent existence, or did they evolve as novel entities? Or did they follow other evolutionary pathways? What are we to make of their great variety, even for a single host? As an illustration, think of all the viruses known to cause disease in humans alone. Unlike bacteria, which evolved independently, viruses did so in association with their hosts. Furthermore, are prions more widespread than presently known, and do they function in basic biological phenomena, such as differentiation? These and many other questions are challenging topics and fertile areas for further investigation.

STUDY QUESTIONS

1. How do viruses differ from cellular forms of life?

2. Which components are common to all viruses?

3. Besides shape, what is an important difference between icosahedral and filamentous viruses?

4. What is the origin of the envelope in the enveloped viruses?

5. Certain viruses carry enzymes in their virions. What kinds of enzymes are these?

6. How does the nucleic acid of retroviruses replicate?

7. Discuss how viral proteins are synthesized during the viral replication cycle.

8. Distinguish between virulent and temperate viruses.

9. How do temperate viruses remain quiescent?

10. In a lysogenic bacterium, how is the decision between a lytic and a lysogenic cycle made?

11. Contrast viroids and prions.

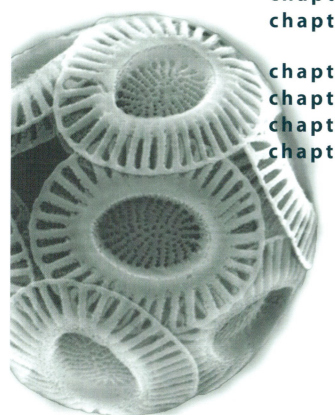

section VII interactions

chapter 18 ecology

OVERVIEW

Ecology is the study of interactions between organisms and their environment—how the chemical, physical, and biological environments affect particular kinds of organisms and how the organisms affect their environments. You might ask, What's new here? We have been discussing this topic throughout the book. That is true, but here we will change a bit. We will put greater emphasis on the grander scheme of the activities of groups of microbes rather than on those of individuals. We will pay particular attention to how microbes have molded Earth's environment and how they maintain it in a balanced, life-sustaining condition.

The fundamental themes of microbial ecology rest on three characteristics of microbes: (i) their **ubiquity**—they are present just about everywhere that liquid water exists; (ii) their **abundance**—they occur in huge numbers; and (iii) their **metabolic power**—they are extremely active, and in astoundingly diverse ways. We have already discussed (see Chapter 4) how some microbes grow under the most extreme conditions of temperature, hydrostatic pressure, and concentrations of salt, and their numbers are staggering. It's estimated that there are over 10^{30} bacteria and archaea on Earth. There are more bacterial cells in our intestines than human cells in our bodies. A milliliter of a bacterial culture or about 10 g of soil contains more microbes than there are humans on Earth.

The microbes' metabolic diversity contributes to their ubiquity. If the components of a thermodynamically feasible reaction are present in a particular environment, a microbe will almost certainly be there to exploit it in order to grow and reproduce. Of course, there must be a lower limit to how much energy a reaction must yield (the change in Gibbs free energy, or ΔG) in order for a microbe to be able to use it successfully, but it is certainly remarkable how low this value can be. In other words, prokaryotes seem able to squeeze every bit of energy out of a chemical compound. Prokaryotes are particularly adept at growing at the expense of low-energy-yield reactions. They do so in one of two ways.

1. They can elicit the help of neighboring microbes, usually by exploiting their neighbor's ability to scavenge reaction products. Such scavenging shifts the reaction's equilibrium and renders it able to support growth. For example, a butyrate-degrading bacterium, when cocultivated with a methanogen, can grow at the expense of a reaction that yields only −4.5 kilojoules per mole of free energy, a minuscule amount when you consider that it takes over seven times that much energy (−33 kilojoules per mole) to convert ADP to ATP.
2. Another strategy for squeezing sustenance out of low-yield reactions is to collect the small bits of energy that each turn of the reaction yields and to save them until sufficient energy has been accumulated to make a molecule of ATP. Microbes can do this by expelling a few protons with each turn of the reaction; eventually, the proton gradient they generate becomes capable of making ATP.

Microbes may be small, but they do things on a grand scale. They have a major impact on forming and setting the concentrations of the major gases in the atmosphere—nitrogen, oxygen, and carbon dioxide. They play critical roles in degrading remains of plants and animals that would otherwise accumulate and eventually sequester enough carbon and other bioelements to make life impossible. However, microbes are also responsible for a long list of other less readily apparent transformations that have profound effects on our environment. For example, by secreting metabolic end products and chelating agents, they modify the mineral compositions of rivers, lakes, and oceans. They catalyze redox reactions of a variety of metal salts, thereby contributing to the formation of Earth's huge ore deposits and causing rocks to weather. They even alter the weather. The Earth sciences have a major microbial component.

As we have noted, microbiologists have focused their greatest attention on the metabolism of aerobically growing chemoheterotrophs. The reasons are that these microbes are the easiest to cultivate, and most pathogenic microbes fall into this class. Some aerobic chemoheterotrophs do play important roles in the environment, but this group probably has a smaller ecological impact than do anaerobes and autotrophs.

We can get a glimpse of the vast variety of microorganisms that exist in nature, their diverse forms of metabolism, and their ability to grow almost anywhere by noting the types of microbes we might encounter if we were to penetrate into the ground below an open field (Fig. 18.1). In the top, well-aerated layers, we encounter largely the familiar chemoheterotrophs that utilize organic compounds by aerobic respiration. In the small anaerobic pockets that exist within this otherwise aerobic region, we find other heterotrophs that ferment carbohydrates. As we go deeper and enter the realm of shallow ground water, a largely anoxic zone, we find microbes that live by various anaerobic respirations. They use a variety of terminal electron acceptors, such as nitrate, ferric, and sulfate ions. Some of these microbes living by anaerobic respiration are heterotrophs that oxidize compounds, such as acetate, that are the trickle-

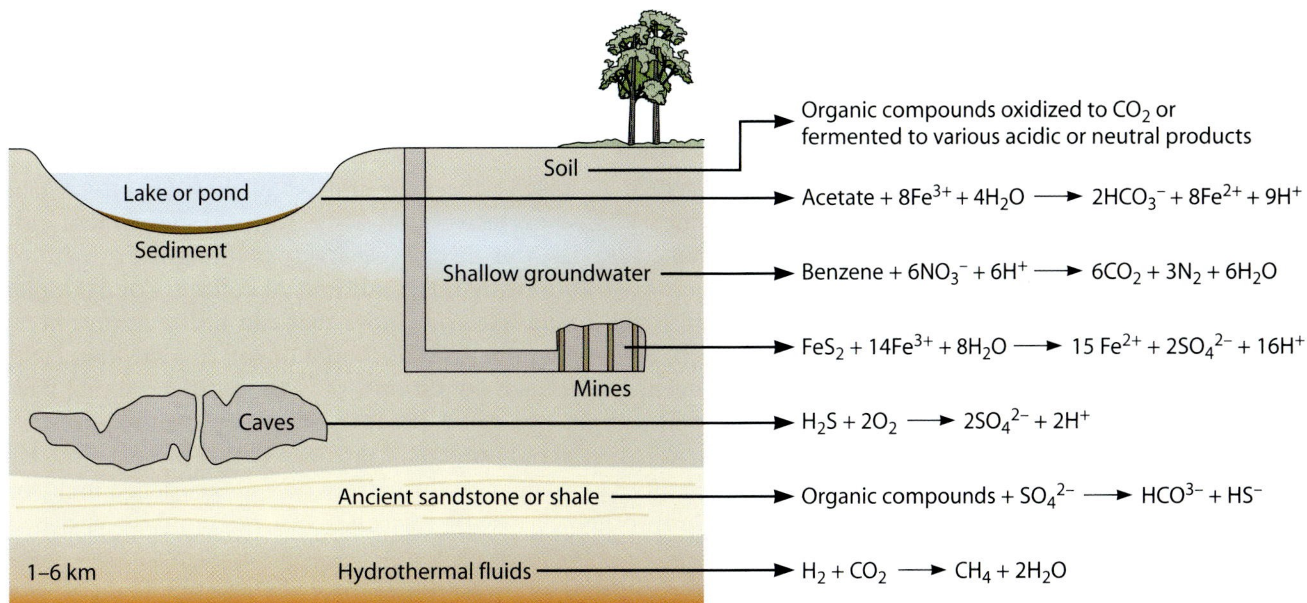

Figure 18.1 Metabolic conversions performed by microbes encountered at various depths in the earth.

down products of fermentations occurring in the soil above them. Others are microbes that oxidize pollutants, such as benzene, trichloroethylene, and various pesticides. We find autotrophs, as well, that also derive energy by anaerobic respiration of inorganic materials, such as sulfide or ferrous ions. As we descend further into the ground, we might be surprised that we continue to encounter microbes. Indeed, we find them at great depths, possibly as deep as 5 or 6 km, as evidenced by their presence in the saline water that leaks into deep mines, water that is known to originate this far down in the Earth's crust.

How can organisms even as metabolically diverse as microbes grow and reproduce at such great depths? Those in the deepest regions are necessarily hyperthermophiles, because their environment is extremely hot, but what is their source of nutrients? Both they and the microbes found at somewhat lesser depths live at the expense of hydrogen-dependent metabolism, coupling its oxidation to the reduction of metal ions or CO_2. Reduction of CO_2 produces methane, which rises to upper regions of the soil, where it is oxidized by methane-metabolizing microbial specialists called **methylotrophs.** Hydrogen gas might seem like an improbable substrate to be encountered at such great depths, but that's not so. Hydrogen is one of the gases that come from Earth's magma; it's formed there by the geochemical reduction of water in that high-temperature, high-pressure environment.

METHODS OF MICROBIAL ECOLOGY

The aim of all ecology is to determine which organisms are present in a particular environment and what they are doing there. In the case of higher organisms, considerable ecological information can be gained just by observation, but that is not so for microbes—although considerable progress is now being made toward that end. Unlike higher organisms, most microbes can't be identified just by looking at them, and because their activities are largely chemical, we seldom can see what they're doing.

Enrichment culture

Until quite recently, to answer the who and what questions, microbial ecology depended almost exclusively on some form of **enrichment culture.** As the name suggests, this means selectively favoring the growth of the desired microbes by adjusting the conditions of culture. For example, to enrich for **nitrogen-fixing bacteria** (those that can utilize atmospheric dinitrogen [N_2]), one might add a small amount of soil to a medium lacking fixed nitrogen. Under such conditions, only the nitrogen-fixing bacteria in that soil sample can grow because only they can acquire this essential nutrient from the only source of nitrogen that is present—the air over the culture. As a consequence, the culture becomes enriched in nitrogen fixers. Many other classes of microbes can be similarly enriched (Table 18.1).

Enrichment culture is a powerful tool for studying microbial ecology. If one suspects that a particular kind of microorganism is carrying out a particular transformation in a particular environment, it is only necessary to inoculate an appropriate medium with material from that environment. If the suspected microbe becomes enriched, it most assuredly is present in that environment. However, quite clearly, relying exclusively on enrichment culture is a risky approach to microbial ecology, because it rests on many unprovable and some incorrect assumptions; among the latter is the assumption that it is always possible to culture the microbes that exert a particular ecological impact, and among the former is the assumption that microbes act in nature as they do in culture.

Studying microbes in the laboratory and in their natural environments

It is a long-recognized and unpleasant fact of microbiological life that most prokaryotes in nature cannot be cultivated. Only recently have the magnitude and impact of this problem been fully appreciated. The num-

Table 18.1 Some examples of enrichment culture

Microbial class	Critical cultural condition	Rationale
Thermophiles	Incubate at a temperature in the thermophilic range, e.g., 55°C.	Only thermophiles can grow at such temperatures.
Endospore formers	Boil soil inoculum before adding to culture medium.	Very few vegetative microbial cells can withstand being boiled; endospores can.
Nitrogen-fixing cyanobacteria	Incubate aerobically in the light in a mineral salts medium lacking fixed nitrogen.	Only certain cyanobacteria can grow aerobically and phototrophically and fix nitrogen.
Sulfate-reducing bacteria	Incubate in the dark, anaerobically, with a nonfermentable carbon source and sulfate ion.	Under such conditions, sulfate-reducing bacteria can derive energy by anaerobic respiration; photosynthesis, aerobic respiration, and fermentation are not possible.
Microbes able to degrade a particular pesticide	Incubate aerobically in the dark in a mineral salts medium with the pesticide as the only source of carbon and energy.	Under such conditions, capability to degrade the pesticide is essential for growth.

ber of live microbial cells in environments such as soil or water determined by counting them under the microscope *exceeds by several orders of magnitude* the number that can be cultivated. However, ingenious but painstaking methods designed to mimic natural environments, such as slowly feeding nutrients at minuscule concentrations or incubating cultures for prolonged periods, are steadily decreasing the gap between the numbers of microbes observed and the numbers cultivated.

Why is it so desirable to obtain pure cultures of microbes? Obviously, most key studies, e.g., determining the genome sequence or conducting physiological and genetic experiments, would be difficult to carry out with single cells, but there are also drawbacks to studying microbes only in culture. It is quite clear that many microbes do not act in pure cultures as they do in nature. Domestication carries a price: some pathogens lose their virulence, others replace their antigens, and yet others are altered in their metabolic activities. In addition, microbes in nature often function as members of communities and not as distinct individuals. As a dramatic example, certain consortia of bacteria are capable of chemical conversions not possible for pure cultures of their members alone. Sometimes, such consortia can be isolated and cultivated as though they were pure cultures. In Chapter 15, we discussed the example of "*Methanobacillus omelianskii,*" a consortium of a bacterium and an archaeon capable of converting ethanol and CO_2 to methane, although in pure culture, neither of its two component prokaryotes can utilize ethanol.

How does one study microbes in their natural environments? With the advent of recombinant DNA technology, genomics, modern immunological methods, and radiochemical techniques, new sets of methods that do not depend on culture became available (Table 18.2). These have become powerful new tools for microbial ecology.

Fluorescence in situ hybridization (FISH) uses synthetic fluorescently labeled DNA probes that hybridize to complementary sequences in the genomes of organisms in a sample. Under a fluorescence microscope, FISH reveals which microbes are present in a particular environment, as well as their relative locations (see Fig. 18.5). Probes can be devised to be used in FISH that are not specific to a particular species but are diagnostic of a larger group of organisms. Thus, for example, FISH can be used to stain selectively either bacteria or archaea. Such information can be invaluable for deciphering microbial interactions in communities. As we have seen repeatedly, if two microbes are found to be closely associated in nature, they probably interact metabolically and might be dependent on one another for growth.

Fluorescent antibody probes are also powerful tools for identifying microbes in nature. This method is highly specific and is used to identify particular species or strains but not classes of microbes (as FISH can). Thus, it cannot be used to answer broad questions regarding the general types of microbes that might be present in a particular environment, although it can provide specific information about one particular type.

Another useful approach is to sequence certain stretches of DNA from a mixture of organisms in a particular environment, e.g., the gene for 16S ribosomal RNA (rRNA). Such information can answer a broad range of questions, from the highly specific to the completely general. For example, is a particular strain representative of a phylum or even a

Table 18.2 Some culture-independent methods of microbial ecology

Method	Procedure	Value
Identification		
FISH	A nucleic acid probe, usually rRNA-encoding DNA, is tagged with a fluorescent dye, hybridized to cells in a natural environment, and viewed by fluorescence microscopy.	Depending on the probe, individual microbial cells can be identified as belonging to a species, a genus, or some larger group, e.g., Bacteria. Using different colored tags and probes, several microbes or kinds of microbes can be identified in the same sample.
Fluorescent antibody probes	The procedure is quite like FISH, except an antibody is used in place of a nucleic acid.	The high specificity of antibodies limits this method to identification of strains or species.
Survey of microbes by sequencing	DNA is extracted from an environment, and either appropriate primers are added to enrich the sample by PCR for its various rRNA genes, which are cloned and sequenced, or all DNA in the sample is cloned and sequenced.	This method makes possible the identification of organisms, including those that cannot be cultured.
Evaluation		
Microscopic evaluation of viability: stains	Certain dye stains discriminate between viable and nonviable cells; fluorescein diacetate, a favored one, is nonfluorescent until taken up and hydrolyzed by nonspecific esterases.	Cells that become fluorescent are probably alive.
Microscopic evaluation of viability: nalidixic acid	Nalidixic acid and a small amount of a nutrient are added to one of two parallel samples; after a period of incubation, the samples are observed microscopically.	Because nalidixic acid inhibits cell division but not elongation, cell types identified as being longer in the treated than in the control sample are presumed to be able to grow by utilizing the added nutrient.
Microradioautography	A radioactive compound is added to a sample; after a few hours, cells are fixed and spread on a photographic-emulsion-coated slide in a light-tight chamber. A few days later, the slides are developed and observed.	Cells surrounded by exposed grains in the photographic emulsion are presumed to be able to metabolize the radioactive compound.

domain present in a sample? The method has the power to detect the more common microbes in a sample, even those that have never previously been encountered. Recently, this approach has been dramatically expanded. Rather than sequencing particular genes from an environmental sample, the complete complement of microbial DNA in a sample is sequenced. The first such technically demanding experiment was done on samples taken from the Sargasso Sea in the North Atlantic Ocean off Bermuda. Samples were filtered to enrich their content of prokaryotic cells; DNA was extracted, and over a billion base pairs were sequenced. Analysis of these voluminous data yielded a wealth of information. An estimated 1,800 microbial species were represented in the samples, and of these, 148 were previously unknown prokaryotic **phylotypes** (species identified solely by sequence similarity [see Chapter 16]). Over 1.2 million previously unknown genes were identified, providing some measure of the degree of microbial diversity in the environment sampled. Studies of this kind are proceeding at a rapid pace.

Being presented with such voluminous information about the numbers and kinds of unknown microbes and genes in the environment is a humbling experience because we are reminded how much more we have to learn. Also, such data offer only marginal information about what these various microbes are doing there. It is possible to make an informed guess about a newly discovered microbe's activities by determining its kinship to known, well-studied microbes and then assuming it might be doing something similar. But the microbe itself must eventually be cultivated and studied before we can know what is really going on. Microscopic evaluation of viability—by adding dyes that differentially stain living and dead cells—adds useful information. In addition, adding radioactive compounds, such as the DNA precursor tritium-labeled thymidine, will label actively metabolizing organisms. One can determine if individual cells have become radioactive by placing them on a microscope slide and covering them with a photographic emulsion. After suitable incubation in the dark, the photographic grains that result from isotope decomposition can be seen under the microscope. Using this technique, called **microradioautography,** and suitable labeled substrates, one can determine whether a particular microbial cell in a sample can utilize those substrates.

BIOGEOCHEMICAL CYCLES

In spite of the continuing challenges of unraveling the roles of the myriad microbes we can see but cannot culture, microbial ecology has progressed impressively. Through it, we have learned that microbes exist in all parts of the biosphere and that they are the sole inhabitants of some ecosystems, for example, deep in the soil, as we discussed earlier. We have also learned that the interconversions of matter in the biosphere are brought about to a large degree by microorganisms, and we have learned that some of these crucial changes are mediated only by prokaryotes. A useful route to comprehending and summarizing all these highly interrelated conversions is to sort them out according to the chemical changes that the various **major bioelements**—carbon, oxygen, nitrogen, sulfur, and phosphorus—undergo in nature. These interconversions are in approximate balance, that is, the total amounts of various forms of the bioelements, for example, nitrogen in the form of atmospheric N_2, do not change much over time, although they are continuously being formed and utilized. Thus, the sum of the reactions that use one particular form of a bioelement is approximately matched by the sum of those that replenish it. Therefore, it is possible to describe the various interconversions of a bioelement as a cycle of utilizations and replenishments. Such cycles are called **biogeochemical cycles.** In the following sections, we will emphasize the roles that microbes play overwhelmingly and uniquely, and we'll touch on the human impacts on these cycles, which have become considerable in recent years.

You will note a generalization that applies to almost all the cycles: *their component steps are oxidation or reduction reactions.* In most cases, the oxidation state of the bioelement changes throughout the cycle. For example, in the carbon cycle, organic matter is *oxidized* to CO_2, which is then *reduced* again to organic matter.

Carbon and oxygen cycles

Figure 18.2 summarizes the essence of the carbon cycle—a cycling between atmospheric CO_2 and fixed carbon, in either organic or inorganic form. Although simple in outline, this cycle encompasses all the

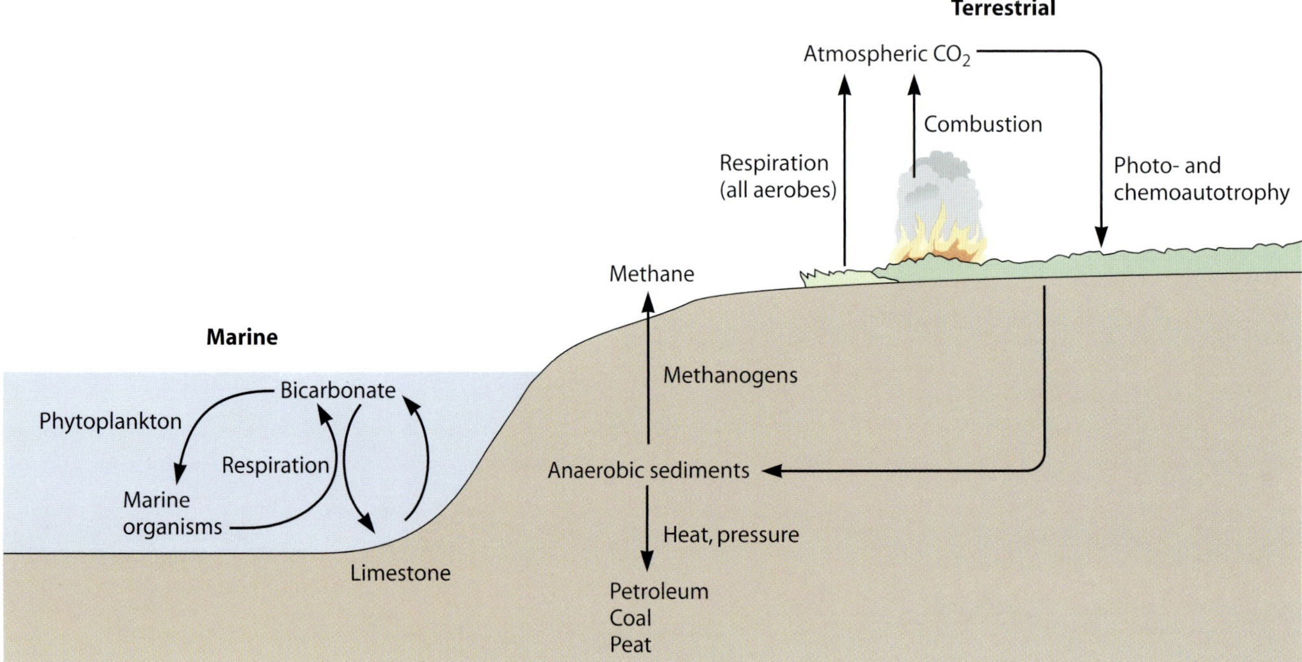

Figure 18.2 The carbon cycle.

complexities of the biological world, many of which we have already discussed, for example, the entrapment of organic carbon in anaerobic environments from which it is only slowly returned to the active cycle (see Chapter 15).

We humans have intervened massively in this cycle with consequences still not fully understood. By burning fossil fuels at an ever-increasing rate, we have markedly increased the concentration of carbon dioxide in Earth's atmosphere, contributing to the global warming we are now experiencing (Fig. 18.3).

Humans have also added a qualitatively new complexity to the cycle. Naturally occurring organic molecules, without exception, are susceptible to microbial attack. Some, such as **humus** (the black, lignin-rich component of soil) and some in anoxic environments, are degraded at quite low rates, but there is overwhelming evidence that *there are no naturally occurring organic compounds that cannot be broken down,* however slowly, by some microorganism (a doctrine sometimes called **microbial infallibility**). However, humans have accomplished what nature could not. Many plastics we have manufactured seem to be completely resistant to microbial attack, and other ecologically dangerous compounds are attacked only extremely slowly. They form a dead-end sump in the carbon cycle. If discarded in nature, they will remain for a very long time (Fig. 18.4).

As we have mentioned, a major route by which degradable organic compounds trapped in anaerobic environments are returned to aerobic environments is through the action of methanogenic Archaea. Worldwide, these microbes (see Chapter 15) produce prodigious quantities of methane from organic material trapped in various anoxic environments—including silt under oceans and lakes and even from the

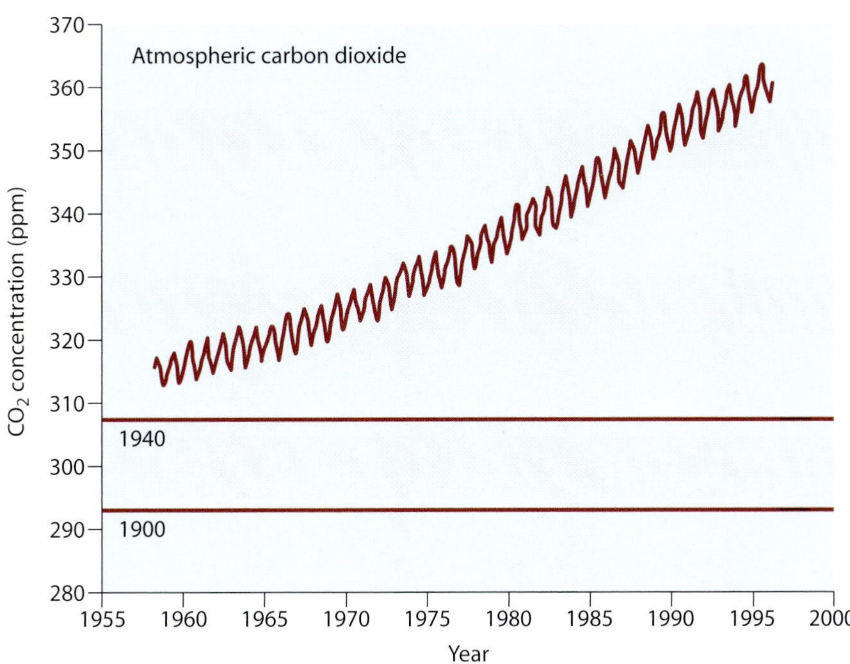

Figure 18.3 Concentration of CO_2 in the atmosphere, measured at Mauna Loa, Hawaii. (The value at the end of 2002 was 373 parts per billion.)

intestinal tracts of ruminant animals and termites. Methanogenesis is a critical link in Earth's carbon cycle because the product, methane, being gaseous and sparingly soluble in water, can escape the trap of its anoxic environment. In addition, huge quantities of methane from biological and nonbiological sources are buried beneath the ocean floor as solid-phase gas hydrates and as free gas.

We might ask what happens to all this methane. Until relatively recently, the only known fate of methane, other than being burned by humans, was oxidation by aerobic bacteria called methylotrophs or escape into the atmosphere, where it acts as a powerful greenhouse gas. (Methane in our atmosphere increased from 1,620 parts per billion in 1984 to 1,750 parts per billion in 2003 but now seems to be leveling off.) It was thought that methane must escape from its anoxic reservoir before it could be degraded. In the 1970s, however, geochemical evidence suggested that large quantities of methane are degraded within anoxic regions of the oceans. In the late 1990s, a remarkable route of disappearance was discovered. It is worth considering this discovery, because it illustrates how much can be learned about microbial ecology by using culture-independent methods.

This route of anaerobic methane utilization proved to be biological, mediated by a two-member microbial consortium consisting of an archaeon and a bacterium that together anoxically oxidize methane to carbon dioxide while reducing sulfate to hydrogen sulfide.

$$CH_4 + SO_4^{2-} \longrightarrow CO_2 + H_2S + 2OH^-$$

These microbes, which consume 80% of the methane produced in marine environments, have not yet been cultivated. The composition and activity of this methane-oxidizing consortium was determined by completely molecular methods. Aggregates of these prokaryotes were examined by

Figure 18.4 Plastic discarded anywhere in the ocean accumulates on beaches, such as this one.

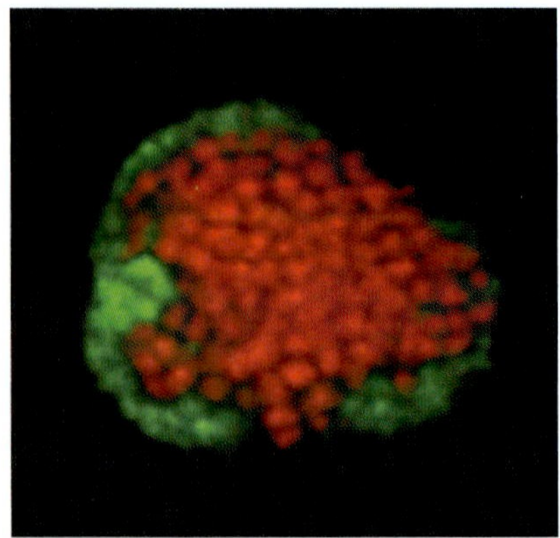

Figure 18.5 A consortium of methane-degrading and sulfate-reducing prokaryotes mediating the anaerobic oxidation of methane. The cells were stained with a green fluorescent RNA probe targeted specifically against a group of sulfate-reducing bacteria (*Desulfosarcina* and *Desulfococcus*) and a red fluorescent RNA probe targeted against a group of archaea called ANME-2. The diameter of the aggregate is approximately 10 μm. The image was taken with a confocal microscope.

Figure 18.6 A mat of methane-degrading prokaryotes from the Black Sea. The preparation is seen under a confocal microscope, which allows one to observe a *section* through the mat. Note that such a section cannot be seen under an ordinary microscope. As in Fig. 18.5, the methane-utilizing archaea are labeled with a red fluorescent probe, and the
sulfate-reducing bacteria are labeled

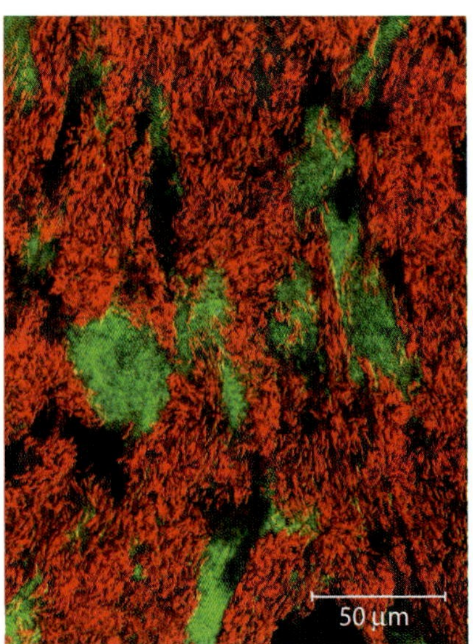

16S rRNA-targeted probes for FISH and were revealed to be packages of about 100 archaeal cells surrounded by a layer of about as many bacterial cells (Fig. 18.5). The sequence of the 16S rRNA of the archaeal cells showed that they constitute a **clade** (related branch) of the methanogenic group of the Archaea. Because of this relationship, it was presumed that these archaea mediate a metabolism similar to methanogenesis, only in reverse, i.e., they convert methane to carbon dioxide and hydrogen.

$$CH_4 + 2H_2O \longrightarrow CO_2 + 4H_2$$

In order for this energetically unfavorable reaction to become feasible and therefore able to support microbial growth, hydrogen must be maintained at a sufficiently low concentration to "pull" the reaction to the right. Efficient removal of hydrogen is accomplished through scavenging by a surrounding bacterial layer consisting of *Desulfosarcina*, which catalyzes oxidation of hydrogen at the expense of reducing sulfate.

$$5H_2 + SO_4^{2-} \longrightarrow H_2S + 4H_2O$$

That the consortium does indeed metabolize methane in its natural environment was also established by chemical methods, exploiting the fact that naturally occurring methane (for complex reasons) always contains a distinctly smaller fraction of ^{13}C than do other organic compounds. Using a technique called secondary ion mass spectrometry in conjunction with FISH, it was shown that both the archaeal and bacterial components of the consortium were extremely depleted in ^{13}C. Therefore, they must have grown at the expense of methane.

Other methane-utilizing Archaea found in the Black Sea occur in consortia with sulfate-reducing bacteria within huge microbial mats, some as high as 4 m and as wide as 1 m (Fig. 18.6). These mats, which are located over methane vents in the floor of the sea, are structurally stabilized by calcium carbonate, which forms as the CO_2 released by the consortia reacts with locally alkaline seawater. A lot has been learned about these organisms, in spite of their never having been cultivated.

The nitrogen cycle

Insufficient nitrogen limits plant growth in many environments. Indeed, the "Green Revolution," a system of improved agriculture that has had a major impact on food production worldwide and decreased starvation, was based principally on providing more nitrogen to crops, principally in the form of ammonia or nitrate.

All the nitrogen that supplies the nutritional needs of plants and indirectly all other life comes from Earth's huge reservoir of atmospheric N_2, and all is returned to it surprisingly rapidly. The half-life of atmospheric N_2 is only about 20 million years, a moment compared with life's 4-billion-year history on Earth. Dinitrogen is a remarkably stable compound, largely because of the huge activation energy required to break its nitrogen-nitrogen triple bond. Prokaryotes are the only organisms capable of breaking this bond, and thereby, of fixing nitrogen (see Chapter 7). Until the early 20th century, when humans learned to convert atmospheric N_2 to ammonia by a chemical method called the Haber process, all fixed nitrogen on Earth (with the exception of small quantities formed by lightning and volcanic activity) was produced by prokaryotes. The importance of microbial activity is illustrated by the calculation that, if microbes ceased to participate in the **nitrogen cycle,** plants would run out of nitrogen in about 1 week. Now, we manufacture about half of the utilizable nitrogen. Other steps of the nitrogen cycle (Fig. 18.7) are also the exclusive provinces of prokaryotes. They include the two steps of **nitrification** by which ammonia is successively converted via nitrite to nitrate ion and the two routes—called **denitrification** and **anammox**—by which fixed nitrogen is returned to gaseous form.

Nitrification is mediated by autotrophs. In the first step of the process, autotrophic bacteria, typified by the genus *Nitrosomonas* (a gammaproteobacterium), oxidize ammonia to nitrate while reducing O_2. In the second step, bacteria typified by the genus *Nitrobacter* (an alphaproteobacterium) oxidize nitrite to nitrate, also reducing O_2 in the process. Curiously, no bacteria are known that oxidize ammonia all the way to nitrate. Nitrifying bacteria are widespread in nature and highly active; therefore, ammonia and nitrite are short-lived in aerobic environments, such as the upper layers of soil. When anhydrous ammonia gas is added as fertilizer to soil (a common practice in modern agriculture), it reacts with water to form ammonium ion, which binds tightly to clay soils and is therefore immobile. However, nitrifying bacteria quickly convert it to nitrate ion, a form that moves freely through soil and is readily used by plants.

Because of the British blockade of Europe during the early 19th century, manure piles were used extensively to supply Napoleon's armies with the nitrate needed to make black powder (a mixture of nitrate, sulfur, and charcoal then used as gunpowder). Such piles are miniature replicas of one arm of the global nitrogen cycle. These piles were turned

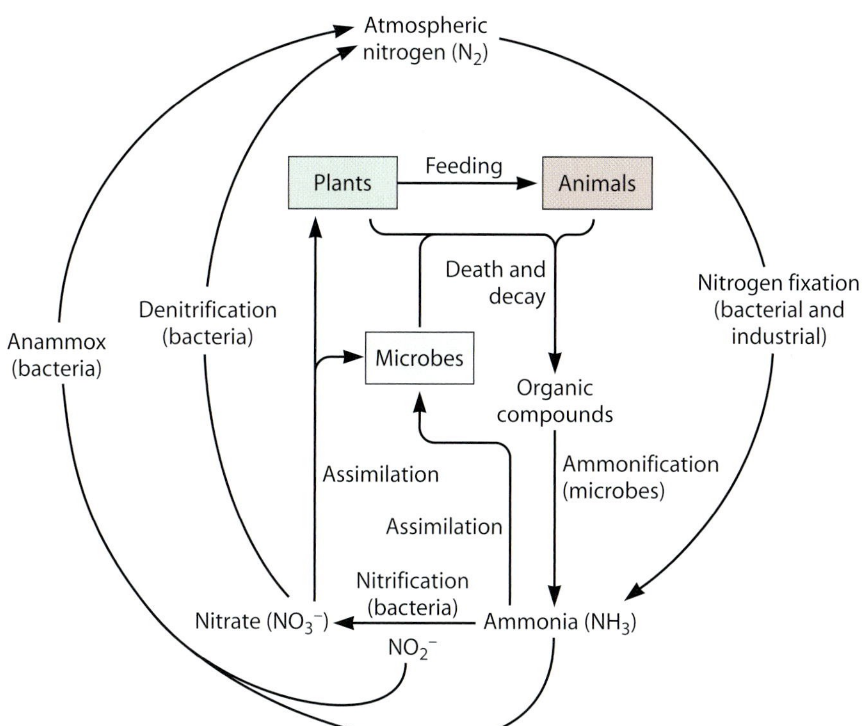

Figure 18.7 The nitrogen cycle.

repeatedly to maintain **aerobiosis.** Within the pile, the complex organic nitrogen in manure was converted to ammonia by a variety of microbes, and then by nitrifying autotrophs successively to nitrite and nitrate. Because denitrification and anammox are obligatorily anaerobic processes, the nitrogen cycle could not be completed in these piles. Nitrogen became trapped in the form of nitrate and accumulated. A similar process occurs in the huge deposits of **guano** (dried excrement of seabirds) on the coastal islands of South America, Africa, and the West Indies.

As seen in Fig. 18.7, organisms use nitrogen in two completely different ways. (i) All organisms use it as a nutrient that they incorporate into cellular constituents, such as proteins, nucleic acids, and phospholipids (see Chapter 7). (ii) Some prokaryotes also use it as a substrate for two types of ATP-generating processes. As we mentioned, some autotrophic bacteria derive energy by oxidizing ammonia or nitrite. Other heterotrophs derive energy by anaerobic respirations, with nitrate, nitrite, NO, or N_2O serving as a terminal electron acceptor. In certain heterotrophs, a particular nitrogen compound, for example, nitrate, can serve either as a nutrient or as a means of generating energy. These two roles are distinguished by the terms **assimilatory** and **dissimilatory.** The process of reducing nitrate to ammonia to serve as a nutrient is called **assimilatory nitrate reduction.** Reduction of nitrate as a consequence of serving as a terminal electron acceptor in an anaerobic respiration is called **dissimilatory nitrate reduction.**

The final step of the nitrogen cycle returns fixed nitrogen to atmospheric N_2. This step is almost exclusively biological. Until recently, it was thought that the only way this was accomplished was through **denitrification,** the cascade of anaerobic respirations through which nitrate is successively reduced to N_2 (see Chapter 6). A wide variety of prokaryotes utilizing a broad spectrum of carbon sources as electron donors can mediate this transformation. Then, in the mid 1990s, a new group of bacteria that can produce N_2 from fixed nitrogen was discovered. These bacteria mediate an anaerobic ammonia oxidation (**anammox**)—a route now recognized as being a major way by which fixed nitrogen (in the form of ammonia or nitrate) is recycled to atmospheric N_2 gas. This newly discovered anammox route was missed because it is not yet possible to cultivate the microbes that mediate the process and because conventional denitrification seemed an adequate explanation for the fixed nitrogen-to-N_2 step of the nitrogen cycle. However, the contribution of anammox has been shown to be huge. From one-third to one-half of Earth's N_2 production occurs in anoxic waters of the ocean, and of that, between 19 and 67% occurs via anammox.

The discovery of anammox employed sophisticated modern methods. It is instructive to consider just how this happened. Two Dutch microbiologists were faced with the following dilemma: ammonia, but not nitrate, was disappearing from an anoxic reactor in a wastewater treatment plant they were studying. Why one but not the other? Then they found that nitrite was also disappearing; thus, they suspected a reaction in which ammonia was oxidized by nitrite, yielding N_2, composed of one N atom from ammonia and the other from nitrite, i.e.:

$$NH_4^+ + NO_2^- \longrightarrow N_2 + 2H_2O$$

They proved their hypothesis by adding $^{15}NH_4^+$ and $^{14}NO_2$ to the reactor and showing that it produced N_2 that contained one atom of ^{15}N and one of ^{14}N. Then, they showed that the process was almost certainly biologically mediated because it was stopped by heat, gamma irradiation, and several **decouplers** (compounds that break down a cell's proton gradient, thereby inhibiting ATP generation via chemiosmosis [see Chapter 6]).

In spite of becoming aware of the anammox reaction, the investigators could not culture the organism that was mediating it, but they could enrich for it in a flowthrough reactor fed with ammonia, nitrite, and bicarbonate. A mixed culture, 70% of which was made up of similar-appearing coccoid cells, developed. By density gradient centrifugation, these coccoid cells were concentrated to more than 90% purity. The cell preparation could convert ammonia and nitrite into N_2 while fixing CO_2. Still, the cells could not be cultured. Sequencing their ribosome-encoding DNA showed that they were bacteria belonging to the phylum Planctomycetes (which we met in Chapter 3, where they were described as having a membrane around their nucleoids). The putative species they comprised was named "*Candidatus* Brocadia anammoxidans" ("*Candidatus*" indicates that it hasn't been cultivated [see Chapter 15]). The discovery of anammox illustrates how much can be learned about the activities of microbes in nature, even if they cannot be cultivated. It also illustrates how difficult it is to cultivate certain microbes, even when you know a lot about them.

The sulfur cycle

The **sulfur cycle** (Fig. 18.8) resembles the nitrogen cycle in that sulfur, like nitrogen, serves two distinct metabolic roles in microbes: for all organisms, it is an essential nutrient (although sulfur is less abundant in the cell than nitrogen), and for some, it enters into both oxidative and reductive pathways that generate ATP. In the latter role, huge amounts of sulfur are processed. The terrestrial reservoirs of sulfur are enormous, and some of them are formed by microbes. Massive deposits of elemental sulfur in the form of gypsum, which occur in the beds of ancient lakes, were formed by the combination of two steps of the sulfur cycle, both anaerobic: sul-

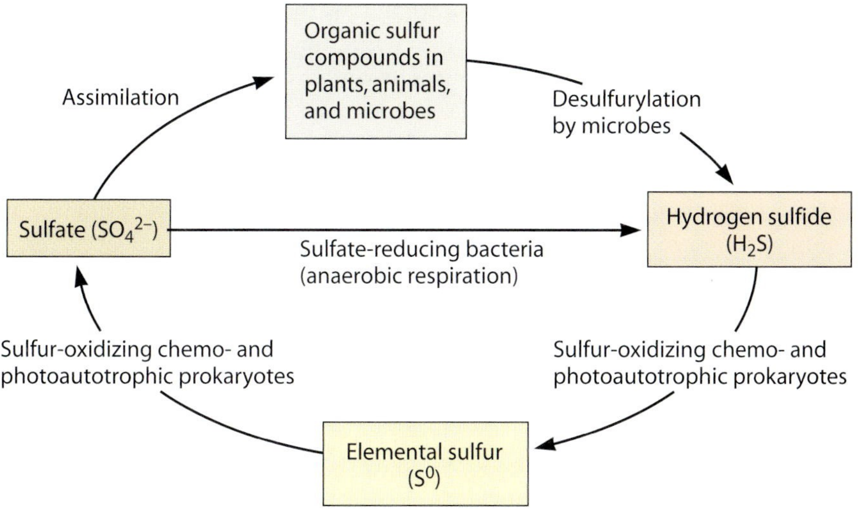

Figure 18.8 The sulfur cycle.

fate-reducing bacteria converted sulfate to hydrogen sulfide, and phototrophic bacteria oxidized it to elemental sulfur. The same two-step process of sulfur formation occurs in some lakes today.

Hydrogen sulfide is also oxidized by a variety of autotrophic bacteria, mostly through aerobic respiration but in a few cases through anaerobic respiration. This conversion occurs abundantly in just about any place hydrogen sulfide is produced by sulfate-reducing bacteria in an underlying anaerobic region—mudflats, for example. Because hydrogen sulfide spontaneously oxidizes in air, chemoautotrophic sulfur oxidizers locate themselves at the interface where rising hydrogen sulfide comes in contact with oxygen—the only place they can have access to both nutrients.

18.1 *Hydrothermal vents*

In most locations, sulfur oxidizers utilize hydrogen sulfide that is supplied by the metabolism of other microbes, but there are spectacular exceptions. Geochemically produced hydrogen sulfide spews out of **hydrothermal vents** in the midocean floors, where Earth's tectonic plates separate and new crust is constantly being formed. A complex community consisting of hundreds of species, both prokaryotic and eukaryotic, thrives there. Unlike the rest of the biosphere, which depends on the primary productivity of photosynthesis, the community in this sunless region depends entirely on chemoautotrophic bacteria that oxidize hydrogen sulfide from the vents at the expense of oxygen diffusing down from the ocean surface. In some cases, this dependence on hydrogen sulfide-oxidizing bacteria is remarkably intimate. The huge tube worms that develop there lack an intestinal system. Instead, they are filled with a spongy tissue, the **trophosome,** consisting principally of sulfur-oxidizing bacteria, which supply the worm with nutrients. The sulfur cycle also has an important atmospheric component. Large amounts of volatile sulfur-containing compounds, chiefly dimethylsulfide, enter the atmosphere and are changed there by sunlight into other forms. This aspect of the sulfur cycle is discussed below (see "Microbes, climate, and the weather").

The phosphorus cycle

The phosphorus cycle (Fig. 18.9) is the simplest of the biogeochemical conversions because phosphorus, for the most part, stays in the same oxidation state, +5 (that is, as phosphate). Phosphate, either organic (as esters, amides, or anhydrides) or inorganic, is the principal form of phosphorus in biological systems. Worldwide cycling of phosphate is extremely slow, because this cycle has no significant gaseous intermediate. Most forms of phosphate, being soluble, are leached from soil and eventually taken to the oceans. Return of phosphate from the oceans to land masses occurs to a minor extent when marine birds that feed on marine animals deposit their feces on land, sometimes forming deposits of guano, as we mentioned in our consideration of the nitrogen cycle. However, most phosphorus from the seas returns to Earth's landmasses only by geological uplift of the ocean floors—a very slow process. The products of these uplifts are phosphate rock, which is mined largely (about 90%) for use in fertilizer.

To a lesser extent, nonphosphate forms of phosphorus do participate in the phosphate cycle. A gaseous form, phosphine (H_3P), which spontaneously ignites in air, has long fascinated humans as a possible cause of "the will o' the wisp," the ghostly pale blue light sometimes seen hover-

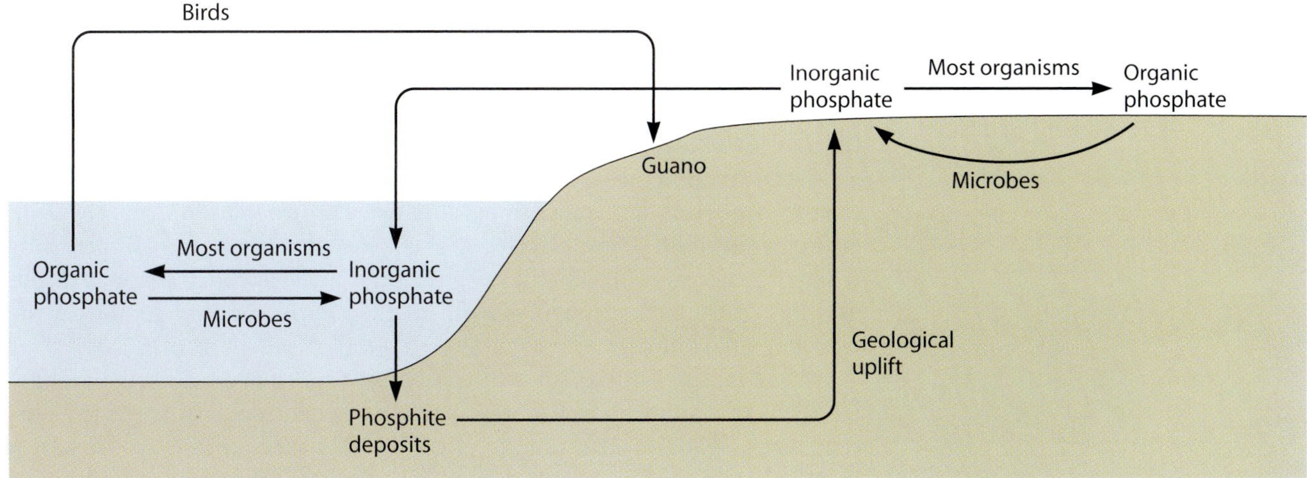

Figure 18.9 The phosphorus cycle.

ing over swamps. Phosphine does exist in nature. It occurs in minuscule quantities (nanograms to micrograms per cubic meter) in the lower atmosphere, but it is not clear whether any of it has a biologic origin. Other reduced forms of phosphorus, including phosphites (PO_3^{3-}), and hypophosphites (PO_2^{3-}) are also found in nature, and some bacteria are capable of oxidizing them for use as sources of phosphate. In some environments, significant quantities of phosphonates ($R\text{-}PO_3^{2-}$) are encountered. These are of clear biologic origin, being made by prokaryotes and eukaryotes (phosphonate antibiotics are made by some streptomycetes). Although the C-P bond of phosphonates is quite stable, certain bacteria can break it and utilize phosphonates as a source of phosphorus by converting them to phosphate.

SOLID SUBSTRATES

Only dissolved nutrients can enter the cells of prokaryotes. Eukaryotes can ingest particulate nutrients by **phagocytosis,** but prokaryotes lack this ability. Nevertheless, prokaryotes can utilize a variety of insoluble nutrients, including starch, cellulose, and even agar. They do so by secreting enzymes that break down these insoluble polymers into soluble subunits, which then enter the cell. Thus, prokaryotic digestion occurs in the cell's immediate external environment.

Recently, a startling exception to this generalization was discovered: insoluble materials, such as iron, magnesium, and uranium oxides, that cannot be broken down into soluble subunits can nevertheless be metabolized by certain bacteria. A genus of bacteria, *Geobacter,* is capable of an anaerobic respiration using one or another of these solid oxides as the terminal electron acceptor. It does so without taking these oxides into the cell. In the case of iron, solid rust-colored pieces around the cell are reduced to pale green ferrous ion. Not surprisingly, in view of their unusual metabolic capacity, geobacters have an unusual morphology. They are comma shaped and bear flagella largely on one side and prominent short pili on the other. The flagella form only in the presence of ferrous ion, apparently an adaptation signaling the cell that the ferric oxide in the particle to which it is attached is nearly depleted and it is time to move on to another piece of rock. Most probably, the pili fulfill a dual role: (i) they are the organelles that attach the cell firmly to the solid surface, and (ii) they are conductors of electrons, rather like the prongs on the plug on an electrical appliance—on the outer surfaces of their tips, they bear electron-transporting molecules, including cytochromes, an unusual location for such enzymes. Thus, electrons, or an electric current if you like, flow from organic nutrients in the cell, through an **electron transport chain,** and out of the cell through the pili to the chunks of iron oxides in the external environment.

A fascinating consequence of *Geobacter*'s nonspecific ability to donate electrons to a solid surface is its capacity to generate a usable electric current, because it can also donate electrons to a piece of metal. Thus, if a metal plate is buried in an ocean floor where *Geobacter* is abundant, a biofilm of *Geobacter* will develop on the surface of the plate and current will flow from it to another plate in the overlying water. Although, the current generated is small—on the order of microamperes—it is sufficient to power small oceanographic sensing devices, and it is being put to use for that purpose. A related bacterium, *Shewanella,* shares these abilities.

MICROBIAL ECOSYSTEMS

We might ask where the microbe-mediated transformations that we've just discussed take place. The simple answer is, just about everywhere. Few places on the planet are too hot, too cold, too acid, too alkaline, too salty, or at too high a hydrostatic pressure for microbes to thrive. We present a few of the major habitats.

Soil

The upper layers of soil are an especially active microbial ecosystem in which many steps of the biogeochemical cycles occur. They teem with microorganisms. A gram of typical soil contains a million to a billion bacterial cells, 10 to 100 m of fungal hyphae, thousands of algal cells, and thousands to millions of protozoa. As we discussed at the beginning of this chapter, microbial growth is not limited to the upper layers. Although the number of microbes decreases in lower layers, microbial activity continues down to the extraordinary depths of 5 or 6 km below the surface of the Earth.

Oceans

The upper regions of the oceans, which cover 71% of Earth's surface, are the other globally important microbial ecosystem. It is estimated that about half of the photosynthesis (the CO_2-fixing and oxygen-producing arm of the carbon cycle) that occurs on the planet is carried out by **phytoplankton** (floating phototrophic microbes) living in the upper layers of the ocean, where enough light penetrates to support their growth (Table 18.3). In addition to being major modulators of the proportions of gases in Earth's atmosphere, phytoplankton comprise the beginning of the food chain for life in the seas. Other marine organisms, from krill (small shrimp-like creatures) through crustaceans and fish to whales, depend directly or

Table 18.3 Abundances of organisms in 1 ml of seawater

Type of organism	No./ml of seawater
Zooplankton	<<1
Phytoplankton	
Algae	3,000
Protozoa	4,000
Photosynthetic bacteria	100,000
Heterotrophic bacteria	1,000,000
Viruses, including phages	10,000,000

indirectly almost exclusively on the primary productivity of the phytoplankton. The most abundant members of the phytoplankton are single-celled cyanobacteria, belonging principally to just two genera, *Synechococcus* (see Fig. 15.4B) and *Prochlorococcus*. Amazingly, the discovery that these cyanobacteria play such import roles in the ocean's contribution to the world's carbon cycle is relatively recent. The abundance of tiny (for a cyanobacterium) *Synechococcus*, which has cells that are about 0.9 μm in diameter, was recognized in 1979. The even smaller *Prochlorococcus* (0.5 to 0.7 μm) was discovered in 1988 (see Chapter 15). Eukaryotic phytoplankton, such as diatoms and algae, are more than 10 times larger, but being more abundant and metabolically active, cyanobacterial phytoplankton exert a greater ecological impact.

Like those in soil, microbes in the ocean are not restricted to the upper regions. Microbes have proliferated all the way down to the regions of darkness, cold, and crushing hydrostatic pressures that exist at the bottom of the seas (which have a median depth of 4 km). We have already discussed the sulfide-oxidizing bacteria that flourish on the ocean floor near hydrothermal vents. Although distinctly layered, so that the water column in the ocean presents different environments at different depths, the seas are subject to stirring, both by the action of winds and tides on the surface and by circulating currents. As a result, the water column becomes somewhat mixed, but not entirely.

Now let us consider the nutritional environment of the ocean at various depths, keeping in mind that the oceans of the world vary enormously. In addition to dissolved nutrients, the oceans contain, among other things, seaweed, crabs, and fish and are greatly affected by their bottom sediments, as well as the solid earth at the coasts. Diverse though it is, some generalizations about the ecology of the ocean are possible. For example, in the open seas, the concentration of available organic material is low but measurable. There is enough organic material to sustain a considerable population of heterotrophic microbes. In fact, the concentration of bacteria in surface seawater is about 1 million/ml, a figure that is relatively constant throughout the world's oceans.

Light is plentiful on the surface, but only about 1% reaches a depth of 100 m. Thus, the zone where phytoplankton flourish is quite narrow. Some bacteria ensure that they remain at this favorable location by changing their buoyant density via gas vacuoles (see Chapter 3). In contrast, the zone favorable to aerobic heterotrophs is much broader. Oxygen, which, like sunlight, enters the ocean at the top, either from that in air or from that produced by phytoplankton, extends to the bottom. Its concentration at depths of 4 km is still half as great as it is on the surface. Phosphorus is scarce on the surface and more abundant below about 1 km. Consequently, considerable microbial life is found in deep waters. Of course, aerobic heterotrophs also need organic nutrients. These are largely supplied by the constant "rain" of organic detritus that slowly settles through the water column. This material, known as **marine snow,** consists of polysaccharide matrices that are readily visible with the naked eye and in which living or dead or dying animals and plants are embedded. Sometimes, marine snow assumes blizzard proportions and can limit the visibility of ocean divers to a few feet. Marine snow sediments at rates in the range of 30 to 70 feet per day and, when it hits the bottom, becomes available for the microbial and nonmicrobial residents of the depths. By far the greatest number of **pelagic** microbes (those in the water column) are attached to these particles, which supply them with nutrients.

The sea floor is a different environment than the water column. The organic nutrients there are steadily deposited by falling organic detritus. The concentration of **benthic** microbes (those in the sediment on the bottom) exceeds that of pelagic microbes by up to 5 orders of magnitude. Oxygen is quickly depleted in the benthic environment. The deeper layers of the sediment, being oxygen depleted, are the home of microbes that use nitrate, sulfate, or ferric iron as their terminal electron acceptor (see Chapter 6). Those that utilize sulfate produce sulfide (H_2S), which in turn is oxidized to sulfur by other anaerobes using nitrate as their terminal electron acceptor. Nitrate is fairly abundant in seawater, but it is depleted in the sulfide-rich sediment where these bacteria proliferate. How do such bacteria acquire their nutrients, which are separated in space: one (nitrate) in the water column and the other (sulfide) in the sediment? Two remarkable bacteria solved the problem in different ways. Cells of one of these organisms, *Thioploca,* become encased in a long mucous **sheath,** which serves as a transport route between the underlying sediment and the overlying water (an example of a sheathed bacterium is shown in Fig. 18.10). *Thioploca* cells travel up and down these sheaths, acquiring sulfide from below and the nitrate needed to oxidize it from above. These organisms take an elevator to work!

The other bacterium, *Thiomargarita,* solves the problem in a different way. Instead of commuting between food and oxidant, its cells remain relatively stationary in strands of cells that are evenly separated by a mucous sheath. There, they sit and wait, accumulating nitrate while waiting for a puff of food in the form of H_2S to pass by. These remarkable organisms have a very large vacuole in which they concentrate nitrate at levels up to 10,000 times higher than in seawater. The vacuole is the equivalent of a tank on the back of a scuba diver, but instead of air, it contains nitrate. This vacuole is huge, making the cells that contain it the largest prokaryotes known: up to three-quarters of a millimeter in diameter (see Chapter 3 for details on gargantuan bacteria). *T. namibiensis* is startlingly bigger than other bacteria: if ordinary bacteria were enlarged to the size of a newborn mouse, *Thiomargarita,* equally enlarged, would be larger than a blue whale. The researchers who found this organism in the waters of Namibia named it *Thiomargarita namibiensis,* which means sulfur pearl of Namibia. Besides nitrate, *T. namibiensis* also stores elemental sulfur, which shines with an opalescent blue-green whiteness, making the string resemble a strand of pearls.

Figure 18.10 A thin section of a sheathed bacterium *(Leptothrix discophora)* examined under an electron microscope. Bar, 1 μm.

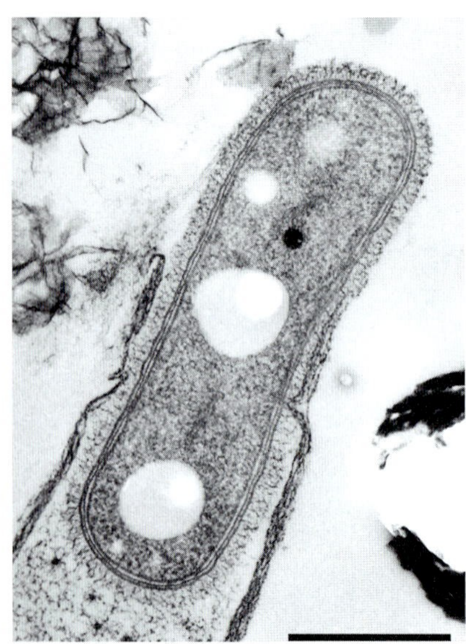

Microbes, climate, and weather

That microbes should affect the climate of our planet is not surprising, because they play critical roles in the turnover of the main gases in the atmosphere: nitrogen, oxygen, and CO_2. The atmospheric reservoirs of O_2 and N_2 are so vast that their turnover times are in the millions of years. The turnover time of CO_2 is considerably shorter: only a few years. However, as we have discussed, atmospheric CO_2 is the greenhouse gas that exerts major effects on Earth's long-term climate. It is surprising that, as is now becoming clear, microbes also affect local weather by causing atmospheric changes that take place over a period of just days.

The story is about cloud formation. It is still an unfolding story, but the general plot is already known. Clouds form when certain compounds in the air act as nuclei for water vapor to condense on, thereby forming fine droplets. Some of these compounds contain sulfur. Where do they come from? A large fraction of them are from nonbiological sources, such as volcanic emissions and the burning of coal and high-sulfur petroleum. These sources produce SO_2, which is oxidized to SO_3 in the atmosphere and, when hydrated, becomes sulfuric acid (H_2SO_4), the principal nucleating form. Soon, in a mater of days, this sulfuric acid is returned to earth in water droplets (as acid rain) or on the surfaces of particulate matter.

However, large amounts of nucleating sulfur compounds do have a biologic origin. This part of the story begins in the top layers of the oceans, where the phytoplankton, including its eukaryotic members, such as the coccolithophores (see Chapter 16), are found in prodigious quantities. On cloudless days, the sun's intense ultraviolet irradiation stresses them. In response, these eukaryotic microbes make huge amounts of a protective compound, dimethylsulfoniopropionate (DMSP). During the periods of algal blooms, DMSP can reach concentrations as high as the millimolar range in the surrounding seawater. A variety of bacteria that are present in the water metabolize DMSP, breaking it down to a volatile compound, dimethylsulfide (DMS), which escapes into the atmosphere. Massive amounts are made: the total annual flux of biogenic DMS to the atmosphere approaches 50 million tons of sulfur per year. In the atmosphere, DMS reacts with oxygen to create various sulfur compounds that act as nuclei of the water droplets that form clouds.

Notice that the process acts as a feedback loop. As the intensity of sunshine increases, algal stress increases, causing more DMSP to be made, forming more clouds, which cause a decrease in the intensity of sunshine. Then, algae become less stressed; they make less DSMP, and fewer clouds form. The cycle repeats.

Do these cycles occur over land? The answer is not yet clear, but it is well established that microbes do make DMS wherever they grow, including in our large intestines. Indeed, the main "rotten-egg" smell in our flatus is due to DMS.

THE FUTURE OF MICROBIAL ECOLOGY

The commanding fact that the overwhelming majority of prokaryotes in most environments have not yet been cultivated announces clearly that microbial ecology remains a work in progress. We can only speculate about its future. Certainly, many new prokaryotes will be discovered—most probably by molecular methods, such as genomic sequencing, fluorescence-labeled probes, and FISH. Some of these newly discovered microbes—probably for some time still a minority—will be cultivated as more sophisticated and imaginative methods are discovered. Certainly, the discovery and cultivation of still unknown microbes will continue to present significant challenges. It is worth noting that molecular methods of discovery depend on comparisons to known microbes. For example, sequences of bases in primers used in polymerase chain reaction (PCR) searches for DNA from new microbes in nature are those in the highly conserved regions of known microbes. Truly unusual microbes would not be discovered by such methods, although random sequencing of DNA from nature (a procedure now in its infancy) would find them.

However, the far greater challenge is discovering the metabolic roles of microbes that resist cultivation. We saw that certain uncultivated microbes were shown to oxidize methane in anoxic environments because their ^{13}C content is low. However, that is a very special case, and few microbial substrates in nature are uniquely labeled in this manner. New imaginative methods are needed. Any ideas?

In spite of the myriad challenges, microbial ecology is a newly revitalized field that promises to reveal much about our environment and how we might protect and restore it.

CONCLUSIONS

In this chapter, we've considered microbial ecology in its most general context—the impact of microbes and their activities on our planet. In subsequent chapters, we shall deal with more specific microbial interactions, including, in the next chapter, **symbioses** (intimate interactions between pairs of species), and then, in the following chapters, **pathogenesis** (disease-causing interactions between microbes and humans, animals, or plants).

STUDY QUESTIONS

1. What are the sources of energy and of carbon that support microbial growth at great depths in the earth?

2. What is enrichment culture? What sorts of ecological questions can it be employed to answer? What ecological questions can't it answer?

3. What leads certain microbes to live as consortia? Give several examples, and discuss their physiological roles.

4. How can culture-independent methods be used to determine which unculturable microbes are in a particular environment and what they might be doing there?

5. What is the doctrine of microbial infallibility?

6. What role do methanogens play in the carbon cycle?

7. What processes balance biological and industrial nitrogen fixation to maintain a relatively constant concentration of dinitrogen in the atmosphere?

8. What microbially mediated processes occur in a manure pile to convert manure to nitrate? How does turning the pile speed the conversion?

9. How were naturally occurring deposits of elemental sulfur formed?

10. List three chemical transformations of matter carried out by microbes that are essential for the continuation of life on Earth, and explain why they are essential.

chapter 19

symbiosis, predation, and antibiosis

SYMBIOSIS

Introduction

Practically no organism lives in isolation. With rare exceptions, to be alive is to live in companionships. This is a powerful reality, because it affects practically all our considerations of biological phenomena, including evolution, ecology, and the functioning of individuals and species. Living in relationship to other organisms defines health and disease. The spectrum of associations, or symbioses, varies from the beneficial, or **mutualistic,** to the damaging, or **parasitic.** (Note that in common parlance, the term symbiosis is used to denote mutualistic associations. In biology, it denotes all forms of association.) The boundaries are not always precise, and definitions can be elusive, because this is a shifting landscape, biased by the vast reactivity of all living things. For human beings, for example, today's benevolent companion may become tomorrow's assailant. A person suffering from AIDS becomes the victim of infections by otherwise-innocuous members of the body's normal microbial flora. The balance between a helpful and a destructive relationship is therefore delicate and can be readily altered by genetic or induced changes in one of the partners. The two share common steps in the way the association is established: encounter, association, and multiplication. The two partners then enter into intense back-and-forth negotiations whose outcome determines the kind of interaction that will prevail.

Symbiosis is a pervasive force in evolution. Through it, organisms develop novel ways to occupy environmental niches, produce energy, acquire nutrients, or defend themselves from predation. Examples are all around us. Some are easily noticed, such as the **lichens** (fungus-alga [sometimes fungus-cyanobacterium] partnerships that adorn rocks and trees) or the root nodules of legumes (whereby bacteria supply plants with a nitrogen source). Other mutualistic relationships are not so readily visible. One would have to travel to the depths of the ocean to see bacteria and worms that have combined into non-sun-based biological communities in thermal vents. The effects of symbiosis are sometimes revealed when the connection between the partners is severed. Thus, the importance of the human intestinal bacteria becomes apparent when they are reduced in number by treatment with antibiotics. People so treated become susceptible to diarrhea caused by yeast, suggesting that, at a minimum, intestinal bacteria play a role in keeping out unwanted intruders.

The scope and variety of symbiotic relationships are illustrated by the examples described below. Each one tells a different story, but it should be kept in mind that the repertoire of symbioses is vast, and any selection can convey only a limited sense of the subject. Further examples of mutualistic symbioses are shown in Table 19.1.

19.1 *Endosymbiosis and the origin of eukaryotes*

Mitochondria, chloroplasts, and the origin of eukaryotic cells

The most wide-ranging and perhaps most enduring symbiotic venture has been the acquisition of mitochondria by animals and plants and of chloroplasts by plants. These organelles were once microbes. The origin of these symbioses, which happened a billion or so years ago, led to a momentous event, the development of the eukaryotic cell. It has taken some time to confirm that cell organelles have a microbial origin because today they hardly resemble their ancestors (Fig. 19.1). Especially convincing has been the finding that the DNA of mitochondria has homologies to that of rickettsiae, the alphaproteobacterial agents that cause typhus and other diseases. Interestingly, rickettsiae remain strict intracellular parasites to this day, although they are a long way from becoming mitochondria. Likewise, the DNA of chloroplasts is similar to that of photosynthetic cyanobacteria.

In both mitochondria and chloroplasts, many of the genes of the original microbes were transferred to the nucleus, and others have been lost. Thus, these organelles have become highly reduced genetically from their original free-living status and are far from being capable of independent existence. Mitochondrial genomes usually code for a few dozen proteins; the chloroplast genome often codes for about 10 times as many. Some of the mitochondrial genes are for the organelle's main function, aerobic respiration. Chloroplast genomes contain genes for photosynthesis.

Figure 19.1 A schematic drawing of a mitochondrion.

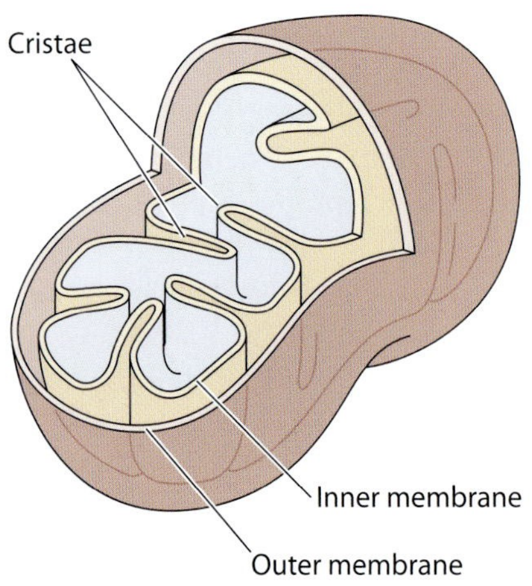

Most of the genes for aerobic respiration are not carried in the mitochondria but in the nucleus, and their protein products must be transported across the cytoplasm to the mitochondria. Although few, mitochondrion-encoded proteins must be especially important because they have been retained through the ages. This does not mean that mitochondrial genes are unchangeable. Several congenital diseases have been attributed to mutations in the mitochondrial genome. In fact, the mutation rate of the mitochondrial DNA is quite high, possibly due to the lack of a proofreading function in their DNA replication machinery. However, the medical importance of such mutations is relatively small because most of these mutations take place in somatic cells and are not inherited. Because there is more than one mitochondrion in most eukaryotic cells, a mutation in one may not have a large genetic effect. (Mitochondrial

Table 19.1 A sampling of mutualistic symbioses not presented in the text

Microbe agent	Host	Nature of interaction
Bacteria		
Bacterium (*Vibrio fischeri*) 19.2	Squid, fish	Bioluminescence, used for camouflage and evasion from predators, perhaps recognition
Bacteria 19.3	Worms, clams in deep-sea hydrothermal vents	Provide food for host by using H_2S oxidation to drive CO_2 fixation to make usable organic compounds
Bacterium (and yeasts) ("*Micrococcus cerolyticus*" and yeasts) 19.4	Bird	Help bird digest beeswax (obtained by birds "guiding" badgers and bears to honeybee hives)
Cyanobacteria 19.5	Sponges, clams	Supply host with food obtained via photosynthesis
Bacterium (*Aeromonas veronii*) 19.6	Leech	Aids in digestion of blood meal? Supplies vitamin B_{12}? Keeps other bacteria from colonizing?
Bacterium (*Anabaena*, a cyanobacterium) 19.7	Fern (*Azolla filiculoides*)	Supplies nitrogen to fern by nitrogen fixation
Protists		
Dinoflagellates 19.8	Clams	Supply host with food obtained via photosynthesis
Algae 19.9	Sea slugs	Chloroplasts of the algae are sucked into slug's gut and incorporated into its cells, where the photosynthesis takes place
Fungi		
Fungi (*Termitomyces* spp.) 19.10	Termites	Serve as direct food for host termites, which cultivate them
Fungi 19.11	Plants	Make mycorrhizae that provide minerals and water to host
Fungi 19.12 19.13	Algae or cyanobacteria[a]	Make lichens, which provide food for the fungi and shelter for the algae or cyanobacteria
Viruses		
Virus (polydnavirus)	Wasp	Inhibits immune defenses of caterpillar (host for wasp's egg maturation and development)[b]

[a]The designation of host is arbitrary here.

[a]See N. Beckage, "The Parasitic Wasp's Secret Weapon," *Scientific American*, November 1997, p. 50–55.

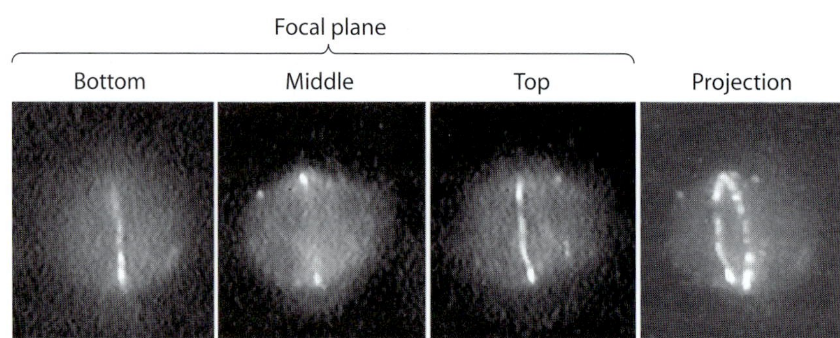

Figure 19.2 The FtsZ protein in a chloroplast of the plant *Arabidopsis*. The protein was visualized by using fluorescent anti-FtsZ antibodies. The three panels on the left are optical sections through the bottom, middle, and top of a chloroplast. In the far right panel, labeled Projection, the images were stacked and rotated 30% to show the FtsZ ring.

genetics is colored by the fact that these organelles are inherited only from the mother—those of the sperm cells are degraded after fertilization. Therefore, inheritance of mitochondrial genes is maternal rather than Mendelian.)

These organelles reveal their bacterial origin in various ways. First, mitochondria and chloroplasts *divide by binary fission*. Some organelles even retain proteins that are typically required for bacterial division. Chloroplasts and certain mitochondria use an FtsZ ring for division and, in some instances, proteins of the Min system (Fig. 19.2) (see Chapter 9). Mitochondrial and chloroplast protein syntheses have several bacterial signatures: both start with formylated methionine rather than with methionine (as in eukaryotes), their ribosomes resemble those of bacteria (they are smaller than eukaryotic ribosomes), and they are sensitive to antibiotics that inhibit bacterial protein synthesis. Mitochondria also have circular chromosomes and no histones.

Mitochondria and chloroplasts are not the only exmicrobes to inhabit eukaryotic cells. Certain protists, such as *Plasmodium* species, the parasites that cause malaria, have structures called **apicoplasts** (Fig. 19.3). These organelles are essential, and their hosts cannot survive without them. They contain only enough DNA to encode about 35 genes. This DNA is related to that of chloroplasts, although apicoplasts reside in nonphotosynthetic organisms. Apicoplasts, then, have functions different than photosynthesis, including the synthesis of fatty acids and the repair, replication, and transcription of DNA. Apicoplasts seem to be the result of an ancient event in which a eukaryotic ancestor acquired a chloroplast but converted it from a photosynthetic factory to one involved in other biochemical activities.

Figure 19.3 A schematic drawing of a protozoon showing the apicoplast.

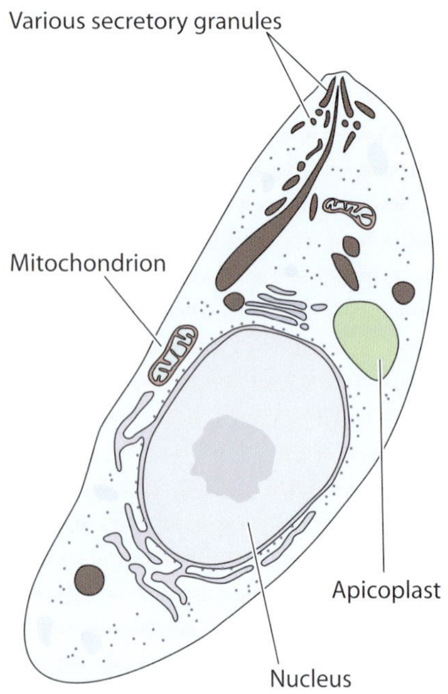

The topic of DNA-containing organelles has further surprises. Certain protozoa, e.g., trypanosomes, the group that contains species that cause sleeping sickness or Chagas' disease, live an Alice-in-Wonderland existence. These organisms have a unique DNA-containing organelle called the **kinetoplast,** which is a highly specialized mitochondrion located at the base of the cilium (Fig. 19.4). Each cell contains a single kinetoplast. Kinetoplasts have a mesh of interwoven DNA circles resembling the chain mail of medieval armor (Fig. 19.5). The circles occur in two sizes: maxi and mini. Maxicircles are fewer—20 to 50 per kinetoplast as opposed to approximately 10,000 minicircles. Maxicircles resemble the DNA of mitochondria, because the proteins they encode are involved in energy production. Minicircle DNAs are much smaller (0.5 to 1.5 kilobases versus

20 to 35 kilobases for the maxicircles) and are heterogeneous in sequence. The DNAs of the minicircles do not encode proteins. Instead, the RNA they encode (guide RNA) is involved in editing the messenger RNA encoded by other genes. The origin of kinetoplasts is under investigation and appears to involve a microbial endosymbiosis.

Bacterial endosymbionts of insects: organelles in the making?

Why do eukaryotic cells have only a few kinds of microbe-derived organelles? Why not more? The advantages of establishing these symbioses are self-evident: in a single step, the recipient cell acquires all the genetic material needed for functions as complex as respiration or photosynthesis and is thus equipped to undertake bold new ventures. The question then arises as to whether symbiotic relationships are limited to mitochondria, chloroplasts, and a few other organelles. The answer is no, because in certain groups of organisms intracellular symbioses are quite common. While the symbiont-host relationship is not as intimate as that of organelles, other kinds of symbiotic partnerships are frequently found in invertebrates such as insects, worms, and clams. In these cases, genes from the symbiont have not been transferred to the host's nucleus but remain in the genome of the symbiont.

For many species of invertebrates, these symbiotic relationships are essential for survival. In contrast, vertebrates have developed few, if any, essential partnerships with microbes. Most intracellular microbes of humans, for example, are pathogenic. However, some intracellular bacteria, such as the tubercle bacillus, can reside in host macrophages for long periods, perhaps for the life of the host. As long as the balance is

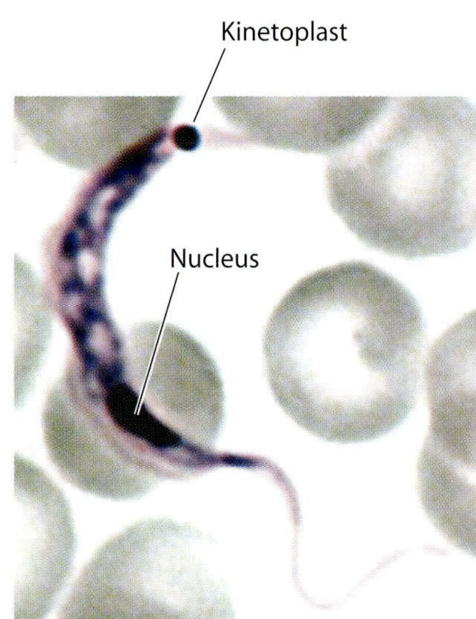

Figure 19.4 The kinetoplast. Shown is a leishmania (a protozoan that causes leishmaniasis), containing a kinetoplast, in a blood smear.

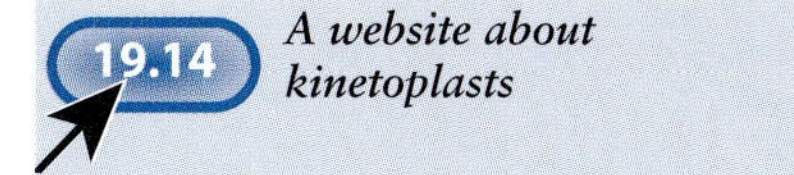

Figure 19.5 Kinetoplast DNA from *Leishmania*. Kinetoplast DNA consists of a giant network of interconnected (catenated) minicircles and maxicircles. There are approximately 10,000 minicircles and 50 maxicircles per network. This DNA makes one of the most unusual structures known.

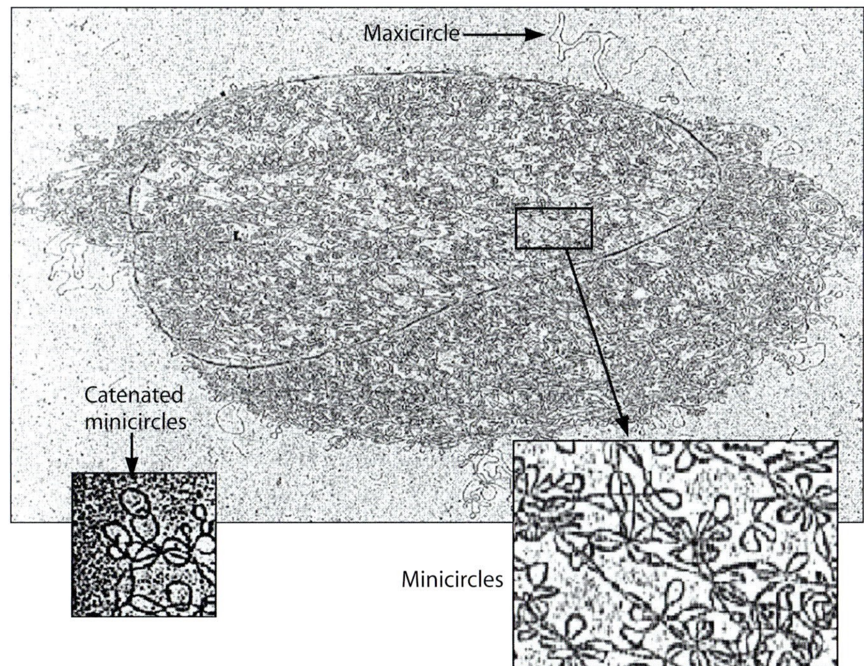

maintained, this is a state of dormancy that affects the host in subtle ways only. Some bacteria, such as rickettsiae (the ancestors of mitochondria) and chlamydiae, are **strict intracellular parasites;** they cannot grow outside host cells. If we had to guess why present-day rickettsiae have not lost more of their genes and evolved as mitochondria did, we would argue that rickettsiae retained many attributes required for the demanding task of being transferred from one host to another.

Rickettsiae and chlamydiae have lost part of their genomes but have also acquired genes for special functions needed for obligate intracellular **parasitism.** Thus, both of these groups of organisms possess genes that encode transport systems for ATP and ADP. Such genes are not found in other bacteria, presumably because they make these nucleotides themselves and are seldom in a position to acquire them from the outside. Rickettsiae and chlamydiae, on the other hand, have a limited ability to produce ATP and need to obtain it from their environment, the host cell's cytoplasm.

A majority of insects have symbiotic partnerships with bacteria. The bacteria are usually **endosymbionts;** they are harbored within large specialized cells called **bacteriocytes.** Bacteriocytes are filled with bacteria, leaving just enough space for the nucleus and other cell components (Fig. 19.6). Most often, these bacteria have lost the ability to be cultured on artificial media and have become totally dependent on their host for survival. In most cases, the host is equally dependent for life on the presence of the microbes. The actual manner in which host and symbiont benefit differs for each partnership. In many cases, symbiosis is nutritional—the host and the symbiont provide needed nutrients to each other. In other cases, the relationship may affect the sex life of the host and lead to changes in speciation and lifestyle. Bacterial endosymbionts are not limited to residing free in their host's cytoplasm. In some cases, the bacteria invade mitochondria (Fig. 19.7) or, in the case of certain rickettsiae, even the nucleus.

A particularly well-studied example of nutritional **mutualism** is that between aphids, small soft-bodied insects that are common pests of plants,

Figure 19.6 A bacteriocyte, an insect cell filled with endosymbiotic bacteria. This electron microscope thin section shows how tightly endosymbionts can be packed within a host cell, here, one in the fat body of a German cockroach. The bacteriocyte is ca. 100 μm in diameter.

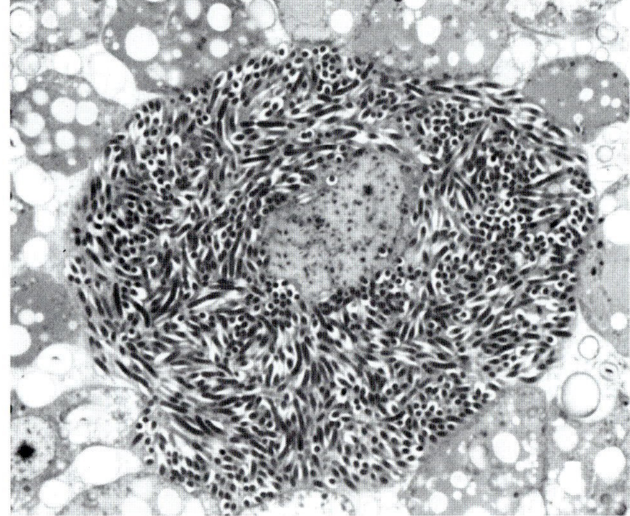

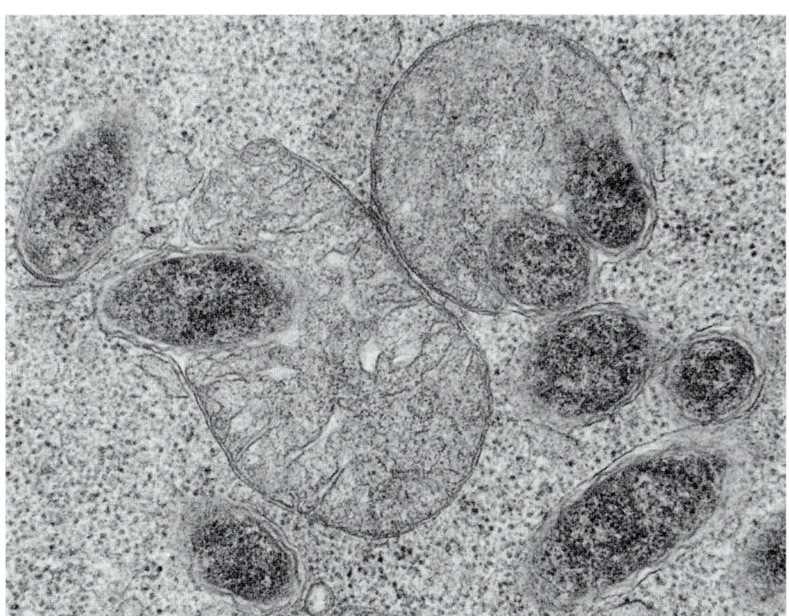

Figure 19.7 Symbiotic bacteria in the mitochondria of a tick. Note that these mitochondria are large enough to readily accommodate bacteria.

and gram-negative bacteria called *Buchnera*. Aphids are sap suckers; the plant sap they feed on is poor in proteins and cannot supply all their needed amino acids. The aphids cannot produce 10 amino acids (similar to humans) and would starve unless they obtained them from another source. That source is their bacterial endosymbionts. Not surprisingly, when exposed to an antibiotic that kills their bacterial endosymbionts, aphids fail to grow and reproduce. The transfer of the essential amino acids can be directly demonstrated by supplying labeled amino acid precursors and showing that the label first appears in the bacteria and is then transferred to the aphids. The other side of the coin is that the bacteria cannot make the other set of amino acids, the ones that the aphids are able to synthesize. The aphids provide precursors, such as glutamate, that the bacteria can use to make their required amino acids. This is a perfect tit-for-tat, each partner feeding the other the missing precursors. Genomic analysis of *Buchnera* has shown that many of the genes required for precursor synthesis are missing. Not surprisingly, these bacteria cannot grow outside the host. Considering the biochemical complexities of amino acid biosynthesis (see Chapter 7), the exact complementary relationship between the host and the symbiont must have required a good number of evolutionary steps. Buchneras are amino acid-making machines: they carry plasmids with multiple copies of some biosynthesis genes, e.g., those for tryptophan.

Buchnera and many other bacterial endosymbionts have small genomes, about one-seventh that of *Escherichia coli*. Phylogenetic analysis of their DNA shows that they appear to have lost most of the genes of their ancestors (who were probably distant relatives of *E. coli*). The genes of this reduced genome are those required for existence within the host. Unlike mitochondria and other organelles, the lost genes do not seem to have been transferred to the nuclei of the host cells. It is possible that such horizontal gene transfers between endosymbiont and host take place all

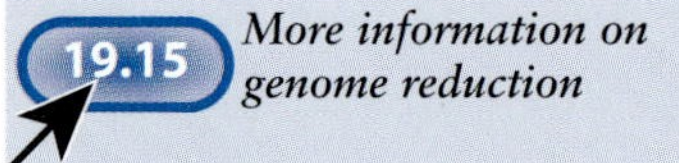

the time but only some become stable. For example, in the case of a certain beetle, a limited number of genes from a bacterial symbiont have been incorporated in the nucleus of the host.

During the evolution of endosymbioses, there is probably a series of steps in the transition from the original free-living bacterium to an organelle whose lost functions must now be provided by the host. The extent to which the bacterial genes have been incorporated into the host nucleus may vary from none to many and may be a defining characteristic of each symbiosis. However, few, if any, intermediate forms between, say, mitochondria and *Buchnera* have been found. A fascinating aspect of *Buchnera* evolution is that it seems to have proceeded synchronously with the evolution of the host aphids (Fig. 19.8), a fine example of coevolution. Because there are fossil records for the aphids, *this concordance can serve as a clock for bacterial evolution.*

Buchneras have retained the capacity to be transmitted vertically among progeny host cells and, in this sense, are organelle-like. Note that this is a difference between endosymbionts, which are generally passed on to the host's progeny by **transovarian transmission,** and **parasites,** which, with some exceptions, must infect each generation of host. Because they are strict intracellular parasites, *Buchnera* cells are shielded from horizontal transmission of genes from other organisms. Consequently, their only obvious source of genetic variation is mutation. How do these organisms cope with the accumulation of lethal mutations? The answer is not really known, but the aphids seem to insure themselves against a catastrophic loss of endosymbionts by having more than one bacterial endosymbiont, each residing in a separate kind of bacteriocyte. Note that such organisms allow the study of the special aspects of evolution in relatively shielded environments.

As we have stated, symbioses between insects and microbes often fulfill nutritional needs, but not always. In other instances, the partnership modifies host reproductive phenomena. An example is the rickettsia *Wolbachia,* an extremely common bacterium that infects between 25 and

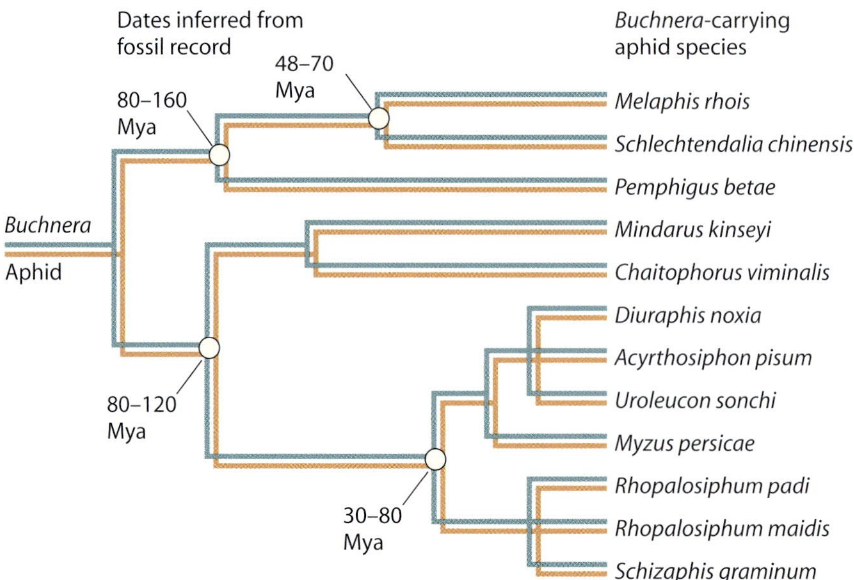

Figure 19.8 Coevolution of aphids and *Buchnera*. The phylogeny of *Buchnera*, derived from genomic relatedness, is shown in turquoise. The phylogeny of the aphid hosts, derived from the fossil record, is shown in orange. Note the remarkable superimposition of the two phylogenies. Mya, millions of years ago.

75% of all insect species, as well as worms and spiders. In different hosts, *Wolbachia* induces a number of diverse phenotypes. These include the determination of sex by selectively killing males, inducing parthenogenesis, and other effects. In colonial insects, such as bees, wasps, and ants, *Wolbachia* infection eliminates males, so that females reproduce parthenogenically and create female-only populations. The selective advantage of eliminating males is to increase the chances of *Wolbachia* being transmitted to the progeny via the eggs. In addition, the female larvae have less competition; they even eat their dead brothers. This widow-making ability of the bacteria puts considerable evolutionary pressure on the host and often alters its behavior. In some butterflies, killing of males is not complete; a few males escape the results of the infection. In these species, the females congregate in groups where they vie for the attention of the scarce males. In this reversal of the often-seen aggregation of males in search of females, it is the males that can be choosy about their mates.

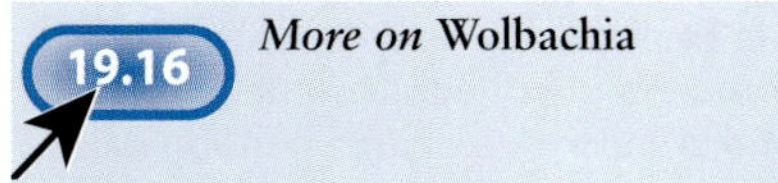

The topic of endosymbiosis is of direct importance to human medicine. The disease **river blindness** (a leading cause of blindness in tropical areas of the world) has long been thought to be caused by a nematode, *Onchocerca*. Until recently, the notion was that clouding of the cornea is caused by an inflammatory response to the worms. It turns out, however, that this may not be so. The disease seems to be the result of an immune response to an essential *Wolbachia* endosymbiont in the worms. In a mouse model, bacterium-free worms do not cause disease, but those that carry *Wolbachia* do. This finding suggests that the disease may be treated by antimicrobial therapy or vaccination.

The world of endosymbioses has many other curiosities. Certain bacteria found in mealybugs themselves harbor other bacteria inside them. The "parent" bacterium benefits the insects by making amino acids, but the role of its own endosymbionts is not yet known. These few examples illustrate the wide range of symbiotic activities promoted by the symbiotic associations between bacteria and eukaryotes.

Nitrogen-fixing bacteria and the legumes

Nitrogen gas is the most abundant molecule in our atmosphere. However, it is relatively chemically inert (unlike oxygen, for example) and cannot be directly used by most living organisms to make their nitrogen-containing compounds, such as amino acids, purines, or pyrimidines. To be biologically available, nitrogen has to be reduced to ammonia, a form in which it can enter biosynthetic pathways. This process, the reduction of the molecule dinitrogen to a utilizable form (**nitrogen fixation**), was introduced in Chapter 7, and the nitrogen cycle in nature was developed further in Chapter 18. *Nitrogen fixation is to nitrogen as phototrophy and lithotrophy are to carbon.*

As is the case with carbon, not all organisms can reduce nitrogen, and those that do not depend on those that do. Biological nitrogen fixation is carried out only by *prokaryotes*, some living free, others in association with plants. Here, we deal only with nitrogen-fixing bacteria that are symbiotic with plants. Such associations are found in trees (alder), shrubs (bayberry), or ferns. The best studied examples are in legumes, such as peas, alfalfa, and beans. When a legume is pulled up from the ground, the roots appear to be decorated with small granules, typically 1 mm in diameter, the so-called

Figure 19.9 Root nodules from an alfalfa plant. The nodules show their pink contents, due to the presence of leghemoglobin, a unique metabolite of this type of symbiosis.

root nodules (Fig. 19.9). Looking at a squashed or sectioned root nodule reveals that they are filled with bacterium-like bodies.

Root nodules are nitrogen-fixing factories. They are symbioses between bacteria and plants. Both partners contribute to nodule formation, and both partners undergo crucial changes in the process. What are the steps in this association? The bacteria involved in nodule formation are soil-dwelling members of the genus *Rhizobium* and related genera. The bacterial species are quite specific for their hosts. In agriculture, crops are often rotated, because the alternating cultivation of legumes and other plants reduces fertilizer needs. Alfalfa and other legumes replace some of the nitrogen removed by corn and other grain crops.

The process of nodule formation begins with the plant roots excreting compounds (flavonoids) that are sensed by nitrogen-fixing bacteria nearby. As a result, the bacterial genes (*nod*) involved in nodulation are induced. The products of several *nod* genes work together to make **nodulation factors,** substances that signal the plant to initiate nodule formation. Nodulation factors are fatty acids plus chitin-like compounds (polymers containing *N*-acetylglucosamine) and are very powerful. When applied at concentrations as low as 10^{-9} M, they can induce nodule formation even in the absence of the bacteria. Interestingly, the *nod* genes are carried on plasmids. The specificity of association between bacterial strains and certain plants is due to the specificity of the products of these genes.

Nodule formation involves the entry of bacteria into the thin filaments that penetrate **root hairs,** the threadlike extensions of certain cells on the root's surface. The invasion process represents a specialized form of bacterial penetration. The first step is binding of a bacterium to a root hair, which requires recognition between a ligand on the bacterial surface and a receptor on the host cell surface. Binding usually takes place at the tip of the root hair, where the bacterium causes localized hydrolysis of the tough plant cell wall. This allows the organism to invade the root hair. Soon, this results in a morphological change in the root hair, which curls up to look like a shepherd's crook (Fig. 19.10). The internalized bacteria reside in intracellular vacuoles, much like certain animal pathogens that

Figure 19.10 Morphological changes leading to a nitrogen-fixing nodule.

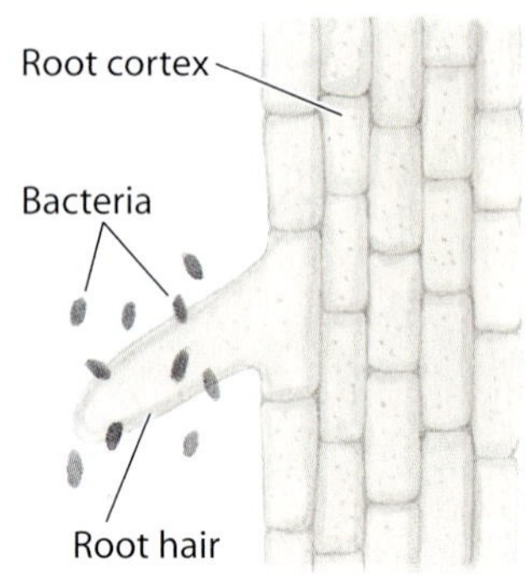

1 Adhesion of *Rhizobium* bacteria to root hair

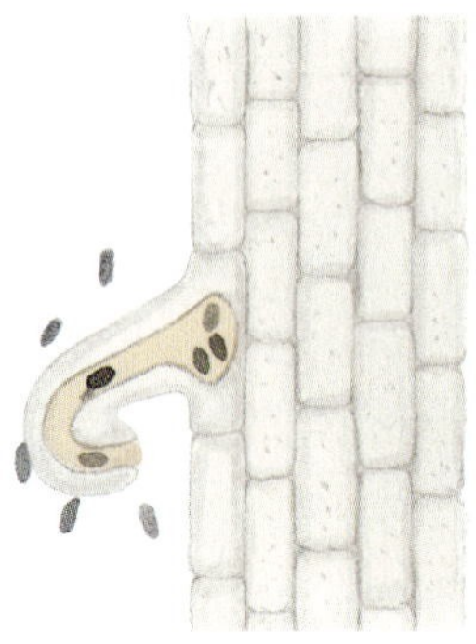

2 Infection: invagination and curling of root hair

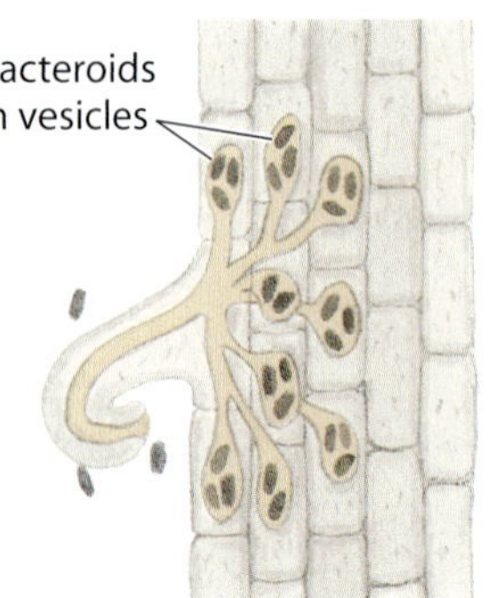

3 Bacteroid development in vesicles in root cortex

4 Bacteroid-induced cell division of cortex cells forming a nodule

survive in the vacuoles of phagocytes. Nodule formation requires that the bacteria move inward, toward the center of the root. This is a difficult trip because the bacteria have to negotiate their way through tough cellulose-encased plant cells. Travel is facilitated when the plant forms a tube (called an **infection thread**) that stretches from the root hair to the root's interior. Bacterial cells travel along this tube. When they arrive at a deep location, they exit the infection thread and begin to multiply. The plant cells respond to bacterial stimuli by rapidly proliferating into a tumorlike structure, the nodule. This host response to the bacteria requires a delicate balance because plants have defense mechanisms that could destroy the invading bacteria. Rhizobia and their relatives make capsules and, being gram negative, possess lipopolysaccharides. We know that both these constituents are required for bacterial survival because mutants lacking them are destroyed after invasion.

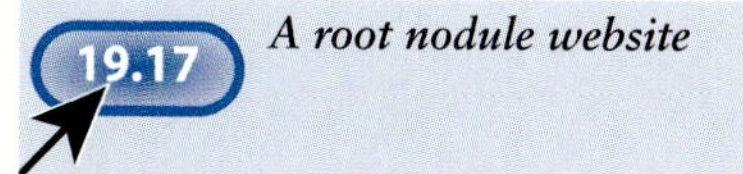

A root nodule website

Having arrived at their ultimate place of business, the bacteria differentiate into nitrogen fixation workshops. They become branched and swollen and are now called **bacteroids** (Fig. 19.11). Before being able to fix nitrogen, the bacteria must address a final problem. Nitrogen fixation is a highly anaerobic process, but rhizobia are aerobes and plant roots are aerobic. Plants synthesize a form of hemoglobin called leghemoglobin that absorbs the oxygen, keeping its concentration at a suitable level for the bacteroids and for the nitrogen-fixing machinery. Bacteroids are incapable of further growth within the nodule and are totally dependent on the plant for nutrients. Thus, for their side of the bargain, bacteroids receive from the host organic acids that provide energy and the reducing power needed for nitrogen fixation. In return, they provide to their host assimilable nitrogen in the form of ammonia. This mutualistic relationship requires that both partners undergo profound biochemical and structural adaptations. As we asked above, what prevented nitrogen-fixing bacteria from becoming cell organelles in parallel with mitochondria and chloroplasts? We cannot answer this question at present.

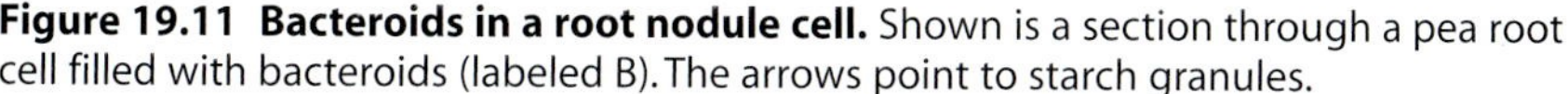
Figure 19.11 Bacteroids in a root nodule cell. Shown is a section through a pea root cell filled with bacteroids (labeled B). The arrows point to starch granules.

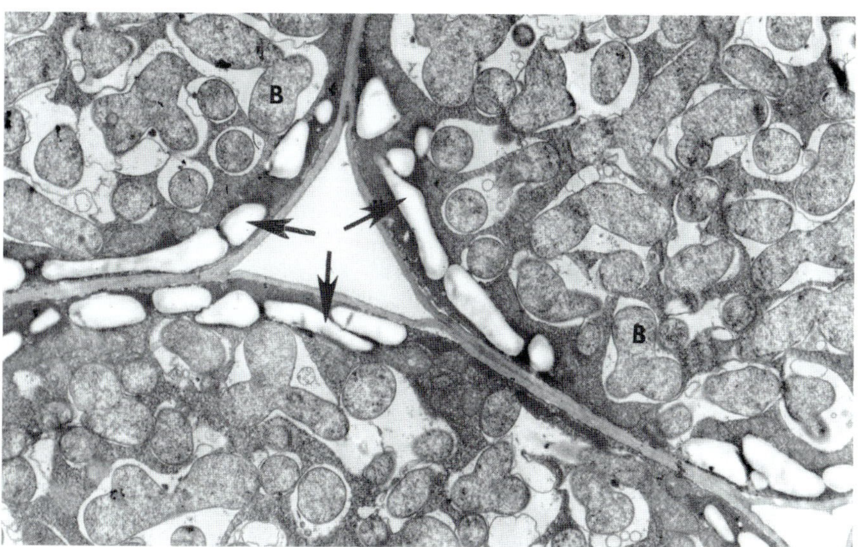

The rumen and its microbes

Animals cannot digest cellulose and some other plant polysaccharides directly, which is a major inconvenience for herbivores. Depending solely on plant material and not being able to use these polymers would be wasteful and inefficient. However, microbial symbionts of plant-eating animals degrade these compounds and transform them into digestible products. Unlike the nodule formation of legumes, this symbiosis does not require major modification of the host by the microbes but depends on the host providing a large nutrient-filled chamber in which the biochemical transformations can take place. There are two general ways to do this. Cattle, goats, and deer have such a large chamber, called the **rumen,** which is why the animals are called **ruminants** (Fig. 19.12). Nonruminants, such as horses, rabbits, and elephants, carry out cellulose digestion in an extralarge cecum in the large intestine. The large intestine of humans is of medium size, and cellulose degradation there is probably not a nutritionally essential process.

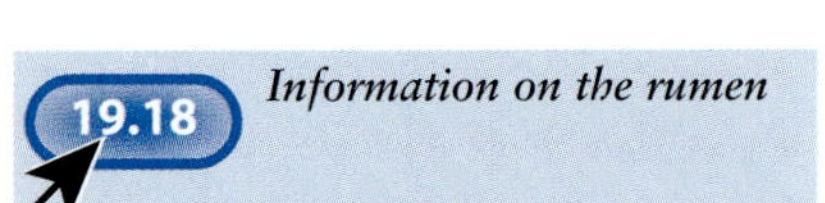

Cellulose and other polymers that are difficult to digest are degraded in steps. The first step is the thorough grinding of plant material into small pieces. Herbivores have teeth with flat opposing surfaces that are well suited to this task. To improve further on this, ruminants chew the cud, meaning that they regurgitate previously eaten food and rechew it in order to reduce it to even smaller bits. This permits these animals to eat rapidly but later to further process the food at their leisure, away from predators.

The well-chewed plant particles then enter the rumen, which in cattle contains about 15 gallons of liquid. Here, diverse groups of **cellulolytic microbes** degrade cellulose into sugars. Other bacteria then ferment the sugars to yield **volatile fatty acids,** such as acetic, propionic, and butyric acids. The rumen is highly anaerobic, and sugars cannot be oxi-

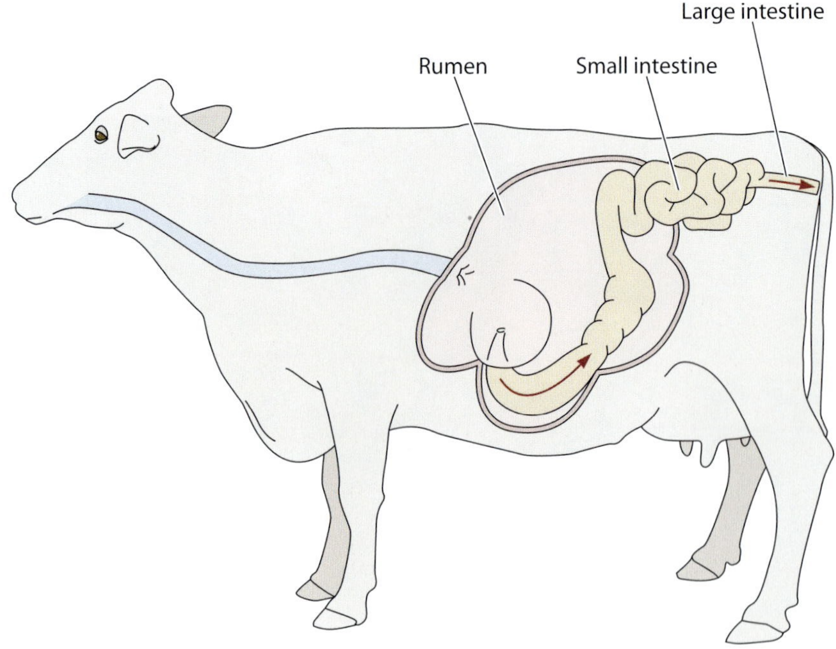

Figure 19.12 A cow's digestive system. Notice the large rumen. The "large" intestine is relatively small, in part because cellulose digestion has already taken place in the rumen. Nonruminant herbivores, such as horses, have a very big large intestine.

dized entirely to carbon dioxide via respiration. Ruminants absorb the fatty acids through the epithelium and use them for their metabolic needs. The pH of the rumen does not drop perceptibly with acid production because ruminants secrete prodigious amounts of well-buffered saliva, 25 gallons and more a day per cow. A large amount of fatty acids is made, supplying the animal with its source of carbon. How do ruminants obtain their required amino acids and other growth factors? The answer is, from the rumen microbes themselves. Once the rumen content is emptied into the next chamber, the stomach, the microbial cells are killed by the acid and are then degraded by digestive enzymes. Many of the bacteria are broken down by a **cell wall-degrading lysozyme** that, uniquely in ruminants and befitting its site of action, is acid resistant. Ruminants make effective use of their feed and their symbionts, and this accounts for the high efficiency of cattle in milk and meat production and the worldwide distribution of ruminant species. It has been said that cattle carry out a fermentation that no industrial microbiologist can match. They use the cheapest substrate (cellulose), gather it themselves, and convert it into valuable products—beef and milk.

In nonruminant herbivores, cellulose degradation takes place in the cecum. Here, the microbes are not recycled as efficiently as in a rumen but are passed into the feces. This accounts for the fact that some of these animals, e.g., rabbits and rats, are **coprophagic,** that is, they eat their highly nutrient-rich feces. Microbial cellulose decomposition also takes place in the gut of termites, again carried out by their microbial partners.

The rumen contains highly diverse populations of bacteria, protozoa, and fungi that are constantly being renewed (Fig. 19.13). A cow could

Figure 19.13 Some bacteria and protozoa from the rumen of a sheep.

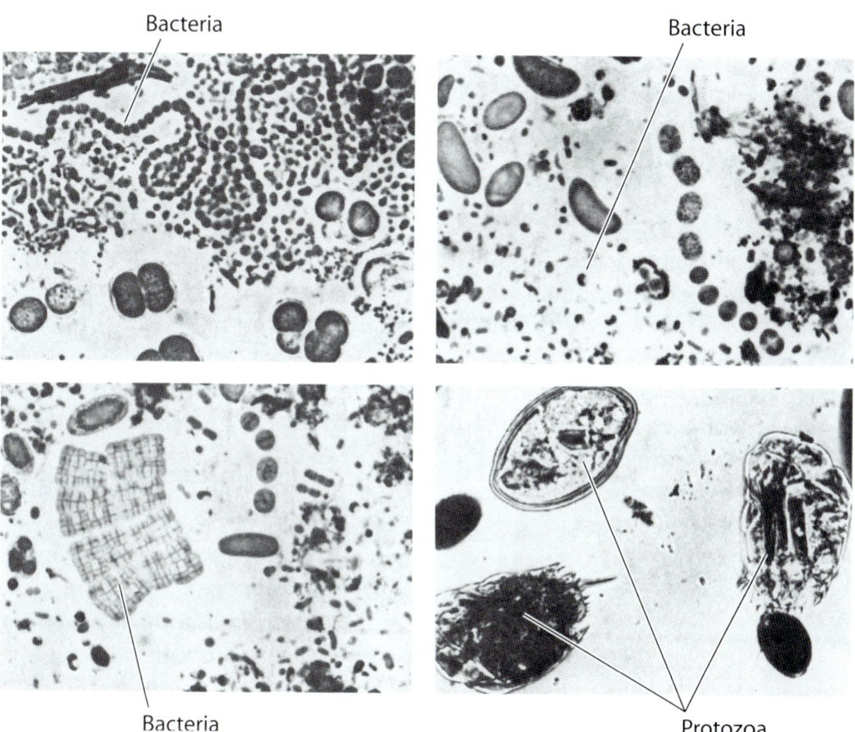

therefore be called a walking continuous-culture device. The rumen microbes include a highly diverse group of organisms. Bacteria are by far the most numerous (as many as 10^{10} per ml) and include over 200 species. Protozoa make up almost half the total microbial load by weight but, being larger, are present in smaller numbers. Although protozoa are also involved in cellulose degradation, they prey on the bacteria and may have a negative influence on the overall fermentation, a still-controversial point.

The biochemical transformation of cellulose to fatty acids and carbon dioxide requires the workings of a **food chain,** starting with cellulose degradation and ending with the fermentation of sugars. The final process, making volatile fatty acids, results in the production of large amounts of hydrogen. If hydrogen were to accumulate under the highly anaerobic conditions of the rumen, it would inhibit further fermentation. Here's why: the formation of acetate from pyruvate is energetically unfavorable and would not take place appreciably unless the concentration of the products of the reaction were decreased. Removing one of the products, hydrogen, allows the reaction to proceed.

How is hydrogen removed from the rumen? The biota of the rumen includes some **methanogens,** archaeal species that can use hydrogen and carbon dioxide to make methane. Note that general aspects of methanogenesis were discussed in Chapter 6 and its ecology in Chapter 18. Being quite insoluble, methane becomes a gas that can be expelled by belching. This can be demonstrated by carefully holding a lit match away from a belching cow and seeing a small flame appear. Removing hydrogen allows fermentation to proceed efficiently. The removal of hydrogen by the methanogens allows more fatty acids to be produced and more ATP to be obtained. This results in the synthesis of more microbial cells, which increases the protein available to the ruminant.

The rumen and its microbes are highly interdependent. This can be expected, because the complexity of the food eaten by ruminants requires a large number of biochemical activities that could not be contained in a single species. No wonder the microbes of the rumen are highly varied and specialized.

Feeding via a murderous partnership: bacteria and nematodes

In some instances, two organisms gang up in order to kill a third. **Nematodes**—small roundworms about 1 mm long—are extremely abundant in soils and varied in their feeding habits. One of them, *Caenorhabditis elegans,* has acquired prominence as a model organism with which to study differentiation and other major biological phenomena. Some nematodes live on a diet of plants, others on fungi or bacteria. Certain kinds of nematodes feed by burrowing into the caterpillars of certain insects. As a result of this invasion, the caterpillars are killed; the nematodes reproduce and eventually leave the caterpillar's carcass. The worms do not feed or reproduce in the soil and must parasitize insects in order to survive. The worms cannot carry on this life cycle alone; they require the help of certain bacteria in a symbiotic relationship. Without the bacteria, the nematodes would be destroyed by the insects. The bacterium *(Xenorhabdus nematophila)* kills the insect host and disables its defense mechanisms.

The bacteria are contained in the nematodes' digestive systems. When a worm penetrates a caterpillar, the bacteria exit the worm and produce toxins that kill the insect. One of these toxins causes apoptosis in the insect's gut epithelial cells, leading to loss of turgor. The caterpillars then look floppy, and the toxin is called **Mcf** (for "makes caterpillar floppy"). The bacteria secrete hydrolases that break down the insect's tissues, which provides a rich soup of nutrients needed by the worms. Once they have fed, the worms mate and reproduce. The nematodes eventually exit from what is left of the caterpillar's body, but not before picking up their symbiotic bacteria. The process may last as long as 2 weeks. Why has the carcass of the caterpillar not putrefied during this period? This calls into play another fiendish-sounding maneuver by the bacteria—they make powerful antibiotics that kill other kinds of bacteria. The dead caterpillar, therefore, becomes a sepulchral chamber that contains the reproducing nematodes and their symbionts.

This conspiracy between worms and bacteria to commit murder is symbiotic indeed because neither partner could subsist in the soil alone. Even though the bacteria involved can grow on ordinary laboratory media, they are not found free in the soil. The worms, then, provide the bacteria with shelter and transportation, and the bacteria make food available to the worms.

As an aside, some of the bacteria involved in this symbiosis (*Photorhabdus luminescens* and others) are bioluminescent. It is not known why they emit light, although it has been speculated that it may serve to attract the worms. Some of these bacteria can also cause wound infection in people. In the darkness of the trenches of World War I, the infected tissues were actually seen to glow. Experienced physicians took this as a good omen, because luminous wounds were likely to heal quickly. Was this due to the production of antibiotics by the bacteria, which kept out more powerful invaders? Given better lighting in modern hospitals, this diagnosis is likely to be missed.

Leaf-cutting ants, fungi, and bacteria

In tropical and subtropical America (from Argentina to the southern United States), there are ants that subsist solely on the fungi they grow. These ants make huge underground nests, to which they bring pieces of leaves, flowers, and other organic materials to construct elaborate fungus gardens. In cross-section, the **fungus gardens** are whitish, irregular masses that almost fill the large cavity of the nest. To grow these gardens, various agricultural chores are divided among members of the different ant castes. The larger workers gather sections of leaves, often half an inch across, and carry them in long and fast-moving processions to their nest. The ants effortlessly carry a burden that to a human would be as heavy and awkward as carrying a large plywood panel (Fig. 19.14). Given this industrious dedication to gathering leaves, it is not surprising that early observers thought that the leaves themselves were the food for the ants. However, the plant material serves only as a growth medium for the real foodstuff, the fungi.

Some species of leaf-cutting ants leave trails of denuded earth between their nest and the source of vegetation. These clear trails can reach astonishing dimensions, over 600 feet in length and 8 inches wide.

Figure 19.14 A leaf-cutting ant carrying its burden.

The amount of coming and going is remarkable, often resulting in traffic jams where the outbound and inbound individuals must literally climb on top of one another (a technique that has yet to be developed for the cars on urban freeways). At the nest, the larger workers carry out an elaborate composting operation. The leaf pieces are cut into smaller sizes, licked thoroughly, and mixed with fecal material. Using their mandibles, legs, and antennae, the ants then knead them into tiny juicy balls. This pulpy material is carefully deposited at the edge of the garden and jabbed into place. It is then "seeded" with the mycelium from the older sections of the garden. The garden rapidly becomes permeated with new filaments. The fungal surface consists of aggregates of hyphal tips that end in roundish bodies.

Leaf-cutting ants have received very bad press because of the damage they cause to vegetation and the threat they constitute to cultivated crops, such as coffee and cacao. They are, in fact, the dominant herbivores in the American tropics. In Brazil, the saúva, the local name for these ants, inspired an old saying: "Either Brazil gets rid of the saúva, or the saúva get rid of Brazil." In partial defense of the ants, it should be mentioned that they do not completely defoliate the plants on which they feed. In addition, the spent litter from the fungus gardens, still rich in organic matter, is returned to the environment, where it serves as plant fertilizer. Typically, this material is carried up the trunk of a tree or over vines and allowed to drop to the ground. The ants of one nest have been observed to carry the waste to a smooth rock and allow it to tumble down the slope.

The relationship appears to be remarkably old, dating perhaps to 50 million years ago. When the ants selected a given fungus as a favorite crop, they remained with it for eons. Comparing the ribosomal RNAs of different species of ants and their fungi, researchers learned that once the symbiotic relationship was established, the two partners remained true to each other over the ages.

There is more to this story. Why do only the desired fungi grow in the fungus gardens? Other fungi can grow on such a substrate, but only one kind predominates—a monoculture. The answer is that these ants carry on their bodies a bacterium of the genus *Streptomyces* that makes antibiotics that inhibit a common fungal invader. This, then, is a three-way symbiosis among ants, fungi, and bacteria. Note that the ants made use of antibiotic agents some 50 million years earlier than humans.

BEHAVIORAL CHANGES DUE TO PARASITISM

Microbial-host interactions range from beneficial mutualism to deleterious forms of parasitism. The chapters on pathogenesis deal with the latter in their classical manifestations. There are, however, other nuances in the interactions among living organisms, some of which are quite startling. For example, the behavior of animals and plants may be altered significantly by their interaction with microbes. A few examples of such manipulations will do better than definitions.

Reckless rats and fatal attraction

Rats infected with a certain parasite lose their fear of cats. This suicidal change from normal behavior is due to infection by the protozoon *Toxoplasma gondii*. This may make little sense for the rat, but it makes wonderful sense for the parasite. Understanding why requires knowing about the life cycle of *T. gondii*. The main hosts are cats, whereas rats, other rodents, and people are incidental hosts. In the cat's intestine, these protozoa reproduce and eventually develop into tough environmental forms called **oocysts** that are eliminated through the feces. Oocysts survive in the soil for long periods, where they may be picked up orally by rodents. Once in the rat, the agents reproduce and induce a strong immune response. To withstand it, *T. gondii* makes resistant forms that remain dormant in the rat's tissue and usually cause no further damage. Similar events take place when people ingest *T. gondii* oocysts. This would seem to be a dead end for the parasite. However, when an infected rat is eaten by a cat, the parasites reproduce in the cat's intestine and are eventually shed in the feces to start the cycle anew. Thus, the capture and ingestion of rodents by cats are essential aspects of the parasite's life cycle (Fig. 19.15).

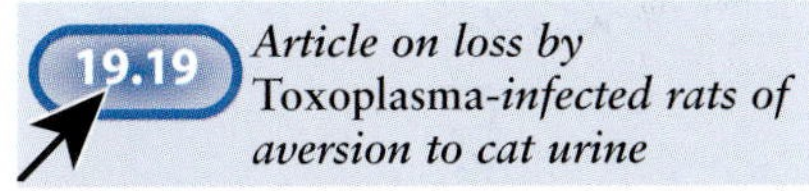

19.19
Article on loss by Toxoplasma*-infected rats of aversion to cat urine*

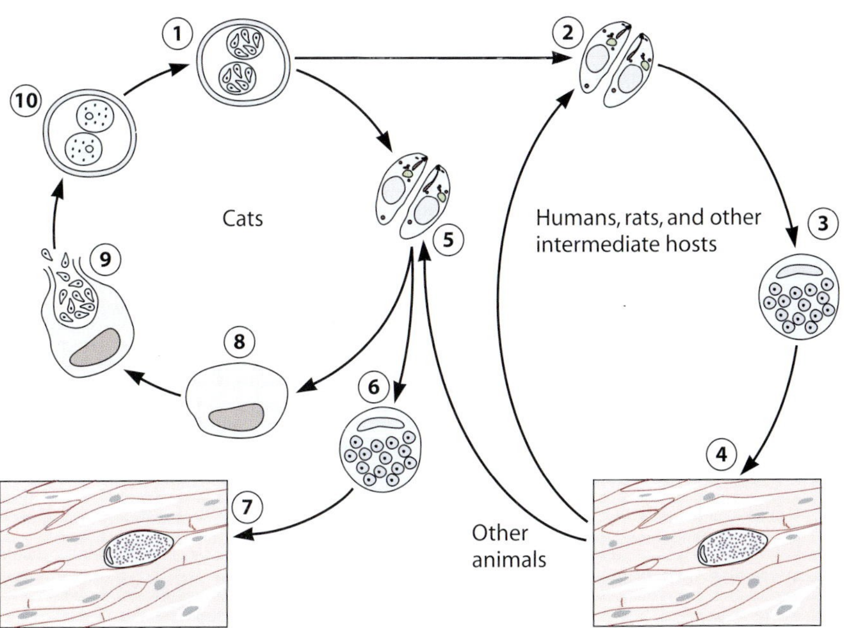

Figure 19.15 The life cycle of *Toxoplasma*. Humans and rats become infected with *Toxoplasma* by ingesting cysts **(1)** in improperly prepared contaminated meat or cat feces. The cysts germinate into active forms **(2)** that enter the bloodstream and disseminate **(3)**. In most hosts, the immune response eliminates the active form and leaves only dormant cysts in tissues **(4)**. The same happens in cats **(5** to **7)**, but here, some parasites progress to finish their life cycle, which includes a sexual stage **(8** to **10)**.

Figure 19.16 "Summit disease." A weevil is firmly attached to the stem of a plant as the result of a fungal infection (*Cordyceps curculionum*) with fruiting bodies of the fungus emerging on stalks.

Using outdoor pens, researchers compared the reactions of normal and infected rats to cats' urine. Healthy rats, not surprisingly, were highly averse to the cats' scent, as if knowing what is good for them. Infected rats, on the other hand, appeared to have lost this inhibition. In fact, they even seemed to be attracted to their nemesis' aroma. Such behavior would hardly be to the rats' advantage, but it certainly helps the parasites complete their life cycle. The significance of this finding has not been confirmed in field studies, but the implications remain intriguing.

The urge to climb

Certain ants that live on the forest floor show drastic changes in their behavior when infected by fungi. The invading fungi develop slowly enough for the infected ants to stay active for some time, but their deportment is altered: they acquire the urge to climb up the stalks of vegetation and trees. When they have reached a certain height, they attach themselves to the plant with their mandibles and remain perched aloft for the rest of their life and thereafter. Other insect groups—grasshoppers, locusts, aphids, and flies—also exhibit this so-called "summit disease." The fungi then grow and develop fruiting bodies filled with spores (Fig. 19.16). The spores can be dispersed from on high, possibly to be carried over great distances.

The reason given for the ants' urge to climb depends on one's tolerance for teleology. "Because it's there" won't do, but claiming that the fungus makes the insect climb for its own benefit also seems suspicious to some researchers. Clearly, remaining on the forest floor decreases the chances for aerial spore dispersal. However, getting off the forest floor means that infected insects are exposed to sunlight and to temperatures deleterious to the fungi. In the words of the entomologist R. A. Humber, "this is a behavior quite analogous to your heading for a warm bed and constant supply of chicken soup when feeling sick." In addition, the infected insect may climb for altruistic reasons, namely, to avoid infecting other members of its colony. Indeed, certain other insects exhibit the opposite behavior when infected by fungi. When infected, larvae of certain butterflies and moths crawl into inaccessible spaces, such as beneath the tree bark, as if to get away from their kin. Such fungi must develop long stalks on their fruiting bodies to spread their spores. Whatever the reason, the interplay of signals between the fungus and the insects is extraordinary. Is there a mechanism that keeps the fungus from growing until the insect reaches a certain distance above the ground? What makes the insects develop the urge to climb up a tree? Who gains and who loses?

Figure 19.17 A pseudoflower. This is caused by the growth of a fungus on a wild mustard plant, transforming leaves into what look like petals.

When is a flower not a flower?

For tweaking the host into making a new and elaborate structure, the prize goes to a fungus, *Puccinia monoica*. This species infects wild plants of the mustard family and induces them to develop dense clusters of leaves at the tips of stems. These rosettes look like the petals of a real flower, all the more so because they become covered with fungal growth (Fig. 19.17). The surface becomes sticky and sweet smelling. These pseudoflowers, as they are called, are a beautiful yellow color, different from the normal flowers of the plant but similar to those of other plants that grow in the same area. Insects arrive, with nectar on their agenda, and poke around the pseudoflower, collecting fungal spores instead of the desired pollen. And off they go, spreading spores to other plants. As can be seen in the photograph, the impersonation is nearly faultless, and at a distance has fooled even professional botanists.

PREDATION

Microbes are usually considered predators, but in reality, they are more often the prey. This is a useful concept, because it explains why many natural microbial populations do not reach the sizes that would be expected based on available food. In oceans, for example, microbes are the center of a food web that includes their predators: phages and protists.

Microbial predation in the environment has been hard to study in the laboratory because more than 99% of microbes have not been cultured. There are, for example, about 1 million bacteria per ml of seawater throughout the world's seas. Even though this is a large number, it is smaller than would be expected based on the amount of food, in the form of dissolved organic carbon, available to marine microbes. This discrepancy is due to the fact that microbes are kept in check by their predators. In fact, without predation, the ocean's microbes could well be 10-fold more abundant.

The major predators of bacteria are protists (e.g., ciliates and dinoflagellates), unicellular eukaryotes that roam the oceans feeding on bacteria, and bacterial viruses, or phages, that prey on marine bacteria. Phages are common in the ocean. As microbiologist Forest Rohwer pointed out, seawater contains about 10^7 phages per ml but only 10^{-19} great white sharks per ml! About half the predation of oceanic microbes is due to ingestion by protozoa, the other half to phage infection. Attempts to study marine phages are hindered by the fact that to study them, one must grow them on a bacterial host, but most of the bacterial hosts have not been cultured. Shotgun sequencing, however, has gotten around this limitation by providing snapshots of the types of phages in marine phage communities without culturing. Using this approach, investigators have been able to show that almost all known phage groups are present in the ocean. These phages are also incredibly diverse, with thousands of different species detectable in 1 liter of seawater.

Most marine phages appear to be lytic rather than lysogenic (see Chapter 17). Infection, then, leads to lysis and the release of cell contents. The bacterial constituents that result from lysis are so small that only other bacteria can feed on them. This means that when a phage kills a bacterium, it actually stimulates other bacteria to grow. In contrast, when protozoa eat bacteria, their contents can enter higher trophic levels of the food chain, including fish. Juvenile cod, for example, eat large quantities of protozoa.

Some bacteria act directly as true parasites of other bacteria. That is, they either graze upon or actually penetrate their prey. The discovery of such predators was as surprising to the discoverers as it may be to the reader. The researchers searched for phages by a tried and true method, looking for "plaques," or holes, in lawns of susceptible bacteria on agar plates (see Chapter 17). They found plaques, all right, but noticed that, unlike those caused by phages, which stop enlarging as the culture stops

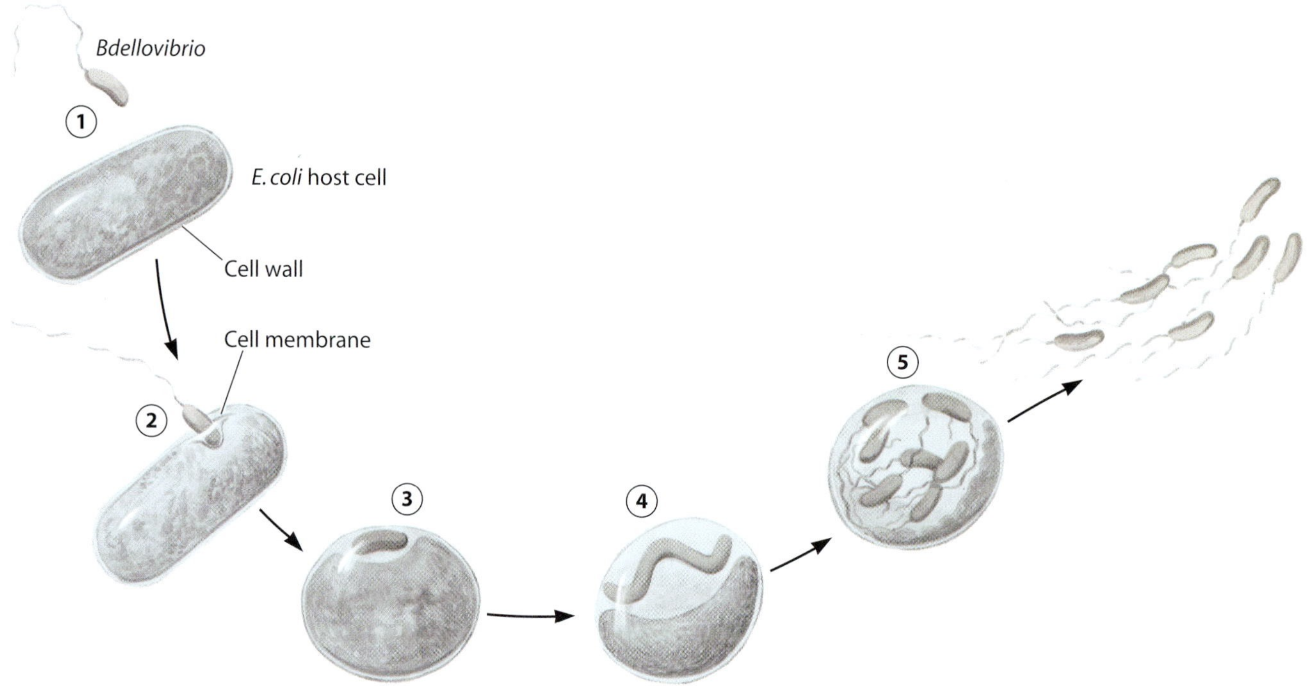

Figure 19.18 The life cycle of *Bdellovibrio.* Attachment to the host cell **(1)** is quickly followed by penetration into the periplasm **(2).** The parasite grows without dividing, making a helical bdellovibrio **(3** and **4).** This divides into small rods that make flagella **(5).** The weakened host cell lyses, releasing progeny bdellovibrios.

growing, these holes grew bigger and bigger even after several days of incubation. When they looked at the material from these plaques under a microscope, they saw very tiny objects that were unmistakably bacteria.

The bacterial predators were given a tongue twister of a name, *Bdellovibrio bacteriovorus* ("bdello" means "leech"). These organisms are about one-fifth the size of an *E. coli* cell but are otherwise typical bacteria. Bdellovibrios occupy a specialized ecological niche (the periplasm of the host or the space between two membranes [Fig. 19.18]). Thus, bdellovibrio infections are confined to gram-negative bacteria. How these predators penetrate the outer membrane is not known. Bdellovibrios are highly motile (among the fastest of all bacteria), but there is little evidence that they home in on the host bacteria by chemotaxis. Once in the periplasm, the invading bacterium (there is only one per host cell) augments its food supply by making the host's inner membrane leaky and introducing hydrolytic enzymes into the host's cytoplasm. Bdellovibrios thrive on degradation products of the host's proteins and nucleic acids. With time, the invading cell elongates and divides into three to five progeny cells that are then released by lysis of the host. The whole process takes about 3 to 4 hours, which is long in comparison to the growth rate of host bacteria under nutritionally abundant conditions but is ample to cause a microepidemic when the host is no longer growing. This explains the gradual increase in size of the plaques on "old" agar plates.

Bdellovibrios are often found in ocean water and soils and are not too choosy about which bacterial species they invade. The question then arises as to how susceptible host bacteria survive in nature. No simple answer is forthcoming, nor is it for other predator-prey interactions. Some tantalizing ideas have been put forth, e.g., that nonsusceptible species act as "decoys" to sop up bdellovibrios or that microbial predators are not able to invade biofilm-forming prey. Such considerations make bdellovibrios unlikely candidates for the control of undesirable microbes, although they may have specialized applications. Another consideration is that bdellovibrios in nature do not grow except when invading their host. As is the case with lions and killer whales, their numbers will therefore wane in the absence of adequate prey.

As with pathogens of more complex hosts, e.g., animals and plants, invasion by *Bdellovibrio* consists of defined steps: encounter, entry, multiplication, and damage. Each of these steps requires the expression of specialized genes and the silencing of unneeded ones. An example is the fact that bdellovibrios on the prowl are highly motile but lose their flagella when in the host. At least 30 proteins are specific to the periplasmic stage. The subject is ripe for detailed analysis. Wild-type bdellovibrios can grow only on host cells, but mutants can be grown on artificial media. However, the faster and longer they grow on artificial media, the less invasive they become.

Bdellovibrios are not the only bacteria that prey on other bacteria. A lot of predators can be found by using a "baiting" technique. A sample of soil on a petri dish is overlayed with a piece of filter paper with pores large enough for microbes to pass through. The filter is seeded with bacteria that act as the bait. In a few days, a large number of predators from the soil will have found their way to this bait. The repertoire of predation is, in fact, vast. Other organisms, such as the myxobacteria, hunt in "wolf packs," swarms of cells that move on surfaces and feed on other kinds of bacteria via the excretion of hydrolytic enzymes (see Chapter 14). An analogous kind of predation relies on the excretion of antibiotics. Other predators display variations on the theme of *Bdellovibrio*: instead of invading the periplasm, they either penetrate the cytoplasm or establish cell-to-cell "bridges" between themselves and the prey. A unique strategy is used by a marine gliding bacterium called *Saprospira grandis*. This organism feeds on other bacteria by first "catching" them by their flagella. The flagella of the prey adhere to the sticky surface of *S. grandis*, and the organisms become trapped, much as flies are trapped on flypaper or insects in a spider web. Flagella, which are otherwise so useful for the growth and survival of many bacteria, here become their downfall.

Predation in the microbial world has been a neglected subject, but in partnership with its converse phenomenon, cooperation among microbial forms, it helps explain complex ecological relationships.

Besides predation, microbial species use subtle strategies to gain an advantage over other species in their environment. Some secrete antibiotics and other antimicrobial products; others use subtle physiological means to compete. One example of a physiological strategy is competition for iron, a metal that, although abundant in nature, is quite scarce in an assimilable soluble form. In the environment, as well as in the bodies of animals, competition for iron is crucial for growth. Basic information on iron acquisition by bacteria was given in Chapter 8. A common tactic bacteria use to acquire the needed iron is to secrete iron-chelating compounds called **siderophores**, small molecules that bind iron with great avidity. The iron-bearing chelates can then be reabsorbed and the iron can be utilized for metabolic purposes, e.g., the synthesis of the cofactors of cytochromes and other oxidative enzymes. As these microbes take up iron, they also reduce its concentration to even lower levels, thus hindering the growth of organisms less adroit at scavenging the mineral.

ANTIBIOTICS AND BACTERIOCINS

Microbes communicate via a "chemical language," that is, by secreting compounds that inhibit the growth of other species. **Antibiotics** are the best known of these chemicals, having gained this ascendancy because they are used to treat human and animal diseases. Paradoxically, we know a great deal about the way antibiotics inhibit bacteria in the laboratory and in infected hosts, but we know relatively little about what they actually do in nature. The concentration of antibiotics in soils is usually below that required to inhibit microbes. However, it is conceivable that in very small niches such concentrations may be achieved, permitting the producers to outgrow species that are inhibited. It is fair to say that we don't know what selective pressures have led to the emergence of antibiotic-producing organisms.

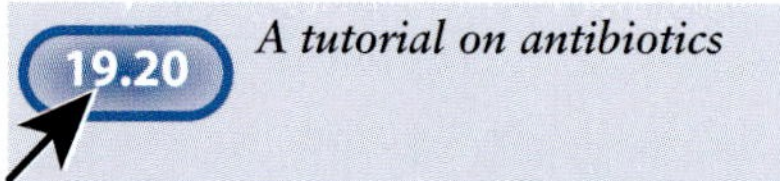

A tutorial on antibiotics

Antibiotics comprise a large array of different organic molecules, generally with molecular masses in the 300- to 1,000-dalton range. The first antibiotics to be used were natural products produced by bacteria and fungi. Over time, some of these compounds have been modified by chemists, who also created altogether new synthetic classes. A list of useful antibiotics and how they work is shown in Table 19.2.

Bacteriocins are peptides, but they are also a varied group. Bacteriocins are made by many bacteria and probably play important roles in the competition among different species in the environment. Some bacteriocins are active only against relatives of the species that produced them, but others have a broad spectrum and affect distantly related species as well. There are even bacteriocins that affect all the gram-positive bacteria tested. Bacteriocins inhibit sensitive bacteria in a variety of ways (Table 19.3). The most common mode of action is to disrupt the cell membrane, making it porous for potassium and phosphate ions. The bacteria then attempt to reaccumulate these ions by using an ATP-dependent transport system. As ATP is utilized, its level falls below that necessary for maintaining the membrane potential, as well as other energy-requiring functions.

Why are bacteriocin-producing strains not sensitive to the actions of their own compounds? The reason for such immunity is that, along with bacteriocins, producing strains make proteins that counteract their activities. Often, synthetic and immunity genes are on the same operon, which allows coordinate regulation of their expression. Losing the immunity genes makes bacteriocin-producing strains sensitive to the actions of these proteins.

Bacteriocins have found practical uses, especially in the dairy industry, where they are used in both a predictable and an unexpected way. The predictable use is to limit the growth of undesirable bacteria, such as *Listeria*, a pathogen that sometimes contaminates dairy products. A

Table 19.2 Commonly used antibacterial antibiotics and how they work

Antibiotic(s)	Mechanism of action
Murein synthesis inhibitors	
β-Lactams (e.g., penicillins, such as penicillin V; cephalosporins, such as cephalexin [Keflex]; and synthetic derivatives, such as ampicillin) Others (vancomycin, aztreonam, imipenem)	Interfere with cell wall biosynthesis, leading to autolysis
Protein synthesis inhibitors	
Aminoglycosides (streptomycin, kanamycin, neomycin, gentamicin, amikacin, tobramycin)	Bind to 30S subunit of bacterial ribosome Cause translational misreading and inhibit elongation of protein chain Kill by blocking initiation of protein synthesis
Others	
Tetracyclines (tetracycline, doxycycline)	Bind to ribosome 30S subunit Inhibit chain elongation step
Chloramphenicol	Binds to ribosome 50S subunit Inhibits chain elongation step
Erythromycins (e.g., azithromycin [Zithromax])	Block the exit of the growing peptide chain from the ribosome
RNA synthesis inhibitors	
Rifampin	Binds to bacterial RNA polymerase and blocks transcription initiation step
DNA synthesis inhibitors	
Nitrofurans Metronidazole	Partially reduced nitro groups give addition products on DNA that lead to strand breakage
Nalidixic acid Novobiocin Ciprofloxacin (Cipro) and other quinolones	Interfere with DNA replication by inhibiting DNA topoisomerases
Folate antagonists	
Sulfonamides (trimethoprim-sulfamethoxazole [Septra])	Block synthesis of tetrahydrofolate and 1-carbon metabolism

handy way to deliver the bacteriocins is to clone their genes into the strains that are used for cheese production. The less obvious use of bacteriocins is to enhance the flavor of cheeses, such as Cheddar. When bacteria used in the production of such cheeses lyse, they release enzymes that improve flavor development and speed up the ripening of the cheese.

CONCLUSIONS

The interactions among living organisms are varied and are central to the survival of many species. No biological entity has evolved without being molded by the presence of other organisms. In this chapter, we have considered several mutually beneficial interactions, as well as others that are more one-sided. In the next two chapters, we will tackle the weighty topic of microbial pathogenesis, which is characterized mainly by interactions that have a deleterious effect on the host.

Table 19.3 Modes of action of some bacteriocins[a]

Forming ion channels in membranes (depolarization)
Breakdown of DNA
Inactivation of ribosomes
Inhibition of cell wall murein synthesis
Hydrolysis of cell wall murein

[a]Some affect eukaryotic cells as well.

STUDY QUESTIONS

1. What has led to the thought that mitochondria and chloroplasts are derived from bacteria?

2. What are key characteristics of bacterial endosymbionts of insects?

3. What is the function of root nodules in legumes, and how do they arise?

4. What are the main microbial activities that take place in the rumen of a cow?

5. Discuss the symbiosis between the bacterium *Xenorhabdus* and nematodes and its effect on certain insect caterpillars.

6. In what way is the relationship between leaf-cutting ants and certain fungi a three-way symbiosis?

7. Present examples for a lay audience of how infectious agents can alter the behavior of their hosts.

8. Discuss examples of bacteria being used as prey by predators.

chapter 20 infection: the vertebrate host

INTRODUCTION

Even without the benefit of a microbiology course, most people have no problem naming a dozen or more infectious diseases. (Try it.) Likely to be included in the list will be some of the terrifying scourges of mankind, such as AIDS, tuberculosis, and smallpox, but also some more benign conditions, such as strep throat, the common cold, and the "stomach flu" (whatever that is—there is no clinically recognized entity by that name). The point here is that infectious diseases are common and widespread. Some are deadly, others nearly trivial, and everyone knows something about them.

Not only have infectious diseases caused untold havoc in the past, they still do now in many parts of the world. They have shaped history in the past and continue to do so even now, but at least one dreaded disease, smallpox, has been eradicated from the globe. Polio is a candidate for imminent eradication, and over a longer timeframe, so are tuberculosis, syphilis, and measles. Meanwhile, the tussle between hosts and parasites goes on, making continuing demands on human ingenuity. We possess three main kinds of external weapons for controlling infectious diseases: **sanitation** (including insect control), **vaccination**, and **antimicrobial drugs.** A moment's reflection will suggest how these can be applied and how they have potential downsides. In any case, by far the most powerful weapon is internal: our own immune system.

The bonds between humans and microbes are forever changing. Some diseases, such as AIDS, appear anew on our horizon; others, like the *Helicobacter* infection of the stomach, have been with us for a long time but have only recently been recognized. In most cases, *human behavior* influences our relationship to the offending microbes by altering the environment, allowing greater chance for transmission of the agents, or overusing antimicrobial drugs (Table 20.1).

Table 20.1 Human behavior that contributes to the emergence of infectious diseases

Factor	Circumstances	Example(s)
Poverty in countries and regions	Poor health services Contaminated food and water Malnutrition	Multidrug-resistant tuberculosis Cholera, many others Enhanced severity of measles
Ecological changes	Reforestation (increase in deer and ticks)	Lyme disease
Personal behavior	Sexual behavior and intravenous drug use	AIDS, hepatitis C
Social behavior	Urbanization	AIDS, dengue, tuberculosis
International travel and trade	Shipping Global food trade	Cholera to South America Food-borne outbreaks (many kinds)
Modern production practices	Mass food production Feeding scraps to cattle Superabsorbent tampons	*Escherichia coli* O157:H7 in hamburgers "Mad cow" disease Toxic shock syndrome
Medical and agricultural use of antimicrobials	Overuse of antibacterial and antiviral agents	Microbial drug resistance

Clearly, there are too many infectious diseases to be able to learn them all. Instead, we will concentrate on what they all have in common. A conceptual framework on which to hang particular facts is this: *the development of all infectious diseases, whether of humans, animals, or plants, proceeds through the same steps.* Before discussing these steps, a word of apology is called for. We shall employ the terminology of conflict and war and use terms such as defense, attack, invasion, weapons, and so forth. This usage is handy but misleading, because it implies a belligerent intent on the part of the agents. Microbes that cause disease simply go about their business. Except for a few cases, causing disease is incidental to their lifestyle.

Encounter

For an infectious disease to happen, the host and the infecting agent must meet. The encounter may take place not only just before getting sick, but also at any time during the host's life. One may "catch a bug" and get sick within a few days, or an agent may reside in the body for a considerable time before causing disease. The encounter may happen earlier than expected: certain human diseases, such as human immunodeficiency virus (HIV) infection, German measles (rubella), and syphilis, can be acquired while in the mother's womb—a dubious bequest.

The symptoms of infection are seldom noticeable right away and are preceded by an **incubation period** that may last days, months, or years (as in leprosy and AIDS, where the symptoms may arise a decade or more after infection). The source of the organisms may be other humans (as for the common cold, the flu, and the sexually transmitted diseases), animals (as in Lyme disease, which is caught from deer ticks), or the inanimate environment (as for *Salmonella* food poisoning and cholera).

One need not look far for infectious agents, because many are common in our own bodies. All vertebrates and most invertebrates are endowed with a large and varied **normal flora** (Table 20.2). (Note that the term **normal biota** is the correct one, because microbes are not plants, but "normal flora" is widely used.) The body's bacterial populations may be very dense, as in the mouth or the large intestine; of moderate size, as on the skin; or virtually absent, as in deep tissues. Although most of these

Table 20.2 Normal microbial biota of humans: an assortment of frequent types

	Gram-positive bacteria		Gram-negative bacteria		
Location	**Cocci**	**Rods**	**Cocci**	**Rods**	**Other**
Skin	Staphylococci	Corynebacteria (diphtheroids)		Enteric bacilli (in some sites)	
Mouth	Streptococci (alpha-hemolytic), other cocci	Corynebacteria (diphtheroids)	Neisseriae	*Haemophilus*	Spirochetes
Large intestine	Streptococci (enterococci)	Lactobacilli		*Bacteroides*, enteric bacilli	
Vagina	Streptococci	Lactobacilli		*Bacteroides*	Mycoplasmas

organisms are **harmless commensals** or even participate in a mutualistic relationship, many of them can turn on the host and cause disease. We call such diseases **endogenously acquired.**

Invariably, the establishment of every infection requires some *breach in the defense mechanisms* of the host. The breach may be as trivial as a cut through the skin, or it may be a catastrophic event such as the breakdown of the entire immune system in AIDS patients. Infectious agents often actively ensure that the host's defenses are breached. Each does it in its own way.

Table 20.3 Modes of entry

Inhalation (influenza virus, hantavirus)
Ingestion, usually fecal-oral route (*Salmonella*, cholera agent, *E. coli*)
Insect bites (malaria and Lyme disease agents)
Sexual contact (sexually transmitted disease agents, HIV)
Wound infection (natural or surgical)
Organ transplants (cornea, blood transfusion)

Entry

To cause an infectious disease, the agent must enter its host (Table 20.3). Entry has two meanings. In the more familiar sense, infectious agents penetrate into the host's tissues. On the other hand, serious diseases can take place without penetration. The agent can enter one of the cavities of the animal body, such as the gastrointestinal, respiratory, or genitourinary tract, all of which are *topologically contiguous with the exterior* (Fig. 20.1). For example, an agent can pass from the mouth to the anus without crossing any epithelial surface. Consequently, the intestinal agents, such as cholera, reside on the outside of the body; they just adhere to epithelial membranes on the body's surface. Examples of serious diseases caused by bacteria that do not enter deep tissues are shown in Table 20.4.

Becoming established

Becoming established in the body means only that the invading agent has breached a certain set of defenses, e.g., the skin or the mucous membranes. Then begins a series of complex interactions between invader and host—a choreography of action and reaction or, to a musician, point and counterpoint. The outcome determines whether there will be symptoms of disease and whether the agent will persist in tissues.

Three factors are generally involved in determining if an infectious agent will become established and cause disease:

- the size of the inoculum (the number of invading microbes)
- the invasive ability of the infectious agent
- the state of the host's defenses

These factors are interrelated. If the invasive ability of the agents is high, fewer of them will be required to establish infection than if the agent were less virulent. The size of the inoculum refers not just to how many organisms are inhaled or ingested, but to how many actually *reach the target tissue or organ*. Much can happen while getting there. For example, bac-

Figure 20.1 The microbe's view of the human body. Our bodies consist of a series of tubes, one of which (the digestive system) has two orifices, others (the respiratory and genitourinary systems) only one. Not drawn are the various sphincters that allow the tubes to be opened and closed. For example, the digestive system has sphincters at the junctions of the esophagus and stomach, stomach and small intestine, etc.

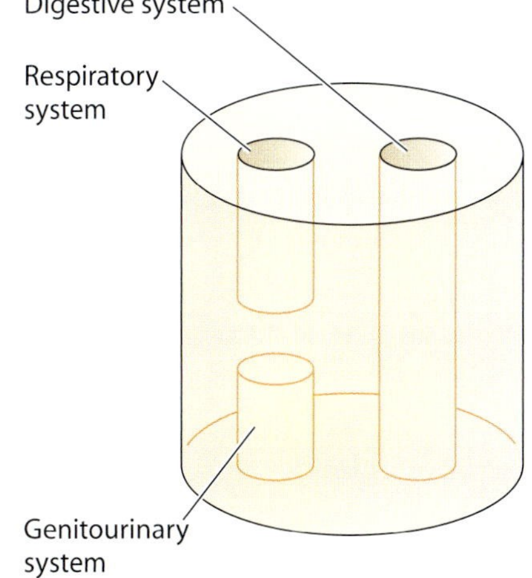

Table 20.4 Bacteria can cause serious diseases without entering the deep tissues of the body: a few examples

Disease	Site	Penetration into tissue
Cholera	Small intestine	Practically none
Bacterial diarrhea (some)	Small intestine	Practically none[a]
Whooping cough	Airways	Practically none
Diphtheria	Throat	Superficial layers only
Bacterial dysentery	Large intestine	Superficial layers only
Cystitis (bladder infections)	Bladder	Superficial layers only
Gonorrhea	Urethra	Superficial layers only[a]
Chlamydia infection	Urethra, eye	Superficial layers only[a]

[a]Uncomplicated cases of the disease only. In some patients, the organisms penetrate into deeper tissue, in which case the symptoms of disease are different.

teria that are swallowed and cause disease in the bowel must withstand the acid in the stomach. It has been shown that fewer cholera bacilli are required to cause disease if the patient is deficient in stomach acid production. Despite such particulars, the lower the defenses of the host, the easier it is for a smaller number of invading organisms to cause disease, and vice versa.

One may then ask, *what is a pathogen*? A tricky term, that. Microbes cause disease in one person but not in another. Patients whose defenses are very low (as in advanced cases of AIDS) may be infected by almost any microbe, including many **commensals.** Thus, the definition of a pathogen is conditional. Given a chance, agents that are harmless commensals in a normal host may act as pathogens and cause disease in a compromised one. The **immune competence** of the host is of paramount importance, but it is not the only factor. Anatomic considerations sometimes make the difference between health and disease. Thus, anaerobic bacteria that live harmlessly in the colon can, if untreated, cause fatal peritonitis if the gut wall is punctured.

Some organisms are host specific, whereas others can infect a wide variety of hosts. The gonococcus, for example, affects only humans, whereas *Pseudomonas aeruginosa*, a bacterium associated with infections of patients with cystic fibrosis and burns, is the least choosy of all pathogens known, causing a variety of diseases in such disparate groups as plants, insects, worms, and vertebrates.

Causing damage

The establishment of an agent in the host will not result in disease unless it damages tissue. Damage occurs in a variety of ways, which is one of the reasons why each infectious disease has its own characteristics.

Damage may result from cell death, either by **lysis** of host cells or by an untimely induction of **apoptosis** (programmed cell death). The effects on the host depend greatly on which tissues are involved and can range from life threatening when vital organs are affected to relatively mild when less essential sites are involved. Tissues may be affected without the death of host cells but rather by *pharmacological action of microbial toxins.* For example, bacterial diarrheal diseases, such as cholera, often result from a derangement of normal cell function, in this case, ion and water exchange in the intestinal cells. The affected intestinal cells remain intact. Rarely, damage results from mechanical action, as occurs when blood or lymph vessels become blocked.

The damage caused by infectious agents may be seen close to the site of entry, for example, when a cut on the skin becomes inflamed with pus. Alternatively, the targeted tissues may be far away from the site of invasion, as happens in tetanus. In this disease, the offending bacteria may be found in a cut on the foot, but their real damage is caused by a toxin that acts on distant neuromuscular junctions. These two examples illustrate another of the great themes of infectious diseases. In tetanus, the damage is caused directly by the agent. In an infected wound, most of the damage is caused by an exuberant response of the host itself. Most infectious diseases involve both kinds of damage (Tables 20.5 and 20.6).

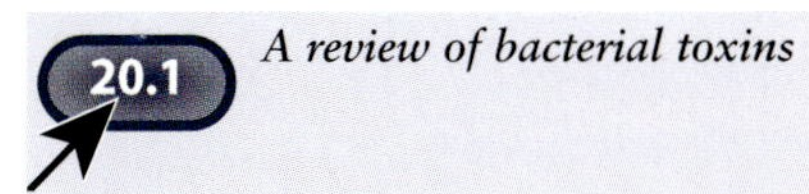

Table 20.5 Damage caused during infections

Lethality (cell death)
Lysis by toxins
Lysis by immune lymphocytes (killer cells)
Programmed cell death (apoptosis)
Pharmacological changes
Tetanus, botulism, cholera
Mechanical
Obstruction of vital passages
Due to host responses
Inflammation
Immunopathology

HOST DEFENSES

The most imposing testimony to the powers of our bodies' defenses is that we live in a "buggy" world but only rarely come down with infections. Our defenses are convoluted, and understanding them is demanding. In any case, a grasp of the immune response is crucial for the understanding of infectious diseases. The importance of the defense mechanisms is best demonstrated by what happens in their absence. Genetic defects or induced failure of the defense mechanisms invariably leads to increased frequency and severity of infection. The consequences can be life threatening.

We can conveniently divide the defenses into two kinds of systems: **innate**, those that are always available, and **adaptive**, those that are set in motion after contact with a particular invading agent (Table 20.7). We inherit the innate defenses from our parents, and with few exceptions, each of us is identical in this regard to all the other individuals of our species. The adaptive defenses, on the other hand, are not inherited but reflect the experience of each individual. Therefore, we all have different repertoires of adaptive responses.

As we will see, these two systems are highly interrelated, and they are presented separately only for convenience.

Table 20.7 Defenses of the body against microbes

Constitutive or innate
Mechanical
Skin, mucous membranes
Chemical (examples only)
Fatty acids on the skin (due to skin bacteria)
Hydrochloric acid in stomach
Antimicrobial peptides in tissues and white blood cells
Lysozymes in cells, tissues, and body fluids
Bile salts in intestine
Complement in circulation and tissues *(major factor)*
Cellular
White blood cells (neutrophils, macrophages, etc.)
Adaptive or induced
Humoral immunity (antibodies)
Cell-mediated immunity

Table 20.6 Examples of symptoms caused by host responses

Disease	Immune response	Symptoms
Pus causing (e.g., gonorrhea, strep throat, acne)	Acute inflammation	Pus, irritation
Brain abscess	Acute inflammation and effects of having a foreign mass in a confined space (the skull)	Neurological symptoms due to compression of vital sites
Tuberculosis	Chronic inflammation	Cough with bleeding, due to ongoing tissue destruction
Respiratory syncytial virus infection	Allergy-like reaction in bronchioles	Wheezing and difficult breathing (asthma-like)
Malaria	Synchronized release of fever-inducing cytokines	Periodic episodes of shaking chills followed by sweats

Innate defenses

External barriers

Few, if any, bacteria or viruses can penetrate the intact skin. Thin as it is, our skin acts like a massive wall surrounding a medieval city, only to be breached upon significant coercion. Additionally, the skin surface contains antimicrobial substances, such as salt in high concentration, certain peptides, and fatty acids. The mucous membranes of the respiratory, digestive, and genitourinary tracts are bathed in antimicrobial substances, such as antibodies and lysozyme (an enzyme that hydrolyzes the cell wall murein of bacteria). In addition, in the tubular organs of the digestive and urinary tracts, liquid currents sweep away, with the vehemence of peristalsis, bacteria not firmly attached to the organ's wall.

An especially effective barrier to microbes is the stomach, where the strong acid kills most organisms ingested. Persons unable to make sufficient hydrochloric acid are at risk of serious intestinal infections. However, the stomach is not a perfect sterilizing chamber: some acid-resistant pathogens, such as the shigellas (agents of dysentery) and cholera bacilli, survive the journey through it. One species of bacteria, *Helicobacter pylori,* even lives in the stomach. This organism causes inflammation and ulcers of the stomach and duodenum and has been implicated in gastric cancer. *H. pylori* survives in this acidic environment by producing large amounts of ammonia, which neutralizes the acid in its immediate environment.

In tissues

The host's most powerful reaction to an invading agent is the **inflammatory response.** It is manifested by **inflammation,** an event common in most infections. A characteristic of inflammation is *local reddening, swelling, pain, and pus.* Reddening is due to increased blood flow at the site, swelling to an outpouring of fluids in the tissues, and pain to the release of chemical mediators and compression of the nerves. The strongest antimicrobial elements in the inflammatory response are the **phagocytes,** white blood cells that ingest foreign particles and, in the case of microbes, try to kill them. Phagocytes must be recruited to the site of infection and become active before they can ingest foreign particles. One way to prod the phagocytes into action is through chemical signals produced in part by a soluble (not cell-bound) system called **complement.**

A central property of the inflammatory response is that *it is normally switched off but is switched on by microbial challenge.* Were this not so, the body would be in a constant state of inflammation, which would lead to serious illness and death. So, what is the master switch? How does the body recognize the presence of microbial invaders? The answer is that invading microbes are recognizable by some of their unique chemical constituents. You may guess what such bacterial visiting cards may be: cell wall murein, the lipopolysaccharide (endotoxin) of gram-negative bacteria, flagellin (the major protein of bacterial flagella), or pilin (the protein of pili). This is correct: the body recognizes these molecular signatures as being exclusively microbial. Thus, a large number of different organisms can be recognized as long as they have *typical microbial components.* Such **microbe-associated molecular patterns** (**MAMPs**) are found in both bacteria and other infectious agents, including surface constituents from fungi and animal parasites and some chemically distinct viral and bacterial nucleic acids. (The misleading term PAMPs, for pathogen-associated molecular patterns, is often used. However, these patterns are shared by pathogens and nonpathogens alike.) It is thought that the innate immune system is able to recognize about 1,000 distinct MAMPs, which ought to include a large number of pathogens. A pathogen lacking all MAMPs would, in theory, be able to escape the body's radar screen, but in fact, our surveillance mechanism is broad enough to detect practically all invaders.

When MAMPs are recognized, the alarm that invaders are present is sounded and the defensive responses are set in motion. Two tracks lead to the development of the innate immune response, one involving soluble components, the other involving cells. We deal first with the main response involving soluble components, the complement system.

The complement system. Complement is a multifunctional defense system consisting of some 30 different proteins. It plays a central role in eliciting the inflammatory response and is essential for health and well-being. The name complement is derived from the early notion that it "complements" the action of antibodies, although it was later determined that it also works in other ways.

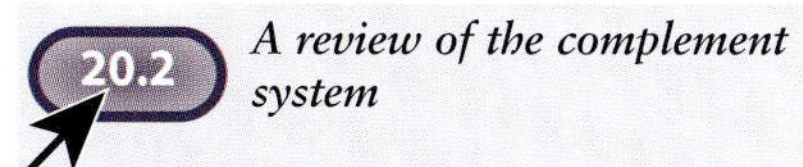

A review of the complement system

Complement is normally dormant and must become *activated* to become a defense mechanism. An important hint about the role of complement comes from observing persons born with hereditary defects in its components. Some people with rare mutations in one of the components can live healthy lives, but others are highly susceptible to bacterial infections, as well as to noninfectious diseases, such as the autoimmune disease lupus erythematosus.

There are various ways to activate complement. A typical one is by the binding of certain complement proteins to MAMPs, such as microbial surface polysaccharides. This sets off a series of sequential proteolytic cleavages—a **proteolytic cascade** reminiscent of that in blood clotting. Normally, when these proteins are circulating freely, they are not susceptible to proteolysis, but they become so once bound to particles. The resulting peptides stimulate the cleavage of other proteins, and so on down the line. Some of the products of these proteolytic events are **pharmacologically active peptides** that strongly promote inflammation (Table 20.8). This is the most common mode of complement activation, but it was given the misleading name of **alternative pathway** (it was not discovered first). Other activation pathways differ in their timing (Fig. 20.2). One of these, known as the **classical pathway,** depends on the presence of antibodies. Because it takes a week or more to make sufficient levels of antibodies after antigenic stimulation, this pathway does not come into play early in infection.

The normal dormancy of the complement system is under the control of several regulatory molecules that monitor the extent of complement activation and switch the system off when the stimulus is gone. The host also avoids setting off complement activation by adding the amino sugar **sialic acid** to the surfaces of its cells. The presence of sialic acid hinders the attachment of the complement protein that sets off the proteolytic cascade.

Table 20.8 Roles of complement

Roles of complement
Helps recruit phagocytes to site of invasion
Makes invading microbes susceptible to phagocytosis (opsonization)
Lyses some microbes directly
Promotes inflammatory response

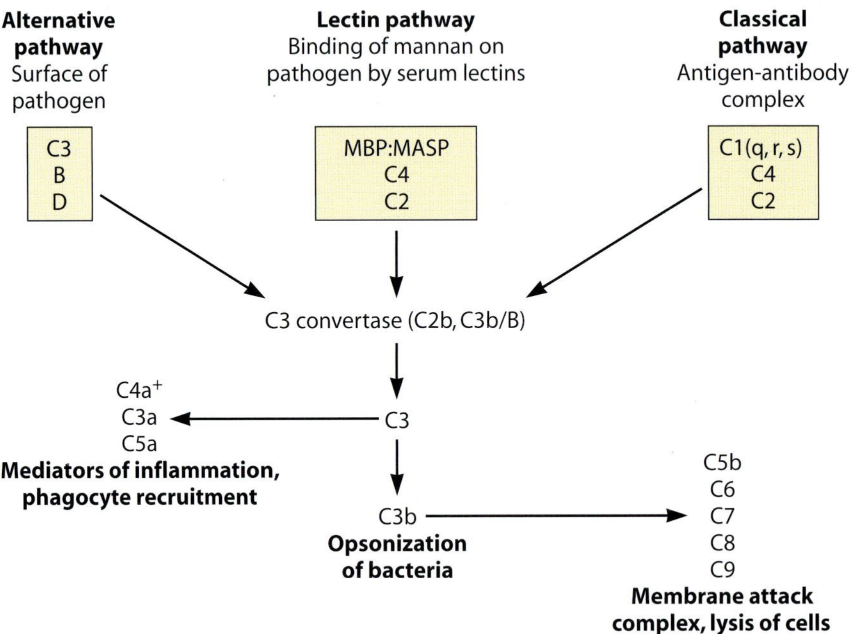

Figure 20.2 Complement activation pathways and their components. The complement system can be activated in three different ways, all of which lead to the same end products. In each case, the end result is the formation of proteins involved in the major activities of complement: mediation of inflammation, phagocyte recruitment, opsonization, and lysis of foreign cells. The difference is in the stimulus that elicits the reaction. The **alternative pathway** is elicited by recognition of surface components of invading organisms. The **lectin pathway** depends on the recognition of mannan residues on bacterial surfaces by serum lectins. The **classical pathway** results from the presence of antigen-antibody complexes. Note that, unlike the other two, the classical pathway must await the formation of specific antibodies, a process that can take 10 to 14 days, unless the host has made antibodies from vaccination or a previous encounter with organisms with the corresponding antigens. The names of the complement proteins often, but not always, start with C (for complement). MBP, mannan-binding protein; MASP, MBP-associated serum protease.

Complement plays an important role in promoting the inflammatory response and, through it, phagocytosis by the white blood cells. It does this in two ways: by recruiting white blood cells to the site of the microbes and by enhancing their phagocytic power.

White blood cells are normally present in the circulation, some free, others sticking to the walls of small blood vessels. **Recruitment** involves the passage of these cells through the blood vessel walls and their massing at the site of the invaders (Fig. 20.3). What sets off the alarm? Complement activation results in the formation of peptides called **complement chemotaxins.** White blood cells sense and move up a concentration gradient of these substances. Thus, if complement is activated by invading microbes, the white blood cells will flock toward them. There is another way phagocytes are drafted to the site of an infection: by sensing **bacterial chemotaxins.** A phenomenon unique to bacteria is the fact that their proteins are synthesized with several extra amino acids at the amino end (see Chapter 8). These are then removed in order for the protein to "mature." Peptides generated by this cleavage are usually only three or four amino acids long and act as attractants (chemotaxins) to recruit white blood cells. Bacterial chemotaxins are remarkably potent. One of

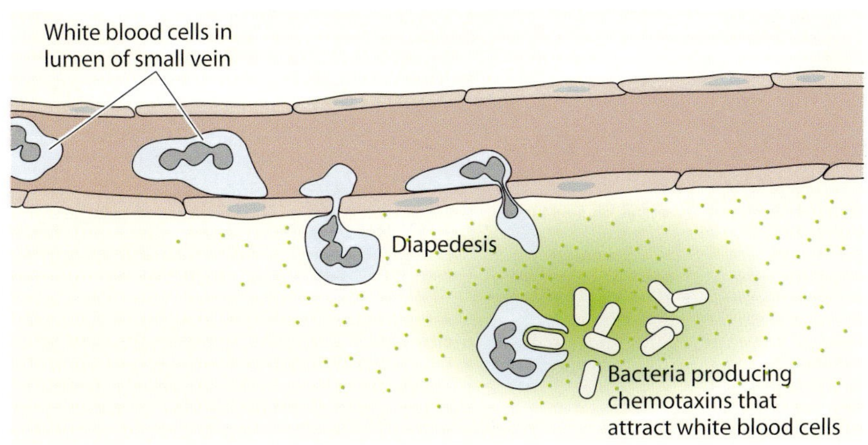

Figure 20.3 Recruitment of white blood cells to a site where microbes are present. Bacteria produce diffusible substances called **chemotaxins** that serve to attract white blood cells from the circulation into the sites in tissues where microbes are present. The passage of white blood cells through the walls of the vessels is called **diapedesis.**

them (*N*-formylmethionyl-leucyl-phenylalanine) works at concentrations of 10^{-11} M! Thus, bacteria loudly advertise their presence, and the body heeds the call.

Unless helped, phagocytes are sluggish eaters and ingest bacteria slowly and inefficiently. Some complement peptides called **opsonins** provide help and turn phagocytic cells into voracious eaters and formidable killers (see below). Complement does not provide the only kind of opsonins; antibodies also function in this way (Fig. 20.4).

Complement also acts as a defense mechanism without acting on phagocytes. Other complement-derived peptides *lyse* bacteria, animal cells, or even enveloped viruses. These peptides assemble into a doughnut-shaped structure, the **membrane attack complex,** which inserts into foreign membranes (Fig. 20.5). This results in holes that make the bacteria or enveloped viruses permeable to ions and water, which leads to lysis. This armor-piercing weapon destroys bacteria and viruses that are not killed by the phagocytes.

Phagocytes. Phagocytosis is the engulfment of foreign particles by cells. It is the most effective of all innate host defenses against bacteria. *The majority of the bacteria that invade humans over their lifetimes are killed or removed by phagocytosis.* The ability of cells to act as phagocytes—to take up particles—is shared by many cells of the body, but only a few do it with gusto. Chief among the cells that may be called professional phagocytes are the white blood cells, yet even such licensed practitioners do not normally perform as phagocytes but await, once again, the signals that invaders are about, the MAMPs.

Phagocytic white blood cells fall into several classes that have different functions. The most numerous are the **neutrophils** (or **polymorphonuclear leukocytes**), which are **terminally differentiated** (Table 20.9). That is, they cannot proliferate, and they live for only a few days or weeks. Their role is to seek out foreign particles, such as bacteria; ingest them; and to try to kill them. What makes them so successful is that they move about very rapidly, like small amoebas, responding effectively to chemotactic gradients. They can reach bacterial targets quickly and in large numbers, which leads to inflammation. Note that if unchecked, inflammation can

Figure 20.4 Opsonization enhances phagocytosis. Microbes and other particles are not readily phagocytized by white blood cells unless they are coated with proteins called opsonins. Several kinds of proteins can act as opsonins. Microbes coated with the complement protein C3b bind to C3b receptors on the surfaces of phagocytic cells by a mechanism that resembles a zipper. An analogous process takes place when microbes are coated with antibodies.

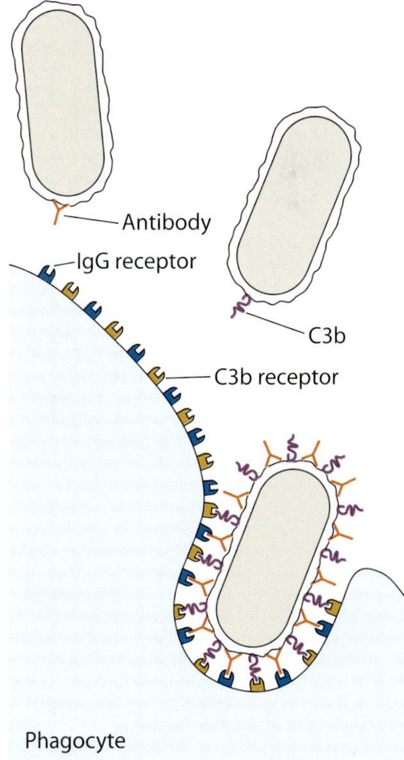

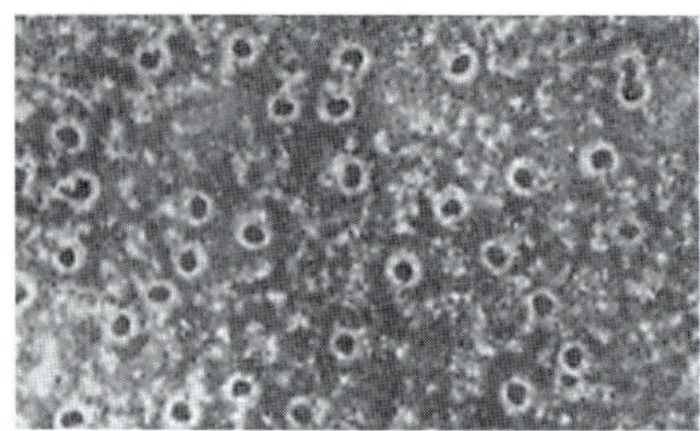

Figure 20.5 The membrane attack complex seen under the electron microscope. These doughnut-shaped MACs are inserted into the membrane of a red blood cell.

Table 20.9 Properties of neutrophils, or polymorphonuclear leukocytes

Short-lived cells (live a few weeks)
Loaded with large lysosomal granules that contain hydrolytic enzymes that can destroy many bacteria and oxidative enzymes that make toxic products, especially hypochlorite (bleach)
Recruited to the site of the microbes by several chemotaxins, some derived from complement and others products of bacterial metabolism
Require opsonins to be effective in killing microbes

cause damage to tissues and symptoms of disease; thus, it is a two-edged sword (Table 20.6).

Phagocytosis starts with **ingestion,** which takes place when the neutrophil membrane enfolds the bacterium. This step depends on the interaction between molecules on the surface of the bacterium and that of the neutrophil. **Opsonins** (complement derived or antibodies) coat the surface of the bacterium. **Opsonin receptors** on the phagocyte membrane interact in zipper-like fashion until the bacterium becomes completely engulfed (Fig. 20.6). Both neutrophil movement and engulfment require the rearrange-

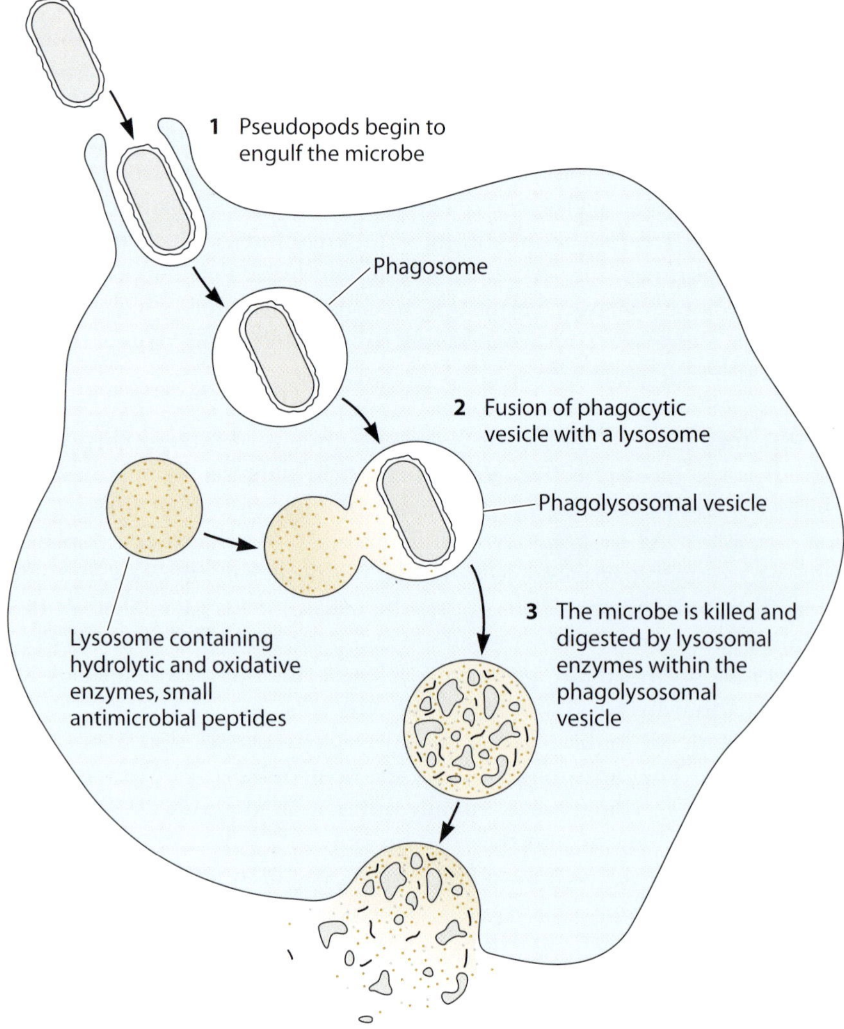
1 Pseudopods begin to engulf the microbe
Phagosome
2 Fusion of phagocytic vesicle with a lysosome
Phagolysosomal vesicle
Lysosome containing hydrolytic and oxidative enzymes, small antimicrobial peptides
3 The microbe is killed and digested by lysosomal enzymes within the phagolysosomal vesicle

Figure 20.6 Steps in phagocytosis. A microbe attaches to a phagocyte. The phagocyte forms pseudopods that surround the (opsonized) microbe **(1),** leading to the formation of a vesicle called a **phagosome (2).** Lysosomes filled with antimicrobial compounds fuse with the phagosome and release their contents within the vesicle, which is now called a **phagolysosomal vesicle.** Microbes within a phagolysosomal vesicle are killed and digested by hydrolytic enzymes **(3).** Note that this is not an inevitable outcome, because many microbes have mechanisms to withstand phagocytosis (Table 20.13).

ment of the cell's cytoskeleton through the actions of both actin and myosin filaments. (Does a neutrophil remind you of an amoeba?)

Engulfment of a particle results in the pinching off and release into the neutrophil's cytoplasm of a vesicle called a phagosome, which contains the bacterium. If nothing further happened, the ingested microbe would not be killed, and it might even grow in the phagosome. That is not usually the case: neutrophils possess large saclike **cytoplasmic granules** (actually gigantic lysosomes) that act as veritable bombs. These granules contain antimicrobial peptides called **defensins** plus **hydrolases** and other enzymes that damage bacteria and, if released, even the cells themselves. Usually, the granules fuse to the phagosomes, assembling a phagolysosomal vesicle. When this happens, the enzymes of the granules come in direct contact with the ingested bacteria. These enzymes fall into two classes: (i) hydrolases and other antibacterial proteins capable of directly affecting the cellular integrity of microbes and (ii) oxidative enzymes that act to form hypochlorite, the same powerful chemical found in laundry bleach. Fusion of the granules to the phagosome may take place before the ingestion of the invaders is complete; thus, they may be killed even before being fully taken up. This resembles the strategy of a poisonous snake that immobilizes its prey before swallowing it.

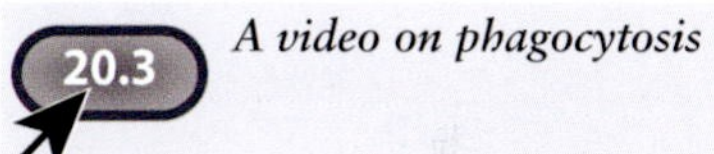

A video on phagocytosis

The arrival of the neutrophils at the scene does not mean that the microbes will necessarily be killed. Rather, the stage is now set for an epic struggle between microbes and phagocytes, the outcome of which is by no means foreseeable. Each combatant has many weapons to overwhelm the other. Even when the neutrophils prevail, there is a cost to the host. The recruitment of white cells to the site where invading microbes are present is not well regulated and may be overdone. This often results in **pus,** the accumulation of live and dead white cells, debris, and tissue fluids. As we have said, pus is a characteristic feature of inflammation, the others being reddening, swelling, heat, and pain at a local site.

The other major class of white blood cells is the **monocytes.** Unlike the neutrophils, these cells are long-lived and capable of differentiating further. Monocytes have the ability to settle in tissues, in which case they are known as **macrophages** (Table 20.10). Monocytes arrive at sites of infection after the neutrophils and are, in fact, called to these sites by substances secreted by the neutrophils. Monocytes and macrophages serve various functions. One is to clean up the debris left at the scene of the battle between neutrophils and microbes, but besides being garbage collectors, they *communicate with the other arms of the immune system.* They do this by making a large assortment of proteins called **cytokines,** some of which activate complement, while others promote inflammation, and

Table 20.10 Properties of macrophages and monocytes

Long-lived cells
Arrive at site of microbes more slowly than neutrophils
Act as "garbage collectors"; clear up microbial debris and remaining microbes
Become activated by cytokines, proteins made in response to microbial invasion
Make cytokines that attract and activate neutrophils, thus contributing to acute inflammation
Help initiate specific innate immune responses, i.e., cell-mediated immunity

yet others set off **adaptive immunity.** Thus, this class of white blood cells plays a complex and central role in all aspects of the immune response.

MAMPs, the signatures of invading microbes, reenter the discussion here. Macrophages and other cells of the immune system have protein receptors that specifically bind certain classes of MAMPs. These receptors are called **TLRs,** for **Toll-like receptors,** because they resemble so-called Toll receptors on *Drosophila* (fruit flies with mutations in these proteins behave in odd ways, causing their German discoverers to call them by the German word *toll,* which means, among other things, peculiar). TLRs are found in many organisms, from fruit flies to humans and even plants, and may be the product of an early coevolution of hosts and parasites. One could argue that such receptors normally play a different role in the host's physiology and that they have been fortuitously appropriated for the purpose of recognizing invading organisms. This is indeed the case for some receptors, but not for all. The notion that some may have a unique microbe recognition function is strengthened by the finding that mice lacking these receptors are normal in all respects other than their ability to fight disease.

TLRs are specific for subsets of MAMPs. Thus, some recognize the murein of gram-positive bacteria, others recognize the lipopolysaccharide of gram-negative bacteria, others recognize the single-stranded RNA of viruses such as influenza virus or mumps virus, and so forth. Some TLRs are located on the cytoplasmic membranes of many cells, others on those of the phagocytic vacuoles. Thus, they have two shots at recognizing MAMPs: first, when the microbes become bound to the host cell membrane, and second, after they are phagocytized and their constituents are released after lysis. Binding of microbial constituents to TLRs sets off a signaling cascade that leads to the activation of a key host cell transcription factor called **nuclear factor κB (NF-κB)** that turns on cytokine genes.

An example of the consequences of TLRs interacting with MAMPs is the activation of macrophages. Macrophages, like other members of the immune response system, are normally quiescent. To reach their greatest capacity for killing microbes, they must be *activated* (become "angry"). Angry macrophages are more effective phagocytes and are also better cytokine producers. Consequently, when macrophages recognize the presence of microbes in tissues, they sound alarms that affect all aspects of the immune response.

Prominent among macrophage-activating stimuli is a paradigm of MAMPs, the lipopolysaccharide (endotoxin) of the outer membrane of gram-negative bacteria (Table 20.11). When lipopolysaccharide is present at a high level, the term "endotoxin" becomes fully justified, because the patient may go into severe shock caused by the widening of blood vessels and a drop in blood pressure. This sometimes lethal condition is seen in patients with infections of the blood caused by gram-negative bacteria (bacterial sepsis). The term "endotoxin" denotes that it is associated with the body of the producing bacteria, unlike soluble toxins, which are known as **exotoxins.**

How do microbes evade the innate defenses?

Working together, complement and the phagocytes would appear to constitute an impenetrable impediment against invading microbes. Most

Table 20.11 Properties of bacterial endotoxins

Consist of lipopolysaccharide
Found only in the outer leaflet of the outer membrane of gram-negative bacteria (thus, not found in gram-positive bacteria)
Act on:
Macrophages to produce cytokines and induce fever
Neutrophils to produce compounds that dilate blood vessels, causing edema and shock
Complement system to induce its activation, causing inflammation

of the time this works, which helps explain why infectious diseases are not a daily norm, at least not in the technologically advanced countries. Still, some infectious agents run this formidable gauntlet and cause disease, because microbes have evolved ways to subvert the host defenses. Individual pathogens differ in their counterdefensive arsenals. Attempts to subvert the host defenses leave practically no corner of the immune response untouched. The breadth of these microbial counterdefenses suggests a possible reason why the immune system is so complex. Evolution here elicits an arms race between the pathogens, who devise new strategic counterdefense initiatives, and hosts, who respond with novel methods to overcome them.

How do microbes defend themselves against complement? The most effective way that microbes defend themselves is to prevent the activation of complement, which can be done in various ways. For example, meningococci *(Neisseria meningitidis) mask the complement-activating components* on their surfaces by coating themselves with a thick **capsule.** Gonococci *(Neisseria gonorrhoeae)* coat their surfaces with sialic acid, the same amino sugar that keeps complement from being activated by the host cells. Salmonellas counteract the lytic effects of complement with a coat of surface components that keep the hole-making proteins from reaching their surfaces. Vaccinia viruses (the viruses used to vaccinate against smallpox) inhibit complement activation by making a protein that mimics one that controls complement activation. Mutants of vaccinia virus lacking this protein are less virulent in laboratory animals.

How do microbes defend themselves against phagocytosis? Again, microbes have a wide repertoire of strategies to defend against phagocytosis (Table 20.13). Some microbes impair the recruitment of phagocytes by inhibiting complement activation, thus impeding the formation of chemotaxins. Others paralyze neutrophils by making toxins that disrupt their cytoskeletons. In yet others, the slimy capsule keeps phagocytes from swallowing them. The capsule-making strategy is widespread and is used by bacteria transmitted via the bloodstream (such as *Haemophilus influenzae* and meningococci).

Other microbes avoid damage by phagocytes *after* being taken up. For example, tubercle bacilli *(Mycobacterium tuberculosis) inhibit the fusion of the lysosomes with the phagosomes.* Other bacteria use a more brutal approach: they *lyse the phagolysosome membrane*, thus releasing its contents into the host cell cytoplasm, which kills the phagocyte. This works best for microbes that are relatively resistant to the action of the lysosomal enzymes themselves. A particularly ingenious technique, used by shigellas and the food-borne pathogen *Listeria,* is to escape into the cytoplasm (Fig. 20.7). This site is a safe haven for the organisms because the cell does not have a way to spill the contents of its lysosomes into its cytoplasm—that would be suicide.

Intracellular life

Living within the cells of the host not only provides a haven from phagocytosis, it also protects the organisms from circulating, damaging substances, e.g., antibodies and antimicrobial drugs that cannot penetrate the host cells. Which agents avail themselves of this opportunity? Viruses *have* to reproduce in cells, but most bacteria, fungi, and protozoa do not. Some bacteria, such as chlamydiae, rickettsiae, and the leprosy bacillus (*Mycobacterium leprae*), cannot be grown outside of cells, but they are the exceptions. However, some organisms perfectly capable of growing on agar plates have chosen an intracellular lifestyle. Tubercle bacilli, for example, are found within macrophages, where they may lie dormant for years and are not disturbed by host defenses.

The intracellular location is no picnic, because it limits opportunities for traveling and for making new encounters. Organisms living inside host cells face the problem of how to reach other target cells. Some viruses and bacteria lyse the host cell and are thus transmitted to others via the circulation, but this exposes them to antibodies, phagocytic cells, and antimicrobial drugs. Certain viruses, such as herpesviruses, solve this problem by inducing the cells in which they have multiplied to fuse with other cells, making giant **multinuclear syncytia.** An elegant solution is that of certain bacteria, such as *Listeria,* which use the cell's cytoskeleton to be pushed from one infected cell into a neighboring one (Fig. 20.7). This is a good example of how infectious agents appropriate host cell functions for their own use, a theme that recurs in microbial pathogenesis. Over time, intercellular existence does not ensure long-range survival because the host has adaptive defense mechanisms to seek out infected cells and destroy them (see below).

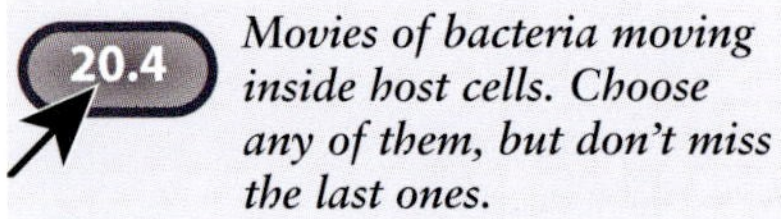

Movies of bacteria moving inside host cells. Choose any of them, but don't miss the last ones.

Adaptive defenses

The main adaptive defenses are **antibodies** and **cell-mediated immunity.** Both require exposure to the agent and, unlike the innate defenses, both are *highly specific* and *vary from individual to individual.* They work extremely well but for one thing: they are not there to protect the host upon first exposure to an agent but require a week or more to become effective. On the other hand, protection may be acquired by previous contact with an agent (whether or not this led to disease symptoms), immunization, or, for the first few months of life, from the transfer of antibodies from mother to fetus across the placenta and via the **colostrum** (high-protein mother's milk secreted for the first few days after giving birth).

Antibodies

Most children in the developed countries are given several vaccines in their early months of life. One of these vaccines, called DTP, is a multiple

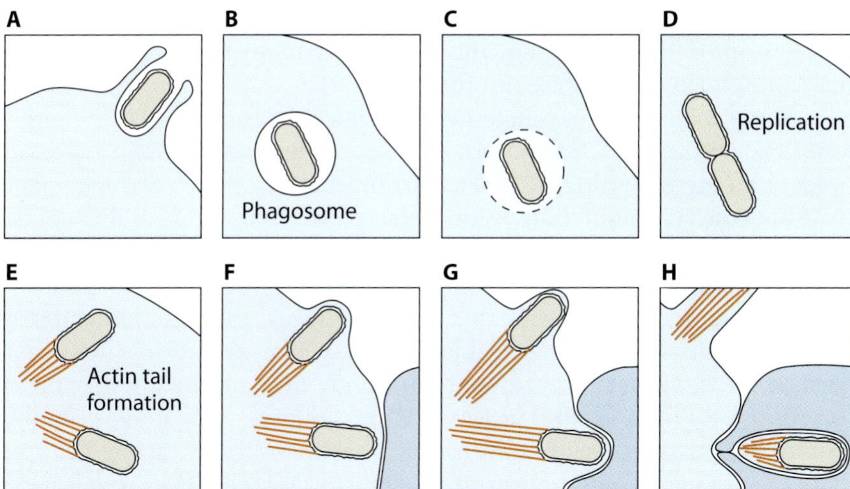

Figure 20.7 How bacteria move within the cytoplasm of host cells. Some bacteria (e.g., *Listeria, Shigella,* and *Rickettsia*), when ingested by phagocytosis **(A),** escape from the phagosome **(B)** into the cytoplasm by lysis of the phagosomal membrane **(C)** and divide **(D).** Actin then polymerizes at one of the poles of the bacterium **(E),** propelling the bacterium in a forward motion **(F).** When a bacterium reaches the plasma membrane, it sometimes causes the formation of a cytoplasmic extrusion **(G).** If this takes place next to an adjacent cell, the bacterium penetrates into it to set up another round of infection **(H).**

vaccine and induces the formation of antibodies against diphtheria toxin (the D), tetanus toxin (the T), and pertussis, or whooping cough (the P). Such vaccination stimulates the formation of antibodies that protect against these diseases.

Here is an example of how antibodies helped fight infections before the advent of antibiotics. Patients with pneumonia caused by pneumococci (*Streptococcus pneumoniae*) were typically quite ill for a period of a week or two, with symptoms increasing in severity. At this point, the "crisis" occurred and patients either died or got better rather suddenly. Some patients rose from their deathbed within hours and demanded a good meal. This miraculous cure occurred because antibodies against the organisms had reached an effective threshold level. The pneumococci are unusually resistant to phagocytosis because they are surrounded by a thick slimy capsule (see Fig. 2.12). When antibodies are bound to the capsule, they serve as opsonins, allowing phagocytes to bind and ingest the bacteria.

Antibodies are proteins called globulins made in response to foreign substances called **antigens.** The emphasis here is on *foreign.* The vertebrate host has an intricate mechanism to differentiate between "self" and "nonself," thus preventing the production of antibodies against its own body components. When this happens on occasion, it leads to so-called **autoimmune diseases.** Many antibodies are amazingly specific and can distinguish between proteins that differ by a single amino acid or polysaccharides that differ by an α or a β linkage, reacting with one and not the other. Normally, antiserum made against a certain antigen is **polyvalent,** that is, it is a mixture of antibodies that recognize different antigenic sites on the same antigen, the **epitopes.** (An ingenious trick permits the production of so-called **monoclonal antibodies,** which are specific against one epitope only. Monoclonal antibodies have many uses in research and, increasingly, in diagnostics and therapy.)

Antibodies come in a huge number of varieties, not because there is one gene encoding each antibody, but because immunoglobulin genes can undergo a large number of DNA rearrangements. These rearrangements (**site-specific recombination** events) take place all the time and engender a large variety of antibody-producing cells, the **B lymphocytes (B cells),** each of which produces a different antibody (Fig. 20.8). It has been calculated that recombination can generate upward of 6 million different kinds of antibody molecules. Additionally, somatic mutations (not occurring in the germ line) contribute to antibody diversity by another factor of 10 to 100, for a total of at least 10^8 possibilities. Such a huge number could not possibly result from having that many genes in the genome. The human genome contains only (some say) 30,000 genes. *Each B-cell clone produces a large amount of a given specific antibody.* The actual amount is about 2,000 molecules per second, an astonishing number.

Normally, the B lymphocytes lie dormant, but when their cognate antigen is present, each proliferates to make a **clone** of cells. They are set off on their path to become antibody-producing factories by specific cytokines made by macrophages and other cells. In this manner, the innate immunity is responsible for signaling the activation of antibody formation. We will see that this is also true for the other arm of this response, cell-mediated immunity. Therefore, the two aspects of immunity are intimately linked with each other.

Antibodies help combat infectious diseases in various ways.

1. They neutralize microbial toxins and render them ineffective.
2. They facilitate the removal of infectious agents by:
   - acting as opsonins—when bacteria are coated with antibodies, they are recognized by receptors on the phagocyte surface, much like complement-derived opsonins
   - clumping bacteria into larger particles, which facilitates their removal
   - utilizing the filtering mechanisms of the body
3. They interact with complement to lyse certain bacteria.

Cell-mediated immunity

Intracellular organisms are shielded from antibodies, complement, and phagocytes. A special mechanism, cell-mediated immunity, deals with those agents that reside inside host cells. Frankly, this is a complex business (Fig. 20.8). It requires killing of infected cells by specialized cells, the **killer** or **cytotoxic T lymphocytes.** As expected, this system must be highly regulated, lest the killer cells attack normal cells of the body (Table 20.12). Briefly, this is how it works. Infected cells carry on their surfaces some of the antigens made by the microbes that reside within them. Most cells of the body possess surface glycoproteins called **MHC** (for **major histocompatibility complex,** a misguided term of historical importance) proteins that help hold foreign antigens in a proper configuration to be recognizable. These cells are then called the **antigen-presenting cells.** Cytotoxic T lymphocytes recognize the MHC-associated surface antigens

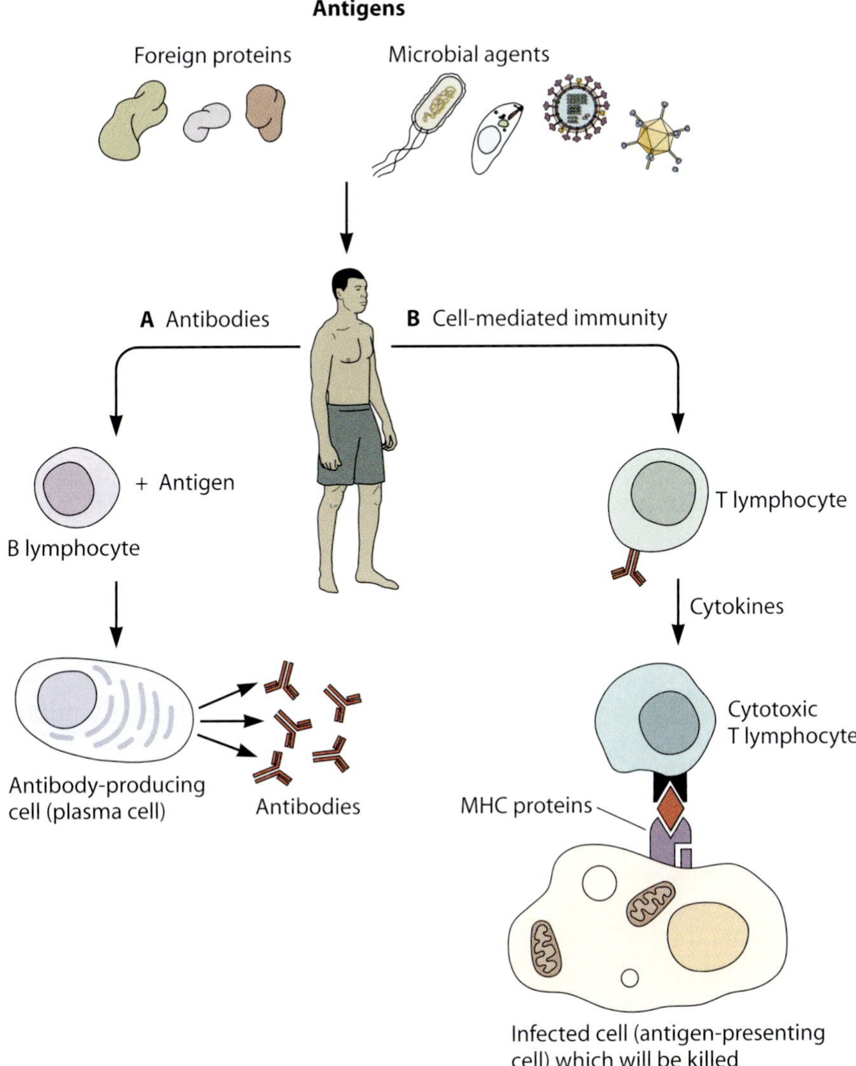

Figure 20.8 The two branches of adaptive immunity. (A) In the presence of an antigen, specific B lymphocytes become antibody-forming cells **(plasma cells)** that proliferate to produce specific antibody-forming clones. **(B)** T cells differentiate into cytotoxic T lymphocytes (CTLs, or killer cells) that recognize MHC-associated antigens on the surfaces of antigen-presenting cells (such as those infected by microbes). As a result, the infected cells are killed by the CTL.

and bind to this complex. Just as with antibody-forming cells, such detection serves as the recognition signal that makes T lymphocytes proliferate into large specific clones. Cell-mediated immunity is therefore as specific as the antibody response. Thus, powerful defense mechanisms operate in infectious diseases caused by intracellular parasites, such as all viruses, the tubercle bacilli, and the salmonellas.

Cell-mediated immunity requires intense communication among various kinds of cells. The language used is chemical, and cytokines deliver the messages (Table 20.13). Some cytokines are made by macrophages, which, to repeat, serve as the sentinels to sound the alarm. One set of lymphocytes, called **T helper cells,** are the master communicators and are

Table 20.12 Some cells involved in adaptive immunity

Cell type	Important function
Macrophages	Antigen presentation; kill microbes directly; kill antigen-presenting cells
Dendritic cells	Antigen presentation
B lymphocytes	Recognize antigens directly; differentiate into antibody-forming cells (plasma cells)
T lymphocytes	Involved in cell-mediated immunity
T helper lymphocytes	Promote differentiation of B lymphocytes; activate macrophages
Cytotoxic T lymphocytes	Kill antigen-presenting cells

involved in activating both branches of adaptive immunity. T helper cells make cytokines that stimulate both B and T lymphocytes, thus serving as middlemen between antibody formation and the establishment of cell-mediated immunity. In addition, T helper cells stimulate the innate response by producing cytokines that result in inflammation. The key role of T helper cells is seen in HIV infection, where the virus specifically targets those that carry a surface protein called CD4, the CD4 T cells. Destruction of CD4 T cells is the primary mechanism by which HIV causes AIDS (see Chapter 21).

Immunological memory

Sometime in life, we are likely to get a booster shot of a vaccine, most likely against tetanus. If we were to measure levels of antibodies or the cell-mediated responses, we would find that they are higher, reach their maximum faster, and are more persistent than after the first vaccination. The same is true after an infection with a microbial agent. A second bout of the same disease is likely to be milder. This effect is called **immunological memory.** How does it come about? After the first exposure to an agent or its antigens in a vaccine, we mount a **primary response** manifested by the proliferation of specific B and T lymphocytes. These cells are short-lived, and most of them disappear after the antigen is eliminated. However, if all of them disappeared, a new encounter with the same agent would require that the immune system start anew. To avoid this, the immune system retains a memory of its past. Some of the B lymphocytes differentiate into **memory cells,** which are not destroyed but become quiescent as long as there is no further antigenic stimulation. When the body encounters the same or a similar antigenic agent, memory cells pro-

Table 20.13 Some important cytokines involved in inflammation and immunity

Name	Major activities
Interleukin-1 (IL-1)	Induces fever; activation of T and B lymphocytes
Interleukin-2 (IL-2)	Induces proliferation of T and B lymphocytes
Interleukin-4 (IL-4)	Activates B lymphocytes to make antibodies
Interleukin-10 (IL-10)	Modulates functions of macrophages and some lymphocytes
Gamma interferon (IFN-γ)	Activates macrophages, other immune cells
Tumor necrosis factor alpha (TNF-α)	Activates macrophages, neutrophils; induces fever

liferate to mount a rapid and efficient immune response, the **secondary response.**

Memory cells are long-lived; they may last for decades. They are poised to rapidly proliferate every time the same antigen is encountered. Immunological memory is a key benefit of being exposed to antigens of pathogens as a child, as it will reduce the risk of the same pathogen causing illness later in life. Thus, the immune response is indeed a gift that keeps on giving.

How do microbes defend themselves against adaptive immunity?

As we have seen, intracellular residence is one way for microbes to try to thwart the host defenses, but there are several others. The most efficient way is to shut off the immune responses altogether. Fortunately, only a few agents can do this, the main one being HIV. By killing cells that direct traffic within the arms of the immune system, HIV causes infected persons, if untreated, to become susceptible to all kinds of infectious agents (see Chapter 21). This collapse of the immune system is the hallmark of the devastating progression of HIV infection to AIDS. Less dramatic changes to the regulation of the immune system also occur. For instance, measles infection leads to a relatively mild immunosuppression, but this is exacerbated by malnutrition, especially in children. Therefore, measles infection is very threatening to children in developing countries with high malnutrition rates, in part because these children become highly susceptible to bacterial infections, such as tuberculosis.

Microbes have evolved other ways of hindering the immune response. A particularly devious one is not to suppress antibody formation but *to actually stimulate it.* The antibodies formed, however, are nonspecific and thus useless to combat infectious agents. This wasteful production of random antibodies is induced by the so-called **superantigens,** microbial products that fool B lymphocytes into making antibodies with a wide range of specificities. There are now more antibodies around, but practically all of them are useless. Superantigen makers include certain streptococci and staphylococci, as well as HIV and other viruses.

Some microbes frustrate both antibodies and the cell-mediated immunity by periodically changing their surface antigens, a phenomenon called **antigenic variation.** The strongest immune response to microbes is usually directed against the surface components, which are the ones most readily sensed by the host. Antigenic variation works this way: once the host has mounted an effective immune response, the agents respond by making a different kind of antigen to which there is no immunity. This switch to a new antigen may be a rare event, but the organisms with the new antigens are immunologically safe. The body tries to play catch-up by mounting an immune response to these new antigens. In some cases, the race is prolonged because the agent can make a large variety of different antigens (Fig. 20.9). The genetic basis for the emergence of new antigenic types depends on some form of genomic rearrangement reminiscent of that used to generate antibody diversity. Agents that undergo antigenic variation are gonococci, protozoa that cause malaria and sleeping sickness, and HIV. No wonder vaccines against these agents are hard to devise and are not yet available.

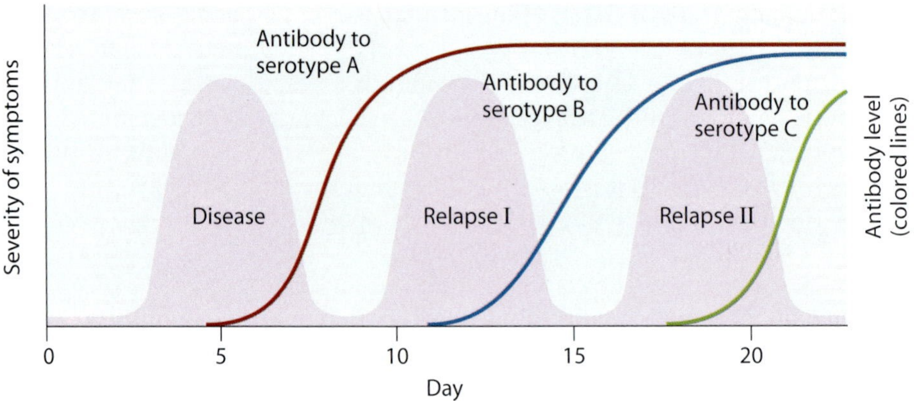

Figure 20.9 Antigenic variation protects microbes from the immune response. An animal infected with a pathogenic microbe (serotype A) becomes ill but recovers when antibodies against the agent reach a sufficient titer. However, this does not lead to the total removal of the microbe, which now mutates to a new antigenic type (serotype B). The host becomes ill again and mounts an antibody response to serotype B. This pattern of appearance of symptoms followed by antigenic variation can repeat itself for many rounds.

Integration of the defense mechanisms

Humans and other vertebrates have evolved a large number of defense mechanisms that ensure the maintenance of health and integrity. The advantage of having so many mechanisms is that should one fail, others can fill the breach. Although these mechanisms differ in specificity, strength, and the time when they come into play, they are interactive and work in a cooperative manner. Thus, neither the adaptive responses nor the innate defenses work independently. We have seen examples of such interactions, e.g., antibodies help activate complement (the "classical" pathway of complement activation), complement greatly enhances the action of phagocytic cells, and macrophages produce cytokines that activate B and T lymphocytes. Once the adaptive responses—antibodies and cell-mediated immunity—reach an effective level, the body can respond effectively to severe challenges. To appropriate what Shakespeare said (in a different context), the body becomes "a fortress built by Nature for herself against infection...."

CONCLUSIONS

We leave this overview of microbial pathogenesis with deference for the numerous and clever-appearing ways that microbes have developed to cause damage in the host and for the equally remarkable responses that hosts have evolved to counter these attacks. In the next chapter, we will discuss a few specific examples of these complex yet fascinating interactions.

STUDY QUESTIONS

1. What are the steps in pathogenesis common to all infections? Can you say a few salient things about each of them?

2. Define these terms: incubation period, inoculum size, normal flora, and endogenously and exogenously acquired infections.

3. What are the main processes that lead to symptoms in infections?

4. What are the main elements of inflammation?

5. What are MAMPs? Give examples.

6. What are some of the consequences of activation of complement?

7. What are the steps in phagocytosis leading to bacterial destruction?

8. How do bacteria evade the innate host defenses?

9. Distinguish between the innate and the adaptive defenses, and give examples.

10. How do antibodies help fend off pathogens?

11. What is cell-mediated immunity, and how is it elicited?

chapter 21 infection: the microbe

INTRODUCTION

A pervasive motif in microbial pathogenesis, yet one that has gained currency only recently, is the idea that microbes do not simply grow in the host; they commandeer host functions for their own purposes. A whole new field known as **cellular microbiology** has arisen around this notion. Research in this field is helping not only to understand the microbes involved, but also to elucidate basic mechanisms in eukaryotic cell biology. Cellular microbiology of viruses is well established, because the only way they can multiply is by appropriating the cell's machinery. Some viruses possess enzymes for nucleic acid replication, but none of them can grow and none of them can generate energy on their own. Most pathogenic bacteria, on the other hand, can grow on agar plates. For a long time, it was thought that bacteria use the body as a giant petri dish and that they simply grow in one or another part of it. There was no compelling reason to believe that bacteria on the plate and inside the host did totally different things. This notion has been turned on its head by newer research. It turns out that bacteria are programmed to do very different things inside the body and in the laboratory. Some of the "housekeeping genes" involved in central metabolism and macromolecular biosynthesis are expressed in both cases. However, analysis of gene expression has revealed that a large number of genes are switched on only when bacteria invade a host. This work is based on two general classes of experimental approaches. One is the use of genetic tools to identify genes that are required for virulence and expressed in the host but not on laboratory media. The other approach is known as transcriptional monitoring, which uses microarray analysis to determine which mRNAs are made under various conditions (see Chapter 13).

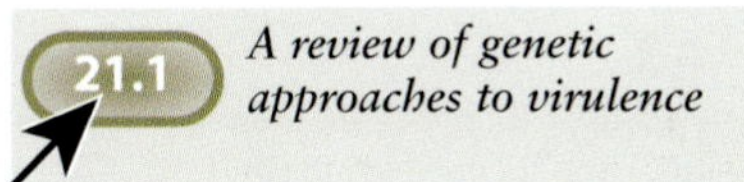

A review of genetic approaches to virulence

Here, we shall consider five examples of infectious diseases. Each one illustrates a number of features, some unique, others common to many infectious agents. As in Chapter 20, we approach the study of these diseases mindful that the universal steps of pathogenesis are encounter, entry, establishment, and damage. We shall address questions such as the following. To what extent is the agent or the host responsible for the damage? To what degree does the agent usurp host functions for its survival and replication? What is the interplay between the host's defenses and the agent's attempts to subvert them? Each of our examples illustrates certain principles, as shown in Table 21.1. To get an idea of the scope of some of the major infectious diseases, see Table 21.2.

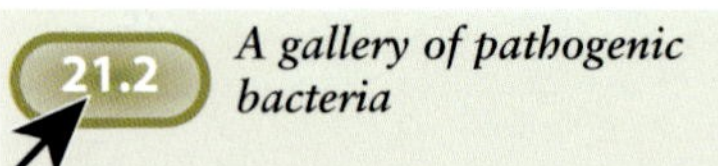

A gallery of pathogenic bacteria

CASE REPORTS

Tetanus, a relatively "simple" infectious disease

A 22-year-old farm worker came to the doctor's office complaining of pain in his jaw for 3 days and inability to open his mouth fully. Ten days before, he had pushed inadvertently against a rusty nail sticking out from a plank in a horse corral. The nail had penetrated deep through the skin, and although the wound hurt and bled, he had not sought medical attention. He had received his tetanus shots as a child but had gotten his last booster more than 10 years before.

The patient was told about the possibility of tetanus and received two doses of human anti-tetanus immunoglobulins intramuscularly into each buttock, as well as an antibiotic. Over a period of several weeks, his symptoms gradually subsided. His wound stopped draining by the next day, and the redness around the wound faded.

To understand what goes on in a patient with tetanus, one must focus on several issues: where the organism causing the infection resides in the environment, how it is encountered, how it enters the body, how it survives the body's local defenses, how the toxin is spread to the target tissue (in this case, the nervous system), and how the toxin acts once it gets there. Tetanus pathogenesis is certainly not simple, but it may appear so in comparison with other infectious diseases.

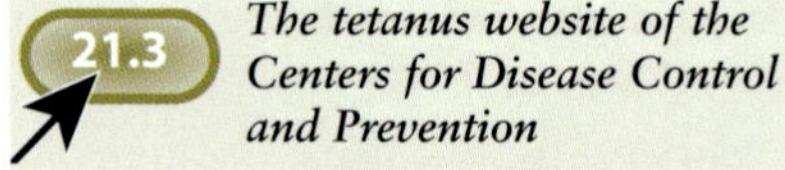

The tetanus website of the Centers for Disease Control and Prevention

Tetanus is a disease caused by the toxin produced by *Clostridium tetani*. This gram-positive bacterium is common in soil and feces. It forms

Table 21.1 Clinical cases in this chapter

Disease	Principles illustrated
Tetanus	The major symptoms of a disease can be caused by a toxin that works far from the site where the bacteria made it.
E. coli hemorrhagic colitis	An infectious agent can appropriate host functions for its own use, in this case to attach to and enter host cells.
Tuberculosis	By a variety of devices, bacteria can survive harmlessly for long periods within host cells. The main symptoms of the disease can be caused as much by the host's response as by the infecting agent.
Infectious mono	Certain viruses can cause a variety of conditions, in this case an acute infection and permanent residence in the host, sometimes leading to cancer.
HIV infection and AIDS	Progressive immunodeficiency caused by the destruction of key cells of the immune system (T helper lymphocytes) results in uncontrollable infections, often by members of the normal flora.

Table 21.2 Examples of infectious diseases

System affected and type of agent	Disease(s)
Upper respiratory and digestive systems	
Bacterial	Strep throat; diphtheria; middle ear infections caused by *Haemophilus influenzae*, pneumococcus
Viral	Common cold (many agents), herpes
Fungal	*Candida* infection (thrush)
Lower respiratory system	
Bacterial	Pneumonias caused by pneumococcus, streptococci, mycoplasmas; pulmonary tuberculosis; legionellosis; plague; anthrax
Viral	Influenza, infection by respiratory syncytial viruses
Lower digestive tract	
Bacterial	Gastritis and ulcers caused by *Helicobacter*; diarrhea caused by cholera bacilli, *Salmonella, Listeria, E. coli* (some strains); dysentery caused by *Shigella, E. coli* (some strains)
Viral	Diarrhea caused by rotavirus, liver infections caused by hepatitis viruses (A, B, C, and D)
Parasitic	Diarrhea caused by *Giardia, Cryptosporidium, Entamoeba*
Urogenital tract	
Bacterial	Gonorrhea, bladder infections caused by *E. coli* (some strains), *Chlamydia* infection, syphilis
Viral	Herpes, warts
Yeast	Candidiasis
Parasitic	*Trichomonas* infection
Nervous system	
Bacterial	Tetanus; botulism; meningitis caused by meningococci, *H. influenzae*, pneumococcus, tubercle bacilli
Viral	Encephalitis caused by herpesviruses, arthropod-borne viruses (e.g., West Nile encephalitis); rabies
Circulatory system	
Bacterial	Endocarditis caused by streptococci, staphylococci, enterococci
Viral	Ebola hemorrhagic fever
Parasitic	Malaria
Skin and soft tissues	
Bacterial	Purulent (pus-forming) infections caused by staphylococci and streptococci, strep fasciitis, gas gangrene
Viral	Herpes, chickenpox, measles, rubella, warts, smallpox
Fungal	*Candida* infections, ringworm, athlete's foot
Bones and joints	
Bacterial	Arthritis caused by staphylococci, streptococci, gonococcus, Lyme disease spirochete
Immune system	
Bacterial	TB, salmonella enteric fever, tularemia
Viral	HIV disease

endospores, which explains its ability to persist for long periods in the environment. *C. tetani* releases a toxin that interferes with neurotransmitters, causing extraordinary tightening of muscles (spastic paralysis), such as the farm worker was experiencing. Left untreated, tetanus may result in death by asphyxiation. The patient did well, possibly because he had some residual protection from his previous tetanus shots.

The toxin of another spore-forming anaerobe, the agent of botulism, *Clostridium botulinum,* may also cause asphyxiation, but it does so by inducing the relaxation of muscles (flaccid paralysis). For this reason, botulinum toxin (Botox) is used in very low doses for cosmetic and occasionally therapeutic purposes. *C. tetani* is closely related to *C. botulinum,* and they share many attributes. Neither organism is highly invasive, and both penetrate via breaks in the body's integrity, such as those caused by wounds or contaminated syringes. Although one thinks of botulism as being caused by eating the contents of improperly sterilized cans of food, the disease can also be acquired from wound contamination. A rare disease in the past, wound botulism has been on the rise among infected drug users. As a result of vaccination, tetanus is relatively infrequent in the United States, but there are still some 50 cases per year. It is quite common in countries with poor hygiene and lack of vaccination.

What must take place for tetanus or botulism to become established? The organisms cause local inflammation at the site of entry. Tetanus and botulism bacilli do not survive very long in tissues and are usually cleared from the site in a matter of days. One reason is that these organisms are *strict anaerobes* and do not grow in the presence of oxygen. However, a few survive, which is possible in small anoxic areas. The severe damage they cause is not a local phenomenon but takes place at a distant site. In fact, the local wound may be so slight that it is not noticed. The reason why a relatively small number of bacterial cells can cause such a potentially devastating disease is that the toxins are enormously powerful and work at minute concentrations. Botulism and diphtheria toxins are among the most powerful poisons known (1 g of toxin could kill some 10 million people!). It has been remarked that purified botulism toxin is a white powder of unknown taste.

An outbreak of hemorrhagic colitis, a complicated infection caused by *E. coli* strain O157:H7

In 1999, a food-borne outbreak of infection took place among people attending the Washington County Fair in New York State. The causative organism was found to be a strain of *Escherichia coli* called O157:H7. The source of the organisms may have been a well on the fairgrounds that was probably contaminated with cow manure. (In other outbreaks, the source of the organism was traced to undercooked hamburger meat.) In the 1999 outbreak, a 3-year old girl and a 79-year-old man died from complications of the infection. Hundreds of others became ill with bloody diarrhea, a condition known as **hemorrhagic colitis.** Seventy-one people had to be hospitalized. Of these, 14 developed a severe complication of *E. coli* O157:H7 infection (hemolytic-uremic syndrome) that can lead to kidney failure.

E. coli O157:H7 is one of a large number of strains of the species. In fact, the limits of the species *E. coli* are very broad and include a large number of pathogenic varieties. They are often designated with a number for "O," the O antigen of their lipopolysaccharide, and one for "H," the flagellin protein of flagella. Most of the pathogenic *E. coli* strains cause intestinal disease, some cause urinary tract infections, and a few invade deep tissues. On the other end of the spectrum is the *E. coli* K-12 strain, an old laboratory workhorse that has not been shown to colonize anyone, let alone cause disease among the innumerable workers who come in contact with it or the volunteers who were fed cultures of the strain.

Although the genome backbones of all *E. coli* strains are similar, many of the pathogenic strains carry upward of 20% more DNA. It seems probable that these extra genes were acquired by horizontal transmission from other bacteria (see Chapter 11).

The *E. coli* strain responsible for this outbreak normally resides in the intestines of cattle, where it causes little, if any, perceptible disease. Animals colonized with this organism become **carriers.** Like many other infectious agents, this organism causes serious disease in one host species and little or none in another. Humans are **accidental hosts** of the organism, which would survive quite well even if it never infected people. Diseases acquired from animals are called **zoonoses** (there are also diseases that animals get from people).

Once ingested, the organisms make their way to the colon, where they cause inflammation (hence, colitis). They are relatively acid resistant, which explains how they survive the trip through the stomach. The organisms do not invade the intestinal epithelium and are confined to its surface, and they must adhere to it in order not to be swept away by liquid currents (Fig. 21.1). Soon, the adhering bacteria destroy the villi of the intestinal epithelial cells and create a lesion that impairs intestinal function and causes inflammation and severe diarrhea. When the lesion becomes deep enough, it breaks through the layer underneath the epithelium (the lamina propria) and affects the underlying blood vessels. This results in hemorrhage and bloody stools. It is thought that the profuse bleeding is brought about by **inflammatory** (also called **proinflammatory**) **cytokines** that are elicited by toxins made by the organisms. Complications of this disease are infrequent, but they can lead to damage to small blood vessels in the kidneys, brain, and other organs. This can cause death.

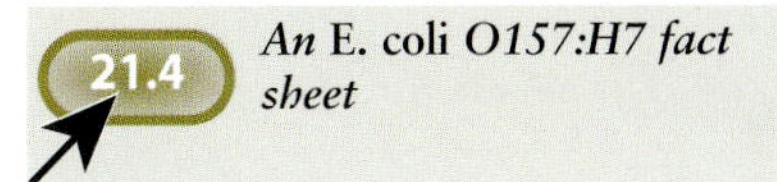

An E. coli *O157:H7 fact sheet*

The toxins made by these organisms are not secreted into the environment; surprisingly, they are introduced directly into the cells of the

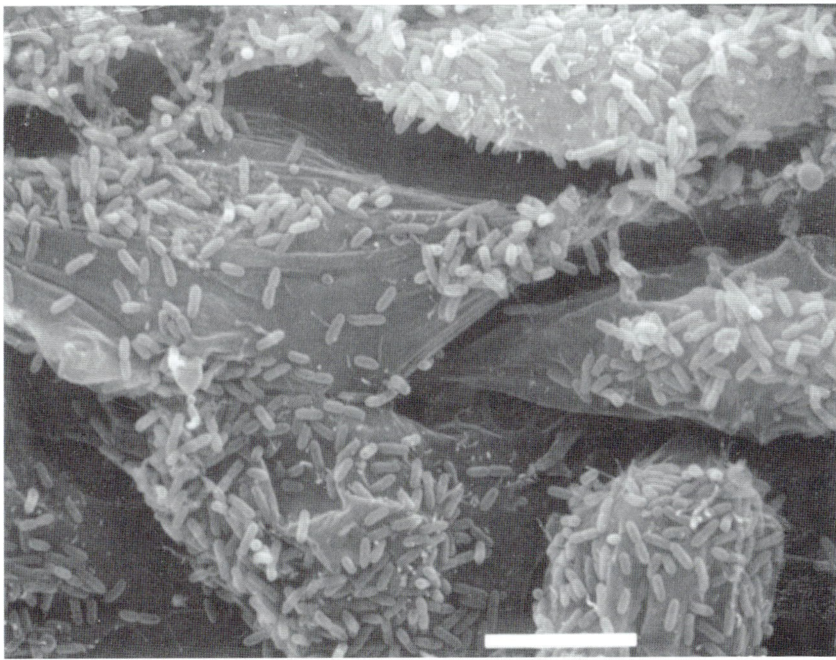

Figure 21.1 Adhesion of bacteria to epithelial cells. A scanning electron micrograph showing *E. coli* adhering to human HeLa cells is shown.

body. Contrast this with the tetanus bacillus, which secretes a toxin that spreads throughout the body. This dilutes the toxin, but here it does not matter, because tetanus toxin is so potent that only a few molecules need to reach their target. *E. coli* is less wasteful: all the toxin molecules reach sensitive cells only. The introduction of bacterial proteins into host cells superficially resembles injection with a syringe. Sticking out from the bacterium is a microscopic apparatus that makes contact with the host cell surface (Fig. 21.2). However, this analogy is misleading, because there is no plunger. This machinery bears the nondescript name of type III secretion (see Chapter 8). The apparatus has several structural proteins that are homologous to proteins of bacterial flagella, suggesting that the two nanomachines may have had a common origin. The type III secretion apparatus is *made only on demand*. The stimuli that set it off include properties of the host environment, such as temperature, ionic conditions, and contact with the surfaces of sensitive cells.

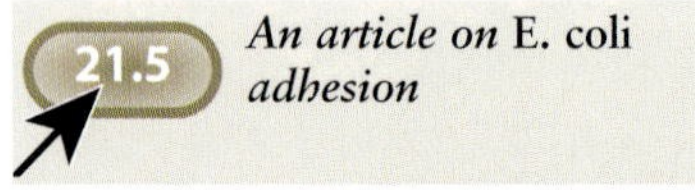

An article on E. coli *adhesion*

Toxins delivered to the host cell by type III secretion alter the cell's cytoskeleton, leading to morphological changes that favor attachment of the bacteria and to the destruction of the intestinal-cell microvilli.

Binding takes place in several steps. The first results in relatively weak binding. To ensure that the bacterium is not dislodged, this step is followed by a stronger binding requiring a receptor on the host cells that is provided by the bacteria themselves. This receptor, called **Tir** (for translocated intimin receptor), is introduced via type III secretion and is initially inactive. In the host's cytoplasm, Tir is activated by phosphory-

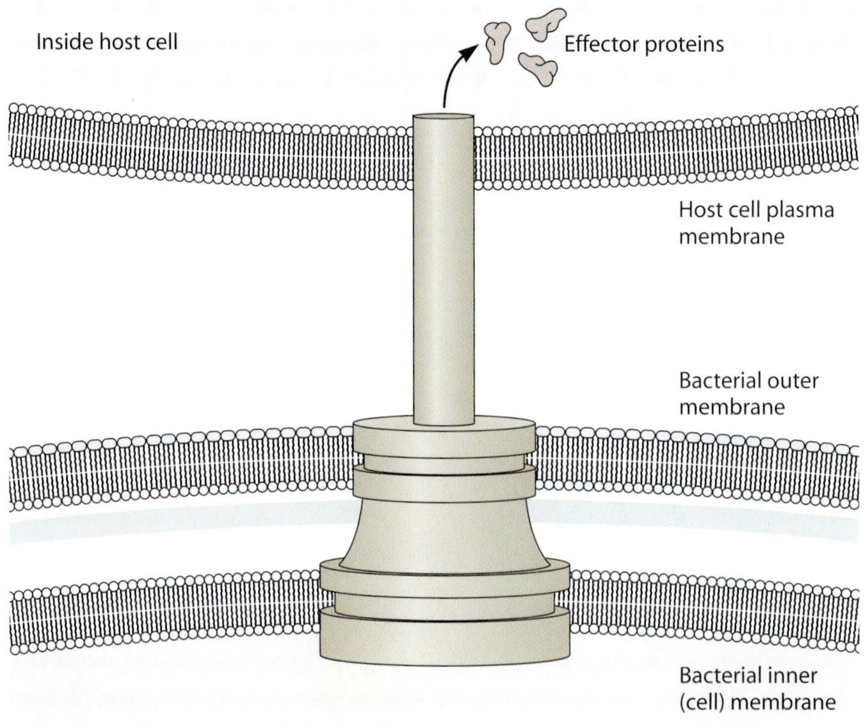

Figure 21.2 The type III secretion apparatus. This schematic drawing shows the structure of the **injectosome,** the device used by a variety of gram-negative bacteria to introduce proteins into eukaryotic host cells. Note the similarity of this structure to bacterial flagella. The "injected" proteins, called **effectors,** may act to impair a vital function of the host cell, such as those associated with the cytoskeleton. The injectosome consists of many proteins, some of which make up the basic structure of the injectosome; others serve to anchor it in the membranes of the bacterium and to pierce the host cell membrane (see Chapter 8).

lation, and in this modified form, it is able to insert into the host cell membrane. In the membrane, Tir acts as a strong receptor for the bacterium, which can bind tightly to it. In this manner, the bacterium uses the host's phosphorylating system to convert one of its own proteins into an effective receptor, another example of how microbes appropriate cell functions for their own use.

Many bacterial species, not just *E. coli*, use type III secretion to introduce toxins and other proteins directly into cells. This mechanism is common among human and plant pathogens, suggesting that it may have evolved once and later been transferred laterally among species. This notion is reinforced by the fact that the genes for type III secretion are often clustered in contiguous regions of the chromosome or plasmids. Such gene clusters are called **genomic** or **pathogenicity islands** and may be distinguishable from the rest of the chromosome, for example, by having a different G+C ratio (Fig. 21.3), suggesting that they have been acquired recently through horizontal gene transfer (see Chapter 11). It seems reasonable that, once evolved, such a sophisticated process as type III secretion should be shared by many pathogenic species.

Some bacteria make use of other protein modifications in the host cell. This discovery has explained long-standing mysteries. For example, until recently, there was no good explanation for how salmonellas, common agents of food poisoning, cause intestinal disease. Other diarrhea-producing bacteria, e.g., the cholera bacillus, make a toxin that is readily found in culture filtrates and even in the contents of a patient's intestine. Despite many attempts, no such soluble toxin was ever found to be produced by salmonellas. It turns out that, as in the case of *E. coli*, a salmonella protein must first be introduced into a host cell's cytoplasm, and there it becomes modified. Once modified, the toxin works enzymatically, adding an ADP-ribosyl group to actin (Fig. 21.4). In the modified state, the actin's cytoskeletal activity is altered, which leads to uptake of the bacteria into intestinal cells, subsequently causing the symptoms in the affected patient. Other bacterial species make toxins with the same enzymatic activity (ADP-ribosylation), but in most cases, the target proteins are different. In the case of diphtheria bacilli and pseudomonads,

Figure 21.3 G+C contents of a pathogenicity island and adjacent regions on the chromosome. A stretch of chromosomal DNA that has a different G+C ratio than the rest is indicative of a pathogenicity, or genomic, island, a region introduced from another organism by horizontal gene transfer.

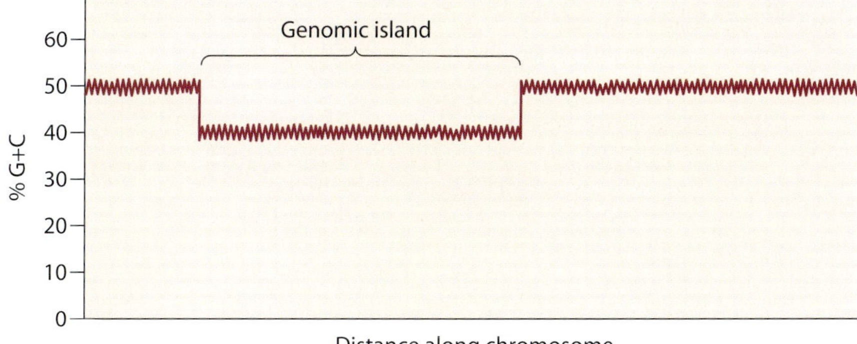

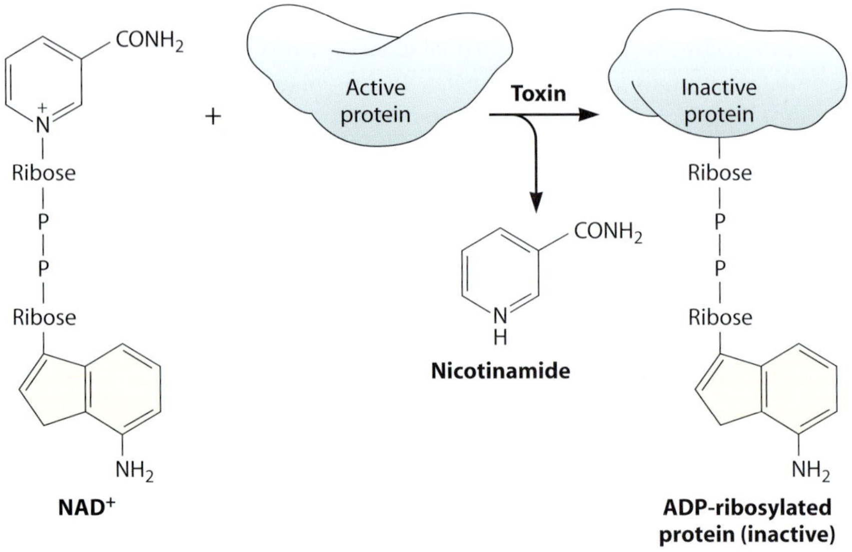

$CONH_2$ — N^+ — Ribose — P — P — Ribose — NH_2

NAD+ + Active protein —**Toxin**→ Inactive protein + **Nicotinamide** ($CONH_2$, N, H)

Ribose — P — P — Ribose — NH_2

ADP-ribosylated protein (inactive)

Figure 21.4 ADP-ribosylating toxins. Many bacterial toxins (e.g., diphtheria, cholera, pseudomonas, and salmonella toxins) modify an essential host protein by adding an adenine-ribose-diphosphate (ADP-ribose) group to a target protein, using nicotinamide adenine dinucleotide (NAD) as the donor.

the protein affected is one required for host cell protein synthesis. In cholera, the target protein regulates the level of cyclic AMP required to maintain the proper ionic balance of host cells. When this protein does not work properly, there is a heavy outpouring of water, resulting in the copious liquid diarrhea characteristic of this disease.

TB, a disease caused mainly by the host response

On a snowy winter evening, a 32-year-old man living on the streets came to a walk-in clinic in Boston complaining of a cough he had had for several months, fever, and night sweats. He appeared slightly drunk and chronically malnourished and had a temperature of 102.6°F. Examination of his chest revealed "rales," crackly lung noises indicating fluid in the air sacs that are suggestive of pneumonia. After getting a chest X ray and depositing a sputum sample, he left abruptly and spent the night in an abandoned building he shared with several friends. Laboratory examination of the sputum revealed the presence of acid-fast bacilli, consistent with the tubercle bacillus, *Mycobacterium tuberculosis* (Fig. 21.5). The X rays added further credence to the diagnosis of tuberculosis (TB). The patient was human immunodeficiency virus (HIV) negative. When he returned to the clinic 4 months later with similar complaints, he was given a cocktail of several antibiotics with the stern and explicit admonition that it was essential that he not skip a single dose. This treatment was to last for 9 months. The personnel at the clinic doubted that the patient would comply with this regimen.

Figure 21.5 TB in a smear of sputum. This smear was stained by the acid-fast technique, in which tubercle bacilli and other acid-fast bacteria retain the original color. The tubercle bacilli appear red because the first dye used to stain them is red (fuchsin).

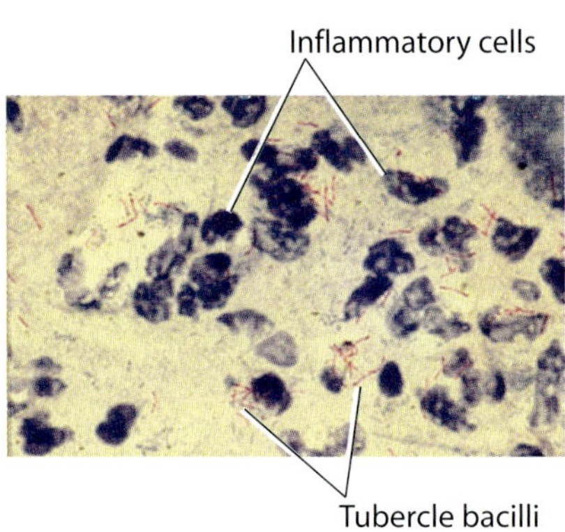

TB conjures up the image of a severe lung disease that can be fatal unless treated. Book lovers will be reminded of Thomas Mann's *Magic Mountain*, and opera fans will remember the dying heroines in *La Traviata* and *La Bohème* ("Che gelida manina...." ["What an icy little hand...."]). Actually, TB is not a single disease but has many manifestations, depending on previous exposure, nutrition, and other health factors. Although now relatively infrequent in developed countries, worldwide TB is still the leading cause of death from a single infectious disease. It was the cause of the "White Plague" of the 17th and 18th centuries in Europe (not to be confused with the Black Plague, caused by the plague bacillus, *Yersinia pestis*). During that period, virtually everyone in Europe was infected, and 25 percent of all adult deaths were due to TB. In the United States, the number of TB cases dropped almost every year since records were kept, but in 1985, the number of cases rose somewhat, to fall again beginning around 1999 (Fig. 21.6). The rise has been largely attributed to the increase in the number of homeless persons (as in the case described). The greater severity of TB cases can be attributed to the emergence of HIV infections, which impair defenses against the disease. The recent fall may be attributed in part to the more successful use of anti-HIV drugs. In recent times, multidrug resistance has emerged among strains of *M. tuberculosis*, making them particularly threatening. A big reason for drug resistance is noncompliance with long-term antibiotic treatment, which leads to the **selection** of drug-resistant mutant strains. Few diseases have as large a social component as TB.

Pulmonary TB typically has two stages, which is a crucial point in understanding the disease. In the first, or **primary**, stage, exposure to the tubercle bacilli, usually by inhalation, leads to a mild, self-limiting disease that may be totally imperceptible or as mild as a cold. In most healthy people, there are no more symptoms. In people whose defense mechanisms are lowered, noticeably or imperceptibly, a much more serious disease, **secondary** TB, emerges later. The time span between the primary and the secondary stages ranges from months to many years. Secondary TB produces symptoms that are classically associated with the deadly image of TB. Between the two stages, the tubercle bacilli lie dormant within macrophages and are little noticed by the host.

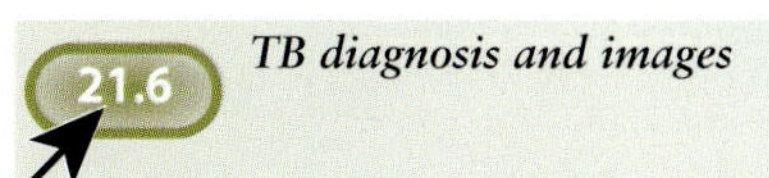

Tubercle bacilli have several distinctive characteristics that help explain the disease process. They are **acid-fast** (see Chapter 2), meaning that once they are stained, they retain the dyes even after treatment with acids. This property reflects the unusual resistance of these organisms to harsh chemicals. Acid-fastness is rare among bacteria. Acid-fast organisms are surrounded by a waxy outer membrane (see "The acid-fast solution" in Chapter 2) that makes them impermeable to many polar molecules, including the common germicides used to "swab the decks" in hospitals. Effective mycobactericidal disinfection requires the use of special compounds. Mycobacteria are singularly preoccupied with the metabolism of lipids, both their synthesis and utilization. A disproportionately large number of genes are dedicated to this metabolism. The mycobacterial outer membrane contains unique components called **mycolic acids,** which

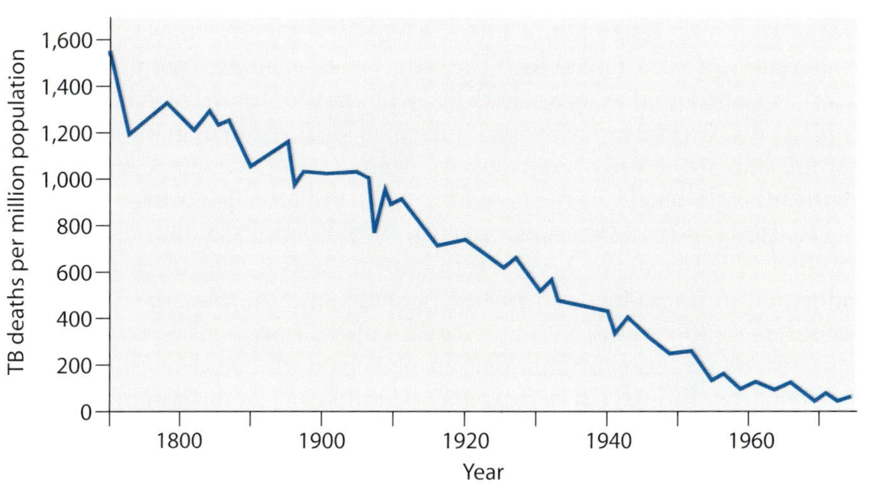

Figure 21.6 Change in incidence of tuberculosis during its period of steady decline. This New Zealand experience is typical of those that occurred in other developed countries.

are long-chain waxes about 80 carbons long (see Chapter 2). Some genes involved in the synthesis and export of lipids are essential for both primary infection by and persistence of the organisms. Others genes, such as the one that encodes the addition of cyclopropane residues to the mycolic acids, are not needed for growth during the acute phase of infection. However, strains carrying null mutations in this gene cannot cause persistent long-term infection.

The outer membrane makes tubercle bacilli resistant to drying. In advanced stages of pulmonary TB, patients cough up huge numbers of the organisms, which persist in the air in aerosols or as dust particles. These two properties, causing a pulmonary disease and being resistant to drying, conspire to increase the chances of the organisms being transmitted to other persons. Another relevant property of tubercle bacilli is that they grow very slowly, doubling about once very 14 hours. Slow growth delays laboratory diagnosis by cultivation, because it can take several weeks to see a colony on agar. Once grown, however, the colonies of tubercle bacilli are distinctive: they appear as yellowish lumps of wax on the agar.

The number of tubercle bacilli required to cause disease is actually quite high. However, when people are crowded together, as in poor housing, jails, or hospitals, the likelihood of contracting the disease increases. It seems plausible that the patient described inhaled a sufficient inoculum from the persons with whom he shared refuge; then he, too, became a source of further spread of the organisms.

What makes people sick with TB? Tubercle bacilli do not make toxins or other products that damage cells. Their slow growth signals no intent to rapidly overwhelm the host—they are nearly innocent bystanders. However, their very presence is noticed by the immune system of the host, which indirectly causes damage to tissues. In fact, it is the **host response** to these organisms that accounts for most of the symptoms. Tissue damage is caused by an uncontrolled, progressive inflammation and eventually severe lesions. It follows that the disease is manifested differently in a virgin host (primary tuberculosis) than in one who has been harboring the organisms (secondary tuberculosis).

Primary TB is readily controlled in most people. However, the organisms are not cleared and remain viable within macrophages. Previous contact with these organisms in a person can be readily demonstrated by the **tuberculin test.** When a mixture of bacterial antigens, collectively called **tuberculin,** is introduced into the skin, a cell-mediated immunity reaction is manifested only in persons who harbor or have harbored the organisms. A positive test is recognized as local reddening and hardening of the skin at the site of inoculation. Many people who have had contact with tubercle bacilli in their youth remain tuberculin positive for years. This attests to the long-range survival of the organisms in the body. The ability of the body to keep the organisms under control is, of course, impaired in immunocompromised patients, and the disease may rapidly progress to an invasion of many tissues of the body and be fatal.

Secondary TB is usually seen in persons with defects in the immune system, even though the defects may be mild or even unrecognized. Here, the balance between microbe and host is tilted in favor of the microbe. The body answers with a vigorous **cell-mediated response** that damages the tissues. Some of the bacterial chemicals that trigger the response are breakdown products of the envelope—the mycolic acids from the outer membrane and muramyl dipeptide from the murein cell wall. These two compounds bind to receptors on macrophages, leading them to release cytokines. One of these compounds, called **tumor necrosis factor alpha,** causes severe inflammation. Damage is also caused by the release of toxic lysosomal components from the macrophages that are trying to kill *M. tuberculosis*. The result is **necrosis,** or cell death. When a lesion becomes sufficiently large, it turns into a cheesy-looking material that contains few host cells but many bacteria. In the lung, such a lesion may break through into the airways. When the contents are coughed up, a cavity is left behind. At this stage, the disease progresses rapidly. It was known as "galloping consumption."

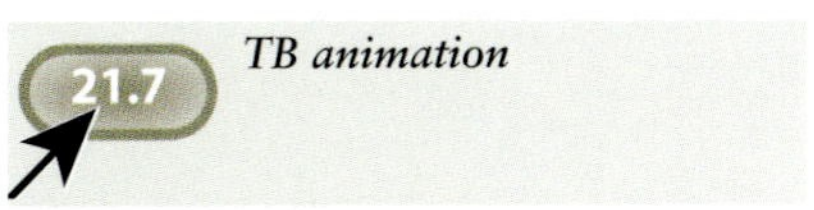

TB animation

Because the immune response is responsible for so much damage, one may ask if it does more harm than good and whether we would be better off without it. Obviously, the immune system is capable of both damaging and healing. The immune system does control TB, making it in most people a slow disease that only accelerates in its last stages. People with TB can live with it for many years, even without treatment. Contrast this with TB in an AIDS patient whose immune system is damaged. Here, the latent tubercle bacilli rapidly cause a severe, fast-progressing, and life-threatening disease. The choice between a functioning immune system and an impaired one is clear.

The interaction between humans and the acid-fast bacteria does not end here. The organism that causes leprosy, *Mycobacterium leprae*, is also acid-fast. It shares some features with the tubercle bacillus, but it has resisted cultivation and can be studied only with a few animals (notably, the armadillo) or, more recently, by cloning its genes into surrogate organisms. There are many other acid-fast bacteria in waters and soil. Most of them are quite benign and rarely cause disease in healthy people. However, they are quite dangerous in immunocompromised people and are a frequent cause of severe infections in AIDS patients.

Infectious mononucleosis: the "kissing disease"

A 19-year-old, healthy male football player developed flu-like symptoms 1 week before a homecoming game. He complained to the doctor of a sore throat, low-grade fever, swollen glands, fatigue, and malaise (feeling ill). A rapid antibody-screening test (costing about $25) was positive. He was told that he probably had infectious mononucleosis and that, to his and his coach's dismay, he should not participate in sports for at least 1 month. A more definitive test (costing about $300) came back positive 2 days later.

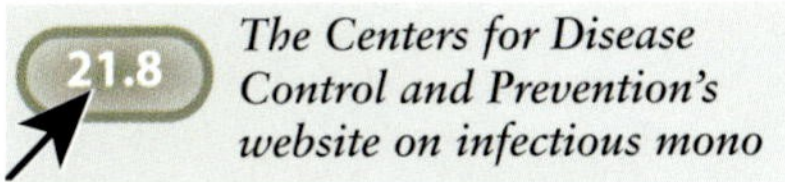

The Centers for Disease Control and Prevention's website on infectious mono

Infectious mononucleosis ("mono") is most common in people 10 to 35 years old, with its highest incidence in 15- to 17-year-olds. Infectious mono is not usually considered a dangerous illness, but it may lead to serious complications. An older but quite descriptive name for this disease is "glandular fever," for the swelling of **lymph nodes.** Infectious mono is caused by **Epstein-Barr virus (EBV).** The virus is usually transmitted though saliva and mucus, hence the nickname "kissing disease." The virus can also be spread by sneezing or sharing a drinking glass or straw with an infected person. The incubation period of the disease is not known with certainty, which makes it hard to trace the initial contact.

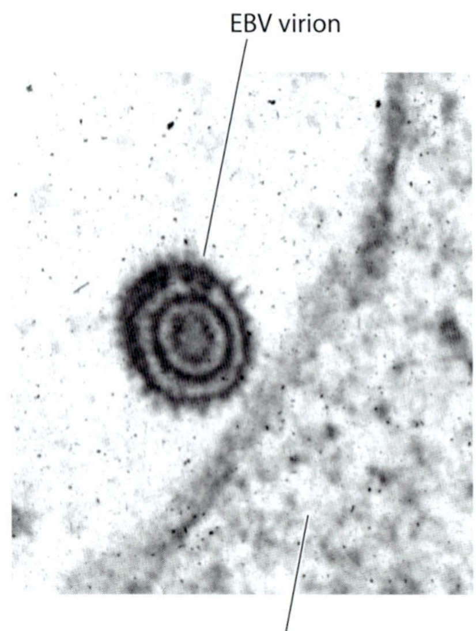

Figure 21.7 Electron micrograph of an EBV virion (thin section). The virion has been released from an infected lymphocyte. Note the complex envelope and the nearly spherical shape that is characteristic of herpesviruses.

That person may not have had any symptoms, because the virus can be carried without signs of illness.

Infectious mono can be caused by either of two DNA viruses, the more common one being EBV and the other being cytomegalovirus. These viruses belong to the family of herpesviruses, which includes the agents that cause cold sores, genital herpes, and chickenpox (Fig. 21.7). About half the people in the United States become infected with EBV sometime during their lives without noticeable consequences. Most often, the infection takes place in childhood, when the infections cause either no symptoms or symptoms indistinguishable from those of other mild illnesses of childhood. By age 40, almost 90% of people in the United States have antibodies against EBV, suggesting that they have the virus in their systems and are immune to further infection. Persons who do not become infected with EBV until they are in their teens or older are more likely to develop the symptoms of infectious mono. Unusually among human viruses, *EBV can persist for life*. Thus, EBV could fairly be considered to be a member of the normal microbial flora.

EBV is usually a fairly benign agent, causing at worst a fairly mild disease. However, this virus has another face: it can also cause serious **human cancers,** called Burkitt's lymphoma, Hodgkin's disease, and nasopharyngeal carcinoma. EBV has three distinct ways of life. It can be:

1. a harmless commensal
2. an agent of a mild disease
3. the cause of serious malignancies

These characteristics of EBV bring up a number of questions.

- What causes the symptoms of infectious mono?
- How does EBV withstand the immune response and persist for so long?
- What role does EBV play in cancer? Is it the direct cause of the malignancies?

These questions are related, and the answers are convoluted. An understanding of the lifestyle of the virus helps here.

EBV first infects the epithelial cells of the mouth and pharynx. The virus then enters the underlying tissues and selectively infects certain cells of the immune system, the **B lymphocytes.** These cells are primarily involved in antibody production, in contrast to the T lymphocytes, which play a major role in cell-mediated immunity. Living in cells designed for host defense may appear to be a dangerous thing for a virus to do. However, many infectious agents have adopted such a strategy, and perversely, they thrive in such unlikely places. We have already considered how the tubercle bacillus resides in macrophages, cells that from a microbial point of view are even more hazardous than B lymphocytes. In time, EBV-infected B lymphocytes reenter the lymphatic vessels and spread to adjacent areas and eventually to the circulation. At any given time, about 20% of infected people have the virus in their saliva, which accounts for its characteristic form of spread, contact with saliva through hands or kissing. Such concentration of the virus in particular fluids suggests that the virus replicates in privileged sites that are protected from T lymphocytes.

Why is EBV so specific for B lymphocytes? These cells carry on their surfaces a **specific receptor** to which EBV virions bind. Interestingly, this receptor is normally used by the cells to bind proteins of the complement system in order to establish communication between that system and the B lymphocytes. This specific binding is yet another example of an infectious agent's appropriating a normal cell function for its own use.

How does EBV cause infectious mono?

Infectious mono occurs in persons not previously infected and having no immunity to EBV. The virus infects a huge number of B lymphocytes, up to 20% of those in the body. By itself, this infection has few direct consequences and leads to no symptoms. In time, however, the body begins to make antibodies and to mount a cell-mediated immune response. T lymphocytes then come into action, including the subclass called killer cells. Killer T lymphocytes know how to seek out cells with viral antigens on the surface and destroy them, a one-sided kind of civil war. The outpouring of cellular constituents from the destroyed B cells causes the symptoms, such as fever and inflammation. Here is how. White blood cells, including lymphocytes, have lysosomes (here called granules) that are replete with materials such as hydrolytic enzymes and substances that cause inflammation; these are potentially damaging to tissues. When the lysosomes are intact, these substances are kept under control. However, when white blood cells and their lysosomes are destroyed, the highly active lysosomal constituents are released, and they damage adjacent cells and tissues. Such events are common in other diseases that result in the death and lysis of white blood cells, such as pus-forming staphylococcus (staph) or streptococcus (strep) infections.

How does EBV persist in the body?

Persistence of any virus (also known as **viral latency**) requires at least two things: (i) that the virus replicate in unison with the cells it infects and (ii) that the infected cells are not destroyed by the immune system. A common way to ensure concurrent replication is for the viral genome to become integrated into that of the host, similar to what happens in lysogenic bacteria (see Chapter 17). Many viruses, e.g., HIV and herpesvirus, persist by such mechanisms, but not EBV; it resides in the nucleus as an **independently replicating DNA entity,** or **plasmid.** Obviously, EBV uses a special mechanism for viral DNA to be replicated in step with the cell's chromosomes. This teamwork relies on the interaction of several proteins, some encoded by the virus and some by the host. In addition, cells normally try to eliminate foreign DNA; thus, the EBV DNA must somehow be protected from eradication. Certain viral proteins bind to the viral DNA and act in consort with several cell proteins to make a stable complex that escapes elimination. Normally, these host proteins play a different stabilizing role: they bind to the ends of the linear human chromosomes, the telomeres, and protect them against degradation. EBV, being circular, has no ends, but its genome has sequences similar to those found in the telomeres. Thus, EBV appropriates the telomere-binding proteins and turns them to its own use. Instead of working as telomere stabilizers, these proteins now act as viral stabilizers. When the formation of telomeric protein-EBV complexes is inhibited, the latent viral genome becomes unstable and is lost from the cell.

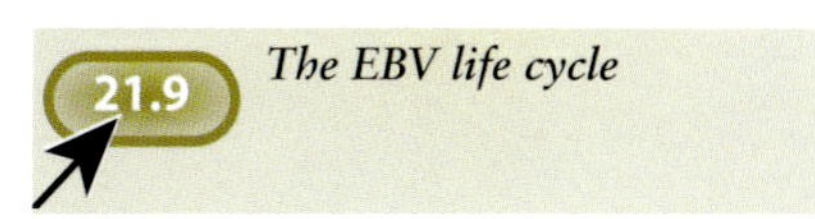

The EBV life cycle

The body has several ways to get rid of virus-infected cells; all must be thwarted for EBV to hang around for the lifetime of the host. One of the reasons for its permanence is that EBV does not reside in every kind of B lymphocyte. Rather, EBV is found preferentially in a subset of B lymphocytes, the **memory cells.** As discussed in Chapter 20, when stimulated with antigens, certain B cells proliferate and make large amounts of a specific antibody. These cells are short-lived, and most of them disappear after the antigen is eliminated. However, to avoid having to start anew when the same or a similar antigen is encountered again, the memory cells are not destroyed but become quiescent until there is further antigenic stimulation. Memory cells then proliferate to mount a rapid and efficient immune response. Immunological memory is the reason why booster shots of vaccines are so efficient and why people infected once become resistant to a second bout of the same infection.

Inside the memory cells, EBV is safe and survives for a long time. Being quiescent, these cells express few viral proteins. Thus, the virus cannot be "seen" by the immune response. One of the amazing things about EBV is that it targets the part of the immune system that lasts for life—the memory cells—and thus can persist for life.

How does EBV contribute to cancer?

How EBV contributes to cancer is not an easy question to answer. What is known with assurance is that virus infection stimulates B lymphocytes to become growth-capable cells called lymphoblasts and that, if unrestrained, lymphoblasts grow into **lymphomas.** In healthy persons, EBV-induced B-cell proliferation is kept in check by the immune system. Because immunodeficient states may develop at various times in the life of a person, the onset of cancer may take place many years after the virus is first acquired. Thus, EBV can act like the "mole" in spy stories, remaining hidden until called into action.

Whether infected B lymphocytes proliferate and eventually make cancers depends on a number of virus-induced proteins. Some prevent **apoptosis,** or **programmed cell death,** and thus "immortalize" the cells. This works by a virus-encoded protein activating a host protein that antagonizes apoptosis. Another clever aspect of EBV is that two of the proteins it expresses late in infection precisely mimic a signaling process required to maintain the long-term survival of the memory cells.

EBV is also associated with cancers of epithelial cells of the nasopharynx. Why does EBV not readily infect other types of cells? Almost certainly because these other cells lack a specific EBV receptor on their surfaces; the receptor is found only on B lymphocytes and certain epithelial cells. Whatever the reason, this can be counted as a good thing.

HIV infection and AIDS

Mr. B., a 28-year-old bisexual male hospital maintenance worker, told his doctor that for the last week he had had fever, swollen lymph glands, and a headache. He related that 3 weeks before he had had unprotected sex with a new male partner and that he considered himself fairly promiscuous. In the past, he had been an intravenous (i.v.) drug user but had ceased that practice over a year before. The physician ordered a laboratory test for the diagnosis of HIV infection.

Mr. B's complaints are typical of many infections and do not suggest a specific condition. The physician, however, was quick to include HIV infection in the list of possibilities because Mr. B. had several relevant risk factors: he had sex with multiple partners, he had been an i.v. drug user, and he worked in a hospital where he could have come in contact with the blood of HIV-infected patients.

Could Mr. B. been infected with HIV by his new partner? In most cases, symptoms such as his are seen within 6 months of HIV infection and are mild, lasting 3 to 14 days. A blood test for antibodies against HIV (an **enzyme-linked immunosorbent assay [ELISA]** or a **Western blot**) would likely be negative at the time of his visit to the physician because antibodies are unlikely to be detectable so early in the infection. More sensitive tests that detect viral antigens of viral RNA are available, but they are expensive and are not usually run in early screening.

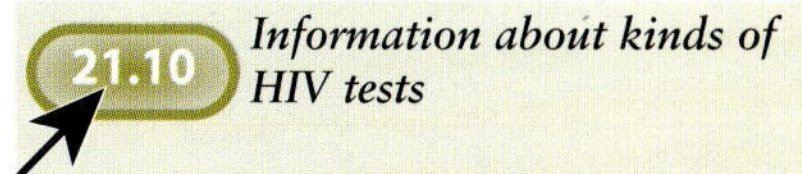

Information about kinds of HIV tests

The laboratory test for HIV antibodies was negative, but a subsequent one for a viral RNA in the blood was positive. Ten days later, when Mr. B. returned to the physician's office and was informed of the test results, he asked, "Does that mean that I have AIDS?" He was told that he did not yet, but if untreated, eventually he would almost certainly develop AIDS. Before going further, he was counseled to abstain from further sexual contacts or to practice safe sex using a condom. He was urged to return to the clinic within 3 months for further tests.

What was going on in Mr. B.'s system? In the typical course of the disease, certain white blood cells, the lymphocytes called **CD4 T cells,** become rapidly infected. The reason for their name is that they carry on their surfaces a protein called CD4 that acts as one of the receptors to which HIV binds. These cells spread to the bloodstream as the virus multiplies within them and infect other organs, e.g., the lymph nodes, liver, and spleen. There is massive killing of CD4 T cells, and symptoms such as those reported by Mr. B. become apparent. This is known as the **acute phase** of the infection.

After a few weeks, the **chronic phase** of HIV infection begins, when the immune system starts to gain control over the infection. The immune response starts to kill virus-producing cells, at which time symptoms disappear but the infection persists. Over a period of 5 to 10 years, the number of CD4 T cells in the blood begins a gradual and continuous decline from its normal level (800 to 2,000 per mm^3). Generally speaking, when the figure reaches 200, people are considered to have AIDS. Below 100, the immune system loses its ability to control not only HIV, but also almost any other infection.

Initially, Mr. B's CD4 T-cell count was normal, and he felt completely well. He was strongly urged to see his physician every 6 months, but he did not keep these appointments. Six years passed, and Mr. B returned to the clinic with a strong rash on his shoulders and upper torso and swollen lymph nodes. He also had a sore throat and white spots in his mouth (indicative of an infection called thrush, caused by the yeast *Candida*). His CD4 T-cell count was now 185 cells/mm^3.

What went on during these 6 years? Although Mr. B's symptoms appeared slowly and gradually during this period, HIV did not remain quiescent. A large number of virions were produced, and a huge number of CD4 T cells were being killed. The body is able to make new CD4 T cells at a prodigious rate, but in time, the virus wins this race.

Why does the depletion of CD4 T cells lead to increased infections by a variety of agents? CD4 T cells play a key role in both branches of adaptive immunity: antibodies and cell-mediated immunity (see Chapter 20). Also known as **T helper cells,** these lymphocytes are essential for the proliferation of cells that make antibodies (B lymphocytes) and cells that are involved in cell-mediated immunity (killer cells, or cytotoxic T lymphocytes). In addition, helper cells release cytokines that result in inflammation. Without CD4 cells, the body becomes defenseless against invading microbes, including those that are virtually devoid of virulence against normal persons. The progression of HIV infection in untreated persons is shown in Fig. 21.8.

The treatment that was available to Mr. B. consists of a combination of drugs that impair the replication of HIV. They fall into three general classes, **reverse transcriptase inhibitors, protease inhibitors,** and inhibitors of viral entry. These drugs work on specific steps required for HIV replication. As discussed in Chapter 17, HIV is a **retrovirus,** an RNA virus that has an obligatory DNA intermediate stage. In order to make DNA from RNA, the virus carries in its virions the enzyme **reverse transcriptase.** Since it is unique to this class of viruses, this enzyme is a ready target for inhibitory drugs. The protease works on a late step in virion maturation, acting within the virion to cleave precursor viral proteins into their mature forms. A number of drugs that inhibit each of these two enzymes are available and are used in combination, usually as a threesome. The reason for the multiple drugs is that HIV mutates very rapidly to become resistant to these drugs if given singly. (High mutability is a general property of RNA viruses and is due to the lack of a proofreading function of enzymes involved in RNA replication.) However, the chance of its becoming simultaneously resistant to three drugs is very low. When should such treatment be initiated? One would think that the earlier, the better, but the evidence to support this is controversial.

Mr. B.'s physician prescribed a cocktail of two reverse transcriptase inhibitors and one protease inhibitor. Mr. B. was told emphatically that he had to take these drugs according to a specific schedule and that failure to do so could cause the virus to become drug resistant. After a few weeks, Mr. B. did not follow this regime, in part because of unpleasant side effects from the drugs, such as sleep deprivation, nausea, and diarrhea. Six months later, he went to an emergency room complaining of a dry cough and shortness of breath. His temperature was 102°F, and he was diagnosed as having pneumonia, for which he was successfully treated with antibiotics. The agent of pneumonia was a fungus called *Pneumocystis*

Figure 21.8 A typical natural history of an untreated HIV infection.

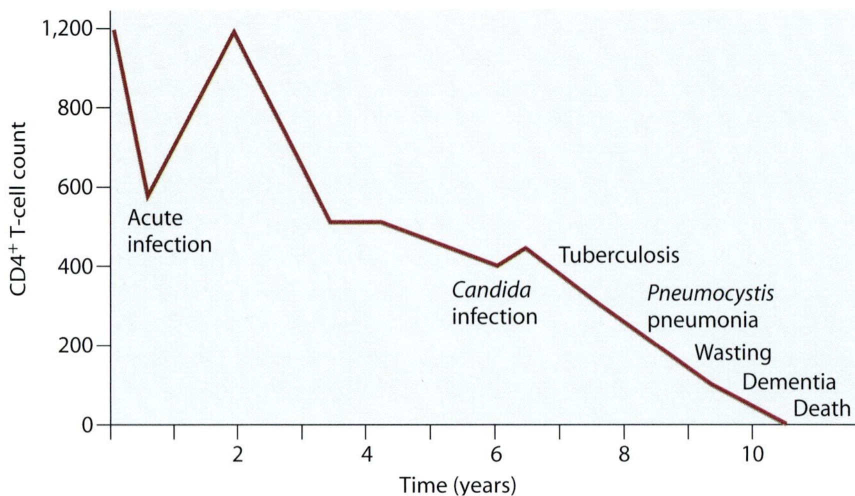

jirovecii (formerly *Pneumocystis carinii*), which is present as a harmless commensal in the lungs of most people.

In people who take these drugs, the virus level in the blood often falls precipitously, and within 6 months, becomes undetectable. The number of CD4 T cells rises, and the patients tend to no longer experience opportunistic infections. As a result of such treatment, death from AIDS in the United States declined by about 20% starting in 1998. It does not appear that prolonged treatment cures the disease, because if treatment is stopped, viral multiplication promptly begins again. Thus, these drugs do not cure HIV infection, they just halt it. There is another downside. Patients must swallow at least 8 and often more than 16 pills a day along with other medicines to control opportunistic infections until their CD4 counts recover. These drugs produce unpleasant side effects ranging from sleep disturbances, rash, nausea, diarrhea, and headaches to anemia, hepatitis, and perhaps diabetes. In addition, the cost of medication is frightening.

After his bout of pneumonia, Mr. B. returned to work. Four months later, he developed a high fever and severe headaches. He returned to his physician, who ordered a spinal tap, which revealed the presence of a fungus, *Cryptococcus neoformans.* Mr. B. responded well to antifungal treatment with fluconazole (a triazole antifungal agent), but he increasingly worried about his weight loss of about 80 pounds in the previous year and an increasing number of skin infections. Two months later, he came to the clinic accompanied by his parents, who noticed that he had become more forgetful and withdrawn. He could no longer follow simple commands. A CAT (computerized axial tomography) scan of the brain showed a profound loss of brain tissue. Two weeks later, Mr. B.'s parents called the physician to tell her that he had died at home that day.

The tragic outcome of Mr. B.'s illness can be multiplied millions of times every year throughout the world. Less than 30 years after it was first recognized, AIDS has threatened to become the most devastating **pandemic** in human history. Over 15 million people have died of AIDS, and some 30 million people are now infected with HIV. The life expectancy in sub-Saharan Africa has dropped to 47 years; it would have been 62 years without AIDS. In Botswana, life expectancy has dropped by almost 35 years. Fourteen million children have lost one or both of their parents. The future appears bleak: many millions more will die from AIDS unless extraordinary measures are instituted. One of the main problems is money, as the drug treatment costs up to $15,000 per patient per year. Compare this with a health expenditure of $10 per person per year in many African countries.

Prevention of HIV infection also depends on powerful social intervention. Note that HIV has three main modes of transmission, all of which must be addressed: (i) by sexual contact with an infected person, leading to the transfer of infected body fluids (blood, semen, or vaginal secretions); (ii) by infected blood from sharing i.v. needles with an infected person or from blood transfusions; and (iii) from mother to infant around the time of birth and in breast milk shortly after birth. Each of these links can be broken with sufficient education and motivation; none of them is easy to achieve. The most powerful weapon would be a vaccine, but HIV's remarkable ability to mutate makes this approach difficult. More fundamentally, as a noted virologist has said, "If the immune system in HIV-infected individuals cannot wipe out the virus, why should a vaccine that activates the same immune response be expected to block infection?" But considering the talent and resources being expended on this effort, we have reason to be hopeful. We have little choice but to have faith that effective prevention and treatment will become available soon, or else we will witness one of humanity's most dreadful experiences.

CONCLUSIONS

We conclude these descriptions of infectious diseases with the awareness that they are merely examples of the vast number of ways in which humans interact with bacteria, fungi, and other parasites, including protozoa and worms. To consider such vast amounts of information, it helps to keep in mind that all infectious diseases share the same steps in pathogenesis and vary only in detail.

The study of infectious diseases is a movable feast because these are not static interactions. With time, both the host and the parasite evolve, changing the picture altogether. In addition, with changing ecology and altered human behavior, new agents emerge. In the next chapter, we will examine some of the factors involved in the changes in the relationship between human beings and pathogenic microbes.

STUDY QUESTIONS

1. Name 20 infectious diseases of humans. Which are caused by bacteria? Viruses? Fungi? Animal parasites?

2. What characterizes bacterial exotoxins? Give examples.

3. How are some bacterial toxins delivered directly into host cells?

4. What are the outstanding characteristics of tubercle bacilli?

5. What does a positive tuberculin skin test tell us?

6. What are the main characteristics of EBV?

7. Describe the main stages in an untreated case of HIV infection.

chapter 22 microbes and human history

INTRODUCTION

Evolution is shaped by interactions between living organisms. Like other species, ours has evolved to survive not only changing physical environments, but also biological challenges, such as mutualism, parasitism, and predation. It is not surprising that we, like all other animals, possess ancient and effective means to respond to such challenges.

To make life in a world of microbes possible, we have evolved powerful defense mechanisms, the **innate** and **adaptive immune responses.** The innate responses, which rely on the skin and mucous membranes, phagocytes, the complement system, and antimicrobial compounds, were the first to evolve. The adaptive response—antibodies and cell-mediated immunity—came later.

Innate immunity has deep roots in the tree of life, appearing before the divergence of invertebrates and vertebrates some 500 million years ago. To this day, both insects and mammals have retained some of these ancient immunological features. Phagocytosis, in fact, had evolved by the time starfish evolved, a discovery made by Elie Metchnikoff in the late 1800s. Both insects and mammals possess alarm mechanisms to alert them to the presence of microbial invaders. The homologs of the vertebrate Toll-like receptors that recognize microbe-associated molecular patterns (see Chapter 20) were first discovered in *Drosophila* cells. Thus, insects and mammals have retained similar mechanisms to sense the presence of invading microbes and to elicit a defensive reaction. Insects, however, have not developed adaptive immune responses.

One can think of several reasons why acquired immunity systems are not generally found in invertebrates. Insects have a brief life span, perhaps too short for acquired immunity to become established, and they are generally smaller than mammals. In humans, acquired immunity is mediated by a diffuse tissue represented in many organs (spleen, lymph nodes, and others). Although much smaller in a mouse than in a human, the immune system still takes up considerable space to accommodate the large diversity of cells programmed for antibody and cell-mediated immunity responses. Perhaps most insects are too small to accommodate such a large number of cells.

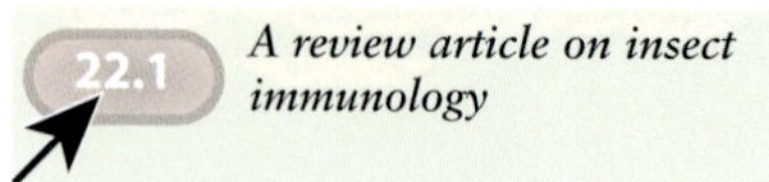

A review article on insect immunology

Infectious diseases are an unfinished business—some wane, new ones emerge, and old ones reappear with altered degrees of severity. There have been times when infectious agents have had a clear ascendancy over humans. We have no record to prove it, but it seems probable that in its early phase of evolution, the human species may have come close to extinction from infectious diseases. Anthropologists point to the existence of other species of the genus *Homo,* some of which overlapped in time with our *Homo sapiens* ancestors and all of which passed into extinction. Even in historical times, isolated groups of our species have disappeared and others have been perilously reduced in number. An example is the sad fate of an Amazon tribe, the Southern Coyapos, who were visited in 1903 by a lone missionary. An epidemic of some kind broke out, reducing the members of the tribe from about 8,000 to 500 by 1913. The 27 members who were still alive by 1927 intermarried with outside groups, and the original tribe disappeared. Another example is what happened in the Hawaiian Islands upon the arrival of Captain James Cook and his crew in 1778. With the introduction of new infectious agents, e.g., those causing tuberculosis, measles, typhoid fever, and other diseases, the population of the islands decreased from 300,000 to 30,000 within one generation. These are not just distant events: today, certain countries in sub-Saharan Africa, such as Botswana, are at risk of a catastrophic loss of population due to AIDS.

The idea that great plagues have shaped our history is seen in the conquest of Mexico by the Spaniards. The native people had not been exposed to many of the infectious diseases common in Europe, where the population had become partly resistant. The introduction of measles, smallpox, and other diseases devastated the Mexican population. In the 100 years after the arrival of the conquistadors, the native population decreased from about 30 million to about 1.6 million. The decline was precipitous and is considered one of the major, if not the main, reasons why a handful of invaders could subjugate a large, powerful, and sophisticated empire. Here are words passed on to us from what is now Guatemala:

> Great was the stench of death. After our fathers and grandfathers succumbed, half the people fled to the fields. The dogs and vultures devoured their bodies. The mortality was terrible. Your grandfathers died, and with them died the son of the king, and his brothers and kinsmen. So it was that we became orphans, oh, my sons! All of us were thus. We were born to die!
>
> A. Recinos et al., translators, *The Annals of Cakchiquels and Title of the Lords of Totonicapan* (University of Oklahoma Press, Norman, 1953), quoted in A. W. Crosby, *The Columbian Exchange,* p. 58 (The Greenwood Press, Westport, Conn., 1972)

The Spaniards' apparent immunity to these diseases led the Native Americans to believe the invaders had supernatural powers. The same thing happened during the conquest of the Inca Empire in South America, which was achieved by Pizarro with just 168 men. Soon other diseases, including measles and probably epidemic typhus, took hold. Not all the new infectious diseases were of European origin. Malaria and yellow fever, for example, came from Africa and soon became established in the Americas.

The question arises as to why diseases were common in the Old World and not the New. A reason proposed by Jared Diamond is that many of the infectious diseases were originally acquired from animals and domestication of large animals was more prevalent in the Old World. In time, purely animal pathogens adapted to human hosts and became increasingly pathogenic. Humans in the Old World necessarily developed some resistance to these diseases. A list of human diseases with animal counterparts is shown in Table 22.1.

HOW INFECTIOUS DISEASES CHANGE

With time, populations come to terms with their microbial invaders, and vice versa. With no changes ensuing, some kind of equilibrium between host and parasite will be established. In general, it's no great selective advantage for parasites to kill off their hosts, and even less to interrupt their transmission to new hosts. These relationships are not always straightforward. For example, the tetanus bacilli live in the soil and need not infect animals or humans to survive. For this organism, the host is incidental to the life cycle of the bacterium, because only a small fraction of its population is produced in infected humans. In the long run, the occasional killing of a host probably makes little difference to these bacteria. This leads to the question of why tetanus bacilli make their toxin. The disappointing answer is that we do not know.

For agents that are not adapted to an existence outside the body of the host, killing one host before transmission to another would clearly be selected against. However, most of the great epidemics of infectious diseases are caused by organisms that have humans or animals as their reservoir and that do not survive for extended periods in the environment. Examples are smallpox, the plague, human immunodeficiency virus (HIV) infection, typhus, diphtheria, syphilis, and polio. It is true that massive outbreaks of typhoid fever, cholera, and bacterial dysentery are caused by the ingestion of contaminated water and foods, but with the exception of cholera, these other causative agents are only occasionally found in the environment. In most other cases, the reservoirs of the agents are humans or animals. As a consequence, *changes in human behavior*

Table 22.1 Deadly gifts from domesticated animals

Human disease	Animals with most closely related pathogens
Measles	Cattle (rinderpest)
Tuberculosis	Cattle
Smallpox	Cattle (cowpox) and other livestock with related poxviruses
Influenza	Pigs and ducks
Pertussis (whooping cough)	Pigs and dogs
Falciparum malaria	Birds (chickens and ducks)

have a profound effect on the establishment of old and new infectious diseases. Examples of such changes include the following.

- Acquisition of a new pathogen by a previously immunologically "naïve" population can be devastating. Striking examples are the introduction of smallpox to the Americas by Europeans, discussed above, and the arrival in Europe from Asia of rats carrying plague bacillus-laden fleas. Modern travel facilitates the rapid spread of infectious agents throughout the world.
- Urbanization, the shift from rural to city existence, affects the spread of infectious diseases. In 1900, about 5% of the world's population lived in cities; in the early 21st century, urban dwellers are a majority. Closer proximity of large numbers of people fosters the transmission of diseases such as tuberculosis.
- Wars, famine, and chronic poverty lead to malnutrition. This results in immunosuppression and heightened susceptibility to digestive and respiratory infections. Children are greatly affected and become more sensitive to severe respiratory and digestive system infections.
- Increases in risky behavior, such as the use of contaminated needles and unprotected sex, are the main causes of many diseases, including the present AIDS epidemic.
- The development of modern methods of meat production, in which large numbers of cattle, poultry, or fish are forced to live in crowded conditions, and large central plants, in which tons of meat and eggs are processed and prepared for national or international distribution, while seemingly economical, has created exactly the circumstances leading to growth and massive dissemination of pathogens (pathogenic strains of *Escherichia coli*, for example).
- Reforestation of land in parts of North America, such as the Northeast and upper Midwest in the United States, combined with the tendency of humans to build houses on wooded lots, has had the unintended consequence of increasing the proximity of humans to a growing deer population. At the same time, there has been an increase in the number of deer ticks that carry the Lyme disease bacterium.
- The widespread use of antibiotics leads to drug resistance. The danger here is that microbial resistance will outpace the effectiveness of the available drug arsenal. Cultures of bacteria that were set aside before the advent of antibiotics have proven to be uniformly drug sensitive.

Not all the alterations in the microbial environment are man-made. The cycles of climate in historical times have affected the spread of diseases such as cholera and malaria. Of course, man-made global warming is of serious concern here.

Environmental changes, weighty or subtle, impose selective pressures on microbes and lead to the emergence of novel forms or the reemergence of old ones. Given the large number of microbes found either in the environment or in an infected host, mutations arise, even if their frequency is

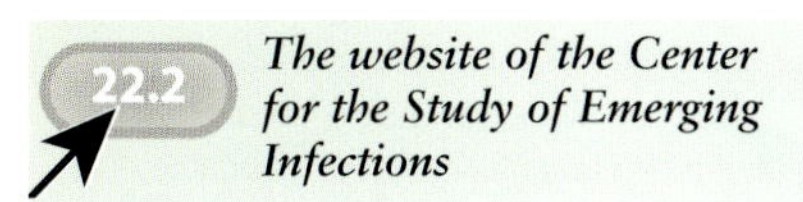

The website of the Center for the Study of Emerging Infections

low. Thus, mutations in bacteria that lead to antibiotic resistance typically occur once every 10^6 to 10^8 cells, but these are not especially large numbers for naturally occurring bacterial populations. However, the acquisition of a more complex function, such as virulence, is not likely to happen by the accumulation of single mutations. Such massive changes are more likely to take place by the wholesale acquisition of genes. As discussed in Chapter 10, bacteria have several mechanisms for transferring large segments of their genomes from cell to cell.

As a consequence of critical changes made in the global environment and the ability of microbes to rapidly respond to them, we cannot predict what new germs will assail us, but we can predict that new ones will emerge. Contrast this realistic view with the tarnished confidence expressed in the 1969 dictum by the Surgeon General of the United States: "It is time to close the book on infectious diseases." Closer to our present perception is a quote from Louis Pasteur: "The microbes will have the last word!"

MICROBIAL AGENTS OF WARFARE

The intentional use of infectious agents to disable an enemy has a long and dishonorable history. One early recorded instance took place in the 14th century during the siege of a city in the Crimean Peninsula. The attackers flung corpses of plague victims over the city walls, and as a result of the epidemic that ensued, the defenders surrendered. In the Americas, both the Spanish and the English provided natives with clothing or blankets infected with smallpox virus with the explicit intent of causing an epidemic. The practice continued well into the 20th century, although the exact scope of these activities has not been fully revealed. It is worth noting that biological warfare preceded the discovery of the microbial origin of infectious diseases, suggesting that the notion of contagion was firmly embedded even if the cause could only be guessed at.

The effectiveness of microbiological warfare is a contentious issue, and not just for ethical reasons. In general, biological agents can be unpredictable, and their dispersal is not readily controllable. For example, the attacking personnel may be placed at risk. The world has come close to deciding to eliminate this threat. In 1969, the United States unilaterally ended its biological weapons program and announced in 1972 that it had destroyed its stockpiles. In 1969, 160 countries signed a treaty banning the use of both biological and chemical weapons. Among the 143 countries that ratified the treaty were the United States, Russia, Iraq, Iran, Libya, and North Korea. Nevertheless, the threat of microbiological warfare continues, presently under the rubric of bioterrorism. Below, we discuss two of the agents that figure prominently in these considerations.

Smallpox

Smallpox has been eradicated from our planet. At least, no cases have been reported anywhere since 1977, and the World Health Organization declared on 8 May 1980 that the disease was conquered. This great triumph of medical science and international cooperation is now clouded by the threat of smallpox virus being used as a bioterrorism agent. Eradication

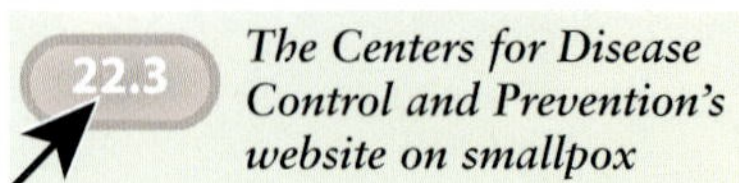
The Centers for Disease Control and Prevention's website on smallpox

has been successful because vaccination is highly effective in stopping the spread of the disease. The reason for the concern is that although the known stocks of the smallpox virus are now in safe repositories in the United States and Russia, it is possible that some live virus had previously found its way to other sites. Reemergence of smallpox would have severe consequences. Since victory over the smallpox virus was declared, the vaccine has not been administered, and the population is now susceptible to the virus.

Smallpox is an ancient disease that was widespread throughout the world. In its most virulent form, the disease had a mortality rate of about 25%, and it was highly contagious. Smallpox (so called to distinguish it from the "great pox," or syphilis) results in a number of severe symptoms: high fever, intense pains, rash, and numerous sores that eventually become filled with pus (pocks).

The smallpox story has stimulated great interest in setting up programs to eradicate other serious infectious diseases, such as polio, measles, leprosy, and some parasitic infections. The success of smallpox eradication points to issues that are pertinent to the suppression of other infectious diseases. Smallpox eradication was possible for the following reasons.

- Humans are the only reservoir of the agent.
- Vaccination is simple and effective.
- The vaccine does not require refrigeration, an important factor in some regions.
- The symptoms of the disease are readily recognizable.
- Subclinical, or persistent, infections were virtually unknown; thus, there was little chance of the virus lurking undetected.

The smallpox virus belongs to a large family, the poxviruses, which includes agents of a number of animal diseases and a few human diseases. Poxviruses are encased in two membranous layers, contain double-stranded DNA, and are among the largest of all viruses, being barely visible with a light microscope. Their genome contains 200,000 bases, over one-third the number in the smallest bacterial genome. They are unusual among the DNA viruses because they replicate in the host cell's cytoplasm, separated from its replication machinery. To accomplish such a feat, their genomes encode all the enzymes needed to make and modify RNA (by methylation and adenylation) and to synthesize DNA. To assert their biosynthetic independence, the RNA polymerase and the RNA-modifying enzymes are packaged in the virions. Functional messenger RNA can then be made soon after infection. The complexity of the macromolecular transactions in the poxviruses suggests that they may be suitable targets for future antiviral drugs.

Anthrax

Before the 2001 scare, most people had barely heard of anthrax. This has not always been the case, because anthrax was a prevalent disease in humans and cattle until the 20th century. It is still important in some parts of the world, e.g., Afghanistan. The disease is ancient and, according to some scholars, may have been the cause of some of the Biblical plagues of Egypt. Anthrax was the first infectious disease to be studied intensively by Robert Koch and Louis Pasteur, both of whom developed a vaccine against it. The disease waned and became a medical and veterinary curiosity due to a combination of good farming practices (proper disposal of contaminated carcasses and vaccination of farm animals). Still, the anthrax bacillus lives in many soils, and occasional outbreaks are reported. Several farms in the upper Midwest of the United States are under quarantine because of anthrax.

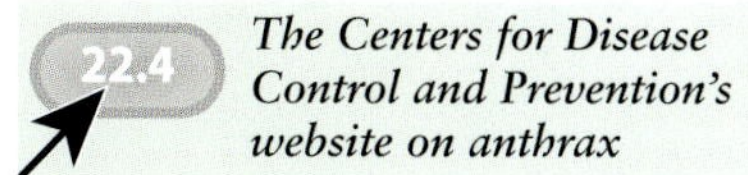

The Centers for Disease Control and Prevention's website on anthrax

As with all infectious diseases, anthrax represents a complex interaction between the parasite (the anthrax bacillus) and the host (us). Anthrax bacilli (*Bacillus anthracis*) penetrate into the bloodstream, where they resist phagocytosis by the white cells that would otherwise dispose of them. Anthrax bacilli are covered with a slimy layer, the capsule, that prevents the white blood cells from ingesting them. These bacteria cause harm by producing powerful toxins inside the body. Anthrax toxin is a complex of **three plasmid-encoded proteins**, two (called LF, for lethal factor, and EF, for **edema** factor) that are directly toxic and one (called PA, for protective antigen) that ushers the other two into cells. LF works by destroying white blood cells, and EF acts by increasing the amount of cyclic AMP in cells, which impairs their energy and water balance. PA, which is itself not toxic, makes a multimeric ring that inserts into the host cell membranes, allowing *specifically* the passage of the two toxic components (Fig. 22.1). Thus, anthrax bacilli have their own special system to deliver toxins to host cells. Knowing this may well help in the design of drugs that neutralize the actions of the toxins. Note that if PA alone were inactivated, the two toxins would cause no harm because they could not penetrate into host cells.

Figure 22.1 How anthrax toxin gets into cells. The toxin consists of three molecules. One of them, called protective antigen, or PA, binds to receptors on susceptible cells and, after being cleaved **(1),** multimerizes into a pore-like structure **(2)** to which the toxic components (edema factor, or EF, and lethal toxin, or LT) bind **(3).** The complex is internalized into a vesicle **(4)** that, upon acidification, releases the toxic EF and LT into the cytoplasm **(5),** where they disrupt the cell function.

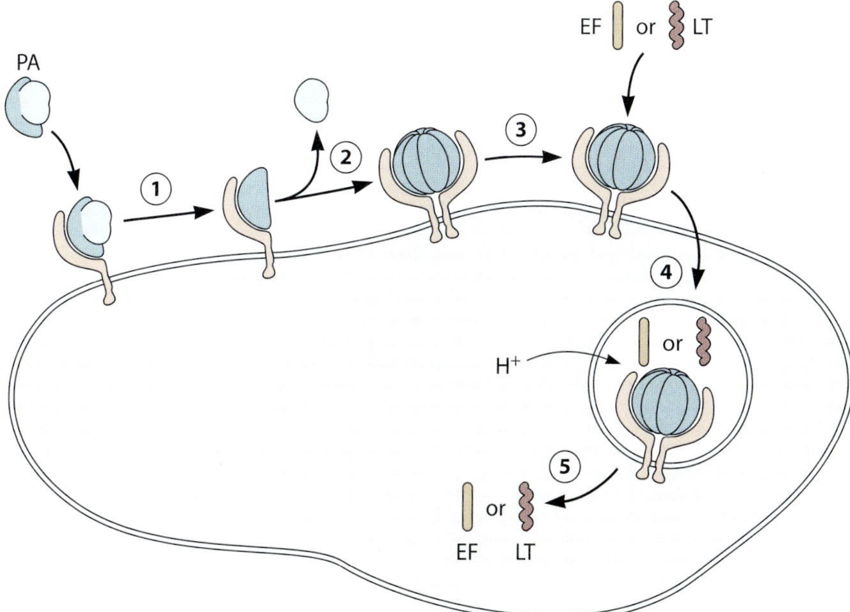

Anthrax bacilli are found in soils worldwide. They are spore-forming gram-positive rods (we look back nostalgically to the days when bacteria were divided into gram-positive and gram-negative bacteria, rather than into "weapons grade" and "non-weapons grade"). The spores of these organisms are typically very resistant to drying and heat and cannot possibly be eradicated from the environment. The most common form of the disease is **cutaneous,** acquired by handling contaminated material. It is thought that the spores enter through cuts or small skin abrasions and germinate there to cause local lesions. Most patients with cutaneous anthrax recover within 10 days, but a few progress to a life-threatening disease. Inhalation of anthrax spores leads to the **pulmonary** form of the disease, in which the organisms spread through the circulation and which has a high mortality rate. It is believed that the spores enter the body after being taken up by macrophages that reside in the lung's alveoli.

Anthrax as a weapon

Several countries developed anthrax as a weapon during the 20th century, and some may have used it for warfare purposes, directed toward farm animals. In 2001, the anthrax scare in the United States illustrated the importance of this organism as a weapon of terror. It can be argued that anthrax is not an ideal biological weapon. The organisms are not particularly pathogenic, because it takes a large number of spores to infect people; they are hardly ever transmitted from person to person; and they are most effective when delivered in the form of a very fine powder. "Weaponizing" anthrax cultures requires grinding the preparation into a fine powder and using anticaking agents to prevent the spores from clumping. For this reason, even though anthrax bacilli are easy to grow, making weapons-grade preparations requires specialized containment facilities and great care by the persons working in them.

Prevention of anthrax is difficult because the available vaccine is not fully effective. Newer vaccines are under development. On the other hand, treatment of patients is effective as long as the diagnosis is made soon after infection. The organisms are generally sensitive to antibiotics, such as **penicillin** and ciprofloxacin (Cipro), although the organisms can be engineered to become resistant to these drugs. Because of the need for early diagnosis, rapid methods to detect the organisms are being developed. Even though anthrax is not a perfect weapon, even small outbreaks caused by terrorists will elicit a strong response, leading to disruption of normal activities and the diversion of critical resources.

A microbiological ruse

There have been times when microbiological quirks have been put to a humanitarian use. Particularly poignant is the story of two physicians living in a small town in Poland during the German occupation in World War II. These two doctors were aware that the Germans were extraordinarily anxious about the spread of epidemic typhus and did not send people suspected of having the disease to labor camps. The test then used (but since superseded) to diagnose typhus relied on measuring antibodies in serum against an unrelated bacterium of the genus *Proteus*. Antigens of these organisms resemble those of the typhus bacterium *(Rickettsia)*, and antirickettsial antibodies cross-react with *Proteus*. Unlike the typhus bacterium, *Proteus* can be readily cultivated even in a simple laboratory. Ingeniously, the Polish doctors made an innocuous vaccine of killed *Proteus* and used it to inoculate the people of several villages. When the Germans screened these persons, they found that they had high antibody titers, which they took to mean that these villages were hotbeds of epidemic typhus. The inhabitants of the villages were spared, thanks to the microbiological acumen of the two physicians.

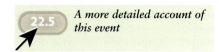

22.5 *A more detailed account of this event*

COPING WITH DANGER IN A MICROBIAL WORLD

Sanitation

We use three main types of intervention to defend ourselves against infectious diseases: sanitation, vaccination, and treatment. The first two are preventive and are therefore the much-preferred options. Of these, sanitation is the oldest and is practiced not only by humans, but also by animals. When animals dispose of their feces at special sites ("latrines"), not only do they avoid the spread of microbes, but they may also distance themselves from their leavings, which may mislead predators. Colony-dwelling insects, such as certain ants and termites, carry their detritus to places far away from their colonies (see Chapter 19). Humans, from ancient times on, have also practiced intricate sanitary measures. The large amount of space devoted in the Old Testament (e.g., Leviticus) to personal cleanliness and care of food attests to the importance given to such preventive actions. Certain simple sanitary measures take relatively little effort. However, the widespread provision of clean water and food requires considerable resources, often beyond the scope of poor people. Because microbes do not respect political boundaries, it is in the self-interest of prosperous countries to be concerned about the sanitation in less-endowed regions.

Vaccination

In the Western world, vaccination began to be used on a wide scale in the 19th century, alongside great improvements in drinking water supplies and sewage disposal. Together, these measures greatly reduced the incidence of food- and waterborne diseases, such as cholera and typhoid fever, as well as those associated with close living, e.g., tuberculosis, typhus, and plague. The spectrum of useful vaccines, great as it is, does not include (at the time of writing) a number of enormously harmful diseases, such as malaria, AIDS, and other sexually transmitted diseases (Table 22.2). The lack of effective vaccines against such important agents is not for lack of trying. For several reasons, many effective vaccines are easy to come by and therefore have been available for over 100 years. The development of vaccines proceeded rather rapidly in the early days of microbiology and immunology only to slow down to a trickle thereafter. In recent years, the pace of development of new vaccines has picked up considerably, but relatively few new ones have become available. There are reasons why developing vaccines is difficult.

- Ethical questions arise when vaccines are field tested in humans. These include the risks inherent in scaling up a new preparation and the choice of who should get the vaccine and who should serve as a control.

Table 22.2 Some vaccines commonly used in the United States

Viral vaccines
Influenza
Measles
Mumps
Polio
Hepatitis A
Hepatitis B
Bacterial vaccines
Diphtheria
Tetanus
Pertussis (whooping cough)
Meningococcus
Haemophilus influenzae type b
Pneumococcus

- Certain agents, such as HIV, gonococci, and the trypanosomes of African sleeping sickness, mutate at a high rate, thus changing their antigenic characteristics. The immune response then becomes a race to catch up with the newest antigen.
- Agents, such as the one that causes malaria, go through a number of stages in their life cycles, each of which looks different to the immune system.
- Antibodies in the circulation may be able to neutralize free-floating agents but not affect them in an intracellular location.
- Cell-mediated immunity, the main acquired defense against intracellular pathogens, is difficult to elicit by vaccination. In general, only live, inactivated agents stand a chance of eliciting cell-mediated immunity. An example is the vaccine against tuberculosis called BCG (which stands for *bacille Calmette-Guérin,* for its developers). However, immunocompromised patients are at risk from such inactivated but live organisms.

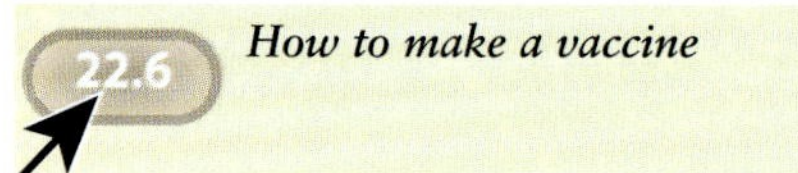

Antimicrobial drugs

The emergence of resistance to antimicrobial compounds is a cogent example of the microbial genome's plasticity. Widespread resistance is now the norm, although luckily, it is not yet universal. So far, the game has been one of catch-up, where microbes develop resistance and humans make novel drugs. At first, in the 1950s, new and powerful drugs became available with great dispatch. However, the efforts of microbiologists and chemists met with relatively modest success thereafter, and only a handful of new clinically useful antibiotics have been made since that early period. Most of these drugs are chemical modifications of previously known compounds; few are the result of discovery of novel classes. The reason may be that microbes, because of their extremely long evolution, have had the time to learn to synthesize most of the effective compounds that ensure their survival. A major exception has been the development of a variety of anti-HIV drugs that have permitted halting the progress of the infection to AIDS. Although these drugs are expensive and cumbersome to administer, they have become a bright spot, because in general, the antiviral drugs have lagged behind the antibacterial ones.

CONCLUSION

This chapter should not end with the vision of microbes as nasty agents of disease. The life of microbes is not dictated by malicious intent, and none should be attributed to them. A society that prizes cleanliness and practices "microbiophobia" risks distorting the reality of the interactions between humans and microbes, sometimes to the detriment of the former. Attempts to sterilize our environment are not only inane, they are counterproductive, because we benefit so much from the close presence of microbes. Our normal microbial flora helps us ward off pathogens and provides important stimuli to our immune systems. A germ-free existence is only possible under conditions of extreme isolation, being born by caesarian section into sterile bubbles and living in them thereafter. This has actually been done with a child diagnosed with a life-threatening deficiency in his immune system. David, the "bubble boy," eventually died in 1984 when exposed to the environment at the age of 12.

For all the threat represented by pathogens, humans and all other forms of life have learned to coexist with microbes. Indeed, the coexistence of humans and microbes, as we have noted throughout this book, has a little appreciated aspect: we cannot live without microbes.

SUGGESTED READING

Diamond, J. 1997. *Guns, Germs, and Steel: the Fates of Human Societies.* W. W. Norton & Co., New York, N.Y.

McNeill, W. H. 1976. *Plagues and Peoples.* Anchor Books, New York, N.Y.

STUDY QUESTIONS

1. Prepare an outline for a lecture to high school students on the historical aspects of infectious diseases.

2. Discuss factors involved in the emergence of new infectious diseases and the reemergence of old ones, and give some examples.

3. What factors contributed to the success of smallpox eradication?

4. Prepare an outline for a lecture to a lay audience on the main tools available to us to combat infectious diseases.

chapter 23

putting microbes to work

INTRODUCTION

Most of our interactions with microbes are involuntary. They do their thing; we do ours. This usually works to our advantage, because microbes are essential participants in this planet's metabolism (see Chapter 1). Other times, the microbes' activities are noxious: they spoil our food, destroy our crops, or make us sick. We reciprocate: certain human pursuits, such as habitat modification, greatly affect the microbial biota, as they do all other living things, but these are usually unintended consequences. On the other hand, we intentionally put microbes to work in a variety of ways and have done so since the beginning of human culture.

VARIOUS USES OF MICROBES

The list of microbial domestications by people is long and varied (Table 23.1). Some of these transactions require deliberate biological engineering (e.g., the production of antibiotics); others happen without sophisticated human intervention (e.g., the production of bread, cheese, pickles, vinegar, sauerkraut, soy sauce, wine, or beer in traditional ways). As microbiology has advanced, we have increased our understanding of how to do this best. Usually, progress has come from recognizing which microbe, acting alone or in a consortium, is involved; this is frequently followed by selecting a better strain or genetically modifying it. Of course, this approach is not unique to microbes and has been used since time immemorial to select for better strains of domestic animals and plants, including wheat, rice, and corn.

We will begin with a story of selecting better strains and go on to discuss a number of other instances where humans have put microbes to work.

Table 23.1 Some uses of domesticated microbes

Purpose	Comments
Making, preserving, or enriching food	
Bread	The carbon dioxide produced when baker's yeast, *Saccharomyces cerevisiae*, ferments sugar causes bread to rise.
Cheese	Lactic acid bacteria cause milk to curdle, the first step in making certain cheeses; many different bacteria, through their proteolytic and lipolytic activities, contribute to the ripening of cheese.
Yogurt	Yogurt is just one of many dairy products made by the fermentation of the lactose in milk by various lactic acid bacteria.
Pickles	Pickles are made by lactic acid bacteria fermenting the sugars that cucumbers contain. Many other vegetables, including olives, are similarly preserved.
Vinegar	Vinegar is made when acetic acid bacteria oxidize ethanol to acetic acid. Starting materials are usually made by yeast fermentation of various fruits, principally grapes and apples.
Sauerkraut	Sauerkraut is the product of the action of lactic acid bacteria on cabbage (see Chapter 1).
Silage	Silage is preserved animal forage made by the action of lactic acid bacteria.
Beer	Beer is the product of a yeast fermentation of grain that has been saccharified.
Wine	Wine is fermented fruit, principally grapes.
Vitamins	Many vitamins, including riboflavin, vitamin C, and vitamin B_{12}, are made by microbial fermentations.
Amino acids	Several amino acids, including lysine, methionine, and monosodium glutamate, are made by microbial fermentations; they are added to human and animal food to increase the nutritional value or to enhance flavor.
Enzymes	Many microbial enzymes are used industrially or therapeutically (Table 23.2).
Medicinal drugs	
Antibiotics	With very few exceptions, antibiotics in use today are made by microbial fermentations.
Probiotics	Various live microbes or mixtures of them are used to treat animal diseases or enhance the growth of crops. For example, mixtures of bacteria approximating a chicken's normal intestinal flora are given to chicks to prevent infection by *Salmonella* spp., and legume seeds are coated with nitrogen-fixing bacteria to ensure the developing plant will be able to fix nitrogen.
Therapeutic proteins	Bacteria into which human genes have been inserted are used to produce certain therapeutic proteins, including insulin and human growth hormone.
Corticosteroids	Microbes are used to mediate certain steps in the chemical synthesis of corticosteroids.
Waste treatment	
Sewage disposal	Sewage treatment involves the oxidation of the organic components of sewage by microbes. Subsequent microbial processes are also employed to reduce the nitrogen and phosphorus content of the effluent.
Composting	Microbial action converts plant waste into enriched material to augment soil.
Bioremediation	Various microbes are used to eliminate toxic waste.
Mining	In certain ores, minerals such as gold are trapped in insoluble sulfides; bacteria are used to oxidize the sulfides, thereby releasing the mineral.
Making fuel	Sewage and animal waste are used as substrates for archaea to make methane, which is used domestically and municipally as fuel. Starch from corn is converted to sugar and fermented to form alcohol, to be used in gasohol.

MAKING BETTER WINES: THE MALOLACTIC FERMENTATION

Enology, the science of wine making, has become a highly sophisticated enterprise. University departments and industrial institutes are dedicated to the basic science that supports this major, worldwide industry—in spite of the fact that making wine is the inevitable microbiological consequence of crushing grapes. The powdery bloom on a grape's surface contains a population of yeasts capable of mediating an alcoholic fermentation; the grape's contents constitute a near-ideal growth medium and a rich source of fermentable sugar. Introduce one into the other by crushing the grape, and wine results. Because yeasts cannot ferment starch, fermenting grains to make beverages, such as beer or sake (or potatoes to make vodka), requires an initial human intervention. The process of converting starch, which grains and potatoes contain, into the fermentable sugars, glucose and maltose, is called **saccharification**. In the Western world, saccharification is usually accomplished by adding **malt** (roasted germinated barley) as a source of the enzyme amylase to hydrolyze the grain's starch. In Asia, mixtures of amylase-producing fungi are encouraged to grow on the grain. In some native cultures, human saliva has been used as a source of amylase. In contrast to grains or potatoes, however, grape juice is ready to go. It contains an equimolar mixture of glucose and fructose, both of which yeasts can ferment, along with all the nutrients yeasts need to grow.

Crushing grapes produces wine, but it does not necessarily produce good wine. Making good wine requires careful attention to myriad details, a good many of which are microbiological. Traditional practices for making wine, developed by trial and error over centuries, have yielded remarkably good results; modern microbiology has made them better and much more reliable. Now, wines from a quality producer vary from year to year because growing conditions change, not the wine maker's luck. Microbiology's best-known contributions to enology are controlling the alcoholic fermentation by favoring the participation of desirable strains of yeasts and preventing spoilage, principally by acetic acid bacteria and lactic acid bacteria. Louis Pasteur developed a moderate heating procedure called pasteurization to eliminate spoilage microorganisms. Pasteurization of milk, which has had major public health benefits, derives from wine making.

In spite of bacteria being identified with wine spoilage, a bacterial fermentation known as the malolactic, or secondary, fermentation of wine is essential to making high-quality red wines. Traditionally, the malolactic fermentation begins spontaneously in midwinter, several months after the grapes are crushed and fermented in the fall. The winemaker may notice a slight activity, possibly a rumbling in the holding tank, or the fermentation might pass unnoticed unless the winemaker were to determine which organic acids were present in the wine. Grape juice contains two organic acids: tartaric acid and malic acid. The primary alcoholic fermentation leaves both of them untouched, but the later secondary fermentation converts all the malic acid to lactic acid via the malolactic fermentation:

$$\underset{\text{Malic acid}}{\text{HOOC-CH}_2\text{-CHOH-COOH}} \longrightarrow \underset{\text{Lactic acid}}{\text{CH}_3\text{-CHOH-COOH}} + \text{CO}_2$$

This conversion yields minuscule amounts of energy, but it is sufficient to support slow growth of the lactic acid bacteria that mediate it (see Chapter 18). The CO_2 that the fermentation produces is immaterial to winemaking except in certain wines from Portugal, called vinhos verdes, which depend on this fermentation to make them sparkling. (Champagne-type sparkling wines depend on the CO_2 produced by yeast fermentation.) It is the conversion of malic acid to lactic acid that is critical to making a high-quality red wine. This transformation of a dicarboxylic acid into a monocarboxylic acid decreases the wine's acidity and softens it. Secondary products of the lactic acid bacteria add to the wine's flavor complexity.

If the malolactic fermentation occurs regularly, albeit mysteriously, why should the winemaker want to control it? There are several reasons. If this fermentation were delayed and occurred after bottling, some volatile components would be trapped, giving the wine a foul taste. A wide variety of different lactic acid bacteria mediate the malolactic fermentation, some conferring much more desirable flavors than others. The solution would seem straightforward, rather like developing a better variety of wheat: isolate the causative organism from a high-quality wine and use it as an inoculum to start a desirable fermentation in other wines. However, that does not seem to work. The reason is related to the delay between the onset of a natural malolactic fermentation and the alcoholic fermentation. Yeasts are sponges for nutrients: when they stop growing, they take nutrients from their environment and store them in their vacuoles. Lactic acid bacteria, including those that mediate the malolactic fermentation, require a long list of growth factors, such as amino acids and vitamins. When this fact was recognized, it became possible to induce a malolactic fermentation with an inoculum selected for quality. Simply add the inoculum near the end of the alcoholic fermentation, before the yeasts have a chance to deplete the wine of nutrients. Natural malolactic fermentations begin in the winter because only by then have a sufficient number of yeast cells autolyzed to enrich the medium with the growth factors required for the growth of malolactic bacteria. Now, malolactic fermentations are routinely induced. A popular inoculum worldwide is the bacterium *Oenococcus oeni* ML34, which was isolated from a premium winery in the Napa Valley of California; it was the strain used to start the first reported induced malolactic fermentation.

PROTECTING PLANTS AND MAKING SNOW: ICE-MINUS BACTERIA

It might sound improbable that bacteria cause frost damage to plants, but it is true. They do not potentiate the chemical effect of low temperature. Instead, they do mechanical damage by causing frost to form. The resulting ice crystals rupture the plant cells, killing them. This frost-forming ability is restricted to a small group of bacteria, including *Pseudomonas syringae* and related species of *Pseudomonas*, *Xanthomonas*, and *Erwinia*—all members of the *Gammaproteobacteria*. These bacteria live on the surfaces of the leaves of plants, and under the right conditions, they produce a protein (InaX) that triggers the formation of ice crystals by serving as a focal point for water molecules to arrange themselves in a regular array. Of course, ice crystals form only at the freezing point (0°C) or below, but low temperature alone is usually not enough. Water remains liquid below 0°C in the absence of a nucleation agent to initiate ice formation. If such an agent is added, ice forms almost instantly. In the case of plant leaves, strawberry leaves for example, in the absence of ice nucleation bacteria, frost does not form until the ambient temperature has dropped below −5°C. In their presence, it forms at 0°C (Fig. 23.1).

The discovery of ice nucleation bacteria makes a fascinating story. It was made in the 1970s by two groups of microbiologists independently on the bases of quite different observations. One group, at the University of Wyoming, investigated an unusual pattern of local weather: rainfall was greater over vegetation-covered areas than over barren ones. The other group, in Wisconsin, pursued an observation that corn seedlings dusted with ground plant material were much more sensitive to frost damage than untreated controls. The group in Wyoming faced the dilemma of cause and effect. Certainly, greater rainfall would favor the growth of vegetation. However, they showed that vegetation stimulated rainfall by establishing that certain bacteria, notably *P. syringae*, on plant leaves were powerful nucleation agents. When swept high into the air, they cause water vapor in clouds to coalesce into raindrops. The group at Wisconsin showed that certain bacteria, notably *P. syringae*, were the frost-inducing component of the ground plant material. Thus, *P. syringae* was revealed to be a powerful nucleation agent, able to cause rain and frost.

Figure 23.1 Two test tubes of water held just below 0°C. The one that contains a bean leaf with wild-type *P. syringae* **(right)** is frozen; the one with an Ice$^-$ strain of *P. syringae* **(left)** is not.

The practical implication of this discovery was immediately apparent. If the naturally occurring microflora of *P. syringae* and other ice nucleation bacteria could be displaced from frost-vulnerable plants, frost damage at temperatures above −5°C could be diminished or even eliminated. Eliminating particular microbes from a natural environment by simple eradication is never feasible, but displacing them with other microbes adapted to that environment sometimes works. For example, a method (called **probiotic;** developed by the U.S. Department of Agriculture) of administering to baby chicks a grand mixture of naturally occurring members of an adult chicken's intestinal microflora dramatically decreases their likelihood of being infected by *Salmonella*: the incidence of contaminated chickens in a flock is reduced from 90 to 10%. Similarly, administering to baby rats strains of the caries-causing bacterium *Streptococcus mutans* that do not produce acid displaces the naturally occurring acid-producing strains and confers lifelong protection against cavities. One day, it might be approved for use to treat humans.

One would predict that the ideal organism to displace frost-inducing bacteria would be a strain of one of them that did not produce the nucleating protein, InaX. The scientists prepared Ice$^-$ strain mutants from which the encoding gene, *inaX*, was deleted. Because these strains were made using recombinant DNA methods, the pathway for obtaining approval to use them in field tests was long and arduous, in spite of the fact that such Ice$^-$ strains can be isolated from nature. However, in the early 1990s, approval was granted; field tests were done, and the results were spectacular. Spraying suspensions of Ice$^-$ strains of *P. syringae* on potato plants decreased their damage during a subsequent overnight frost by up to 80%. Still, because of the public's concerns about the release into nature of genetically engineered microbes, use of Ice$^-$ bacteria has not been commercialized. Instead, a naturally occurring strain of *Pseudomonas fluorescens* is used. It displaces Ice$^+$ strains as effectively as Ice$^-$ strains do. About 50,000 pounds of freeze-dried bacterial cells were sold in 2004 for use on various crops in the United States. At 10^{10} cells per g, that is a lot of bacteria. It probably constitutes the largest example of the use of a biological control agent.

Of course, there are situations in which the reverse is true and one would want ice crystals to form quickly and abundantly. In such cases, Ice^{+} strains of *P. syringae* might be helpful. One such application is making artificial snow for ski runs. Natural snow forms when tiny particles of water form ice crystals around nucleating particles, such as small particles of dust. Then, as the particles begin to fall, water vapor deposits more crystals on the surface of the ice by the process of **deposition** (direct transition from the gaseous to the solid state, bypassing the liquid phase—the reverse of sublimation); slowly, as the snowflake falls, its size increases. Snowmaking machines, called snow guns, cannot accommodate an equivalent leisurely process.

These machines send a spray of water, which, because it is propelled by compressed air, disperses into a fine mist, only about 20 or 30 feet into the air. If nucleation did not occur immediately, water would fall to the ground, forming ice—a skier's nemesis. In order to make snow under such conditions, cultured *P. syringae*, which has been lyophilized, made into pellets, and sterilized (some strains of *P. syringae* are plant pathogens), is added to snow guns. This works very well.

USING MICROBES TO MAKE PROTEIN DRUGS: INSULIN AND HUMAN GROWTH HORMONE (hGH)

From the mid-20th century, microbes have been major contributors to the manufacture of medicinal drugs. Antibiotics, with few exceptions (including chloramphenicol and those that are chemically modified, such as tetracycline), are made by microbes, and microbes are used to carry out difficult steps in the otherwise chemical synthesis of some drugs, for example, certain steroids. These uses of microbes all exploit their natural capacities. For example, antibiotic-producing microbes produce those antibiotics when they are first isolated from nature. Subsequent genetic and cultural manipulations necessary for commercialization are just to induce them to synthesize larger quantities of the antibiotics. In the 1970s, with the advent of recombinant DNA technology, the uses of microbes in drug manufacture changed completely: the therapeutic agents that could be produced by microbes were no longer limited by their natural abilities. Rather than searching for a microbe that could produce a useful agent, a microbe could be genetically altered so that it would produce that particular agent. At least in theory, genes from any organism could be introduced into a microbe and its remarkable biosynthetic capacity would churn out quantities of that gene's product. However, there are hurdles in the track.

Even before starting, one has to decide which compound to make. There are obvious guidelines. Without doubt, the compound has to be of sufficient value to justify the costs of development and manufacture. Better still, it should not be available from another source. Therapeutically valuable human proteins seemed to be obvious choices. Insulin, used for treating patients with diabetes, was the first therapeutic agent made by a microbe into which a human gene had been cloned. Insulin had been available for therapeutic use since shortly after its discovery in the 1920s. It was purified from the pancreases of animals—largely from cows in the United States and from pigs in Europe. Although most human patients who received repeated doses of these foreign proteins tolerated them well, there were certain complications. Some patients developed insulin resistance because animal proteins, although very similar, are not antigenically identical to the human one. Also, the supply of insulin became problematic because feedlot cattle, the major source of bovine pancreases, produce smaller amounts of insulin than range cattle, the traditional source. The solution to the problem was to produce insulin in *Escherichia coli* into which insulin-encoding human genes had been introduced. Because it was not a foreign protein, there was no immune response against it, and the supply became unlimited.

hGH presented an even more compelling need. The only source of this hormone, which is essential for children whose pituitary glands make insufficient amounts of it (who, if untreated, are destined to become pituitary dwarfs), was the pituitary glands of human cadavers. This source presented clear dangers, as there was evidence that a few children had contracted the prion disease Creutzfeldt-Jakob disease from such treatments.

The methods used to clone and express the hGH gene in bacteria are typical of those used to make other human proteins. All human cells carry the gene that encodes hGH, but how do you separate that gene from all others and clone it? One way is to start with messenger RNA (mRNA) instead of DNA. Human pituitary gland tissue makes the body's supply of hGH, so its cells contain large amounts of the corresponding mRNA. These mRNA molecules can be identified with an appropriate **probe** (a short DNA molecule corresponding to part of the hGH-encoding gene). The probe, being complementary to its corresponding mRNA, will hybridize (form hydrogen-bonded base pairs) with it and thereby identify it. There is a second and more compelling reason for isolating mRNA rather than isolating the gene directly. Genes from eukaryotes, including humans, contain **introns** (stretches of DNA that do not encode the protein because they are cut out of mRNA as it matures). Because prokaryotes lack the enzymes to eliminate introns, they would synthesize an incorrect protein from an intron-carrying eukaryotic gene.

How do you make the probe? First, you have to know the sequence of amino acids in the hGH protein. Then, the sequence of bases for the probe can be determined by referring to the genetic code. However, the genetic code is redundant—most amino acids are encoded by several different codons. The genetic code states precisely which amino acid sequence is encoded by a particular sequence of bases in DNA, but not the reverse. Because of the redundancy of the code, all the DNA sequences that designate the desired amino acid sequence must be synthesized. This mixture is used as a probe.

Once the correct mRNA has been isolated, **reverse transcriptase** (the enzyme from retroviruses that uses RNA as a template to make DNA [see Chapter 17]) is used to make DNA. This DNA product is called complementary DNA (cDNA) to indicate that it is a copy of mRNA, not the DNA in the gene itself. Then, the complementary DNA is cut with restriction endonucleases, ligated into a bacterial plasmid, and inserted into the bacterial host by transformation. In the host, the recombinant DNA molecule is replicated, and its genes, including the hGH gene, are expressed. The transformed cell becomes an hGH factory.

The method used to clone hGH is not the only way it could be done. Many variations are possible, and new ones are developed almost daily. The most fundamental changes are in the **cloning vector** and the host cell. A plasmid can be used to clone hGH, but viral DNA can also be used. Introducing native or recombinant DNA into a host is called **transformation.** Almost any kind of cell or intact organism can now be used as a host for the recombinant DNA molecule. However, no matter which host cell is finally selected to make the protein product, bacteria are almost always used in the cloning process.

After the hGH-producing bacterial strain is obtained, the job of producing hGH commercially has just begun. It is not enough to obtain a strain to produce some hGH. The plasmid must be modified and cultural conditions must be developed to produce commercially feasible amounts of hGH. For example, because large amounts of hGH, and other foreign proteins as well, are detrimental to the cell, methods have been devised to trigger *E. coli* to make hGH only after the culture achieves a high density of cells. Methods have also been devised to purify properly folded hGH from the bacterial culture and to prevent its being destroyed by the host's proteolytic enzymes, whose function is to destroy abnormal proteins.

MICROBIAL ENZYMES: SWEETENERS FROM CORN

Enzymes are the biocatalysts that make life possible, but they are also articles of commerce. Microbial enzymes are produced industrially in huge quantities and used for myriad purposes, from making candies to medical uses to large-scale production of sweeteners from corn (Table 23.2). Some of these uses are intriguing. For example, invertase, which

Table 23.2 A sampling of microbial enzymes that are used commercially

Enzyme	Microbial source	Activity	Use
Invertase	*Saccharomyces cerevisiae*	Hydrolyzes sucrose into glucose and fructose	Baking, stabilizing syrup, candy making
β-Glucanase	*Bacillus subtilis, Aspergillus niger, Penicillium emersonii*	Hydrolyzes β-glucans	Clarifying beer
Lactase	*Saccharomyces lactis, Aspergillus niger, Aspergillus oryzae, Rhizopus oryzae*	Hydrolyzes lactose into glucose and galactose	Digestive aid to those with lactose intolerance; dairy industry
Pectinase	*Aspergillus niger, Aspergillus oryzae, Rhizopus oryzae*	Hydrolyzes pectin	Clarification of fruit juices and wines
Rennin	*Escherichia coli* with cloned gene from cattle	Hydrolyzes a bond in the milk protein, casein, causing it to coagulate	Cheese making
Neutral protease	*Bacillus subtilis, Aspergillus niger*	Hydrolyzes proteins at neutral pH	Enhancing flavor of meats and cheese
Alkaline protease	*Bacillus licheniformis*	Hydrolyzes proteins at an alkaline pH	An additive to detergents that removes protein-based stains
Lipase	*Aspergillus niger, Aspergillus oryzae, Rhizopus oryzae*	Hydrolysis of ester bonds in fats and oils	Dairy industry; an additive to detergents that removes fat-based stains
Cellulase	*Trichoderma konigi*	Hydrolyzes cellulose	A digestive aid
α-Galactosidase	*Lactobacillus plantarum, Lactobacillus fermentum, Lactobacillus brevis, Lactobacillus buchneri*	Hydrolyzes α-galactosides	Treatment of legumes to decrease their ability to cause flatulence

hydrolyzes sucrose to glucose and fructose, is added to the filling of chocolate-covered maraschino cherry candies to make their centers liquid: the filling is solid as it is being covered with hot chocolate, and it becomes liquid only later as the less soluble sucrose it contains is converted to a more highly soluble mixture of glucose and fructose. The enzyme α-galactosidase is used to decrease the flatulence-causing properties of beans, which is caused by their content of α-galactoside sugars, principally raffinose and stachyose. Because humans and other monogastric animals do not produce α-galactosidase, these sugars pass undigested through the duodenum. When they reach the large intestine, they are fermented by species of *Clostridium* and *Bacteroides* that reside there to produce gaseous products, H_2 and CO_2, which cause flatulence. Adding microbial α-galactosidase while soybeans are being processed or as a digestive aid when beans are eaten minimizes the problem. Microbiologically, economically, and socially, the microbial enzymes used to produce sweeteners from cornstarch illustrate the general principles of making and using microbial enzymes commercially.

Corn contains starch, which is not sweet, yet it is the source, annually, of over 10 million tons of sweetener, called **high-fructose corn syrup (HFCS)**, which is added to just about everything sweet or almost sweet that we eat. This extensive list includes most carbonated drinks; fruit drinks; candied fruits; canned fruits; dairy desserts, such as flavored yogurts; most baked goods; many cereals; and most jellies. All the steps in the process of converting cornstarch to HFCS are enzymatic. There are three of them, each catalyzed by different enzymes (Table 23.3):

$$\text{Starch} \xrightarrow[\alpha\text{-Amylase}]{} \text{Maltodextrans} \xrightarrow[\text{Glucoamylase and pullulanase}]{} \text{Glucose} \xrightarrow[\text{Glucose isomerase}]{} \text{HFCS}$$

The enzymes that catalyze the first two steps have been produced industrially since the 1950s, but glucose isomerase, the critical third enzyme of the process, is a more recent innovation. Although the product of the first two steps, glucose, is a sugar, it is not very sweet. If table sugar (sucrose) were assigned a sweetness value of 100, glucose's sweetness value would be only 70, and fructose's would be 130. Moreover, glucose is relatively insoluble, so it tends to precipitate from concentrated syrups, making

Table 23.3 Enzymes used to convert cornstarch to HFCS

Enzyme	Microbial sources	Activity
α-Amylase	*Bacillus licheniformis, Bacillus subtilis*	Cuts starch (a mixture of amylose, a straight-chain polymer of α-D-glucopyranose, and amylopectin, a branched form that contains some α-1,6 bonds) to maltodextrins, which are short chains
Glucoamylase	*Aspergillus niger, Aspergillus oryzae*	Splits glucose from maltodextrans by cleaving α-1,4 bonds
Pullulanase	*Klebsiella aerogenes, Bacillus* spp.	Cleaves α-1,6 bonds in maltodextrans
Glucose isomerase	*Bacillus coagulans, Actinoplanes missouriensis, Streptomyces* spp.	Converts D-glucose into D-fructose

them difficult to store and use industrially. Fructose is twice as soluble as glucose. As a result, converting about half the glucose in a syrup to fructose (producing HFCS) solves both problems: the product is as sweet as sucrose, and it is a stable solution that does not readily precipitate if chilled.

In retrospect, the discovery of glucose isomerase seems improbable. The enzyme's metabolic role in the bacteria that produce it has nothing to do with glucose or fructose. Instead, it catalyzes the first step in the catabolism of xylose, an abundant source of carbon in nature because it is the hydrolysis product of hemicellulose, a major constituent of the cell walls of plants. The enzyme, more properly called xylose isomerase, converts xylose to xylulose. In 1957, it was discovered to have about 160-fold less activity for glucose (a homologue of xylose, differing only in possessing an additional $—CH_2OH$ group). It was soon discovered that the enzyme's specificity for glucose is increased in the presence of cobalt ion and magnesium ion, which are required for activity. Intensive research on this enzyme, which is widely distributed among bacteria, has led to its practical use in producing HFCS. Such research involved mutational improvement of the enzyme, optimization of the means of producing it, and improved conditions for using it. Among the many hurdles, the requirement for expensive xylose as its inducing substrate had to be overcome. Also, immobilizing the enzyme on a column, through which the glucose solution was passed, rather than adding it to batches solved several problems: less of this relatively expensive enzyme was needed, and the cobalt and magnesium cofactors could be immobilized on the column and kept out of the product.

Glucose isomerase catalyzes an equilibrium reaction in which approximately 41% of the final mixture is fructose at moderate temperatures. Because about 55% fructose is needed to give sufficient sweetness for carbonated drinks, some fructose is purified from the mixture by column chromatography and used to adjust other batches to this level. Because the equilibrium between glucose and fructose is shifted toward fructose as the temperature increases, intense research has been done to increase the heat stability of glucose isomerase so that the reaction temperature could be increased and more fructose could be made. With an enzyme that could withstand 95°C, the concentration of fructose would be high enough for it to crystallize, thereby making it possible to produce granular fructose, a direct competitor with sucrose for use as table sugar. This goal has not yet been attained.

We might ask why HFCS is needed, in view of the availability of sucrose as a sweetener. The answer is principally economic. HFCS is cheaper than sucrose. For this reason, production of HFCS continues to rise and to displace sucrose. Already, more HFCS than sucrose is used worldwide. Its principal use is in the United States, but overseas use is growing rapidly. Unlike sucrose, fructose does not cause dental caries, because it cannot be used by the caries-inducing *S. mutans* to produce glucans on the surfaces of teeth, an essential step in the process of forming cavities. Glucan traps the lactic acid produced by *S. mutans* close to the tooth, where it eats holes in the dental enamel, causing cavities. However, some see ominous health implications associated with our rising consumption of HFCS. Its introduction into our food supply in the late 1970s coincides with the onset of the present epidemic of obesity and accompanying childhood diabetes. Furthermore, the consumption of HFCS and the incidence of obesity have grown together. However, there is more evidence than mere coincidence for the presumed causal relationship between HFCS and obesity. Owing to its low cost, portion sizes of soft drinks in restaurants, in which HFCS is the sole sweetener, have increased severalfold, and there seems to be a physiological rational for our high consumption of fructose in this form. In response to glucose, the hydrolysis product of much of our food, the pancreas produces insulin, which stimulates the release of **leptin**, the hormone that signals satiety, telling us we have had enough to eat. Because the pancreas lacks a transport system for fructose, it cannot initiate the physiological cascade leading to a feeling of satiety. But let us not blame the microbe; it was only trying to break down products from hemicellulose in wood.

BIOLOGICAL INSECTICIDES: *Bt*

Bacillus thuringiensis is one of a number of spore-forming soil bacteria that have become useful in agriculture as pesticides. This bacterium makes a set of powerful protein toxins that are used widely by farmers to control a variety of insects. Known as *Bt* toxins, they kill over 150 different species of noxious insects. The toxins are proteins made by the bacteria during sporulation and are produced in such large amounts that they make crystals readily seen under the electron microscope (Fig. 23.2).

There are a large variety of *Bt* toxins, each specific for a different insect. When *Bt* toxin-forming bacteria are ingested, the toxins bind specifically to receptors in insects' intestines. The receptors differ with the insect species, which accounts for the specificity of the various toxins. Humans and other vertebrates do not have these receptors; hence, the

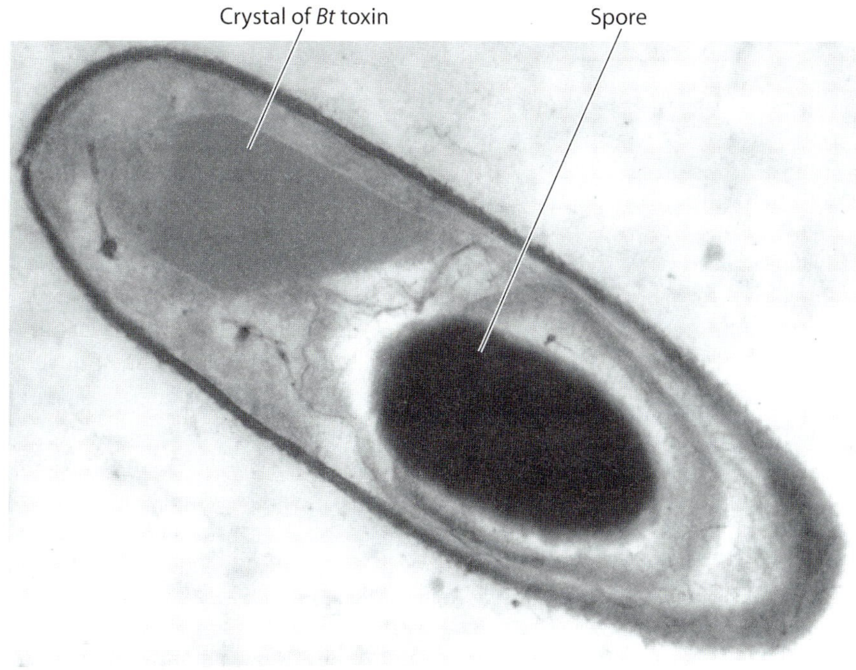

Figure 23.2 A spore of *B. thuringiensis* with an associated crystal of *Bt* toxin.

toxins are inactive in larger animals and are considered safe for food production.

More accurately, perhaps, the protein as it is made in the bacterium should be called a protoxin, because in itself it is not toxic. It is converted into lethal form only when ingested by the insect. The protein crystals are solubilized in the alkaline, reducing conditions of the midguts of susceptible insect larvae. The solubilized protein, which is still inactive, is converted into an active form by proteolysis. The toxins then insert into the membranes of the cells lining the midgut to form cation-selective channels, which leads to cell death and eventually to the insect's death.

We know little about the role of these toxins in nature. What would be the selective advantage for a soil-dwelling bacterium to kill insects? In nature, *B. thuringiensis* rarely causes epidemics among insects, and it occurs in sites where there are no susceptible hosts. However, *Bt* toxins can also kill nematodes, little worms that are abundant in soil and feed on bacteria. Is *Bt* toxin a mechanism, perhaps, of converting a potential predator into a source of nutrients? If so, its insect-killing activity might be accidental. Such speculation is plausible, because the toxins kill nematodes and insects by the same mechanism.

The insecticidal properties of *Bt* toxins are widely used for insect control. Commercial preparations of several varieties are available. Farmers have to match the particular *Bt* toxin protein with the insect to be controlled. Thus, beneficial insects, such as bees, are not usually harmed by agricultural use of *Bt*. Because once applied, *Bt* toxins do not persist, repeated applications are usually necessary. However, the targeted species may develop resistance, which usually occurs as a result of mutational changes in the insect's intestinal-cell receptors. For these reason, *Bt* toxins must be applied with care and foresight.

Genetic engineers have inserted the genes for toxin production directly into plants, such as corn, making it unnecessary to apply toxins to the field. Also, these genes have been inserted into **rhizosphere bacteria** (bacteria that are abundant in the soil surrounding a plant's roots). The bacteria then deliver the toxins to the plant, protecting it from attacking insects. Appealing though these alternative methods of delivering *Bt* may seem, they may lead to increased risk of resistance, because the toxin would persist for longer periods. Also, because most plant parts, including pollen, contain toxin, some desirable insects may be killed.

UNDOING POLLUTION: BIOREMEDIATION

The introduction, whether unintentional or intentional, of toxic and radioactive chemicals into the environment causes biological, physical, and financial damage of massive proportions (see "Bioremediation" in Chapter 5). Such pollution occurs in many ways, including dumping of contaminants into rivers, lakes, or soils; maritime disasters; and leakage from the millions of buried tanks that contain toxic chemicals. Cleaning up toxic-waste sites is a formidable task that challenges the limits of our societal capabilities and resolve. Generally speaking, it is difficult and expensive. Storage of contaminated material is costly and carries with it the danger of leakage. Incineration causes air pollution; other methods are no less dangerous and costly.

Fortunately, microbes, with their enormous repertoire of chemical activities (see "Overview of Fueling Reactions" and "Summary" in Chapter 6), are well suited to rid the environment of the toxic materials we add to it. A tremendous amount of microbial activity takes place spontaneously, although often at a pace too slow to provide adequate relief. However, we can improve the microbes' effectiveness by studying natural ways in which microbes help clean the environment. The ecological principle that the chemical and physical properties of the environment dictate the physiology of the organisms applies here. For example, microbes that may help clean up sites contaminated with radioactive materials can be expected to be present in effluents from nuclear power plants. In such waters, one indeed finds bacteria that are not only resistant to high levels of radiation (like the world champion, *Deinococcus radiodurans* [see Chapter 15]), but that can also make radioactive compounds insoluble and easier to dispose of. Such organisms can be cultured and then genetically modified to become even more effective.

Cleanup operations that purposefully involve microbes or plants are called **bioremediation** (see "Bioremediation" in Chapter 5). This process is being used at some of the 50,000 or so hazardous-waste sites in the United States, with the focus on the 1,200 designated **Superfund sites.** These sites have been selected for special attention by the Environmental Protection Agency because of the grave threat they represent. There are several routes to bioremediation. In some cases, the aim is to lower the level of pollutants at the site of contamination; in others, the polluted material is extracted and removed to special sites, such as containment tanks or lagoons, there to be treated biologically.

Bioremediation has been moderately successful at the federal government's 310-square-mile Savannah River site in South Carolina, where radioactive materials for nuclear weapons were made for over four decades. This is one of the most polluted tracts of land in the United States. Here, solvents used in the process were transported through buried pipes and stored in underground tanks. The pipes eventually leaked, and the solvents seeped underground. Among the most abundant of these solvents is trichloroethylene (TCE), which is quite toxic and, like other chlorinated solvents, difficult to clean up. TCE is one of the most abundant toxic solvents because it is widely used for dry cleaning and other industrial applications. Microbiologists thought of a way to increase the number of bacteria that can oxidize TCE. They enlisted the help of soil bacteria called methanotrophs (see Chapter 15) that oxidize methane but can also utilize other hydrocarbons, including TCE.

Methanotrophs produce an enzyme called methane mono-oxygenase that is not highly specific and can degrade TCE and other compounds in addition to methane. Normally, these indigenous organisms are present in relatively small amounts. How can we enrich for the organisms? To enhance the growth of methanotrophs, methane was pumped into the contaminated soils. Because hydrocarbon oxidation requires oxygen, air was pumped into the soil along with the methane. Measurements showed that at some sites the number of methanotrophs increased by 7 orders of magnitude. These bacteria then proceeded to reduce the concentration of TCE. The process is not quite as simple as it sounds, because considerable engineering is required for the effective delivery of the gases

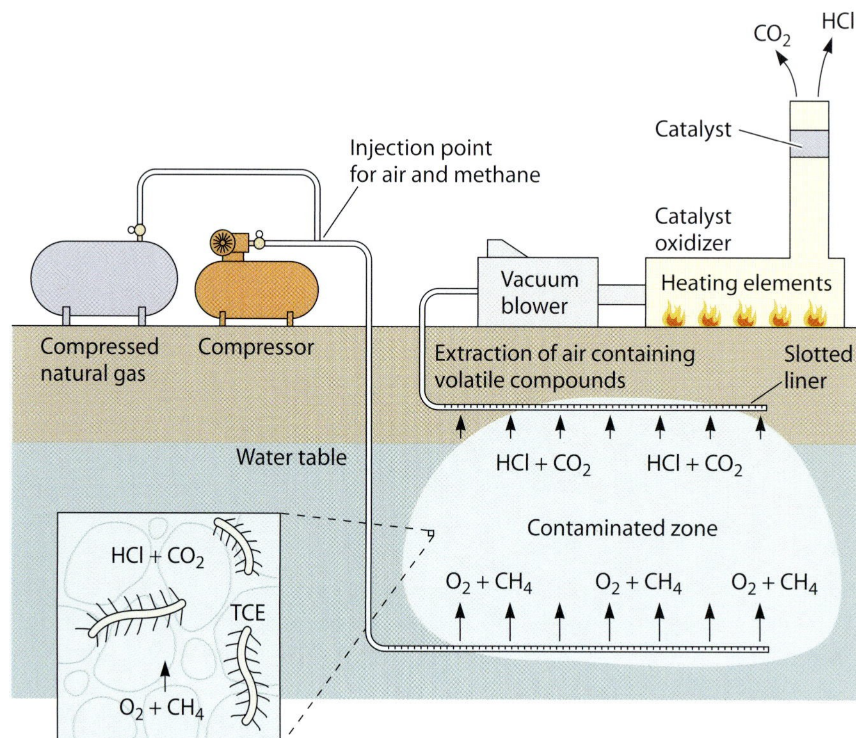

Figure 23.3 Diagram of the arrangement for bioremediation of the TCE-polluted Savannah River site in South Carolina.

Figure 23.4 Extent of the 1989 *Exxon Valdez* oil spill drawn to scale on a map of New England. Note that the oil spill covered an area the size of several states.

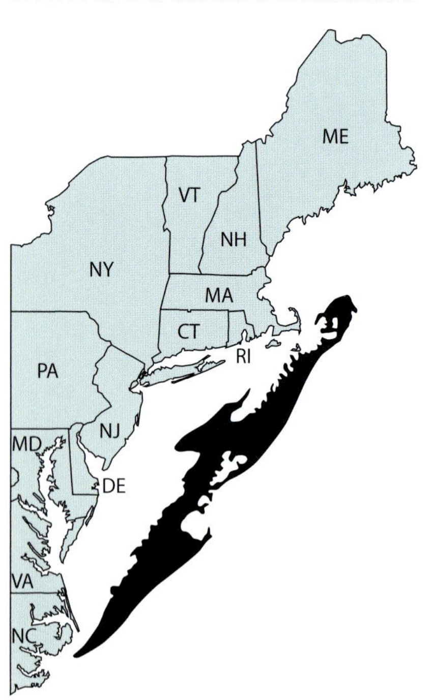

(Fig. 23.3). One advantage of this process is that its biology is reversible. When methane is no longer pumped, the methanotrophs return to their natural level.

This experience illustrates that enrichment for desired microbes requires an understanding of their nutritional requirements. In fact, methane and oxygen were not sufficient to obtain maximal degradation of TCE. The controlled addition of sources of phosphorus and nitrogen helped to further enhance the growth of the methanotrophs. Another example of the effect of such nutritional supplementation was the oil spill caused by the tanker *Exxon Valdez* in Alaska (Fig. 23.4). There, the addition of commercial fertilizer supplied needed phosphorus and nitrogen sources and permitted the enhanced degradation of petroleum pollutants by resident microbes.

Of course, the approach used at the Savannah site reduced the levels of only certain pollutants. The site remains contaminated with heavy metals (lead, chromium, mercury, and cadmium), radioactive compounds (tritium, uranium, fission products, and plutonium), and others. Bioremediation techniques have been developed for the removal of some of these pollutants, including the radioactive nuclides. The same can be said for pesticides, PCBs (polychlorinated biphenyls), and other environmentally damaging substances. Consequently, one can be optimistic for the future of bioremediation. With further research, it seems likely that this approach could become helpful in solving one of humanity's most daunting problems.

CONCLUSION

In this chapter, we have barely sampled the long history of the many ways humans have enlisted the cooperation of microbes for useful purposes. Others abound, and new uses are constantly being discovered. This field of microbiology will undoubtedly expand as new microbes are discovered and our ability to genetically engineer known microbes improves.

STUDY QUESTIONS

1. How do bacteria contribute to improving the quality of wines? What do they get in return?

2. What do you think of the public concern for the use in the field of genetically engineered *P. syringae* for the purpose of delaying ice formation on cultivated plants? Why is this not a concern in the case of using bacteria for the opposite effect, making snow?

3. What are the steps in engineering a bacterium to produce a human protein for industrial purposes?

4. How do bacterial enzymes contribute to the production of high-fructose sweeteners used in industry?

5. *B. thuringiensis* produces toxins that kill insects. However, the bacterium inhabits the soil, where there are not many insects. How do you think the ability to make this toxin evolved?

6. Discuss how physiological attributes of bacteria may suggest a strategy for bioremediation.

Coda

We are becoming increasingly aware that microbes are essential for all forms of life on Earth, including humans. Microbes are our heritage because they are the ancestors of all other living things. And right now, as we live and breathe, our lives are still critically dependent on microbial activities. As we have seen, microbes process the nutrients and elements necessary for life, they influence our climate and weather, and they sculpt our planet's rocks and bodies of water. Not all of our interactions with microbes are benign, and certain microbes pose a threat to human health and to the health of plants and animals. To us, this may seem like yin and yang, but this is an anthropocentric point of view. The microbes are simply trying to make a living, indifferent to whether they help or hinder human beings.

Microbes, then, are the foundation of the biosphere and major determinants of human health. Hence, the study of microbes, microbiology, is pivotal to the study of all living things. Microbiology is therefore a fundamental subject and is essential for the study and understanding of all life on this planet. We hope that your excursion into the microbial world will serve as the groundwork for further studies and, more broadly, in your sojourn on this planet.

This book ends with a wish that what we have learned so far and will continue to learn about the microbial world will, for the sake of this planet and all its inhabitants, be put to a sane and prudent use. This requires not only good intentions but also a deepening of our understanding of all that microbes do.

Glossary

Abortive transduction An event in which the fragment of DNA introduced by transduction fails to be recombined into the recipient chromosome and, lacking an origin of replication, is inherited by only one of the daughter cells at each cell division.

Accommodation (chemotactic) A gradual decrease in responsiveness to a chemotactic stimulus brought on by the methylation of protein sensors.

Acid-fast Resistant to destaining by mild acids; a property exhibited by mycobacteria and some actinomycetes.

Acquired immunodeficiency syndrome A deadly disease syndrome that develops after infection with human immunodeficiency virus. Abbreviated AIDS.

Actinomycetes A large group of gram-positive bacteria capable of differentiation into aerial hyphae and spores; producers of many antibiotics.

Activator A gene product (usually protein) that positively regulates transcription. Activators may either increase binding of RNA polymerase to the promoter (closed complex formation) or stimulate RNA polymerase to begin transcription (open complex formation).

Active site The region of an enzyme responsible for catalysis.

Adaptive immunity The portion of the immune system that is activated by exposure to antigens and that differs among individuals.

Addiction module A two-gene array of toxin and antitoxin involved in the maintenance of plasmids and in bacterial programmed cell death.

Adenosine triphosphate An adenine-ribose-triphosphate compound that conserves metabolic energy in its two high-energy phosphate bonds. Abbreviated ATP.

Adventurous motility A form of gliding motility that enables individual cells to move out from a group of cells that are moving together as a raft.

Aerobic In the presence of oxygen. Often used to describe an organism that requires oxygen for growth or growth conditions with oxygen.

Aerobic respiration A form of respiration in which oxygen gas is the terminal electron acceptor.

AIDS *See* Acquired immunodeficiency syndrome.

Alkaliphiles Organisms that thrive in alkaline environments.

Allosteric effector A small molecule that binds to and changes the activity of an allosteric protein.

Allostery The stereospecific modification of a protein by an effector to influence the activity of another site of the protein.

Alternative pathway The complement activation cascade induced by some microbe-associated molecular patterns, such as bacterial lipopolysaccharide.

Aminoacyl-tRNA synthetases Enzymes that charge (add amino acids to) molecules of transfer RNA (tRNA).

Anaerobic In the absence of oxygen. Often used to describe an organism that is sensitive to oxygen or that requires growth conditions without oxygen.

Anaerobic respiration A respiratory process with a terminal electron acceptor other than oxygen (O_2).

Anammox reaction The anaerobic formation of nitrogen gas from ammonia and nitrite.

Annealing Formation of double-stranded nucleic acid from complementary single strands.

Annotation The process of assigning functions to DNA sequences.

Anoxygenic photosynthesis The type of photosynthesis carried out by certain prokaryotes that does not produce oxygen as a product.

Antibiotic A substance that interferes with a particular step of cellular metabolism, causing either bactericidal or bacteriostatic inhibition; the term is sometimes restricted to those having a natural biological origin.

Antibody A protein complex that specifically interacts with an antigen.

Anticodon triplet A sequence of 3 bases on a molecule of transfer RNA that is complementary and that binds to a codon on messenger RNA.

Antigen A molecule that is recognized by the immune system.

Antigenic variation The ability of some microbes to change their surface antigens periodically.

Antigen-presenting cells Cells that display on their surface degradation products of ingested antigens.

Antiport Countermovements, one in and one out, of two molecules across the cytoplasmic membrane.

Antiterminator A protein that allows RNA polymerase to continue transcription through a transcription termination site.

Apicoplast A special DNA-containing organelle of some protists.

Apoptosis Cell death due to an intracellular developmental program or induced by other cells or infectious agents.

Archaea One of the three domains of living organisms: Archaea, Bacteria, and Eukarya. Although they share a basic morphology with bacteria and they are also prokaryotes (i.e., they lack a true nucleus), in many molecular details they resemble eukaryotes more than bacteria. Previously called Archaebacteria or Archaeobacteria.

Artificial transformation The process by which cells take up exogenous DNA following chemical or physical treatment.

Assembly modification The addition of a chemical moiety to a protein to make it functional.

Assimilation The taking up and utilization of a nutrient.

Assimilatory nitrate reduction Reducing nitrate to ammonia to serve as a nutrient.

Atomic force microscope A scanning microscope that uses the attractive and repulsive forces between atoms to generate images.

ATP *See* Adenosine triphosphate.

Attenuation A mechanism for regulating the level of transcription by interfering with messenger RNA elongation. Slowed translation through a regulatory region allows formation of an RNA secondary structure that promotes transcription termination. The process depends on coupled transcription and translation and thus is restricted to prokaryotes.

Autoinducer A compound produced by an organism that, when it accumulates, induces a response in that organism.

Autoregulation The control of synthesis of a gene product (usually a protein) by itself.

Autotroph An organism that can derive all its carbon from CO_2.

Auxotroph A mutant that will grow only when a particular nutritional requirement (e.g., an amino acid, nucleotide, or vitamin) is provided.

Baby machine A device for obtaining cells that have recently divided. It relies on attaching cells to a surface, such as a surface to which the "mother cells" adhere; after division, one "baby" remains adhered and the other becomes detached.

Bactericidal Having a lethal effect on bacteria.

Bacteriocin A small protein made by bacteria that is toxic to other bacteria.

Bacteriocyte A cell in an insect that carries endosymbiotic bacteria.

Bacteriophage A virus that infects a bacterium. Also called phage.

Bacteriorhodopsin A rhodopsin-like pigment that occurs in certain archaea and mediates a primitive form of photosynthesis.

Bacteroid A differentiated form of nitrogen-fixing, nodule-producing bacteria.

Balanced growth The condition of growth in which all cell constituents increase by the same factor over a period of time. Usually synonymous with exponential or steady-state growth.

Barophiles Organisms that grow best at pressures higher than 1 atmosphere.

Basal body A structure of the flagellum that is inserted into the bacterial cell envelope and acts as a "stator," allowing the flagellar filament to turn.

B cell *See* B lymphocyte.

Bergey's Manual of Systematic Bacteriology A widely used book with a classification scheme of prokaryotes.

β-Galactosidase An enzyme that catalyzes the cleavage of lactose into glucose plus galactose. In *Escherichia coli*, this enzyme is encoded by the *lacZ* gene. Often used as a reporter for assaying gene expression.

β-Lactam antibiotics Antibiotics that contain a β-lactam ring and act by inhibiting peptidoglycan synthesis, for example, penicillins, cephalosporins, and related antibiotics.

β-Lactamase An enzyme that cleaves the β-lactam ring of β-lactam antibiotics, thus inactivating the antibiotics. The ampicillin resistance encoded by many common plasmids is due to a secreted β-lactamase.

Bidirectional replication The process by which two DNA replication forks proceed in opposite directions from the same origin of replication.

Binding proteins Proteins in the periplasm that bind to specific nutrients and facilitate their passage through the cell membrane.

Biofilm A layer of microbes embedded in an extracellular slime.

Biological species A group of actually or potentially interbreeding populations.

Bioremediation The use of microbes to detoxify or eliminate toxic materials.

B lymphocyte A lymphocyte involved mainly in antibody production. Also called B cell.

Bubble (1) In replicating DNA, a region of the DNA in which the strands are separated to initiate replication. (2) In RNA synthesis, a region of DNA in which the strands are separated to permit transcription.

Calvin cycle The pathway, used by most autotrophs, by which CO_2 is incorporated into cellular constituents.

cAMP *See* Cyclic AMP.

CAP *See* Catabolite gene activator protein.

Capsid The protein layer that encloses the nucleic acid of a phage or virus and protects it from the environment.

Capsule A diffuse outermost layer, usually carbohydrate, surrounding many microbes.

Carbon cycle The cycle of interconversions of carbon-containing compounds in nature.

Carboxysome A protein-bound vesicle of autotrophic bacteria containing ribulose bisphosphate carboxylase.

Carriers Healthy individuals who are reservoirs of infectious agents.

Cassette A DNA fragment that can be cloned into a site to confer some property, for example, antibiotic resistance.

Catabolite gene activator protein A protein whose interaction with cyclic AMP modulates many aspects of catabolite repression in enteric bacteria. Abbreviated CAP.

Catabolite repression system A global regulatory system resulting in decreased expression of many genes due to the addition of an efficient carbon source, such as glucose; mediated in part by a complex between the CRP protein and cyclic AMP in enteric bacteria.

Catalase Enzyme that catalyzes decomposition of hydrogen peroxide to water and oxygen.

cDNA *See* Complementary DNA.

Cell-mediated immunity Immune responses mediated by T cells.

Cell membrane A phospholipid bilayer that surrounds all cells. Also called cytoplasmic membrane or plasma membrane.

Cellular microbiology The study of effects of microbes on the cell biology of their hosts.

Cell wall The tough envelope surrounding many cells, including nearly all bacteria and archaea, located outside the cytoplasmic membrane.

Central dogma The biological dictum stating that DNA makes RNA, which makes protein.

CFP *See* Cyan fluorescent protein.

Channel proteins Proteins that form channels through membranes.

Chaperone A protein that facilitates the folding of other proteins or assembly of multiprotein complexes. Some heat shock proteins are chaperones.

Chemoautotroph *See* Lithotroph.

Chemostat *See* Continuous-culture device.

Chemotaxis Migratory process directed by a chemical concentration gradient.

Chitin A polymer of *N*-acetylglucosamine.

Chloroplast An intracellular organelle in which photosynthesis occurs in phototrophic eukaryotes.

Chromosome (1) A self-replicating DNA molecule that carries essential genetic information for growth and replication of a cell or virus. (2) The DNA organized into a tightly packaged structure by associated histone-like proteins in bacteria and histones in eukaryotes.

Ciliates A class of protozoa that bear cilia.

Cistron A DNA sequence encoding a single polypeptide.

Clade A phylogenetically related branch of organisms.

Clamp loader A complex of polypeptides that places the β-clamp, or sliding bracelet, subunit of polymerase III (Pol III) in position ahead of replication on each strand of DNA.

Classical pathway Antibody-activated complement activation cascade.

Clone A set of cells derived from a single cell and thus expected to be genetically identical.

Cloning The production of multiple, genetically identical molecules of DNA, cells, or organisms.

Cloning vector A DNA molecule that is capable of replication in a suitable host cell, that has a suitable site(s) for the insertion of DNA fragments by recombinant DNA techniques, and that has genetic markers that allow selection for the vector in a host cell.

Closed complex An early stage of transcription in which RNA polymerase has bound to DNA but the strands have not yet separated.

Coccus A spherical prokaryotic cell. Plural, cocci.

Coding region A sequence of DNA that encodes a polypeptide.

Coding strand The strand of DNA used as a template to transcribe RNA.

Codon The three consecutive nucleotides (triplets) in DNA or RNA that encode a particular amino acid or signal the termination of polypeptide synthesis.

Coenzyme An organic molecule required for certain enzymes to be active.

Cofactor An inorganic molecule required for certain enzymes to be active.

Cold shock (1) Cell death that results from the sudden chilling of a rapidly growing bacterial culture. (2) Sudden chilling and dilution of the osmolarity of a culture of gram-negative bacteria resulting in leakage of periplasmic contents.

Colony A visible group of cells arising from a single cell plated on a solid medium.

Commensals Organisms in a symbiotic relationship in which neither partner is harmed.

Complement A complex of more than 30 blood proteins that functions as a nonspecific defense against infection.

Complementary Describing two polynucleotide chains that can base pair to form a double-stranded molecule.

Complementary DNA DNA synthesized from the messenger RNA using reverse transcriptase. Abbreviated cDNA.

Concatenate A DNA molecule that contains multiple, tandem, head-to-tail repeats of a sequence.

Confocal microscope A light microscope that uses the computer-processed output of a photodetector to generate an image of slices through a specimen.

Conjugation In bacteria, the transfer of DNA from a donor to a recipient cell in direct contact.

Conjugative plasmid A bacterial plasmid that encodes functions required for conjugation.

Consensus sequence A sequence derived from comparing a set of related sequences and incorporating the most commonly occurring base at each site.

Consortium An interdependent group of individuals of different types. Plural, consortia.

Continuous-culture device A device for the culture of microorganisms in liquid medium under controlled conditions, with continuous additions of fresh medium and removal of culture medium over a lengthy period of time. Also called a chemostat.

Copy number The number of molecules or plasmids present in a cell.

Cortical inheritance The inheritance of surface properties without the participation of nuclear genes.

Covalent modification The posttranslational chemical alteration of a protein by addition of a functional group that becomes linked to the protein by a covalent bond.

Crystalline surface layer A surface layer of some bacteria and archaea consisting of protein arrays, usually quite resistant to chemicals and proteases. Also called S-layer.

Cyan fluorescent protein A variant of green fluorescent protein that emits bluish light upon ultraviolet irradiation. Abbreviated CFP.

Cyanobacteria Oxygen-producing photosynthetic bacteria.

Cyclic AMP An adenosine monophosphate molecule with the phosphate covalently attached to both the 3′ and 5′ carbons of the ribose. Abbreviated cAMP.

Cyclic photophosphorylation Generation of ATP by phototrophs whereby an activated electron is passed through a membrane-located electron transport chain to chlorophyll in its ground state.

Cytokine A messenger protein (e.g., tumor necrosis factor or interleukin) released by white blood cells that is responsible for communication between cells of the immune system and the rest of the body.

Cytoplasm The cell contents excluding cell envelopes plus nuclei in eukaryotes and nucleoids in prokaryotes.

Cytoplasmic membrane *See* Cell membrane.

Cytoskeleton The intracellular structure of eukaryotic cells, composed of microtubules, microfibrils, and intermediate filaments.

Dam methylase The enzyme that transfers methyl groups into the adenine of GATC sites.

Decoating The release of viral nucleic acids from virions within a host cell.

Deconvolution microscopy A technique that applies algorithms to a stack of images of an object along its optical axis to enhance signals specific for a given image plane. Typically, deconvolution analysis is used to sharpen an image by removing out-of-focus light from a particular focal plane by using fluorescence excitation and emission.

Deletion The loss of one or more bases or base pairs from a molecule of DNA.

Dendrogram A tree-like representation of the relationships among taxonomic units.

Denitrification Bacterially mediated cascade of anaerobic respirations that converts nitrate ion to nitrogen gas.

Deoxyribonuclease An enzyme that degrades DNA. Abbreviated DNase.

Deoxyribonucleic acid A macromolecule usually made up of two antiparallel polynucleotide strands held together by weak hydrogen bonds and with deoxyribose as the component sugar. Abbreviated DNA.

Development The progression of changes occurring in an organism during maturation.

Differentiation The structural and functional changes that occur during development of cells and tissues.

Dikaryon An organism composed of cells each containing two genetically distinct nuclei. Typical of certain fungi.

Dimorphism The characteristic of some fungi to be able to switch between a yeast and a mycelial phase of growth.

Dinitrogenase The enzyme that catalyzes the conversion of nitrogen gas to ammonia.

Dinitrogenase reductases The homodimeric proteins that donate electrons to dinitrogenase.

Diploid An organism that contains pairs of each chromosome.

Dissimilatory nitrate reduction Reduction of nitrate by serving as a terminal electron acceptor in anaerobic respiration.

Divisome A complex of proteins involved in cell division found at the bacterial division site.

DNA *See* Deoxyribonucleic acid.

DnaA boxes Regions of DNA where the protein DnaA binds to initiate DNA replication.

DNA bending A crook that occurs in a DNA sequence as a result of binding to particular proteins.

DNA gyrase A topoisomerase that removes supercoils from DNA by first producing double-strand breaks and then sealing them.

DNA polymerases Enzymes that polymerize deoxyribonucleotides onto an existing polynucleotide chain using the complementary strand of DNA as a template.

DNA repair A variety of different mechanisms that remove or correct damaged DNA.

DNase *See* Deoxyribonuclease.

Domain (1) A discrete, independently folded region of a protein. Different functions of a multifunctional protein are usually localized in separate domains. (2) One of the three major taxons: Bacteria, Archaea, or Eukarya.

Downstream region A region of DNA that is further away in the direction of replication or transcription.

Driving force Collective term for energy and reducing power.

Duplication A region of DNA that is present in two copies. If present in an adjacent direct repeat, it is called a tandem duplication. It is also possible for the duplicated DNA to be present at distant sites on a chromosome.

***D* value** Decimal reduction time. The time required for a lethal treatment to kill 90% of a microbial population.

Eclipse period The period following viral infection, when no intact virions are present.

Edema Swelling of tissue by the outpouring of fluid in spaces between cells.

Electron acceptor A compound or atom that becomes reduced by taking up electrons.

Electron donor A compound or atom that becomes oxidized by losing electrons.

Electron transport chain The sequential oxidation-reduction of compounds embedded in a membrane that creates a proton gradient across the membrane.

Electrophoresis Movement, and thereby separation, of charged molecules in an electric field.

Electroporation A method for introducing DNA (or other large molecules) into cells by exposure to rapid pulses of high voltage, which causes the transient formation of pores in the cell membrane.

ELISA *See* Enzyme-linked immunosorbent assay.

Elongation cycle The portion of translation in which a peptide chain is elongated by the addition of a single amino acid residue.

Elongation factor A protein that participates in the elongation cycle of translation.

Embden-Meyerhof-Parnas pathway The catabolic pathway between a molecule of glucose and two molecules of pyruvate. Also called glycolysis. Abbreviated EMP pathway.

EMP pathway *See* Embden-Meyerhof-Parnas pathway.

Endocytosis Engulfment by a cell of extracellular material.

Endonuclease An enzyme that cuts DNA at sites within the molecule.

Endospore A metabolically inactive, nonreplicating form of certain bacteria, including those in the genera *Bacillus* and *Clostridium*. Endospores tend to be highly resistant to physical and chemical damage.

Endosymbiont A bacterium that resides within a host cell, usually in an obligate relationship.

Endotoxin The lipopolysaccharide in the outer leaflet of the outer membranes of gram-negative bacteria that is harmful to humans and other animals; most of the toxicity is mediated by lipid A.

End product inhibition *See* Feedback inhibition.

Enhancer A *cis*-acting regulatory sequence that can increase transcription from an adjacent promoter.

Enrichment culture A method of cultivating microbes designed to favor growth of a particular microorganism or type of microorganism from a large, complex natural population.

Enterochelin A particular siderophore (iron-binding compound).

Enterosome A vesicle resembling a carboxysome containing enzymes involved in special aspects of metabolism of some enteric bacteria and others.

Entner-Doudoroff pathway A two-enzyme pathway joining the pentose phosphate pathway to glycolysis that generates no ATP, reducing power, or precursor metabolites.

Enzyme-linked immunosorbent assay A diagnostic immunological test that uses an enzyme linked to an indicator antibody. Abbreviated ELISA.

Epigenetics The inheritance of a particular trait that is not encoded in the nucleotide sequence; for example, methylation of certain DNA sequences can influence gene expression.

Epitope A portion of an antigen recognized by an antibody-binding site. For protein antigens, an epitope is typically 5 to 8 amino acids. Gene fusions to add epitopes to a protein can be used to tag that protein.

Eukaryote An organism with a nuclear membrane and membrane-bound organelles (e.g., mitochondria) and a mitotic apparatus. The taxonomic name for the eukaryotes is Eukarya.

Evolutionary distance The quantitative expression of the phylogenetic relatedness of organisms.

Excisionase An enzyme involved in the excision of prophage DNA from the host cell genome.

Excision repair A DNA repair system that removes nucleotides from a damaged strand of DNA and then replaces them with a new tract of DNA synthesized using the undamaged complementary strand as a template.

Exergonic Yielding energy.

Exosporium A loosely fitting envelope surrounding the outer coat of a bacterial endospore.

Exotoxins Toxic proteins produced by certain bacterial pathogens.

Exponential growth The growth phase during which the number of cells in the population doubles repeatedly over the same time interval. Also called logarithmic growth. *See also* Balanced growth.

Exponential phase *See* Log phase.

Extremophiles Microorganisms that inhabit environments with extreme values of temperature, pressure, pH, or salinity.

Facilitated diffusion Movement of molecules across a membrane from higher to lower concentration mediated by proteins that permit the passage of specific molecules only.

Facultative organism An organism that can grow under different conditions. Commonly used to describe bacteria that can grow oxidatively under aerobic conditions and fermentatively under anaerobic conditions.

Fastidious Requiring more than the ordinary number of nutrients for growth.

F^- cell A cell that does not contain the F factor and hence is able to act as a recipient (female) in a conjugative DNA transfer in matings with F^+ or Hfr strains.

F^+ cell A cell that carries an F factor as an autonomous plasmid, which enables the cell to act as a donor (male) to transfer the F factor to a recipient (female) cell.

Feedback inhibition Regulation of a biosynthesis pathway through allosteric inhibition, usually of the first enzyme in the pathway by the product of the pathway. Also called end product inhibition.

Feeder pathways Pathways that convert a source of carbon into intermediates of central metabolism.

Fermentation (1) Scientific: an anaerobic process in which ATP is generated exclusively by substrate level phosphorylation. (2) Industrial: any microbial transformation, either aerobic or anaerobic.

F factor An *Escherichia coli* plasmid that encodes conjugative-transfer ("fertility") functions. The F plasmid may exist as an autonomous plasmid in the cytoplasm or may integrate into specific sites in the chromosome, producing an Hfr cell.

Filament The helical portion of the flagellum that rotates to provide motility.

Fimbriae Thin, proteinaceous filaments that extend from the cell surfaces of microbial cells and facilitate adhesion to solid surfaces or other cells. Singular, fimbria. Also called pili.

FISH *See* Fluorescence in situ hybridization.

5′ terminus The end of a polynucleotide that carries the phosphate group attached to the 5′ position of the sugar.

Flagella Long, flexible, helical protein structures that extend from the surface of a cell and cause motility by rotation. Singular, flagellum.

Flagellin The principal protein component of the bacterial flagellar filament.

Flow cytometry A technique for determining the distribution of a fluorescent chemical (naturally present or artificially introduced) in a population of cells; usually accompanied by a cell sorter function that separates cells with certain characteristics.

Fluorescence in situ hybridization Method of identifying microorganisms in natural environments by hybridization to fluorescence-labeled nucleic acid probes, usually fragments of DNA encoding ribosomal RNA. Abbreviated FISH.

Fluorescence microscopy Visualization of cells or their constituents when they are labeled with a fluorochrome and usually illuminated with ultraviolet light.

Fluorochrome A fluorescent chemical that absorbs light of one wavelength and gives off light of a higher wavelength.

Folding The posttranslational process through which a protein assumes its three-dimensional shape.

Forespore A developmental stage in the formation of an endospore at which a spore-like structure is first discernible.

Frameshift A mutation that adds or deletes 1 or 2 base pairs (or any nonmultiple of 3) from a coding sequence in a molecule of DNA, so that the genetic code is read out of phase. All codons translated downstream of a frameshift mutation will be misread, and frequently an out-of-frame stop codon will prematurely terminate translation.

Fruiting body A structure bearing or containing spores.

Fueling pathways Metabolic pathways that generate precursor metabolites, ATP, and reducing power.

Gel electrophoresis Electrophoretic separation of molecules using a gel, usually composed of agarose or acrylamide.

Gene The genetic unit of function. A gene may encode a polypeptide or a molecule of nontranslated RNA (e.g., ribosomal RNA, transfer RNA, or a regulatory RNA).

Gene dosage The number of copies of a particular gene. The position of a gene along a replicating circular chromosome affects gene dosage. In cells undergoing replication, genes closer to the origin will be present in higher numbers than those closer to the terminus.

Gene expression The process by which a gene product is produced. For genes that encode proteins, the gene must be transcribed into messenger RNA and then translated into protein. For genes that encode structural RNAs (ribosomal RNA, transfer RNA, etc.), the gene must be transcribed into RNA.

Generalized transduction A phage-mediated process of genetic exchange that occurs when reproductive errors result in incorporation of chromosomal genes into a phage head.

Generation time The amount of time required for one cell to divide, forming two daughter cells.

Genetic code The assignment of each of the triplet codons of messenger RNA to amino acids and translation stop signals.

Genetic competence The transient physiological state required for a bacterial cell to take up transforming DNA.

Genetic engineering The use of molecular techniques to produce DNA molecules containing new genes or new combinations of genes.

Genetic recombination The process by which a fragment of DNA from one molecule (chromosome, plasmid, or phage genome) is exchanged with or integrated into another molecule to produce a recombinant molecule(s).

Genome The complete genetic content of a cell or organism, including chromosomes, plasmids, and prophages.

Genomic island A large region of DNA that is present on the chromosome of an organism but absent from closely related organisms.

Genomics The analysis of genomic DNA sequences from one or several organisms. Genomic analysis can provide information about the evolution of genes and can make predictions about their function and the metabolism of an organism.

Genomic species *See* Phylotype.

Genotype A description of the genetic constitution of an organism. For simplicity, usually only relevant parts are described.

Germination The transition of a resting structure, such as a spore, to a vegetative cell.

GFP *See* Green fluorescent protein.

Gliding motility A form of nonflagellar motility on solid surfaces exhibited by some prokaryotes.

Global control system A system mediating global regulation.

Gluconeogenesis The metabolic synthesis of glucose from other nutrients. The term is usually applied to reverse flow through glycolysis.

Glycerol ether A lipid characteristic of the Archaea, containing isoprenoid lipids that are ether linked to glycerol.

Glycolysis *See* Embden-Meyerhof-Parnas pathway.

Glyoxylate cycle A modification of the tricarboxylic acid cycle in which one intermediate, isocitrate, is linked through glyoxylate to another intermediate, malate.

Gram-negative bacteria A group of bacteria with a cell envelope composed of an outer membrane surrounding a thin peptidoglycan layer.

Gram-positive bacteria A group of bacteria with a cell envelope composed of a thick peptidoglycan layer and no outer membrane.

Green fluorescent protein An intrinsically fluorescent protein from the jellyfish *Aequorea victoria*. Abbreviated GFP. Gene fusions with GFP-encoding DNA are commonly used for determining protein localization by fluorescence microscopy. *See also* Cyan fluorescent protein, Yellow fluorescent protein.

Greenhouse gas A gas that allows sunlight to enter the atmosphere but traps in the atmosphere infrared radiation (heat) that is reflected from Earth back to space. Water vapor, carbon dioxide, methane, and nitrous oxide are naturally occurring greenhouse gases; gases used in aerosols are human-made greenhouse gases.

Group translocation Entry of a compound into a cell with a simultaneous chemical alteration, such as phosphorylation.

Growth metabolism The aspect of metabolism primarily directed toward growth.

Growth rate The rate of increase of a cellular population per unit of time.

Half-life The time in which half the amount of a substance decomposes or is modified or half the population of an organism is killed.

Halophile An organism that grows in the presence of high concentrations of salt.

Haploid Having only one form of each chromosome in each cell. (Prokaryotes are generally haploid, although at higher growth rates more than one copy of a chromosome may be present.)

Heat shock response A global regulatory response resulting in increased or decreased expression of a number of genes in response to injury by heat, osmotic change, and certain other forms of stress.

Helical filament A viral capsule in which capsomeres are arranged as a spiral with a hollow core.

Helicase A protein that unwinds the DNA double helix.

Helper phage A phage that is introduced into a host cell in order to provide functions needed for replication, morphogenesis, or packaging of a mutant (defective) phage.

Hemimethylation Methylation of only one strand at a particular site in double-stranded DNA.

Heterocysts Specialized, oxygen-impermeable cells of some cyanobacteria in which nitrogen fixation occurs.

Heterotroph *See* Organotroph.

Hfr For "high frequency of recombination." A cell in which the F factor has integrated into a specific location in the chromosome, causing it to act as a high-frequency donor of chromosomal genes in crosses with F^- cells.

High-energy phosphate bond An anhydride or enol linkage, which is easily ruptured, releasing a phosphoryl group that readily enters into other reactions, thereby driving them.

Holdfast A structure that attaches an organism to a solid surface, for example, the structure that attaches stalked *Caulobacter* cells.

Homologous recombination The physical exchange of DNA between two homologous DNA molecules. It requires the RecA protein in enteric bacteria.

Homology (1) Sequence identity between two nucleotide sequences. (2) Genetic relatedness between two sequences of common ancestry. Commonly confused with sequence similarity, for example, 85% similarity means that 85 nucleotide positions out of 100 are identical in two polynucleotides.

Hook The structural component of a bacterial flagellum that couples the filament to the basal body.

Horizontal gene transfer Transfer of genes other than by reproduction. Also called lateral gene transfer.

Hormogonium A motile, chemotactic fragment of a filamentous cyanobacterium.

Hydrogenosome A membrane-bound organelle that converts pyruvate to acetate, CO_2, and H_2.

Hydroxypropionate pathway A CO_2-fixing pathway found in certain green phototrophic bacteria.

Hyperthermophile An organism that grows at extremely high temperatures.

Hypha A multicellular filament formed during the vegetative reproduction of fungi or actinomycetes. Plural, hyphae.

Icosahedron A regular geometric polyhedron with 20 equilateral triangular faces and 12 corners. This is a particularly stable structure. The capsids of many phages and viruses are icosahedral.

Immune competence The capability to function fully in the body's defense.

Immunity (1) The resistance of a lysogen to superinfection by a phage with a similar regulatory mechanism. (2) An adaptive antibody or cellular response against specific microbial infections.

Immunological memory Persistence of immunity long after exposure to an antigen.

Inclusion body A visible localized intracellular accumulation of a compound, such as polyphosphate, polyhydroxyalkanoate, or a protein made by hyperproduction in a genetically modified organism.

Incubation period The time between acquisition of a pathogenic agent and the appearance of symptoms.

Inducer A chemical or physical agent that turns on gene expression. Usually refers to an agent that alters repressor-operator interactions, often by decreasing the extent of DNA binding.

Inducible Describing a regulatory system in which the genes are expressed only under appropriate conditions (e.g., when the substrate is present or under specific environmental conditions).

Induction (1) The switching on of transcription in a repressed system due to the interaction between the inducer and a regulatory protein. For example, the *lac* operon is induced by adding lactose or isopropyl-β-D-thiogalactopyranoside. (2) A condition that causes a lysogen to begin lytic growth, as occurs when the *c*I repressor of phage lambda is inactivated following DNA damage.

Inflammation *See* Inflammatory response.

Inflammatory cytokine A hormone-like protein made by immune cells that stimulates the inflammatory response. Also called proinflammatory cytokine.

Inflammatory response The body's nonspecific reaction to injury or infection, consisting of redness, pain, heat, swelling, and sometimes loss of function. Also called inflammation.

Initiation complex The complex of proteins that promote the binding of ribosomes and the initiator transfer RNA to messenger RNA to begin the process of translation.

Initiation factors Proteins that promote the binding of ribosomes and the initiator transfer RNA to messenger RNA to begin the process of translation.

Innate immunity The portion of the immune system that is naturally present and does not require prior exposure to an antigen to be active.

Inoculum The cells used initially to start a culture or to infect a host.

Insertion element A transposable nucleotide sequence that encodes only the functions required for its own transposition. Insertion elements are typically less than 5 kilobases long. Also called insertion sequence; abbreviated IS.

Insertion sequence *See* Insertion element.

In situ A Latin phrase meaning "in the original place." Commonly used to describe a process that visualizes the position of a biological molecule in a cell.

In situ hybridization A technique for gene detection involving hybridization of a labeled sample of a cloned gene to a large DNA molecule (usually a chromosome), often within a cell.

Integrase A viral enzyme that promotes the integration of a proviral genome into a chromosome.

Introns Noncoding regions common within eukaryotic genes but also found (if rarely) in prokaryotes.

Inverted repeat A DNA or RNA sequence where the sequence of nucleotides along one strand of DNA is repeated in the opposite physical direction along the other strand. Inverted repeats are commonly separated by a tract of nonrepeated DNA.

In vitro A Latin phrase meaning "in glass." Commonly refers to a process that takes place in a test tube, outside the cell.

In vivo A Latin phrase meaning "in life." Commonly refers to a process that takes place inside the cell.

IS *See* Insertion element.

Isoelectric point The pH at which a molecule carries no net charge.

kb *See* Kilobase.

Kilobase 1,000 nucleotides of DNA or RNA. Abbreviated kb.

Kinase An enzyme that transfers phosphate from ATP to another molecule.

Kinetoplast A highly specialized mitochondrion located at the base of the cilium in ciliates.

Krebs cycle *See* Tricarboxylic acid cycle.

Lactoferrin A kind of iron-binding protein.

Lagging strand The strand of DNA that is copied discontinuously during replication.

Lag phase The phase of a microbial growth cycle before a culture begins to grow.

Lambda (λ) A temperate phage that infects *Escherichia coli*. Derivatives of lambda are widely used as cloning vectors.

Lateral gene transfer *See* Horizontal gene transfer.

Leader peptide The peptide that is encoded in front of operons that are regulated by attenuation.

Leader sequence The DNA sequence encoding a leader peptide.

Leader transcript The transcript of a leader sequence.

Leading strand The strand of newly replicated DNA that is synthesized continuously in the same direction as the replication fork. DNA synthesis proceeds in the 5′-to-3′ direction.

Leukocyte A white blood cell.

Lichen A symbiotic association between a fungus and an alga or a cyanobacterium.

Ligand A small molecule that binds to a molecule or a cell surface.

Ligase (DNA) An enzyme that joins a 3′ OH residue of a deoxyribonucleotide to the 5′ phosphate residue of an adjacent deoxyribonucleotide.

Linkage group A set of genes that tend to be coinherited.

Lipid A A phosphorylated glycolipid common to all bacterial lipopolysaccharides.

Lipopolysaccharide A major component of the outer layer of the outer membrane in gram-negative bacteria. Abbreviated LPS.

Lipoprotein A protein containing covalently bound fatty acids.

Lithotroph An organism that derives its ATP from energy obtained by oxidizing inorganic nutrients, such as ferrous ion, hydrogen sulfide, or hydrogen gas. Also called chemoautotroph.

Loading factor DnaC, a protein that helps bind the helicase (DnaB) to the origin of DNA replication.

Logarithmic growth *See* Exponential growth.

Logarithmic phase *See* Log phase.

Log phase The phase of the microbial growth cycle when growth is exponential. Also called logarithmic or exponential phase.

Low-energy phosphoryl bond A relatively stable ester linkage of a phosphate group to an organic molecule.

LPS *See* Lipopolysaccharide.

Lumen A space within a tube or tubelike structure.

Lymph nodes Small bean-shaped organs located along lymphatic vessels throughout the body.

Lymphocyte A white blood cell; various types participate in immune responses.

Lysate The contents of cells released after lysis.

Lysis Disruption of cells with release of their contents.

Lysogen A bacterial cell carrying a phage genome as a repressed prophage.

Lysogenic cycle The pattern of phage infection that involves integration of the phage DNA into the host chromosome.

Lysogeny The ability of a temperate bacteriophage to maintain itself as a quiescent prophage until it is induced into the lytic cycle.

Lysozyme An enzyme that hydrolyzes murein.

Lytic cycle The development of bacteriophage particles, either after infection of a host bacterium or after induction of a prophage, resulting in production and release of free progeny phage particles and lysis of the host cell.

Lytic phage A phage that can enter a lytic cycle only when it infects a sensitive bacterial cell.

Macronucleus The large nucleus of ciliates, containing repeated copies of selected genes.

Macrophages Large phagocytic cells found in many tissues that are part of the innate immune system and are derived from a monocyte.

Magnetosome A small particle found in magnetotactic bacteria consisting of magnetite or other iron-containing substances.

Magnetotaxis Motility of microbes along magnetic lines of force, such as Earth's magnetic field.

Major histocompatibility complex A class of proteins on the surfaces of mammalian cells that present antigens derived from intracellular pathogens. Abbreviated MHC.

MAMPs *See* Microbe-associated molecular patterns.

Membrane attack complex A cylinder-like protein complex of complement proteins that makes a hole through the cytoplasmic membrane and causes cell lysis.

Memory cell A lymphocyte involved in a specific immune response that remains quiescent as long as there is no further antigenic stimulation.

Mesophile An organism that grows best at moderate temperatures, around 37°C.

Messenger RNA The transcript of a segment of chromosomal DNA that is a template for protein synthesis. Abbreviated mRNA.

Methanogen A methane-producing archaeon.

Methylotrophs Microorganisms that utilize methane or other C_1 compounds.

Methyltransferase An enzyme that catalyzes the transfer of methyl groups from one molecule to another.

MHC *See* Major histocompatibility complex.

Microaerophilic Pertaining to microbes that grow best at lower concentrations of oxygen than are present in air.

Microarray An ordered set of DNA fragments fixed to solid surfaces. The DNA fragments may represent all the open reading frames in a genome, a particular gene family, or any other subset of genes. Sometimes called a gene chip or microchip.

Microbe-associated molecular patterns Unique microbial molecules, such as proteins, murein, or certain nucleic acids, that are recognized by specific host cell receptors. Abbreviated MAMPs.

Microbial infallibility The doctrine which postulates that there exists no naturally occurring organic compound that cannot be broken down by some microbe.

Microchip *See* Microarray.

Micronucleus The nucleus of ciliates that contains the entire genome.

Microradioautography A technique that reveals the presence of a radioactive isotope as photographic grain on a microscopically visible structure. This is done by overlaying the object with photographic emulsion, incubating it in the dark, and developing it.

Minimal medium A defined medium that provides only those nutrients needed to support the growth of a particular microbe.

Mismatch repair system A DNA repair system that detects and corrects base pairs other than A-T or G-C. Abbreviated MMR system.

Mitosis The normal process of nuclear division in a eukaryote, whereby nuclear division occurs on a spindle structure without reduction in the chromosome number.

MMR system *See* Mismatch repair system.

Mobile element A sequence of DNA that is able to promote its own transposition; an insertion sequence or a transposon.

Modulator A protein or nucleic acid that changes the activity of an enzyme by reversibly associating with it.

Modulon A group of independent operons subject to a common regulator even though they are members of different regulons.

Molecular clock The measurement of presumed regular, clocklike changes that occur in a sequence of DNA.

Monoclonal antibody An antibody molecule produced by a specific lymphocyte clone that reacts with a single epitope.

Monocyte A large phagocytic cell with an oval or horseshoe-shaped nucleus found in the circulation and capable of differentiating into a macrophage.

mRNA *See* Messenger RNA.

Multifork replication The mode of replication seen in fast-growing bacteria, in which a new round of replication starts before a previous round has finished.

Murein The form of peptidoglycan found in the walls of bacteria.

Mutagen A chemical or physical agent that increases the frequency of mutation.

Mutagenesis The induction of mutations.

Mutant An organism with an altered base sequence in one or several genes. Usually refers to an organism with a mutation that causes a phenotypic difference from the wild type.

Mutation Any heritable alteration in the base sequence of the genetic material.

Mutation rate The number of mutations per cell division. Sometimes the mutant frequency is used instead of the mutation rate.

Mutualism A symbiotic relationship beneficial to both partners.

Mycelium A mass of hyphae produced by many fungi and actinomycetes.

Mycolic acid A long-chain organic acid found in the waxy cell envelope of mycobacteria.

Mycoplasmas A group of small, wall-less bacteria that includes some pathogens of animals and plants.

Myxospores Spores formed by myxobacteria.

NAD *See* Nicotinamide adenine dinucleotide.

Nematodes Roundworms, a phylum of worms or helminths.

Neo-Lamarckism Recent proposals that certain environmental stresses direct formation of compensatory mutations.

Neutrophil *See* Polymorphonuclear leukocyte.

Nicotinamide adenine dinucleotide A chemical reservoir of reducing power that acts by accepting and donating pairs of hydrogen atoms. Abbreviated NAD.

Nitrification The successive conversion of ammonia via nitrite to nitrate ion.

Nitrogenase The enzyme that converts (fixes) nitrogen gas to ammonia.

Nitrogen cycle Conversions of the nitrogen-containing compounds that occur in nature.

Nodulation factor A substance that signals a plant to initiate nodule formation.

Nomarski differential interference contrast microscope A microscope that exploits differences in refractive index within a specimen to form a pseudo-three-dimensional image.

Noncyclic photophosphorylation Generation of ATP from light energy by passage of an excited electron from chlorophyll through an electron transport chain to DNA.

Nonsense codon A codon that does not code for any amino acid but signals a termination of translation, or punctuation. The three nonsense codons are UAG (amber), UAA (ochre), and UAG (opal).

Nonsense mutation A mutation that replaces an amino acid-encoding codon with a nonsense codon.

Normal biota *See* Normal flora.

Normal flora The microbial population normally present on animals. Properly, normal biota.

Nuclease An enzyme that cleaves phosphate-deoxyribose bonds within (endonuclease) or at the end of (exonuclease) a nucleotide sequence.

Nucleoid The condensed organization of a prokaryote chromosome inside the cell.

Nucleosome The basic structural unit of eukaryotic chromosomes, consisting of nearly two turns of DNA wound around a core of four kinds of histones.

Nucleotide A nucleoside (ribose-bearing purine or pyrimidine) with an attached phosphate group.

Numerical taxonomy A system of taxonomy that involves measuring a large number of traits and weighing them equally.

O antigen A polysaccharide side chain on the lipopolysaccharide of gram-negative bacteria. It serves as the receptor for some types of phage.

Okazaki fragments Short DNA fragments of about 1,000 to 2,000 nucleotides formed during DNA replication of the lagging strand by discontinuous replication of DNA and later joined together by ligation.

Oncogene A gene responsible for eukaryotic cell transformation, a cancer-related process.

Open complex A complex that forms during transcription in which the DNA double helix has been opened, forming a bubble within which RNA polymerase is attached to one of the strands.

Open reading frame A stretch of DNA that potentially codes for protein. Abbreviated ORF.

Operator The DNA sequence in which a repressor protein reversibly binds to regulate the activity of one or more closely linked structural genes; the term is of less significance now that the great variety of potential sites for regulators is appreciated.

Operon A sequence of adjacent genes read as a single, polycistronic messenger RNA. Changes in the level of transcription thus affect all of the genes in an operon, so such genes are often coordinately regulated.

Opines Unusual amino acids produced by plants with a gall caused by the crown gall-forming *Agrobacterium tumefaciens.*

Opsonin A substance that facilitates phagocytosis.

ORF *See* Open reading frame.

Organotroph An organism that utilizes organic nutrients as a source of carbon and energy. Also called heterotroph.

Origin of replication The nucleotide sequence where DNA replication is initiated (*ori*).

Outer membrane The outer lipid bilayer of gram-negative bacteria, consisting of an outer lipopolysaccharide leaflet and an inner phospholipid leaflet plus proteins.

Oxidation-reduction reaction A reaction in which one molecule loses electrons and another gains them. Also called redox reaction.

Oxidative phosphorylation ATP formation resulting from aerobic respiration.

Oxygenic photosynthesis Photosynthesis carried out by plants, algae, and cyanobacteria that produces oxygen as a product.

Palindromic sequence A nucleotide that reads the same in both directions, in analogy to language palindromes (such as "Madam, I'm Adam").

Pandemic A worldwide epidemic.

Parasitism A symbiotic relationship in which the host is harmed and the parasite apparently benefits.

Passive transport Diffusional passage of a compound across a membrane.

PBP *See* Penicillin-binding protein.

PCR *See* Polymerase chain reaction.

Pelagic Resident within a water column. *See also* Planktonic.

Penicillin An antibiotic that inhibits cross-linking of peptidoglycan chains in the cell walls of bacteria. A large variety of penicillin derivatives are available, e.g., ampicillin.

Penicillin-binding protein An enzyme that binds penicillin and also links murein strands. Abbreviated PBP.

Pentose phosphate pathway Catabolic pathway that produces pentose phosphates from glucose.

Peptide bond formation Formation of the bond that links amino acids in a peptide.

Peptidoglycan A polymer consisting of glycan and peptide. The term is commonly used as a substitute for murein, the common constituent of bacterial walls.

Periplasm The region between the cytoplasmic membrane and the outer membrane in gram-negative bacteria.

Permease An enzyme system functioning in the transport of specific substances, usually nutrients, through the cytoplasm membrane.

Phage *See* Bacteriophage.

Phagocytosis Engulfment of a particle or a cell by another cell.

Phagolysosomal vesicle A vesicle arising from the fusion of a phagosome (phagocytic vesicle) with a lysosome.

Phagosome A membrane-bound vesicle of a phagocytic cell containing phagocytized material.

Phase-contrast microscope A compound microscope that creates contrast in transparent objects by exploiting differences in refractive index within the specimen.

Phenotype The appearance or other observable characteristics of an organism.

Pheromone A secreted chemical that affects the behavior of other cells or organisms.

Phosphodiester bond A diester bond between phosphate and a hydroxyl group of a sugar molecule. The linkage (—O—PO_2—O—) between nucleotides in nucleic acids.

Phospholipid bilayer A membrane consisting of two leaflets, each composed of phospholipid.

Phosphorylation Chemical modification of a molecule by addition of a phosphate group.

Phosphorylation cascade A series of reactions in which a phosphate group is passed from one compound (usually a protein) to the next.

Photoautotrophs Organisms that derive ATP and reducing power from light energy and carbon from CO_2.

Phototroph An organism that generates ATP and reducing power from light energy.

Phylogeny The evolutionary history of species.

Phylotype A group with at least 94% DNA sequence identity. Also called genomic species.

Phytoplankton Phototrophic microbes floating in oceans.

Pigment antenna A complex of pigments in phototrophs that gathers light energy to be fed to chlorophyll.

Pili *See* Fimbriae. Singular, pilus.

Pilin A structural protein component of pili (fimbriae).

Pinocytosis Cellular uptake of soluble material by engulfment.

Planktonic Free-floating; usually refers to microbial cells in bodies of water. *See also* Pelagic.

Plasma cell An antibody-producing cell.

Plasma membrane *See* Cell membrane.

Plasmid A molecule of extrachromosomal DNA existing as an autonomous replicon in the cytoplasm. Most plasmids are covalently closed circular DNA, although examples of linear plasmids are known.

Point mutation A mutation involving the substitution, addition, or deletion of a single base pair.

Polycistronic Referring to a stretch of DNA or messenger RNA that encodes several discrete gene products.

Polyhydroxyalkanoate A carbon storage polymer of alpha hydroxycarboxylates.

Polymerase chain reaction A method for amplifying a particular region of DNA by a repeated sequence of denaturation, annealing of specific primers, and synthesis. The concentration of the amplified DNA fragment increases exponentially with each cycle. Abbreviated PCR.

Polymerization Production of a macromolecule by joining of monomers.

Polymorphonuclear leukocyte The most abundant class of white blood cells in the circulation, with lobate nuclei. They are short-lived and phagocytic.

Polyribosome A messenger RNA molecule being transcribed by more than one ribosome. Also called polysome.

Polysome *See* Polyribosome.

Precursor metabolites The 13 compounds from which all carbon constituents of a cell can be synthesized.

Primase An enzyme that synthesizes ribonucleotide primers for lagging-strand DNA synthesis of Okazaki fragments.

Primer A short oligonucleotide complementary to a strand of DNA or RNA that is used to initiate synthesis of the complementary DNA strand (i.e., to extend the primer). The primer provides a 3′-OH end, which is required by DNA polymerases to initiate synthesis of the complementary DNA.

Prion An infectious agent composed only of protein.

Probe A fragment of DNA labeled with radioactivity or with a fluorescent moiety used to hybridize to another DNA molecule to identify complementary base sequences.

Progenote A putative ancestral organism from which all extant organisms are derived.

Proinflammatory cytokine *See* Inflammatory cytokine.

Prokaryote An organism lacking a nuclear membrane and certain organelles, such as mitochondria. Includes both Bacteria and Archaea.

Proofreading The process by which errors during protein and DNA synthesis are detected and corrected.

Prophage A temperate phage genome whose lytic functions are repressed and which replicates in synchrony with the bacterial chromosome.

Prosthecate Bearing an appendage (prostheca).

Protease An enzyme that degrades proteins.

Protein domain *See* Domain.

Protein export Transport of protein out of the cell.

Protein families Groups of phylogenetically related proteins.

Protein secretion Transport of a protein across the cell membrane.

Protein superfamilies Groups of phylogenetically related protein families.

Protein-synthesizing system The collection of enzymes, cofactors, and ribosomes needed for protein synthesis. Abbreviated PSS.

Proteobacteria The largest phylum in the domain Bacteria.

Proteome The complete array of proteins made by an organism.

Protists Common (nontaxonomic) collective term for protozoa and microscopic algae.

Proton motive force Potential energy stored in the form of a proton gradient.

Protoplast A cell bound by the cytoplasmic membrane and a portion (usually unknown) of the envelope material exterior to it. If significant amounts of envelope remain, the structure is known as a spheroplast.

Prototroph A strain of a microorganism that has the same nutritional requirements as it did when first isolated from nature.

Protozoa Nonphotosynthetic unicellular eukaryotes.

Pseudomurein The form of peptidoglycan found in the cell walls of some archaea.

PSS *See* Protein-synthesizing system.

Psychrophile An organism that grows at low temperatures.

Purple membrane A patch of the cell membranes of halophilic archaea in which a primitive form of photosynthesis occurs.

Pus A mixture of dead white blood cells, tissue fluids, and microbes.

Quorum sensing The chemical process by which populations of microbes sense their density as a result of each cell secreting a small amount of a certain compound.

Reading frame The way in which a sequence of nucleotides that code for a polypeptide are read as consecutive triplets.

RecA The protein, encoded by the *recA* gene, that is essential for homologous recombination. The RecA protein is also involved in the induction of the SOS response and the induction of lambda prophage in response to DNA-damaging agents.

Recombinant DNA A molecule of DNA in which DNA fragments from different sources are covalently joined.

Recombinase An enzyme that mediates recombination.

Recombination Genetic exchange resulting from a crossover between two different DNA molecules or different regions of a DNA molecule. *See also* Homologous recombination, Illegitimate recombination, Site-specific recombination.

Redox reaction *See* Oxidation-reduction reaction.

Reducing power The sum of reduced nicotinamide adenine dinucleotide (NADH) and reduced nicotinamide adenine dinucleotide phosphate (NADPH).

Reductive poise The fraction of pyridine nucleotides that is in the reduced state, i.e., (NADH + NADPH)/(NAD^+ + $NADP^+$).

Reductive tricarboxylic acid cycle The oxaloacetate-to-2-oxoglutarate portion of the tricarboxylic acid cycle flowing in that direction.

Regulatory site A region of the genome that mediates a regulatory effect.

Regulon A group of genes or operons located at different positions on the chromosome that respond to a common regulatory protein.

Release factor A protein that facilitates separation of a nascent polypeptide from the ribosome that synthesized it. Abbreviated RF.

Repair synthesis DNA synthesis that fills gaps formed during DNA repair.

Replica plating A technique for transferring a particular pattern of bacterial colonies from one agar-containing petri dish (the master plate) to one or more other plates.

Replication The process of duplicating a DNA molecule.

Replication fork The region on a replicating double-stranded DNA molecule where synthesis of new DNA is taking place. The replicating fork produces a Y-shaped region in the DNA molecule where the two strands have separated and replication is taking place.

Replicon A DNA molecule that is able to initiate its own replication. A replicon must have an origin of replication and usually also has the regulatory information required for the proper initiation of DNA replication.

Replisome A complex comprising the elements of the DNA-replicating machinery.

Repression Switching off the expression of a gene or a group of genes in response to a chemical or other stimulus.

Repressor A gene product that negatively regulates gene expression. Usually refers to a DNA-binding protein that inhibits transcription under certain conditions.

Resolving power In microscopy, the ability to tell that two objects are distinct.

Response regulator The member of a two-component regulatory system that senses the state of the other component and mediates regulation.

Restriction The cleavage of double-stranded DNA by an endonuclease (restriction enzyme). The restriction enzyme distinguishes between self and foreign DNA based upon the modification of its DNA-binding site (for example, by methylation).

Restriction endonuclease An endonuclease that cuts double-stranded DNA by binding to specific sites, many of which are arranged in palindromes. Also called restriction enzyme.

Restriction enzyme *See* Restriction endonuclease.

Retroviruses A family of viruses whose reverse transcriptase converts their RNA genome into DNA as an obligatory part of their life cycle.

Reverse transcriptase An enzyme that can synthesize a strand of DNA complementary to an RNA template and that is used to make complementary DNA from messenger RNA.

Reversion A mutation that restores the wild-type phenotype of a mutant.

RF *See* Release factor.

R factor *See* R plasmid.

Rhizosphere The region of soil surrounding a plant's roots.

Ribonuclease An enzyme that hydrolyzes RNA molecules.

Ribosomal RNA An RNA molecule that forms part of the structure of a ribosome. Abbreviated rRNA.

Ribosome An RNA-protein complex responsible for the correct positioning of messenger RNA and charged transfer RNAs, allowing proper alignment of amino acids during protein synthesis.

Ribosome binding site *See* Shine-Dalgarno sequence.

Ribozyme A catalytic molecule of RNA; analogous to an enzyme, which is a catalytic molecule of protein.

Ribulose bisphosphate carboxylase The enzyme in the Calvin cycle that mediates CO_2 fixation. Abbreviated RuBisCo.

RNA polymerase An enzyme responsible for the synthesis of RNA from its constituent ribonucleotides (NTPs), using one strand of DNA as a template.

RNA replicase A virus-encoded RNA-dependent RNA polymerase.

Rolling-circle replication A type of DNA replication in which a replication fork moves around a circular DNA molecule, producing a single-stranded concatamer (much as toilet paper peels off the roll). The resulting single-stranded DNA may become double stranded by the synthesis of a complementary strand.

Root nodule A swelling on the root of a plant within which symbiotic nitrogen-fixing bacteria are housed.

R plasmid A transmissible plasmid that carries genes that code for resistance to one or several different antibiotics. Also called R factor.

rRNA *See* Ribosomal RNA.

RuBisCo *See* Ribulose bisphosphate carboxylase.

Rumen The first vessel of the multichambered stomach of ruminants, such as cattle, sheep, and goats, where much of the plant food is degraded and fermented by microbes into volatile fatty acids that can be used by the animal.

Sacculus The three-dimensional bag made up of murein (peptidoglycan). Plural, sacculi.

Scalar reaction A reaction that adds to a cell's proton gradient by utilizing protons on the inside or producing them on the outside, in contrast to vectorial reactions that form a gradient by transporting protons across a membrane.

Scanning electron microscope An electron microscope in which an image is generated by scanning the surface of the specimen with an electron beam.

Scanning tunneling microscope A microscope that views surfaces of semiconducting materials by processing a signal generated by a flow of electrons to a probe held close to the specimen.

Secondary active transport Active transport of substances using proton motive force.

Secondary metabolite A metabolic end product that an organism produces after growth has ceased.

Secreted protein A protein that is exported through the cytoplasmic membrane. Most secreted proteins have specific signal sequences that promote interaction with the export apparatus in the membrane.

Selection Use of a condition in which only mutant or recombinant cells with a particular phenotype will grow and divide.

Selective medium A growth medium that allows growth of only certain species, strains, or mutants.

Semiconservative replication DNA replication in which each new double helix is composed of one new and one old (conserved) strand.

Sense strand The strand of DNA that has the same nucleotide sequence as the messenger RNA (except that the DNA has T residues where the RNA has U residues).

Sensor The protein component of a two-component regulatory system that detects a regulatory signal.

Sequencing The process of determining the sequence of monomers in a macromolecule, e.g., a protein or a nucleic acid.

Sessile A term describing an organism that adheres to a surface and is not free to move about (derived from Latin "to sit").

Sheath A long transparent polysaccharide tube produced by some bacteria.

Shine-Dalgarno sequence A sequence of bases in a messenger RNA molecule that determines where a ribosome will bind to initiate translation.

Shuttle vector A vector that can replicate in the cells of more than one organism (e.g., in both *Escherichia coli* and yeast).

Siderophore A small molecule of bacterial production that binds iron with great avidity.

Sigma factor A protein that functions as a subunit of bacterial RNA polymerases and is responsible for recognition of promoters. Different sigma factors allow recognition of different promoter sequences.

Signal peptide A sequence of amino acids on the N terminus of a protein that targets it for secretion. Also called signal sequence.

Signal sequence *See* Signal peptide.

Signal transduction Alteration in the conformation of a protein that regulates expression of other genes. Initially, signal transduction was used to refer to extracellular conditions that alter the conformation of a membrane protein and cause it to relay the regulatory signal inside the cell, but more recently, the term has been broadly applied to a variety of regulatory cascades.

Simple diffusion Diffusion that is not regulated or mediated by a protein.

Single-hit kinetics Kinetics characterized by linear decay of the logarithm of the concentration of a compound or of the number of live organisms with time.

Single-strand displacement A recombination process in which one strand of DNA double helix is replaced by a single strand of DNA.

Single-stranded DNA-binding protein A small basic protein that has a high affinity for single-stranded DNA. This protein protects single-stranded DNA from nuclease attack and inhibits it from reannealing into double-stranded DNA. Abbreviated SSB protein.

Site-directed mutagenesis A method for introducing specific mutations at a defined site in a nucleotide sequence.

Site-specific recombination Genetic exchange that occurs between particular short DNA sequences not requiring RecA. Each site-specific recombination system requires unique enzymes that catalyze the genetic exchange but little sequence homology between the two DNA molecules.

S-layer *See* Crystalline surface layer.

Slime layer A thin slimy or gummy layer that surrounds many prokaryotic cells.

Small RNA One of several RNA molecules with regulatory properties. Abbreviated sRNA.

Social motility Coordinated motility of cells of myxobacteria.

Sodium motive force A concentration gradient of sodium ions across the cell membrane.

Solute transport Movement of solutes across the cell membrane.

SOS response The coordinate induction of many genes in response to certain types of DNA damage. Many of the induced gene products facilitate repair of the damaged DNA, but the repair processes result in a high frequency of mistakes in the repaired DNA, a process often called error-prone repair.

Specialized transduction A method of gene transfer between bacteria in which a specific region of the bacterial donor DNA is carried by a phage. The host DNA carried by a specialized transducing phage arises by aberrant excision of a prophage. Thus, only regions of DNA adjacent to an integrated phage can be transferred by this method.

Specific growth rate Expression of growth rate as the differential rate constant.

Spheroplast An osmotically sensitive cell whose cell wall has been partially removed. *See also* Protoplast.

Spirillum A corkscrew-shaped microbe.

Spore coat The outer layer of an endospore.

Sporulation The process of forming a spore.

sRNA *See* Small RNA.

SSB protein *See* Single-stranded DNA-binding protein.

Stationary phase The postgrowth stage of a microbial culture.

Stimulon A group of genes whose expression responds to the same stimulus.

Stop codon *See* nonsense codon.

Strains Individual clones of a species that are distinguishable genetically or by having been isolated separately from nature.

Stringent response The ability of a bacterium to limit the synthesis of transfer RNA and ribosomal RNA during amino acid starvation. The compounds ppGpp (guanosine 3′-diphosphate 5′-diphosphate) and pppGpp (guanosine 3′-diphosphate 5′-triphosphate) are at least partially responsible for the stringent response.

Structural gene A gene encoding a polypeptide or an RNA molecule, in contrast to a regulatory gene.

Substrate A chemical utilized by an enzyme or an organism.

Substrate level phosphorylation ATP formation from the reaction between ADP and a metabolic intermediate with a high-energy phosphate bond.

Sulfur cycle The chemical conversion of sulfur compounds that takes place in nature.

Supercoiled Pertaining to double-stranded circular DNA in which either overwinding or underwinding of the duplex makes the circle twist; the conformation of a covalently closed circular DNA molecule, which is coiled by torsional strain into the shape taken by a wound-up elastic band.

Superoxide dismutase The enzyme that catalyzes the breakdown of superoxide (O_2^-) into hydrogen peroxide and oxygen.

Surfactin A lipopeptide wetting agent. Some are synthesized and secreted by swarming prokaryotic species.

Swarming Coordinated movement of cells of certain bacteria.

Symbiosis The living together of two different kinds of organism.

Symport The moving together of two molecules across the cell membrane. Usually the concentration of one of them drives the movement of the other.

TCA cycle *See* Tricarboxylic acid cycle.

Teichoic acid A molecule composed of glycerol or ribitol units linked by phosphate groups. Found in the walls of gram-positive bacteria.

Temperate phage A phage that is capable of becoming a prophage in the bacterial host (i.e., that can maintain itself in a quiescent state).

Temperature-sensitive mutant A mutant with a mutation that results in functional expression of a gene within a certain temperature range (e.g., at less than 30°C) but that is nonfunctional at a different temperature (e.g., at 42°C).

Template A single-stranded polynucleotide (or region of a polynucleotide) that can be copied to produce a complementary polynucleotide.

Terminal electron acceptor A compound at the end of an electron transport chain, e.g., oxygen for aerobic respiration.

Termination codon *See* Nonsense codon.

Terminator A DNA sequence that results in termination of transcription.

Terminator utilization substance A protein that binds to *ter* sequences at the termination site of bacterial DNA replication and inhibits DnaB helicase. Abbreviated Tus.

Terminus The region of DNA sequences where DNA replication terminates.

Ternary complex A complex composed of three molecules.

Tetracycline An antibiotic that inhibits protein synthesis by preventing aminoacyl transfer RNA from binding to ribosomes.

T helper cell A T lymphocyte that activates the functions of other lymphocytes.

Thermoacidophile An organism that grows at high temperature and low pH.

Thermophile An organism that grows at high temperatures.

3′ terminus The end of a polynucleotide that carries the hydroxyl group attached to the 3′ position of the sugar.

Thylakoid A membrane structure of chloroplasts where light energy is captured and converted into chemical energy.

TLR *See* Toll-like receptor.

Toll-like receptor A receptor on cells of the immune system that recognizes microbe-associated molecular patterns (MAMPs). Different receptors are specific for various kinds of MAMPs.

Topoisomerase An enzyme which affects supercoiling of the DNA circular duplex by causing a nick, rotating the strands, and then religating them.

Toxin A poisonous compound, commonly a protein, produced by some microbes.

Transamination The transfer of an amino group from one molecule to another.

Transcript A strand of RNA transcribed from a DNA template.

Transcription The synthesis of RNA from a DNA template catalyzed by RNA polymerase.

Transcriptional unit A region of DNA (a gene or an operon) transcribed into a single RNA molecule.

Transcriptome The complete set of RNA transcripts made by a cell under a particular condition. Typically determined by microarray analysis.

Transducing particle An aberrant phage virion that contains bacterial DNA and is thereby capable of mediating transduction by attacking another cell.

Transduction A method of gene transfer between bacteria in which the bacterial donor DNA is carried by a phage. There are two types of transduction: generalized and specialized.

Transfer RNA Adaptor molecules that translate the triplet code from the messenger RNA sequence into the corresponding chain of amino acids. Abbreviated tRNA.

Transformation Genetic exchange resulting from transfer of naked DNA from one cell to another.

Transhydrogenases Enzymes that transfer hydrogens between the reduced and oxidized forms of nicotinamide adenine dinucleotide and nicotinamide adenine dinucleotide phosphate (NAD^+ and $NADP^+$).

Translation The assembly of amino acids into polypeptides using the genetic information encoded in the molecules of messenger RNA.

Translocation Movement from one place to another.

Transposase An enzyme (or enzyme complex) required for the transposition of a particular transposable element. A transposase must recognize specific sites on the ends of a transposon, cut the transposon out of the original site, and insert the transposon into a new site.

Transposition The movement of a discrete segment of DNA from one location in the genome to another.

Transposon A genetic element that, in addition to encoding the proteins required for its own transposition, confers one or more new observable phenotypes (often resistance to one or more specific drugs) on the host cell.

Tree of life A dendrogram showing the phylogenic relationships among all groups of organisms.

Tricarboxylic acid cycle A cyclic metabolic pathway that oxidizes acetate, generates ATP, and forms four precursor metabolites. Also called Krebs cycle. Abbreviated TCA.

Trigger factor A ribosome-associated chaperone encountered by a newly made polypeptide; trigger factor is a multidomain protein with a peptidyl proline isomerase activity. It mediates *cis-trans* conversions of proline peptidyl bonds in the growing polypeptide.

tRNA *See* Transfer RNA.

Tuberculin test A skin test carried out by injecting a small amount of tuberculin, a mixture derived from tubercle bacilli. A positive reaction is seen after 12 to 48 hours as a hard swelling of the skin due to a delayed hypersensitivity reaction. It indicates previous exposure to the agent, not necessarily active disease.

Tus *See* Terminator utilization substance.

Twitching motility A form of pilus-dependent motility through which some bacteria can move by pulling themselves across a solid surface.

Two-component regulatory system A regulatory system consisting of a sensor that detects a signal and a response regulator that effects a control.

Type III secretion A contact-dependent secretion system found in some bacterial pathogens by which proteins are injected (translocated) directly into the cytosol of eukaryotic host cells.

Uniport A secondary active-transport system through which a single cation is driven into the cell or an anion is driven out of the cell by the transmembrane proton gradient.

Universal ancestor The presumed ancestor of all extant organisms.

UP element *See* Upstream region.

Upstream region A region in front of (in the opposite direction from that of transcription) the promoter. Also called UP element.

Vector A replicon that is useful for cloning DNA fragments so that they can be amplified or transferred to other cells; commonly a derivative of a plasmid, phage, or virus.

Vectorial reaction A reaction that drives a reactant into or out of a cell, as when a proton is transported.

Vertical gene transfer Transfer of genes by replication and cell division.

Viability The ability of an organism to grow and divide.

Viable count The determination of the number of living cells in a population, usually carried out by placing aliquots of dilutions on agar plates and counting the resulting colonies.

Viral induction Formation of viable virus particles by excision of a provirus from a chromosome.

Viral latency The persistence of a viral genome in a host without symptoms of disease.

Virion A virus particle.

Viroid A small circular molecule of single-stranded RNA without a capsid; viroids cause many plant diseases and hepatitis D.

Virulence The relative ability of an organism to cause disease.

Virulence factor Any gene product that enhances the ability of an organism to cause disease.

Virulent phage A bacteriophage that always grows lytically.

Virus A small obligate intracellular parasite that comes apart into its component proteins and nucleic acid during its replication.

Western blot A test in which proteins are transferred from an acrylamide gel onto a membrane filter for detection with specific antibodies.

YAC *See* Yeast artificial chromosome.

Yeast A unicellular form of fungi. Because the species is so widely used in fermentation and bread making, the term often refers to *Saccharomyces cerevisiae.*

Yeast artificial chromosome A cloning vector that contains sequences from a yeast chromosome required for DNA replication and segregation. Often used for cloning very large fragments of DNA. Abbreviated YAC.

Yellow fluorescent protein A variant of green fluorescent protein that emits bluish-yellowish light upon ultraviolet irradiation. Abbreviated YFP.

YFP *See* Yellow fluorescent protein.

Zoonoses Human diseases caused by pathogens that normally reside in animals. Singular, zoonosis.

Zygote A cell arising from the fusion of gametes.

Answers to Study Questions

Chapter 1

1. Consider using the following as guiding principles. All life forms are derived from microbes. Microbes are more phylogenetically diverse than plants and animals, are enormously abundant, and are found in virtually every place on earth where there is liquid water. In addition, microbes are involved in all the cycles of matter that are essential for life, transform the geosphere, and even affect the climate. Microbes participate in countless symbiotic relationships with animals, plants, and other microbes, and they can cause disease.

2. You may point out the following. Microbes are wonderful model systems for learning about all living things, including humans. Microbes carry out chemical activities of major industrial importance and can be engineered for production of useful proteins, e.g., vaccines and drugs, or for enhancing food production and preservation. They are used for bioremediation of polluted sites. Microbes can be used for malevolent purposes, such as biological warfare and bioterrorism. Moreover, the ones that inhabit our bodies when we are healthy serve as part of our protection against pathogenic microbes.

3. The larger the cells, the smaller the surface/volume ratio, which suggests that uptake of nutrients by diffusion and excretion of spent metabolites is slower than in small cells. Some of the largest bacteria can be expected to possess special characteristics for efficient transport of solutes or to carry out this process slowly. Size also dictates how many cells can be packed in a given volume; the smaller the cells, the greater the number, and vice versa.

4. It is generally accepted that eukaryotic cells arose by the ingestion of endosymbiotic prokaryotic cells, which became the present-day organelles mitochondria and chloroplasts. The many helpful and hurtful interactions between prokaryotes and eukaryotes have affected and will continue to affect the evolution of both kinds of organisms. This reflects the intimate connection between microbes and all other forms of life.

Chapter 2

1. Gram-positive bacteria protect their cytoplasmic membranes with a tough cell wall that consists of many layers of a peptidoglycan (murein) capable of excluding many kinds of toxic compounds. Gram-negative bacteria have a thin cell wall, but outside it they have a special protective outer membrane whose outer bilayer leaflet consists of lipopolysaccharides instead of phospholipids. In addition, attached to this layer is a large amount of polysaccharide chains (O antigens) that exclude possibly toxic hydrophobic compounds.

2. The fact that these two groups differ in the chemical compositions of their envelopes invites speculation that this situation may reflect the organisms' environmental preferences. However, we do not know enough about the envelopes of the two groups to make definitive statements. Extremophiles are represented more by archaea than bacteria, but many bacteria thrive in extreme environments as well. The ether-linked lipids of the archaea may be more heat resistant than the ester-linked ones of the bacteria. In addition, some archaea have monolayer membranes, which may also reflect greater thermal stability.

3. The entry of compounds smaller than 700 daltons into gram-negative bacteria occurs through hydrophilic channels. The pores are made by trimers of special proteins called porins.

4. The outer layer of acid-fast bacteria consists of waxes known as mycolic acids arranged in a lipid bilayer. This layer is highly resistant to organic solvents and acids.

5. Crystalline S-layers are widely distributed among Bacteria and Archaea. In some Archaea, they are the only layer outside the cell membrane. They constitute an outer layer of these cells, made up of a single kind of protein molecule that is highly resistant to protein-denaturing agents and proteolytic enzymes.

6. Capsules and their more loosely knit cousins, slime layers, protect microbes from phagocytosis, desiccation, and other noxious events. In some species, they facilitate the adhesion of the organisms to surfaces, such as tooth enamel.

7. Bacterial flagella consist of filaments made up of a single kind of protein (flagellin, which varies in composition between organisms) plus a hook structure that connects the filament to a basal body. The basal body is inserted into the cell membrane of gram-positive bacteria, and in the gram-negative bacteria, into their two membranes. Flagella are organs of locomotion in many prokaryotes, although some can move by mechanisms that do not involve these structures.

8. Pili (fimbriae) consist of straight protein filaments, thinner and often shorter than flagella. They are involved in attachment of bacteria to host cells and other surfaces, resistance to phagocytosis, the transfer of proteins and nucleic acids to other cells, and motility. Not all pili do all these things, and most are specialized. They are also antigenic and elicit an immune response in the host. They are found mainly in gram-negative bacteria. Unlike flagella, they grow from the inside of the cell outward. Although pili have a central channel, it is too small for pilins (the pilus proteins) to pass through it.

Chapter 3

1. The two major components inside prokaryotic cells are the nucleoid and the cytoplasm. Missing are the mitochondria, nucleus, Golgi apparatus, endoplasmic reticulum, plastids, and mitotic apparatus—elements that are almost universal among eukaryotic cells. The cytoplasm of prokaryotes appears to be denser than that of eukaryotes, and molecules diffuse faster in the former than in the latter.

2. Nucleoids lack a nuclear membrane and the nucleosome organization of chromosomes (they lack histones). Most prokaryotes contain a single chromosome, which is usually circular. Some prokaryotes have linear chromosomes like those of eukaryotes, but their ends lack the characteristic structure of telomeres. Nucleoids look like irregular blobs of DNA and do not divide using the machinery of a mitotic apparatus.

3. It can become supercoiled (which can be used to regulate gene expression) and has no need to protect ends from exonucleases or to solve the problems connected with their replication (as in linear chromosomes).

4. By having gas vesicles that fill with gas under some circumstances. This permits photosynthetic bacteria during daytime to reach upper layers in the water column, where there is more light.

5. Gas vesicles, photosynthetic vesicles, carboxysomes, and enterosomes. Gas vesicles function to change the buoyant density of the cells. The functions of carboxysomes and enterosomes are not well understood, but they may function to facilitate biochemical reactions by dense packing of enzymes.

Chapter 4

1. Direct counts (with a microscope counting chamber or an electronic particle counter) tell the total number of cells. Viable counts (plate counts) tell the number of living cells.

2. The growth rate of a bacterial culture can be determined by measuring over time any relevant parameter, such as cell mass by turbidimetry, viable count by plate count, and total cell number by particle count, dry weight, or any chemical cell constituent. These measurements will be the same as long as the culture is in balanced growth.

3. A culture in balanced growth is in a reproducible steady state, whereas all other conditions are transient and hard to reproduce. As a consequence, any property of cells in balanced growth will increase by the same factor as any other, allowing the statement in the answer to the previous question, that to determine the growth rate of such a culture, one can use any property that can be conveniently measured.

4. In the lag phase, the cells adjust to a new medium. In the exponential phase, growth is unhindered. In the stationary phase, growth slows down due to either the accumulation of toxic metabolites or the exhaustion of needed nutrients.

5. A chemostat, or continuous-culture device, consists of a culture vessel kept at a desired temperature and aerated in the case of the growth of aerobes. Fresh medium is added at a rate set by a metering pump or a valve that regulates the flow rate. The volume is kept constant by removing medium through an overflow device at the same rate as fresh medium is added. For a chemostat to function properly, the bacterial density must not exceed the one that allows balanced growth in a batch culture. This condition is achieved by making an essential nutrient limiting, e.g., by reducing the concentration of some essential nutrient, such as glucose, ammonia, phosphate, or a required amino acid.

6. Some Bacteria and Archaea have evolved to tolerate and thrive in environments at extremes of temperature, pH, hydrostatic pressure, or concentration of salt (see Table 4.2). Some even tolerate more than one of these environmental challenges at the same time.

Chapter 5

1. There are various acceptable answers based on different assumptions about the other planet and its alien life form. A few of the many possibilities: (i) shortage of nitrogen may favor the exclusive use of RNA (or inorganic metal complexes) rather than proteins as catalysts; (ii) the life form may be so primitive, and the environment so rich in organic molecules, that only spontaneous polymerization reactions make up their metabolic activity; (iii) energy input to the planet from its sun may be harvested solely by photochemical reactions external to the life form, which then utilizes these "high-energy" compounds directly in metabolism; (iv) life on the planet may be acellular, and all metabolism may occur on surfaces other than the membranes familiar to us Earthlings.

2. The biosphere is described as the sum of all places on Earth in which living things exist and (presumably) grow. All these locations are occupied by prokaryotes, and in many of these sites, prokaryotes are the only life form.

3. Kinds of starting materials and products are shown in the table.

Metabolic process	Starting materials	Products
Fueling	Environmental inorganic and organic molecules; light	NAD(P)H ATP 13 precursor metabolites
Biosynthesis	NAD(P)H ATP 13 precursor metabolites Sources of nitrogen and sulfur	Amino acids Purine and pyrimidine nucleotides Vitamins and cofactors Sugars and other carbohydrates Fatty acids
Polymerization	Amino acids Purine and pyrimidine nucleotides Vitamins and cofactors Sugars and other carbohydrates Fatty acids ATP	Proteins RNA DNA Carbohydrates Phospholipids
Assembly	Proteins RNA DNA Carbohydrates Phospholipids	Membranes Wall Nucleoid Pili Flagella Polyribosomes
Cell division	Membranes Wall Nucleoid Pili Flagella Polyribosomes	Two cells from one

Of these processes, the only one that conceivably could be dispensable in whole or in part is biosynthesis, assuming that its products can be supplied preformed in the environment.

Chapter 6

1. The ability of microbes to utilize light as energy, and inorganic as well as organic molecules as sources of energy and carbon, far exceeds the fueling capability of animals and plants (see all the tables in Chapter 6).

2. Reducing power and energy are equivalent in cell biology because they are interconvertible through forward and reverse electron transport. Given this interconvertibility, reducing power and energy can be regarded collectively as the driving force of the cell.

3. In substrate level phosphorylation, a "low-energy" phosphoryl bond in an organic metabolite is converted into a "high-energy" one by oxidation of a metabolite, involving the transfer of electrons, usually to NAD^+. The phosphoryl group can then be used to convert ADP to ATP. Transmembrane ion gradients can be used to produce ATP by the action of the enzyme F_1F_0 ATP synthase, which uses energy from the passage of protons (or other ions) down a concentration gradient. Most commonly, the gradient is established from the passage of electrons down an energy gradient in respiration (using membrane electron transfer systems) or photosynthesis (using light energy to strip electrons from water or other compounds. Less commonly, the gradients are established by ion pumps or scalar reactions (using the energy of decarboxylation or other chemical reactions not involving separate electron carriers).

4. Microbes have evolved at least four mechanisms to establish transmembrane ion gradients: (i) ion (usually proton) secretion using the energy of electrons as they are passed from carrier to carrier in membrane-bound electron transport systems; (ii) ion secretion using the energy of electrons activated by light; (iii) direct secretion of ions by membrane-bound enzymes that utilize the energy of certain chemical reactions, such as decarboxylation (ATP hydrolysis by F_1F_0 ATP synthase is a special case of this sort); and (iv) scalar reactions that, for example, consume protons within the cell. The chief mechanism whereby a transmembrane ion gradient generates ATP is through the agency of the integral membrane enzyme complex, F_1F_0 ATP synthase.

5. Though there are many variations, involving different photosensitive pigments, different pathways, and different structures for ensuring efficiency of the whole process, photosynthesis involves sunlight-induced change in a pigment molecule, raising the energy level of its electrons from a basal state to an activated state, following which the activated electron can reduce some oxidized molecule, thereby creating chemical energy.

6. The forms of autotrophic and heterotrophic fueling reactions among the prokaryotes far exceed in number and chemical diversity those in plants and animals. This greater diversity is summarized in Table 6.1 and illustrated in all the tables of Chapter 6.

7. Some of the diversity has to do with the nature of the transported solute, but the great variety must also reflect the variety of selection pressures on prokaryotes in their diverse environmental niches. One example is particularly instructive: the group transfer reaction costs no extra energy because the ATP expended in bringing glucose and other sugars into the cell as their phosphorylated derivatives would have had to be expended in the first steps of glycolysis anyway. Thus, one finds this mode of entry prevalent among fermenting anaerobes, which have a harder time obtaining metabolic energy than do aerobes.

8. In one sense, the central pathways are certainly more similar among all the prokaryotes than are the other areas of metabolic fueling, but one must recognize that even among the seemingly constant reactions of glycolysis, the pentose phosphate pathway, and the TCA cycle, there is much flexibility and variety of function. These pathways are constructed in ways that make them reversible and able to function in partial segments. Examples are shown in Fig. 6.16.

Chapter 7

1. Biosynthetic pathways are costly. Genetically, they require structural genes for the enzymes and regulatory genes (or other mechanisms) to control them. In terms of energy, biosynthesis costs both ATPs and reducing power to make most of the building blocks. Environments that consistently supply preformed building blocks will favor the growth of organisms that dispense with the redundant machinery to make these compounds.

2. The longest is that leading to tryptophan (12 enzymes); the shortest is that leading to alanine (1 enzyme).

3. Your answer could use any of several pathways (leading to an amino acid or nucleotide or another building block) to illustrate (i) a precursor metabolite(s) as starting material, (ii) use of ATP and NADPH, (iii) a linear path or a branched one leading to multiple products, (iv) a building block(s) as a product(s), and (v) feedback to one or more early steps controlling flow through the pathway.

4. Most likely it was the need to achieve two opposite goals: the need to maintain low concentrations of the oxidized form for use as a hydrogen acceptor in fueling reactions and the need to maintain high concentrations of the reduced form for use in biosynthetic reactions.

5. All biosynthesis of nitrogenous compounds depends on the fixation of atmospheric nitrogen to make ammonia available, a uniquely prokaryotic process.

6. Glutamate is key in the assimilation of nitrogen. Glutamate and, to a lesser extent, glutamine donate their nitrogen to the various pathways of biosynthesis of building blocks. Quantitatively, glutamate is by far the more important: about 90% of the cell's nitrogen flows through it. Cysteine is key in the assimilation of sulfur. L-Cysteine serves directly or indirectly as the source for most other sulfur-containing compounds in the cell (for example, L-methionine, biotin, thiamine, and CoA).

7. Two features distinguish the assimilation of phosphorus: the process occurs in fueling rather than biosynthesis reactions, and it does not involve oxidation or reduction.

8. A major difference is that nucleic acid building blocks are synthesized as deoxynucleoside triphosphates (deoxyadenosine triphosphate [dATP], deoxyguanosine triphosphate [dGTP], deoxycytidine triphosphate [dCTP], and ribosylthymine triphosphate [TTP]); amino acids, on the other hand, must be activated by ATP after synthesis before being assembled as polypeptides.

Chapter 8

1. Several remarkable characteristics can be cited. (i) The structure of the translation apparatus is a model of speed and efficiency: a polycistronic mRNA coated with ribosomes making protein while the mRNA itself is still being made by RNA polymerase. (ii) There is no need to transport the mRNA from one cell compartment (nucleus) to another (cytoplasm). (iii) All RNA synthesis is performed by a single RNA polymerase. Two other streamlining features, not emphasized in Chapter 8, can be pointed out: (i) the ribosomal subunits are smaller than those of eukaryotes, and (ii) prokaryotic mRNAs have no poly(A) tails.

2. Similarities include identical building blocks (dATP, dGTP, dCTP, and TTP); semiconservative replication, with each chain acting as a template for synthesis of its complement; similar needs for a replication fork, primer synthesis, and formation and ligation of Okazaki fragments; and existence of multiple growing points. Differences include the need in eukaryotes to unpackage the DNA by separating it from histone before replication and much larger chromosomes and much slower progression of replication forks in eukaryotes, which necessitate many more replication forks than in prokaryotes and which initiate from multiple points.

3. DNA methylation functions in replication initiation. The enzyme Dam methylase places a methyl group on the A of GATC sequences, but it takes several minutes for this to occur after synthesis of a new strand, during which time the DNA is hemimethylated. The hemimethylated segments become bound by the protein SeqA, thereby preventing inappropriate initiations at the origin. DNA methylation also functions in the marking ("branding") of a cell's DNA. The cell protects its own chromosomal DNA by methylating an adenine or a cytosine residue in the recognition site

of its own restriction endonuclease. This modification by methylation marks the DNA as self and protects it from destruction.

4. The trick is accomplished by having multiple replication forks at high cell growth rates. Each cell inherits a partially replicated chromosome, thereby shortening the time needed to prepare for the next cell division.

5. In essence, the prokaryotic cell wipes out its blueprint for protein synthesis every few minutes, enabling it to completely change which proteins will be made in the next interval. For cells that are subject to rapid, frequent, and massive changes in environmental conditions, this feature helps ensure adaptability.

6. Prokaryotes have no single promoter sequence, either within one organism or among different species. Instead, there are general patterns of sequences rather than unique nucleotide sequences that spell "promoter" to RNA polymerase. Promoters can be said to have a core flanked by upstream and downstream regions. The upstream region is the site of action of regulatory proteins; the downstream region includes the start of the coding sequences of the gene. The core has three parts: a −35 hexamer and a −10 hexamer (the minus signs indicate distance upstream from the codon initiating translation of the gene) separated by a 17-bp spacer.

7. Constant conflict is a characteristic of the prokaryotic chromosome. The DNA replication forks collide head-on with RNA polymerase molecules moving in the opposite direction and overtake and collide with the rear of RNA polymerase molecules moving in the same direction but much more slowly than the replisome. Head-on collisions are the worst; they halt replication, if only briefly, and abort transcription. Same-direction collisions have less impact; they only slow replication and permit continued transcription.

8. Prokaryotic transcripts encode information for making more than one polypeptide and hence are said to be polycistronic. The significance of polycistronic messages is the opportunity they provide for efficiency and speed in making proteins that have related functions (such as enzymes in a single pathway) and for coregulating their syntheses.

9. An authentic start AUG codon is preceded approximately 10 nucleotides upstream by a sequence of 4 to 6 bases that is complementary to the 3′ end of the 16S rRNA of the 30S ribosomal subunit. These bases, called the Shine-Dalgarno sequence, are believed to help position the 30S ribosomal subunit at the proper site by hydrogen bonding with the 16S RNA. Internal AUG codons, of course, lack this positioning function.

10. The polyribosomes would be of different sizes and with different numbers of ribosomes, because transcription and translation are coupled, and translation begins before transcription is completed (Fig. 8.14).

11. The rate of synthesis of mRNA is approximately 45 nucleotides per second (in *E. coli*). Since ribosomes advance along the mRNA at a speed of 15 codons per second, and since a codon consists of 3 nucleotides, the ribosomes move along the message at the same rate that the message is being made. Thus, each mRNA molecule is coated with a parade of ribosomes busily translating the growing mRNA, and the ribosome leading the parade is keeping pace with RNA polymerase at the transcription bubble. One consequence of this synchrony is that, as long as translation is proceeding, there is never a large segment of mRNA exposed to nucleases. A faster RNA polymerase could cause great difficulties for the cell, since mRNA would be subject to premature degradation.

12. Many features of protein export demand that the solution be complex and diverse. Proteins differ in their structures and susceptibilities to entering membranes and in their ultimate locations in the cell envelope. The gram-positive and gram-negative body plans provide very distinct challenges to protein translocation. Finally, the fact that type III protein secretion results in direct injection of the protein into a host eukaryotic cell provides an elegant example of a highly specialized process.

Chapter 9

1. The most often used ones consist of examining a single cell during its growth, determining the distribution of a given property of single cells in a population (e.g., by flow cytometry), and synchronizing a population (e.g., by using a baby machine).

2. DNA replication proceeds (bidirectionally) from the origin to the terminus. In *E. coli*, the process takes about 40 min at 37°C. In fast-growing cells, initiation begins once per cell cycle, leading to multifork replication (see the answer to question 8.5).

3. Bacterial cell division involves invagination of the cell membrane and the cell wall and, in the case of gram-negative bacteria, the outer membrane as well. Only when these processes are completed can progeny cells separate. Most gram-negative bacteria divide by making a constriction of their envelope layers at midcell, which is similar to what is seen with most animal cells. In some gram-positive cocci and rods, cell division proceeds without apparent constriction of their girth, which resembles cell division in plants.

4. Cell septum formation in most prokaryotes involves a protein called FtsZ, which makes a constricting ring where the cell septum will eventually form. This structure, called the Z-ring, closes gradually as the septum forms. Other proteins are involved in this process as well.

5. Cell division and DNA replication in bacteria are linked in time in the sense that one does not take place when the

other is inhibited. The two systems "talk" to each other via molecular signals. Thus, when DNA is damaged and cannot replicate, a protein called SulA is made and inhibits the formation of the FtsZ ring needed for cell division.

6. The mechanism whereby the Z-ring is formed precisely in the middle of the cell is not well understood. The choreography of various Min proteins helps explain why the septum is normally not formed elsewhere along the cell length. Models have been proposed to explain how the oscillation of Min proteins may serve as a kinetic measuring device to define the cell middle.

7. Chromosome segregation in growing bacteria begins before replication has ended, unlike what is seen in eukaryotic cells. Segregation requires the migration of the replicative origin from the cell center to a location toward the poles, where the septum for the future cell division will be made.

Chapter 10

1. **a.** Such a strain would be able to serve as a donor but not as a recipient. No recombinational event participates in the formation of a generalized transducing particle, which involves fragmentation of the host cell chromosome and insertion of the pieces into a phage head. However, recombination is required in order to act as a recipient, because as is the case with most types of genetic exchanges among prokaryotes, only a small bit of DNA enters the recipient, and such a small piece is extremely unlikely to be a replicon. To become a replicon and thereby become a part of the recipient's genome, it must be inserted by recombination into one of the recipient's preexisting replicons.

b. Such a strain would not be able to serve as either a donor or a recipient of chromosomal genes. In order for chromosomal genes to be transferred, they must be attached to the F plasmid by a recombinational event, and (for the reasons stated above) a second recombinational event is necessary for them to become a part of the recipient's genome.

2. There are many possible pairs of phenotypes and genotypes. One, for example, might be described as having a phenotype of an auxotroph or a tryptophan auxotroph. Such a strain would have suffered a mutation that inactivated one of the genes, for example, *trpA*, encoding one of the several steps of tryptophan biosynthesis. That would be its genotype, which could be described at various levels of precision, from identifying the mutated gene to identifying the actual change of bases of the encoding DNA.

3. Smaller colonies on the plate might be suspected to have developed from an abortive transductant, but the proof would be to show that only one cell in a colony would be able to grow on the minimal medium. Thus, if the entire colony on the plate were picked and plated on a fresh plate of minimal medium and only one colony developed, that colony certainly would have developed from an abortive transductant.

4. Yes. By shifting all reading frames, all downstream codons will change. With high probability, some of these will be nonsense codons.

5. A mutagenized culture would be grown in a medium in which ribose is the sole source of carbon, and then penicillin would be added. Because penicillin kills only growing cells and the desired mutant cannot grow in this medium, it will survive. Wild-type cells can grow and will be killed. As a result, the desired mutant will be enriched.

Chapter 11

1. Biological evolution is the appearance of new species as a consequence of natural selection.

2. **a.** The sequence of bases in the DNA of extant organisms provides more information about relationships among microbes because the fossil record of microbes is so sparse. Also, the fossils reveal only morphology, and the morphology of microbes is inadequate to distinguish species.

b. Probably not, because at least so far, only sequences of extant organisms can be determined.

c. The fossil record of microbes establishes that microbes were present on Earth by a certain early date, and it establishes the morphologies of these early microbes.

3. The gene must be present in all the organisms to be compared, and it must evolve slowly enough so that some similarities between even distantly related organisms are preserved.

4. If one set of genes were transferred horizontally within the group and the other were not, comparing the similarities of their sequences would suggest different evolutionary relationships.

5. Because RNA can serve as a repository for accumulated genetic information, as well as catalyze chemical reactions, one can conceive of a primitive cell existing during an early era of evolution when DNA and protein had not yet evolved.

6. Cyanobacteria and chloroplasts of plants produce oxygen, and there is strong evidence that chloroplasts are derived from endosymbiotic cyanobacteria. Similarly, mitochondria confer on eukaryotes the ability to utilize oxygen, and there is strong evidence that mitochondria are derived from endosymbiotic oxygen-utilizing bacteria.

7. Particular activities of proteins are mediated by specific domains. Reshuffling preexisting domains would be a more rapid process than evolving together the various domains within a single protein.

Chapter 12

1. (i) Economy. Bacteria cannot afford to make useless or redundant enzymes, nor can they survive without producing essential enzymes; both situations call for rapid adjustment of enzyme synthesis. (ii) Rapid growth. By turning off the synthesis of an enzyme, a bacterial cell will halve the level of that protein with each doubling of cell mass. If the generation time is 20 minutes, the level can be reduced eightfold in an hour. (iii) mRNA degradation. Bacterial mRNAs turn over rapidly, and hence, the blueprints for protein synthesis are completely renewed every few minutes, affording a marvelous opportunity to control enzyme synthesis rapidly at the transcriptional level. (iv) Multicistronic operons. The multicistronic nature of bacterial mRNA facilitates unitary control of whole pathways or other groups of functionally related proteins. By a single adjustment of the synthesis of one mRNA, the cell can change the rate of synthesis of an entire pathway.

2. Modulation of protein activity by allosteric interactions is the faster control; it can occur almost instantaneously. If all control were exerted by changing the activity of existing enzymes, cells would lose the great economy of avoiding synthesis of redundant or unnecessary enzymes.

3. (i) Covalent modification affords the opportunity to modulate protein activity in cases where there might not be an appropriate organic compound (metabolite) to serve as the allosteric ligand. (ii) Covalent modification makes it possible to make quantitative adjustments (rather than simple on-off switching) in activity, as in the case of multiple adenylylation of glutamine synthetase; the adenylylated forms are less sensitive to allosteric control. (iii) In the case of phosphorylation of proteins, the phosphate group can be transferred from one protein to another to create signal transduction, as in the chemotaxis pathway or the scores of two-component sensor-response systems.

4. A useful index of a cell's energy status is the energy charge of the cell, defined as follows:

$$\text{Energy charge} = \frac{([\text{ATP}] + [\text{ADP}]/2)}{([\text{ATP}] + [\text{ADP}] + [\text{AMP}])}$$

In general, ATP-replenishing (fueling) pathways are inhibited by high levels of energy charge, and ATP-utilizing (biosynthesis and polymerization) pathways are stimulated.

5. (i) Attenuation, in which the ability of ribosomes to translate a leader sequence affects further transcription, and (ii) translational repression, in which unassembled ribosomal proteins block initiation of translation.

6. The most expensive ways are those that regulate the cellular level by varying the rate of degradation of a protein, rather than its synthesis. The least expensive ways are those that block the earliest step of gene expression: initiation of transcription.

7. At a minimum, transcriptional repression involves at least one regulatory gene (encoding the repressor protein) and a site in the DNA for it to bind. Attenuation requires no regulatory genes, simply the small leader sequence.

8. Control regions located distant from the operon they control are called enhancers, because they usually stimulate initiation. A DNA sequence can control a distant promoter by the bending of DNA to bring the bound regulatory protein to the promoter region, where it can influence initiation. Bending does not happen spontaneously, but is facilitated by DNA-bending proteins.

9. Most, if not all, regulatory proteins (repressors and activators) are allosteric proteins that can be modulated in activity by the binding of specific ligands (inducers or corepressors).

10. (i) In some cases, the number of genes to be coregulated is too great to be accommodated within a single workable operon. (The 150 or so genes that encode parts of the translation machinery are a good example.) (ii) Some processes involve genes that must be independently regulated and also be subject to an overriding, coordinating control. (Genes for utilizing different carbon sources, for example, must be under specific control, depending on the presence of the particular growth substrate, and be subject overall to catabolite repression.)

11. A regulon is a group of independent operons governed by the same regulator, usually a protein repressor or activator. A modulon is also a group of independent operons governed by the same regulator, but a modulon is a higher-order regulatory unit because its group of independent operons includes some that are parts of different regulons (illustrated in Fig. 12.10).

12. Modulons that are very large (i.e., consist of many operons and regulons) and that affect very many metabolic functions are referred to as global regulatory systems. Catabolite repression and the stringent-response system are two good examples cited in the text. One might add the RpoS stationary-phase cascade as well.

Chapter 13

1. Stress is a technical term referring to the effect on a microbial cell of any significant change in the physical (temperature, pressure, etc.), chemical (presence of inorganic and organic nutrients and toxic substances, etc.), or biological (presence of interacting or competing organisms) environment.

2. If the bacterial species is thermophilic, the inoculation will provoke a cold shock response. If the bacterial species is psychrophilic, a heat shock response will ensue. (Note: if you answered this question by stating that it matters at what temperature the inoculum had been grown, you deserve extra credit. Consult your instructor for a reward; the authors are not in a position to provide rewards.)

3. (i) Sensor. The usual sensor is an allosteric protein, commonly an integral membrane protein, since its function is to sense the external environment. (ii) Response regulator. The regulator is commonly a cytoplasmic allosteric protein that modulates (adjusts) transcription by binding to the DNA that governs the expression of a set of target operons tied together in a regulon. This regulon produces the responding proteins that bring about the cellular response. (iii) Feedback controls match the cellular response to the problem, i.e., to the magnitude and length of the period of stress.

4. Both proteomic and transcriptional monitoring can reveal the components of a stimulon. The former involves recognizing the responding proteins displayed on two-dimensional polyacrylamide gels; the latter involves recognizing the mRNA that produces these proteins as displayed on gene-loaded microarrays.

5. UP elements are special sequences upstream of several of the seven rRNA operons of *E. coli* that provide especially strong binding for RNA-P. The protein Fis (factor for inversion stimulation) binds to sites near the UP sequences and attracts RNA-P to the promoter; it is a positive controlling factor. The protein H-NS (histone-like nucleoid structuring protein) affects DNA bending and is a negative controlling factor because it inhibits rRNA promoter activity. The guanosine triphosphate derivative ppGpp (guanosine tetraphosphate) is the effector molecule of the stringent-response modulon and has the ability to totally inhibit the synthesis of rRNA and tRNA. When a nutritional restriction slows growth, ppGpp accumulates quickly and changes RNA-P in such a way that it cannot function at promoters of rRNA, perhaps by favoring the association of RNA-P with sigma (σ) factors other than σ^{70}. Also, conditions that limit the cells' ability to synthesize nucleoside triphosphates, the substrates of RNA-P, will of course also restrict RNA synthesis. Complex and redundant controls can be expected because of the importance of a regulatory system that controls the bulk of macromolecule synthesis. (In *E. coli* during fast growth in rich media, rRNA transcripts alone constitute over half the RNA made by the cell. In addition, there are 100 or so proteins constituting the PSS, including initiation and elongation factors, aminoacyl-tRNA synthetases, and naturally, the ribosomal proteins.)

6. Cells in stationary phase are smaller than growing cells. They are also tougher than growing cells and harder to lyse. Metabolism is altered in ways that promote survival rather than synthesis for growth. The overall chemical composition of stationary-phase cells is different, and there are changes in the chemical and/or physical nature of each cellular component (the appendages, outer membrane, and periplasm of gram-negative bacteria, and the murein wall, cell membrane, nucleoid, and ribosomes). To a first approximation, it matters little what caused the cells to stop growing; the end result seems almost the same. This approximation, however, is far from the whole story; a cell in stationary phase as a result of depletion of glucose will not be identical to one that has been poisoned by hydrogen peroxide or a heavy metal, but the differences will be reflected largely in the profile of enzymes made under the two conditions, not the appearance and overall chemical toughening of the cell.

7. A regulatory circuit of enormous complexity guides the process of turning a growing cell into a nongrowing stationary-phase cell. Part of the process involves sRNA molecules regulating translation initiation. These sRNA molecules (DsrA, RprA, and OxyS) act as integrators; they process signals from different stresses and integrate them into a single, coherent action. One critical action is to stimulate the synthesis of σ^s, the sigma factor directing RNA-P to operons having a distinctive promoter. σ^s is the major player in directing entry into the stationary phase. Association of σ^s with core RNA-P diminishes the amount associated with σ^{70} and in this way diverts the polymerase from transcribing operons related to growth to those related to survival in the nongrowth state. (σ^s is not the only regulator of stationary phase. Other global regulators, such as H-NS, CRP, Lrp, and FhlA, modulate and fine-tune the expression of individual operons of the σ^s-dependent stationary-phase regulon. A network of this sort, in which regulators govern regulators, which in turn may govern other regulators, is called a regulatory cascade.)

8. Genome size generally reflects the extent to which an organism faces environmental stresses. The more stresses an organism faces, the larger its genome and the greater the fraction of its genome devoted to regulatory genes.

9. Usually not all bacteria in a given population may be killed by a particular deleterious treatment. If a large (99%) or even very large (99.99%) fraction of a microbial population is killed, the survivors may still be a large number. Killing 99% of 10^8 cells leaves a million (10^6) cells still alive, ready to contaminate other surfaces or to multiply and quickly restore the original number if conditions are favorable.

10. Bacteria sense gradients by continually comparing the concentration of an attractant solute with what their memory tells them it was a short time ago. Binding of an attractant at a receptor near the cell surface alters the endogenous routine schedule of runs and tumbles. It does so by interrupting a phosphorylation cascade that governs the direction of rotation of the flagellar motors, which has the effect of prolonging the run. Accommodation by a methylation system restores the endogenous schedule and resets the cell's sensitivity to the attractant to require a higher concentration to prolong the run. (Accommodation accounts for many properties of human perception, such as our gradual failure to smell an odor after being in its presence for an extended time.) In effect, accommodation con-

stitutes a molecular memory that enables change in attractant concentration to be sensed and ensures progress toward it.

11. Many bacterial species possess the ability of quorum sensing. The cells produce and secrete a small, diffusible molecule called an autoinducer. (Many autoinducers are acyl-homoserine lactones.) In simple cases, the secreted autoinducer simply diffuses back into the cell. When a sufficient number of cells (a quorum) are growing in a confined space, the concentration of autoinducer diffusing back into the cells reaches a threshold amount and triggers certain responses. In many cases, the autoinducer, rather than diffusing back into the producing cell, is sensed by a typical sensor kinase on the cell surface, triggering the transmission of a phosphorylation signal to some response regulator protein. Quorum sensing is used by bacteria to determine when to initiate processes that benefit from large populations. Preparation for invasion of pathogenic bacteria is one example of a quorum-triggered process.

12. Biofilms are natural communities of bacterial cells of one or more species tightly adhering to each other, and usually to some solid surface. These sessile bacteria form multicellular associations, which may develop complex channels, compartments, and other structures of some architectural complexity. Biofilms facilitate the capture of nutrients that are in extremely low concentration; they afford protection from deleterious agents, such as antibiotics; and they enable the cells to colonize areas and not be washed away by liquid currents.

Chapter 14

1. Differentiation to form these two structures is irreversible; neither can dedifferentiate again to become vegetative cells.

2. Exospores form by cell division at the tips of filaments; endospores form within cells. Exospores are primarily agents of dispersal; endospores are primarily agents of survival.

3. Two: its own and that of the spore mother cell.

4. The kinase activity, which is the first component of the cascade, consists of three separate proteins, each regulated by a different environmental signal. The level of Spo0A~P, the active product of the cascade, as well as those of phosphate intermediates of the cascade, is modulated by three phosphatases, each of which responds to different environmental signals.

5. Swarmer and stalked cells, the differentiated cell types of *Caulobacter*, each contribute to the bacterium's ability to compete in nature. By their ability to stick to solid surfaces, stalked cells prevent being swept away in moving currents of their aqueous environment; the motility of swarmer cells provides a mechanism to move to new environments.

6. Yes. Experiments utilizing protein labeling show that certain regulatory proteins are restricted to certain regions of the cell.

7. They allow myxobacteria to kill their prey.

8. Being packaged in sporangioles increases the probability that a group of myxospores will germinate in the same spot and be able to participate in group feeding, which is critical to their survival.

Chapter 15

1. The most dramatic difference is that eukaryotic, but not prokaryotic, species constitute breeding groups; the two types of species share the property of being groups of similar organisms with few intermediates to related species.

2. Their huge population, rapid multiplication, and prevalence in a diversity of environments lead to immense genetic variation among prokaryotes, which indicates that a vast number of prokaryotic species must exist.

3. The term is used to indicate that the species has not been cultivated and, thus, to warn that it does not meet the requirements for designating a species.

4. The many profound biochemical differences between Bacteria and Archaea and the similarity, in certain respects, of the latter to eukaryotes suggest that these two groups are not closely related. Therefore, sharing a prokaryotic cell organization does not necessarily imply relatedness.

5. Archaea are ecologically unique in the abilities of some of their members to grow in extremes of high concentrations of salt and high temperature; they are biochemically unique in respect to the composition of the lipids that form their cytoplasmic membranes.

6. Most probably the bubbles are methane, and an anaerobic layer at the bottom of the pond contains methanogens.

7. *D. radiodurans* has an extraordinary ability to repair the double-strand breaks in DNA that radiation causes.

8. The Ti plasmid carries genes that cause infected plants to synthesize nutrients that *A. tumefaciens* can utilize and a protective housing.

Chapter 16

1. Yes and no. Fungi are a broad but reasonably well-defined group, but protists include a wide variety of totally distinct organisms (e.g., protozoa, algae, and diatoms). Examples of fungi are yeast, *Candida*, mushrooms, and molds. Examples of protists are small unicellular algae (e.g., *Chlamydomonas*, *Chlorella*, and *Euglena*), amoebas, malarial parasites, *Giardia*, *Paramecium*, red tide dinoflagellates, and diatoms. (See Table 16.2.)

2. No chlorophyll and no photosynthesis; thus, fungi depend on other life forms for food. No phagocytosis. They are the main decomposers of vegetable material. Some are pathogens, others are mutualistic symbionts of plants (mycorrhiza).

3. Yeast cells are easy to work with, grow fast, have a relatively small genome for a eukaryote (15 megabase pairs), and can be mated, and their products of meiosis can be analyzed, have stable haploid and diploid phases, and readily allow the introduction of foreign DNA and plasmids.

4. The micronucleus is the repository of the genome and divides by mitosis. The macronucleus carries many copies of genes needed for growth but lacks others. It does not divide by mitosis but simply pulls apart into two macronuclei.

5. The *Plasmodium* parasites growing in red blood cells deform them and make them susceptible to lysis in the spleen and anemia. Uninfected red blood cells are also lysed, suggesting an autoimmune mechanism. Cell lysis leads to the release of inflammatory compounds.

6. Diatoms have a hard shell made of silica; coccolithophores have one made of carbonates. Together, these eukaryotic microbes contribute much of the photosynthesis in the oceans. Diatom shells make large silica deposits, and coccolithophore shells sequester large amounts of carbonates, an important part of the carbon cycle.

Chapter 17

1. Unlike all cells, viruses "fly apart" into their component proteins and nucleic acids during replication. Viruses lack ribosomes and most of the protein-synthesizing machinery; thus, they depend on the host for the synthesis of their proteins.

2. Viruses possess one kind of nucleic acid, RNA or DNA, and a protein shell, the capsid. Many viruses have an envelope, and some contain enzymes required for various aspects of their replication cycle.

3. Generally speaking, icosahedral viruses are limited in size by the self-assembly properties of the capsid proteins, the capsomeres. Filamentous viruses, on the other hand, are not necessarily of defined length and thus can accept foreign nucleic acids, a property useful for making proteins by genetic engineering.

4. Viral envelopes are derived from host cell membranes (e.g., those of the nucleus, Golgi membranes, the endoplasmic reticulum, or the plasma membrane). Envelopes are formed when the virus "buds" through a host cell membrane during maturation.

5. Most of the enzymes found in virions are involved either in attachment to host cells (e.g., the influenza virus hemagglutinins) or in nucleic acid replication (negative-strand RNA viruses depend on their RNA replicase for making the positive, or mRNA, strand). Some viruses, e.g., the herpesvirus group, contain other enzymes necessary for DNA replication.

6. They first copy their RNA into DNA via a virion-carried reverse transcriptase. The viral DNA is now integrated into the host cell genome and can be transcribed by a host RNA polymerase to make new viral RNA.

7. Making viral proteins represents a series of well-timed and interrelated events. They are synthesized using the host cell machinery, and this takes place in the cytoplasm (which is where the host keeps its ribosomes). The source of mRNA is either the viral RNA in the virion (as in positive RNA viruses) or a copy of the virion RNA made by a virus-encoded RNA replicase. Viral proteins are not all made at once: "early" proteins redirect the cell's biosynthetic machinery toward making viral nucleic acids and protein; "late" proteins are mainly the structural components of the virions.

8. Virulent viruses usually kill their host cells by inducing lysis. Although temperate viruses can do this too, they can also live in harmony with their hosts by stably integrating into the host genome or as extrachromosomal elements. Viruses of both eukaryotes and prokaryotes (phages) can be virulent or temperate.

9. A virus remains a quiescent prophage because most of its genes are silenced by a viral repressor. The repressor is usually the only viral protein that is made in lysogenic cells. The level of repressor is carefully balanced to allow repression of the genes that would lead to viral replication but to permit induction of the prophage under certain conditions.

10. The decision between a lytic and a lysogenic cycle relies on the properties of the prophage repressor. In lambda, the repressor is unstable and is broken down by a protease (Hfl). Hfl senses the cellular level of cAMP. At high cAMP levels, as when the glucose concentration in the medium is low, less protease is made and the repressor will function to keep the lysogenic state. The converse is also true, and in high-glucose media, the repressor is cleaved and the cell undergoes a lytic cycle. Thus, the decision about which cycle will dominate reflects the environment of the bacteria.

11. Viroids are composed of capsidless naked RNA molecules. They replicate using host RNA polymerases and produce long copies of the RNA by a mechanism called rolling-cycle replication. These molecules are cleaved into viroid size either by the ribozyme activity of the viroid RNA or by a host nuclease. Prions are devoid of nucleic acids and are made up entirely of protein. Like other proteins, the prion protein can fold into a number of different three-dimensional structures. In one configuration, it is a normal constituent of the cells of the central nervous system and causes no harm. In another configuration, it becomes a prion and gains the unusual ability to act as a template that can convert molecules of the normal proteins into prions.

Chapter 18

1. Hydrogen gas, which is formed by the geochemical reduction of water in the high-temperature, high-pressure environment of the Earth's magma.

2. Enrichment culture is the setting up of particular environmental and nutrient conditions that favor the growth of particular kinds of microorganisms and inoculating the medium with a sample from a particular environment. Such culture answers whether organisms capable of a particular conversion are present in a particular environment, but it does not reveal how many such organisms are present.

3. Consortia permit microbes to cooperate in utilizing certain substrates. For example, one member of the consortium might scavenge the products produced by another member, thereby reducing its concentration sufficiently to render the producing reaction thermodynamically feasible.

4. Hybridizing with specific DNA probes reveals which organisms are present; radioautography gives some indication of what they are doing there.

5. The doctrine holds that all naturally occurring organic compounds can be metabolized by some microbe.

6. By producing a gaseous product, methane, they allow carbon that might otherwise be trapped in an anaerobic environment to rise to an aerobic one.

7. Two microbe-mediated processes, denitrification and anammox, convert fixed nitrogen to dinitrogen.

8. Nitrification, of which two steps, oxidation of ammonia to nitrite and then to nitrate, are brought about by distinct autotrophs. Because both mediating groups of microbes are aerobes, turning the pile, which aerates it, speeds the process.

9. In the bottoms of ancient lakes, sulfate, which is abundant on Earth's surface, was reduced to H_2S by sulfate-reducing bacteria and was subsequently oxidized by photoautotrophs to elemental sulfur.

10. Denitrification and anammox allow nitrogen to recycle through the atmosphere and prevent its accumulation in the oceans. Plant life is dependent on nitrogen fixation. Nitrification produces nitrate, which is mobile in soils and thus is dispersed to support plant growth.

Chapter 19

1. Their genomes have a significant degree of homology to those of present-day bacteria (rickettsiae in the case of mitochondria and cyanobacteria for chloroplasts). These organelles divide by binary fission, some with the aid of an FtsZ ring, which is typically bacterial. Their ribosomes are closer to those of bacteria than to those found in the cytoplasm of eukaryotic cells. Mitochondria are sensitive to certain antibacterial antibiotics, mainly those that inhibit protein synthesis.

2. Most of the known bacterial insect endosymbionts have a reduced genome, i.e., lack some of the functions necessary for living freely and thus rely on the host for some of these functions. They retain functions that are useful to the host, thus contributing to a mutualistic existence.

3. Root nodules are nitrogen-fixing factories that supply legumes with utilizable nitrogen. Nodule formation occurs after plants excrete compounds sensed by nodule-forming bacteria in the soil. Bacterial genes are then turned on to make nodulation factors that induce nodule formation in the plant roots. Nodule formation involves making structures that allow the bacteria to penetrate into the roots. The plant then responds by making tumorlike nodules in which the bacteria differentiate into the nitrogen-fixing cells, the bacteroids.

4. Some bacteria break down cellulose into sugars. These are fermented by other bacteria to make the volatile fatty acids that the cow absorbs and uses as its main energy source. In the process, much hydrogen is made, some of which is changed into methane by methanogenic Archaea.

5. Certain soil nematodes feed solely by invading the caterpillars of certain insects. The caterpillars are killed, and the nematodes reproduce. The worms require the help of certain bacteria, e.g., *Xenorhabdus nematophila*, which kills the insect host and disables its defense mechanisms. The bacterium makes a toxin that induces apoptosis in the intestinal cells of the insect, as well as hydrolases that break down its tissues. In order to avoid putrefaction of the insect's carcass, the bacteria make antibiotics that kill other kinds of bacteria. Neither the worms nor the bacteria subsist in the soil alone.

6. These ants make fungus gardens in their underground nests. These structures are composed of fungal growth on chopped-up leaves the ants bring into the nest and are their sole source of food. To ensure that only the "right" kinds of fungi are grown, the ants carry on them a bacterium that makes antifungal antibiotics that do not affect the desired species but act on undesirable ones. Thus, these ants discovered both agriculture and the use of antibiotics some 50 million years ago.

7. Suitable examples are the loss of aversion to cat urine by *Toxoplasma*-infected rats, which permits the parasite to complete its life cycle; "summit disease," in which insects infected with fungi climb on plant stalks, perhaps to better spread the fungal spores; the formation of "pseudoflowers"

on fungus-infected plants, which fools insects into "pollinating" the spores and thus helping their propagation; and more familiarly, the behavioral change in humans when affected by the common cold.

8. In many habitats, bacteria are at the bottom of the food chain and are prey to other organisms and viruses. Predation appears to regulate their population density. The major agents in bacterial predation are phages; bacterium-eating bacteria, such as *Bdellovibrio*; and antibiotic and bacteriocin production.

Chapter 20

1. Encounter, entry, establishment of the agent, and damage are the nearly universal steps common to all infections. Encounter may happen before birth, at birth, or thereafter. Symptoms appear at various times after the encounter (incubation time). Encounters with agents that are part of the normal flora took place earlier than the onset of the disease, unlike encounters with exogenous agents. Entry of agents into the body refers both to crossing epithelial layers, thus entering the deep layers of the body, and to causing disease by adhering to outer surfaces, such as those of the respiratory and digestive systems. The establishment of the agent depends on factors such as the size of the inoculum, the invasive ability of the agent, and the state of the host defenses. Damage may be caused by cell damage (lysis or apoptosis), pharmacological action (many exotoxins), or mechanical damage (obstruction of blood vessels or lymph vessels).

2. The incubation period is the time between the acquisition of an infectious agent and the appearance of symptoms. The inoculum size is the number of agents acquired during an infection. The normal flora (properly, "normal biota") consists of the microbes usually present in a healthy individual. Endogenous infections are those caused by infectious agents present in the body. Exogenous infections are those acquired from the outside (e.g., via insect bites, ingestion of contaminated food, or inhalation).

3. The main processes are inflammation, cell death by lysis or apoptosis, pharmacological action of toxins, and mechanical blockage.

4. The main elements of inflammation are local reddening, swelling, pain, and pus.

5. They are microbial constituents that are recognized by certain receptors called Toll-like receptors. Typical MAMPs are bacterial lipopolysaccharide, murein, flagellin, other surface components of fungi and protozoa, certain viral nucleic acids, and some sequences in bacterial DNA.

6. Formation of chemotaxins and opsonins, leading to recruitment and enhanced function of phagocytes, inflammation via pharmacologically active peptides, and killing of cells by the insertion of pores in their membranes (membrane attack complex).

7. Attachment of particles to surface receptors on phagocytic cells, engulfment via cytoskeletal rearrangements, formation of phagocytic vacuole, fusion with lysosomes, and outpouring of bactericidal lysosomal enzymes.

8. Different microbes can evade each of various aspects of the innate defenses. Some defend themselves against complement by masking their surface complement-activating constituents or altering complement-regulatory proteins. Defenses against phagocytosis include inhibiting the action of phagocyte chemotaxins, avoiding uptake by being coated by capsules, impeding phagolysosome formation, escaping into the cytoplasm, killing the phagocyte, or being intrinsically resistant to lysosomal bactericidal compounds.

9. Innate defenses are present in all healthy persons and act rapidly. They are represented by antimicrobial chemicals in tissue, the complement system, and the phagocytic cells. Adaptive defenses are particular to each individual, reflecting everyone's history of exposure to foreign antigens. On first exposure, adaptive defenses take time to be effective. On repeated exposure, these defenses also come into play rapidly. There are two kinds of adaptive responses: soluble antibodies and cell-mediated immunity.

10. By neutralizing toxins, facilitating removal of invading microbes by clumping them into readily filterable particles, acting as opsonins, interacting with complement to lyse certain pathogens, and enhancing cell-mediated immunity.

11. Cell-mediated immunity depends on the action of cytotoxic T lymphocytes that recognize and lyse cells that present microbial antigens on their surfaces.

Chapter 21

1. We assume that every reader will be able to come up with the names of 20 infectious diseases. For help, see Table 21.2.

2. Bacterial exotoxins are poisonous proteins that act either close to where they are made in the body or at distant sites. They are secreted by bacteria, often as growth stops. Exotoxins act in various general ways: lysing host cells by making pores in the plasma membrane; altering their function, e.g., by elevating the level of cyclic AMP; or influencing neurotransmission to produce either spastic or flaccid paralysis.

3. By using a type III secretion mechanism, which consists of a syringe-like apparatus on the bacterial surface that penetrates into the host cell membrane. This allows delivery of the toxin directly into target cells instead of being secreted into the medium.

4. They are acid-fast because they are surrounded by a waxy envelope. They are resistant to harsh chemicals and to drying, which facilitates their spread from patients with pulmonary tuberculosis. They grow slowly, doubling about once in 24 hours.

5. An indication that the person has been in contact with the tubercle bacillus and has developed cell-mediated immunity to antigens of the organism. It does not indicate that the person has or has had active tuberculosis. If a person in a developed country converts from tuberculin negative to tuberculin positive, a medical examination and even antibiotic treatment are warranted.

6. EBV is a DNA virus in the herpesvirus family that infects B lymphocytes. These cells are destroyed by T lymphocytes, and the outpouring of cell constituents causes damage to tissues. EBV can persist in the memory cells as a plasmid, later to become reactivated or contribute to cancer development.

7. Upon first infection with HIV, patients develop a mild infection characterized by flu-like symptoms. In time, usually a few years, a variety of more serious symptoms appear, often including skin rashes, swollen glands, sore throat, and thrush. Pneumonia often ensues, often caused by an otherwise innocuous commensal, *Pneumocystis*. This is followed by infections due to a large number of agents with a large number of manifestations. Death then follows.

Chapter 22

1. Include in your discussion the early appearance in evolution of defense mechanism and the antiquity of certain infectious diseases. Give examples of their effects on unexposed populations. Discuss the appearance of new infectious agents.

2. Discuss changes in human behavior, such as travel, urbanization, wars and their consequences, increased risky sexual behavior, industrialization of food production, reforestation of parts of North America, widespread and imprudent use of antibiotics, and naturally occurring and anthropogenic environmental changes.

3. Several factors have contributed to the successful eradication of this disease. They include the following: humans are the only reservoir of the agent, vaccination is simple and effective, the vaccine does not require refrigeration, and the symptoms of the disease are readily recognizable. In addition, subclinical or persistent infections are virtually unknown; thus, there is little chance of the virus lurking undetected.

4. Include in your discussion sanitation, vaccination, and treatment. Sanitation is largely a social and engineering problem that requires resources and political will. Vaccination is highly effective for many diseases but not yet available for others. The reasons for the lack of certain vaccines are difficulties in testing candidate vaccines, the ability of many agents to change antigens, and difficulties in eliciting cell-mediated immunity. Antimicrobial drugs are highly effective, especially against bacteria, but become ineffective when the agents mutate to become resistant. The prudent use of antimicrobial drugs has become imperative.

Chapter 23

1. The taste of red wines is improved by converting malic acid into lactic acid, the so-called malolactic fermentation. The bacteria responsible for this fermentation derive a small amount of energy from the reaction, but it is sufficient to sustain their growth.

2. Concern for the field use of any altered species is reasonable, whether the change was brought about by genetic engineering or by standard breeding manipulations. Relevant to this case is the existence of naturally occurring ice-retardant bacterial strains. For snowmaking, the bacteria are killed and thus present no danger, real or perceived.

3. The gene for the desired protein must be cloned in an industrially suitable organism. This requires identifying the gene and its mRNA with appropriate hybridization probes, copying the mRNA into clonable DNA, introducing it into a suitable organism, ensuring that the gene is expressed at suitable levels and under controllable conditions, setting up an industrial-scale production plant, and purifying the protein.

4. The reason why a soil bacterium makes insecticidal proteins is not obvious because the two organisms do not meet often. However, *Bt* toxins also affect soil nematodes by the same mechanism, suggesting that this is how the ability to make toxins was selected.

5. The toxin affects not only insects but also nematodes, which are predators of bacteria. That the toxins also work on insects may be a fortuitous coincidence.

6. Some bacteria may degrade pollutants by pathways similar to those used for their usual nutrients. For example, methane-utilizing bacteria can degrade not only methane but also certain pollutants, such as trichloroethylene. Growing these bacteria in contaminated sites requires knowledge of all their nutritional requirements, including the need for sources of nitrogen and phosphorus.

Figure and Table Credits

Chapter 1

Figure 1.4 Courtesy of E. Angert.

Chapter 2

Figure 2.5 Courtesy of M. R. J. Salton.

Figure 2.6 Reprinted from Y.-L. Shu and L. Rothfield, *Proc. Natl. Acad. Sci USA* **100**:7865–7870, 2003.

Figure 2.7 Courtesy of C. Weibull. Reprinted from F. C. Neidhardt, J. L. Ingraham, and M. Schaechter, *Physiology of the Bacterial Cell* (Sinauer Associates, Inc., Sunderland, Mass., 1990), p. 37.

Figure 2.10 Courtesy of M. A. Wells.

Figure 2.11 Courtesy of R. G. E. Murray.

Figure 2.13 Courtesy of T. J. Beveridge.

Figure 2.14 Courtesy of T. J. Beveridge.

Chapter 3

Figure 3.1 Courtesy of E. Kellenberger.

Figure 3.2 Reprinted from M. R. Lindsay, R. I. Webb, M. Strous, M. S. Jetten, M. K. Butler, R. J. Forde, and J. A. Fuerst, *Arch. Microbiol.* **175**:413–429, 2001, with permission.

Figure 3.4 (A) Courtesy of H. Schulz; (B) courtesy of E. Angert.

Figure 3.5 Courtesy of S. S. DasSarma.

Figure 3.6 Reprinted from G. C. Cannon, C. E. Bradburne, H. C. Aldrich, S. H. Baker, S. Heinhorst, and J. M. Shively, *Appl. Environ. Microbiol.* **67**:5351–5361, 2001, with permission.

Figure 3.7 Courtesy of J. F. Wilkinson. Reprinted from F. C. Neidhardt, J. L. Ingraham, and M. Schaechter, *Physiology of the Bacterial Cell* (Sinauer Associates, Inc., Sunderland, Mass., 1990), p. 58.

Chapter 4

Figure 4.8A Courtesy of C. L. Woldringh.

Chapter 5

Table 5.2 Reprinted from M. Riley and B. Labedan, p. 2118–2202, *in* F. C. Neidhardt, R. Curtiss III, J. L. Ingraham, E. C. C. Lin, K. B. Low, B. Magasanik, W. S. Reznikoff, M. Riley, M. Schaechter, and H. E. Umbarger (ed.), Escherichia coli *and* Salmonella: *Cellular and Molecular Biology*, vol. 2 (ASM Press, Washington, D.C., 1996), with permission.

Table 5.3 Modified from F. C. Neidhardt, J. L. Ingraham, and M. Schaechter, *Physiology of the Bacterial Cell* (Sinauer Associates, Inc., Sunderland, Mass., 1990), p. 4.

Chapter 6

Figure 6.3 Redrawn from F. C. Neidhardt, J. L. Ingraham, and M. Schaechter, *Physiology of the Bacterial Cell* (Sinauer Associates, Inc., Sunderland, Mass., 1990), p. 162.

Figure 6.7 Redrawn from material submitted by Robert Gunsalus.

Figure 6.8 Redrawn from F. C. Neidhardt, J. L. Ingraham, and M. Schaechter, *Physiology of the Bacterial Cell* (Sinauer Associates, Inc., Sunderland, Mass., 1990), p. 167.

Figure 6.9 Redrawn from F. C. Neidhardt, J. L. Ingraham, and M. Schaechter, *Physiology of the Bacterial Cell* (Sinauer Associates, Inc., Sunderland, Mass., 1990), p. 169.

Figure 6.12 Based on A. G. Moat, J. W. Foster, and M. P. Spector, *Microbial Physiology*, 4th ed. (Wiley-Liss Inc., New York, N.Y., 2002), p. 390.

Table 6.9 Reprinted from D. Gutnick, J. M. Calvo, T. Klopotowski, and B. N. Ames, *J. Bacteriol.* **100**:215–219, 1969, with permission.

Chapter 8

Figure 8.1 Redrawn from F. C. Neidhardt, J. L. Ingraham, and M. Schaechter, *Physiology of the Bacterial Cell* (Sinauer Associates, Inc., Sunderland, Mass., 1990), p. 75.

Figure 8.2 Redrawn from F. C. Neidhardt, J. L. Ingraham, and M. Schaechter, *Physiology of the Bacterial Cell* (Sinauer Associates, Inc., Sunderland, Mass., 1990), p. 77.

Figure 8.3 Derived from A. G. Moat, J. W. Foster, and M. P. Spector, *Microbial Physiology*, 4th ed. (Wiley-Liss Inc., New York, N.Y., 2002), p. 44.

Figure 8.5 Derived from A. G. Moat, J. W. Foster, and M. P. Spector, *Microbial Physiology*, 4th ed. (Wiley-Liss Inc., New York, N.Y., 2002), p. 39.

Figure 8.10 Based on Fig. (p. 796) of M. T. Record, Jr., W. S. Reznikoff, M. L. Craig, K. L. McQuade, and P. J. Schlax, p. 792–820, *in* F. C. Neidhardt, R. Curtiss III, J. L. Ingraham, E. C. C. Lin, K. B. Low, B. Magasanik, W. S. Reznikoff, M. Riley, M. Schaechter, and H. E. Umbarger (ed.), Escherichia coli *and* Salmonella: *Cellular and Molecular Biology*, vol. 1 (ASM Press, Washington D.C., 1996).

Figure 8.15 Redrawn from F. C. Neidhardt, J. L. Ingraham, and M. Schaechter, *Physiology of the Bacterial Cell* (Sinauer Associates, Inc., Sunderland, Mass., 1990), p. 90.

Figure 8.16 Redrawn from F. C. Neidhardt, J. L. Ingraham, and M. Schaechter, *Physiology of the Bacterial Cell* (Sinauer Associates, Inc., Sunderland, Mass., 1990), p. 91.

Figure 8.18 Redrawn from F. C. Neidhardt, J. L. Ingraham, and M. Schaechter, *Physiology of the Bacterial Cell* (Sinauer Associates, Inc., Sunderland, Mass., 1990), p. 107.

Figure 8.19 Derived from A. G. Moat, J. W. Foster, and M. P. Spector, *Microbial Physiology*, 4th ed. (Wiley-Liss Inc., New York, N.Y., 2002), p. 78.

Figure 8.21 Redrawn from F. C. Neidhardt, J. L. Ingraham, and M. Schaechter, *Physiology of the Bacterial Cell* (Sinauer Associates, Inc., Sunderland, Mass., 1990), p. 104.

Figure 8.22 Redrawn from F. C. Neidhardt, J. L. Ingraham, and M. Schaechter, *Physiology of the Bacterial Cell* (Sinauer Associates, Inc., Sunderland, Mass., 1990), p. 115.

Table 8.1 Reprinted from F. C. Neidhardt, J. L. Ingraham, and M. Schaechter, *Physiology of the Bacterial Cell* (Sinauer Associates, Inc., Sunderland, Mass., 1990), p. 294.

Table 8.2 Reprinted from F. C. Neidhardt, J. L. Ingraham, and M. Schaechter, *Physiology of the Bacterial Cell* (Sinauer Associates, Inc., Sunderland, Mass., 1990), p. 73.

Chapter 9

Figure 9.2 Reprinted from K. Skarstadt, E. Boye, and H. B. Steen, *EMBO J.* **5**:1711–1717, 1986, with permission.

Figure 9.5 Courtesy of D. E. Caldwell. Previously published on the cover of the 2003 issues of *International Microbiology*; reprinted with permission.

Figure 9.6 Courtesy of T. J. Beveridge.

Figure 9.7 Courtesy of T. J. Beveridge.

Figure 9.8 Courtesy of J. Pogliano.

Figure 9.9 Courtesy of Y. Hirota. Reprinted from F. C. Neidhardt, J. L. Ingraham, and M. Schaechter, *Physiology of the Bacterial Cell* (Sinauer Associates, Inc., Sunderland, Mass., 1990), p. 407.

Figure 9.13 Courtesy of D. Sherratt.

Chapter 11

Figure 11.1 Courtesy of J. W. Schopf.

Chapter 12

Figure 12.2 Redrawn from F. C. Neidhardt, J. L. Ingraham, and M. Schaechter, *Physiology of the Bacterial Cell* (Sinauer Associates, Inc., Sunderland, Mass., 1990), p. 308.

Figure 12.3 Redrawn from F. C. Neidhardt, J. L. Ingraham, and M. Schaechter, *Physiology of the Bacterial Cell* (Sinauer Associates, Inc., Sunderland, Mass., 1990), p. 309.

Figure 12.4 Redrawn from F. C. Neidhardt, J. L. Ingraham, and M. Schaechter, *Physiology of the Bacterial Cell* (Sinauer Associates, Inc., Sunderland, Mass., 1990), p. 311.

Figure 12.5 Redrawn from F. C. Neidhardt, J. L. Ingraham, and M. Schaechter, *Physiology of the Bacterial Cell* (Sinauer Associates, Inc., Sunderland, Mass., 1990), p. 333.

Figure 12.6 Redrawn from F. C. Neidhardt, J. L. Ingraham, and M. Schaechter, *Physiology of the Bacterial Cell* (Sinauer Associates, Inc., Sunderland, Mass., 1990), p. 329.

Figure 12.7 Redrawn from F. C. Neidhardt, J. L. Ingraham, and M. Schaechter, *Physiology of the Bacterial Cell* (Sinauer Associates, Inc., Sunderland, Mass., 1990), p. 341.

Figure 12.8 Redrawn from F. C. Neidhardt, J. L. Ingraham, and M. Schaechter, *Physiology of the Bacterial Cell* (Sinauer Associates, Inc., Sunderland, Mass., 1990), p. 343.

Figure 12.9 Redrawn from F. C. Neidhardt, J. L. Ingraham, and M. Schaechter, *Physiology of the Bacterial Cell* (Sinauer Associates, Inc., Sunderland, Mass., 1990), p. 346.

Figure 12.11 Redrawn from F. C. Neidhardt, J. L. Ingraham, and M. Schaechter, *Physiology of the Bacterial Cell* (Sinauer Associates, Inc., Sunderland, Mass., 1990), p. 363.

Table 12.1 Reprinted from F. C. Neidhardt, J. L. Ingraham, and M. Schaechter, *Physiology of the Bacterial Cell* (Sinauer Associates, Inc., Sunderland, Mass., 1990), p. 315.

Table 12.2 Modified from F. C. Neidhardt, J. L. Ingraham, and M. Schaechter, *Physiology of the Bacterial Cell* (Sinauer Associates, Inc., Sunderland, Mass., 1990), p. 354–356.

Chapter 13

Figure 13.4 Based on T. Nyström, *Mol. Microbiol.* **54**:855–862, 2004.

Figure 13.5 Based on Fig. 1 (p. 1673) of G. W. Huisman, D. A. Siegele, M. M. Zambrano, and R. Kolter, p. 1672–1682, *in* F. C. Neidhardt, R. Curtiss III, J. L. Ingraham, E. C. C. Lin, K. B. Low, B. Magasanik, W. S. Reznikoff, M. Riley, M. Schaechter, and H. E. Umbarger (ed.), Escherichia coli *and* Salmonella: *Cellular and Molecular Biology,* vol. 2 (ASM Press, Washington, D.C., 1996).

Table 13.2 Reprinted from F. C. Neidhardt, J. L. Ingraham, and M. Schaechter, *Physiology of the Bacterial Cell* (Sinauer Associates, Inc., Sunderland, Mass., 1990), p. 379. Based on J. B. Stock, A. J. Ninfa, and A. N. Stock, *Microbiol. Rev.* **53**:450–490, 1989.

Table 13.6 Based on Table 1 (p. 4) of G. M. Dunny and S. C. Winans, p. 1–5, *in* G. M. Dunny and S.C. Winans (ed.), *Cell-Cell Signaling in Bacteria* (ASM Press, Washington, D.C., 1999).

Chapter 14

Figure 14.3B Courtesy of T. J. Beveridge.

Figure 14.5 Reprinted from H. Reichenbach, p. 13–62, *in* M. Dworkin and D. Kaiser (ed.), *Myxobacteria II* (American Society for Microbiology, Washington, D.C., 1993), with permission.

Chapter 15

Figure 15.2 Courtesy of the Image Science and Analysis Laboratory, NASA Johnson Space Center.

Figure 15.3 Courtesy of A.Vasilenko and M. J. Daly.

Figure 15.4 (A) Courtesy of J. E. Frias; (B) reprinted from J. McCarren, J. Heuser, R. Roth, N. Yamada, M. Martone, and B. Brahamsha, *J. Bacteriol.* **187**:224–230, 2005, with permission.

Figure 15.5 Reprinted from J. L. Ingraham and C. A. Ingraham, *Introduction to Microbiology,* 2nd ed. (Brooks/Cole, Pacific Grove, Calif., 2000), p. 281. Source: D. A. Glave, Biological Photo Service.

Figure 15.6 Courtesy of T. J. Beveridge.

Figure 15.7A Courtesy of K. J. McDowall and K. Jolly.

Chapter 16

Figure 16.1 Courtesy of A. Wheals.

Figure 16.3 Courtesy of K. Johnson.

Figure 16.6 Courtesy of G. Grimes and S. L'Hernault.

Figure 16.7 Redrawn from a figure provided by P. Rotkiewicz.

Figure 16.9 Courtesy of R. Edgar and the Center for Diatom Informatics.

Figure 16.11 Courtesy of M. Y. Cortés.

Figure 16.12 Courtesy of S. Carty.

Chapter 17

Figure 17.1 Adapted from S. J. Flint, L.W. Enquist, V. R. Racaniello, and A. M. Skalka, *Principles of Virology: Molecular Biology, Pathogenesis, and Control of Animal Viruses,* 2nd ed. (ASM Press, Washington, D.C., 2004).

Figure 17.7 Modified from S. Wold et al., *in* D. P. Nayak (ed.), *Molecular Biology of Animal Viruses,* vol. 2 (Marcel Dekker, Inc., New York, N.Y., 1978).

Figure 17.11 Courtesy of E. C. Klatt.

Chapter 18

Figure 18.4 Courtesy of Heal the Bay.

Figure 18.5 Reprinted from A. Boetius, K. Ravenschlag, C. J. Schubert, D. Rickert, F. Widdel, A. Gieseke, R. Amann, B. B. Jørgensen, U. Witte, and O. Pfannkuche, *Nature* **407**:623–626, 2000, with permission.

Figure 18.6 Reprinted from W. Michaelis, R. Seifert, K. Nauhaus, T. Treude, V. Thiel, M. Blumenberg, K. Knittel, A. Gieseke, K. Peterknecht, T. Pape, A. Boetius, R. Amann, B. B. Jørgensen, F. Widdel, J. Peckmann, N. V. Pimenov, and M. B. Gulin, *Science* **297**:1013–1015, 2002, with permission.

Figure 18.10 Courtesy of T. J. Beveridge.

Chapter 19

Figure 19.2 Reprinted from S. Vitha, R. S. McAndrew, and K. W. Osteryoung, *J. Cell Biol.* **153**:111–119, 2001, with permission.

Figure 19.4 Courtesy of M. F. Wiser.

Figure 19.5 Courtesy of M. F. Wiser.

Figure 19.6 Courtesy of L. Sacchi.

Figure 19.7 Courtesy of L. Sacchi.

Figure 19.8 Courtesy of N. A. Moran.

Figure 19.9 Courtesy of D. Gage.

Figure 19.11 Courtesy of D. A. Phillips. Reprinted from F. C. Neidhardt, J. L. Ingraham, and M. Schaechter, *Physiology of the Bacterial Cell* (Sinauer Associates, Inc., Sunderland, Mass., 1990), p. 450.

Figure 19.13 Courtesy of J. Smiles and M. J. Dobson. Reprinted from F. C. Neidhardt, J. L. Ingraham, and M. Schaechter, *Physiology of the Bacterial Cell* (Sinauer Associates, Inc., Sunderland, Mass., 1990), p. 478.

Figure 19.14 Source: Online image gallery, Agricultural Research Service, U.S. Department of Agriculture. Photograph by Scott Bauer.

Figure 19.16 Courtesy of J. Beach.

Figure 19.17 Courtesy of B. A. Roy.

Chapter 20

Figure 20.5 Courtesy of J. Tranum-Jensen.

Chapter 21

Figure 21.1 Courtesy of R. R. Isberg. Reprinted from F. C. Neidhardt, J. L. Ingraham, and M. Schaechter, *Physiology of the Bacterial Cell* (Sinauer Associates, Inc., Sunderland, Mass., 1990), p. 469.

Figure 21.5 Source: ASM Microbe Library (http://www.microbelibrary.org). © G. Delisle and L. Tomalty.

Figure 21.7 Courtesy of E. Kieff.

Chapter 22

Table 22.1 Reprinted from Table 11.1 of J. Diamond, *Guns, Germs, and Steel: the Fates of Human Societies* (W. W. Norton & Co., Inc., New York, N.Y., 1997). Copyright 1997 by Jared Diamond. Used by permission of W. W. Norton & Co., Inc.

Chapter 23

Figure 23.1 Courtesy of S. E. Lindow.

Figure 23.2 Courtesy of D. Lereclus.

Figure 23.3 Redrawn from an illustration provided by T. Hazen.

Figure 23.4 Redrawn from an illustration provided by the Alaska Wilderness League.

Index

D

E

H

I

J

K

L

M

N

O

Q

R

S

T

U

V

W

X

Y

Z